WITHDRAWN

ANNUAL REVIEW OF EARTH AND PLANETARY SCIENCES

EDITORIAL COMMITTEE (1987)

ARDEN L. ALBEE
ROBIN BRETT
WILLIAM W. HAY
RAYMOND JEANLOZ
CARL KISSLINGER
GERARD V. MIDDLETON
FRANCIS G. STEHLI
GEORGE W. WETHERILL

Responsible for the organization of Volume 15
(Editorial Committee, 1985)

ARDEN L. ALBEE
JOHN C. HARMS
WILLIAM W. HAY
CARL KISSLINGER
V. RAMA MURTHY
FRANCIS G. STEHLI
DONALD L. TURCOTTE
GEORGE W. WETHERILL
GORDON E. BROWN (Guest)
MICHAEL H. CARR (Guest)

Production Editor   KEITH DODSON
Subject Indexer   CHERI D. WALSH

ANNUAL REVIEW OF EARTH AND PLANETARY SCIENCES

VOLUME 15, 1987

GEORGE W. WETHERILL, *Editor*
Carnegie Institution of Washington

ARDEN L. ALBEE, *Associate Editor*
California Institute of Technology

FRANCIS G. STEHLI, *Associate Editor*
DOSECC Science Advisory Committee

ANNUAL REVIEWS INC 4139 EL CAMINO WAY P.O. BOX 10139 PALO ALTO, CALIFORNIA 94303-0897

ANNUAL REVIEWS INC.
Palo Alto, California, USA

COPYRIGHT © 1987 BY ANNUAL REVIEWS INC., PALO ALTO, CALIFORNIA, USA. ALL RIGHTS RESERVED. The appearance of the code at the bottom of the first page of an article in this serial indicates the copyright owner's consent that copies of the article may be made for personal or internal use, or for the personal or internal use of specific clients. This consent is given on the condition, however, that the copier pay the stated per-copy fee of $2.00 per article through the Copyright Clearance Center, Inc. (21 Congress Street, Salem, MA 01970) for copying beyond that permitted by Sections 107 or 108 of the US Copyright Law. The per-copy fee of $2.00 per article also applies to the copying under the stated conditions, of articles published in any Annual Review serial before January 1, 1978. Individual readers, and nonprofit libraries acting for them, are permitted to make a single copy of an article without charge for use in research or teaching. This consent does not extend to other kinds of copying, such as copying for general distribution, for advertising or promotional purposes, for creating new collective works, or for resale. For such uses, written permission is required. Write to Permissions Dept., Annual Reviews Inc., 4139 El Camino Way, P.O. Box 10139, Palo Alto, CA 94303-0897, USA.

International Standard Serial Number: 0084-6597
International Standard Book Number: 0-8243-2015-8
Library of Congress Catalog Card Number: 72-82137

Annual Review and publication titles are registered trademarks of Annual Reviews Inc.

Annual Reviews Inc. and the Editors of its publications assume no responsibility for the statements expressed by the contributors to this *Review*.

TYPESET BY AUP TYPESETTERS (GLASGOW) LTD., SCOTLAND
PRINTED AND BOUND IN THE UNITED STATES OF AMERICA

CONTENTS

HOW I BECAME AN OCEANOGRAPHER AND OTHER SEA STORIES, *R. Revelle*	1
THE CORE-MANTLE BOUNDARY, *Christopher J. Young and Thorne Lay*	25
APPLICATION OF STABLE CARBON ISOTOPES TO EARLY BIOCHEMICAL EVOLUTION ON EARTH, *M. Schidlowski*	47
PHYSICAL PROCESSES IN VOLCANIC ERUPTIONS, *L. Wilson, H. Pinkerton, and R. Macdonald*	73
GEOARCHAEOLOGY, *George Rapp, Jr.*	97
THREE-DIMENSIONAL SEISMIC IMAGING, *Clifford H. Thurber and Keiiti Aki*	115
INTEGRATED DIAGENETIC MODELING: A Process-Oriented Approach for Clastic Systems, *Ronald C. Surdam and Laura J. Crossey*	141
THE ATMOSPHERES OF VENUS, EARTH, AND MARS: A CRITICAL COMPARISON, *Ronald G. Prinn and Bruce Fegley, Jr.*	171
TECTONICS OF THE TETHYSIDES: Orogenic Collage Development in a Collisional Setting, *A. M. Celâl Şengör*	213
TERRESTRIAL IMPACT STRUCTURES, *Richard A. F. Grieve*	245
ORIGIN OF THE MOON—THE COLLISION HYPOTHESIS, *D. J. Stevenson*	271
EXPERIMENTAL AND THEORETICAL CONSTRAINTS ON HYDROTHERMAL ALTERATION PROCESSES AT MID-OCEAN RIDGES, *W. E. Seyfried, Jr.*	317
TECTONICS OF THE SOUTHERN AND CENTRAL APPALACHIAN INTERNIDES, *Robert D. Hatcher, Jr.*	337
ORGANIC GEOCHEMISTRY OF BIOMARKERS, *R. Paul Philp and C. Anthony Lewis*	363
THERMOMETRY AND BAROMETRY OF IGNEOUS AND METAMORPHIC ROCKS, *Steven R. Bohlen and Donald H. Lindsley*	397
MODELS OF LITHOSPHERIC THINNING, *Horst J. Neugebauer*	421
ARCHITECTURE OF CONTINENTAL RIFTS WITH SPECIAL REFERENCE TO EAST AFRICA, *B. R. Rosendahl*	445

(*continued*)

CONTENTS (*continued*)

MARINE MAGNETIC ANOMALIES—THE ORIGIN OF THE STRIPES, *C. G. A. Harrison*	505
STRESS NEAR THE SURFACE OF THE EARTH, *D. I. Gough and W. I. Gough*	545
POLAR WANDERING AND PALEOMAGNETISM, *Richard G. Gordon*	567
INDEXES	
Subject Index	595
Cumulative Index of Contributing Authors, Volumes 1–15	604
Cumulative Index of Chapter Titles, Volumes 1–15	607

SOME RELATED ARTICLES IN OTHER *ANNUAL REVIEWS*

From the *Annual Review of Astronomy and Astrophysics*, Volume 24 (1986)

　The Quiet Solar Transition Region, John T. Mariska

From the *Annual Review of Biochemistry*, Volume 55 (1986)

　Genetics of Mitochondrial Biogenesis, Alexander Tzagoloff and Alan M. Myers

　The Heat-Shock Response, Susan Lindquist

From the *Annual Review of Ecology and Systematics*, Volume 17 (1986)

　Ecology of Tropical Dry Forest, Peter G. Murphy and Ariel E. Lugo

　The Socioecology of Primate Groups, J. Terborgh and C. H. Janson

　Nutrient Cycling in Moist Tropical Forest, P. M. Vitousek and R. L. Sanford, Jr.

　The Ecology of Coralline Algal Crusts: Convergent Patterns and Adaptive Strategies, Robert S. Steneck

　Information, Economics, and Evolution, Niles Eldredge

　Risk and Foraging in Stochastic Environments, Leslie Real and Thomas Caraco

　Phenetic Taxonomy: Theory and Methods, Robert R. Sokal

　Molecular Aspects of the Species Barrier, Michel Krieber and Michael R. Rose

　Sexual Selection and the Evolution of Song, William A. Searcy and Malte Andersson

　Genetic Polymorphism in Heterogeneous Environments: A Decade Later, Philip W. Hedrick

　Vascular Plant Breakdown in Freshwater Ecosystems, J. R. Webster and E. F. Benfield

　Rates of Molecular Evolution, John H. Gillespie

　The Evolution of Phenotypic Plasticity in Plants, Carl D. Schlichting

(*continued*)

RELATED ARTICLES (*continued*)

From the *Annual Review of Fluid Mechanics*, Volume 19 (1987)

Tsunamis, S. S. Voit

Isolated Eddy Models in Geophysics, G. R. Flierl

From the *Annual Review of Materials Science*, Volume 16 (1986)

Materials Modification and Synthesis Under High-Pressure Shock Compression, R. A. Graham, B. Morosin, E. L. Venturini, and M. J. Carr

From the *Annual Review of Microbiology*, Volume 40 (1986)

Microorganisms in Reclamation of Metals, S. R. Hutchins, M. S. Davidson, J. A. Brierley, and C. L. Brierley

Microbial Ecology and Evolution: A Ribosomal Approach, Gary J. Olsen, David J. Lane, Stephen J. Giovannoni, Norman R. Pace, and David A. Stahl

From the *Annual Review of Nuclear and Particle Science*, Volume 36 (1986)

Nuclear Reactions in Stars, B. W. Filippone

From the *Annual Review of Physical Chemistry*, Volume 37 (1986)

Thermodynamic Behavior of Fluids Near the Critical Point, J. V. Sengers and J. M. H. Levelt Sengers

ANNUAL REVIEWS INC. is a nonprofit scientific publisher established to promote the advancement of the sciences. Beginning in 1932 with the *Annual Review of Biochemistry*, the Company has pursued as its principal function the publication of high quality, reasonably priced *Annual Review* volumes. The volumes are organized by Editors and Editorial Committees who invite qualified authors to contribute critical articles reviewing significant developments within each major discipline. The Editor-in-Chief invites those interested in serving as future Editorial Committee members to communicate directly with him. Annual Reviews Inc. is administered by a Board of Directors, whose members serve without compensation.

1987 Board of Directors, Annual Reviews Inc.

Dr. J. Murray Luck, Founder and Director Emeritus of Annual Reviews Inc.
  Professor Emeritus of Chemistry, Stanford University
Dr. Joshua Lederberg, President of Annual Reviews Inc.
  President, The Rockefeller University
Dr. James E. Howell, Vice President of Annual Reviews Inc.
  Professor of Economics, Stanford University
Dr. William O. Baker, *Retired Chairman of the Board, Bell Laboratories*
Dr. Winslow R. Briggs, *Director, Carnegie Institution of Washington, Stanford*
Dr. Sidney D. Drell, *Deputy Director, Stanford Linear Accelerator Center*
Dr. Eugene Garfield, *President, Institute for Scientific Information*
Dr. Conyers Herring, *Professor of Applied Physics, Stanford University*
Mr. William Kaufmann, *President, William Kaufmann, Inc.*
Dr. D. E. Koshland, Jr., *Professor of Biochemistry, University of California, Berkeley*
Dr. Gardner Lindzey, *Director, Center for Advanced Study in the Behavioral Sciences, Stanford*
Dr. William D. McElroy, *Professor of Biology, University of California, San Diego*
Dr. William F. Miller, *President, SRI International*
Dr. Esmond E. Snell, *Professor of Microbiology and Chemistry, University of Texas, Austin*
Dr. Harriet A. Zuckerman, *Professor of Sociology, Columbia University*

Management of Annual Reviews Inc.

John S. McNeil, Publisher and Secretary-Treasurer
William Kaufmann, Editor-in-Chief
Mickey G. Hamilton, Promotion Manager
Donald S. Svedeman, Business Manager

ANNUAL REVIEWS OF		SPECIAL PUBLICATIONS
Anthropology	Materials Science	
Astronomy and Astrophysics	Medicine	Annual Reviews Reprints:
Biochemistry	Microbiology	Cell Membranes, 1975–1977
Biophysics and Biophysical Chemistry	Neuroscience	Immunology, 1977–1979
Cell Biology	Nuclear and Particle Science	
Computer Science	Nutrition	
Earth and Planetary Sciences	Pharmacology and Toxicology	Excitement and Fascination
Ecology and Systematics	Physical Chemistry	of Science, Vols. 1 and 2
Energy	Physiology	
Entomology	Phytopathology	Intelligence and Affectivity,
Fluid Mechanics	Plant Physiology	by Jean Piaget
Genetics	Psychology	
Immunology	Public Health	
	Sociology	Telescopes for the 1980s

A detachable order form/envelope is bound into the back of this volume.

Roger Revelle

HOW I BECAME AN OCEANOGRAPHER AND OTHER SEA STORIES

R. Revelle

Program in Science, Technology and Public Affairs, University of California, San Diego, La Jolla, California 92093

> *Let us now praise famous men*
> *Men of little showing*
> *For their work continueth*
> *Broad and deep continueth*
> *Great beyond their knowing*
>
> Rudyard Kipling, *Stalky and Co.*

BECOMING AN OCEANOGRAPHER

When I went to Pomona College as a 16-year-old freshman in 1925 I thought I wanted to be a journalist. While attending high school in Pasadena I had been on the editorial staff of the student paper and had had a good time. But in my sophomore year of college I took a course in elementary geology from an inspired teacher, Alfred O. Woodford. On the first or second session of the class he took us all up to Indian Hill, a flat-topped erosional remnant about 2 miles north of the college. It rises some 50 feet above the surrounding alluvial fan that debouches from the nearby San Gabriel Mountains. When the class had gathered around him he asked, "How did this hill get here?" This was a question none of us had thought to ask, let alone answer. Then he said, "Look around you, and tell me what you see and think."

Actually, Woody's question was by no means simple. Its precise answer is not clear even today, although a fairly reasonable explanation of the origin of Indian Hill and other similar features was given by Rollin Eckis

(Woody's all-time prize student) in his classic paper, "Alluvial Fans of the Cucamonga District, Southern California" (Eckis 1928), which was written while Rollin was still a Pomona College undergraduate. After we had puzzled over Woody's question for half an hour he gave us several partial alternative answers, explaining that he was not sure himself as to which of these, if any, was correct.

I still remember the final examination in this course. It consisted of a series of quotations from a book entitled *The New Geology* by a religious fundamentalist, George Price. We students were requested to comment on these quotations in the light of what we had learned. I remember I answered that it all dependend on whether one accepted Lyell's doctrine of uniformitarianism—that the laws of physics have always applied everywhere, and consequently the present is the key to the past. From that time onward I gave up all thought of becoming a journalist.

Woody took a sabbatical leave during my junior year, leaving me to the tender mercies of two cold-blooded professors of physics and mathematics. As a hangover from my previous attachment to journalism I had become editor of Pomona College's so-called funny magazine, the *Sagehen*, and in the middle of the fall semester I was drafted to become co-editor of the student newspaper, together with a fraternity brother of mine, Murray Putnam. The paper was in the process of being turned from a weekly into a daily; after a few months the previous editor had had a nervous breakdown. Murray and I usually got to bed about 3:00 A.M. My physics class was at 7:30 in the morning. Needless to say, I did very badly, and no better in the calculus class during the next period.

I learned many years later that when Woody got back from his sabbatical he found the physics professor pushing hard to flunk me out of college. Through heroic efforts, he managed to avoid this. But it resulted in a permanent break between him and the physics professor.

Woody was the sole member of the Pomona Geology Department, and he could teach only a limited number of geology courses. He traditionally sent his students to Berkeley for graduate work in geology; usually they were given a teaching assistantship. This happened to me in 1930/31. There were lots of graduate students in Bacon Hall, the old brick geology building next to the Campanile. Nearly all of them had taken their undergraduate work at Berkeley. Many did not have teaching assistantships, even though they had obviously learned more geology at Berkeley than Woody could teach me at Pomona. I brooded for some time about the injustice of this situation. But I finally decided that I deserved that teaching assistantship, for Woody had taught me three things: a knowledge of the vocabulary of geology, so that I was able to read the literature easily; a recognition that in spite of that literature, little was actually known about the Earth and its history, but that it was possible through research to find out a good deal; and that geological research was wonderfully exciting and very good fun. My fellow students at Berkeley had learned so many geological facts that they had missed these three fundamentals.

In those days, Berkeley was—as it still is—a marvelous place. I have worked for the University of California during most of my adult life, but mostly in La Jolla, about as far from Berkeley as one can get and still be in the state. Thus, whenever I go to Berkeley a chill runs up and down my spine from the sheer joy of being there. The best parts were the old redwood faculty club and the chemistry buildings, both very close to Bacon Hall. The great thermodynamicist Gilbert Newton Lewis was head of chemistry; he spent part of nearly every afternoon in the faculty club playing cribbage or kriegspiel while we students watched in awe.

In the first half of this century a significant number of professors of chemistry in the United States, including Harold Urey, Joseph Mayer, Frank Long, and many others, had taken their PhD's at Berkeley under giants like Lewis, Latimer, Hildebrand, Eastman, Giauque, Branch, and others. I took the elementary course in chemistry. The lectures were given by Joel Hildebrand, and the other full professors acted as laboratory assistants. This was one of the great experiences of life. Hildebrand never mentioned the peculiar properties of arsenic or lead, or any other element. He was concerned entirely with the principles of chemistry, such as the laws of mass action and of ionic dissociation.

Several of us graduate students had desks in the large library room of Bacon Hall. Off this library was a private office occupied by the pioneering Berkeley geologist Andrew Cowper Lawson—usually called, for good reason, "the King." With his walrus moustache and fierce looks he seemed to us the very model of a curmudgeon. Several times a day he would stomp in and out of his office without saying a word to any of us. He was said to have a high-pitched voice, and there were many stories about him, mostly slightly indecent, among California geologists. He had recently married a woman 50 years his junior, the daughter of his old friend, Dr. Collins, the Canadian Government's chief geologist. For all I knew, he might have been a gentle soul underneath his crust.

The chairman of the Geology Department was George D. Louderback, usually called "Uncle George." Just after World War I he had played a major part in the faculty revolt at Berkeley that established the almost omnipotent role of the Academic Senate in the University of California. Uncle George was a native of San Francisco; he spoke a New Englandish dialect that was said to be nearly San Franciscan. He had a large office with many tables, all of which were covered by about a foot of papers, with a stratification going down to the early Tertiary, but he never had any trouble finding a paper when he wanted it. He faced life and science with a cool skepticism that we graduate students tried our best to emulate. I remember that his seminar in marine sedimentation was a superb example of what a seminar should be, given in a style that I still try to follow in my own classes.

In the spring of 1931 T. Wayland Vaughan, the Director of the Scripps Institution of Oceanography, came to Bacon Hall in search of a graduate student who was willing to spend a year in La Jolla looking at muds collected from the Pacific and Atlantic deep-sea floors by the nonmagnetic yacht *Carnegie*. The ship, which had belonged to the Department of Terrestrial Magnetism of the Carnegie Institution of Washington, had been on a round-the-world combined oceanographic and geomagnetic expedition. Before the voyage was half-finished, she had been destroyed by an explosion in the harbor of Apia, Samoa. (Many oceanographic ships in those days came eventually to a similar bad end.) Fortunately, the samples and data had been shipped back to the United States before the accident, and the mud samples were resting safely in La Jolla.

I applied and was accepted for the job of studying them. My fiancée, Ellen Clark, had been born in La Jolla, and she and I had spent a good deal of time there in her mother's beach house (where, by the way, we live today). Hence, we thought it would be nice to spend the first year of our marriage in these familiar surroundings. In preparation for the work at Scripps, I spent both the summer- and inter-session at Berkeley, twelve weeks in all, taking the four elementary physics courses that ordinarily took two years during the regular college term. These helped to make up for my dismal physics performance at Pomona.

Ellen and I went down to La Jolla in September 1931 to live in the upper story of one of the small wooden cottages on the Scripps campus. John Wells, today a famous specialist on fossil corals, lived downstairs, and a well-behaved but noisy family of skunks lived under the house. This was during the middle of the Depression, and I remember our rent was about $10 a month. We and the couple next door (Horace Byers, later to become a well-known meteorologist, and his wife Frances), shared a maid. We each paid her $6 a month.

After a few weeks Dick Fleming, another research assistant (now Professor Emeritus R. H. Fleming of the University of Washington), came to my lab in what is now the Old Scripps Building and said, "You're the new boy here. Tomorrow morning we have to go to sea. I'll pick you up in front of your house at about 2:30 A.M." Sure enough, he appeared in pitch darkness at the appointed time, accompanied by Maynard Harding, also a research assistant. We drove down to the San Diego Yacht Club at Point Loma, where the Scripps Institution's so-called oceanographic ship—a 64.5-foot ex-purse seiner named *Scripps*—was waiting for us, with her one-man crew, a former railroad engineer named Murdy Ross. (Murdy was used to keeping a steam locomotive in tip-top shape, and that meant covered with grease. He tended to apply the same principles to *Scripps*.)

We sailed out through San Diego Harbor, past the bell and the whistling buoys to an area over 1000 meters deep, some 15 miles west of Point Loma, which we later called the San Diego Trough. Here we stopped the engine and hove to. I learned that this same trip was to be taken every week as part of a year-round study of the ocean conditions off San Diego. We lowered a set of Nansen bottles equipped with reversing thermometers into the water. When the proper depth had been reached a small brass weight, called a messenger, was attached to the wire, dropped through the water to "trip" the bottles, the whole "cast" was retrieved, and the process was repeated for a series of different depths. Along about 11:30 Dick Fleming said "You're the new boy, so you get to cook lunch." I dutifully went down to the galley and cooked what seemed to me a wonderful meal of steak and boiled potatoes, with a salad of lettuce and tomatoes. The three others came down to the galley, bolted their meal in absolute silence, as sailors do, and announced "We're liable to get seasick down here. We'll go back up topside and continue the sampling, and you can stay down here and do the dishes."

About 4:30 the station was completed; we headed back to San Diego, arriving well after dark. I thought it was one of the finest days I had ever spent. I believe I decided then and there, more or less subconsciously, without actually saying so, that I would spend the rest of my life as an oceanographer. Being at the same time a sailor and a scientist just seemed too good to be true. Of course this meant that I would have to spend my life at an oceanographic institution, of which in those days there were only three—the Woods Hole Oceanographic Institution, the Oceanographic Laboratories of the University of Washington, and the Scripps Institution of Oceanography. For a California boy, Scripps was the obvious choice, although Dr. Vaughan and the faculty of course didn't realize it, and I never actually told them. I just stayed.

I believe part of the reason for my decision was the result of a field trip I had taken a couple of years before with Rollin Eckis in the Santa Rosa Mountains at the eastern edge of San Diego County. Like most good geologists, Rollin climbed cliffs and walked along precipices as casually as a mountain goat. I tried my best to follow him but it was often an ordeal, for I was afraid of heights. One of the great things about being an oceanographer was that the only heights to climb were the rope ladders attached to the masts of ships, and on these you could hang on with your hands as well as your feet.

One day Dr. Vaughan (whom we called T. Willey) gave me some calcareous mud from the Bahamas to look at. It consisted mainly of small spherulites, called "oolites," and many small needles, which I was able to identify as aragonite and to describe in my first scientific paper. Thus began my lifelong love affair with calcium carbonate and carbon dioxide. Erik Moberg, Scripps' sole chemical faculty member at that time, had been working with David Greenberg, a biochemist at Berkeley, on the role of carbon dioxide in the buffer mechanism of seawater. Together with a chemical technician, Esther Allen, they had found a discrepancy between the theoretical relationships of carbonate, bicarbonate, free CO_2, and hydrogen ion concentrations and their experimental results. Moberg asked me to look into the reasons why, and after a considerable time I was able to show that the various dissociation products of boric acid played a minor but important part in the buffering of seawater. The four of us published a long paper, "The Buffer Mechanism of Sea Water" (Moberg et al 1934), but we were pretty much preempted by a paper published by four European marine chemists, named Buch, Harvey, Wattenberg and Gripenberg, at about the same time that the paper by Moberg, Greenberg, Revelle, and Allen appeared. Both papers had identical conclusions. What a wonderful law firm those eight names would have made!

During this period, Dick Fleming and I undertook experiments to determine the solubility of calcium carbonate in seawater. We found, as is now well known, that ocean surface waters nearly everywhere are vastly oversaturated with calcium carbonate. (I was later able to show, as did Wattenberg, that the deep ocean waters are undersaturated because of the effect of hydrostatic pressure on the solubility.)

By this time I had become "captain" of the *Scripps*, having taken out a proper small-craft operator's license, and we ventured out beyond Catalina Island 100 miles or so on two- to three-week-long cruises. Besides Murdy Ross, we had acquired another crew member, a cook and general handyman named Frank. In spite of our primitive marine facilities we were also beginning to learn something about the deep North Pacific. Dick Fleming spent several months on the US Naval hydrographic ship *Hannibal* in 1933, and I went on a cruise off northern California in the Coast and Geodedic Survey ship *Pioneer*. In the summer of 1934 I went on another cruise, to the Gulf of Alaska and across the Pacific from the Aleutian Islands to the Hawaiian Islands, on the USS *Bushnell*, the flagship of what was then called the Submarine Force. This was a major naval exercise under the command of a rear admiral, with six submarines (the entire submarine fleet of the United States at that time) and two submarine tenders. My part of it was to take a series of oceanographic stations between the Aleutian Islands and Hawaii. The admiral and I shared a common weakness—we both tended to get seasick when *Bushnell* was pitching into a head sea. The place of minimum motion was right at the fantail, and we spent a good many hours together there. I got along well with the other officers on *Bushnell*, and her captain, later Rear Admiral A. T. Bidwell, urged me to join the Naval Reserve, which I did in 1936. From this, many consequences followed. The most important was that I spent seven years, from July 1941 to early 1948, on active duty as a reserve officer, mostly in Washington, DC, but also in the Pacific. Though I never heard a shot fired in anger, I became to a very considerable extent the Navy's oceanographer, involved (in the role of project officer) with many research and development programs in applied oceanography. [See my article "The Age of Innocence and War in Oceanography" (Revelle 1969).]

HARALD SVERDRUP FINDS A SHIP

After considerable, somewhat rude prodding by T. Willey and Jno A. Fleming, the director of the Carnegie Institution's Department of Terrestrial Magnetism, I finally finished my study of the Carnegie muds and was given a PhD degree in the spring of 1936 (Revelle 1944). Having given up on the idea that I should go away, the faculty offered me a position as an instructor at the munificent salary of $150 a month, a considerable improvement over my previous salary of $100 a month. During that summer Dr. Vaughan retired; he was succeeded by the great Norwegian physical oceanographer and geophysicist Harald Sverdrup. Unlike the common notion of Norwegians as blond vikings, Sverdrup was a small, dark, quiet man, tough as nails, but gentle at the same time. He had spent seven years frozen into the Arctic ice on Roald Amundsen's ship, *Maud*, together with four Norwegian shipmates. Amundsen's idea was to repeat Fridjof Nansen's earlier attempt on his ship, *Fram*, to drift across the polar sea with the sea ice, which was believed to move at relatively high velocity. In both cases the ice proved uncooperative, and the ships remained fairly close to where they had started. In *Maud*'s case this was in the Chukchi Sea, west of the Bering Strait. Sverdrup spent most of one year with the Chukchis, a kind of eskimo. He learned what he thought was the Chukchi language. But it turned out that the Chukchis have two languages, one for women and one for men; because he had spent most of his time in the village he had learned the women's language, causing great merriment among the Chukchi men when they came home.

Sverdrup was at that time (and probably still should be considered) the greatest oceanographer of our century. His coming to the Scripps Institution worked a transformation. Before his coming, none of the faculty at Scripps had had much training or even experience as real oceanographers.

Sverdrup was one of the leading products, together with Carl Gustaf Rossby and Jack Bjerknes, of the Scandinavian school—mostly Norwegians, but with some Swedes, Danes, and Finns mixed in—that had literally invented physical oceanography and dynamical meteorology in the modern sense of these terms.

As a young "postdoc" I would have done much better, scientifically, to stay at Scripps and work with Harald. But I was already committed to spend the year of 1936–37 in Norway at the Geophysical Institute of Bergen, headed by another famous Norwegian oceanographer, Björn Helland Hansen.

In September 1936 my wife Ellen and I and our two daughters, one a babe in arms, the other nearly four years old, sailed across the Atlantic to London on a moderately slow boat. Of course there were no trans-Atlantic planes in those days. From London we went by train to Edinburgh, where there was a general assembly of the International Union of Geodesy and Geophysics. I presented a series of papers, detailing some of our work on the distribution of nutrients and other properties in the North Pacific, and met many famous scientists, including Darcy Thompson, Harold Jeffreys, Jack Bjerknes, George Wüst, Joseph Proudman, Björn Helland Hansen (whom I had already met in La Jolla several months previously), Martin Knudsen, Walfried Ekman, Colonel R. B. Seymour Sewell, J. D. H. Wiseman, and Columbus Iselin, who was beginning his remarkable career as director of the Woods Hole Oceanographic Institution. Edinburgh at that time was a dirty gray, depressing city, filled with the acrid smell of burning coal, quite unlike its present charming face. But we were entertained splendidly by the Lord Provost, and Darcy Thompson convinced us that Edinburgh really was the "Athens of the North." I never shall forget the sight of Edinburgh Castle illuminated with searchlights, rising out of the darkness above our heads.

We sailed to Norway from Newcastle; in Newcastle we saw many pigeon-breasted, obviously malnourished young men with very poor teeth, wearing black shirts, followers of the British fascist Oswald Mosley. Those were bad days in Europe, with Adolf Hitler screaming his bloody nonsense over the radio, the on-going tragedy of the Spanish civil war, and some fascists in every country.

Bergen was nevertheless a wonderful place for us. We learned to ski and became very good friends with several of the scientists at the Geophysical Institute and their families. I didn't learn much about oceanography, but I did learn a good deal about people.

During the year we were in Bergen our little boat *Scripps* blew up while loading fuel. Frank, the cook, was killed, and Murdy Ross critically maimed. Harald Sverdrup mourned Frank's death and Murdy's injury.

But at the same time he must have thought of the accident as a blessing, for he was able to persuade Robert P. Scripps, the son of E. W. Scripps and nephew of Ellen Browning Scripps, both of whom had contributed greatly to the Institution in its early days, to buy a new ship for us. This was a Gloucester-type, two-masted, topsail schooner named *Serena*, which had belonged to the movie actor Lewis Stone. Sverdrup renamed her *E. W. Scripps* after Bob's father, cut off the topmasts but kept the other sails and rigging, and installed a more powerful diesel engine. (After seven years frozen in the Arctic ice, Sverdrup had little patience for the romantic side of sailing. During her career as our research vessel, *E. W. Scripps*'s masts became progressively shorter, her sails smaller, and her engines more powerful.) With this ship we were able to go far out to sea and to stay at sea for a month or more without refueling or resupplying. After the first step of obtaining Harald Sverdrup as Director, the acquisition of *E. W. Scripps* was the second great leap forward in the evolution of the Scripps Institution toward a genuine world oceanographic institution.

When we returned to La Jolla I found that another marine geologist, Francis Parker Shepard, then of the University of Illinois, was spending more and more time at Scripps. Though he was interested in every aspect of submarine geology, Fran was primarily a geomorphologist. He had little interest in or aptitude for theory, but he was a keen and tireless observer. By poring over hydrographic charts, crude as they were in those days, he had discovered or rediscovered the great abundance, large size, and extreme depths of the submarine canyons that cut into the continental shelves off every continent. He hoped to make detailed echo-sounding surveys of the canyons off California, collecting samples of their rocky walls and bottom sediments and measuring water currents within them, in order to understand their origin and mode of formation. During 1938 and 1939 he made a prolonged visit to La Jolla, bringing with him two graduate students from the University of Illinois, Kenneth Emery and Robert Dietz, who developed in later years into famous marine scientists, considerably exceeding the accomplishments of their teacher. More immediately important from Harald Sverdrup's viewpoint was a grant of $12,000 that Fran had obtained from the Geological Society of America—the largest grant the Society had made up to that time—which was mainly to be used to pay most of the costs of operating *E. W. Scripps* for a year. (Today one of the large Scripps vessels costs more than $10,000 *per day* at sea.) With this windfall of money, Shepard, Dietz, Emery, and I, together with several others, embarked on an extensive study of what Fran called the "Continental Borderland," the series of undersea troughs, basins, canyons, ridges, and banks, laced with islands, off southern California.

We found a great variety of bottom types, from the varved sediments of the virtually anoxic Santa Cruz Basin to rocky phosphorite on the flanks of the outer ridges. Because of our crude sampling instruments, we could learn relatively little about the underlying geology of the borderland. In contrast, today's ability to drill for considerable depths into the seafloor might make it possible to learn whether the borderland is a mosaic of shards from distant continents similar to some other parts of the California Coast Ranges.

My fondest memory of this period concerns my first close encounter with Walter Munk, my friend and colleague for nearly 50 years. Scripps had begun a summer fellowship program for undergraduates, and Walter, then an undergraduate at Caltech, was one of our first fellows. Fran Shepard was interested in measuring the bottom currents in the submarine canyon that lies northwest of Scripps. This was accomplished by anchoring a rowboat over the canyon and making current measurements from it with an instrument that had to be raised and lowered, from the bottom to the surface, about 70 meters for each reading. We divided into three teams, each making measurements for four hours in the rowboat and then off for eight hours. Walter and I formed one team. Because he was so cheerful, willing, and enthusiastic, this was a very pleasant experience. He retains those same qualities today.

In 1939 and again in 1940, *E. W. Scripps* made two cruises to the Gulf of California, the longest voyages we had ever undertaken in our own ship. The Gulf was then a wild, remote place, sparsely inhabited by a few Mexican and Indian fishermen. Harald Sverdrup led the first of these expeditions, the second was a geological cruise, led by Charles Anderson of the US Geological Survey, Fran Shepard, and me, together with J. Wyatt Durham of Berkeley, Ken Emery, Bob Dietz, and others. We found that the waters of the Gulf are very rich biologically. (Hernán Cortés in the sixteenth century had called it the "Vermillion Sea" because of the red color of the waters during springtime. We recognized that this color results from extraordinarily heavy spring blooms of phytoplankton.) There was a heavy fallout of organic remains on the seafloor during the spring and summer, including numerous tests of diatoms. In the Gulf's many deep basins, where the bottom waters were virtually stagnant and very low in oxygen, this resulted in the rapid accumulation of thickly varved, highly siliceous green and black sediments, stinking of hydrogen sulfide and rich in organic matter. We concluded that we had discovered the conditions of deposition of the famous diatomaceous shales of the Miocene Monterey Formation in California. But we did not suspect what we now believe— that the Gulf was being formed by the splitting of the Earth's crust along an embryonic spreading center.

ATOLLS AND ATOM BOMBS

In the United States, nearly all oceanographers during World War II were deeply involved in research and development related to problems of the war—development of underwater sound equipment for detecting and tracking submarines; development of devices for submarines to help them avoid anchored mines and to protect them against acoustic detection, studies of the effects of variations in ocean temperature, salinity, and bottom conditions on the behavior of underwater sound; methods for predicting surface waves in the deep sea and surf in shallow water; methods for determining shallow-water depths from aerial photographs of shallow-water surf; use of smoke to protect surface ships from kamikazes; search-and-rescue methods for downed aviators, using what was known about surface currents and the drift of life rafts; and studies of the noises made by marine mammals, fishes, and marine invertebrates that interfered with underwater acoustic devices. These various activities had brought bright graduate students, as well as more senior physicists, chemists, biologists, mathematicians, and engineers into the marine sciences, where many of them stayed after the war.

In 1946 the newly created Department of Defense decided to carry out tests of the effect of atom bombs on naval ships. This "Operation Crossroads" was to be conducted at Bikini Atoll, a remote location in the Central Pacific. One aspect of the tests was a study of the possible effects of nuclear weapons on the atoll itself and its animal and plant inhabitants. I was assigned to the staff of the joint task force commander, Admiral W. H. P. Blandy, to organize a survey of the atoll, which could serve as a baseline for evaluation of the potential effects. This task expanded to include all aspects of the oceanographic problems associated with the tests, particularly the waves that would be generated by the explosions, and the diffusion and fate of the radioactivity that would be released. Most of the oceanographers in the United States spent a good many months in 1946 at Bikini on these various tasks. John Isaacs, Jeffrey Holter, Alexander Forbes, and Allyn Vine were given responsibility for different kinds of wave measurements, ranging from automatic photography from huge towers erected on Bikini Island to rugged underwater pressure devices that were placed on the bottom of the lagoon, directly underneath the location of the bomb blast. Marston Sargent was responsible for the biologists and geologists who spent several months in the spring and summer of 1946 on the old hydrographic survey ships *Bowditch* and *Sylvania* studying the biology and geology of the atoll and the surrounding seafloor. John Lyman, then a lieutenant commander and later chief oceanographer of the old Naval Hydrographic Office, was in charge of the six small ships that were assigned to study the diffusion of radioactivity outside the atoll. Walter Munk, Gifford Ewing, W. S. von Arx, and William Ford studied the currents and diffusion within the lagoon. Kenneth Emery worked on the bottom topography and geology. He discovered a flat-topped submerged mountain, 2000 meters deep, close by the atoll, which he called Sylvania Seamount (Emery et al 1954). (The existence of this seamount, far beneath the sea surface, with its flat, wave-cut summit virtually free of sediment, next to an atoll, is a mystery that becomes harder to understand the more one thinks about it.)

One of the principal scientific consequences of Crossroads was the proof of the hypotheses of Charles Darwin and James Dwight Dana about the origin and history of coral atolls. One hundred and ten years earlier these two scientists had proposed that atolls have a sunken volcanic core that had been eroded by rain, wind, and rivers but not by ocean waves because it was first protected by a fringing coral reef and later, as it began to sink, by a barrier reef. Finally, when the volcano became completely submerged it was covered with an upward-growing platform of shallow-water coral and coralline algae perhaps thousands of meters thick. Such structures could form only where the ocean waters were warm enough for shallow-water reef-forming corals and coralline algae to live and flourish while the volcanic core sank slowly and continuously beneath the sea. Darwin and Dana supposed that high volcanic islands such as the Marquesas and Tahiti must be geologically very young and had not had time to sink beneath the waves. Darwin postulated the existence of submerged, wave-eroded—and hence flat-topped—seamounts outside the tropic seas where reef corals could not grow. We now know these as Harry Hess's guyots. The theorems about coral atolls were vigorously disputed over the succeeding century.

In 1947 I organized a resurvey of Bikini, working from my desk in the Navy's Bureau of Ships. One aspect of this resurvey was a program of drilling into the atoll, hopefully down to what we supposed to be its volcanic core. The work was done under the supervision of Harry S. Ladd and Joshua Tracey of the US Geological Survey, together with Gordon Lill, who was attached to my other office in the Geophysics Branch of the Office of Naval Research. It turned out that coring is extraordinarily difficult in the limestone material of the reef; nonetheless, Harry Ladd and his colleagues were able to find successively older strata of shallow-water coral-reef limestone down to the Eocene, some 40 million years ago. Although the volcanic core was not reached on this expedition, it was clear that the atoll was very old, and that Darwin and Dana were essentially right about its origin and history. Seismic reflection studies in 1946 had already shown a marked discontinuity at somewhat greater depths than the drill penetrated, which we assumed to represent the surface of the volcanic core. Later studies by Russell Raitt on our Mid-Pacific Expedition in 1950 showed this surface to be quite irregular, apparently eroded only by wind and running water above sea level, just as Darwin and Dana would have expected (Emery et al 1954).

THE NEW AGE OF EXPLORATION

At the end of World War II not much more was known about the floor of the deep sea than had been described in Sverdrup, Johnson, & Fleming's *The Oceans*, written in the 1930s but published in 1942. The general distribution of deep-sea sediments, basically according to the *Challenger* classification of organic oozes, red clays, and "muds," was known. But it was generally believed that the deep-sea floor was mainly a flat and featureless plain, somewhat shallower in the middle of the Atlantic—the Mid-Atlantic Ridge—and somewhat deeper around the borders of the Pacific—the Mindanao and other deep-sea trenches. One new discovery had been made—the existence of "guyots," or flat-topped seamounts at depths of several thousand meters. Harry Hess, on active duty as a reserve naval officer in command of a naval transport, had found these by watching and listening to his echo sounder as his ship crisscrossed the Pacific between Hawaii and the Philippines. Because of the prevailing notion that the oceans were a permanent feature of the Earth's surface, Harry thought these seamounts must have been islands eroded by waves several billion years ago. Their apparent sinking was probably due to the filling up of the ocean by thousands of meters of deep-sea sediments over billions of years.

In contrast to the stagnation of knowledge about the Earth beneath the sea, the development of instruments and techniques that could be used for deep-sea research had proceeded rapidly during the war. A great deal had been learned about the behavior of sound in seawater, including Maurice Ewing's discovery that a comparatively small signal, emitted at the right depth (for example, the explosion of a 1-lb block of TNT at a depth of 500 meters), could be heard over distances of thousands of kilometers. One of the most important developments was an accurate recording echo sounder that could give a continuous profile of the ocean depth along the ship's track. Sensitive, rapidly responding devices for magnetic field measurements had been developed for submarine detection. Another useful invention was a relatively simple one, the rubber O-ring, which made it possible to encase instruments in pressure-resistant cylinders that became more and more tightly sealed as they were lowered to greater and greater depths.

The wartime research and development effort had brought many physicists into research on underwater sound and other marine phenomena, using the many new developments in electronics. (In the 1930s, electronic equipment was quite unreliable; we oceanographers had a saying that there should be less than one vacuum tube per instrument.) Before the war, the Federal Government had provided only sporadic and meager support for marine science; now the Navy's Bureau of Ships and Office of Naval Research, and later the National Science Foundation, were prepared to finance oceanographic research on generous and liberal terms. And finally, many surplus naval ships were available that were capable of extended high-seas voyages at relatively modest operating costs, of the order of $500 to $800 per day at sea. The combination of circumstances soon made it possible for oceanographers and marine geologists in the United States to join in a worldwide scientific enterprise. This was nothing less than a new age of exploration, comparable in scope and intensity to the great ages of exploration of the fifteenth to the eighteenth centuries. The difference was that the new explorers concentrated on the world that lay beneath the surface of the sea.

Ships and scientists of several nations took part in this great undertaking, including Soviet oceanographers with their large, multipurpose ships—the names of Gleb Udintsev and Vladimir Kort stand out in my memory; Cambridge University geophysicists led by Teddy Bullard and Maurice Hill; and geophysicists of the British National Institution of Oceanography at Wormley, with Tony Laughton. In the United States a highly productive effort was made by Maurice Ewing, with his colleagues Bruce Heezen, Frank Press, and Joe Worzel, at the Lamont Geological Observatory. However, in this account I concentrate on what was most familiar to me, the role of the Scripps Institution near the beginning of the new age of exploration.

Three expeditions were organized by Europeans in the late 1940s: the Swedish *Albatross* expedition, led by Hans Pettersson on board an ancient freighter that had been renamed *Albatross* after Alexander Agassiz' nineteenth century exploring ship; the Danish *Galathea* expedition aboard a similar converted freighter, led by the biologist Anton Bruun; and the British *Challenger* expedition, led by the British geophysicists Tom Gaskell, Maurice Hill, and John Swallow.

The work of *Albatross* and *Galathea* was centered around the use of a single giant winch containing 12,000 meters of tapered steel wire rope, on which sampling and measuring instruments could be lowered to and retrieved from the Pacific trenches, the greatest depths of the sea. *Albatross* used this contraption to collect long cores of deep-sea sediments, employing a new type of "piston corer" invented by the Swedish oceanographer Bjorn Kullenberg. Under the right circumstances, this instrument was capable of taking cores 10 to 15 meters long, more than five times as long as any previously recovered. *Galathea* had used the great winch for a different purpose—to dredge the trench bottoms for animals and bacteria that might be living there. *Challenger*, a much smaller ship, about the same size as its nineteenth century prototype, did not have a great winch, but she used a recording echo sounder and other underwater acoustic devices to good advantage.

At the close of World War II Harald Sverdrup felt that he should return to Norway to help in the postwar reconstruction of the country. This feeling was compounded by his chagrin at not sharing the years of German occupation with his fellow countrymen. In 1948 he resigned as director of the Scripps Institution. Carl Eckart, the great theoretical physicist who had turned his attention from quantum theory and irreversible thermodynamics to the problems of underwater sound propagation in the sea, became the Scripps director, and I returned from my prolonged tour in the Navy as Associate Director.

The directorship was not really Carl's cup of tea; after a couple of years he resigned, and I became Acting Director. One of my first projects in this new position was to organize a deep-sea exploring expedition, the first ever undertaken by the Scripps Institution of Oceanography. This expedition was made possible because Sverdrup had acquired an ocean-going tug, which we called *Horizon*, from the Navy as one of the ships to be used in the California Cooperative Fisheries investigation that he had organized in 1946.

The expedition we had in mind would require two ships to carry out seismic refraction studies of the oceanic crust and the upper portions of the underlying mantle. One ship was to lie still in the water, deploying a series of cable-connected hydrophones. The other, starting about 50 miles away, would set off large charges of TNT as sound signals at regular intervals; it would come close by the listening ship and then proceed for an equal distance in the opposite direction. This work was to be done by Russell Raitt of the Scripps Institution faculty and his associates.

Captain (later Rear Admiral) Rawson Bennett was the Director of the US Naval Electronics Laboratory (NEL) in Point Loma, where we kept our ships. The NEL oceanographers were enthusiastic about participating in the proposed expedition, and it was not hard to convince Rawson that the laboratory's research vessel (which had no name but only some letters and a number—*EPCER 857*) should be the second ship for our expedition.

Our proposed track led from San Diego southwest to the equator, north to the Hawaiian Islands, then southwest to Bikini Atoll, and homeward along a track at about 40°N latitude. The operation was to be called the "Mid-Pacific Expedition," thus starting the Scripps Institution custom of giving each one of its expeditions a distinctive name.

After many months of preparation, the two ships started off bravely toward the equator in June of 1950. Among the scientists on *EPCER 857* were H. W. Menard, Robert S. Dietz, Ed Hamilton, and Robert Dill. On *Horizon* were Russell Raitt, Arthur Maxwell, Kenneth Emery, Jeff Frautschy, and I. Jim Faughn, one of the unsung heroes of the new age of exploration, was captain of *Horizon*.

After about four days at sea the clutch on one of *EPCER 857*'s engines broke down. It was strongly hinted by her officers that the ship should return to San Diego. Fortunately, *Horizon* carried enough fuel for both ships to go all the way to Hawaii and beyond, while *EPCER 857* had only enough fuel for a few days' sailing. The two ships pulled alongside each other so that the Navy ship could be refueled, but Jim was careful to provide them with only three days' worth of fuel. Thus they had no choice but to proceed with us toward the equator. For the next refueling stop three days later, he provided them with enough for five days' sailing, still not enough to reach Point Loma. Thus we limped along for some three weeks till we arrived at Pearl Harbor in Hawaii. Being familiar with the usual leisurely pace of Navy shipyards, the officers and crew of *EPCER 857*, anticipating a stay of at least two months, sent for their families. I went to see the commandant of the yard, told him that we were on a high-priority expedition for the Office of Naval Research, and pleaded with him to expedite the *EPCER 857* repairs. He promised to give the ship top priority in the yard. *Horizon* took aboard the entire scientific party of both ships and proceeded westward to an area where Harry Hess had discovered several guyots. We found that the flat-topped seamounts were the eroded summits of a giant undersea mountain range, extending westward from Necker Island in the Hawaiian chain for at least 1500 kilometers. (It was later shown that this mountain range, which we called the Mid-Pacific Mountains, extends all the way to Wake Island in the central North Pacific.)

After an echo-sounding survey of one of the first guyots, we lowered a rock dredge in order to obtain a sample of the material on the mountain summit. The dredge came up partly filled with manganese hydroxide crusts, evidently very similar in composition to the familiar manganese nodules that had been dredged by HMS *Challenger* and other early expeditions from the deep-sea floor. But among these black, irregular fragments was a white object, a fossil shallow-water coral, which Ed Hamilton, our paleontologist, was able to identify tentatively as Cretaceous in age, i.e. approximately 80 million years old. Harry Hess's idea that the guyots had lain undisturbed for billions of years while sediments accumulated in the deep sea around them was clearly wrong. The Mid-Pacific Mountains could not be much older than their flat-topped summits.

The next surprise came from Russell Raitt's seismic observations. The top layer of the seafloor, about 100 meters thick, had the expected sound velocity for marine sediments; beneath it was a material with much higher velocity that was most likely volcanic rock, and beneath that layer, one or two kilometers thick, was denser material with still higher velocity. (Russ called it the second layer.) Underneath the second layer at a depth of 4 to 5 kilometers, the sound velocity jumped abruptly to more than 8 kilometers per second. This high velocity corresponded to that found, from seismic measurements in continental areas, for the mantle material below the "Moho," the Mohorovičić discontinuity that lies at a depth of 35 to 50 kilometers or more under the continents.

The surprising result of the seismic studies was the thinness of the sedimentary layer. It was generally believed that the rate of sedimentation in the red clay areas of the deep sea that surround the Mid-Pacific Mountains was at least one millimeter per thousand years, or one meter per million years. If the sediments were only 100 meters thick, the entire sedimentary column could have been laid down in only 100 million years. The guyots and the bottom stratum of the sediments in the deep sea surrounding the mountains were of about the same age. The supposed column of sediments extending far back into Precambrian time simply did not exist.

The first successful measurements of heat flow from the interior of the Earth through the seafloor were made by Art Maxwell on the Mid-Pacific Expedition. Thousands of similar measurements have been made since 1950, but at the time we thought of these measurements as very difficult. The measuring instrument consisted of a spear containing two thermistors spaced two meters apart, connected to an electronic recording device in a sealed, pressure-proof cylinder. (The seal was formed by rubber O-rings, which we have already mentioned as one of the great wartime discoveries.) This device had been developed in the Scripps instrument shops during the previous two years by Teddy Bullard (later Sir Edward Bullard) and Art Maxwell. It was based on the fact that the temperature of the ocean water overlying the deep-sea floor is virtually constant over periods of many years, and hence the temperature gradient in the subsea clays or oozes must directly reflect the flux of heat from below, unlike the situation on land, where seasonal temperature changes near the Earth's surface mask the temperature gradient.

Art Maxwell was able to obtain six apparently satisfactory measurements with his instrument. The basic problem was to pay out enough wire rope so that the spear would sit firmly in place in the bottom mud without wobbling, and yet not to pay out so much that the wire would kink on the bottom with the accompanying danger of losing the instrument. To accomplish this we attached a weighted tripping mechanism containing a glass ball—an ordinary glass fishnet float—to the wire just below the spear, following a design made by John Isaacs. This ball broke and imploded when the instrument hit the bottom; the resulting sound signal could easily be heard on shipboard, and we could stop paying out wire.

We had a pretty good idea of the thermal conductivity of the red clay. Multiplying this value by the measured temperature gradient, we calculated that the heat flow through the seafloor was about the same as that which had been measured previously in mines and deep wells on land. This was a puzzling result, because the basaltic material that was believed to lie under the ocean was known to contain a far smaller amount of radioactive, heat-producing elements than the granitic rocks of the continental crust. Harold Jeffreys, in *The Earth*, had speculated that the heat flow through the seafloor would be not much more than 30% of the heat coming from the continental rocks.

Bullard, Maxwell, and I explained our result by proposing that the heat in the Earth's interior deep beneath the sea was carried to the ocean floor by slow convective motions of the mantle rocks (Bullard et al 1956). This conclusion seemed particularly reasonable after Art Maxwell, on a later Scripps expedition, had made a series of measurements across the East Pacific Rise to the Middle America Trench, in which he found high heat flow near the rise, progressing regularly to low flow in the neighborhood of the trench. We were not courageous enough, or perhaps not smart enough, to conclude what is now widely believed—that the rocks of the Pacific ocean floor are carried along by convective motion and the force of gravity from the ridges (where they originate as lava) to the trenches (where they are subducted as relatively cool lithospheric crust). This of course is the reason why the sediments of the deep-sea floor are so thin and the seamounts are so young—the older ocean crust has long since disappeared in the process of seafloor spreading, first described in the mid-1960s by Harry Hess and Robert Dietz. It should be pointed out, however, that Maxwell might easily have obtained just the opposite result, because we know now that hydrothermal circulation within a few hundred kilometers of the mid-ocean ridges results in both very high and low heat flows at closely spaced intervals.

After about 10 days, during which *Horizon* worked alone, we were rejoined by *EPCER 857*, and the two ships proceeded together to Bikini Atoll. Some of us camped on the beach to study reef erosion processes in the intertidal zone, while Russ Raitt and his assistants on the ships made the seismic survey I have already referred to of the volcanic rock surface underlying the reef limestone. One thing I remember about this episode was how quickly the works of man, in the form of the massive debris of

Operation Crossroads—the Quonset huts and furniture, the bulldozers and the jeeps, the refrigerators and other food containers—had rotted away under the action of the tropical Sun and the salty air or had been covered over by the lush vegetation of the atoll.

On the return voyage the two ships separated. Bill Menard and Jeff Frautschy sailed along the 40th parallel while most of the rest of us went to Hawaii and flew home. Menard and Frautschy made another great discovery. They found a huge, undersea cliff extending east–west for several thousand kilometers and more than 1000 meters high. They followed it into the California coastal waters near Cape Mendocino, where previous charts had shown a marked break in depth from south to north. They called the entire structure the Mendocino Escarpment. Later, Menard and his colleagues found half a dozen features approximately parallel to the Mendocino Escarpment, separated by several hundred kilometers from each other, which they called "fracture zones." We now know that they are the surface expressions of several of the transform faults, first explained in the late 1960s by Tuzo Wilson, that cut across the mid-ocean spreading centers.[1]

Some two years after the return of the Mid-Pacific Expedition the Scripps Institution undertook another long voyage of geologic discovery, the "Capricorn Expedition" of 1952/53. This time we used our own two ships, *Horizon* and *Spencer F. Baird*, both converted sea-going tugs with powerful engines and large fuel capacity. *Baird* was equipped with a newly constructed giant winch, similar to the one used successively by *Albatross* and *Galathea*. As with all our expeditions in those days, "Capricorn" was a multipurpose affair, participated in by geologists, geophysicists, meteorologists, biologists, and physical oceanographers. Among the scientific members of the expedition were Gustaf Arrhenius, Willard Bascom, Richard Blumberg, Milton Bramlette, Robert Dill, Rhodes Fairbridge, Robert L. Fisher, Ted Folsom, Don Hilleary, John Isaacs, Philip Jackson, Martin Johnson, Alan Jones, Ronald Mason, Arthur Maxwell, H. W. Menard, Walter Munk, J. R. Nicholson, Willard North, Russell Raitt, William Riedel, Henri Rotschi, Stanley Ruttenberg, Maxwell Silverman, Harris Stewart, Edward Taylor, Richard von Herzen, and I. Robert Livingstone was our expedition physician as well as one of the scuba divers.

The two ships first went to Bikini and Eniwetok, where they took part in the awesome first tests of fusion weapons. In November they sailed together to the Fiji Islands and thence to spend Christmas in TongaTabu, the capital island of the kingdom of Tonga. Here we were royally enter-

[1] The principal results of the "Mid-Pacific Expedition" were described in my article "The Earth Beneath the Sea—Geophysical Exploration Under the Ocean," published in what turned out to be an obscure book edited by Louis Ridenour (Revelle 1954).

tained by Prince Tungi, then the hereditary Prime Minister for his mother, Queen Salote, and now King Taufa'ahau Toupou IV of the Tonga Islands. From TongaTabu we sailed through the Tonga archipelago to American Samoa, eastward through the Cook Islands to Tahiti and the archipelago of the Tuamotus, and then to the Marquesas. And finally a long voyage home eastward across the East Pacific Rise, which was then called the Albatross Plateau, and northward across the empty equatorial Pacific. (We never saw another ship in the open sea on either the "Mid-Pacific" or the "Capricorn Expeditions.")

For the first six weeks of "Capricorn" the monster winch behaved very badly. It had been built specifically to explore the great Tonga Trench east of the islands. On our first attempt to use it in the trench we had paid out about 10,000 meters of wire when an appalling kink developed on the huge reel. We spent the next 24 hours making a long splice in the wire while 10,000 meters of it, with a Kullenberg coring device on the end, altogether weighing many tons, dangled over the side secured only by a wire clamp called a "comealong." After this frightening experience, in which *Baird*'s crew behaved with cool skill, we successfully lowered several sampling instruments down to the trench floor. But these came up scratched and battered with very little in them.

In fact, this was one of our most important discoveries about the trench—it was nearly empty of sediments. We found also that it was V-shaped in cross section. On subsequent expeditions, Bob Fisher of the Scripps staff showed that scarcity of sediments, exposures of hard rocks, and V-shaped cross sections were characteristic of most of the Pacific trenches (Fisher & Revelle 1955).

We found from careful echo-sounding that the trench was close to 10,800 meters deep, within 100 to 200 meters of the same depth as the two trenches that were then thought to be the deepest in the world, the Mindanao and Marianas trenches of the North Pacific (Fisher & Revelle 1954).[2] Apparently a depth of about 10,000 meters is about the limit for a hole in the bottom of the sea. More important was our discovery that on the seaward flank of the trench there was a flat-topped guyot, tilted downward toward the abyss. If we had had the sense to realize it, this was clear evidence of the process of subduction, which, as we now believe, slowly consumes the deep-sea floor.

The other principal geological and geophysical results of the "Capricorn Expedition" extended over a very wide area the discoveries of the "Mid-

[2] Fisher's most recent data from repeated soundings indicate that the Marianas Trench is $10,915 \pm 10$ meters deep, about 115 meters deeper than the Tonga Trench, while the maximum depth of the Mindanao, or Philippine, Trench is $10,057 \pm 5$ meters, about 750 meters shallower than the Tonga Trench.

Pacific Expedition"—particularly the thinness of the sediments overlying the bottom volcanic rocks and the high heat flow through the seafloor. As Gustaf Arrhenius, one of our shipmates, had expected from his experience on the Swedish *Albatross* Expedition, Russell Raitt's seismic survey showed that the sediments in the equatorial zones of high biological productivity were several hundreds of meters thick. This did not mean that they had been deposited over a longer time interval than 100–150 million years but simply that the rate of deposition of the siliceous and calcareous remains of pelagic organisms was much greater in the equatorial regions. In contrast, near the crest of the East Pacific Rise there was very little sediment. Also in this region, Russell Raitt's seismic refraction studies did not show the shallow "Moho" and the simple pattern of rock layers of increasing sound velocity that we had come to expect from elsewhere in the deep Pacific. This different seismic structure may be explainable, as has been suggested in recent years, by the presence of great chambers of molten or nearly molten rock underlying the crest of the rise, which serve as the source of the volcanic rock that flows out of the mid-ocean ridges and continually renews the seafloor.

One outcome of "Capricorn" was the breaking of the Scripps taboo against women on oceanographic ships. Helen Raitt, Russell Raitt's wife, had met us at the Grand Pacific Hotel in Fiji, and she found an island freighter that would take her to our next stop in TongaTabu. But after we and she arrived there, it turned out that she might have to stay for several weeks in Tonga before finding a boat back to Fiji, where she could get an airplane. The obvious solution was to ask her to join us on board *Spencer F. Baird* with Russ and the rest of us. She proved to be a first-rate shipmate, standing regular watches and keeping up all our spirits. After our return she wrote a popular account of the expedition, *Exploring the Deep Pacific* (Raitt 1964), which was translated into Russian and Japanese and widely read among oceanographers and their friends.

Victor Vacquier, who later joined our Scripps faculty, had developed during World War II a sensitive instrument for measuring variations in the Earth's magnetic field. This was adapted for oceanographic use by fitting it into a torpedo-like casing, which could be towed far enough behind a research vessel so that the ship's own magnetic field would be negligible. We towed this contraption continually on "Capricorn," obtaining, among other records, interesting but difficult to interpret maps of the magnetic fields of seamounts. Away from the seamounts the magnetic record also showed many fluctuations, presumably reflecting the remanent magnetism of the materials beneath the seafloor.

On "Capricorn," the magnetometer was the particular charge of Ronald Mason, a young geophysicist from Imperial College in the London University. He found it hard to explain the "wiggles" in the record taken along the ship's track. He supposed that a two-dimensional map of the remanent magnetism would be easier to understand.

After the return of "Capricorn," Ron found that the Coast and Geodetic Survey ship, *Pioneer*, was about to undertake a topographic survey of the seafloor off the California and Oregon coasts out to several hundred miles. He proposed to me that he should go along on this survey with his towed magnetometer in order to make his dreamed-of magnetic map. He estimated that the total cost would be about $100,000. I tried my best to find this money in Washington, DC, but the magnetic experts of the US Geological Survey and other Washington scientific bureaus thought that Ron's proposed map would be useless and advised against any funds for the project. In the end, I had to use my Director's Contingency Fund to pay for the magnetic survey and the working up of the data.

Ronald produced a map showing the magnetic striations on the seafloor, which we now know represent a time series of reversals of the Earth's magnetic field. He found that the striations were displaced along Bill Menard's fracture zones. This result was later confirmed by Victor Vacquier and Arthur Raff, who found an apparent lateral displacement of fourteen hundred kilometers along the Mendocino Fracture Zone. These apparent displacements represent Tuzo Wilson's transform faults, and the entire ocean floor beyond the continental slopes exhibits magnetic striations like those first discovered by Ronald Mason.

The "Mid-Pacific" and "Capricorn Expeditions" were the first stages of the program of oceanwide exploration of the Earth beneath the sea undertaken by Scripps scientists over the following 25 years. During this period our ships sailed nearly two million miles through the Indian, Pacific, and Atlantic oceans, from the Arctic to the Antarctic, in more than 25 major geological/geophysical expeditions to study the seafloor and what lies beneath it. H. W. Menard, R. L. Fisher, Victor Vacquier, Arthur Raff, Russell Raitt, and George Shor were the leaders of this effort, but many others participated. As for me, "Capricorn" was my last long voyage. After 1953 the problems of directing Scripps, and at the same time trying to create a new univeristy, took all my time.

Literature Cited

Bullard, E. C., Maxwell, A. E., Revelle, R. 1956. Heat flow through the deep-sea floor. *Adv. Geophys.* 3: 153–81

Eckis, R. P. 1928. Alluvial fans of the Cucamonga District, southern California. *J. Geol.* 36: 224–47

Emery, K. O., et al. 1954. Geology of Bikini and nearby atolls, Marshall Islands. *US Geol. Surv. Prof. Pap. 260.* 265 pp., 64 pl.

Fisher, R. L., Revelle, R. 1954. A deep sounding from the Southern Hemisphere. *Nature* 174: 469–70

Fisher, R. L., Revelle, R. 1955. The trenches of the Pacific. *Sci. Am.* 193: 36–41

Moberg, E. G., Greenberg, D. M., Revelle, R., Allen, E. C. 1934. The buffer mechanism of sea water. *Scripps Inst. Oceanogr. Bull.* 3: 231–78.

Raitt, H. 1964. *Exploring the Deep Pacific.* Denver: Sage. 272 pp. 2nd ed.

Revelle, R. 1944. Marine bottom samples collected in the Pacific Ocean by the *Carnegie* on its seventh cruise. *Carnegie Inst. Washington Publ. No. 556.* 180 pp.

Revelle, R. 1954. The earth beneath the sea—geophysical exploration under the ocean. In *Modern Physics for the Engineer*, ed. L. Ridenour, pp. 306–29. New York: McGraw-Hill

Revelle, R. 1969. The age of innocence and war in oceanography. *Oceans* 1969 (Jan): 6–16

THE CORE-MANTLE BOUNDARY

Christopher J. Young and Thorne Lay

Department of Geological Sciences, University of Michigan, Ann Arbor, Michigan 48109

INTRODUCTION

The largest compositional discontinuity within the Earth is the core-mantle boundary (CMB), at a depth of 2889 km. This boundary and the adjacent transition zones in the lowermost mantle and outermost core play a critical role in the Earth's thermal and chemical evolution. Estimates of the heat flux out of the core indicate that the 200-km-thick D'' region at the base of the mantle is a major thermal boundary layer. The D'' region, like its counterpart boundary layer in the lithosphere, is laterally heterogeneous and possibly chemically stratified. This complexity obscures the exact dynamic behavior of the D'' region. However, it is clear that the large density contrast (4.3 g cm^{-3}) across the CMB provides favorable conditions for density stratification on both sides of the boundary. The D'' region may be the refuse pile for "heavy" mantle heterogeneities, while the outermost core may have a concentration of "light" components. Boundary-layer instabilities in D'' have been invoked as sources of both localized mantle plumes and major upwelling currents in whole-mantle convection systems. Dynamically supported topography on the CMB and lateral temperature gradients in D'' may produce coupling between the core and mantle, influencing the excitation of the geodynamo. The outermost core is generally believed to be radially and laterally homogeneous, although chemical stratification has not been ruled out. Seismological, geodynamical, and geomagnetic investigations are unveiling the three-dimensional complexity of the deep Earth in the vicinity of the CMB.

SEISMOLOGICAL CONSTRAINTS

Several geophysical disciplines provide information about the CMB; however, seismological investigations continue to provide the best resolution of the region. A variety of seismic phases have been used to investigate the density and elastic velocity structure, anelastic properties, and the lateral heterogeneity of the lowermost mantle and outermost core. However, the history of seismological models, particularly for the D'' region, has been a checkered one, with many incompatible results being published. Present-day quantitative modeling procedures are yielding somewhat more consistent models, although there is still no consensus on the detailed characteristics of the D'' region. A new generation of three-dimensional models that explicitly include lateral heterogeneity may reconcile the outstanding inconsistencies. In this review, contemporary seismological models for the CMB are considered first and then integrated with constraints from other disciplines. Particular emphasis is placed on the evidence for a thermal or chemical boundary layer at the base of the mantle.

Radial Models Based on Body-Wave Travel Times and Free Oscillations

While recognizing the importance of lateral heterogeneity near the CMB, we still find it useful to consider radially symmetric "average" models for the region. The average properties, such as velocity gradients in D'', are important boundary conditions for thermal modeling of the CMB. The classic procedure for determining radially symmetric velocity models is inversion of large data sets of body-wave travel times. As early as 1939, seismic velocity models (Gutenberg & Richter 1939, Jeffreys 1939) indicated that the P- and S-wave velocity gradients in the lowermost mantle are anomalously diminished relative to the overlying mantle. The primary evidence for this is changes in the slopes of the travel-time curves at a distance of about 95°. In an important review, Cleary (1974) compared the early velocity models for D'' based on travel-time inversions and showed that most of the models suggest decreases in the velocity gradients about 200 km above the CMB. Some models, notably that of Gutenberg, actually predict decreasing velocities with depth. The most recent Earth model, PREM (Dziewonski & Anderson 1981), based on global travel-time and free oscillation measurements, shows similar smooth decreases in velocity gradients above the core (Figure 1). These decreases in velocity gradients above the CMB led Bullen (1949) to define the D'' region, which he concluded was an inhomogeneous zone. Such inhomogeneity could result from either compositional changes or nonadiabatic thermal structure within D''. Nonadiabatic conditions could arise as a result of the presence of a thermal boundary layer, which requires a large heat flux from the core into the mantle. Alternatively, compositional stratification or a phase change could account for the seismic velocity behavior, although a smooth model such as PREM would require gradational changes.

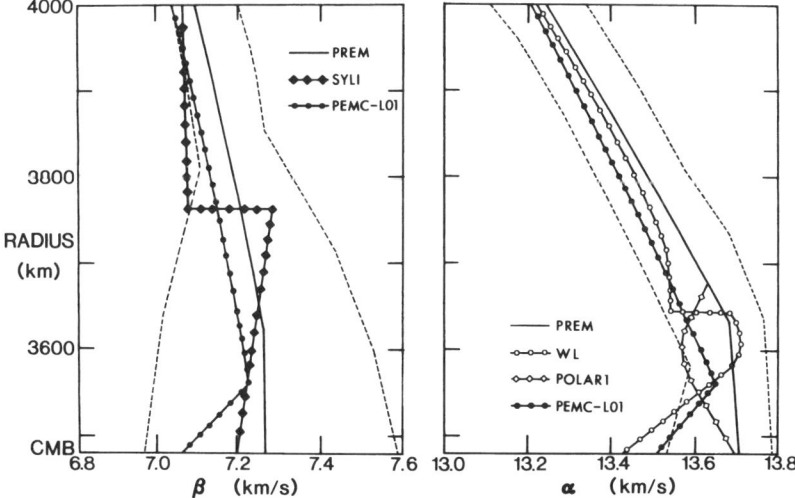

Figure 1 Recent S-wave (*left*) and P-wave (*right*) velocity models proposed for the D'' region; PREM (Dziewonski & Anderson 1981); PEMC-L01 (Doornbos & Mondt 1979b); SYLI (Young & Lay 1986); WL (Wright & Lyons 1981); POLAR1 (Ruff & Helmberger 1982). The dashed lines are the extremal velocity bounds from Lee & Johnson (1984a,b); bounding velocity values are consistent with travel-time inversions for radially symmetric models.

Models for the D'' region based on travel-time inversions alone have limited resolution for several reasons. Foremost among these is that the travel-time slope, $dT/d\Delta$, is nearly constant at distances beyond about 92°. A constant slope is expected for either a diffracted wave (traveling parallel to the CMB) or for a velocity structure with a critical velocity gradient, $dv/dr = v/r$. In either case travel-time inversions are unreliable. Frequency-dependent effects due to diffraction are not accounted for in ray-theory inversions, leading to potential biases in the resulting structure. A critical velocity gradient over a 100-km thickness of D'' requires a P velocity decrease of 0.4 km s^{-1} or an S velocity decrease of 0.2 km s^{-1}. In regions of the Earth where seismic velocities decrease with depth, travel-time inversions tend to be unstable. Direct measurement of the travel-time slopes using array techniques should improve the accuracy of the travel-time inversions; however, typical array measurements (e.g. Chinnery 1969, Johnson 1969, Corbishley 1970, Wright & Cleary 1972) show substantial scatter at distances beyond 85°, which limits their usefulness in global inversions [see Cleary (1974) for a summary of such measurements]. An indication of the limitations of travel-time inversions for resolving D''

velocity structure is given by the extremal bounds in Figure 1, taken from a study of 108,000 P-wave travel times and 75,000 S-wave travel times by Lee & Johnson (1984a,b). These bounds indicate the range of radially symmetric Earth models consistent with the travel-time observations. The average bound widths in the central mantle are 0.13–0.14 km s^{-1}. The splaying of the bounds in D'' indicates both increased variation in the data and decreased resolution of the inversions. These bounds do not necessarily bracket lateral velocity variations. Part of the increase in the S-wave velocity bound widths can be attributed to the fact that at distances greater than 80° the core-traversing phase SKS arrives ahead of direct S and contaminates the travel-time measurements. As is discussed later, relaxing the requirement of a radially symmetric model allows further information about D'' to be obtained from large travel-time data sets.

The seismic properties of the outermost core are also difficult to extract from travel-time measurements. The principal reason is that the core behaves as a liquid for seismic wave frequencies and thus does not transmit shear waves, while the P-wave velocities are much lower than in the lower mantle, producing a major low-velocity zone. The resulting downward deflection of P waves crossing the CMB prevents PKP-type phases from bottoming in the outermost core; therefore, these phases only provide integral constraints on the structure. The P velocities in the outer core are fortunately higher than the fastest S-wave velocities in the mantle, so that phases such as SKS and $SKKS$ do turn in the outermost core. However, these phases are often difficult to time accurately, and most travel-time studies extrapolate the outermost core velocities from the directly determined velocities at greater depths. None of the travel-time studies have resulted in particularly complex outermost core P velocity structures. The P velocity just below the CMB has been estimated by SKS travel-time studies to lie between 7.9 (Hales & Roberts 1971) and 8.1 km s^{-1} (Jeffreys 1939), a range that spans the extremal bounds on outer core velocities found by Johnson & Lee (1985). The low-velocity estimate of Hales & Roberts (1971) results in a steep outermost core velocity gradient, suggestive of a transition zone mirroring the D'' region.

In addition to travel-time data, observations of free oscillations, particularly higher modes, have been used to develop radial models for the CMB region. While intrinsically insensitive to fine velocity characteristics, free oscillations do provide uniform global averaging that is difficult to attain even with large International Seismological Centre travel-time data sets. Utilization of the normal modes to determine deep Earth structure is complicated by strong trade-offs between density and shear velocity structure and depth of the CMB (Masters 1979). However, joint analyses of travel-time and free oscillation measurements have helped to constrain

D'' properties. Differential travel times $PcP-P$ and $ScS-S$ (Engdahl & Johnson 1974, Jordan & Anderson 1974) are particularly useful in conjunction with normal-mode measurements, since they place relatively tight constraints on the depth to the CMB. The currently preferred depth is 2889 ± 4 km (Dziewonski & Anderson 1981), corresponding to an outer core radius of 3482 ± 4 km, although this value is dependent upon the accuracy of the mantle velocity models [See Dziewonski & Haddon (1974) for a review of this problem.] The velocity models for D'' obtained by joint inversion of travel-time and free oscillation measurements tend to result in positive velocity gradients that are usually slightly diminished relative to the shallower mantle (Dziewonski & Gilbert 1972, Jordan & Anderson 1974, Dziewonski et al 1975, Gilbert & Dziewonski 1975, Anderson & Hart 1976, Dziewonski & Anderson 1981). The limited resolution of the combined data sets is indicated by the fact that models with strong negative velocity gradients in a thin zone above the CMB cannot be ruled out (Jordan & Anderson 1974). It is not possible to unambiguously distinguish between thermal and compositional explanations for D'' inhomogeneity using normal-mode data (Masters 1979). The velocity models of the outermost core derived by the same joint inversions tend to agree well with the travel-time models, having P velocities just below the CMB of 7.98–8.06 km s^{-1}. The PREM model has a steep velocity gradient of 0.17 km s^{-1} per 100 km at the top of the outer core, compared with a gradient of about 0.12 km s^{-1} per 100 km at a depth 600 km below the CMB. This decrease in gradient, albeit only marginally resolved, is evidence for inhomogeneity of the outermost core. As is discussed later, a region of thermally stratified, stable outermost core has been suggested on other grounds, and this possibility cannot be ruled out by the travel-time and normal-mode data (Masters 1979). However, Choy (1977) performed detailed waveform modeling of S_nKS phases, which is the most accurate procedure for modeling the outermost core, and he found that a relatively low gradient, like that in Jeffreys' model, fits the data best. More studies of this type are needed to resolve this issue. In order to place tighter constraints on the velocity structure above the CMB, detailed analyses of a variety of body-wave amplitudes and waveforms have been performed.

Radial Models of D'' Based on Body-Wave Analyses

Many of the early attempts to constrain the properties of the CMB used core-reflected phases PcP and ScS. In theory, by determining the reflection coefficients of PcP and ScS, bounds on the density and P-wave velocity contrast, as well as the outer core viscosity, can be determined (Kanamori 1967a,b). However, in practice, substantial unexplained amplitude variations of the short-period phases have precluded definitive results. In general, the efforts to interpret the unstable short-period PcP observations led to complex transition zones with strong decreases in velocities above the core (Buchbinder 1968a,b, Ibrahim 1971, 1973, Berzon et al 1972, Buchbinder & Poupinet 1973). Spectral studies of ScS have suggested finite outer core rigidity (Sato & Espinosa 1967, Suzuki & Sato 1970). However, the scatter of the data and the intrinsic instability of the spectral techniques applied in the early studies have weakened confidence in these analyses (Frasier & Chowdhury 1974). Quantitative waveform modeling of more stable long-period phases also gives mixed results, with PcP data suggesting a mild negative gradient in D'' (Müller et al 1977) and ScS data requiring at least some regions either to have a strong positive gradient or to be anisotropic (Mitchell & Helmberger 1973, Lay & Helmberger 1983b). The core reflections also intrinsically sample restricted regions, so it is difficult to develop radially averaged models.

Perhaps the most straightforward analysis of D'' velocity structure involves body waves that are diffracted along the CMB and observed at distances in the core shadow (>95°). These phases usually have long pathlengths in D'', providing laterally averaged (but not necessarily globally representative) structures. Many of the studies of these phases have involved simply measuring the apparent velocity of the diffracted signals. In conjunction with an estimate of the core radius, a geometric optics interpretation of the waves yields a direct estimate of the velocity at the base of the mantle. This procedure typically results in quite low velocities at the CMB for both P waves (Sacks 1967, Bolt 1970, 1972) and S waves (Cleary 1969, Bolt et al 1970), requiring negative velocity gradients in D''. However, this simple analysis may lead to incorrect models for two reasons. First, diffraction is a frequency-dependent phenomenon that cannot be rigorously described by geometrical ray theory. Second, the presence of a low-velocity zone at the base of the mantle could simply refract energy to large distances, where it may mistakenly be interpreted as having been diffracted. In either case, the signals are sensitive to velocity structure throughout the D'' region, rather than only at the CMB.

Frequency-dependent effects of the diffracted signals are most readily accounted for by synthesizing waveforms for a given Earth model. This allows the frequency-dependent amplitude effects of diffraction as well as the travel-time measurements to be exploited in determining the D'' structure. Since the pioneering study by Alexander & Phinney (1966), it has been standard practice to analyze the amplitude decay coefficients. Diffraction theory [see Chapman & Phinney (1972) for a review] predicts that the amplitude decay into the shadow zone is directly related to the velocity gradients in D''. Positive velocity gradients produce faster rates of amplitude decay, and high-frequency signals decay much more rapidly than long periods. While long-period signals are less sensitive to D'' structure, they are less influenced by other propagation effects, so most quantitative studies have analyzed the amplitude decay of long-period (8–64 s) P waves (Alexander & Phinney 1966, Sacks 1966, Phinney & Cathles 1969, Chapman & Phinney 1972). More recent studies of larger data sets of both long-period P and SH waves by Mondt (1977), Doornbos & Mondt (1979a,b), Mula & Müller (1980), Mula (1981), and Doornbos (1983) have presented radially symmetric models for D'', but despite similarities in the procedures, these studies have inconsistent results. The models preferred by Doornbos & Mondt (1979b) are shown in Figure 1, where the PEMC model (Dziewonski et al 1975) has been modified to have negative velocity gradients for both P and S waves in the lowermost 75 km of the D'' region. Mula (1981) prefers a model with slightly positive gradients that is essentially the same as PREM. These studies have shown that the S-wave decay coefficients differ substantially from the P-wave coefficients, and that in the long-period band it is principally the S-wave observations that require decreased or negative velocity gradients. The major explanation for the inconsistencies in the models is that the S-wave decay coefficients, as well as the P-wave decay coefficients at the shortest periods, show substantial scatter. The various authors have chosen different ways of weighting and linearly combining the scattered values in determining a "best" model. The very presence of the large scatter in the observations casts doubt on the validity of the resulting models as global averages. These studies explored a limited class of smoothly varying structures, and the evidence for radial discontinuities discussed below indicates that more complex models may need to be considered.

The recent quantitative modeling of diffracted phases has shown that the diffracted ray parameters are more sensitive to average velocity levels in D'' than to velocity gradients (Okal & Geller 1979, Mula & Müller 1980). Synthetic models for long-period body waves also indicate that measured apparent velocities can be lower than the true CMB velocities by as much as 0.4 km s^{-1} for P waves and 0.12 km s^{-1} for S waves as a result of the effects of dispersion. Thus, even the PREM model may be consistent with the early measurements of low apparent velocities of diffracted waves. However, the evidence for strong dispersive effects in the diffracted waves has been disputed by Doornbos (1983) and Bolt & Niazi (1984). Doornbos (1983) found that for SH waves, low-frequency dispersion is close to zero, indicating a near critical velocity gradient. Bolt & Niazi (1984) argue that impulsive, high-frequency onsets of the diffracted waves are usually picked in determining the travel-time slopes, and the high frequencies should "see" the deepest part of D''. Both arguments have merit and suggest that short-period diffracted signals hold the key to determining D'' velocity structure.

The amplitude decay of diffracted short-period P-wave signals has been studied by several workers (Sacks 1966, Carpenter et al 1967, Bolt 1972, Ansell 1974, Booth et al 1974). These studies had limited success because of the large amount of scatter in the P-wave data. This scatter can be only partially attributed to the D'' region, since receiver and source effects are known to be very strong. Ruff & Helmberger (1982) devised an experiment to suppress the shallow effects by taking the ratio of amplitude patterns at North American stations for events at several Soviet test sites. They found an abrupt onset of short-period amplitude decay near 95° and a subsequent, rather steep falloff with distance that they attribute to a localized positive velocity gradient at the base of D''. One of the proposed models (POLAR1) that fits these data is shown in Figure 1. Doornbos (1983) and Bolt (1972) have presented observations of high-frequency P-wave arrivals at large distances into the shadow zone that would not be consistent with a positive gradient of this type, which suggests that such a model is only appropriate for a limited portion of D''. Recently, Ruff & Lettvin (1984) have observed very different amplitude behavior for short-period P waves sampling other regions of D'', which indicates that lateral variations are in fact significant.

A fundamentally different class of models based on short-period P waves has been suggested in a series of detailed array studies (Wright 1973, Wright & Lyons 1975, 1981, Wright et al 1985). In these studies, P-wave apparent velocity measurements between 75 and 95° have been interpreted as resulting from a lower-mantle triplication produced by a sharp 1.5–3% P-wave velocity increase about 180 km above the CMB. Figure 1 shows one of the proposed models (WL; Wright & Lyons 1981), which has a strong negative velocity gradient below the 1.5% discontinuity. This model differs dramatically from the other P-wave models proposed for D'', but a 1.5% discontinuity could easily go undetected by travel-time, long-period diffracted wave, and free oscillation studies. The detection of such a discontinuity is very complicated because the associated changes in ray parameter around 85° are small, and the later branches of the triplication are difficult to recognize because of strong interference with the first arrivals. The possibility of crustal contamination of the array measurements has prevented general acceptance of the P-wave velocity discontinuity model, and an array study by Schlittenhardt (1984) has failed to find supporting evidence for this model.

The evidence for a velocity discontinuity in D'' has been strongly bolstered by the studies of Lay & Helmberger (1983a), Zhang & Lay (1984), and Young & Lay (1986). These investigations have established that in four separate regions of the lowermost mantle (beneath Eurasia, Alaska, Central America, and India), a 2.75% shear velocity discontinuity exists about 280 km above the core. Unlike the evidence for the *P*-wave discontinuity, the *S*-wave structures (an example of which is shown in Figure 1) are based on long-period observations in the distance range 70 to 95°. The observations show a systematic arrival preceding the core reflection, *ScS*, throughout this range. This arrival results from a triplication produced by the abrupt velocity increase, the size of which was constrained by detailed waveform modeling. The relatively low *S*-wave velocities in D'' separate the triplication arrivals more than for the *P* waves, enabling the use of stable long-period signals in the analysis. The long-period data that were modeled are not particularly sensitive to the velocity gradient below the discontinuity (Lay & Helmberger 1983b), but analysis of diffracted waves in the presence of such a discontinuity strongly favors the presence of at least a mild negative velocity gradient (Lay 1985). The discontinuities in the proposed *P*- and *S*-wave models are separated in depth by about 100 km, but it may be possible to reconcile the models if they are constrained to have similar velocity gradients below the discontinuities. It appears that a global *P*-wave velocity discontinuity of greater than 1.5% is incompatible with some diffracted *P* observations (Schlittenhardt et al 1985); however, a smaller discontinuity cannot be detected with these phases. Alternate interpretations of the *S*-wave triplication arrivals as results of *SKS* scattering, receiver reverberations, or source multipathing have been ruled out by detailed analysis (Lay & Young, 1986, Lay 1986). The existence of a globally stratified D'' would have profound impact upon the thermal and compositional models for the CMB, so intensive efforts are needed to establish the validity and extent of these structures. In particular, it is important to appraise the significance of these models in the light of the evidence for strong lateral heterogeneity in D''.

Lateral Heterogeneity of D'' *and the Outermost Core*

Radial characterizations of the average velocity structure near the CMB are required for thermal, chemical, and dynamical analyses of the region. However, the previous discussion and the perplexing array of recent radial seismic models shown in Figure 1 do not provide a unified model. In fact, the significance of all of these models is placed in question because of clear evidence for lateral heterogeneity in D''.

Travel-time studies provided the first indication of large-scale lateral heterogeneity in D'' (e.g. Chinnery 1969, Julian & Sengupta 1973). With the accumulation of large travel-time data sets, it has been possible to invert directly for the three-dimensional configuration of the lower-mantle velocity variations (Sengupta & Toksöz 1976, Dziewonski et al 1977, Sengupta et al 1981, Clayton & Comer 1983, Dziewonski 1984). Each of these global inversions has indicated the presence of greater heterogeneity in D'' than in the overlying mantle. These large travel-time data sets cannot resolve detailed velocity layering but do extract the long-wavelength component of the velocity variations in D''. In a recent analysis of 500,000 travel-time residuals, Dziewonski (1984) found that the low-order (degrees 2 to 6) spherical harmonic components of the heterogeneity have 1.0–1.5% velocity variations in D'', three to four times greater than variations in the central mantle. This is considered to be a lower bound on the actual range of variations, given the smoothing effect of the low-order expansion. In the higher-resolution models (Clayton & Comer 1983), some of the heterogeneities extend upward from the CMB into the mantle in a manner suggestive of rising plume structures.

While the global inversions indicate the presence of very long wavelength velocity variations in D'', many scales of heterogeneities are indicated by different seismic waves. Long-period diffracted signals indicate coherent large-scale variations in the velocity gradients in D'' (Alexander & Phinney 1966, Bolt & Niazi 1984) that may account for the scatter found in diffracted S-wave decay coefficients. Diffracted short-period P waves indicate similar lateral variations (Ruff & Lettvin 1984). In numerous studies, short-period precursors to PKP phases have been attributed to lateral heterogeneities in D'' with scale lengths of 10 to 150 km and mean velocity variations of about 1% (Cleary & Haddon 1972, Doornbos & Vlaar 1973, Haddon & Cleary 1974, King et al 1974, Wright 1975, Husebye et al 1976). Comparison of the $PKPab$ and $PKPdf$ branches led Sacks et al (1979) to conclude that various portions of D'' produce different amounts of focusing and defocusing. Strongly scattering regions have scale lengths of 150 km with 1% mean velocity changes, whereas other regions have much smaller lateral variations. Haddon (1982) suggests that the heterogeneity has both short (10–20 km) and longer (500–1000 km) scale lengths, with the large-scale features concentrated in preferred directions.

Some of the scattering attributed to D'' heterogeneity may instead be due to topography on the CMB itself. Doornbos (1978) showed that the PKP precursors may be explained by topography of a few hundred meters. Precursors to $PKKP$ phases and off-azimuth anomalies of $PKKP$ have been attributed to backscattering from the underside of a rough CMB (Doornbos 1974, 1980, Chang & Cleary 1978, 1981). The variability of the $PKKP$ precursors is indicative of considerable lateral variation in the topography on the CMB. Greater relief of up to 5 km has been suggested by recent three-dimensional models of the outer core obtained by inverting large PKP travel-time data sets (Creager & Jordan 1986, Morelli & Dziewonski 1986). The amount of inferred topography trades off with the degree of heterogeneity attributed to the D'' region; however, comparable excess ellipticity of the CMB may also be required to account for anomalous splitting of normal modes that are sensitive to the outermost core (Ritzwoller et al 1985). Any topography on the CMB can be sustained only by dynamic processes because of the large density contrast between the core and mantle, and hence such topography may be crucial to understanding these processes. It may also affect the interpretation of core reflections and diffracted waves.

Lateral variations in the outermost core are difficult to detect for the same reasons that radial models have limited resolution. In general, however, the outer core has usually been considered to be laterally homogeneous, principally on the basis of the small scatter of travel times of $PmKP$ phases (Engdahl 1968, Buchbinder 1972). In a recent global travel-time inversion, Creager & Jordan (1986) suggested substantial lateral heterogeneity of the outermost core, but this is a preliminary result. As a result of the low viscosity of the outer core, it is unlikely that lateral heterogeneity of sufficient magnitude to be observed seismically can be sustained, even dynamically.

Seismic Constraints on Anelastic Properties Near the CMB

Anelastic processes in the Earth are thermally activated; thus, a detailed model of the attenuating properties near the CMB would help to resolve whether thermal or compositional effects dominate in this region. Many studies have suggested that anomalously high attenuation of seismic waves occurs in the D'' region, while the outer core appears to transmit seismic body waves with almost no anelastic loss. Radial models of the quality factor Q (which is inversely proportional to the amount of anelastic loss) based on normal-mode data tend to have low-Q values in D'' relative to the rest of the deep mantle (Anderson & Hart 1978a,b, Anderson & Given 1982); however, it is not clear that the mode data alone require this low-Q zone (Sailor & Dziewonski 1978, Masters & Gilbert 1983). Anderson & Given (1982) presented a frequency-dependent absorption-band model for Q variations in the Earth in which the absorption band in the D'' region shifts to higher frequencies in order to match relatively high Q values for the low-order spheroidal modes $_0S_2$ to $_0S_4$. This model predicts low-Q values in the range 100 to 150 for periods less than 500 s, with increasing Q for longer periods.

The question of whether body-wave observations are consistent with the presence of a low-Q zone in D'' has been strongly contested. Many early studies indicated the presence of a low-Q zone at the base of the mantle by spectral ratio measurements of body waves (Mikumo & Kurita

1968, Teng 1968, Shore 1984); however, these studies have not convincingly accounted for frequency-dependent effects due to diffraction. Separating the effects of Q and diffraction is difficult, because both strongly modify the spectral content, and accurate correction for the diffraction effects requires detailed knowledge of the D'' velocity structure. Doornbos & Mondt (1979a) considered this problem in detail and concluded that the observed S-wave spectral decay coefficients are inconsistent with a thick low-Q zone at the base of the mantle, although a thin zone cannot be excluded. Mula (1981) explored this issue further, finding that Q values of less than about 250 lead to P-wave amplitude decay predictions that are inconsistent with the data. Velocity models with positive velocity gradients can match the observations well with no attenuation at all (i.e. infinite Q). If the Q structure in D'' is similar to that in the overlying mantle (Q values near 300–500), models with slightly negative velocity gradients in D'' will fit the data best. Recent high-frequency body-wave studies have also failed to detect any low-Q zone in D'' (Ruff & Helmberger 1982, Choy & Cormier 1986).

Body-wave studies of the outer core indicate high-Q values of 5000 to 10,000 (Qamar & Eisenberg 1974), 3000 to 10,000 (Sacks 1969), 4000 (Buchbinder 1971), and 10,000 (Cormier & Richards 1976). Q values of several hundred were proposed by Suzuki & Sato (1970), but these results may be in error as a result of the spectral procedure employed and the assumption of frequency independence of the reflection transmission coefficients of SKS. There has been no clear indication of radial variation of Q in the outermost core.

GEODYNAMIC, GEOMAGNETIC, AND COMPOSITIONAL CONSTRAINTS ON THE CMB

While seismology directly probes the velocity structure near the CMB, other geophysical studies place general constraints on the region. It is important to recognize that many of these studies explicitly utilize inferences from seismology; therefore, the ambiguity in the current seismic models must affect the reliability of the conclusions drawn from other disciplines.

Evidence for a Thermal Boundary Layer

The general decrease in velocity gradients in the D'' region has often been attributed to the presence of a thermal boundary layer. A thermal boundary layer would exist if there is a strong contrast in temperature between the lower mantle and the outer core, which would lead to heat flux out of the core. The existence of such a thermal boundary layer is of paramount importance to models of lower-mantle dynamics because of the difference in convection geometries resulting from heating from below versus internal heating only. Systems with heating from below tend to have much stronger upwelling thermal plumes arising from boundary-layer instabilities. Unfortunately, estimating the heat flux out of the core is fraught with uncertainty, as is estimating the temperature at the base of the mantle (Jeanloz & Richter 1979). The calculations require assumptions about whether whole-mantle convection occurs, as well as whether the D'' region and the outermost core are stratified, all of which are unresolved problems.

Several calculations do predict a strong temperature contrast of 650 to 1300° across the CMB, which would produce a major thermal boundary layer (Verhoogen 1973, Jones 1977, Jeanloz & Richter 1979, Stacey & Loper 1983, Zharkov 1985). These estimates indicate boundary-layer thicknesses of 75 to 100 km over which seismic velocity gradients would be expected to decrease and possibly become negative (Doornbos 1983). If the D'' region is dynamically separate from the overlying mantle, a second thermal boundary layer may exist at the top of D'', although all of the estimates of deep Earth temperature contrasts may be in error if this is the case (Jeanloz & Richter 1979). If the D'' region is not dynamically stratified, the boundary layer should have reduced viscosities at the base, giving rise to strong horizontal flow and thermal plumes of relatively small dimensions (on the order 20 km in diameter) that would rise into the mantle (Loper & Stacey 1983, Stacey & Loper 1983, Loper 1984, Zharkov et al 1985). The time scale for the growth of these instabilities is on the order of 1 Myr (Yuen & Peltier 1980). The localized thermal plumes may be entrained in a larger-scale lower-mantle circulation (Boss & Sacks 1985). If viscosity in the Earth increases with depth, the surrounding large-scale circulation may overturn slowly, while the small-scale plumes will ascend rapidly. If D'' includes patches of material that are too dense to rise, the convective overturn in D'' may be irregularly distributed around them. In this case the D'' region may resemble the lithosphere, where oceans and continents have very different participation in the upper-mantle convective system (Jordan 1979, Doornbos et al 1986). The strong horizontal shear flow at the base of D'' predicted by the thermal boundary layer calculations may result in observable seismic anisotropy (Lay & Helmberger 1983b, Doornbos et al 1986).

The three-dimensional heterogeneity structures for the lower mantle obtained by global travel-time inversions indicate a complex configuration of velocity and, presumably, density variations. Over the long time scales operating in the Earth, any density heterogeneity must drive viscous flow. The resulting large-scale flow in the deep mantle should produce dynamically supported topography on the CMB with a total excursion of about 3 km (Hager et al 1985). Larger relief may exist in regions with stronger heterogeneity.

Geomagnetic Constraints on the Outermost Core Structure

The Earth's magnetic field is produced by convective motion in the outer core; hence, it is plausible that some characteristics of these motions should be reflected in measurements of the magnetic field at the surface of the Earth. Several efforts have been made to extract information about the actual radial convection currents in the outermost core by analyzing the secular variation of the magnetic field (Roberts & Scott 1965, Backus 1968, Benton et al 1979, Whaler 1980, 1982, Gubbins 1982, 1983, Le Mouël et al 1985, Gire et al 1986); however, Backus (1982) has applied singular perturbation theory to demonstrate the nonuniqueness and limited resolution of this technique. Even if the outermost core is actively convecting, the very low viscosities should not allow significant lateral heterogeneity that would be seismically observable to persist. The core convection will also maintain the CMB at a nearly constant temperature.

A thermally stratified outermost core, which would inhibit large-scale radial motions, has been proposed on the basis of various thermal and chemical arguments (Higgins & Kennedy 1971, Gubbins et al 1982). This stable layer may serve as a reservoir for light materials segregated out as the inner core grows. As discussed previously, the seismological evidence for such a stabilized zone is marginal, and efforts to directly observe inertia-gravity waves due to such a region have yielded ambiguous results (Masters 1979, Yukutake 1981, Crossley 1984).

Correlation of changes in the geomagnetic and gravitational fields with changes in the length of day and minima in the Earth's rotation rate suggests that there is a coupling between the outer core and mantle. The coupling is thought to be due to either of two causes: electromagnetic coupling caused by motions of the convective systems on either side of the CMB, or topographic coupling caused by irregularities on the CMB (Hide 1969, Anufriyev & Braginski 1975, Moffatt & Dillon 1976, Moffatt 1978, Le Mouël & Courtillot 1981, Le Mouël et al 1981, Hassan & Eltayeb 1982). Topography of several kilometers would be sufficient to couple the mantle and core. This level of topography is consistent with that suggested in the recent global travel-time inversions and the geodynamic calculations. As the seismic models improve, it will be possible to test the competing coupling hypotheses.

Compositional Models for the CMB Region

The composition of the deep mantle is not precisely known, as only rough constraints are provided by seismic velocities and densities. In general, it is believed to consist primarily of magnesiowustite $(Mg,Fe)O$ and silicates $(Mg,Fe)SiO_3$. Iron enrichment of the D'' region, brought about by exsolution from magnesiowustite and resulting in a general decrease in seismic velocity, has been suggested by Anderson & Jordan (1970) and Jeanloz & Ahrens (1980). The seismic discontinuity models for D'' are suggestive of a compositional boundary, but it is not clear whether this is of global extent, nor is it known what compositions are involved. Jeanloz & Ahrens (1980) suggested that a phase change in $(Mg,Fe)O$ could explain some of the anomalous properties of D''. No other candidate phase changes that might account for the discontinuity models have been proposed.

The composition of the outer core is principally iron, but both density and bulk modulus constraints require a light alloying component. The cosmochemically plausible light constituents are Si, C, S, K, and O (Ahrens 1982). The possibility of K being a significant component is particularly important because of the large heat release resulting from radioactive decay. This heat source could increase the temperature contrast across the CMB, enhancing the likelihood of a thermal boundary layer (Elsasser et al 1979). Ruff & Anderson (1980) have suggested that a heterogeneous layer in the outermost core involving an uneven distribution of Al, Ca, U, and Th could have accumulated as the core evolved. Uneven heating in this layer could drive the geodynamo. The only seismological support for such a layer is in the recent core travel-time inversions of Creager & Jordan (1986).

CONCEPTUAL MODELS OF DYNAMICS NEAR THE CMB

The CMB clearly represents a compositional and geodynamical discontinuity in the Earth. The exact role of the D'' region in lower-mantle dynamics has not yet been established by seismological or thermal modeling. Several plausible scenarios are schematically portrayed in Figure 2. Perhaps the simplest model of the CMB is that it is a major thermal boundary layer at the base of a convecting lower-mantle system, with no significant stratification. In this case (Figure 2a), a thin thermal boundary layer will exist from which small-scale plumes may rise as a result of boundary-layer instabilities. The mass flux will be balanced by slow downward flow from the surrounding mantle, and large-scale upwelling plumes may also be present. The density heterogeneity resulting from temperature variations will produce viscous flow resulting in dynamically supported relief of several kilometers on the CMB: Some of the plumes originating in the D'' region may penetrate to the surface, producing hotspots.

An alternative model, which is also consistent with seismological observations, is one in which the D'' layer is a stably stratified, compositionally distinct layer in the lower mantle. This model is supported by the evidence for widespread, if not global, velocity discontinuities at the top of D''.

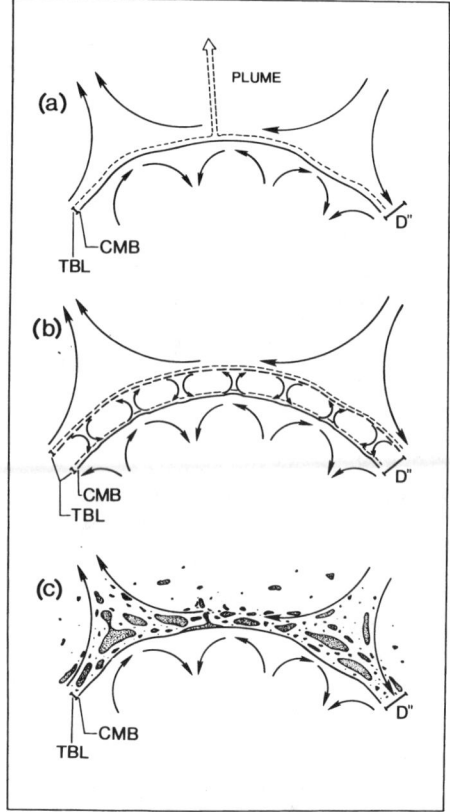

Figure 2 A schematic diagram of three possible dynamic models for the CMB region. (*a*) A model with a single-layer convective system and a thermal boundary layer (TBL) above the CMB. A plume is shown emanating from the TBL. (*b*) A compositionally stratified model with a two-layer convective system. (D'' is separated from the overlying mantle.) Note the second TBL at the top of D'' and the different scale lengths of the two convective layers. (*c*) A mixed model with a single-layer convective system in a laterally heterogeneous mantle. The heterogeneities (shaded) are concentrated in D'' as a result of an increase in viscosity with depth.

These discontinuities are presumably embedded within or on top of the laterally varying D'' zone, with the lateral variations arising from thermal and compositional gradients associated with small-scale convection in the layer. This possibility has not received much consideration in recent thermal modeling, but it would presumably result in a double boundary-layer system (Figure 2b), with a layered convection system of very different scale lengths. While dynamically supported topography may exist on the CMB, the actual sign of the deflections depends on the rheology of D'' (Hager et al 1985).

A third conceptual model, similar to one proposed by Davies (1984), lies between these two extremes. This model is one in which significant compositional heterogeneity exists in the lowermost mantle, although the D'' region is not strictly a compositionally distinct layer (Figure 2c). The chemical heterogeneities exist on many scales, some of which are stably stratified and do not participate in the D'' dynamics, while others are entrained in narrow thermal plumes as well as any larger-scale upwelling currents rising out of D''. Some of the heterogeneities may represent material subducted from the upper mantle, including oceanic crust (Hofmann & White 1982). In this model, dynamically supported topography of the CMB would exist as a result of the viscous flow driven by the mantle heterogeneities. If viscosities are sufficiently high, the residence time for these heterogeneities may be on the order of billions of years, providing the reservoir of primordial material needed to satisfy geochemical isotopic observations. Strong concentrations of heterogeneities could give rise to reflections that have been interpreted as seismic discontinuities in D''. Though complex, this model seems to be the most readily reconciled with the current constraints on the CMB.

CONCLUDING REMARKS

Seismological investigations have not yet reached a consensus on the velocity structure near the CMB, apparently because of the strong lateral heterogeneity of the region. The D'' region is particularly complex and may be both a chemical and a thermal boundary layer. While the majority of the seismic analyses indicate the presence of a general decrease in velocity gradients in D'', it is not clear whether or not the velocities actually decrease with depth. There is evidence for stratification of D'' as well as lateral variations in the velocity gradients, which indicates that the use of a single smoothed velocity model in geothermal analysis is unjustified. Topography on the CMB is likely to result from convection in the overlying mantle. This relief is dynamically supported and provides coupling between the mantle and core. The possibility of a chemically or thermally stratified zone in the outermost core with a concentration of light components cannot be ruled out. Better resolution of the seismic structure of the CMB region is needed before confidence can be placed in either the chemical or dynamical studies that have been published.

ACKNOWLEDGMENTS

C. J. Young was supported by an NSF Graduate Fellowship. This research was supported by NSF Grants EAR-8407792 and EAR-8451715, a Shell Faculty Career Initiation Grant, and the Alfred P. Sloan Foundation. We appreciate comments on the manuscript by Susan Schwartz, Christopher Lynnes, and Elizabeth Finkel.

Literature Cited

Ahrens, T. J. 1982. Constraints on core composition from shock wave data. *Philos. Trans. R. Soc. London Ser. A* 306: 37–47

Alexander, S. S., Phinney, R. A. 1966. A study of the core-mantle boundary using P waves diffracted by the Earth's core. *J. Geophys. Res.* 71: 5943–58

Anderson, D. L., Given, J. W. 1982. Absorption band Q model for the earth. *J. Geophys. Res.* 87: 3893–3904

Anderson, D. L., Hart, R. S. 1976. An Earth model based on free oscillations and body waves. *J. Geophys. Res.* 81: 1461–75

Anderson, D. L., Hart, R. S. 1978a. Q of the Earth. *J. Geophys. Res.* 83: 5869–82

Anderson, D. L., Hart, R. S. 1978b. Attenuation models of the Earth. *Phys. Earth Planet. Inter.* 16: 289–306

Anderson, D. L., Jordan, T. H. 1970. The composition of the lower mantle. *Phys. Earth Planet. Inter.* 3: 23–35

Ansell, J. H. 1974. Observation of the frequency-dependent amplitude variation with distance of P waves from 87° to 119°. *Pure Appl. Geophys.* 112: 683–700

Anufriyev, A. P., Braginski, S. I. 1975. Influence of irregularities on the boundary of the Earth's core on the velocity of the liquid and on the magnetic field. *Geomagn. Aeron.* 15: 1075–82

Backus, G. E. 1968. Kinematics of geomagnetic secular variation in a perfectly conducting core. *Philos. Trans. R. Soc. London Ser. A* 263: 239–66

Backus, G. E. 1982. The electric field produced in the mantle by the dynamo in the core. *Phys. Earth Planet. Inter.* 28: 191–214

Benton, E. R., Muth, L. A., Stix, M. 1979. Magnetic contour maps at the core-mantle boundary. *J. Geomagn. Geoelectr.* 31: 615–26

Berzon, I. S., Kogan, S. D., Passechnik, I. P. 1972. The character of the mantle-core boundary from observations of PcP waves. *Earth Planet. Sci. Lett.* 16: 166–70

Bolt, B. A. 1970. PdP and $PKiKP$ waves and diffracted PcP waves. *Geophys. J. R. Astron. Soc.* 20: 367–82

Bolt, B. A. 1972. The density distribution near the base of the mantle and near the Earth's center. *Phys. Earth Planet. Inter.* 5: 301–11

Bolt, B. A., Niazi, M. 1984. S velocities in D'' from diffracted SH-waves at the core boundary. *Geophys. J. R. Astron. Soc.* 79: 825–34

Bolt, B. A., Niazi, M., Somerville, M. R. 1970. Diffracted ScS and the shear velocity at the core boundary. *Geophys. J. R. Astron. Soc.* 19: 299–305

Booth, D. D., Marshall, P. D., Young, J. B. 1974. Long and short period P-wave amplitudes from earthquakes in the range 0°–114°. *Geophys. J. R. Astron. Soc.* 39: 523–37

Boss, A. P., Sacks, I. S. 1985. Formation and growth of deep mantle plumes. *Geophys. J. R. Astron Soc.* 80: 241–55

Buchbinder, G. G. R. 1968a. Amplitude spectra of PcP and P phases. *Bull. Seismol. Soc. Am.* 58: 1797–1819

Buchbinder, G. G. R. 1968b. Properties of the core-mantle boundary and observations of PcP. *J. Geophys. Res.* 73: 5901–23

Buchbinder, G. G. R. 1971. A velocity structure of the Earth's core. *Bull. Seismol. Soc. Am.* 61: 429–56

Buchbinder, G. G. R. 1972. Travel times and velocities in the outer core from $PmKP$. *Earth Planet. Sci. Lett.* 14: 161–68

Buchbinder, G. G. R., Poupinet, G. 1973.

Problems related to PcP and the core-mantle boundary illustrated by two nuclear events. *Bull. Seismol. Soc. Am.* 63: 2047–70

Bullen, K. E. 1949. Compressibility-pressure hypothesis and the Earth's interior. *Mon. Not. R. Astron. Soc.* 5: 355–68

Carpenter, E. W., Marshall, P. D., Douglas, A. 1967. The amplitude-distance curve for short period teleseismic P-waves. *Geophys. J. R. Astron. Soc.* 13: 61–70

Chang, A. C., Cleary, J. R. 1978. Precursors to *PKKP*. *Bull. Seismol. Soc. Am.* 68: 1059–79

Chang, A. C., Cleary, J. R. 1981. Scattered *PKKP*: further evidence for scattering at a rough core-mantle boundary. *Phys. Earth Planet. Inter.* 24: 15–29

Chapman, C. H., Phinney, R. A. 1972. Diffracted seismic signals and their numerical solution. In *Methods in Computational Physics*, ed. B. A. Bolt, 12: 165–230. New York: Academic

Chinnery, M. A. 1969. Velocity anomalies in the lower mantle. *Phys. Earth Planet. Inter.* 2: 1–10

Choy, G. L. 1977. Theoretical seismograms of core phases calculated by frequency-dependent full wave theory, and their interpretation. *Geophys. J. R. Astron. Soc.* 51: 275–312

Choy, G. L., Cormier, V. F. 1986. Direct measurement of the mantle attenuation operator from broadband P and S waveforms. *J. Geophys. Res.* 91: 7326–42

Clayton, R. W., Comer, R. P. 1983. A tomographic analysis of mantle heterogeneities from body wave travel time data. *Eos, Trans. Am. Geophys. Union.* 64: 776 (Abstr.)

Cleary, J. R. 1969. The S velocity at the core-mantle boundary, from observations of diffracted S. *Bull. Seismol. Soc. Am.* 59: 1399–1405

Cleary, J. R. 1974. The D'' region. *Phys. Earth Planet. Inter.* 19: 13–27

Cleary, J. R., Haddon, R. A. W. 1972. Seismic wave scattering near the core-mantle boundary: a new interpretation of precursors to *PKIKP*. *Nature* 240: 549–51

Corbishley, D. J. 1970. Multiple array measurements of the P wave travel time derivative. *Geophys. J. R. Astron. Soc.* 19: 1–14

Cormier, V. F., Richards, P. G. 1976. Comments on "The damping of core waves," by A. Qamar and A. Eisenberg. *J. Geophys. Res.* 81: 3066–68

Creager, K. C., Jordan, T. H. 1986. Large-scale structure of the outermost core from P'_{DF} and P'_{AB} travel times. *Eos, Trans. Am. Geophys. Union.* 67: 311 (Abstr.)

Crossley, D. 1984. Oscillatory flow in the liquid core. *Phys. Earth Planet. Inter.* 36: 1–16

Davies, G. F. 1984. Geophysical and isotopic constraints on mantle convection: an interim synthesis. *J. Geophys. Res.* 89: 6017–40

Doornbos, D. J. 1974. Seismic wave scattering near caustics: observations of *PKKP* precursors. *Nature* 247: 352–53

Doornbos, D. J. 1978. On seismic-wave scattering by a rough core-mantle boundary. *Geophys. J. R. Astron. Soc.* 53: 643–62

Doornbos, D. J. 1980. The effect of a rough core-mantle boundary on *PKKP*. *Phys. Earth Planet. Inter.* 21: 351–58

Doornbos, D. J. 1983. Present seismic evidence for a boundary layer at the base of the mantle. *J. Geophys. Res.* 88: 3498–3505

Doornbos, D. J., Mondt, J. C. 1979a. Attenuation of P and S waves diffracted around the core. *Geophys. J. R. Astron. Soc.* 57: 353–79

Doornbos, D. J., Mondt, J. C. 1979b. P and S waves diffracted around the core and the velocity structure at the base of the mantle. *Geophys. J. R. Astron. Soc.* 57: 381–95

Doornbos, D. J., Vlaar, N. J. 1973. Regions of seismic wave scattering in the Earth's mantle and precursors to *PKP*. *Nature Phys. Sci.* 243: 58–61

Doornbos, D. J., Spiliopoulos, S., Stacey, F. D. 1986. Seismological properties of D'' and the structure of a thermal boundary layer. *Phys. Earth Planet. Inter.* 41: 225–39

Dziewonski, A. M. 1984. Mapping the lower mantle: determination of lateral heterogeneity in P velocity up to degree and order 6. *J. Geophys. Res.* 89: 5929–52

Dziewonski, A. M., Anderson, D. L. 1981. Preliminary reference Earth model. *Phys. Earth Planet. Inter.* 25: 297–356

Dziewonski, A. M., Gilbert, F. 1972. Observations of normal modes from 84 recordings of the Alaskan earthquake of 1964 March 28. *Geophys. J. R. Astron. Soc.* 27: 393–446

Dziewonski, A. M., Haddon, R. A. W. 1974. The radius of the core-mantle boundary inferred from travel time and free oscillation data; a critical review. *Phys. Earth Planet. Inter.* 9: 28–35

Dziewonski, A. M., Hales, A. L., Lapwood, E. R. 1975. Parametrically simple earth models consistent with geophysical data. *Phys. Earth Planet. Inter.* 10: 12–48

Dziewonski, A. M., Hager, B. H., O'Connell, R. J. 1977. Large-scale heterogeneities in the lower mantle. *J. Geophys. Res.* 82: 239–55

Elsasser, W. M., Olson, P., Marsh, B. D. 1979. The depth of mantle convection. *J. Geophys. Res.* 84: 147–55

Engdahl, E. R. 1968. Seismic waves within the Earth's outer core: multiple reflection. *Science* 161: 263–64

Engdahl, E. R., Johnson, L. E. 1974. Differential *PcP* travel times and the radius of the core. *Geophys. J. R. Astron. Soc.* 39: 435–56

Frasier, C. W., Chowdhury, D. K. 1974. Effect of scattering on *PcP/P* amplitude ratios at Lasa from 40° to 84° distance. *J. Geophys. Res.* 79: 5469–77

Gilbert, F., Dziewonski, A. M. 1975. An application of normal mode theory to the retrieval of structural parameters and source mechanisms from seismic spectra. *Philos. Trans. R. Soc. London Ser. A* 278: 187–269

Gire, C., Le Mouël, J. L., Madden, T. 1986. Motions at the core surface derived from *SV* data. *Geophys. J. R. Astron. Soc.* 84: 1–30

Gubbins, D. 1982. Finding core motions from magnetic observations. *Philos. Trans. R. Soc. London Ser. A* 306: 247–54

Gubbins, D. 1983. Geomagnetic field analysis—I. Stochastic inversion. *Geophys. J. R. Astron. Soc.* 73: 641–52

Gubbins, D., Thomson, C. J., Whaler, K. A. 1982. Stable regions in the Earth's liquid core. *Geophys. J. R. Astron. Soc.* 68: 241–51

Gutenberg, B., Richter, C. F. 1939. On seismic waves. *Beitr. Geophys.* 54: 94–136

Haddon, R. A. W. 1982. Evidence for inhomogeneities near the core-mantle boundary. *Philos. Trans. R. Soc. London Ser. A* 306: 61–70

Haddon, R. A. W., Cleary, J. R. 1974. Evidence for scattering of seismic *PKP* waves near the mantle-core boundary. *Phys. Earth Planet. Inter.* 8: 211–34

Hager, B. H., Clayton, R. W., Richards, M. A., Comer, R. P., Dziewonski, A. M. 1985. Lower mantle heterogeneity, dynamic topography and the geoid. *Nature* 313: 541–45

Hales, A. L., Roberts, J. L. 1971. The velocities in the outer core. *Bull. Seismol. Soc. Am.* 61: 1051–59

Hassan, M. H. A., Eltayeb, I. A. 1982. On the topographic coupling at the core-mantle interface. *Phys. Earth Planet. Inter.* 28: 14–26

Hide, R. 1969. Interaction between the Earth's liquid core and solid mantle. *Nature* 222: 1055–56

Higgins, G., Kennedy, G. C. 1971. The adiabatic gradient and the melting point gradient of the core of the Earth. *J. Geophys. Res.* 76: 1870–78

Hofmann, A. W., White, W. M. 1982. Mantle plumes from ancient oceanic crust. *Earth Planet. Sci. Lett.* 57: 421–36

Husebye, E. S., King, D. W., Haddon, R. A. W. 1976. Precursors to *PKIKP* and seismic wave scattering near the mantle-core boundary. *J. Geophys. Res.* 81: 1870–82

Ibrahim, A. K. 1971. The amplitude ratio *PcP/P* and the core-mantle boundary. *Pure Appl. Geophys.* 91: 114–33

Ibrahim, A. K. 1973. Evidences for a low-velocity core-mantle transition zone. *Phys. Earth Planet. Inter.* 7: 187–98

Jeanloz, R., Ahrens, T. J. 1980. Equations of state of FeO and CaO. *Geophys. J. R. Astron. Soc.* 62: 505–28

Jeanloz, R., Richter, F. M. 1979. Convection, composition, and the thermal state of the lower mantle. *J. Geophys. Res.* 84: 5497–5504

Jeffreys, H. 1939. The times of *P*, *S*, and *SKS* and the velocities of *P* and *S*. *Mon. Not. R. Astron. Soc.* 4: 498–533

Johnson, L. R. 1969. Array measurements of *P* velocities in the lower mantle. *Bull. Seismol. Soc. Am.* 59: 973–1008

Johnson, L. R., Lee, R. C. 1985. Extremal bounds on the *P* velocity in the Earth's core. *Bull. Seismol. Soc. Am.* 75: 115–30

Jones, G. M. 1977. Thermal interaction of the core and mantle and long term behavior of the geomagnetic field. *J. Geophys. Res.* 82: 1703–9

Jordan, T. H. 1979. Structural geology of the Earth's interior. *Proc. Natl. Acad. Sci. USA* 76: 4192–4200

Jordan, T. H., Anderson, D. L. 1974. Earth structure from free oscillations and travel times. *Geophys. J. R. Astron. Soc.* 36: 411–59

Julian, B. R., Sengupta, M. K. 1973. Seismic travel time evidence for lateral inhomogeneity in the deep mantle. *Nature* 242: 443–47

Kanamori, H. 1967a. Spectrum of *P* and *PcP* in relation to the mantle-core boundary and attenuation in the mantle. *J. Geophys. Res.* 72: 559–71

Kanamori, H. 1967b. Spectrum of short-period core phases in relation to the attenuation in the mantle. *J. Geophys. Res.* 72: 2181–86

King, D. W., Haddon, R. A. W., Cleary, J. R. 1974. Array analysis of precursors to *PKIKP* in the distance range 128 to 142°. *Geophys. J. R. Astron. Soc.* 37: 157–73

Lay, T. 1985. Analysis of diffracted *S* waves traversing a region with a lower mantle shear velocity discontinuity. *Eos, Trans. Am. Geophys. Union.* 66: 310

Lay, T. 1986. Evidence for a lower mantle shear velocity discontinuity in *S* and *sS* phases. *Geophys. Res. Lett.* In press

Lay, T., Helmberger, D. V. 1983a. A lower mantle *S*-wave triplication and the shear velocity structure of D''. *Geophys. J. R. Astron. Soc.* 75: 799–838

Lay, T., Helmberger, D. V. 1983b. The shear-wave velocity gradient at the base of the mantle. *J. Geophys. Res.* 88: 8160–70

Lay, T., Young, C. J. 1986. The effect of *SKS* scattering on models of the shear velocity-structure of the D'' region. *J. Geophys.* 59: 11–15

Lee, R. C., Johnson, L. R. 1984a. Tau estimates for mantle *P*- and *S*-waves from global travel-time observations. *Geophys. J. R. Astron. Soc.* 77: 655–66

Lee, R. C., Johnson, L. R. 1984b. Extremal bounds on the seismic velocities in the Earth's mantle. *Geophys. J. R. Astron. Soc.* 77: 667–81

Le Mouël, J.-L., Courtillot, V. 1981. Core motions, electromagnetic core-mantle coupling and variations in the Earth's rotation: new constraints from geomagnetic secular variation impulses. *Phys. Earth Planet. Inter.* 24: 236–41

Le Mouël, J.-L., Madden, T. R., Ducruix, J., Courtillot, V. 1981. Decade fluctuations in geomagnetic westward drift and earth rotation. *Nature* 290: 763–65

Le Mouël, J.-L., Gire, C., Madden, T. 1985. Motions at core surface in the geostrophic approximation. *Phys. Earth Planet. Inter.* 39: 270–87

Loper, D. E. 1984. The dynamical structures of D'' and deep plumes in a non-Newtonian mantle. *Phys. Earth Planet. Inter.* 34: 57–67

Loper, D. E., Stacey, F. D. 1983. The dynamical and thermal structure of deep mantle plumes. *Phys. Earth Planet. Inter.* 33: 304–17

Masters, G. 1979. Observational constraints on the chemical and thermal structure of the Earth's deep interior. *Geophys. J. R. Astron. Soc.* 57: 507–34

Masters, G., Gilbert, F. 1983. Attenuation in the Earth at low frequencies. *Philos. Trans. R. Soc. London Ser. A* 308: 479–552

Mikumo, T., Kurita, T. 1968. *Q* distribution for long-period *P* waves in the mantle. *J. Phys. Earth* 16: 11–29

Mitchell, B. J., Helmberger, D. V. 1973. Shear velocities at the base of the mantle from observations of *S* and *ScS*. *J. Geophys. Res.* 78: 6009–20

Moffatt, H. K. 1978,. Topographic coupling at the core-mantle interface. *Geophys. Astrophys. Fluid Dyn.* 9: 279–88

Moffatt, H. K., Dillon, R. F. 1976. The correlation between gravitational and geomagnetic fields caused by interaction of the core fluid motion with a bumpy core-mantle boundary. *Phys. Earth Planet. Inter.* 13: 67–78

Mondt, J. C. 1977. *SH* waves: theory and observations for epicentral distances greater than 90 degrees. *Phys. Earth Planet. Inter.* 15: 46–59

Morelli, A., Dziewonski, A. M. 1986. 3D structure of the Earth's core inferred from travel time residuals. *Eos, Trans. Am. Geophys. Union.* 67: 311 (Abstr.)

Mula, A. H. 1981. Amplitudes of diffracted long-period *P* and *S* waves and the velocities and *Q* structure at the base of the mantle. *J. Geophys. Res.* 86: 4999–5011

Mula, A. H., Müller, G. 1980. Ray parameters of diffracted long period *P* and *S* waves and the velocities at the base of the mantle. *Pure Appl. Geophys.* 118: 1272–92

Müller, G., Mula, A. H., Gregersen, S. 1977. Amplitudes of long-period *PcP* and the core-mantle boundary. *Phys. Earth Planet. Inter.* 14: 30–40

Okal, E. A., Geller, R. J. 1979. Shear-wave velocity at the base of the mantle from profiles of diffracted *SH* waves. *Bull. Seismol. Soc. Am.* 69: 1039–53

Phinney, R. A., Cathles, L. M. 1969. Diffraction of *P* by the core: a study of long-period amplitudes near the edge of the shadow. *J. Geophys. Res.* 74: 1556–74

Qamar, A., Eisenberg, A. 1974. The damping of core waves. *J. Geophys. Res.* 79: 758–65

Ritzwoller, M., Masters, G., Gilbert, G. 1985. Observations of anomalous splitting and their interpretation in terms of aspherical structure. *Eos, Trans. Am. Geophys. Union.* 66: 966 (Abstr.)

Roberts, P. H., Scott, S. 1965. On analysis of the secular variation. *J. Geomagn. Geolectr.* 17: 137–51

Ruff, L. J., Anderson, D. L. 1980. Core formation, evolution, and convection: a geophysical model. *Phys. Earth Planet. Inter.* 21: 181–201

Ruff, L. J., Helmberger, D. V. 1982. The structure of the lowermost mantle determined by short-period *P*-wave amplitudes. *Geophys. J. R. Astron. Soc.* 68: 95–119

Ruff, L. J., Lettvin, E. 1984. Short period *P*-wave amplitudes and variability of the core shadow zone boundary. *Eos, Trans. Am. Geophys. Union.* 65: 999 (Abstr.)

Sacks, I. S. 1966. Diffracted wave studies of the Earth's core, 1. Amplitudes, core size, and rigidity. *J. Geophys. Res.* 71: 1173–81

Sacks, I. S. 1967. Diffracted *P*-wave studies of the Earth's core, 2. Lower mantle velocity, core size, lower mantle structure. *J. Geophys. Res.* 72: 2589–94

Sacks, I. S. 1969. Anelasticity of the outer core. *Carnegie Inst. Washington Yearb.* 69: 414–16

Sacks, I. S., Snoke, J. A., Beach, L. 1979. Lateral heterogeneity at the base of the mantle revealed by observations of amplitudes of *PKP* phases. *Geophys. J. R. Astron. Soc.* 59 : 379–87

Sailor, R., Dziewonski, A. M. 1978. Measurements and interpretation of normal mode attenuation. *Geophys. J. R. Astron. Soc.* 53 : 559–82

Sato, R., Espinosa, A. F. 1967. Dissipation in the Earth's mantle and rigidity and viscosity in the Earth's core determined from waves multiply reflected from the mantle-core boundary. *Bull. Seismol. Soc. Am.* 57 : 829–56

Schlittenhardt, J. 1984. *Array-Untersuchungen von reflektierten und diffraktierten Kernphasen.* PhD thesis. Univ. Frankfurt, West Ger.

Schlittenhardt, J., Schweitzer, J., Müller, G. 1985. Evidence against a discontinuity at the top of D''. *Geophys. J. R. Astron. Soc.* 81 : 295–306

Sengupta, M. K., Toksöz, M. N. 1976. Three dimensional model of seismic velocity variation in the Earth's mantle. *Geophys. Res. Lett.* 3 : 84–86

Sengupta, M. K., Hassell, R. E., Ward, R. W. 1981. Three-dimensional seismic velocity structure of the Earth's mantle using body wave travel times from intra-plate and deep-focus earthquakes. *J. Geophys. Res.* 86 : 3913–34

Shore, M. J., 1984. A seismic Q-profile for the lower mantle from short-period P waves. *Nature* 310 : 399–401

Stacey, F. D., Loper, D. E. 1983. The thermal boundary-layer interpretations of D'' and its role as a plume source. *Phys. Earth Planet. Inter.* 33 : 45–55

Suzuki, Y., Sato, R. 1970. Viscosity determination in the Earth's outer core from *ScS* and *SKS* phases. *J. Phys. Earth* 18 : 157–70

Teng, T. 1968. Attenuation of body waves and the Q structure of the mantle. *J. Geophys. Res.* 73 : 2195–2208

Verhoogen, J. 1973. Thermal regime of the Earth's core. *Phys. Earth Planet. Inter.* 7 : 47–58

Whaler, K. A. 1980. Does the whole of the Earth's core convect? *Nature* 287 : 528–30

Whaler, K. A. 1982. Geomagnetic secular variation and fluid motion at the core surface. *Philos. Trans. R. Soc. London Ser. A* 306 : 235–46

Wright, C. 1973. Array studies of P phases and the structure of the D'' region of the mantle. *J. Geophys. Res.* 78 : 4965–82

Wright, C. 1975. The origin of short period precursors to *PKP*. *Bull. Seismol. Soc. Am.* 65 : 765–80

Wright, C., Cleary, J. 1972. P wave travel time gradient measurements for the Warramunga seismic array and lower mantle structure. *Phys. Earth Planet. Inter.* 5 : 213–30

Wright, C., Lyons, J. A. 1975. Seismology, $dT/d\Delta$ and deep mantle convection. *Geophys. J. R. Astron. Soc.* 40 : 115–38

Wright, C., Lyons, J. A. 1981. Further evidence for radial velocity anomalies in the lower mantle. *Pure Appl. Geophys.* 119 : 137–62

Wright, C., Muirhead, K. J., Dixon, A. E. 1985. The P wave velocity structure near the base of the mantle. *J. Geophys. Res.* 90 : 623–34

Young, C. J., Lay, T. 1986. Evidence for a shear velocity discontinuity in the lowermost mantle beneath India and the Indian Ocean. *Eos, Trans. Am. Geophys. Union.* 67 : 311 (Abstr.)

Yuen, D. A., Peltier, W. R. 1980. Mantle plumes and thermal stability of the D'' layer. *Geophys. Res. Lett.* 7 : 625–28

Yukutake, T. 1981. A stratified core motion inferred from geomagnetic secular variations. *Phys. Earth Planet. Inter.* 24 : 253–58

Zhang, J., Lay, T. 1984. Investigation of a lower mantle shear wave triplication using a broadband array. *Geophys. Res. Lett.* 11 : 620–23

Zharkov, V. N. 1985. Thermal state and thermal regime of the Earth's interior. *Phys. Earth Planet. Inter.* 41 : 133–37

Zharkov, V. N., Karpov, P. B., Leontjev, V. V. 1985. On the thermal regime of the boundary layer at the bottom of the mantle. *Phys. Earth Planet. Inter.* 41 : 138–42

APPLICATION OF STABLE CARBON ISOTOPES TO EARLY BIOCHEMICAL EVOLUTION ON EARTH

M. Schidlowski

Max-Planck-Institut für Chemie (Otto-Hahn Institut),
D-6501 Mainz, West Germany

INTRODUCTION

The emergence and early evolution of life on Earth have long been intriguing subjects for scientific inquiry and a continuous source of intellectual excitement. Current work at this frontier proceeds along a number of different avenues and comprises a variety of disciplinary approaches that together strongly suggest that the Earth was inhabited by life as early as almost 4 Gyr ago. The present state of the field as inferred from the results of a broad interdisciplinary approach has recently been summarized by J. W. Schopf and coworkers (cf. Schopf 1983).

It is with the advantage of such a synoptic view that the present review addresses a single facet of the theme—namely, the potential isotopic inferences of the presence of CO_2-fixing ("autotrophic") organisms for the carbon cycle of the early Earth. With biologically derived (organic) carbon as well as carbonates (limestone, dolomite, etc.) abundantly preserved in sediments, relevant isotopic evidence should have been encoded in the geologic record from its onset about 3.8 Gyr ago.

It has been known since the pioneering studies by Nier & Gulbransen (1939), Murphey & Nier (1941), and Rankama (1948) that the transformation of inorganic carbon into living matter entails a marked bias in favor of the light isotope (^{12}C), with the heavy species (^{13}C) retained in the inorganic reservoir. Subsequent work performed by many investigators (e.g. Craig 1953, Park & Epstein 1960) has confirmed that all common pathways of autotrophic (specifically photosynthetic) carbon fixation discriminate against ^{13}C, principally as a result of a kinetic isotope effect inherent in the first irreversible enzymatic CO_2-fixing reaction. This leads to a preferential accumulation of ^{12}C in all forms of biogenic (reduced) carbon as compared with the inorganic (oxidized) carbon pool of the surficial environment, which is mainly composed of atmospheric carbon dioxide and dissolved marine bicarbonate (Figure 1). Since the isotopic difference thus established is largely retained when biogenic materials and

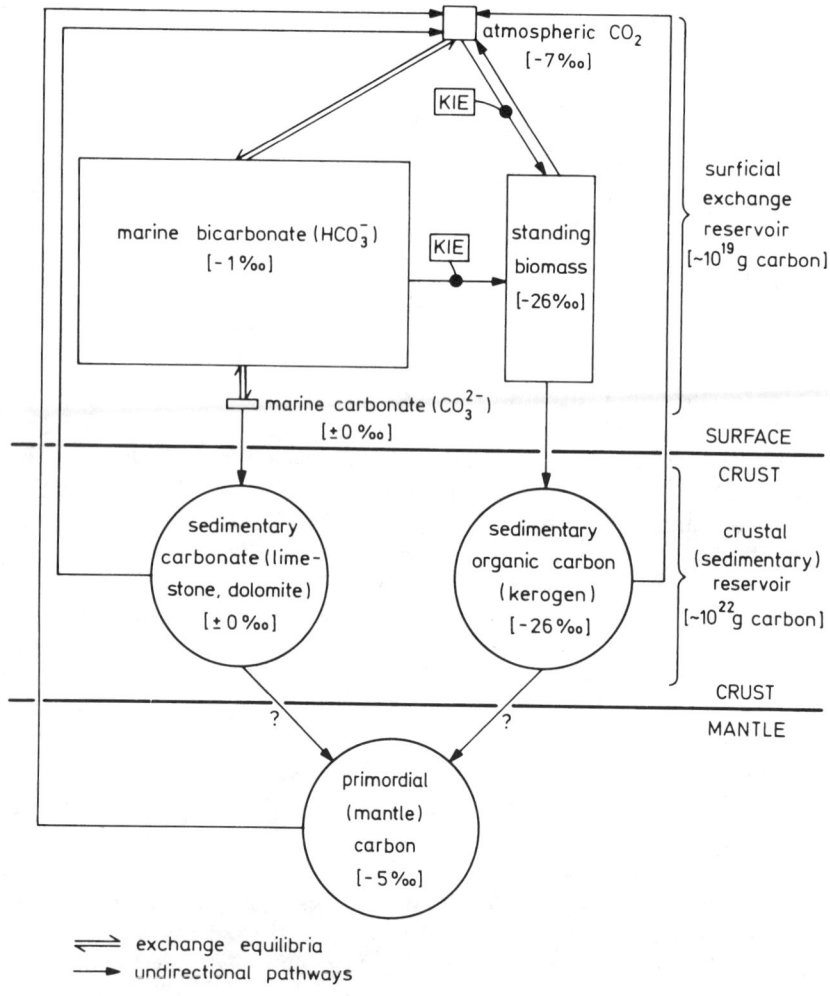

carbonate are incorporated in newly formed sediments, the kinetic isotope effect associated with autotrophic carbon fixation is propagated from the surficial exchange reservoir into the rock section of the carbon cycle. Over geologic time, this effect has ultimately brought about a conspicuous isotopic disproportionation of the planet's original endowment of primordial carbon into a "light" and a "heavy" crustal carbon reservoir (Figure 1).

ISOTOPIC BIOGEOCHEMISTRY OF AUTOTROPHIC CARBON FIXATION

Autotrophic carbon fixation is primarily the assimilation of carbon dioxide (CO_2) and bicarbonate ion (HCO_3^-) by plants and microorganisms; the process proceeds by a limited number of biochemical pathways and carbon-fixing reactions (Table 1). Imposed on these pathways are thermodynamic and kinetic isotope effects that cause the biosynthesized materials to be isotopically distinctive (^{12}C enriched) relative to the feeder substrate. With biochemical reactions largely enzyme controlled, and living systems constituting dynamic states undergoing rapid cycles of anabolism and catabolism, it has become widely accepted that most biological isotope fractionations are due to kinetic rather than equilibrium effects (Vogel 1980, O'Leary 1981).

Kinetic isotope fractionations reflect differences in either the reaction or transport rates of the heavy and light isotope species and are known to be principally imposed on two steps in the primary metabolism of autotrophs. These are (*a*) the diffusion of external CO_2 to the assimilatory reaction centers, and (*b*) the first irreversible enzymatic fixation of CO_2 in the carboxyl (COOH) group of a carboxylic acid. Thermodynamically controlled *equilibrium fractionations*, on the other hand, determine the isotope

---

Figure 1 Simplified box model of the terrestrial carbon cycle, with $\delta^{13}C$ averages of individual reservoirs given in brackets. Surficial reservoirs (rectangular boxes) are drawn approximately to scale; rock reservoirs are represented as circles. Carbon fluxes into the biosphere are beset with kinetic isotope effects (KIE) responsible for a preferential accumulation of ^{12}C in organic matter, with the heavy complement (^{13}C) basically retained in marine bicarbonate (the largest carbon reservoir at the Earth's surface). Organic carbon and carbonate from the surface are continuously entering newly formed sediments, thereby propagating the isotopic difference between the two carbon species into the crustal cycle, where the species are apt to reside with a half-life of several 100 million years. Biological processing of carbon in the surficial compartment has thus, over the ages, brought about a large-scale isotopic disproportionation of the element between a heavy (carbonate) and a light (kerogen) crustal reservoir. Note that the amount of carbon stored in the crust exceeds that residing on the surface by three orders of magnitude.

Table 1 Principal CO_2-fixing reactions utilized in common autotrophic pathways (abridged from Schidlowski et al 1983)[a]

---

(1) CO_2 + ribulose-1,5-bisphosphate → phosphoglycerate
  Operated by: *C3 plants, *algae, *cyanobacteria, *purple photosynthetic bacteria (Chromatiaceae), chemoautotrophic bacteria

(2) CO_2/HCO_3^- + phosphoenolpyruvate/pyruvate → oxaloacetate
  Operated by: *C4 plants,[b] *CAM plants,[b] anaerobic and facultatively anaerobic bacteria

(3) $CO_2 + CO_2$ → acetate/acetyl coenzyme A
  Operated by: *Green photosynthetic bacteria (Chlorobiaceae),[c] anaerobic bacteria, *methanogenic bacteria[d]

(4) CO_2 + acetyl coenzyme A → phosphoenolpyruvate/pyruvate
  Operated by: *Green photosynthetic bacteria (Chlorobiaceae),[e] autotrophic sulfate-reducing bacteria, *methanogenic bacteria[d]

---

[a] Reduction of CO_2 primarily gives rise to C3 compounds (possessing 3-carbon skeletons such as phosphoglycerate and pyruvate), C4 compounds (oxaloacetate), or C2 compounds (acetate, acetyl coenzyme A). The geochemically most important enzymatic carboxylation proceeds via the RuBP carboxylase reaction of the Calvin cycle operative in C3 photosynthesis (reaction 1). Asterisks mark groups of organisms for which carbon isotope fractionations are known (cf. Figure 3).
[b] C4 and CAM plants combine this carboxylation with reaction (1).
[c] Primary CO_2 fixation via succinyl coenzyme A and α-ketoglutarate.
[d] Primary CO_2 fixation probably via C1 acceptors. [Details of the assimilatory pathway of methanogens are as yet poorly known, but the presence of both α-ketoglutarate and pyruvate synthases suggests the involvement of reactions (3) and (4).]
[e] Combined with reactions (3) and (2).

exchange between CO_2 and HCO_3^- in bicarbonate-utilizing pathways, such as C4 and CAM photosynthesis, and also have been claimed to be essential in inter- and intramolecular isotope exchange among different metabolites or classes of organic substances, such as lipids, carbohydrates, and proteins (Galimov 1985).

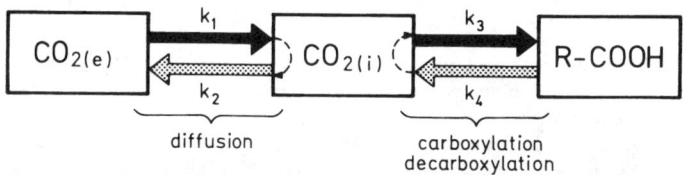

Figure 2 The principal isotope-discriminating steps in the assimilatory pathway of autotrophic organisms (black: assimilatory reactions; stippled: dissimilatory and other reverse processes). $CO_{2(e)}$ and $CO_{2(i)}$ stand for external and internal CO_2, respectively, and R-COOH represents the product of the first CO_2-fixing carboxylation; component reactions have been assigned rate constants k_1–k_4. In sum, these processes lead to a preferential enrichment of light carbon (^{12}C) in organic substances relative to the feeder substrate ($CO_{2(e)}$); the largest single effect is associated with the enzymatic carboxylation step.

Figure 2 summarizes the principal isotope-discriminating steps in autotrophic (notably photosynthetic) carbon fixation and is an elaborated version of the scheme first proposed by Park & Epstein (1960). In sum, these fractionations result in a sizable shift of the $\delta^{13}C$ values[1] of biosynthesized matter in a negative direction relative to the feeder substrate (CO_2); this shift can be conveniently expressed as the difference between the isotopic composition of cells and the isotopic composition of the substrate, i.e.

$$\Delta\delta = \delta^{13}C_{cells} - \delta^{13}C_{CO_2}. \qquad (1)$$

As for the kinetic isotope effect inherent in the initial diffusion step (Figure 2, k_1 and k_2), this discriminates only slightly against ^{13}C, attaining at best the value for CO_2 diffusion in air ($-4.4‰$). With the gaseous effect usually modulated by processes such as dissolution, hydration, and liquid transport of CO_2, actual fractionations tend to stay well below this maximum. As liquid diffusion of CO_2 is considerably slower than the gaseous process, the associated isotope effect is believed to be very small (-1.6 to $-3.2‰$; Vogel 1980) or possibly to approach unity within a few tenths of a per mill (O'Leary 1981). Such minor fractionations are particularly seen in the case of aquatic autotrophs, whose assimilatory pathways are largely diffusion limited (Benedict et al 1980, Schidlowski et al 1984).

In contrast, fractionations in the subsequent enzymatic carboxylation step (Figure 2, k_3) are usually much larger but highly variable in detail. For the quantitatively most important ribulose-1,5-bisphosphate (RuBP) carboxylase reaction (Table 1, No. 1), isotope effects were shown to lie mainly in the range -20 to $-40‰$, the wide range observed being due to the fact that fractionations in enzymatic reactions tend to vary widely with pH, metal cofactor, temperature, and other variables (Winkler et al 1982). Since the carboxylation product generated in this reaction is a compound with a 3-carbon skeleton (phosphoglycerate) that is immediately fed into the reductive pentose phosphate (or "Calvin") cycle, the corresponding pathway has been termed C3 (or Calvin cycle) photosynthesis. The Calvin cycle is utilized by all green plants (those relying on it exclusively are called C3 plants), eukaryotic algae, and the bulk of photoautotrophic and

---

[1] Differences in carbon isotope composition are commonly expressed in terms of the conventional δ-notation giving the per-mill deviation of the isotope ratio of a sample (sa) relative to the PDB standard (st) that defines zero per mill on the δ-scale, i.e.

$$\delta^{13}C = \left[\frac{(^{13}C/^{12}C)_{sa}}{(^{13}C/^{12}C)_{st}} - 1\right] \times 1000 \qquad (‰, PDB).$$

chemoautotrophic bacteria, and it constitutes the principal contrivance for the biologically mediated reduction of CO_2 to the carbohydrate level that channels most of the carbon transfer from the nonliving to the living realm.

Another common (though quantitatively less important) carboxylation reaction that fixes CO_2 as a 4-carbon compound (oxaloacetate) is catalyzed by the enzyme phosphoenolpyruvate (PEP) carboxylase (Table 1, No. 2). This carboxylation constitutes the initial carbon-fixing reaction in the C4 dicarboxylic acid (or Hatch-Slack) pathway and entails a minor fractionation of -2 to $-3‰$ relative to bicarbonate ion that serves as feeder substrate in this pathway. The isotope effects associated with some ferredoxin-linked carboxylations of minor quantitative importance relying on CO_2 acceptors such as succinyl coenzyme A and acetyl coenzyme A (Table 1, Nos. 3 and 4) are as yet poorly known, but fractionations exercised by organisms that have these reactions incorporated in their pathways are usually smaller than those of others utilizing the more common enzymatic reactions.

Finally, enzymatic decarboxylations and related dissimilatory processes (Figure 2, k_4) may also contribute to the overall isotopic composition of autotrophs. Although conflicting results have been reported with regard to the magnitude and even the direction of the isotope effect in these processes, *in vivo* decarboxylations seem to release CO_2 that is lighter than its source material, with isotope discriminations thus running counter to those of the assimilatory pathway. Discrepancies between fractionations predicted by the isotopic discrimination properties of specific enzymes and the actual observed compositions of plant matter could thus be accounted for by the antagonistic effect of respiratory decarboxylations (such as photorespiration in C3 plants). Although such processes seem potentially capable of making substantial contributions to the overall isotope budget of plants, the small number of detailed investigations so far available leaves important aspects of the issue open (O'Leary 1981).

ISOTOPIC COMPOSITION OF EXTANT AUTOTROPHIC ORGANISMS

Depending on which of the isotope-discriminating steps shown in Figure 2 and/or which of the carbon-fixing reactions listed in Table 1 become rate controlling in the specific instance, the isotopic compositions of plants and autotrophic microorganisms may vary over a considerable range. Figure 3 shows the $\delta^{13}C_{org}$ spreads reported for the principal groups of extant primary producers as compared with the compositions of the surficial feeder pools of atmospheric carbon dioxide and marine bicarbonate.

CARBON ISOTOPES AND EVOLUTION 53

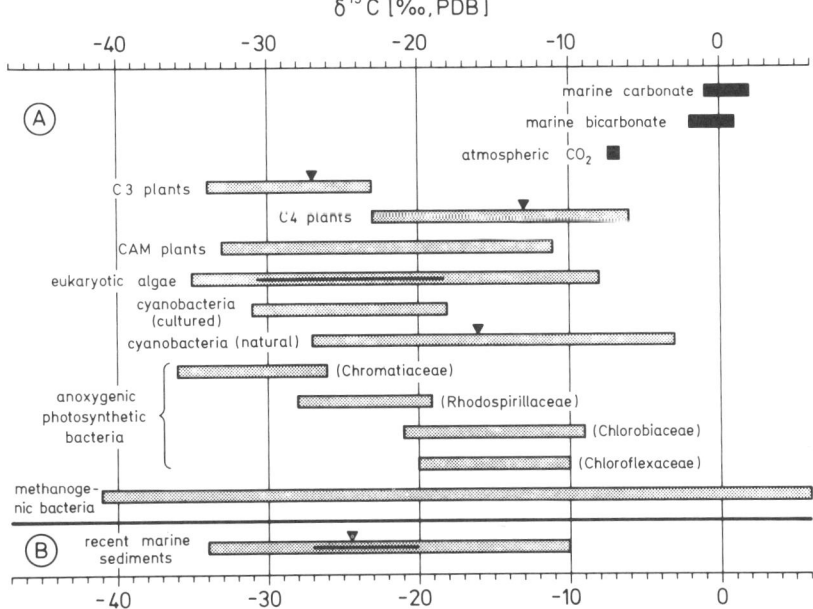

Figure 3 (*A*) Carbon isotope spreads of major groups of higher plants and autotrophic microorganisms compared with the isotopic compositions of the principal forms of oxidized carbon (CO_2, HCO_3^-, CO_3^{2-}) in the environment. Triangles indicate approximate means; black insert in band for eukaryotic algae shows main spread for marine plankton according to Sackett et al (1965). Fractionations obtained in culture experiments with artificial feeder substrates have been normalized to an atmospheric CO_2 source ($\delta^{13}C = -7‰$). The positive segment of the spectrum for methanogens was obtained for cultures grown at very low CO_2 pressures when competition between CO_2 conversion to CH_4 (consuming >90% of the available substrate with an extreme bias for ^{12}C) and CO_2 assimilation was apt to give rise to unusually heavy cell material (Fuchs et al 1979). (*B*) Corresponding isotope spread for organic carbon from recent marine sediments; black insert covers >90% of some 1600 data points represented by the band (adapted from Deines 1980)

C3 PLANTS   Operating C3 (Calvin cycle) photosynthesis and making up the bulk of higher plants, this group basically covers the $\delta^{13}C$ range between -23 and $-34‰$, with an average close to $-27‰$ (Smith & Epstein 1971, Benedict 1978). The most important single contribution toward total fractionation in this group is made by the sizable isotope effect of the RuBP carboxylase reaction. Since discrimination against ^{13}C in this reaction is often larger (-20 to $-40‰$) than the average observed for C3 plants, we must necessarily infer either that diffusion becomes partially rate limiting, or that fractionations in the carboxylation step are partially offset by an inverse effect associated with photorespiration, which is particularly pronounced in C3 species due to the "oxygenase" activity of RuBP carboxylase (cf. Walker 1979).

C4 PLANTS $\delta^{13}C$ values for this group range from -6 to about $-23‰$, with an average between -12 and $-14‰$ (Smith & Epstein 1971, Benedict 1978). These small gross fractionations have often been ascribed to the poor isotope selection properties of the PEP carboxylase reaction that is assigned the initial CO_2 scavenging function in the C4 pathway. However, with PEP carboxylase working on bicarbonate instead of carbon dioxide, the equilibrium fractionation between HCO_3^- and CO_2 of 7–8‰ (Mook et al 1974) is necessarily superimposed on this pathway. If we assume diffusion were fast and the HCO_3^-–CO_2 interchange kept close to equilibrium, the $\delta^{13}C$ value of external CO_2 ($-7‰$) would change to about $+1‰$ in the internal bicarbonate substrate, whose subsequent carboxylation would yield a value between -1 to $-2‰$ for the resulting oxaloacetate. The very fact that the $\delta^{13}C$ values of C4 plants do not come close to this value, but rather approach averages of the range -12 to $-14‰$ (Figure 3), indicates that actual control of isotope fractionation in C4 photosynthesis is exercised by the initial diffusion step (Figure 2, k_1), with fractionations around $-4‰$ relative to atmospheric CO_2. However, the wider range of values observed suggests that neither diffusion nor PEP carboxylation is entirely limiting. Hence, the second carboxylation step in this pathway (by RuBP carboxylase), though largely cryptic as a result of "closed-system" conditions imposed by the confinement of the reaction to the inner leaf structure, may at least partly contribute to the total fractionation in C4 species. Nevertheless, there is virtually no overlap between C3 and C4 plants (Figure 3).

CAM PLANTS   Succulents entertaining the Crassulacean Acid Metabolism (CAM) are capable of operating, albeit at different times, both the RuBP and PEP carboxylase reactions as initial carbon-fixing steps; thus, their compositions consequently span the combined spreads of C3 and C4 plants (Benedict 1978). The rather divergent $\delta^{13}C$ averages found in CAM species seem to primarily stem from the fact that one of the possible carboxylation reactions occasionally succeeds in completely overbalancing the other (O'Leary 1981).

EUKARYOTIC ALGAE   Operating the C3 pathway, marine and freshwater algae span the range from about -8 to $-35‰$ (Smith & Epstein 1971, Seckbach & Kaplan 1973, Wong & Sackett 1978, Estep 1984). Eukaryotic algae constitute the bulk of marine phytoplankton, thus determining—either directly or via the food chain—the isotopic composition of the total oceanic plankton population (-9 to $-31‰$; cf. Sackett et al 1965, Deuser et al 1968). Since the positive part of the spectrum ($\delta^{13}C > -20‰$) shown in Figure 3 is untypical of fractionations obtained in C3 (Calvin cycle) photosynthesis, diffusional control of CO_2 supply in the aquatic realm seems to become rate limiting in specific environmental settings.

CYANOBACTERIA Benthic communities of prokaryotic cyanobacteria (formerly termed blue-green algae, or "cyanophytes") were shown to cover the $\delta^{13}C$ range from -3 to $-27‰$, with an average of about $-16‰$ (Barghoorn et al 1977, Behrens & Frishman 1971, Calder & Parker 1973, Smith & Epstein 1971, Estep 1984, Schidlowski et al 1984, 1985). Fractionations reported for cultured taxa are either coincident with or in excess of the negative extremes of the natural spectrum (Pardue et al 1976), with maximum values approaching $-31‰$ (normalized to an atmospheric CO_2 source with $\delta^{13}C = -7‰$). With the group as a whole known to rely on C3 photosynthesis, there is little doubt that the superheavy cyanobacterial biomass encountered in brine habitats (Schidlowski et al 1984) has been generated in a diffusion-limited pathway. Under conditions of extreme CO_2 depletion due to substantially reduced CO_2 solubilities in hypersaline waters, it is not surprising that the sizable fractionations of the RuBP carboxylase reaction appear to be suppressed.

ANOXYGENIC PHOTOSYNTHETIC BACTERIA Photosynthetic sulfur bacteria that presumably operate the Calvin cycle have yielded fractionations consistent with isotope discriminations in the RuBP carboxylase reaction. Autotrophically grown cultures of the purple sulfur bacterium *Chromatium vinosum* were found to range between -30 and $-36‰$ (Wong et al 1975), while single values reported by Sirevag et al (1977) and Quandt et al (1977) for this species were -29.5 and $-26.6‰$, respectively (original data normalized to an atmospheric CO_2 source of $-7‰$). Corresponding figures obtained for purple nonsulfur bacteria (*Rhodospirillum rubrum*) were $-27.5‰$ (Sirevag et al 1977) and $-19.4‰$ (Quandt et al 1977). On the other hand, green photosynthetic sulfur bacteria like *Chlorobium* sp. that rely on the reductive carboxylic acid cycle [a pathway unique to photosynthetic prokaryotes utilizing reaction (3) of Table 1] have furnished consistently smaller fractionations between -9 and $-21‰$ (Barghoorn et al 1977, Quandt et al 1977, Sirevag et al 1977; values normalized to atmospheric CO_2). For naturally occurring pure layers and mats of *Chloroflexus* sp., Estep (1984) has reported values between -10.7 and $-19.6‰$.

CHEMOAUTOTROPHIC BACTERIA Isotopic data for this group (in which CO_2 fixation is powered by chemical rather than light energy) are hitherto confined to methanogens, notably to *Methanogenium* sp. that grows anaerobically on CO_2 and H_2O as its only carbon and energy resources. Normalized to an atmospheric CO_2 source, the data obtained span the range between $+6$ and $-41‰$ (Fuchs et al 1979, Belyaev et al 1983).

SUMMARY FOR THE TERRESTRIAL BIOSPHERE As can be seen from Figure 3, the $\delta^{13}C$ values of the Earth's standing biomass are, on average, 20–30‰ more negative than those of marine bicarbonate, the most abundant inorganic carbon species in our environment (cf. Figure 1). It should be noted that this conspicuous enrichment in ^{12}C derives, for the most part, from the isotope selection properties of the key enzyme of the Calvin cycle, RuBP carboxylase. Carrying the main burden of the carbon transfer to the living world, the RuBP carboxylase reaction has consequently stamped its isotopic signature on the terrestrial biomass as a whole, imparting to it a principal $\delta^{13}C$ range of about $-26\pm 7‰$.

BIOLOGICALLY MEDIATED CARBON ISOTOPE FRACTIONATIONS IN THE GEOLOGIC PAST

Impact on the Isotopic Geochemistry of Terrestrial Carbon

Certainly the strong bias in favor of isotopically light carbon exercised by the common assimilatory pathways was apt to bring about a large-scale redistribution of the stable carbon isotopes during the long-term operation of the carbon cycle. From the advent of autotrophic life on Earth, any input of CO_2 into the atmosphere-ocean system was bound to undergo an isotopic disproportionation into a heavy and a light fraction as a result of the continuous processing of surficial carbon by the terrestrial biosphere. Enrichment of ^{12}C in biogenic substances was necessarily paralleled by a corresponding accumulation of ^{13}C in the residual reservoir that shifted the $\delta^{13}C$ values of inorganic carbon progressively in the positive direction, in direct proportion to the fraction of total carbon ending up as organic matter. The bulk of the residual (^{13}C enriched) carbon dioxide subsequently came to be stored in the ocean in hydrated form (as dissolved bicarbonate ion, HCO_3^-), which in the present world exceeds the gaseous (atmospheric) species by almost two orders of magnitude (see Figure 1).

It can be readily shown that the isotopic compositions of organic carbon (C_{org}) and carbonate carbon (C_{carb}) are coupled with their relative proportions in the exogenic reservoir by the constraints of an isotope mass balance. If a primary carbon flux into the atmosphere-ocean system with the isotopic composition of primordial mantle carbon ($\delta^{13}C_{prim} \cong -5‰$; cf. Deines & Gold 1973) were partitioned between C_{org} and C_{carb} with a fractionation of $-25‰$ imposed on the formation of biogenic carbon, resulting $\delta^{13}C_{carb}$ values around 0‰ would indicate a ratio $C_{org}/C_{carb} = 1/4$ in the surficial exchange reservoir. This follows from the relation

$$\delta^{13}C_{prim} = R\delta^{13}C_{org} + (1-R)\delta^{13}C_{carb}, \tag{2}$$

rendering $R = C_{org}/(C_{org}+C_{carb}) = 0.2$ and, consequently, $C_{org}/C_{carb} = 0.2/0.8 = 1/4$ for the above numerical parameters (implicitly, $\delta^{13}C_{org} = \delta^{13}C_{carb} - 25‰$). If the surficial carbon reservoir were flushed at a rapid rate as it is today (with a residence time of oceanic HCO_3^- of some 10^5 yr; cf. Holland 1978, p. 156), a steady state complying with Equation (2) would be established within a geologically short time interval.

Accordingly, any biologically induced isotopic fractionation of the Earth's original endowment of primordial carbon should be monitored by the $\delta^{13}C$ values of both exogenic carbon species, i.e. organic and carbonate carbon. As biological fractionations give rise to a wide scatter of $\delta^{13}C_{org}$ values (Figure 3), the state of the system as a whole can be expected to be best reflected by the narrow mean of oceanic bicarbonate that is subsequently preserved in marine carbonate with a minor change of about $+1‰$. Because of the rapid mixing of the oceans, which store about 80% of the total carbon residing on the Earth's surface in the form of HCO_3^-, the bicarbonate mean is likely to integrate most faithfully over the surficial compartment of the carbon cycle, conveying a reliable signal of the state of the system at any given time. In terms of the constraints of Equation (2), any shift of $\delta^{13}C_{carb}$ from the primordial value of $-5‰$ in the positive direction would constitute testimony that part of the total carbon flushed through the surficial exchange reservoir had ended up as organic carbon, its relative proportion increasing with increasing magnitude of that shift.

Postdepositional Changes of Isotopic Compositions of Carbonate and Organic Carbon in Sedimentary Rocks

The sedimentary reservoirs of organic and carbonate carbon have been built up, and are still being fed, by carbon from the surficial exchange reservoir (Figure 1). Minor fractions of both surficial carbon species are continuously transferred to newly formed sediments, which in the case of C_{org} amounts to a flux of some 10^{14} g C yr^{-1} corresponding to between 10^{-2} to 10^{-3} of the annual rate of primary production (Holland 1978, p. 218; Woodwell et al 1978). In a similar way, precipitation of marine bicarbonate steadily adds to the formidable pile of carbonate rocks stored in the crust. A conclusive interpretation of the isotopic evidence thus encoded in the record is necessarily contingent on a full understanding of diagenetic and metamorphic overprints that may have subsequently affected the primary isotopic signatures of both carbon species.

Sedimentary carbonates commonly preserve the isotopic composition of their parent carbonate muds within $\pm 1‰$ of the original value, which in turn is inherited from a bicarbonate precursor with a minor shift of about $+1‰$ due to the equilibrium fractionation between HCO_3^- and CO_3^{2-} (Vogel 1961; see also Figure 3). In contrast, isotopic changes during burial and subsequent diagenetic reconstitution of *sedimentary organic matter* are decidedly more pronounced, spanning the range from a few to several per mill over the maturation pathway of these substances in the sediment. This maturation ultimately leads to the formation of kerogen, the polycondensed, acid-insoluble end-product of the diagenetic alteration of primary biogenic substances (Durand 1980). Concomitant isotope changes tend to make the resulting high-rank kerogens (and derivative graphites) slightly heavier relative to their precursor materials; the respective effect integrates over the fractionations inherent in single discrete steps of the maturation pathway, such as

1. preferential loss of isotopically light lipids and hydrocarbons during progressive dehydrogenation of maturing kerogens,
2. preferential loss of isotopically distinctive functional groups (such as "heavy" carboxyl groups) during microbiological and/or thermal degradation,
3. selective scission of $^{12}C-^{12}C$ bounds in response to increasing thermocatalytic stress, with subsequent removal of the isotopically light crack products (mostly in the form of methane), and
4. exposure of kerogen constituents to oxidizing conditions at elevated temperatures with an ensuing preferential volatilization of isotopically light carbon (e.g. $2C + 2H_2O \rightarrow CH_4 + CO_2$, or $C + 6Fe_2O_3 \rightarrow 4Fe_3O_4 + CO_2$).

There is evidence that the above processes are capable of bringing about an enrichment of ^{13}C in the residual kerogen (Chung & Sackett 1979, Galimov 1985, Hayes et al 1983, Peters et al 1981, Sackett et al 1968, and others), but a calculation of the ^{12}C depletion potentially linked to a near-complete dehydrogenation of the primary kerogenous materials as well as actually observed $\delta^{13}C_{org}$ shifts in kerogens suggest an isotope effect on the order of $+3‰$ as a most probable upper limit. Larger shifts could possibly be obtained in process (4), but the pronounced scarcity of oxidizing enclaves in the diagenetic realm (which is typically reducing) is likely to place limits on the quantitative importance of oxidation processes for promoting isotopic changes of this type.

However, both organic carbon and carbonate may suffer sizable alterations in $^{13}C/^{12}C$ when subjected to metamorphism. Except for occurrences of pure marbles or thick coal measures that retain virtually unchanged isotopic compositions throughout successive metamorphic regimes, isotope effects due to metamorphism are pronounced in the more common forms of sedimentary carbon—notably, impure carbonates and finely dispersed kerogenous substances. Siliceous carbonates are bound to undergo a sequence of decarbonation reactions in which the primary carbonate constituents (calcite, dolomite) react with silica to give Ca-Mg-silicates (tremolite, diopside, etc), with a concomitant liberation of CO_2, e.g.

$$5 \text{dolomite} + 8 \text{qtz} + H_2O \rightarrow \text{tremolite} + 3 \text{calcite} + 7 CO_2.$$

Decarbonation processes have been shown to commence in the greenschist facies (300–450°C) and to become more pronounced with increasing metamorphic grade. They produce residual carbonates enriched in ^{12}C, as equilibrium fractionations between CO_3^{2-} and gaseous CO_2 at elevated temperatures favor the accumulation of heavy carbon in the volatile phase (Bottinga 1969). As a result, the $\delta^{13}C_{carb}$ values of metamorphosed carbonates that have been subjected to extensive decarbonation processes with concomitant formation of calc-silicates are usually shifted by 2–5‰ in the negative direction, a phenomenon first noted in contact metamorphic rocks (Deines & Gold 1969, Shieh & Taylor 1969). It has been found, furthermore, that the magnitude of the observed isotope shift is often proportional to the modal abundances of newly formed Ca-Mg-silicates (Sheppard & Schwarcz 1970).

In contrast to the relatively moderate changes recorded for carbonates, metamorphic alteration of the kerogen fraction has been shown to entail formidable shifts of $\delta^{13}C_{org}$ from original means between -25 and -27‰ to about -10‰ and less (Eichmann & Schidlowski 1975, Hayes et al 1983, Hoefs & Frey 1976, McKirdy & Powell 1974, Schidlowski et al 1979). Since calculation of the isotope effect associated with preferential removal of ^{12}C during dehydrogenation of the primary kerogens indicates a maximum shift of about $+3$‰ as an upper limit, there is little doubt that high-temperature $^{13}C/^{12}C$ exchange between kerogenous substances and coexisting sedimentary carbonates must be ultimately responsible for the formation of excessively "heavy" kerogens and corresponding graphites from metamorphic terranes (Valley & O'Neil 1981). As can be inferred from Figure 4, isotopic reequilibration between C_{org} and C_{carb} with a concomitant decrease in fractionation in response to increasing metamorphic temperature starts below 400°C, with thermodynamic equilibrium usually attained above 650°C (granulite facies; cf. Figure 5). It is reasonable to assume that isotopically heavy CO_2 released by decarbonation processes serves as the principal vehicle of isotope exchange. Data assembled for successive metamorphic facies from a host of metasedimentary suites show characteristic distribution patterns for each facies, with average fractionations progressively shifted to smaller values (Figure

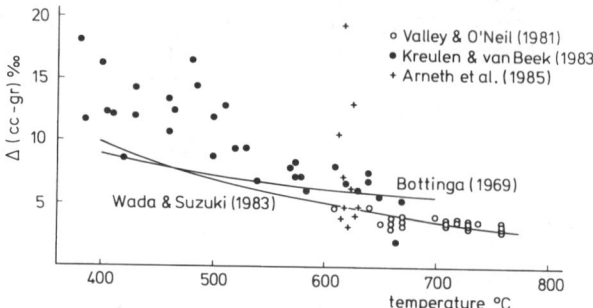

Figure 4 Decrease of isotopic fractionation $\Delta(cc\text{-}gr) = \delta^{13}C_{cc} - \delta^{13}C_{gr}$ between coexisting calcite (cc) and organic carbon [occurring as graphite (gr) at higher metamorphic grades] in response to increasing metamorphic temperature. The changes are due to isotopic reequilibration of the two carbon phases, with $^{13}C/^{12}C$ exchange promoted by the release of isotopically heavy CO_2 during decarbonation of sedimentary carbonates. Fractionations reported by several investigators from different metamorphic terranes scatter around both Bottinga's (1969) function of thermodynamically calculated isotope equilibria and an empirical fractionation curve by Wada & Suzuki (1983) calibrated by dolomite-calcite solvus temperatures. Note that single discordant values may persist up to 600–650°C, which indicates that isotopic reequilibration between organic and carbonate carbon is not complete even in the upper amphibolite facies.

5). These findings are crucial for a conclusive interpretation of carbon isotope data from metamorphic terranes such as the Isua Supracrustal Belt (West Greenland), which stores the oldest (~3.8 Gyr) segment of the carbon record.

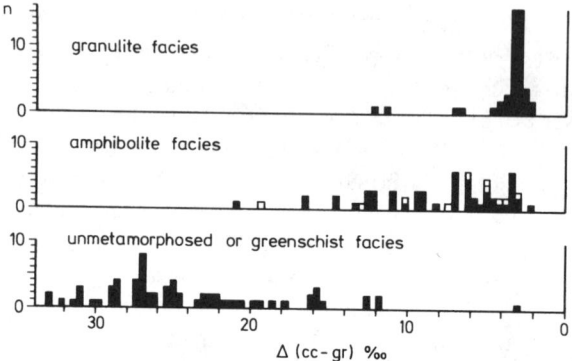

Figure 5 Distribution patterns of $\Delta(cc\text{-}gr)$ for different metamorphic facies [Valley & O'Neil (1981), with additional data for the amphibolite facies (open squares) from Arneth et al (1985)]. Note the systematic decrease of average fractionations with increasing metamorphic grade (greenschist facies: 300–450°C; amphibolite facies: 450–650°C; granulite facies $\geq$ 650°C). The emergence of a narrow peak in the granulite facies indicates that thermodynamic equilibrium is virtually attained at this stage.

Summing up, we may conclude that diagenetic and metamorphic processes are capable of imposing distinct (if not sizable) overprints on the primary isotopic compositions of both sedimentary carbon species. However, these secondary changes usually just blur, and never completely obliterate, the isotopic signal primarily encoded in the record. This even holds for the extremely positive $\delta^{13}C_{org}$ values of some graphites from high-grade metasediments which, because of the possibility of $^{13}C/^{12}C$ exchange with carbonate, cannot be taken as evidence for an abiogenic origin of their host phase.

Isotopic Composition of Organic Carbon and Carbonate Carbon Through Time

As kerogenous materials and carbonates are preserved over almost 4 Gyr of recorded Earth history, and with the secondary changes of their isotopic compositions during diagenesis and metamorphism basically understood, we may attempt to establish an isotope record of both sedimentary carbon species. Figure 6 summarizes the principal features of the currently known age functions of $\delta^{13}C_{org}$ and $\delta^{13}C_{carb}$ that cover some 10,000 measurements assembled over the time interval from Recent to ~ 3.8 Gyr ago.

In the isotope age curve for sedimentary *carbonates*, the best-documented part is that of the last 0.8 Gyr, which is based on 3,016 data points collected by Veizer et al (1980). It seems reasonable to conjecture that the small-scale oscillations of the Phanerozoic ($t < 0.56$ Gyr) record (between a minimum of -0.8 in the Ordovician and a maximum of $+2$ in the Permian) are representative of the amplitudes of the carbonate curve as a whole. In any case, the mean of $+0.5 \pm 2.6‰$ yielded by 2,916 Phanerozoic samples contained in the data base of Veizer et al (1980) compares well with a corresponding average of $+0.4 \pm 2.7‰$ reported by Schidlowski et al (1975) for the largest single investigation (260 samples) of Precambrian carbonates. Apart from variations of the above magnitude, the $\delta^{13}C_{carb}$ function as a whole shows an astounding degree of time invariance, with the overall mean closely tethered to the zero per mill line. As marine carbonates have been shown to preserve the isotopic composition of their HCO_3^- precursor with a slight change of just about $+1‰$, the basic information to be retrieved from the $\delta^{13}C_{carb}$ record is that the isotopic composition of oceanic bicarbonate has varied only slightly through geologic time (Schidlowski 1983, Schidlowski et al 1975, 1983, Veizer & Hoefs 1976).

The isotopic record of *organic carbon* ("kerogen") displays a considerably larger scatter resembling that of the supposed progenitor materials (Figure 6), but the average seems to be constrained by fairly narrow limits (-24 to $-28‰$; the mainstream of the envelope shown for

$\delta^{13}C_{org}$ basically lies within the confines $-26\pm7‰$). Both Welte et al (1975) and Galimov et al (1975) have defined tentative age trends for Phanerozoic kerogens (with a minimum close to $-29‰$ during Ordovician-Silurian and a maximum around $-24‰$ in the Carboniferous) that were subsequently confirmed by other investigators (e.g. Arneth 1984). These curves show a marked covariance with the carbonate age curve, which would be a necessary consequence of a coupling of the isotopic compositions of both carbon species to their relative amounts in the surficial exchange reservoir in terms of Equation (2). Accordingly, the observed oscillations certainly reflect moderate changes in the relative proportions of C_{org} and C_{carb} through geologic time.

The isotopic composition of organic carbon from Precambrian rocks virtually does not differ from that of geologically younger kerogens, with the notable exception of some extremely negative spikes imposed on the mainstream of the record between 2.1 and 2.8 Gyr ago (Figure 6). Kerogen constituents from the ~ 2.6 Gyr old Ventersdorp Supergroup of South Africa and from the 2.8 Gyr old Fortescue Group of Australia were found to preferentially range between -35 and $-52‰$ (Hayes et al 1983, Schidlowski et al 1983), while Schoell & Wellmer (1981) had noted a lower extreme of $-44‰$ for near-coeval graphite-bearing sediments from the Superior Province of the Canadian Shield. Another isotopically anomalous kerogen occurrence ($\delta^{13}C_{org}$ to about $-47‰$) of considerably younger age (~ 2.1 Gyr) has been reported by Weber et al (1983) from the Francevillian Series of Gabon, West Africa. Since these ^{12}C-enriched carbonaceous materials are coeval with a host of isotopically "normal" kerogen provinces, and in view of the absence of a complementary signal in the contemporaneous $\delta^{13}C_{carb}$ record, there is no doubt that these anomalies reflect changes in the local (rather than global) environment in which the organic precursor materials had formed. We may reasonably assume that the bacteria of the methane cycle (methanogens and methylotrophs) were involved in the genesis of these anomalous kerogens, since negative $\delta^{13}C$ values of the above range have hitherto only been reported in connection with methane (see below). It seems worth noting that a single odd occurrence of extremely light, nonvolatile organic matter has also been described from geologically young sediments (Kaplan & Nissenbaum 1966).

As for the conspicuous break in the $\delta^{13}C_{org}$ age curve between the 3.8 Gyr old metasedimentary suite from Isua, West Greenland, and the rest of the record (Schidlowski et al 1979, 1983), there is by now ample evidence that the displacement of the Isua values in the positive direction is due to the amphibolite-grade metamorphism to which these oldest terrestrial sediments have been subjected during their postdepositional history. Both the direction and magnitude of the observed change are fully consistent with the expected results of an isotopic reequilibration between coexisting sedimentary carbonate and organic carbon (and its graphitic derivatives) under the conditions of the amphibolite facies (cf. Figures 4 and 5).

The $\delta^{13}C_{org}$ Record as an Index of Autotrophic Carbon Fixation: Implications for the Origin and Evolution of the Terrestrial Biosphere

Decoding the vast body of information stored in the sedimentary carbon isotope record has posed a continuous challenge to geochemists (Broecker 1970, Hayes et al 1983, Junge et al 1975, Schidlowski et al 1975, 1979, 1983). As carriers of both biochemical and geochemical information, carbon isotopes are potentially capable of providing crucial clues to early organic evolution as well as to the evolution of the long-term biogeochemical cycles. Ever since the accumulation of the first $\delta^{13}C_{org}$ values from Precambrian rocks (Rankama 1948, Hoering 1967), students of the carbon cycle have attempted to utilize isotope data for constraining the antiquity of photoautotrophy as the quantitatively most important biochemical process on Earth. With the prolific accumulation during the last two decades of new data—notably for the oldest carbon record (Eichmann & Schidlowski 1975, Hayes et al 1983, Oehler et al 1972, Perry & Ahmad 1977, Schidlowski et al 1976, 1979, and others)—inferences have been based on increasingly firmer grounds.

With the background information previously presented, the implications of the isotope age functions shown in Figure 6 are fairly straightforward. There is little doubt that the consistent enrichment in ^{12}C relative to carbonate displayed by fossil organic matter is nearly identical to respective enrichments observed in the extant biosphere. Obviously, the bias in favor of ^{12}C inherent in the common autotrophic pathways (notably photosynthesis) has been propagated from the surficial environment into the rock record with very little change; the isotope spread of organic matter from recent marine sediments virtually holds over 3.5 Gyr of recorded Earth history (Figure 6).

Isotopic features of the present biosphere that appear to be poorly reflected by the record are the small fractionations often obtained (a) in the diffusion-limited pathways of some aquatic autotrophs (notably benthic cyanobacteria from hypersaline habitats; cf. Schidlowski et al 1984) and (b) in C4 photosynthesis. This situation is, however, by no means surprising. The generation of isotopically "superheavy" microbial biomass is clearly confined to side stages of the carbon cycle that contribute little to the global budget. The C4 pathway, on the other hand, is a late achievement in angiosperm evolution based on a specific (compartmented) leaf anatomy and operated by a limited number of families (Smith 1976, Walker 1979)

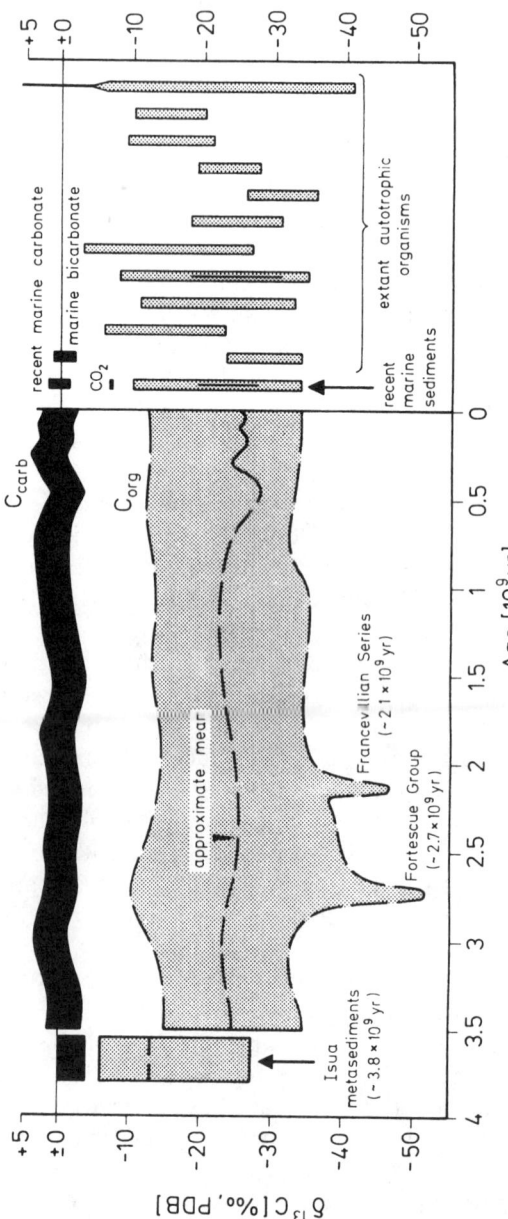

Figure 6 Isotopic composition of sedimentary carbonate (C_{carb}) and organic carbon (C_{org}) over 3.8 Gyr of Earth history. (Right part of graph summarizes spreads of their progenitor materials as shown in Figure 3.) The envelope for fossil organic carbon is an updated synopsis of the data base presented by Schidlowski et al (1983) that comprises the means of some 150 Precambrian kerogen provinces as well as the currently available Phanerozoic record [Phanerozoic data from Degens (1969) and others]. Note that the isotope spreads of the principal contributors to the extant biomass are transcribed into the record with just the extremes eliminated because of negligible contributions from carbon sources of unusual isotopic composition. (Notable exceptions are extremely negative $\delta^{13}C_{org}$ values in the Precambrian that indicate the involvement of methane in the formation of the respective kerogen precursors.) The band for carbonate (dark) gives the mean and standard deviation of $\delta^{13}C_{carb}$ through time. The approximate constancy in both the $\delta^{13}C_{org}$ and $\delta^{13}C_{carb}$ records holds over 3.5 Gyr before values are reset by metamorphism in the Isua suite (presently the oldest data point of the isotope age curve).

that cannot predate the appearance (toward the end of the Cretaceous) of flowering plants as a group. Hence, the isotopically distinctive organic materials produced in this pathway have necessarily failed to leave a significant imprint in the carbon record.

As a whole, the fossil $\delta^{13}C_{org}$ record appears to offer a fair transcription of the average isotopic composition of the extant biomass, integrating over the total range of biological carbon isotope fractionations in the present environment. Accordingly, it is almost impossible to escape the conclusion that sedimentary $\delta^{13}C_{org}$ values constitute signals to autotrophic carbon fixation that date back to at least 3.5, if not 3.8, Gyr ago. Moreover, the glaring uniformity through time of this isotopic signal suggests an extreme degree of evolutionary conservatism in the principal biochemical mechanism that underlies biological CO_2 fixation. The mainstream of the envelope for $\delta^{13}C_{org}$ shown in Figure 6 can, in fact, be best explained as the geochemical expression of the isotope-discriminating properties of ribulose-1,5-bisphosphate carboxylase, the key enzyme of the Calvin cycle (and supposedly the most abundant protein in the present world). Since the RuBP carboxylase reaction constitutes the principal avenue of carbon transfer from the nonliving to the living realm, we may indeed expect the record of fossil organic matter to exhibit the isotopic signature of this process.

A feature of the record that presently lacks a well-documented precedent in the contemporary biosphere is the presence of negative offshoots of $\delta^{13}C_{org}$ typical of some Precambrian (notably Archean) kerogen provinces. As pointed out by several authors (Hayes et al 1983, Schidlowski et al 1983, Schoell & Wellmer 1981), such kerogens most probably derive from isotopically "superlight" organic substances generated with at least partial involvement of the biological methane cycle. Alternatively, CH_4 could have been either directly utilized by methylotrophic bacteria in a methane-based assimilatory pathway entailing the oxidation sequence $CH_4 \rightarrow CH_3OH \rightarrow HCHO \rightarrow HCOOH \rightarrow CO_2$ [of which the HCHO (formaldehyde) intermediate is subsequently metabolized], or isotopically light CO_2 stemming from the oxidation of a methane precursor could have been processed in the common autotrophic pathways.

Since both processes are contingent on the presence of an oxidant, the first abundant occurrence of isotopically light kerogens between 2.7 and 2.8 Gyr ago has been tentatively equated with the rise of oxygenic conditions in the surficial environment (Hayes 1983); alternative explanations might invoke various mechanisms of anaerobic methane oxidation as have been shown to operate in present-day reducing environments (Alperin & Reeburgh 1984, Reeburgh 1983). Since, however, the widespread occurrence of Precambrian banded iron formation is widely believed to indicate at least transient oxygen levels in the ancient seas (Cloud 1973), it would not seem improbable that some of the O_2 flux into the iron oxides was set aside for sustaining the activities of methane-oxidizing bacteria in suitable habitats. Such interpretation of the "superlight" kerogens appears to be the most plausible way to account for the extremes at the lower fringe of the envelope for $\delta^{13}C_{org}$ depicted in Figure 6. In any case, it is not surprising that negative anomalies of the above type are primarily documented in the Archean and early Proterozoic ($t > 2.0$ Gyr), when methanogenesis as a key process of anaerobic decomposition of organic matter was certainly more important than it is today.

As for the reset values of the 3.8 Gyr old Isua Suite [$\delta^{13}C_{carb} = -2.3 \pm 2.2‰$ ($n = 187$); $\delta^{13}C_{org} = -13.0 \pm 4.9‰$ ($n = 78$)], Schidlowski and coworkers were the first to propose that the observed changes had been caused by the amphibolite-grade metamorphism suffered by these rocks (Schidlowski et al 1979); this prediction has been borne out by a wealth of additional evidence (cf. Hayes et al 1983, Schidlowski et al 1983). What lends particular support to this interpretation is that both $\delta^{13}C_{carb}$ and $\delta^{13}C_{org}$ have responded to increasing metamorphism in a predictable manner consonant with the thermodynamics of high-temperature $^{13}C/^{12}C$ exchange and with a host of corresponding observations from geologically younger metamorphic terranes (cf. Figures 4 and 5). It is, therefore, reasonable to conjecture that the premetamorphic Isua sediments had contained carbonate and organic carbon of about "normal" isotopic composition. Hence, the uniform trends of the sedimentary $\delta^{13}C_{carb}$ and $\delta^{13}C_{org}$ records had, originally, almost certainly extended to Isua times, with the isotopic signature of biological activity thus covering the time interval from 3.8 Gyr ago to the present.

If we take the currently documented isotope age function of sedimentary organic carbon as a whole (Figure 6), a reasonable case can be made that autotrophy—and notably photoautotrophy as the quantitatively most important process of CO_2 fixation—had been extant as a biochemical process and as a geochemical agent since at least 3.8 Gyr ago, which implies that biologically mediated photoinduced uphill electron transfer had evolved as a means of creating reducing power for assimilatory reactions (cf. Krasnovsky 1981) *before* that time. Such a conclusion is also consistent with the observation that the average C_{org} content of Precambrian sediments does not basically differ from that of Phanerozoic formations, amounting to means around 0.6% in the case of organic-rich rocks (cf. Cameron & Garrels 1980, Schidlowski 1982). Such formidable C_{org} contents of the oldest terrestrial sediments (inclusive of the Isua suite) are, however, by no means astounding, since microbial communities are known to figure among the most productive ecosystems; notable examples are benthic prokaryotes such as cyanobacteria, which are capable of sustaining rates of primary production of more than 8 g C_{org} m^{-2} day^{-1} (Cohen et al 1980, Krumbein & Cohen 1977). Moreover, it is generally accepted that the Precambrian was the time when prokaryotic ecosystems held dominion over the Earth (Cloud 1976, Schopf 1983). If such high rates of productivity can be sustained by microbial photoautotrophs operating at the prokaryotic level, photosynthesis may have gained little in quantitative importance since the establishment of the first microbial communities on the ancient Earth.

Incidentally, an implication of the largely time-invariant $\delta^{13}C_{carb}$ and $\delta^{13}C_{org}$ functions of Figure 6 in terms of an isotope mass balance (Equation 2) would be that organic carbon has always accounted for about 20% of total carbon in the exogenic system from the start of the sedimentary record. Such inferences lend strong support to the concept of "biotic plenitude" (or "fullness") based on the intrinsic property of life to proliferate to ultimate limits set by environmental resources, notably critical nutrients such as nitrogen and phosphorus. A state of global biotic saturation may be preferentially envisioned for the Precambrian, when flourishing microbial ecosystems monopolized the biosphere, having the aquatic realm essentially to themselves.

These ancient microbial communities (preserved in the form of impressive stromatolite carpets) can be expected to have proliferated to limits set by the nutrient supply, and the perception of habitats saturated with microorganisms to a degree approaching that of artificially eutrophicated environments of the present world certainly does not strain the imagination (Stanley 1981, p. 194). Since critical nutrients such as phosphate were not retained by any land biota because the ancient continents were virtually barren of life, the Precambrian seas were probably eutrophicated by at least a factor of two as compared with the present ocean. Acceptance of the role of phosphorus as the ultimate determinant for the size of the terrestrial biomass (cf. Broecker 1973, Junge et al 1975, Holland 1978, p. 213) would, in fact, imply the existence of a state of global biotic saturation or "plenitude" from the time of the first appearance of life on Earth. There is, moreover, little doubt that the relative constancy of the $\delta^{13}C_{carb}$ and $\delta^{13}C_{org}$ records over almost 4 Gyr of Earth history primarily stems from the fact that the C_{org}/C_{carb} ratio in the atmosphere-ocean-crust system has been fixed by a roughly uniform rate of phosphorus supply to the exogenic compartment throughout geologic time.

The early appearance of sedimentary organic matter bearing the isotopic signature of biological processes is, furthermore, consistent with the paleontological record of apparently autotrophic life (Awramik 1982, Schopf 1983). Available evidence includes relics of cellular microfossils as well as biosedimentary structures deriving from mat-building prokaryotic microbenthos (stromatolites). Lithified microbial colonies of the stromatolitic type have been shown to extend back to ~ 3.5 Gyr ago (Lowe 1980, Walter et al 1980), while the biogenicity of the presumably oldest cellular fossils (Awramik et al 1983, Pflug & Jaeschke-Boyer 1979) is still under discussion.

To invalidate the interpretation of the Archean $\delta^{13}C_{org}$ record as an indicator of early biological activity, we would have to postulate an inorganic process operating on a global scale that was capable of mimicking both the direction and magnitude of autotrophic carbon fixation with a remarkable degree of precision. Potential candidates fail, however, to meet the requirement of a near-perfect isotopic mimicry. Fractionations between reduced and oxidized carbon in *Fischer-Tropsch reactions* have been shown to largely exceed those of photosynthetic processes, amounting to between -50 and $-100‰$ (Lancet & Anders 1970). *Miller-Urey-type spark discharge syntheses*, on the other hand, entail relatively small isotope effects ($-10‰$ and less in the case of their bulk reaction products; see Chang et al 1983). Although these latter fractionations marginally overlap the biological range, they are untypical of average organic matter and should lend themselves to a quick detection if prevailing in the organic fraction of unmetamorphosed sediments. Their virtual absence from the record would exclude substantial contributions from such sources to the reduced carbon content of the Earth's oldest sediments.

SUMMARY AND CONCLUSIONS

With the currently known sedimentary carbon isotope record at hand, it can be stated with fair confidence that biologically mediated carbon isotope fractionations have persisted rather uniformly throughout 3.5 Gyr of recorded Earth history. Since changes of $\delta^{13}C_{org}$ and $\delta^{13}C_{carb}$ observed in the oldest terrestrial sediments are due to an isotopic reequilibration between organic carbon and carbonate in response to amphibolite-grade metamorphism, it is virtually certain that the biological signature of the record had originally extended to ~ 3.8 Gyr ago. There is no doubt that the enrichment of light carbon in fossil organic matter principally derives from the bias in favor of ^{12}C during photosynthesis (notably the RuBP carboxylase reaction of the Calvin cycle). Hence, the mainstream of the sedimentary $\delta^{13}C_{org}$ record can best be interpreted as geochemical evidence of the isotope-discriminating properties of the principal CO_2-fixing enzymatic reaction of the assimilatory pathway, suggesting an extreme degree of evolutionary conservatism in the biochemistry of autotrophic carbon fixation. As a consequence, biological control of the geochemical carbon cycle was established very early in the geologic past and was fully operative by the time of formation of the Earth's oldest sediments.

Literature Cited

Alperin, M. J., Reeburgh, W. S. 1984. Geochemical observations supporting anaerobic methane oxidation. In *Microbial Growth on C1 Compounds*, ed. R. L. Crawford, R. S. Hanson, pp. 282–89. Washington, DC: Am. Soc. Microbiol.

Arneth, J. D. 1984. Stable isotope and organo-geochemical studies on Phanerozoic sediments of the Williston Basin. *Isot. Geosci.* 2: 113–40

Arneth, J. D., Schidlowski, M., Sarbas, B., Goerg, U., Amstutz, G. C. 1985. Graphite content and isotopic fractionation between calcite-graphite pairs in metasediments from the Mgama Hills, Southern Kenya. *Geochim. Cosmochim. Acta* 49: 1553–60

Awramik, S. M. 1982. The pre-Phanerozoic fossil record. In *Mineral Deposits and Evolution of the Biosphere*, ed. H. D. Holland, M. Schidlowski, pp. 67–81. Berlin: Springer-Verlag

Awramik, S. M., Schopf, J. W., Walter, M. R. 1983. Filamentous fossil bacteria from the Archaean of Western Australia. In *Developments and Interactions of the Precambrian Atmosphere, Lithosphere and Biosphere. Dev. Precambrian Geol.*, ed. B. Nagy, R. Weber, J. C. Guerrero, M. Schidlowski, 7: 249–66. Amsterdam: Elsevier

Barghoorn, E. S., Knoll, A. H., Dembicki, H., Meinschein, W. G. 1977. Variation in stable carbon isotopes in organic matter from the Gunflint iron formation. *Geochim. Cosmochim. Acta* 41: 425–30

Behrens, E. W., Frishman, S. A. 1971. Stable carbon isotopes in blue-green algal mats. *J. Geol.* 79: 94–100

Belyaev, S. S., Wolkin, R., Kenealy, W. R., DeNiro, M. J., Epstein, S., Zeikus, J. G. 1983. Methanogenic bacteria from the Bondyuzhskoe Oil Field: general characterization and analysis of stable carbon isotopic fractionation. *Appl. Environ. Microbiol.* 45: 691–97

Benedict, C. R. 1978. The fractionation of stable carbon isotopes in photosynthesis. *What's New Plant Physiol.* 9: 13–16

Benedict, C. R., Wong, W. W. L., Wong, J. H. H. 1980. Fractionation of the stable isotopes of inorganic carbon by seagrasses. *Plant Physiol.* 65: 512–17

Bottinga, Y. 1969. Calculated fractionation factors for carbon and hydrogen isotope exchange in the system calcite–carbon dioxide – graphite – methane – hydrogen – water vapor. *Geochim. Cosmochim. Acta* 33: 49–64

Broecker, W. S. 1970. A boundary condition on the evolution of atmospheric oxygen. *J. Geophys. Res.* 75: 3553–57

Broecker, W. S. 1973. Factors controlling CO_2 content in the oceans and atmosphere. In *Carbon and the Biosphere*, ed. C. M. Woodwell, E. V. Pecan, pp. 32–50. Washington, DC: US At. Energy Comm.

Calder, J. A., Parker, P. L. 1973. Geochemical implications of induced changes in ^{13}C fractionation by blue-green algae. *Geochim. Cosmochim. Acta* 37: 133–40

Cameron, E. M., Garrels, R. M. 1980. Geochemical compositions of some Precambrian shales from the Canadian Shield. *Chem. Geol.* 28: 181–97

Chang, S., Des Marais, D., Mack, R., Miller, S. R., Strathearn, G. 1983. Prebiotic organic synthesis and the origin of life. See Schopf 1983, pp. 53–92

Chung, H. M., Sackett, W. M. 1979. Use of stable carbon isotope compositions of pyrolytically derived methane as maturity indices for carbonaceous materials. *Geochim. Cosmochim. Acta* 43:1979–88

Cloud, P. E. 1973. Paleoecological significances of the banded iron formation. *Econ. Geol.* 68: 1135–43

Cloud, P. E. 1976. Beginnings of biospheric evolution and their biogeochemical consequences. *Paleobiology* 2: 351–87

Cohen, Y., Aizenshtat, Z., Stoler, A., Jørgensen, B. B. 1980. The microbial geochemistry of Solar Lake, Sinai. In *Biogeochemistry of Ancient and Modern Environments*, ed. J. B. Ralph, P. A. Trudinger, M. R. Walter, pp. 167–72. Berlin: Springer-Verlag

Craig, H. 1953. The geochemistry of stable carbon isotopes. *Geochim. Cosmochim. Acta* 3: 53–92

Degens, E. T. 1969. Biogeochemistry of stable carbon isotopes. In *Organic Geochemistry*, ed. G. Eglinton, M. T. J. Murphy, pp. 304–29. Berlin: Springer-Verlag

Deines, P. 1980. The isotopic composition of reduced organic carbon. In *Handbook of Environmental Isotope Geochemistry*, ed. P. Fritz, J. C. Fontes, 1: 329–406. Amsterdam: Elsevier

Deines, P., Gold, D. P. 1969. The change in carbon and oxygen isotopic composition during contact metamorphism of Trenton limestone by the Mt. Royal pluton. *Geochim. Cosmochim. Acta* 33 : 421–24

Deines, P., Gold, D. P. 1973. The isotopic composition of carbonatite and kimberlite carbonates and their bearing on the isotopic composition of deep-seated carbon. *Geochim. Cosmochim. Acta* 37 : 1709–33

Deuser, W. G., Degens, E. T., Guillard, R. R. L. 1968. Carbon isotope relationships between plankton and sea water. *Geochim. Cosmochim. Acta* 32 : 657–60

Durand, B., ed. 1980. *Kerogen—Insoluble Organic Matter From Sedimentary Rocks*. Paris : Editions Technip. 519 pp.

Eichmann, R., Schidlowski, M. 1975. Isotopic fractionation between coexisting organic carbon–carbonate pairs in Precambrian sediments. *Geochim. Cosmochim. Acta* 39 : 585–95

Estep, M. L. F. 1984. Carbon and hydrogen isotopic compositions of algae and bacteria from hydrothermal environments, Yellowstone National Park. *Geochim. Cosmochim. Acta* 48 : 591–99

Fuchs, G., Thauer, R., Ziegler, H., Stichler, W. 1979. Carbon isotope fractionation by *Methanobacterium thermoautotrophicum*. *Arch. Microbiol.* 120 : 135–39

Galimov, E. M. 1985. *The Biological Fractionation of Isotopes*. Orlando, Fla : Academic. 261 pp.

Galimov, E. M., Migdisov, A. A., Ronov, A. B. 1975. Variation in the isotopic composition of carbonate and organic carbon in sedimentary rocks during the Earth's history. *Geochem. Int.* 12 : 1–19

Hayes, J. M. 1983. Geochemical evidence bearing on the origin of aerobiosis. A speculative hypothesis. See Schopf 1983, pp. 291–301

Hayes, J. M., Kaplan, I. R., Wedeking, K. W. 1983. Precambrian organic geochemistry : preservation of the record. See Schopf 1983, pp. 93–134

Hoefs, J., Frey, M. 1976. Isotopic composition of carbonaceous matter in a metamorphic profile from the Swiss Alps. *Geochim. Cosmochim. Acta* 40 : 945–51

Hoering, T. C. 1967. The organic geochemistry of Precambrian rocks. In *Researches in Geochemistry*, ed. P. H. Abelson, pp. 89–111. New York : Wiley

Holland, H. D. 1978. *The Chemistry of the Atmosphere and Oceans*. New York : Wiley. 351 pp.

Junge, C. E., Schidlowski, M., Eichmann, R., Pietrek, H. 1975. Model calculations for the terrestrial carbon cycle : carbon isotope geochemistry and evolution of photosynthetic oxygen. *J. Geophys. Res.* 80 : 4542–52

Kaplan, I. R., Nissenbaum, A. 1966. Anomalous carbon isotope ratios in nonvolatile organic material. *Science* 153 : 744–45

Krasnovsky, A. A. 1981. Evolution of uphill electron transfer in photosynthesis. In *Science and Scientists—Essays by Biochemists, Biologists and Chemists*, ed. M. Kageyama, K. Nakamura, T. Oshima, T. Uchida, pp. 47–55. Tokyo/Dordrecht : Jpn. Sci. Soc. Press/Reidel

Kreulen, R., van Beek, P. C. J. M. 1983. The calcite-graphite isotope thermometer ; data on graphite bearing marbles from Naxos, Greece. *Geochim. Cosmochim. Acta* 47 : 1527–30

Krumbein, W. E., Cohen, Y. 1977. Primary production, mat formation and lithification chances of oxygenic and facultative anoxygenic cyanophytes (cyanobacteria). In *Fossil Algae*, ed. E. Flügel, pp. 37–56. Berlin : Springer-Verlag

Lancet, M. S., Anders, E. 1970. Carbon isotope fractionation in Fischer-Tropsch synthesis and in meteorites. *Science* 170 : 980–82

Lowe, D. R. 1980. Stromatolites 3.400-Myr old from the Archaean of Western Australia. *Nature* 284 : 441–43

McKirdy, D. M., Powell, T. S. 1974. Metamorphic alteration of carbon isotopic composition in ancient sedimentary organic matter : new evidence from Australia and South Africa. *Geology* 2 : 591–95

Mook, W. G., Bommerson, J. C., Staverman, W. H. 1974. Carbon isotope fractionation between dissolved bicarbonate and gaseous carbon dioxide. *Earth Planet. Sci. Lett.* 22 : 169–76

Murphey, B. F., Nier, A. O. 1941. Variations in the relative abundance of the carbon isotopes. *Phys. Rev.* 59 : 771–72

Nier, A. O., Gulbransen, E. A. 1939. Variations in the relative abundance of the carbon isotopes. *J. Am. Chem. Soc.* 61 : 697–98

Oehler, D. Z., Schopf, J. W., Kvenvolden, K. A. 1972. Carbon isotopic studies of organic matter in Precambrian rocks. *Science* 175 : 1246–48

O'Leary, M. H. 1981. Carbon isotope fractionation in plants. *Phytochemistry* 20 : 553–67

Pardue, J. W., Scalan, R. S., Van Baalen, C., Parker, P. L. 1976. Maximum carbon isotope fractionation in photosynthesis by blue-green algae and a green alga. *Geochim. Cosmochim. Acta* 40 : 309–12

Park, R., Epstein, S. 1960. Carbon isotope fractionation during photosynthesis. *Geochim. Cosmochim. Acta* 21 : 110–26

Perry, E. C., Ahmad, S. N. 1977. Carbon isotope composition of graphite and carbonate minerals from 3.8-AE metamorphosed sediments, Isukasia, Greenland. *Earth Planet. Sci. Lett.* 36: 280–84

Peters, K. E., Rohrback, B. G., Kaplan, I. R. 1981. Carbon and hydrogen stable isotope variations in kerogen during laboratory-simulated thermal maturation. *Bull. Am. Assoc. Pet. Geol.* 65: 501–8

Pflug, H. D., Jaeschke-Boyer, H. 1979. Combined structural and chemical analysis of 3.800-Myr-old microfossils. *Nature* 280: 483–86

Quandt, L., Gottschalk, G., Ziegler, H., Stichler, W. 1977. Isotope discrimination by photosynthetic bacteria. *FEMS Microbiol. Lett.* 1: 125–28

Rankama, K. 1948. New evidence of the origin of Pre-Cambrian carbon. *Geol. Soc. Am. Bull.* 59: 389–416

Reeburgh, W. S. 1983. Rates of biogeochemical processes in anoxic sediments. *Ann. Rev. Earth Planet. Sci.* 11: 269–98

Sackett, W. M., Eckelmann, W. R., Bender, M. L., Bé, A. W. H. 1965. Temperature dependence of carbon isotope composition in marine plankton and sediments. *Science* 148: 235–37

Sackett, W. M., Nakaparksin, S., Dalrymple, D. 1968. In *Advances in Organic Geochemistry 1966*, ed. G. D. Hobson, G. C. Speers, pp. 37–53. Oxford: Pergamon

Schidlowski, M. 1982. Content and isotopic composition of reduced carbon in sediments. In *Mineral Deposits and the Evolution of the Biosphere*, ed. H. D. Holland, M. Schidlowski, pp. 103–22. Berlin: Springer-Verlag

Schidlowski, M. 1983. Biologically mediated isotope fractionations: biochemistry, geochemical significance, and preservation in the Earth's oldest sediments. In *Cosmochemistry and the Origin of Life*, ed. C. Ponnamperuma, pp. 277–322. Dordrecht: Reidel

Schidlowski, M., Eichmann, R., Junge, C. E. 1975. Precambrian sedimentary carbonates: carbon and oxygen isotope geochemistry and implications for the terrestrial oxygen budget. *Precambrian Res.* 2: 1–69

Schidlowski, M., Eichmann, R., Fiebiger, W. 1976. Isotopic fractionation between organic carbon and carbonate carbon in Precambrian banded ironstone series from Brazil. *Neues Jahrb. Mineral. Monatsh.* 1976: 344–53

Schidlowski, M., Appel, P. W. U., Eichmann, R., Junge, C. E. 1979. Carbon isotope geochemistry of the 3.7×10^9 yr old Isua sediments, West Greenland: implications for the Archaean carbon and oxygen cycles. *Geochim. Cosmochim. Acta* 43: 189–99

Schidlowski, M., Hayes, J. M., Kaplan, I. R. 1983. Isotopic inferences of ancient biochemistries: carbon, sulfur, hydrogen and nitrogen. See Schopf 1983, pp. 149–86

Schidlowski, M., Matzigkeit, U., Krumbein, W. E. 1984. Superheavy organic carbon from hypersaline microbial mats: assimilatory pathway and geochemical implications. *Naturwissenschaften* 71: 303–8

Schidlowski, M., Matzigkeit, U., Mook, W. G., Krumbein, W. E. 1985. Carbon isotope geochemistry and ^{14}C ages of microbial mats from the Gavish Sabkha and the Solar Lake. In *Hypersaline Ecosystems: The Gavish Sabkha. Ecol. Stud.*, ed. G. M. Friedman, W. E. Krumbein, 53: 381–401. Berlin: Springer-Verlag

Schoell, M., Wellmer, F. W. 1981. Anomalous ^{13}C depletion in early Precambrian graphites from Superior Province, Canada. *Nature* 290: 696–99

Schopf, J. W., ed. 1983. *Earth's Earliest Biosphere: Its Origin and Evolution*. Princeton, NJ: Princeton Univ. Press. XXV + 543 pp.

Seckbach, J., Kaplan, I. R. 1973. Growth pattern and $^{13}C/^{12}C$ isotope fractionation of *Cyanidium caldarium* and hot spring algal mats. *Chem. Geol.* 12: 161–69

Sheppard, S. M. F., Schwarcz, H. P. 1970. Fractionation of carbon and oxygen isotopes and magnesium between coexisting metamorphic calcite and dolomite. *Contrib. Mineral. Petrol.* 26: 161–98

Shieh, Y. N., Taylor, H. P. 1969. Oxygen and carbon isotope studies of contact metamorphism of carbonate rocks. *J. Petrol.* 10: 307–31

Sirevag, R., Buchanan, B. B., Berry, J. A., Troughton, J. H. 1977. Mechanisms of CO_2 fixation in bacterial photosynthesis studied by the carbon isotope fractionation technique. *Arch. Microbiol.* 112: 35–38

Smith, B. N. 1976. Evolution of C4 photosynthesis in response to changes in carbon and oxygen concentrations in the atmosphere through time. *Biosystems* 8: 24–32

Smith, B. N., Epstein, S. 1971. Two categories of $^{13}C/^{12}C$ ratios for higher plants. *Plant Physiol.* 47: 380–84

Stanley, S. M. 1981. *The New Evolutionary Timetable*. New York: Basic Books. 222 pp.

Valley, J. W., O'Neil, J. R. 1981. $^{13}C/^{12}C$ exchange between calcite and graphite: a possible thermometer in Grenville marbles. *Geochim. Cosmochim. Acta* 45: 411–19

Veizer, J., Hoefs, J. 1976. The nature of $^{18}O/^{16}O$ and $^{13}C/^{12}C$ secular trends in sedimentary carbonate rocks. *Geochim. Cosmochim. Acta* 40: 1387–95

Veizer, J., Holser, W. T., Wilgus, C. K. 1980. Correlation of $^{13}C/^{12}C$ and $^{34}S/^{32}S$ secular variations. *Geochim. Cosmochim. Acta* 44: 579–87

Vogel, J. C. 1961. Isotope separation factors of carbon in the equilibrium system CO_2-HCO_3^--CO_3^{2-}. In *Summer Course on Nuclear Geology, Varenna, 1960*, ed. F. G. Houtermans, E. E. Picciotto, E. Tongiori, pp. 216–21. Pisa: Lischi

Vogel, J. C. 1980. Fractionation of the carbon isotopes during photosynthesis. *Sitzungsber. Heidelb. Akad. Wiss. Math.-Naturwiss. Kl.* 3: 111–35

Wada, H., Suzuki, K. 1983. Carbon isotopic thermometry calibrated by dolomite-calcite solvus temperatures. *Geochim. Cosmochim. Acta* 47: 697–706

Walker, D. 1979. *Energy, Plants and Man*. Chichester, Engl: Packard. 31 pp.

Walter, M. R., Buick, R., Dunlop, J. S. R. 1980. Stromatolites 3.400–3.500 Myr old from the North Pole area, Western Australia. *Nature* 284: 443–45

Weber, F., Schidlowski, M., Arneth, J. D., Gauthier-Lafaye, F. 1983. Carbon isotope geochemistry of the lower Proterozoic Francevillian Series of Gabon (Africa). *Terra Cognita* 3: 220

Welte, D. H., Kalkreuth, W., Hoefs, J. 1975. Age trend in carbon isotopic composition in Paleozoic sediments. *Naturwissenschaften* 62: 482–83

Winkler, F. J., Kexel, H., Kranz, C., Schmidt, H. L. 1982. Parameters affecting the $^{13}CO_2/^{12}CO_2$ isotope discrimination of the ribulose-1,5-bisphosphate carboxylase reaction. In *Stable Isotopes. Anal. Chem. Symp. Ser.*, ed. H. L. Schmidt, H. Förstel, K. Heinzinger, 11: 83–89. Amsterdam: Elsevier

Wong, W. W., Sackett, W. M., Benedict, C. R. 1975. Isotope fractionation in photosynthetic bacteria during carbon dioxide assimilation. *Plant Physiol.* 55: 475–79

Wong, W. W., Sackett, W. M. 1978. Fractionation of stable carbon isotopes by marine phytoplankton. *Geochim. Cosmochim. Acta* 42: 1809–15

Woodwell, G. M., Whittaker, R. H., Reiners, W. A., Likens, G. E., Delwiche, C. C., Botkin, D. D. 1978. The biota and the world carbon budget. *Science* 199: 141–46

PHYSICAL PROCESSES IN VOLCANIC ERUPTIONS

L. Wilson, H. Pinkerton and R. Macdonald

Department of Environmental Science, University of Lancaster, Lancaster LA1 4YQ, United Kingdom

INTRODUCTION

Recent advances in the theoretical interpretation of volcanic phenomena have resulted largely from studies in which volcanic mechanisms are modeled in terms of dynamic physico-chemical processes. In this respect, volcanology shares with much of Earth science in general the benefits that accrue from the application of the methods and concepts of other disciplines to its problems. Basic concepts from the fields of thermodynamics, fluid mechanics, and rheology have been particularly important in fostering new insights. Recent developments have been so numerous as to make it necessary to limit the scope of this review, and thus we concentrate here on the basic physical processes involved in the rise of magma to the surface and the distribution of eruption products on the ground or in the atmosphere. The perspective of planetary volcanism (Wilson & Head 1983) underlines the fact that volcanoes can be regarded as probes of the composition and structure of a planetary interior as a function of position, depth, and time.

MAGMA AVAILABILITY AND ERUPTION ONSET

Melt Formation

The wide range of tectonic settings, and thus geothermal gradients, and the associated horizontal and vertical variations in mantle composition on Earth ensure that there is no single depth or even a small range of depths at which partial melting takes place (Yoder 1976) and no unique melt composition that is formed (Thompson 1984). Extreme situations for partial melting depths would be 10–20 km for mid-ocean ridge basalts to about 200 km for kimberlites. The same is probably true of the other planets that have been volcanically active, for despite the apparent absence of plate tectonic processes, the majority show evidence of strong lateral variations in structure (Basaltic Volcanism Study Project 1981).

Volumetrically, most melts are probably formed by adiabatic decompression of rocks moving to lower pressure sites by large-scale convection. Other melting mechanisms include direct heating of rocks being carried into higher temperature regimes, e.g. into subduction zones, where viscous dissipative heating may also be a factor; volatile-induced melting related to lowering of solidus temperatures; and local concentration of radiogenic heat production.

Normally, melting takes place along grain-edge contacts, forming a three-dimensional interconnected network of channels (Beeré 1975, Watson 1982). The efficiency of melt segregation from the grain boundaries of the unmelted components of a rock and collection into larger pockets of fluid depend on the local pressure gradients, the effective permeability of the matrix, and the viscosity of the melt (McKenzie 1984, 1985, Richter & McKenzie 1984). To generate larger magma bodies, melt segregation must be rapid enough to minimize melt infiltration into the dry, unmelted source rocks (Watson 1982).

Diapirism

Once sufficiently large bodies of melt have been formed by filter-pressing, their buoyancy relative to the unmelted residues and the surrounding rock becomes significant. The melt bodies will rise at a speed dictated by the density contrast and the deformation mode of the surroundings, the latter actually having more control than the rheological properties of the magmas. At depths greater than several kilometers, the main deformation mode is plastic creep, and the melt bodies are classified as diapirs. This mode of transport probably dominates magma movement in the mantle and possibly the lower crust, although Weertman (1971), Shaw (1980), and Spera (1980) have considered the possibility of brittle fractures extending to the base of the crust and providing magma pathways in this region.

Fedotov (1977), Marsh & Kantha (1978), and Marsh (1982, 1984) have studied the motions of diapiric bodies over a wide range of sizes. The efficiency of heat loss from a diapir largely determines how far it can ascend into the crust. The internal temperature distribution determines whether convection (which increases the rate of heat loss) takes place within the diapir. The ascent of new diapirs may be dependent on the passage of earlier melt bodies that have effectively preheated the surrounding rocks. Cooling leads to the formation of crystals that are generally denser than the residual liquid, and these may sink through the liquid to be lost at the base of the melt zone. The enhanced buoyancy due to the lower density of the liquid helps to offset the effects of cooling.

Magma Chamber Evolution

As diapiric bodies approach the surface, the ability of the surrounding rocks to deform plastically decreases, with a corresponding decrease in the ascent velocity. The velocity becomes negligible on a geological time scale, and the diapir effectively becomes a fixed magma reservoir. In other cases, magma chambers may be fed by vertically propagating dikes that are arrested by changing stress regimes. Continued chemical and thermal evolution commonly involves the injection of new magma into the base of the original body, so that the chemical system is not closed (O'Hara & Mathews 1981). The important phenomena in magma chambers include fractional crystallization (where crystals form against the walls, roof, or floor of the chamber, depending on the heat-transfer modes, or form within the liquid and either sink or float); the partial melting and assimilation of the walls and roof of the chamber; absorption of volatiles from, or loss of volatiles to, the surrounding rocks; mixing of newly injected magma batches with preexisting material (McBirney 1980, Huppert & Sparks 1980, Sparks et al 1980); and separation of the liquid within the chamber into subhorizontal layers of differing composition and physico-chemical properties by double-diffusive convection (Huppert & Sparks 1984).

Most of these processes lead to volume changes in the reservoir which increase the stress on the surrounding rocks. These rocks eventually fail in a brittle fashion (Blake 1981) when the applied stress exceeds some function of their short-term tensile and shear strengths. Further magma movement is accomplished by the propagation of new pathways in the form of dikes or sills.

Dike Propagation

Studies of the stress conditions necessary for the propagation of magma-filled fractures have generally approximated dikes as planar fractures having a width specifiable as a function of depth and extending to infinity in the horizontal direction at right angles to the width vector (Weertman 1971, Johnson & Pollard 1973, Pollard & Muller 1976, Pollard 1976). The stress distribution around the opening tip of the crack is always great enough to continue the fracture process once initiated. Depending on the available magma volume and the distance between the magma source and the planetary surface, the fracture may be open for the whole of the distance or may almost completely close off behind a magma body of finite size (Stevenson 1982a).

Spence & Turcotte (1985) have shown that the propagation speed of a crack is normally limited not by the properties of the host rocks, but by the ability of the magma to flow into the advancing cavity. Magma movement is determined by the rheological (deformation) properties of the magma, the size and shape of the cavity being formed, and the magnitude of the driving stresses. The stresses may be caused by the buoyancy of the magma, by an excess pressure in the magma source region (Wilson & Head 1981a), or by stress variations due to the topographic shape of a volcanic edifice (Wadge 1977).

These models of dike shape assume that a constant buoyancy force drives the motion. However, the local density difference between a rising magma and the surrounding rocks may vary considerably as the nature of the surrounding rocks changes with depth and, particularly, as volatiles exsolve from the magma. Local variations in the density difference constrain new magma to occupy only certain parts of a growing volcanic edifice (Walker 1986). A more exact treatment of dike stability (yet to be formulated) would consider the integrated effects of density differences and stress gradients over the whole region occupied by the moving magma. The sill formation models of Johnson & Pollard (1973) begin to address problems of this kind, and it is quite possible that some magma reservoirs represent regions where numerous vertically propagating dikes have been converted into sills by the changing stress conditions.

SUBSURFACE MAGMA MOTION

Rheology of Magma

Magma probably exists in the lithosphere as a disperse system, a liquid matrix containing solids (in the form of crystals or accidental lithic debris stripped from the walls of the conduit) and bubbles of exsolving gases. Depending on the magma temperature and ambient pressure conditions, the solid or the gas bubble content may be zero. Disperse systems may exhibit a finite strength (the yield strength) that must be overcome by the applied stresses before motion can occur, and the subsequent motion may involve a nonlinear dependence of strain rate on the applied stress, so that the apparent viscosity (defined as the instantaneous ratio of shear stress to strain rate) is itself a function of the strain rate and, in some cases, of the stress and thermal histories of the material (McBirney & Murase 1984).

The rheological properties of a disperse system depend on the ratio of the length scale at which it is sheared to the size of the solid or gas bubble inclusions. The size range of the solid inclusions is not likely to change rapidly with position or time in a volcanic system, but the sizes of gas bubbles will change substantially with decompression (Sparks 1978), and the width of the conduit system may also change significantly with depth (Wilson & Head 1981a). Thus, the apparent viscosity may be a complicated function of position within the system. Certain lava flows containing millimeter-size crystals show good evidence for the crystals having settled through the liquid as though it were Newtonian, whereas the meter-scale morphology of the flow is dependent on the nonlinear rheological properties (Rowland & Walker 1986).

The rheological properties of many volcanic liquids may be empirically related to chemical composition and temperature (Bottinga & Weill 1972, Shaw 1972). However, these representations take no account of gas bubbles and do not apply at temperatures sufficiently below the liquidus that the crystal content becomes appreciable. No general model of the rheology of disperse systems in terms of the size, shape, rigidity, and number-density distributions of the inclusions currently exists, although treatments such as that of Gay et al (1969) address parts of the problem. It is necessary to rely mainly on empirical determinations of lava rheology to represent the general properties of magmas. These indicate that the behavior of many lavas can usefully be approximated by a Bingham model (Shaw et al 1968, Pinkerton & Sparks 1978). McBirney & Murase (1984) have developed a potentially useful model to calculate the viscosity of magma at subliquidus temperatures.

Magma Motion in Conduits

The velocity of a fluid moving steadily in a pipe may be obtained from the relation between the wall shear stress and the frictional energy losses (Knudsen & Katz 1958). For magma moving through a volcanic conduit, the wall shear stress can be written in terms of the pressure gradient driving the motion. The pressure gradient is in turn related to a combination of the density difference between the magma and its surroundings [in situations where the strength of the conduit walls is negligible (Wilson & Head 1981a)], any excess pressure that may exist in the magma source reservoir [when the conduit and reservoir walls have a finite strength (Druitt & Sparks 1984)], and the stress distribution within the region containing the magma source and conduit system (Solomon 1975, Wadge 1977). Representative magma velocities have been calculated for Newtonian or Bingham plastic rheological models of magmas moving in conduits with quite simple shapes (Johnson & Pollard 1973, Wilson & Head 1981a, Wilson et al 1980), although more complex fluids could be treated using methods such as those given in, for example, Wilkinson (1960). Ivey & Blake (1985) and Blake & Ivey (1986) stress the importance of the time variation of magma properties that occurs when an eruption taps a compositionally (or thermally) stratified magma chamber.

Volatile Exsolution and Pyroclast Formation

Volatile expansion is the major energy source driving volcanic fluid motions at shallow depths in the lithosphere (Wilson et al 1980, Wilson & Head 1981a, Kieffer 1982). The variations of volatile availability and solubility with depth (Holloway & Jakobsson 1986) suggest that CO_2 and SO_2 are most important below depths of a few kilometers (Moore 1979, Macpherson 1984), whereas H_2O, CO_2, SO_2, and halogens may all be significant at shallower depths. When a magma is slightly supersaturated in a volatile, gas exsolves into freshly nucleating bubbles or diffuses into existing bubbles that expand as they decompress (Sparks 1978, Williams & McBirney 1979, Watson et al 1982). In high-viscosity magmas, bubbles cannot migrate appreciably through the magma in the time needed to reach the surface, and they form a relatively uniform foam. If the fractional volume occupied by the bubbles exceeds about 75%, disruption of the fluid into vesicular pyroclasts (pumice or scoria) and released gas can occur (McBirney & Murase 1971, Sparks 1978) at depths of hundreds of meters to a few kilometers, and an explosive eruption takes place.

In low-viscosity magmas, larger bubbles can overtake and assimilate some of the smaller bubbles (Wilson & Head 1981a), leading to a range of eruption styles depending on the size distribution of the larger bubbles relative to the diameter of the conduit (Vergniolle & Jaupart 1986). The intermittent emergence of discrete bubbles at the surface of a lava lake to generate strombolian explosions (Blackburn et al 1976, Wilson & Head 1981a) is favored in the case of low-viscosity magma rising at low velocity (and hence low mass flux) in a narrow conduit. At higher mass fluxes in low-viscosity magmas, intermittent to relatively steady fire fountains (characteristic of Hawaiian basaltic eruptions) are generated in which the extent of the unsteady motions is governed by the combination of released gas content and conduit size. The basic principles of two-phase flow (Wallis 1969) must be applied to these systems (Vergniolle & Jaupart 1986). It is already established that the effective acoustic speed of a gas-pyroclast mixture [commonly of order 100 m s^{-1} (Kieffer 1977, Kieffer 1982, Wilson & Head 1981a)] is a critical limit to explosive eruption velocities in vents unless the conduit system develops into a quite specific outward-flaring shape. In long-lived eruptions, the required geometry may be "manufactured" by the pattern of pyroclast accumulation around the vent (Wilson et al 1980), but the details of this process are critically affected by the pressure of the atmosphere into which the eruption takes place (see discussion below).

The above treatments of magma motion assume negligible time rates of change of velocity at a given location. The more complex case of accelerated flow is likely to be of importance in the opening phases of any eruption (Bennett 1974) and will dominate transient eruptions in which either water gains intermittent access to the vent (Sheridan & Wohletz 1983) or the available volume of magma is so small compared with the volume of the conduit system that steady flow is never established. Numerical models of transient explosive eruptions have been developed by Self et al (1979), L. Wilson (1980), Kieffer (1981), Eichelberger & Hayes (1982), and Wohletz et al (1984). Kieffer & Sturtevant (1984) have used volatile fluids in laboratory simulations of volcanic systems driven by excess source pressures.

Cooling and Heating of Magma in Conduits

Rising magma cools by conduction into the surrounding rocks. Heat loss is maximized if turbulent motion occurs in a continuous liquid (as distinct from a gas-pyroclast mixture). Excessive cooling results in intrusive bodies when magma cannot reach the surface and in intermittent explosive activity when magma motion stops very close to the surface. Fedotov (1978) and Wilson & Head (1981a) have used steady-state models (Carslaw & Jaeger 1959) to establish maximum travel distances for magma in terms of the rise speed and conduit width. More exact treatments (Delaney & Pollard 1982) deal with the time dependence of the temperature distribution in both magma and surrounding rocks; thermal effects may be as important as reductions in driving stresses in terminating eruptions.

At sufficiently high mass fluxes in low-viscosity magmas (such as almost certainly characterized some eruptions of komatiite magmas during the Archean on Earth and some of the late-stage lunar eruptions), heat transfer to conduit walls may have been great enough, after an initial equilibration period (Hulme 1982), to cause wall melting (Huppert & Sparks 1985a). This process is enhanced if significant heating due to viscous dissipation occurs in the rising magma. By reducing magma viscosity, viscous heating can lead to high lava eruption speeds at the surface during nonexplosive eruptions (Hardee & Larson 1977).

ERUPTION STYLES

A classification scheme for styles of volcanic activity is clearly needed. However, even the basic distinction between effusive and explosive activity is not trivially made when many lava flows are formed by the coalescence of explosively generated pyroclasts falling from fire fountains, and when the term explosive is used to refer both to the discrete, transient release of a poorly coupled mixture of pyroclasts and gas in a strombolian or vulcanian explosion and to the steady discharge into the atmosphere of an intimately mixed jet of gas and clasts in a plinian or ignimbrite-forming eruption [see Williams & McBirney (1979) for definitions of these terms].

Given that gas release is an extremely common aspect of volcanism on Earth, the main problem is that of classifying explosive events. Walker's (1973a) scheme based on grain-size variations in the resulting deposits, together with later additions summarized by Fisher & Schmincke (1984), may be used on Earth and is equally applicable to modern or prehistoric eruptions. Classification of active explosive eruptions on Io (McEwen & Soderblom 1983), the products of ancient explosive eruptions on the Moon (Wilson & Head 1979) and Mars (Mouginis-Mark et al 1982), and probably the products of some kinds of explosive eruptive activity on Venus (Head & Wilson 1986) raises many problems.

The predicted ranges of grain sizes are very different from those in otherwise similar eruptions on Earth, with a tendency toward smaller clasts on planets with low atmospheric pressures, where even the smallest gas bubbles attempt to expand to a greater volume than they could attain on Earth (Wilson & Head 1983). Also, the greater degree of gas expansion dictated by the low external pressures will lead to greater velocities for the released gas streams and hence most of the small pyroclasts. Finally, low atmospheric pressure is coupled with low gravity for all the solar system objects known to have had any form of volcanic activity, and the dispersal of pyroclasts of a given size is likely to be much greater than on Earth to varying degrees for all the volcanically active planets and satellites except Venus, where the much higher atmospheric pressure would reduce pyroclast dispersal (Head & Wilson 1986). A classification scheme should, therefore, be based only on process, rather than on the dispersal of eruption products. Such a scheme would probably be parameterized by mass flux and exsolved magma volatile mass fraction (Wilson & Head 1983).

EFFUSIVE ERUPTIONS

Controlling Factors

The motion of a lava flow is defined by the size and shape of the fluid source vent, the rheological properties of the fluid, the mass flux at which it emerges, the topography of the surrounding terrain, and the acceleration due to gravity (Hulme 1974). It is of little importance whether a flow forms by direct overflow of magma from a vent or by coalescence of sufficiently hot pyroclastic fragments falling from a fire fountain or eruption cloud. Problems of access to active vents mean that few reliable measurements of vent size and shape or mass flux exist. Mass fluxes are commonly assessed from the width, thickness, bulk density, and advance rate of lava flows.

Confined Flows

At one extreme, the topography near a vent may contain one or more depressions in which the new lava is completely captured if each channel is sufficiently large relative to the mass flux to be accommodated. The balance between the down-slope gravitational force component and the frictional resistance then determines both the velocity distribution within the flow and its thickness. A common feature of Bingham plastic flows in confined channels is the presence of a rigid, moving plug (within which no shear takes place) surrounded by a region into which the shearing is concentrated (Johnson 1970). The motions of fluids with other rheological properties (e.g. Skelland 1967) could be deduced by similar methods.

Unconfined Flows

A single preexisting channel may not be able to accommodate the available mass flux. When the excess is small, lava will simply form separate streams in two or more depressions. If the mass flux is so large that, given the regional slope of the ground, the thickness of a single flow unit would be much greater than the typical depth of the depressions in the preexisting topography, it is relevant to consider the lava as spreading out onto a plane with a rough surface.

Hulme (1974) considered the motion of lava spreading out in a laminar fashion onto an inclined smooth plane. Apart from the effects of surface tension (which can be shown to be of vanishing importance for most lavas), a Newtonian liquid would spread indefinitely both down and across the slope. The fact that lava flow lobes commonly have a well-defined cross-sectional shape is evidence for a finite yield strength in essentially all lavas. By considering the down-slope and transverse stress fields and by slightly modifying Skelland's (1967) treatment of the motion of a Bingham plastic between parallel plates, Hulme (1974) related the volume fluxes and centerline thicknesses of flows, and the widths and thicknesses of the stationary banks (levees) at the sides of flows, to the lava yield strength, to plastic viscosity and density, and to the gravity and the slope of the ground.

Hulme & Fielder (1977), Moore et al (1978), Zimbelman (1985), and Head & Wilson (1986) used Hulme's model to interpret the shapes of flow deposits in terms of the lava yield strength and plastic viscosity. Moore et al (1978) noted some of its limitations. In that Hulme (1974) uses a solution for fluid motion between parallel plates, his model is applicable only to flows whose width greatly exceeds their depth. In addition, it was developed for constant, high-effusion-rate flows in which cooling during emplacement is unimportant. Both cooling and effusion rate changes can significantly alter the dimensions of confining levees (Sparks et al 1976) and the morphology of the resulting flow field (Wadge 1978). The progressive change in shape of a cooling lava as it flows away from a vent has been modeled by Park & Iverson (1984) and Pieri & Baloga (1986). Special considerations appear to apply to the motions of very high viscosity lavas, which may take a very long time to come to equilibrium with the stresses induced by gravity; Huppert et al (1982) successfully applied a Newtonian model to the growth of a lava dome.

The lengths of lava flows are of fundamental importance. Effusion rate is the dominant variable affecting flow length (Walker 1973b), although order of magnitude differences between lengths produced at similar effusion rates show that other factors, such as lava rheology, are important. For flows on Hawaii, Malin (1980) found that erupted volume was the most important factor in determining flow length. The rate and amount of cooling of flows (Hulme & Fielder 1977) can be expressed in terms of the dimensionless Grätz number (Prandtl 1952). When the Grätz number decreases to less than about 300, flow fronts stop advancing and breakouts commonly occur (Pinkerton & Sparks 1976). The continued advance of some flows for hours to days after vent activity has ceased indicates that lengths attained by many lava flows are not limited by marginal cooling. Only those eruptions that continue for a sufficient time to permit their Grätz numbers to fall below 300 will have lengths controlled by effusion rate; others will be limited by the volume of erupted lava, thus explaining the trend observed by Malin (1980).

Formation of lava tubes reduces heat loss and hence increases the potential length of a lava flow (Swanson 1973). Lava tubes commonly form a few days after the start of eruption of fluid lava (Greeley 1986). Tubes are less common in long-lived aa flows, where they form by roofing over the proximal channels (Guest et al 1980).

Lava tubes are also important in the formation of submarine lava flows (Greeley 1986), as are several other factors. The buoyancy of submerged lava effectively reduces the gravitational force acting on it, and the presence of an insulating sheath of steam at shallow water depths (Mills 1984) reduces heat loss rates; both of these factors contribute to the formation of fields of pillow lavas.

The surface texture of a lava flow is traditionally regarded as being a function of viscosity (Macdonald 1953, Fink & Fletcher 1978). Recent crystal-settling measurements on Hawaiian lava flows have confirmed that the viscosity of pahoehoe lava is less than that of proximal aa; that proximal aa has a comparable viscosity to that of toothpaste lava; and that they in turn have a viscosity that is significantly less than that of distal aa (Rowland & Walker 1986). The transition from pahoehoe to toothpaste to aa lava appears to be controlled by the strain rate generated within the flow (Peterson & Tilling 1980, Kilburn 1981, Rowland & Walker 1986).

Turbulent Flows and Sinuous Rilles

Long (> 50 km), sinuous channels occur near the boundaries of the basaltic lava-flooded mare basins on the Moon (Oberbeck et al 1969) and on Mars (Carr 1981). Some of these features have the characteristics of collapsed lava tubes (Oberbeck et al 1969), but Hulme (1973) and Carr (1974) proposed that in many other cases long-duration lava flows had melted the preexisting ground surface, progressively excavating trenches by thermal erosion of their central channels. High-effusion-rate turbulent flows would do this most effectively because of their nearly uniform temperature profile (Hulme 1973). Turbulent, as opposed to laminar, behavior was related to small differences in topographic slope (Hulme & Fielder 1977). The onset of melting typically required about two weeks (Hulme 1982), and detailed interpretations of rille length imply that sinuous rille-forming eruptions on the Moon (Wilson & Head 1981b) and Mars (Wilson & Mouginis-Mark 1984) involved mass fluxes of 10^8 to 10^9 kg s^{-1} continuing for times of the order of several months and leading to discharges of up to 1000 km^3 per eruption. Such large volumes and discharge rates are presumably related to the difficulty of driving eruptions through thick, cool, strong lithosphere.

Thermal erosion channels were also formed by komatiite lava flows on Earth (Huppert & Sparks 1985b). Turbulence in these flows was encouraged by low viscosities related to high temperatures and ultramafic compositions.

PYROCLAST DISPERSION

By definition, the production of pyroclastic materials in explosive eruptions is accompanied by the release of gas (mainly magmatic gas except in phreatomagmatic eruptions, where near-surface volatiles are incorporated). The dispersal of pyroclasts depends on the amount of gas expansion, on the mass ratio of gas to pyroclasts, on the nature of gas-pyroclast interactions (a function of the pyroclast grain-size distribution), and on the way that the gas and pyroclasts interact with whatever planetary atmosphere may be present.

Dispersion of Pyroclasts in vacuo

On Io and other bodies without atmospheres, the negligible atmospheric pressure permits essentially indefinite gas expansion, so that from both transient explosions and long-duration, steady explosive events, relatively high gas speeds are reached. A limiting speed, controlled by the initial gas temperature and molecular weight, is approached asymptotically as the pressure goes to zero (Kieffer 1982, Wilson & Head 1981a). As gas density decreases with expansion, drag between particles and gas quickly decreases to a negligible value, and clasts eventually follow ballistic trajectories (Strom et al 1982, Wilson & Head 1981a, Kieffer 1982).

The motions of pyroclasts and gas are complicated by two phenomena. First, the expansion of gas from a finite pressure in a vent into a vacuum is accomplished (Kieffer 1982) by the formation of a series of shock waves. Similar phenomena occur when volcanic materials are erupted explosively into a finite atmospheric pressure, unless the vent has had time to develop into a very specific Laval nozzle shape (Wilson & Head 1981a, Kieffer 1982). The velocity flow field of gas and particles emerging through a vent of any other shape is sufficiently complex (Kieffer 1981) that it can only be described in terms of empirical information on jets containing shocks obtained from studies of, for example, rocket engine exhausts (JANNAF 1975). Second, most magmatic (or phreatomagmatic) volatiles do not remain gases or vapors at low pressures; condensation to the liquid or solid form eventually occurs, releasing latent heat. Applying basic thermodynamic arguments, Kieffer (1982) has shown that the order in which (and the pressures at which) phase changes will occur depend on the initial entropy of the magma-volatile system, determined by the pressure and temperature from which the gas or vapor expansion begins. The current activity on Io, where it seems very probable that silicate magma interacts with both sulfur and sulfur dioxide at shallow levels in the crust, almost certainly provides illustrations of all the possible permutations of these phenomena (Kieffer 1982, McEwen & Soderblom 1983).

Smith et al (1979), Strom et al (1982), and Wilson & Head (1981a) have studied the dispersal of pyroclasts from steady eruptions in vacuo. Wilson & Head (1981a) emphasize the significance of the optical depth of pyroclast clouds in connection with heat retention. For the ancient lunar eruptions, pyroclast sizes were probably 100 to 1000 μm (Heiken & McKay 1977), and the released gas weight fractions were of order 0.05 wt% (Housley 1978). Steady explosive lunar eruptions should have been capable of producing opaque eruption clouds over high-mass-flux vents from which clasts at near-magmatic temperatures accumulated into lava ponds up to at least 5 km in diameter. Thermal erosion of the substrate of such ponds produced the source depressions of sinuous rilles (Wilson & Head 1981b). For phreatomagmatic eruptions on Io, the need to share magmatic heat with a comparable or larger mass of volatiles should result in relatively low temperatures for these systems (Kieffer 1982). The eruption clouds in

Voyager images are seen by reflected sunlight, and the optical depth concept can be used to investigate the particle packing density as a function of position in the clouds, leading to estimates of mass fluxes greater than 10^7 kg s^{-1} for the larger Io eruption clouds (Wilson & Head 1981c).

Transient, strombolian explosions apparently took place in lava lakes on the Moon and presumably currently do so on Io, where various permutations of silicate lava lakes and liquid sulfur lakes underlain by magmatic intrusions probably exist (Lunine & Stevenson 1985). The mass fraction of gas in the mixture of gas and pyroclasts in a strombolian eruption can be much higher than the average in the magma. Blackburn et al (1976) and Wilson & Head (1979) proposed that the most widespread lunar dark-mantling deposits were produced in this way. Some of the widespread and roughly circular deposits on Io may have a similar origin. Discrete vulcanian explosions may have occurred on the Moon when gas pressure built up under cooled plugs of basaltic lava chilled by incorporation of fragmental surface regolith materials. Head & Wilson (1979) found no other way of interpreting the masses and extents of very localized dark halo deposits on the floor of the crater Alphonsus. Finally, analogues of all the above eruption styles can probably be found on many of the small, icy bodies in the solar system, where water containing low-molecular-weight volatiles may play the equivalent role to silicate magma (Stevenson 1982b, Wilson & Head 1984).

Atmospheric Influences

The amount of gas expansion in an explosive eruption, and hence the highest gas speed reached, is determined by the ratio of the pressure at which magma disrupts into pyroclasts and the planetary atmospheric pressure (L. Wilson 1980). Pyroclast motion is determined by the ratio of the gravitational force acting on a pyroclast and the mutual drag force between the pyroclast and the surrounding gas. When gas and clasts are essentially locked together, the motion of the mixture can be treated (Wilson 1976) as that of a dusty gas (Saffman 1967). Much of the subsurface motion of pyroclasts in explosive eruptions on all planets takes place in this mode (Wilson & Head 1981a), although Wilson et al (1980) have studied the decoupling of gas and coarse pyroclasts near the level of the vent and shown how the pressure regime will adjust with time so as to reestablish good coupling and prevent the vent from becoming clogged with coarse debris.

The bulk densities of all pyroclast–magmatic gas mixtures are greater than the surrounding atmospheric densities at the same pressure as long as the gas mass fraction does not exceed about 20% (Wilson 1976). So, except in the cases of some phreatomagmatic explosions, all erupting "dusty gases" will initially be negatively buoyant. However, as pyroclasts and gas emerge through a vent, atmospheric gases are entrained into the mixture. The volcanic materials decelerate and the atmospheric component accelerates as momentum is shared via friction. As heat is exchanged, the atmospheric gas becomes less dense and a buoyancy flux is effectively created (Morton et al 1956). If enough air can be entrained, the final mixture will be positively buoyant, and a convecting eruption cloud will form over the vent; if it cannot, the gas-particle mixture will collapse back to the surface to form pyroclastic flows or surges (Sparks & Wilson 1976, Sparks et al 1978). Since the general motion is then largely horizontal rather than vertical, gravitational settling of clasts in the gas eventually dominates all other processes, and pyroclastic flow deposits of various kinds are formed (Walker 1985).

Plinian Eruptions

These events are characterized by a relatively steady magma discharge for at least a few tens of minutes (Walker & Croasdale 1971) and by the presence of a high enough released magma volatile content to ensure that eruption speeds are high [200–600 m s^{-1} typically on Earth (Wilson 1976)] and that most pyroclasts are initially dynamically locked to the gas. Enough air is entrained into the volcanic jet to ensure that the bulk density becomes low enough for convection to dominate the upper part of the resulting eruption cloud. The mean upward motion of cloud material can be modeled in terms of the laws of conservation of momentum, mass, and thermal energy (Morton et al 1956, Briggs 1969, Turner 1979) with due allowance for air entrainment at the edge of the cloud and progressive loss of pyroclasts as the upward gas velocity and the gas density decrease with height. Progressively more detailed dynamic models of eruption clouds have been developed by Wilson (1976), Wilson et al (1978), Suzuki (1983), Sparks & Wilson (1982), Sparks (1986), Carey & Sparks (1986), and L. Wilson & G. P. L. Walker (submitted for publication). Versions have been developed for the atmospheres of Mars (Wilson et al 1982) and Venus (Head & Wilson 1986).

These models predict that the ultimate height of maintained eruption clouds from steady eruptions should be relatively independent of all factors except the rate of heat injection at the base, which is determined by the magma mass flux from the vent with some allowance for the difficulty of extracting heat from very large pyroclasts that quickly fall out of the cloud. The functional relationship is that cloud height is proportional to the fourth root of the mass eruption rate (Wilson et al 1978, Settle 1978). There is an analogous relationship between maximum height and the total amount of heat injected for clouds from discrete volcanic explosions (Wilson et al 1978). The models of Carey & Sparks (1986) and L. Wilson & G. P. L. Walker can predict the maximum height to which pyroclasts of a given density are carried in a cloud and also the variation of cloud radius with height. This information, together with the ambient wind speed profile, allows prediction of the maximum ranges to which pyroclasts can be carried, both downwind and at right angles to the wind direction. Although the two models use different horizontal velocity profiles within the cloud and make differing simplifying assumptions, they yield similar eruption conditions, cloud heights, and wind speed profiles when applied to the analysis of both modern and ancient eruption deposits.

Pyroclastic Flows

Dense, ground-hugging, high-velocity clouds of gas and entrained particles may form from laterally directed, discrete, explosive eruptions; from the disruption of a lava flow or dome; from the sudden collapse of a convectively unstable eruption cloud; or from the steady outflow of material from a vent in conditions where a stable eruption cloud has never formed. Discrete explosions commonly produce block and ash flows, whereas collapsing eruption clouds or near-steady discharges tend to produce much more voluminous ignimbrites—flows dominated by the proportion of vesicular pumice rather than by other components such as poorly vesiculated juvenile magma, old vent wall material, or debris scoured from the ground over which the flow moves (Wilson 1986). The juvenile material in all types of pyroclastic flow undergoes mixing with the atmosphere everywhere on the contact surface.

Near the vent, segregation of gaseous and clastic components under gravity is determined by the ratio of the gravitational weight of the clasts and the mutual drag force between clasts and gas. The coarsest clasts are almost completely decoupled from all other components and follow near-ballistic trajectories to form a lag breccia deposit near the vent (Walker 1985). Clasts with a wide range of intermediate grain sizes, together with some of the erupted gas and entrained air, form the dense, ground-hugging, pyroclastic flow proper, the solids in which are incipiently fluidized by the gaseous components (Sparks 1976, C. J. N. Wilson 1980, 1984). The finest particles, together with the bulk of the gases, escape from the top of the pyroclastic flow to form a dilute, convecting cloud, the clasts from which are eventually deposited as an air-fall mantle over the flow and the surrounding topography (Sparks & Walker 1977).

The main body of an advancing pyroclastic flow is postulated to ingest air, the heating of which leads to strong fluidization and turbulence at the flow front (Wilson & Walker 1982) and the formation of various types of thin, basal deposits in large flows (Wilson 1985, Walker et al 1980, 1981) and lateral deposits in small flows (Wilson & Head 1981d). The deposit from the main body of the flow can be separated into a thin, lower boundary layer and a much thicker, upper zone in which motions were probably laminar and settling of dense clasts and flotation of light pumice fragments occurred in a fine-grained matrix having a density of order 1000 kg m^{-3} (Sparks 1976, Wilson 1985, Wilson & Walker 1982). Where deposition occurs from flows traveling on steep slopes (while crossing topographic obstacles or overshooting sudden bends in valleys), a much thinner version of this layer, termed a *veneer deposit*, may be formed (Walker et al 1981, Wilson 1985, Wilson & Walker 1982).

The dispersal of material in pyroclastic flows should be a function of both the released magma volatile weight fraction and the mass discharge rate (Sparks et al 1978). Incipient instability in convecting eruption clouds or episodic discharge from a vent is likely to produce relatively low-volume pyroclastic flows at correspondingly low mass fluxes, whereas high-mass-flux eruptions are more likely to produce single, high-volume flows. High gas contents in nonconvecting eruption products lead to high fountain structures over the vent. The conversion of potential energy to kinetic energy as material in a fountain reaches the ground will lead to high velocities in the pyroclastic flows, maximizing their chances of crossing topographic barriers—a common feature of ignimbrites (Miller & Smith 1977, Wilson & Walker 1985). Inferred velocities approaching 300 m s^{-1} (Wilson & Walker 1982) imply descent from fountains several kilometers high. High gas contents and consequent high velocities will tend to be linked with high mass fluxes, since pyroclastic flow formation is only preferred over convecting cloud formation for a given gas content if the mass flux exceeds a critical value (Sparks & Wilson 1976).

These trends are apparently reflected in the morphologies of flow deposits, with evidence (Wilson & Walker 1981, Wilson 1986) that relatively thick, multiple flow unit deposits emplaced at relatively low speeds and found largely confined to topographic depressions were produced from low-mass-flux eruptions, whereas relatively thin, extensive, topography-crossing, single flow unit deposits derive from high-mass-flux events. The complex variations of internal welding and chemical alteration seen in flow deposits (Ross & Smith 1960, Sheridan 1970, Schmincke 1974, Sheridan & Ragan 1977) are dictated by the emplacement temperature (which is strongly influenced by the amount of air entrainment) and, in multiple flow unit sequences, the time intervals between the arrival of successive units. Both these factors are also controlled ultimately by eruption conditions at the vent (Sparks & Wilson 1976).

Pyroclastic Surges

These horizontally directed, high-velocity clouds are dilute mixtures of gases (partly entrained air) and solid particles. In surges associated with phreatomagmatic eruptions, the gas is largely water of meteoric or surface origin (Fisher & Waters 1970, Schmincke et al 1973). Surges commonly develop in series as a result of partial instability of eruption clouds or intermittent access of water into the system (Freundte & Schmincke 1984). Surge deposits tend to be poorly sorted, with changes in facies and bed forms related to load and velocity. Characteristically, bed forms show wavy or low-angle cross-bedding (Fisher & Schmincke 1984).

Other types of surges are the hot, dry deposits associated with eruption column collapse (Fisher 1979) or turbulent processes at the margins of larger-scale pyroclastic flows (Wilson & Walker 1982). Fisher & Heiken (1982), Sparks (1983), and Walker & McBroome (1983) discuss the difficulties in distinguishing between the various formation mechanisms.

Fire Fountains and Near-Vent Deposits

Basaltic fire fountains on Earth represent the other extreme of gas-particle coupling from that displayed by plinian eruptions: Most pyroclasts are coarse enough to decouple from the gas motions almost immediately on leaving the vent (Wilson & Head 1981a) and subsequently follow nearly ballistic paths. On reaching the ground, their fate will depend critically on the opacity of the fire fountain in the same way as was discussed earlier for eruption in a vacuum: They may form an unwelded ash or scoria cone, an incipiently or totally welded deposit (Sparks & Wright 1979), or may coalesce to form a lava flow. In some cases, the buildup of stress in a growing deposit may cause deformation or flow to occur only after a sufficient mass has accumulated to overcome an effective yield strength (Wolff & Wright 1981).

By no means do all the pyroclasts in a basaltic fire fountain follow ballistic paths. There is generally observed to be a convecting plume containing the finer grain-size components above the incandescent fountain, and these fines settle out to form an air-fall deposit that is the equivalent, apart from the small size of its source region, of a co-ignimbrite air-fall mantle. The details of the motions of volcanic gas, entrained air, and small particles in fire fountains have not yet been studied in detail, and the influence of planetary atmospheric pressure on determining the nature of near-vent deposits from basaltic fire fountain eruptions is poorly understood from a theoretical viewpoint. Recently, it has become clear that clouds from basaltic eruptions with elongate, fissure-type vents may be important sources of sulfur compounds that have significant effects on climatic conditions. R. B. Stothers et al (submitted for publication) have shown that the rise heights of clouds from such sources should be proportional to the cube root of the mass eruption rate per unit fissure length.

SUMMARY

We have attempted to convey the success of the application of physical principles to volcanology. We have also noted that not all the major physical processes have been identified, much less understood. Further progress in volcanology requires continued interdisciplinary collaboration in order that the remaining problems can be more clearly defined and appropriate methods devised to solve them. This will ensure that the relevant field and laboratory observations and measurements are made to test the validity of the increasingly complex theoretical models of volcanic systems.

Literature Cited

Basaltic Volcanism Study Project. 1981. *Basaltic Volcanism on the Terrestrial Planets.* New York: Pergamon. 1286 pp.

Bennett, F. D. 1974. On volcanic ash formation. *Am. J. Sci.* 274: 648–61

Beeré, W. 1975. A unifying theory of the stability of penetrating liquid phases and sintering pores. *Acta Metall.* 23: 131–38

Blackburn, E. A., Wilson, L., Sparks, R. S. J. 1976. Mechanisms and dynamics of strombolian activity. *J. Geol. Soc. London* 132: 429–40

Blake, S. 1981. Volcanism and the dynamics of open magma chambers. *Nature* 289: 783–85

Blake, S., Ivey, G. N. 1986. Magma-mixing and the dynamics of withdrawal from stratified chambers. *J. Volcanol. Geotherm. Res.* 27: 153–78

Bottinga, Y., Weill, D. F. 1972. The viscosity of magmatic silicate liquids: a model for calculation. *Am. J. Sci.* 272: 438–75

Briggs, G. A. 1969. *Plume Rise.* Washington, DC: US At. Energy Comm. 80 pp.

Carey, S. N., Sparks, R. S. J. 1986. Quantitative models of the fall-out and dispersal of tephra from volcanic eruption columns. *Bull. Volcanol.* 48: 109–25

Carr, M. H. 1974. The role of lava erosion in the formation of lunar rilles and martian channels. *Icarus* 22: 1–23

Carr, M. H. 1981. *The Surface of Mars.* New Haven, Conn: Yale Univ. Press. 232 pp.

Carslaw, H. S., Jaeger, J. C. 1959. *Conduction of Heat in Solids.* Oxford: Oxford Univ. Press. 510 pp.

Delaney, P. T., Pollard, D. D. 1982. Solidification of basaltic magma during flow in a dike. *Am. J. Sci.* 282: 856–85

Druitt, T. H., Sparks, R. S. J. 1984. On the formation of calderas during ignimbrite eruptions. *Nature* 310: 679–81

Eichelberger, J. C., Hayes, D. B. 1982. Magmatic model for the Mount St. Helens blast of May 18, 1980. *J. Geophys. Res.* 87: 7727–38

Fedotov, S. A. 1977. Mechanism of deep-seated magmatic activity below island-arc volcanoes and similar structures. *Int. Geol. Rev.* 6: 671–80

Fedotov, S. A. 1978. Ascent of basic magmas in the crust and the mechanism of basaltic fissure eruptions. *Int. Geol. Rev.* 20: 33–48

Fink, J. H., Fletcher, R. C. 1978. Ropy pahoehoe: surface folding of a viscous fluid. *J. Volcanol. Geotherm. Res.* 4: 151–70

Fisher, R. V. 1979. Models for pyroclastic surges and pyroclastic flows. *J. Volcanol. Geotherm. Res.* 6: 305–18

Fisher, R. V., Heiken, G. 1982. Mt. Pelée, Martinique: May 8 and 20, 1902 pyroclastic flows and surges. *J. Volcanol. Geotherm. Res.* 13: 339–71

Fisher, R. V., Schmincke, H.-U. 1984. *Pyroclastic Rocks.* Berlin: Springer-Verlag. 472 pp.

Fisher R. V., Waters, A. C. 1970. Base surge bed forms in maar volcanoes. *Am. J. Sci.* 268: 157–80

Freundte, A., Schmincke, H.-U. 1984. Hierarchy of facies of pyroclastic flow deposits generated by Laacher See–type eruption. *Geology* 13: 278–81

Gay, E. C., Nelson, P. A., Armstrong, W. P. 1969. Flow properties of suspensions with high solids concentration. *AIChE J.* 15: 815–22

Greeley, R. 1986. The role of lava tubes in Hawaiian volcanoes. In *Reports of the Planetary Geology Program—1985*, NASA TM 88383, ed. S. J. Bougan, pp. 309–10

Guest, J. E., Underwood, J. R., Greeley, R. 1980. Role of lava tubes in flows from the Observatory Vent, 1971 eruption on Mount Etna. *Geol. Mag.* 117: 601–6

Hardee, H. C., Larson, D. W. 1977. Viscous dissipation effects in magma conduits. *J. Volcanol. Geotherm. Res.* 2: 299–308

Head, J. W., Wilson, L. 1979. Alphonsus-type dark-halo craters: morphology, morphometry and eruption conditions. *Proc. Lunar Planet. Sci. Conf., 10th*, pp. 2861–97. Houston: Lunar Planet. Inst.

Head, J. W., Wilson, L. 1986. Volcanic processes and landforms on Venus: theory, predictions, and observations. *J. Geophys. Res.* 91: 9407–46

Heiken, G., McKay, D. S. 1977. A model for the eruption behavior of a volcanic vent in Eastern Mare Serenitatis. *Proc. Lunar Sci. Conf., 8th*, pp. 3243–55. Houston: Lunar Planet. Inst.

Holloway, J. R., Jakobsson, S. 1986. Volatile solubility in magmas: transport of volatiles to planet surfaces. *J. Geophys. Res.* 91: 505–8

Housley, R. M. 1978. Modelling lunar eruptions. *Proc. Lunar Planet. Sci. Conf., 9th*, pp. 1473–84. Houston: Lunar Planet. Inst.

Hulme, G. 1973. Turbulent lava flow and the formation of lunar sinuous rilles. *Mod. Geol.* 4: 107–17

Hulme, G. 1974. The interpretation of lava flow morphology. *Geophys. J. R. Astron. Soc.* 39: 361–83

Hulme, G. 1982. A review of lava flow processes related to the formation of lunar sinuous rilles. *Geophys. Surv.* 5: 245–79

Hulme, G., Fielder, G. 1977. Effusion rates and rheology of lunar lavas. *Philos. Trans. R. Soc. London Ser. A* 285: 227–34

Huppert, H. E., Sparks, R. S. J. 1980. Restrictions on the compositions of mid-ocean ridge basalts: a fluid dynamical investigation. *Nature* 286: 46–48

Huppert, H. E., Sparks, R. S. J. 1984. Double-diffusive convection due to crystallization in magmas. *Ann. Rev. Earth Planet. Sci.* 12: 11–37

Huppert, H. E., Sparks, R. S. J. 1985a. Cooling and contamination of mafic and ultramafic magmas during ascent through continental crust. *Earth Planet. Sci. Lett.* 74: 371–86

Huppert, H. E., Sparks, R. S. J. 1985b. Komatiites I: eruption and flow. *J. Petrol.* 26: 694–725

Huppert, H. E., Shepherd, J. B., Sigurdsson, H., Sparks, R. S. J. 1982. On lava dome growth, with application to the 1979 lava extrusion of the Soufriere of St. Vincent. *J. Volcanol. Geotherm. Res.* 14: 199–222

Ivey, G. N., Blake, S. 1985. Axisymmetric withdrawal and inflow in a density-stratified container. *J. Fluid Mech.* 161: 115–37

JANNAF. 1975. *Joint Army, Navy, NASA, Air Force Handbook of Rocket Exhaust Plume Technology*, Chem. Propul. Inf. Agency Publ. 263, Ch. 2. 237 pp.

Johnson, A. M. 1970. *Physical Processes in Geology*. San Francisco: Freeman. 577 pp.

Johnson, A. M., Pollard, D. D. 1973. Mechanics and growth of some laccolithic intrusions in the Neary Mountains, Utah: I, Field observations, Gilbert's model, physical properties and the flow of the magma. *Tectonophysics* 18: 261–309

Kieffer, S. W. 1977. Sound speed in liquid-gas mixtures: water-air and water-steam. *J. Geophys. Res.* 82: 2895–2904

Kieffer, S. W. 1981. Blast dynamics at Mount St. Helens on 18 May 1980. *Nature* 291: 568–70

Kieffer, S. W. 1982. Dynamics and thermodynamics of volcanic eruptions: implications for the plumes on Io. In *Satellites of Jupiter*, ed. D. Morrison, pp. 647–723. Tucson: Univ. Ariz. Press

Kieffer, S. W., Sturtevant, B. 1984. Laboratory studies of volcanic jets. *J. Geophys. Res.* 89: 8253–68

Kilburn, C. R. J. 1981. Pahoehoe and aa lava: a discussion and continuation of the model of Peterson and Tilling. *J. Volcanol. Geotherm. Res.* 11: 373–82

Knudsen, J. G., Katz, D. L. 1958. *Fluid Dynamics and Heat Transfer*. New York: McGraw-Hill. 576 pp.

Lunine, J. I., Stevenson, D. J. 1985. Physics and chemistry of sulfur lakes on Io. *Icarus* 64: 345–67

Macpherson, G. J. 1984. A model for predicting the volumes of vesicles in submarine basalts. *J. Geol.* 92: 72–82

Macdonald, G. A. 1953. Pahoehoe, aa, and block lava. *Am. J. Sci.* 251: 169–91

Malin, M. C. 1980. Lengths of Hawaiian lava flows. *Geology* 8: 306–8

Marsh, B. D. 1982. On the mechanics of igneous diapirism, stoping and zone melting. *Am. J. Sci.* 282: 808–55

Marsh, B. D. 1984. Mechanics and energetics of magma formation and ascension. In

Explosive Volcanism: Inception, Evolution and Hazards, Geophysics Study Committee, pp. 67–83. Washington, DC: Natl. Acad. Press. 176 pp.

Marsh, B. D., Kantha, L. H. 1978. On the heat and mass transfer from an ascending magma. Earth Planet. Sci. Lett. 39: 435–43

McBirney, A. R. 1980. Mixing and unmixing of magmas. J. Volcanol. Geotherm. Res. 7: 357–71

McBirney, A. R., Murase, T. 1971. Factors governing the formation of pyroclastic rocks. Bull. Volcanol. 34: 372–84

McBirney, A. R., Murase, T. 1984. Rheological properties of magmas. Ann. Rev. Earth Planet. Sci. 12: 337–57

McEwen, A. S., Soderblom, L. A. 1983. Two classes of volcanic plumes on Io. Icarus 55: 191–217

McKenzie, D. 1984. The generation and compaction of partially molten rock. J. Petrol. 25: 713–65

McKenzie, D. 1985. The extraction of magma from the crust and mantle. Earth Planet. Sci. Lett. 74: 81–91

Miller, T. P., Smith, R. L. 1977. Spectacular mobility of ash flows around Aniachak and Fisher calderas, Alaska. Geology 5: 173–76

Mills, A. A. 1984. Pillow lavas and the Leidenfrost effect. J. Geol. Soc. London 141: 183–86

Moore, H. J., Arthur, D. W. G., Schaber, G. G. 1978. Yield strengths of flows on the Earth, Mars and Moon. Proc. Lunar Planet. Sci. Conf., 9th, pp. 3351–78. Houston: Lunar Planet. Inst.

Moore, J. G. 1979. Vesicularity and CO_2 in mid-ocean ridge basalt. Nature 282: 250–53

Morton, B. R., Taylor, G., Turner, J. S. 1956. Turbulent gravitational convection from maintained and instantaneous sources. Proc. R. Soc. London Ser. A 234: 1–23

Mouginis-Mark, P. J., Wilson, L., Head, J. W. 1982. Explosive volcanism on Hecates Tholus, Mars: investigation of eruption conditions. J. Geophys. Res. 87: 9890–9904

Oberbeck, V. R., Quaide, W. L., Greeley, R. 1969. On the origin of sinuous rilles. Mod. Geol. 1: 75–80

O'Hara, M. J., Mathews, R. E. 1981. Geochemical evolution in an advancing, periodically replenished, periodically tapped, continuously fractionated magma chamber. J. Geol. Soc. London 138: 237–77

Park, S., Iverson, J. D. 1984. Dynamics of lava flow: thickness growth characteristics of steady two-dimensional flow. Geophys. Res. Lett. 11: 611–44

Peterson, D. W., Tilling, R. I. 1980. Transition of basaltic lava from pahoehoe to aa, Kilauea volcano, Hawaii: field observations and key factors. J. Volcanol. Geotherm. Res. 7: 271–93

Pieri, D., Baloga, S. 1986. Lava flow profiles: time dependent solutions of the Jeffreys' equation. In Reports of the Planetary Geology Program—1985, NASA TM 88383, ed. S. J. Bougan, pp. 315–19

Pinkerton, H., Sparks, R. S. J. 1976. The 1975 sub-terminal lavas, Mount Etna: a case history of the formation of a compound lava field. J. Volcanol. Geotherm. Res. 1: 167–82

Pinkerton, H., Sparks, R. S. J. 1978. Field measurements of the rheology of lava. Nature 276: 383–85

Pollard, D. D. 1976. On the form and stability of open hydraulic fractures in the Earth's crust. Geophys. Res. Lett. 3: 513–16

Pollard, D. D., Muller, O. H. 1976. The effects of gradients in regional stress and magma pressure on the form of sheet intrusions in cross section. J. Geophys. Res. 81: 975–84

Prandtl, L. 1952. The Essentials of Fluid Mechanics. London: Blackie & Son. 452 pp.

Richter, F. M., McKenzie, D. 1984. Dynamical models for melt segregation from a deformable matrix. J. Geol. 92: 729–40

Ross, C. S., Smith, R. L. 1960. Ash-flow tuffs: their origin, geologic relation and identification. US Geol. Surv. Prof. Pap. 366, pp. 1–77

Rowland, S. K., Walker, G. P. L. 1986. Toothpaste lava: characteristics of a lava structural type distinct from pahoehoe and aa. Geology. In press

Saffman, P. G. 1967. On the stability of flow of a dusty gas. J. Fluid Mech. 13: 120–28

Schmincke, H.-U. 1974. Volcanological aspects of peralkaline silicic welded ash-flow tuffs. Bull. Volcanol. 38: 594–636

Schmincke, H.-U., Fisher, R. V., Waters, A. C. 1973. Antidune and shute and pool structures in base surge deposits of the Laacher See area, Germany. Sedimentology 20: 553–74

Self, S., Wilson, L., Nairn, I. A. 1979. Vulcanian eruption mechanisms. Nature 277: 440–43

Settle, M. 1978. Volcanic eruption clouds and the thermal power output of explosive eruptions. J. Volcanol. Geotherm. Res. 3: 309–24

Shaw, H. R. 1972. Viscosities of magmatic liquids: an empirical method of prediction. Am. J. Sci. 272: 870–93

Shaw, H. R. 1980. The fracture mechanisms

of magma transport from the mantle to the surface. In *Physics of Magmatic Processes*, ed. R. B. Hargraves, pp. 201–64. Princeton, NJ: Princeton Univ. Press. 585 pp.

Shaw, H. R., Wright, T. L., Peck, D. L., Okamura, R. 1968. The viscosity of basaltic magma: an analysis of field measurements in Makaopuhi lava lake, Hawaii. *Am. J. Sci.* 266: 225–64

Sheridan, M. F. 1970. Fumarolic mounds and ridges of the Bishop Tuff, California. *Geol. Soc. Am. Bull.* 81: 851–68

Sheridan, M. F., Ragan, D. M. 1977. Compaction of ash-flow tuffs. *Dev. Sedimentol.* 18B: 677–713

Sheridan, M. F., Wohletz, K. H. 1983. Hydrovolcanism: basic considerations and review. *J. Volcanol. Geotherm. Res.* 17: 1–29

Skelland, A. H. P. 1967. *Non-Newtonian Flow and Heat Transfer*. New York: Wiley. 469 pp.

Smith, B. A., Shoemaker, E. M., Kieffer, S. W., Cook, A. F. 1979. The role of SO_2 in volcanism on Io. *Nature* 280: 738–43

Solomon, S. C. 1975. Mare volcanism and lunar crustal structure. *Proc. Lunar Sci. Conf., 6th*, pp. 1021–42. Houston: Lunar Planet. Inst.

Sparks, R. S. J. 1976. Grain size variations in ignimbrites and implications for the transport of pyroclastic flow. *Sedimentology* 23: 147–88

Sparks, R. S. J. 1978. The dynamics of bubble formation and growth in magmas: a review and analysis. *J. Volcanol. Geotherm. Res.* 3: 1–37

Sparks, R. S. J. 1983. Mont Pelée, Martinique: May 8 and 20, 1902, pyroclastic flows and surges—discussion. *J. Volcanol. Geotherm. Res.* 19: 175–84

Sparks, R. S. J. 1986. The dimensions and dynamics of volcanic eruption columns. *Bull. Volcanol.* 48: 3–15

Sparks, R. S. J., Walker, G. P. L. 1977. The significance of vitric-enriched air-fall ashes associated with crystal-enriched ignimbrites. *J. Volcanol. Geotherm. Res.* 2: 329–41

Sparks, R. S. J., Wilson, L. 1976. A model for the formation of ignimbrite by gravitational column collapse. *J. Geol. Soc. London* 132: 441–51

Sparks, R. S. J., Wilson, L. 1982. Explosive volcanic eruption—V. Observations of plume dynamics during the 1979 Soufriere eruption, St. Vincent. *Geophys. J. R. Astron. Soc.* 69: 551–70

Sparks, R. S. J., Wright, J. V. 1979. Welded air-fall tuffs. *Geol. Soc. Am. Spec. Pap. No. 180*, pp. 155–66

Sparks, R. S. J., Pinkerton, H., Hulme, G. 1976. Classification and formation of levees on Mount Etna, Sicily. *Geology* 4: 269–71

Sparks, R. S. J., Wilson, L., Hulme, G. 1978. Theoretical modelling of the generation, movement and emplacement of pyroclastic flows by column collapse. *J. Geophys. Res.* 83: 1727–39

Sparks, R. S. J., Meyer, P., Sigurdsson, H. 1980. Density variation amongst mid-ocean ridge basalts: implications for magma mixing and the scarcity of primitive basalts. *Earth Planet. Sci. Lett.* 46: 419–30

Spence, D. A., Turcotte, D. L. 1985. Magma driven propagation of cracks. *J. Geophys. Res.* 90: 575–80

Spera, F. J. 1980. Aspects of magma transport. In *Physics of Magmatic Processes*, ed. R. B. Hargraves, pp. 265–324. Princeton: Princeton Univ. Press. 585 pp.

Stevenson, D. J. 1982a. Migration of fluid-filled cracks: applications to terrestrial and icy bodies. *Lunar Planet. Sci. XIII*, pp. 768–69. Houston: Lunar Planet. Inst.

Stevenson, D. J. 1982b. Volcanism and igneous processes in small, icy satellites. *Nature* 298: 142–44

Strom, R. G., Schneider, N. M., Terrile, R. J., Cook, A. F., Hansen, C. 1982. Volcanic eruptions on Io. *J. Geophys. Res.* 86: 8593–8620

Suzuki, T. 1983. A theoretical model for dispersion of tephra. In *Arc Volcanism: Physics and Tectonics*, ed. D. Shimozuru, I. Yokoyama, pp. 95–113. Dordrecht: Reidel

Swanson, D. A. 1973. Pahoehoe flows from the 1969–1971 Mauna Ulu eruption, Kilauea volcano, Hawaii. *Geol. Soc. Am. Bull.* 84: 615–26

Thompson, R. N. 1984. Dispatches from the basalt front. I. Experiments. *Proc. Geol. Assoc.* 95: 249–62

Turner, J. S. 1979. *Buoyancy Effects in Fluids*. Cambridge: Cambridge Univ. Press. 367 pp.

Vergniolle, S., Jaupart, C. 1986. Two-phase flow and volcanic eruptions. *J. Geophys. Res.* In press

Wadge, G. 1977. Storage and release of magma on Mount Etna. *J. Volcanol. Geotherm.* 2: 361–84

Wadge, G. 1978. Effusion rate and the shape of lava flow fields on Mount Etna. *Geology* 6: 503–6

Walker, G. P. L. 1973a. Explosive volcanic eruptions: a new classification scheme. *Geol. Rundsch.* 62: 431–46

Walker, G. P. L. 1973b. Lengths of lava flows. *Philos. Trans. R. Soc. London Ser. A* 274: 107–18

Walker, G. P. L. 1985. Origin of coarse lithic breccias near ignimbrite source vents. *J. Volcanol. Geotherm. Res.* 24: 157–70

Walker, G. P. L. 1986. Koolau dike complex, Oahu: intensity and origin of a sheeted-dike complex high in a Hawaiian volcanic edifice. *Geology* 14: 310–13

Walker, G. P. L., Croasdale, R. 1971. Two plinian-type eruptions in the Azores. *J. Geol. Soc. London* 127: 17–55

Walker, G. P. L., McBroome, L. A. 1983. Mount St. Helens 1980 and Mount Pelée 1902—flow or surge. *Geology* 11: 571–74

Walker, G. P. L., Wilson, C. J. N., Froggatt, P. C. 1980. Fines-depleted ignimbrite in New Zealand—the product of a turbulent pyroclastic flow. *Geology* 8: 245–49

Walker, G. P. L., Self, S., Froggatt, P. C. 1981. The ground layer of the Taupo ignimbrite: a striking example of sedimentation from a pyroclastic flow. *J. Volcanol. Geotherm. Res.* 10: 1–11

Wallis, G. B. 1969. *One Dimensional Two-Phase Flow.* New York: McGraw-Hill. 408 pp.

Watson, E. B. 1982. Melt infiltration and magma evolution. *Geology* 10: 236–40

Watson, E. B., Sneeriger, M. A., Ross, A. 1982. Diffusion of dissolved carbonate in magmas: experimental results and applications. *Earth Planet. Sci. Lett.* 61: 346–58

Weertman, J. 1971. Theory of water-filled crevasses in glaciers applied to vertical magma transport beneath oceanic ridges. *J. Geophys. Res.* 76: 1171–83

Wilkinson, W. L. 1960. *Non-Newtonian Fluids: Fluid Mechanics, Mixing and Heat Transfer.* New York: Pergamon. 138 pp.

Williams, H., McBirney, A. R. 1979. *Volcanology.* San Francisco: Freeman. 397 pp.

Wilson, C. J. N. 1980. The role of fluidisation in the emplacement of pyroclastic flows: an experimental approach. *J. Volcanol. Geotherm. Res.* 8: 231–49

Wilson, C. J. N. 1984. The role of fluidisation in the emplacement of pyroclastic flows, 2: experimental results and their interpretation. *J. Volcanol. Geotherm. Res.* 20: 55–84

Wilson, C. J. N. 1985. The Taupo eruption, New Zealand II. The Taupo ignimbrite. *Philos. Trans. R. Soc. London Ser. A* 314: 229–310

Wilson, C. J. N. 1986. Pyroclastic flows and ignimbrites. *Sci. Prog. (Oxford)* 70: 171–207

Wilson, C. J. N., Walker, G. P. L. 1981. Violence in pyroclastic flow eruptions. In *Tephra Studies,* ed. S. Self, R. S. J. Sparks, pp. 441–48. Dordrecht: Reidel

Wilson, C. J. N., Walker, G. P. L. 1982. Ignimbrite depositional facies: the anatomy of a pyroclastic flow. *J. Geol. Soc. London* 139: 581–92

Wilson, C. J. N., Walker, G. P. L. 1985. The Taupo eruption, New Zealand, I. General aspects. *Philos. Trans. R. Soc. London Ser. A* 314: 199–228

Wilson, L. 1976. Explosive volcanic eruptions—III. Plinian eruption columns. *Geophys. J. R. Astron. Soc.* 45: 543–56

Wilson, L. 1980. Relationships between pressure, volatile content and ejecta velocity in three types of volcanic explosion. *J. Volcanol. Geotherm. Res.* 8: 297–313

Wilson, L., Head, J. W. 1979. Lunar volcanic cones and dark mantling deposits: consequences of patterns of volatile release. *Lunar Planet. Sci. X,* pp. 1353–55. Houston: Lunar Planet. Inst.

Wilson, L., Head, J. W. 1981a. Ascent and eruption of basaltic magma on the Earth and Moon. *J. Geophys. Res.* 86: 2971–3001

Wilson, L., Head, J. W. 1981b. Lunar sinuous rille formation by thermal erosion: eruption conditions, rates and durations. *Lunar Planet. Sci. XII,* pp. 427–29. Houston: Lunar Planet. Inst.

Wilson, L., Head, J. W. 1981c. Io volcanic eruptions: mass eruption rate estimates. *Lunar Planet. Sci. XII,* pp. 1191–93. Houston: Lunar Planet. Inst.

Wilson, L., Head, J. W. 1981d. Morphology and rheology of pyroclastic flows and their deposits, and guidelines for future eruptions. *US Geol. Surv. Prof. Pap. 1250,* pp. 513–24

Wilson, L., Head, J. W. 1983. A comparison of volcanic eruption processes on Earth, Moon, Mars, Io and Venus. *Nature* 302: 663–69

Wilson, L., Head, J. W. 1984. Aspects of water eruptions on icy satellites. *Lunar Planet. Sci. XV,* pp. 924–25. Houston: Lunar Planet. Inst.

Wilson, L., Mouginis-Mark, P. J. 1984. Martian sinuous rilles. *Lunar Planet. Sci. XV,* pp. 926–27. Houston: Lunar Planet. Inst.

Wilson, L., Sparks, R. S. J., Huang, T. C. Watkins, N. D. 1978. The control of volcanic column heights by eruption energetics and dynamics. *J. Geophys. Res.* 83: 1829–36

Wilson, L., Sparks, R. S. J., Walker, G. P. L. 1980. Explosive volcanic eruptions—IV. The control of magma properties and conduit geometry on eruption column behavior. *Geophys. J. R. Astron. Soc.* 63: 117–48

Wilson, L., Head, J. W., Mouginis-Mark, P. J. 1982. Theoretical analysis of martian volcanic eruption mechanisms. *ESA SP-185,* pp. 107–13

Wohletz, K. H., McGetchin, T. R., Sandford, M. T. II, Jones, E. M. 1984. Hydrodynamic aspects of caldera-forming eruptions: numerical models. *J. Geophys. Res.* 89: 8269–85

Wolff, J. A., Wright, J. V. 1981. Rheomorphism of welded tuffs. *J. Volcanol. Geotherm. Res.* 10: 13–34

Yoder, H. S. 1976. *Generation of Basaltic Magma.* Washington, DC: Natl. Acad. Sci. 265 pp.

Zimbelman, J. R. 1985. Estimation of rheological properties for flows on the martian volcano Ascraeus Mons. *J. Geophys. Res.* B90: D157–62 (Suppl.)

GEOARCHAEOLOGY

George Rapp, Jr.

Geology Department and Archaeometry Laboratory, University of Minnesota, Duluth, Minnesota 55812

INTRODUCTION: HISTORY OF DEVELOPMENT

This review is written with an Earth sciences audience in mind. It concentrates on four aspects of geoarchaeology (or archaeological geology) most likely to be of interest to Earth scientists: (*a*) a brief history of the development of geoarchaeology, (*b*) archaeological sediments, (*c*) paleogeomorphic reconstructions, and (*d*) provenance studies. Such coverage leaves as much unsaid as said, since geoarchaeology is nearly as broad a field as geology itself. Geophysics (e.g. Weymouth & Huggins 1985), geochemistry (e.g. Wehmiller & Belknap 1978), mineralogy (e.g. Littmann 1980), geochronology (e.g. Bada & Finkel 1982), paleontology (e.g. Klein 1979), economic geology (e.g. Patterson 1971), petrography (e.g. Dickinson & Shutler 1979), petrology (e.g. Kempe & Harvey 1983), and nearly every other subdiscipline in geology have concepts and methods applicable to the solution of archaeological problems.

From approximately 1830 to 1930 the young disciplines of geology and archaeology were often united in a common goal—a study of the evidence for the "antiquity of man." Sir Charles Lyell, one of the founders of modern geology, in his book *The Geological Evidence for the Antiquity of Man* (1870) clearly established the role of geology in archaeological inquiry. The history and development of geoarchaeology are detailed in Rapp & Gifford (1982a) and Gifford & Rapp (1985a). The dynamic interactions between geology and archaeology in late nineteenth-century North America involved many of the best-known American geologists of that era: T. C. Chamberlin, J. D. Dana, W. H. Holmes, W. J. McGee, J. W. Powell, H. S. Washington, N. H. Winchell, and G. F. Wright (Gifford & Rapp 1985b). John Wesley Powell was instrumental in the birth and early development of both the United States Geological Survey and the Bureau of American Ethnology and became director of both. In the late

1930s, after a somewhat slack period in the early twentieth century, well-known geologists such as Kirk Bryan, Ernst Antevs, Frederick Zeuner, and E. H. Sellards led the way for increased participation of geologists in archaeological investigations. Currently there are hundreds of geologists involved worldwide in archaeology and additional hundreds of archaeologists routinely using geologic concepts and methods in their research.

Butzer (1982) draws a distinction between *geoarchaeology*, which implies archaeological research using the methods and concepts of the Earth sciences but not necessarily linked directly to geology, and *archaeological geology*, which implies geologic research undertaken by geologists to aid in the solution of archaeological problems. He asks that a fundamental distinction be made between technique and goal. Rapp & Gifford (1982a) draw a somewhat similar distinction based on which disciplinary designation is the noun and which is the modifier. If one uses these distinctions, this review covers both archaeological geology and geoarchaeology but leans toward the former. Other major discussions of the role of the geologist in archaeology or of the nature of geoarchaeology can be found in Rapp (1975), Davidson & Shackley (1976), Butzer (1977), Gladfelter (1977, 1981), Bullard (1978), Hassan (1979), and Rapp & Gifford (1985).

Archaeological geology also grades imperceptibly into environmental geology, depending primarily on when the human/environment interaction took place (Folk 1975, Moss & Walker 1978).

For two decades or more, archaeology has been in a transformational phase, partly because of a major dedication to the development of a theoretical basis for archaeological studies and partly because of the increasing realization that the whole context of an archaeological deposit must be understood before ancient lifeways can be reconstructed from incomplete remains. This "whole context" includes the geologic as a critical component. However, views remain divided on what constitutes archaeology. As recently as 1958 two eminent archaeologists opined that archaeology is anthropology or it is nothing (Willey & Phillips 1958). The active interface between a natural and a social science can be strained by perspectives as well as paradigms.

Archaeological geology has recently coalesced as an independent subdiscipline. In 1977 the Archaeological Geology Division of the Geological Society of America was formally organized. Earlier that same year archaeological geology was added to the list of specialties reviewed yearly in an issue of *Geotimes*.

As one can see from the literature cited, publications on geoarchaeology and archaeological geology are spread among hundreds of journals and site reports, and this literature has been expanding geometrically since 1960. The recent founding of the journal *Geoarchaeology* may provide a focused outlet for significant papers in this subdiscipline. Established disciplines generate a means of training practitioners. The Archaeometry Laboratory at the University of Minnesota, Duluth, publishes an annual *Directory of Graduate Programs in Archaeological Geology and Geoarchaeology*. The 1985 edition lists 17 universities with graduate programs.

ARCHAEOLOGICAL SEDIMENTS

Excavation archaeologists have become good at resolving complex stratigraphies (Figure 1). However, the science of archaeological sedimentology has been slow in developing. The sedimentary record of an archaeological site can provide critical environmental, stratigraphic, and cultural information because these sediments are a mix of geologic, biogenic, and anthropogenic components.

The lithology, micropaleontology, structure, texture, spatial contexts, geochemistry, taphonomy, and pedology all contain unique information on the human element in the formation of these deposits. Yet only in the last three or four decades have systematic sedimentological studies been undertaken as a significant component of excavation. Many of the "early" systematic studies of sedimentary records have been in the excavation of caves and rock shelters.

Three American geologists have recently been instrumental in furthering the systematics of archaeological cave and rock-shelter sedimentology. W. R. Farrand's work at Abri Pataud in France (Farrand 1975a,b) and at various sites in the Levant (Farrand 1979) has shown that sedimentological studies can recover critical information on climate, provenance of debris, stratigraphy, human and animal occupation, and chronology of the site. P. Goldberg, particularly using the techniques of sediment micromorphology, has extended the types of data available to reconstruct the human occupation of caves (Goldberg 1979a,b, Goldberg & Nathan 1975).

K. W. Butzer (1973, 1981) has shown that appreciable quantities of lithic debris in caves are brought in by human occupants. He has used analyses of the mineral soil, organic colloids, clay-humus components, organic carbon, pH, and numerous soluble ions to reconstruct the settlement history of cave sites.

In nearly all excavation reports the treatment of the sedimentary matrix has been cursory at best, without even detailed petrography available. Only recently have efforts been made to provide a conceptual basis for archaeological sedimentation. Stein & Rapp (1985) have suggested a framework for studying sedimentation on habitation sites based on the traditional geological construct of sediment source, transport and deposition agents, depositional locations, and postdepositional alterations.

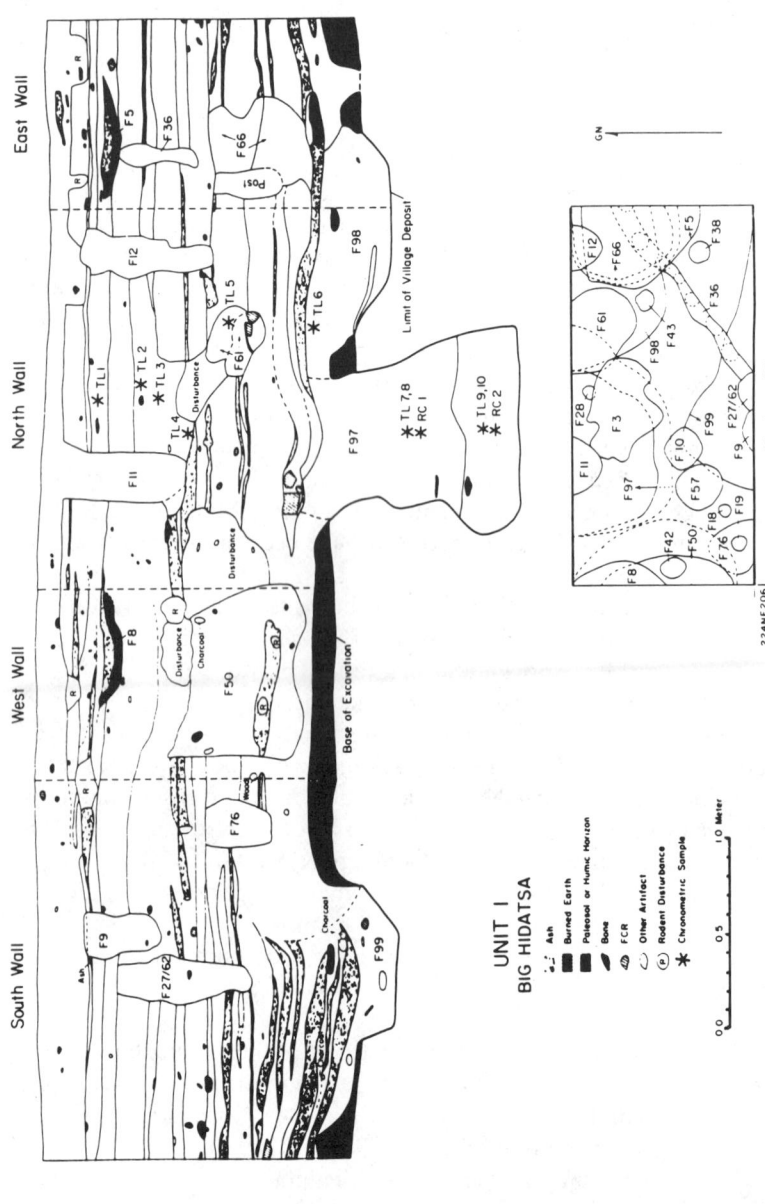

Figure 1 Archaeologists often deal with laterally complex microstratigraphy. This section is taken from *Test Excavations at Big Hidatsa Village* (North Dakota) by S. A. Ahler and A. A. Swenson (by permission of the senior author).

Using the concept of facies would allow excavators to better explain the spatial distribution of contrasting deposits.

As archaeologists consider more fully the reciprocal impact of humans on their environment and environmental constraints on human settlements, the role of interpreting sediments in cultural settings becomes increasingly important. To aid in the study of site formation processes, Stein & Farrand (1985) have edited a volume entitled *Archaeological Sediments in Context*. This collection presents nine papers covering the role of sedimentology in a broad range of settlement and environmental settings that include fluvial and alluvial settings, coastal change, and the special circumstances of arid lands and cold regions. These papers, coupled with a dramatic increase in the quantity and quality of sediment studies that are components of regional (Limbrey 1976, Gladfelter 1977, Hassan 1978) or excavation (Bullard 1970, Shackley 1976, Fedele 1976, Stein & Rapp 1978) archaeology, augur well for the vitality of this field of research.

Diagenetic, taphonomic, and related changes can disturb the sedimentary structures in any deposit. Conspicuous disturbances in archaeological sites are easily detected during excavation. However, there are processes such as earthworm activity (Stein 1983) that redistribute the sediments vertically, thereby playing havoc with the stratigraphy and structures. The presence of 0.5-mm-sized granules of matrix or burrows approximately 10 mm in diameter are indicative of earthworm activity.

Human activities alter the normal processes of sedimentation, pedogenesis, and the postdepositional physical and chemical properties of deposits. Eidt (1977, 1984, 1985) has been instrumental in the study and identification of anthrosols as diagnostic indicators of abandoned settlements. He has identified three principal types of inorganic phosphate in settlement soil phosphate fractionation. Based on the total amounts of phosphate in these three fractions, Eidt has been able to identify ranching, farming, dwelling, manufacturing, burial, refuse, and other areas in archaeological contexts.

Increasing attention is being paid to detailed analyses of archaeological sediments. Macphail & Courty (1985) studied the Roman to Medieval anthropogenic deposits of London and Exeter using pedologic and micromorphologic techniques. Their investigations were able to identify areas used for military ditch digging at Exeter and accumulations of partially or fully reworked materials derived from the destruction and collapse of insubstantial buildings in London. Micromorphology proved to be the best analytical technique.

Courty & Nornberg (1985) were able to distinguish between buried uncultivated and cultivated Iron Age soils in Denmark. Cultivated soils contained an abundance of fine charcoal and silt integrated with the top horizon of the soil. Courty & Nornberg's physical and chemical data showed that ancient cultivation affected particle size distribution, carbon/nitrogen ratio, organic phosphorus content, and distribution of organic matter.

In a related study, Nornberg & Courty (1985) found a very thick layer of occupation material in Iron Age villages in Denmark that seemed to come from the accumulation of grass turf walls from the houses in the villages. The boundary between a brownish and a more gray horizon in the occupation material could be explained as a redox boundary not related to any archaeological event.

Following the lead of geology, in the context of the International Geological Correlation Program, a workshop group has proposed (Gasche & Tunca 1983) some basic definitions and principles for archaeostratigraphic classification and terminology. An equivalent effort is needed for archaeological sedimentology.

The silt fraction of archaeological sediments often contains phytoliths that can indicate the paleoecology of the area as well as be the only material remains of agricultural plants. Phytolith literally means "plant stone." These opaline silica bodies are particularly characteristic of the Gramineae (grasses). Many dietary staples are grasses—for example, rice, corn, wheat, barley, rye, and sugarcane. Grass species also have been used in building construction. Phytoliths have been recovered from a wide range of geological and archaeological sediments (Rovner 1983), but research is still in its infancy to determine the extent to which these silica bodies can be used to reconstruct paleoenvironments and ancient agricultural practices.

Phytolith studies require high-resolution polarizing microscopy equipped with a Nomarski system and the use of scanning electron microscopy (Figure 2). As a guide to environmental discrimination, studies of these fine-grained sedimentary materials and the geoarchaeological study of grain surface textures (Krinsley & Doornkamp 1973, Krinsley 1978) by scanning electron microscopy provide excellent research opportunities.

RECONSTRUCTING THE PALEOGEOMORPHOLOGY

Current geoarchaeological research encompasses a wide range of paleoenvironmental studies, from the paleoclimatic to the paleogeomorphic. This section deals exclusively with the latter type of study, especially Holocene coastal change in regions of important archaeological sites. Sediment infilling of harbors and, conversely, the transgression of the sea into the land were noted even by ancient writers. Yet historians, archaeologists, and historical geographers have few, if any, paleo-

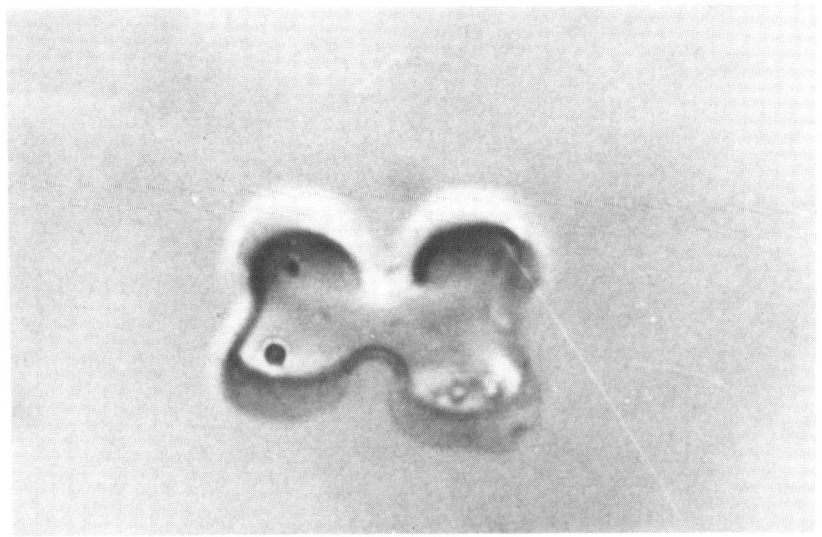

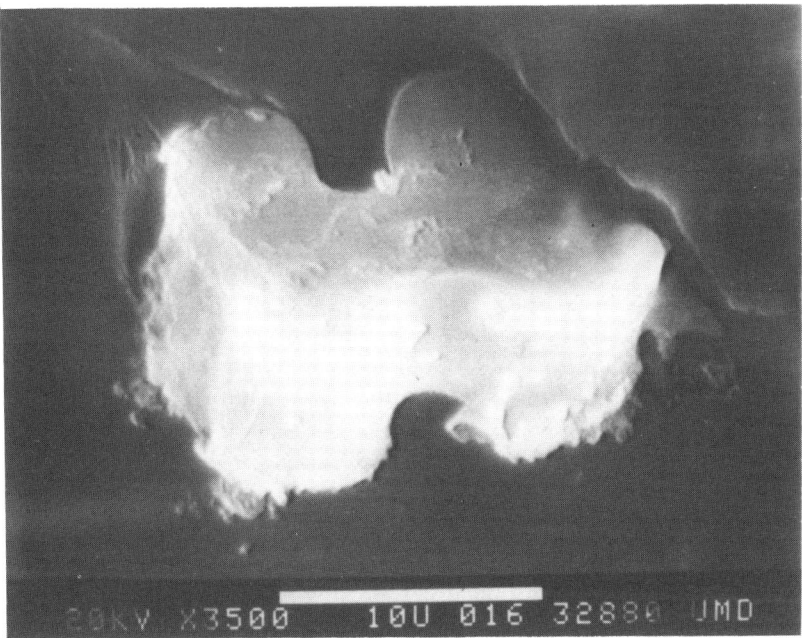

Figure 2 Dumbbell-shaped phytoliths of *Arundo donax* L. from Greece. (*Top*) using light microscopy with Nomarski attachment; (*bottom*) using scanning electron microscope.

geographic maps covering the last 10,000 years that delineate with any accuracy the shift of coastlines (vertical as well as horizontal) in geomorphically dynamic areas. Without such maps, spatial/environmental relationships cannot be adequately described or presented.

Relative sea-level changes in any local area are due to a mix of three separate geologic processes: (*a*) rise or fall of eustatic sea level (relative to a point at the center of the Earth), (*b*) vertical tectonic movements of the regional land mass, and (*c*) addition to or removal of coastal sedimentary deposits. Many examples in the Aegean area have been found where sediment infill has extended the land seaward even during the periods of eustatic sea-level rise in the late Holocene (Kraft et al 1975, 1977).

Three great river systems—the Nile, the Tigris-Euphrates, and the Indus—served as cradles of the great early civilizations. As with all such river systems, their Holocene evolution is marked by dynamic change in the lower reaches of the river as lateral migration overruns or moves away from habitation sites and in the mouth of the system, where fertile deltas have formed from the deposition of new sediment. Greek cities of Alexander's time lie undiscovered under the coastal sediment of the Indus River.

A century and a half ago, C. T. Beke (1835) presented geologic evidence for a theory of delta advance at the mouth of the Tigris-Euphrates system. Since then classicists, historians, and geologists have disputed the location of the ancient shorelines in this region, citing textual, conjectural, and geologic evidence. Cuneiform texts imply that the Sumerian city of Ur was a port, possibly on a large freshwater lake in the Euphrates delta. The site of Ur is now more than 100 km from the sea. Research interest in the question remains unflagging. C. Larsen (1975) has written a good summary of current geologic knowledge of the paleogeography of the Mesopotamian plain.

The modern Nile has only two main branches in its delta, but Herodotus reported that the Nile delta had five branches and Ptolemy listed eight. Aggradation in deltaic regimes causes frequent major shifts in the drainage pattern, and it should be noted that the Roman port at Alexandria lies 6 m below the modern port.

As part of a larger study of the archaeological geology of ancient Troy (Rapp & Gifford 1982b), Kraft et al (1982) undertook a core drilling program to reconstruct the paleogeography of the Trojan plain. The debate concerning the geography of Homeric Troy has involved scholars for over two millennia. The core drilling provided sufficient evidence to reconstruct the geography of the Trojan plain throughout the Holocene.

Troy lies inland some 5 km from the mouth of the Dardanelles on a bluff overlooking a plain at the confluence of the Scamander and Simois rivers. Currently broad and flat, this plain can easily be imagined as the location of battles attending the (Homeric) siege of Troy (3250 yr B.P.). However, the detailed paleogeographic reconstruction of Kraft et al (1982) shows a major marine embayment extending well up the Scamander past Troy and flanked by marshes (Figure 3). Any chariots operating on the "plain" opposite Troy in 3250 yr B.P. would have required pontoons.

Kraft, Rapp, and colleagues have recently (to be published) reconstructed the paleogeography of the famous battle between the Persians and the Spartans at the pass at Thermopylae, Greece. The narrow coastal pass is now part of a broad agricultural plain formed by sedimentation since the fifth century B.C. In order to determine the details of the sequence of geologic environments and to reconstruct the ancient topography at Thermopylae, the group drilled and analyzed the sedimentary materials from seven core holes in the sediments infilling the Gulf of Malia. Massive

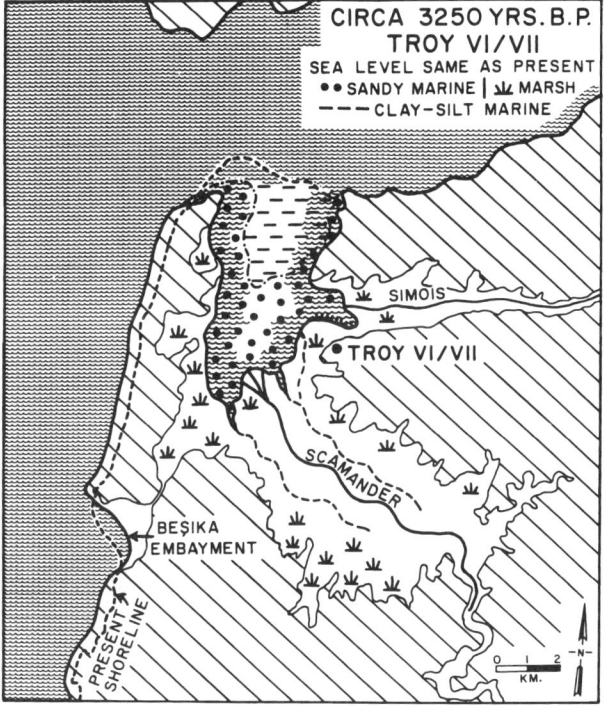

Figure 3 Paleogeographic reconstruction of the vicinity of Troy circa 3250 yr B.P., the time of the Homeric "Trojan War." The Blegen excavation uncovered evidence of a siege and major battle at this time, when an extensive marine embayment still existed northwest of the city. (Reproduced with permission from Kraft et al in Rapp & Gifford 1982b).

sedimentation of the Sperchios River delta has created a delta-plain extending beyond Thermopylae (Figure 4). In addition, a large travertine fan is building upward and seaward from Thermopylae.

The physiography of the pass at Thermopylae has varied considerably through time. Most of this variation is the result of interaction between local relative sea-level changes and sediment infill. When the sea extended to its westernmost limits, the "pass" would have been a sea cliff, impassable to vehicle traffic. During periods of relatively lower sea level, the pass widened sufficiently to allow regular human traffic.

The results of the core drilling program highlight the importance of the third dimension in reconstructing paleogeographies. The battle during the fifth century B.C. would have taken place many meters below the present surface, thus erasing the value of many geomorphic observations.

Two recent international conferences (Thompson 1980, Schwartz 1980) have been devoted exclusively to archaeology and coastal change. Papers in these two volumes cover sites or regions in Japan, Greece, Turkey, China, the United States, South America, Costa Rica, the USSR, Pakistan, Tunisia, Chile, Canada, Great Britain, and the Netherlands. Studies reported in these papers exhibit clearly the major consequences of human occupation of the coastal zone.

Remote sensing in conjunction with high-resolution geomorphic studies and core drilling can provide the data for detailed paleogeographic maps showing the sequence of Holocene geologic environments. Without this context, many interpretations derived from archaeological excavations are not anchored to their environmental setting, an unfortunately frequent state of affairs.

"Inland seas" such as the Great Lakes of North America are also subject to dynamic changes that affect human coastal settlements. The Great Lakes are subject to the differential vertical movements of glacial unloading and the incision of outlet channels as well as normal high-energy erosional processes and sediment influx. C. E. Larsen (1985a,b) has detailed Holocene lake-level stands and the episodic chronology of the fluctuations in the upper Great Lakes region. Neoglacial and pollen records suggest that many fluctuations were climate related. Fluctuations on the order of a few meters are dramatic for coastal settlements. Detailed studies such as these have important implications for modern coastal-zone planning.

There have also been geoarchaeological studies to elucidate the evolution of landscapes away from the coasts. Through a detailed mapping of Quaternary alluvium and soils in the southern Argolid (Greece), Pope & van Andel (1984) identified seven periods of alluviation, each of short duration relative to long intervening periods of stability and soil formation.

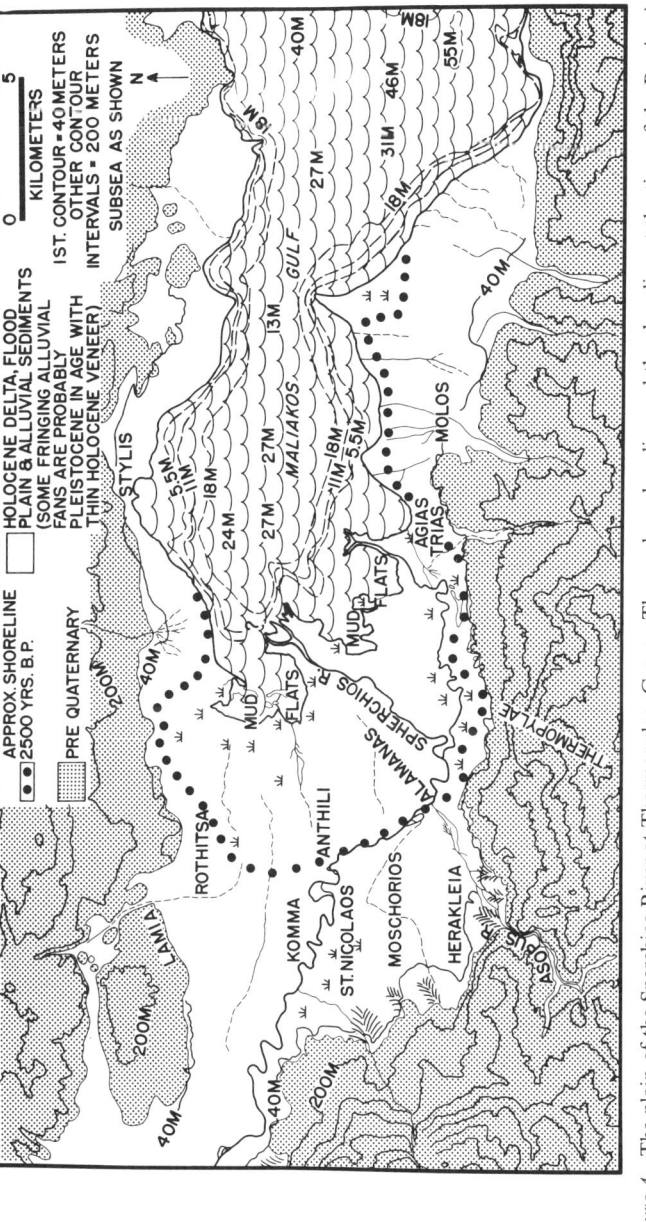

Figure 4. The plain of the Sperchios River at Thermopylae, Greece. The modern shoreline and the shoreline at the time of the Persian invasion (2500 yr B.P.) are indicated.

There was no alluviation accompanying the last glacial maximum. A stable landscape persisted from then until about 4500 yr B.P., when widespread aggradation in the valleys resulted from slope destabilization and soil erosion possibly related to land clearance in the early Bronze Age. These authors found that the nature and chronology of the soil-forming and alluviation events do not have simple correlations with climatic events. After about 4500 yr B.P., human activity seemed to provide the dominant control for soil formation and/or alluviation.

The general relationships between the properties of alluvial sediments and the geomorphologic bases for reconstructing past landscapes in an archaeological context are presented by Gladfelter (1977, 1985). He argues that interpretations must be based upon a composite picture drawn from many properties of the sediment as seen in vertical and horizontal distributions. These properties include carbon/nitrogen ratio, phosphate content, free oxides, pH, Eh, clay mineralogy, pedogenic structure, and organic and mineral content.

PROVENANCE STUDIES

Provenance studies involve a comparison between the trace-element, isotope, or other pattern in an object of interest and the known pattern in potential source deposits. In recent years two geochemical techniques have been used successfully to determine the geographic and geologic sources of exotic materials recovered from archaeological excavations. Such provenance studies seek to identify the geographic source of the deposits that provided the raw materials for the manufacture of a set of artifacts. Although the same analytical techniques can be used to "fingerprint" manufacturing sites, the geoarchaeological studies discussed here do not seek to address the question of where the artifact was manufactured but only the source of the raw material(s).

Trace-element concentrations and stable isotope ratios, combined with rigorous statistical treatment, have proved to be successful in sourcing artifact raw materials. It would be fortuitous for two minerals (or metals) from geologically distinct and geographically separated ore deposits to have coincident trace-element concentrations of a dozen or more geochemically independent elements. A wide variety of rock and mineral materials are amenable to such trace-element "fingerprinting."

The earliest and the most widespread trace-element sourcing has been for obsidian artifacts. Obsidian is not a common rock type, and therefore the number of possible sources is usually limited. In common with other lithic materials, the manufacture of obsidian objects does not alter the chemical composition of the material. Since the early work of Renfrew and his colleagues (Cann & Renfrew 1964) on Mediterranean obsidians and Gordus and his colleagues (Gordus et al 1968) in the United States, obsidian provenance studies have increased in scope until successful applications are now nearly worldwide.

Gale (1981) has demonstrated that the most powerful discrimination of eastern Mediterranean obsidian deposits is made by plotting the ratios of ^{87}Sr to ^{86}Sr against the trace-element concentrations of rubidium.

Following the successful sourcing of obsidian came major provenance studies of chert and soft-stone artifacts. Trace-element determinations of the provenance of chert artifacts have been made successfully by numerous researchers (e.g. Sieveking et al 1970, de Bruin et al 1972, Luedtke 1979).

Soft-stone (soapstone, serpentine, etc.), turquoise, and sanukite have also proved amenable to trace-element fingerprinting (see Rapp 1985). On the other hand, trace-element concentrations do not provide discrimination among sources of Mediterranean marble (Conforto et al 1975). Because smelting destroys the correspondence between trace-element patterns in the ore mineral and the resulting metal or alloy, this technique also is not successful for metal artifacts formed by pyrometallurgical methods.

In North America, the indigenous cultures did not derive their copper from smelting ores. Unlike other regions, North America has large native copper deposits that provided sufficient copper for the needs of these native peoples. Copper ornaments and implements were made by hammering, which did affect the trace-element patterns.

Rapp et al (1980, 1984) have been able to discriminate between discrete sources (deposits) of native copper. The Archaeometry Laboratory at the University of Minnesota, Duluth, has a data base of over 1000 trace-element analyses of native copper from North American deposits. To get a statistically sound trace-element fingerprint of a deposit, one must analyze at least 20 samples if 15–20 chemical elements are used in the determination. It should be noted that attempts to use only one or two elements to distinguish among sources are highly likely to lead to incorrect results.

In a manner analogous to provenance studies of native copper, the determination of the source of archaeological ceramics should be amenable to trace- (or minor-) element characterization of clay sources from the vicinity of a site. However, the little reported work that has been done has not been very encouraging. Adan-Bayewitz & Perlman (1985) have presented a clear statement of the problem and suggest circumstances where clay analysis can have a crucial role in pottery provenance determinations. Their study of Roman and Byzantine pottery in the Galilee region of Israel was able to trace local manufacture and overland trade.

Fortunately, stable isotope techniques have proved successful in situations where trace-element techniques have failed. Using $\delta\ ^{13}C/\delta\ ^{18}O$, N. Herz (Herz 1985, Herz & Dean 1986) has been able to discriminate among Mediterranean marble quarries and to associate broken fragments of marble artifacts. As in the case of native copper, it has been necessary to build up a large data base of analyses from the sources (Greek, Aegean, Turkish, and Roman quarries). Current research on the problem of marble provenance includes the use of electron spin resonance spectroscopy (Lloyd et al 1985).

Stable isotopes also are proving successful in sourcing lead, silver, and copper. In the Mediterranean region, Gale & Stos-Gale (1981, 1982) have used lead isotope signatures to trace the origin of artifacts made from these three metals. Unlike trace elements, which are redistributed in smelting processes, lead isotopes remain associated with the metal phase. Lead isotope ratios have also been used to trace the source of galena in Egypt (Hassan & Hassan 1981) and in North America (Farquhar & Fletcher 1980).

The large data bases needed for many provenance studies require sophisticated data analysis procedures. Following the use of multivariable instrumental techniques to collect the analytical data, various mathematical and statistical methods must be applied to the analyses and to the evaluation of the experimental data. Various statistical procedures have been used (Perlman & Asaro 1969, Rapp et al 1984), but it has become apparent that only the use of the most powerful multivariate statistics and numerical taxonomy methods will adequately reveal the complexities of the data set (Ward 1974, de Bruin et al 1976). Vitali & Franklin (1986) have presented an incisive approach to dealing with chemical characterization in provenance studies. The field of provenance determinations in geoarchaeology will continue to expand with the advent of new instrumental techniques.

EPILOGUE

Archaeology began as a component of antiquarian interest in the human past and developed through humanities and social science scholarship. Currently, archaeology may be on the verge of merging the humanities, the social sciences, and the natural sciences into a more coherent paradigm for studying the material component of human history and prehistory. Archaeology should become more scientific without losing its humanistic component. The synergy of humanistic and scientific scholarship is necessary to assess all the potential information available for reconstructing human history.

Excellent and challenging research opportunities often lie at the interface between traditional disciplines. The current vigor of geoarchaeology attests to this proposition. Unfortunately, this short review leaves unmentioned the excellent work of C. Vance Haynes, Jonathan Davis, and others in the New World as well as vital contributions lying outside the thrust of this paper. Readers are referred to the selective bibliography in the appendix to Rapp & Gifford (1985) for a broader view of the sweep of this field.

Also lamentable is the fact that a short review paper cannot give a good sense of the structure of a field of inquiry unfamiliar to the reader. Serving archaeology and dependent on geology, geoarchaeology has yet to be fully embraced by archaeology and remains a fringe element in geology. Yet it will be driven forward by the unique contributions it can make to the Quaternary history of human/environment reciprocal impacts.

Literature Cited

Adan-Bayewitz, D., Perlman, I. 1985. Local pottery provenience studies: a role for clay analysis. *Archaeometry* 27: 203–17

Bada, J. L., Finkel, R. 1982. Uranium-series ages of the Del Mar man and the Sunnyvale skeletons. *Science* 217: 755–56

Beke, C. T. 1835. On the geological evidence of the advance of the land at the head of the Persian Gulf. *London Edinburgh Philos. Mag. J. Sci. Ser. 3* 7: 40–46

Bullard, R. G. 1970. Geological studies in field archaeology: Tel Gezer, Israel. *Biblical Archaeol.* 32: 98–132

Bullard, R. 1978. Geology in field archaeology. In *A Manual of Field Excavation*, ed. W. G. Dever, H. D. Lance, pp. 197–235. New York: Hebrew Union Coll.

Butzer, K. W. 1973. Geology of Nelson Bay Cave, Robberg, South Africa. *S. Afr. Archaeol. Bull.* 28: 97–110

Butzer, K. W. 1977. Geo-archaeology in practice. *Rev. Anthropol.* 4: 125–31

Butzer, K. W. 1981. Cave sediments, upper Pleistocene stratigraphy, and Mousterian facies in Cantabrian Spain. *J. Archaeol. Sci.* 8: 133–83

Butzer, K. W. 1982. *Archaeology as Human Ecology*. Cambridge: Cambridge Univ. Press. 364 pp.

Cann, J. R., Renfrew, C. 1964. The characterization of obsidian and its application to the Mediterranean region. *Proc. Prehist. Soc.* 30: 111–23

Conforto, L., Felici, M., Monna, D., Serva, L., Taddeucci, A. 1975. A preliminary evaluation of chemical data (trace element) from classical marble quarries in the Mediterranean. *Archaeometry* 17: 201–13

Courty, M. A., Nornberg, P. 1985. Comparison between buried uncultivated and cultivated Iron Age soils on the west coast of Jutland, Denmark. *Nordic Conf. Appl. Sci. Methods Archaeol., 3rd*, pp. 57–69. Vammala, Finl: ISKOS 5

Davidson, D. A., Shackley, M., eds. 1976. *Geoarchaeology: Earth Science and the Past*. London: Duckworth. 408 pp.

de Bruin, M. P., Korthoven, P. J. M., Bakels, C. C., Groen, F. C. A. 1972. The use of non-destructive activation analysis and pattern recognition in the study of flint artefacts. *Archaeometry* 14: 55–63

de Bruin, M., Korthoven, P. J. M., Steen, A. J. v. d., Houtman, J. P. W., Duin, R. P. W. 1976. The use of trace element concentrations in the identification of objects. *Archaeometry* 18: 75–83

Dickinson, W. R., Shutler, R. Jr. 1979. Petrography of sand tempers in Pacific Islands potsherds. *Geol. Soc. Am. Bull.* 90: 1644–1701

Eidt, R. C. 1977. Detection and examination of anthrosols by phosphate analysis. *Science* 197: 1327–33

Eidt, R. C. 1984. *Advances in Abandoned Settlement Analysis*. Milwaukee: Cent. Lat. Am., Univ. Wisc.-Milwaukee. 156 pp.

Eidt, R. C. 1985. Theoretical and practical considerations in the analysis of anthrosols. See Rapp & Gifford 1985, pp. 155–90

Farquhar, R. M., Fletcher, I. R. 1980. Lead isotope identification of sources of galena from some prehistoric Indian sites in Ontario, Canada. *Science* 207: 640–43

Farrand, W. R. 1975a. Analysis of the Abri Pataud sedimenta. In *Excavation of the*

Abri Pataud, ed. H. L. Movius, Jr., pp. 27–68. Cambridge, Mass: Peabody Mus., Harvard Univ.

Farrand, W. R. 1975b. Sediment analysis of a prehistoric rockshelter: the Abri Pataud. *Quat. Res.* 5: 1–26

Farrand, W. R. 1979. Chronology and paleoenvironment of Levantine prehistoric sites as seen from sediment studies. *J. Archaeol. Sci.* 6: 369–92

Fedele, F. G. 1976. Sediments as palaeo-land segments: the excavation side of study. See Davidson & Shackley 1976, pp. 23–48

Folk, R. L. 1975. Geologic urban hind-planning: an example from a Hellenistic Byzantine city, Stobi, Jugoslavian Macedonia. *Environ. Geol.* 1: 5–22

Gale, N. H. 1981. Mediterranean obsidian source characterization by strontium isotope analysis. *Archaeometry* 23: 41–51

Gale, N., Stos-Gale, Z. A. 1981. Lead and silver in the ancient Aegean. *Sci. Am.* 245: 176–91

Gale, N., Stos-Gale, Z. A. 1982. Bronze Age copper sources in the Mediterranean: a new approach. *Science* 216: 11–19

Gasche, H., Tunca, O. 1983. Guide to archaeostratigraphic classification and terminology: definitions and principles. *J. Field Archaeol.* 10: 325–35

Gifford, J. A., Rapp, G. Jr. 1985a. The early development of archaeological geology in North America. In *Geologists and Ideas: A History of North American Geology*, ed. E. T. Drake, W. M. Jordan, pp. 409–21. Boulder, Colo: Geol. Soc. Am.

Gifford, J. A., Rapp, G. Jr. 1985b. History, philosophy and perspectives. See Rapp & Gifford 1985, pp. 1–23

Gladfelter, B. G. 1977. Geoarchaeology: the geomorphologist and archaeology. *Am. Antiq.* 42: 519–38

Gladfelter, B. G. 1981. Developments and directions in geoarchaeology. In *Advances in Archaeological Method and Theory*, ed. M. B. Schiffer, 4: 343–64. New York: Academic

Gladfelter, B. G. 1985. On the interpretation of archaeological sites in alluvial settings. In *Archaeological Sediments in Context*, ed. J. K. Stein, W. R. Farrand, pp. 41–52. Orono, Maine: Cent. Study Early Man

Goldberg, P. 1979a. Micromorphology of sediments from Hayonim Cave, Israel. *Catena* 6: 167–81

Goldberg, P. 1979b. Micromorphology of Pech-de-l'Aze II sediments. *J. Archaeol. Sci.* 6: 17–47

Goldberg, P., Nathan, Y. 1975. The phosphate mineralogy of et-Tabun Cave, Mount Carmel, Israel. *Mineral. Mag.* 40: 253–58

Gordus, A. A., Wright, G. A., Griffin, J. B. 1968. Obsidian sources characterized by neutron activation analysis. *Science* 161: 382–84

Hassan, A. A., Hassan, F. A. 1981. Source of galena in predynastic Egypt at Nagada. *Archaeometry* 23: 77–82

Hassan, F. A. 1978. Sediments in archaeology: methods and implications for paleoenvironmental and cultural analysis. *J. Field Archaeol.* 5: 197–213

Hassan, F. A. 1979. Geoarchaeology: the geologist and archaeology. *Am. Antiq.* 44: 267–70

Herz, N. 1985. Isotopic analysis of marble. See Rapp & Gifford 1985, pp. 331–51

Herz, N., Dean, N. E. 1986. Stable isotopes and archaeological geology: the Carrara marble, northern Italy. *Appl. Geochem.* 1: 139–51

Kempe, D. R. C., Harvey, A. P., eds. 1983. *The Petrology of Archaeological Artefacts*. New York: Oxford Univ. Press. 374 pp.

Klein, R. 1979. Stone Age exploitation of animals in southern Africa. *Am. Sci.* 67: 151–60

Kraft, J. C., Aschenbrenner, S. E., Rapp, G. Jr. 1977. Application of Holocene stratigraphy to paleogeographic reconstructions of coastal Aegean archaeological sites. *Science* 195: 941–47

Kraft, J. C., Kayan, I., Erol, O. 1982. Geology and paleogeographic reconstructions of the vicinity of Troy. See Rapp & Gifford 1982b, pp. 11–41

Kraft, J. C., Rapp, G. Jr., Aschenbrenner, S. E. 1975. Late Holocene paleogeography of the coastal plain of Messenia, Greece, and its relationships to archaeological settings and coastal change. *Geol. Soc. Am. Bull.* 86: 1191–1208

Krinsley, D. H. 1978. The present state and future prospects of environmental discrimination by scanning electron microscopy. In *Scanning Electron Microscopy in the Study of Sediments*, ed., W. B. Walley, pp. 169–80. Norwich, Engl: Geo Abstr., Univ. East Anglia

Krinsley, D. H., Doornkamp, J. C. 1973. *Atlas of Quartz Sand Surface Textures*. London: Cambridge Univ. Press. 91 pp.

Larsen, C. E. 1975. The Mesopotamian delta region: a reconsideration of Lees and Falcon. *J. Am. Orient. Soc.* 95: 43–57

Larsen, C. E. 1985a. Lake level, uplift, and outlet incision, the Nipissing and Algona Great Lakes. In *Quaternary Evolution of the Great Lakes, Geol. Assoc. Can. Spec. Pap. 30*, ed. P. F. Karrow, P. E. Calkin, pp. 63–77

Larsen, C. E. 1985b. A stratigraphic study of beach features on the southwestern shore of Lake Michigan: new evidence of Holocene lake level fluctuations. *Environ.*

Geol. Notes 112, Ill. State Geol. Surv. 31 pp.
Limbrey, S. 1976. Tlapacoya: problems of interpretation of lake margin sediments at an early occupation site in the basin of Mexico. See Davidson & Shackley 1976, pp. 213–25
Littmann, E. R. 1980. Maya blue—a new perspective. *Am. Antiq.* 45: 87–100
Lloyd, R. V., Smith, P. W., Haskell, H. W. 1985. Evaluation of the manganese ESR method of marble characterization. *Archaeometry* 27: 108–16
Luedtke, B. E. 1979. The identification of the sources of chert artifacts. *Am. Antiq.* 44: 744–57
Lyell, C. 1870. *The Geological Evidence for the Antiquity of Man.* Philadelphia: J. B. Lippincott. 526 pp. 2nd ed.
Macphail, R. I., Courty, M. A. 1985. Interpretation and significance of urban deposits. *Nordic Conf. Appl. Sci. Methods Archaeol., 3rd,* pp. 71–83. Vammala, Finl: ISKOS 5
Moss, A. J., Walker, P. H. 1978. Particle transport by continental water flows in relation to erosion, deposition, soils and human activities. *Sediment. Geol.* 20: 81–139
Nornberg, P., Courty, M. A. 1985. Standard geological methods used on archaeological problems. *Nordic Conf. Appl. Sci. Methods Archaeol., 3rd,* pp. 107–17. Vammala, Finl: ISKOS 5
Patterson, C. C. 1971. Native copper, silver and gold accessible to early metallurgists. *Am. Antiq.* 36: 286–321
Perlman, I., Asaro, F. 1969. Pottery analysis by neutron activation. *Archaeometry* 11: 21–52
Pope, K. O., van Andel, T. H. 1984. Late Quaternary alluviation and soil formation in the southern Argolid: its history, causes and archaeological implications. *J. Archaeol. Sci.* 11: 281–306
Rapp, G. Jr. 1975. The archaeological field staff: the geologist. *J. Field Archaeol.* 2: 229–37
Rapp, G. Jr. 1985. The provenance of artifactual raw materials. See Rapp & Gifford 1985, pp. 353–75
Rapp, G. Jr., Allert, J., Henrickson, E. 1984. Trace element discrimination of discrete sources of native copper. *Adv. Chem. Ser.* 205: 273–93
Rapp, G. Jr., Gifford, J. A. 1982a. Archaeological geology. *Am. Sci.* 70: 45–53
Rapp, G. Jr., Gifford, J. A., eds. 1982b. *Troy: The Archaeological Geology, Suppl. Monogr. 4.* Princeton, NJ: Princeton Univ. Press. 209 pp.
Rapp, G. Jr., Gifford, J. A., eds. 1985. *Archaeological Geology.* New Haven, Conn: Yale Univ. Press. 435 pp.
Rapp, G. Jr., Henrickson, E., Miller, M., Aschenbrenner, S. E. 1980. Trace-element fingerprinting as a guide to the geographic sources of native copper. *J. Met.* 32: 35–45
Rovner, I. 1983. Plant opal phytolith analysis: major advances in archaeobotanical research. In *Advances in Archaeological Methods and Theory,* ed. M. Schiffer, 6: 225–66. New York: Academic
Schwartz, M. L., ed. 1980. *Proceedings of the Commission on the Coastal Environment Field-Symposium, Coastal-Archaeology Session, Shimoda, Japan.* Bellingham: West. Wash. Univ.
Shackley, M. L. 1976. The Danebury project: an experiment in site sediment recording. See Davidson & Shackley 1976, pp. 9–20
Sieveking, G. de G., Craddock, P. T., Hughes, M. J., Bush, P., Ferguson, J. 1970. Characterization of prehistoric flint mine products. *Nature* 228: 251–54
Stein, J. K. 1983. Earthworm activity: a source of potential disturbance of archaeological sediments. *Am. Antiq.* 48: 277–89
Stein, J. K., Farrand, W. R. 1985. *Archaeological Sediments in Context.* Orono, Maine: Cent. Study Early Man. 147 pp.
Stein, J. K., Rapp, G. Jr. 1978. Archaeological geology of site. In *Excavations at Nichoria in Southwest Greece: Site, Environs, Techniques,* ed. G. Rapp, Jr., S. E. Aschenbrenner, 1: 234–57. Minneapolis: Univ. Minn. Press
Stein, J. K., Rapp, G. Jr. 1985. Archaeological sediments: a largely untapped reservoir of information. In *Contributions to Aegean Archaeology,* ed. N. Wilkie, W. D. E. Coulson, pp. 143–59. Minneapolis: Kendall/Hunt.
Thompson, F. H., ed. 1980. *Archaeology and Coastal Change.* London: Soc. Antiq. 154 pp.
Vitali, V., Franklin, U. M. 1986. New approaches to the characterization and classification of ceramics on the basis of their elemental composition. *J. Archaeol. Sci.* 13: 161–70
Ward, G. 1974. A systematic approach to the definition of sources of raw material. *Archaeometry* 16: 55–63
Wehmiller, J. F., Belknap, D. F. 1978. Alternative kinetic models for the interpretation of amino acid enantiomeric ratios in Pleistocene mollusks. *Quat. Res.* 9: 330–48
Weymouth, J. W., Huggins, R. 1985. Geophysical surveying of archaeological sites. See Rapp & Gifford 1985, pp. 191–235
Willey, G. R., Phillips, P. 1958. *Method and Theory in American Archaeology.* Chicago: Univ. Chicago Press. 269 pp.

THREE-DIMENSIONAL SEISMIC IMAGING

Clifford H. Thurber

Department of Earth and Space Sciences, State University of New York at Stony Brook, Stony Brook, New York 11794

Keiiti Aki

Department of Geological Sciences, University of Southern California, Los Angeles, California 90089-0741

INTRODUCTION

One of the esoteric beauties of seismic waves is their ability to sample the Earth's elastic (and anelastic) properties as they propagate. As a result, seismograms contain an integrated picture of the structure of the Earth along the wave propagation path. Seismologists have begun to unravel this information in order to construct three-dimensional images of the Earth's interior. In so doing, we are able to extend our geologic knowledge of the Earth deep into its interior.

An analogy is often drawn between seismic imaging and medical tomography (hence the term "seismic tomography"). The two bear strong similarities: A source generates a disturbance that travels through the medium of interest and is recorded by the receiver(s); a suite of such measurements are used to map the spatially varying propagation characteristics of the medium—X-ray attenuation in the case of medical tomography, and seismic-wave velocity in the case of seismic imaging. The medical case is somewhat simpler, however. Only controlled sources of known location and time of occurrence are employed, the propagation path is known, and receivers can be situated with relative freedom. The seismic case does not have these advantages, because the use of natural sources of uncertain location (earthquakes) is often essential, propagation paths are imprecisely known (as they are affected by heterogeneities), and receiver locations are restricted by cultural and geographic constraints.

Seismic imaging has one added feature that can be both an advantage and a complication: a multiplicity of wave types. Three major classes of seismic waves are body waves (P and S), surface waves (Rayleigh and Love), and free oscillations. For body waves, we can also distinguish direct, reflected, refracted, and diffracted waves. Typical characteristic wavelength scales can be associated with each wave class—hundreds of meters to several kilometers for body waves, hundreds of kilometers for surface waves, and thousands of kilometers for free oscillations. These length scales roughly correspond to the resolution length for each wave, indicating the minimum size of features that could be imaged or distinguished within the Earth.

Three-dimensional seismic imaging has the potential to address many basic and crucial issues faced by geoscientists today. Global seismic tomography shows promise for helping to map the flow of mantle convection (Dziewonski & Anderson 1984). Mantle density anomalies indicated by body-wave tomography may help explain the geoid (Hager et al 1985). Teleseismic imaging is beginning to reveal the structure of subducting oceanic lithosphere (e.g. Hirahara & Mikumo 1980) as well as many areas of thermal anomalies and "hotspots" (e.g. Zandt 1978). Local studies can address specific questions regarding subsurface geology (e.g. Thurber 1983).

We review here the methodologies used for three-dimensional seismic imaging, introducing a spectrum of body-wave and surface-wave studies from local to global scale and highlighting examples of the kinds of applications in which seismic imaging has been successful. The advent of state-of-the-art local and global digital seismic arrays promises to open the realm of seismic imaging to an ever-widening suite of geologic problems.

METHODOLOGY

The manifestation of heterogeneous Earth structure depends on the wave type considered. For example, variations are observed in travel times for body waves, dispersion for surface waves, and resonance frequencies for free oscillations. Seismologists employ two complementary procedures for interpreting these observed variations in terms of a model for the Earth's structure: computation of variations in the observables given a hypothetical Earth model (the "forward problem") and, conversely, estimation of an appropriate Earth model, or at least improvements to the current model, given the observations (the "inverse problem").

Immediately we encounter two profound difficulties shared by all seismic-imaging techniques. Existing data are finite in number and contain measurement errors, so that one cannot hope to determine the Earth's structure uniquely and precisely via the inverse problem. This issue has been addressed at length in the geophysical literature (Backus & Gilbert 1967, Jackson 1972, Aki & Richards 1980, Menke 1984), and numerous inversion schemes have been devised. Computers also have finite capacity, so an a priori form for representing the Earth's structure generally must be chosen, in particular in a fashion that allows the forward problem to be carried out efficiently. Although all seismic-imaging techniques are in a sense conceptually the same, the factors of inverse method, structure representation, and of course wave type tend to disguise the underlying similarities.

The following sections introduce the methodologies for seismic imaging that have been developed for body waves and surface waves. These methods evolved quite rapidly through the past decade and appear to have reached a plateau in development at the present time. Current efforts are directed more toward applying imaging techniques to existing data as well as planning for future instrumentation and experiments. However, the next generation of methods, which are now beginning to be developed, clearly will make fuller use of the broadband waveform data promised by the modern seismic network instrumentation of GSN, PASSCAL, Geoscope, and other arrays.

Body-Wave Travel Times

Within the limits of ray theory, the body-wave travel time T can be expressed as a path integral

$$T = \int_{\text{path}} u \, ds, \tag{1}$$

where u is slowness (reciprocal of velocity) along the ray path, and ds is the differential of path length. Seismic imaging via body-wave travel times involves reconstructing the three-dimensional slowness field using data from many natural and/or artificial sources. Generally, the misfit (residuals) between the observed travel times and travel times calculated from a trial model for the structure is used to estimate improvements to the structure model. This problem is confounded by the fact that the true path for the integral in (1) cannot be known until the true structure is determined, and furthermore for earthquake sources the hypocenter location and origin time are similarly uncertain. Fermat's principle is universally appealed to in order to justify the approximation

$$\delta T \cong \int_{\text{initial path}} \delta u \, ds, \qquad (2)$$

where δT and δu are perturbations to the travel time and slowness field, respectively. The limitations of this approximation have been investigated by only a few studies (Thurber 1981, Gubbins 1981, Pavlis & Booker 1983, Koch 1985a).

Three-dimensional ray tracing is a necessity for determining the seismic ray paths in (1) or (2). For imaging problems treated by a one-step linear procedure, standard one-dimensional ray-tracing methods are applicable, assuming that the initial model is laterally homogeneous (Aki & Lee 1976, Aki et al 1977, Roecker 1982). If an iterative solution is desired, several choices are available: retention of the ray path from the original, unperturbed model (Spencer & Gubbins 1980, Comer & Clayton 1984), approximate ray-tracing methods (Thurber & Ellsworth 1980, Thurber 1983, Buland 1982, Eickemeyer & Prothero 1985, Um & Thurber 1986), or exact ray tracing (Julian & Gubbins 1977, Pereyra et al 1980). In the latter case, the type of ray-tracing scheme must correspond with the manner in which the Earth's structure is represented. Examples are constant-velocity blocks (Whitcombe 1982, Koch 1985b), laterally varying layers (Hawley et al 1981), and parameterized orthogonal functions (Firbas 1981).

TELESEISMIC METHODS The most widely used body-wave imaging methods employ P-wave arrival times from large sets of distant earthquakes (teleseisms) recorded by a network of seismic stations located over a region of geologic interest (Figure 1). It is assumed that the incident wavefronts can be approximated by plane waves, a justifiable assumption given the long wave paths through the relatively smooth lower mantle, so that deviations from expected arrival times arise from slowness perturbations beneath the network. In practice, relative arrival-time residuals for each event are computed to remove unwanted systematic effects due to source mislocation or anomalous structure along the common portions of the ray paths (i.e. outside of the modeled volume). As a direct consequence, though, only relative slowness perturbations across each model layer can be determined, as opposed to the actual slowness value. Mathematically, the teleseismic inversion problem can be represented by the equation

$$r_{ij} - \bar{r}_j \cong \sum_{k=1}^{M} (d_{ijk} u_{ijk} - \bar{d}_{jk} \bar{u}_{jk}) \Delta f_k, \qquad (3)$$

where r_{ij} is the arrival time residual at station i for event j; M is the total number of blocks; d, u, and Δf_k are the partial path length, initial slowness, and fractional slowness perturbation, respectively, within the blocks k hit by ray ij; and an overbar indicates an average over i.

Two principal methods for the solution of (3) have been developed: the "ACH" (Aki, Christofferson, & Husebye) method of direct damped least-squares inversion (Aki et al 1977) and algebraic reconstruction (Humphreys et al 1984). The direct solution has the advantage of allowing the computation of the resolution and covariance matrices, which are valuable measures of the solution quality. The algebraic reconstruction technique processes rays sequentially, so no storage or manipulation of large matrices is required, and therefore finely discretized modeling is possible. Either method can be computed iteratively, but several studies have indicated that iteration is often unnecessary for the ACH method (Ellsworth 1977, Koch 1985a). It is vital to have rays with a good distribution of azimuth and incidence angles to assure a well-resolved image. Christofferson & Husebye (1979) describe some ideas that give added flexibility to the ACH method. An important paper by Haddon & Husebye (1978) demonstrates possible ambiguities in the teleseismic method.

This idea can be extended to include a global distribution of stations and teleseismic events in order to model the structure of the entire mantle of the Earth. Early studies by Dziewonski et al (1977) and Sengupta et al (1981) divided the Earth into relatively large cells (blocks) to obtain very

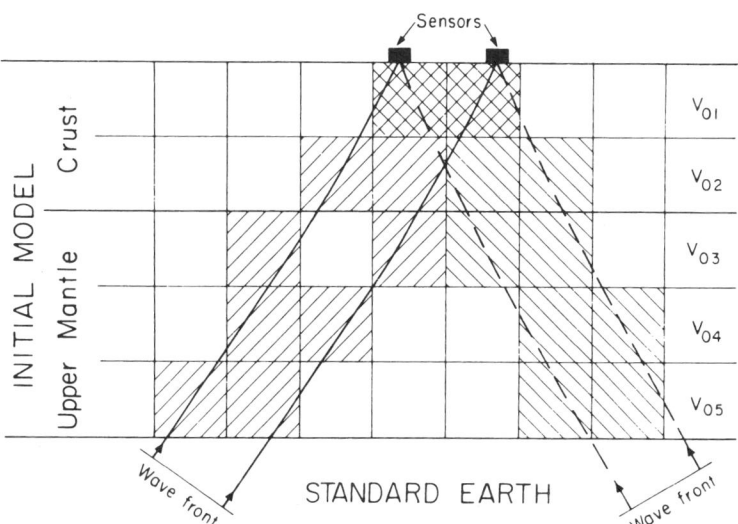

Figure 1 Geometry of the ACH teleseismic inversion method. Velocity perturbations to an initial layered model are derived from relative arrival-time residuals of teleseismic earthquakes. From Aki et al (1977), with permission.

coarse models of the Earth's laterally heterogeneous structure. The Dziewonski study used direct least-squares inversion, while Sengupta used a successive approximation method. Comer & Clayton (1984) extended this to model 50,000 cells using over one million data. An iterative back-projection technique was required, since inversion of a 50,000 by 50,000 element matrix is virtually impossible with present computers. An alternative approach taken by Dziewonski (1984) involves representing velocity perturbations with a spherical harmonic expansion in latitude and longitude and polynomials for depth, so that coefficients for the continuous functions are determined in a damped least-squares inversion. The degree and order of the expansion can obviously be tailored to suit the available data.

Two subtle aspects of these global methods deserve mention. Station corrections are vital for absorbing near-station heterogeneity that could not be resolved in the global model. These corrections can either be constant delay times or functions of azimuth. Source mislocation must also be considered, either through explicit inclusion in the inversion or some other approximate means. This problem is discussed more fully in the following section.

LOCAL AND REGIONAL METHODS   If the seismic sources for an arrival-time study are located within the volume being modeled (Figure 2), then it is essential that the locations and origin times of the sources be considered, and two-point seismic ray tracing is required. For controlled source (i.e. seismic refraction) studies the locations are of course known, but for local earthquake studies these new parameters add significant complexity to the problem, known as simultaneous inversion. Fortunately, clever schemes for handling this complexity have been developed. A further complication is the possibility of multiple P phase arrivals, which can make selection of the actual first arrival path ambiguous. These phases generally provide vital information for refraction studies, but they pose difficulties in terms of phase identification for local earthquake inversions. Eventually, however, we can expect simultaneous waveform inversion methods to be developed to image both the seismic source and heterogeneous Earth structure on a local scale.

Seismic refraction studies now are routinely used to construct images of two-dimensional sections of the Earth's crust and upper mantle (Phinney & Odom 1983). Detailed data interpretation and modeling are typically done manually, but formal travel-time inversion methods for laterally heterogeneous structure are gradually being developed (Wesson 1971, Firbas 1981, Kanasewich & Chiu 1985, Elbring & Braile 1985, Huang et al 1986, Pavlis 1986). For inversion, travel-time residuals r_{ij} are related to model perturbations by the set of equations

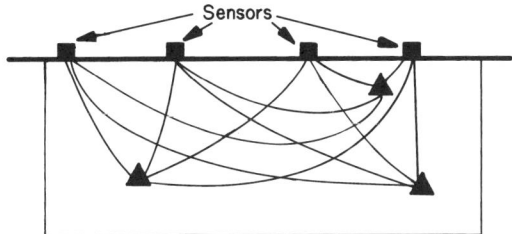

Figure 2 Geometry of the local earthquake simultaneous inversion method. In contrast to the teleseismic method, sources (triangles) are contained within the modeled volume.

$$r_{ij} \cong \sum_{k=1}^{M} \partial T_{ij}/\partial p_k \, \Delta p_k, \qquad (4)$$

where i and j again represent the events and stations, respectively, M is the number of model parameters, T_{ij} is the travel time, and p_k is the kth parameter of the velocity model. Parameters can include variable refractor velocity or block velocities, layer velocity gradients, boundary coordinates, or coefficients of a set of basis functions. The same methods can be applied to earthquake data if source time terms are included (Hearn 1984) or a master-event scheme is employed (Shedlock & Roecker 1985).

If we represent the velocity model as a continuous function, we can write (4) more formally as

$$r_{ij} \cong \int_{\text{path}} G_{ij} \delta m \, ds, \qquad (5)$$

where G_{ij} represents the Frechet derivative (data kernel), and δm is the perturbation to the model. Two approaches to the inversion for the continuous function δm in (5) are via the Backus-Gilbert formalism (Chou & Booker 1979) or through the use of an a priori spatial covariance function (Tarantola & Nercessian 1984). Chou & Booker developed the concept of "ideal averaging volumes" to compute the three-dimensional spatial variations in velocity in a manner that optimizes the resolution of structure. Tarantola & Nercessian's nonlinear algorithm derives the velocity model by minimizing a weighted sum of the data misfit and model perturbation, with the weighting given by the a priori data and model covariance operators. They suggest the use of a spatial covariance function with a length scale defining the smoothness of the model. Both of these methods have the appeal of limiting the bias in the form of the velocity model that is inherent in the parameterized approaches, but obviously they are somewhat inappropriate for modeling structures with major velocity discontinuities.

Returning to the simultaneous inversion problem, we can write the expression parallel to (4) for discrete parameterizations as

$$r_{ij} \cong \sum_{k=1}^{M} \partial T_{ij}/\partial p_k \, \Delta p_k + \Delta t_{0i} + \sum_{l=1}^{3} \partial T_{ij}/\partial x_{li} \, \Delta x_{li}, \qquad (6)$$

where the x_{li} are the coordinates of the ith event, and t_{0i} is its origin time. It is clear that the hypocenter parameters of the various events are coupled via the velocity parameters, which are basically common to all the equations. If the problem were attacked directly, the size for (6) would become unmanageable for large data sets, even if a least-squares inverse is employed. However, an elegant method is available to decouple the hypocenter parameters from the velocity part of the problem (Pavlis & Booker 1980, Spencer & Gubbins 1980, Rodi et al 1980). This method, commonly known as parameter separation, exploits the ability to construct an orthogonal transformation for each event that "annuls" the hypocenter components while transforming but still preserving the velocity part of the problem. If we write the set of equations (6) for one event as

$$\mathbf{r}_i = \mathbf{A}_i \Delta \mathbf{p} + \mathbf{H}_i \Delta \mathbf{h}, \qquad (7)$$

where $\mathbf{A}_i$ and $\mathbf{H}_i$ contain the velocity and hypocenter partial derivatives, and $\Delta \mathbf{p}$ and $\Delta \mathbf{h}$ the respective perturbations, then there always exists an orthogonal transformation matrix $\mathbf{U}_0$ with the property

$$\mathbf{U}_0^T \mathbf{H}_i = 0, \qquad (8)$$

so that

$$\mathbf{r}'_i = \mathbf{U}_0^T \mathbf{r}_i = \mathbf{U}_0^T \mathbf{A}_i \Delta \mathbf{p} = \mathbf{A}'_i \Delta \mathbf{p}. \qquad (9)$$

In this manner, a set of equations (9) is composed that only contain the velocity parameters as unknowns, permitting the solution of problems with very large data sets and many model parameters.

A wide range of discrete velocity parameterizations are available, including constant-velocity blocks (Aki & Lee 1976, Roecker 1982, Koch 1985b), laterally varying layers (Hawley et al 1981), three-dimensional grids (Thurber 1983), and parameterized functions (Spencer & Gubbins 1980). Although the theory for determining truly continuous three-dimensional structure is established, as discussed above, no successful applications to simultaneous inversion yet exist.

For the future, simultaneous inversion with multiple phases should be a primary goal. Engdahl & Gubbins (1985) have demonstrated the value of using reflected phases for resolving subduction zone structure. Inclusion of S waves greatly improves the constraint on the hypocentral parameters, as the focal depth and origin time become less coupled (Roecker 1982). The modernization of existing networks and establishment of the PASSCAL array (Smith 1986) should permit more routine and accurate treatment of S-wave arrival-time data in these studies.

Surface-Wave Dispersion

The nature of surface waves effectively guarantees a diminished resolution of laterally varying structure compared with the resolution obtained with body waves, but at the same time surface waves sample the depth variation of structure in a different manner. The "skin depth" for sampling increases with increasing wavelength, which combined with the tendency for seismic velocity to increase with depth (and the fact that the Earth is not flat) gives rise to surface-wave dispersion, with velocity generally increasing with wavelength.

Part of the power of surface-wave imaging techniques comes from the existence of the two major surface-wave types—Rayleigh and Love waves. The latter arise from trapped SH waves (horizontally polarized shear waves), while the former are generated by the interaction of P and SV (vertically polarized shear) waves. Love and Rayleigh waves can provide strong constraints, particularly on shear velocity and its anisotropy (transverse or azimuthal) in the crust (in the case of regional studies) and upper mantle.

The frequency-dependent propagation velocity of surface waves introduces an additional level of complexity to the imaging problem. In the standard approach, phase or group travel-time anomalies δT are related to perturbations in velocity δv (here either phase or group) using the assumption of "pure-path" dispersion, equivalent to the use of Fermat's principle for body-waves, by the expression

$$\delta T(\omega) = -v(\omega)^{-2} \int_{\text{initial path}} \delta v(\omega) \, ds, \tag{10}$$

where ω is frequency and $v(\omega)$ is the initial dispersion curve. Unlike the case for body waves, the determination of lateral variations in $\delta v(\omega)$ does not really complete the problem—vertical variations in seismic velocity and density must be inferred from $\delta v(\omega)$.

Two types of source-receiver configurations, the "single station" and "great circle" methods (Brune et al 1960, Sato 1958), are used most commonly for surface-wave imaging. The great circle method uses multiple-circuit waves, i.e. waves that propagate from the event to the station

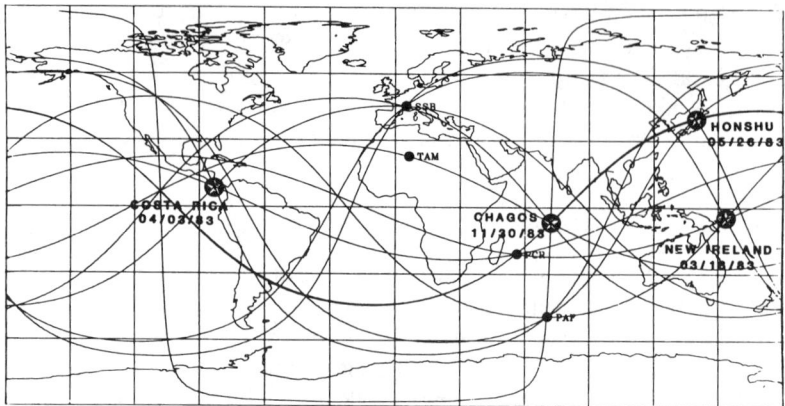

Figure 3 Great circle paths corresponding to four large events (stars) recorded at four stations of the GEOSCOPE network (circles), illustrating the geometry of global surface-wave imaging. From Roult & Romanowicz (1984), with permission.

and then continue around the entire globe to be recorded again at the same station (Figure 3). Differential dispersion between the first and second passes is measured, so that unwanted source effects can be eliminated. Obviously, this method is only appropriate for global modeling. Disadvantages of the great circle approach are the need for sufficiently large events to excite observable multiple-circuit waves, and the difficulty of determining odd-order terms in the spherical harmonic expansion (Backus 1964). For the single-station method, the surface-wave dispersion is measured along the path directly from the source to the station (Figure 4), so that regional or global modeling is possible. The shorter path lengths permit the use of moderate-sized events, so in general many more data are available. A drawback, however, is the need to know the source properties (origin time, epicenter, depth, mechanism, and rupture duration) quite well to correct for effects on the radiated waves (Knopoff & Schwab 1968, Weidner 1974). These corrections can be minor for regional group velocity studies (Feng & Teng 1983). On the other hand, the accuracy of the group velocity measurement is intrinsically poorer than that of the phase velocity measurement, because the group arrival time has an uncertainty comparable to one period, whereas the phase arrival time can be measured as accurately as the signal-to-noise ratio permits (Aki & Richards 1980). A third approach, the two-station method, is popular for modeling laterally homogeneous structure on a regional basis, but to our knowledge it has not been applied to three-dimensional imaging problems.

Procedures for the inversion of $\delta v(\omega)$ for vertical structure date back

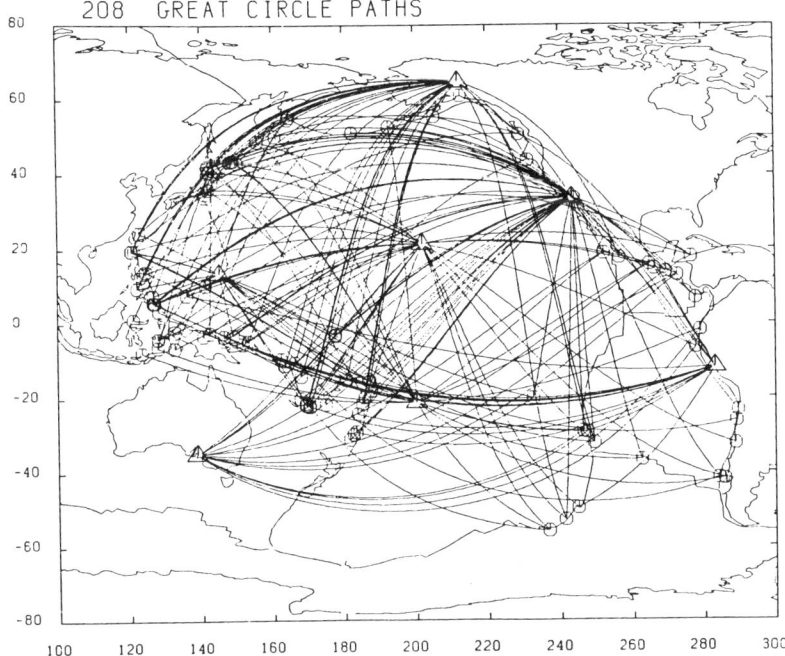

Figure 4 Example of the source-station geometry for regional surface-wave imaging. From Suetsugu & Nakanishi (1985), with permission.

more than two decades (Dorman & Ewing 1962, Brune & Dorman 1963), but detailed studies of the resolution and uncertainty properties of these methods were not accomplished until the early 1970s (Wiggins 1972, Der & Landisman 1972). Recent studies are applying the Backus-Gilbert formalism to study the resolution of the lateral variations in $\delta v(\omega)$ (Tanimoto 1985, Suetsugu & Nakanishi 1985).

Parameterization of the Earth's structure is again a fundamental issue. Three common choices are block regionalization (Feng & Teng 1983), tectonic regionalization (Santo 1961, Patton 1980), or spherical harmonic expansion (Nakanishi & Anderson 1983, Tanimoto & Anderson 1985). The former two are generally applied to regional studies, while the latter is only appropriate for global modeling. In the regionalization methods, (10) is rewritten as

$$\delta T(\omega) = v(\omega)^{-2} \sum_{k=1}^{M} \delta v_k(\omega) \Delta s_k \tag{11}$$

where M is the number of model regions (blocks or tectonic provinces), and δv_k and Δs_k are the velocity perturbation and partial path length, respectively, in the kth model region. The dispersion curves are determined from (11) and then are inverted for vertically varying structure within each region. For the spherical harmonic case, even and odd global dispersion curves $A(\omega)$ and $B(\omega)$ are determined for each degree and order via the expansion (Dziewonski & Anderson 1984)

$$\delta v(\theta, \phi, \omega) = \sum_l \sum_m [A_{lm}(\omega) \cos m\phi + B_{lm}(\omega) \sin m\phi] P_{lm}(\cos \theta), \quad (12)$$

and the corresponding vertically varying structure is computed for each A_{lm} and B_{lm}.

Alternatively, a correlation-length approach may be adopted following the ideas of Tarantola & Nercessian (1984). As in the body-wave application described above, no a priori form is assumed for the structure. Instead, a finely discretized model is used that would be underdetermined but for the introduction of an a priori smoothing filter, the spatial covariance function (Montagner 1986). Generally, this function will have distinct lateral and vertical behavior.

Dziewonski & Steim (1982) and Woodhouse & Dziewonski (1984) developed a fundamentally different approach that bypasses the step of inverting dispersion curves. They formally related very long period seismograms directly to the parameters of a spherical harmonic representation of the Earth's global density and anisotropic shear velocity structure. A path-integral approximation was devised so that the seismogram partial derivatives for each observation would only depend on a subset of the parameters; otherwise, the problem would be computationally intractable. Similarly, a regional surface-wave waveform inversion scheme has been developed by Yomogida & Aki (1985) employing a block (actually gridpoint) regionalization. Their technique has the advantage of directly incorporating focusing and multipath interference effects. After deriving a suitable initial model with traditional methods, they calculated synthetic surface-wave seismograms using a Gaussian beam method (Yomogida 1985). Amplitude and phase residuals were then inverted for lateral variations in phase velocity using the nonlinear approach of Tarantola & Valette (1982).

As in the case of arrival-time inversion, nonlinearity is certainly an important issue for the surface-wave inversion methods. Surface-wave propagation is affected by regional variations in structure, so that surface-wave ray tracing and iterative solutions may be essential. Lay & Kanamori (1985) and Schwartz & Lay (1985) have begun the study of ray-path perturbations in heterogeneous global models. Interestingly, the independent models of Nakanishi & Anderson (1984) and Woodhouse &

Dziewonski (1984) display similar anomalies in amplitude and phase, but travel-time errors from neglecting path perturbations may be significant in both models (Schwartz & Lay 1985).

APPLICATIONS

Three-dimensional seismic-imaging methods have received wide application through the past decade. In the Earth's crust, seismologists are striving to extend geology into the third dimension, depth, to explore the nature of fault zones, volcanic regions, and other features. In the mantle, a major focus is on features related to geodynamics, such as subduction-zone structure and mantle flow. For the first time, we are able to "see" how the Earth operates deep in its interior. Our hope is that this new vision will aid scientists in answering some fundamental geological questions, but as in the opening of any new scientific frontier, there is the possibility that more new questions will be raised than old ones answered.

Fault Zones

Many basic questions concerning fault zones, such as transform plate boundaries or intraplate faults, can be addressed through seismic imaging. For a complex transform fault like the San Andreas system, the enigmatic relation between crustal faults and the configuration of the plate boundary at depth can be explored by imaging of upper-mantle structure. Realistic crustal models allow accurate location of earthquakes, which permits the detailed study of the fault surface. Heterogeneities or asperities along a fault may also be detected in images of the three-dimensional structure, and these could provide clues to the mechanical behavior of the fault.

A number of body-wave imaging studies have explored different parts of the San Andreas fault system. Two regional-scale teleseismic studies covered the north-central (Zandt 1981) and southern sections (Humphreys et al 1984). Both detected strong lateral variations in crustal velocities, locally reaching 10% or greater. Zandt (1981) was able to relate the zones of low velocity, most lying between parallel faults, to the presence of thick sedimentary material, whereas areas of high velocity corresponded to exposed basement. Results for the uppermost mantle revealed a narrow linear zone of low velocity trending north-northwest, oblique to the orientation of the northern San Andreas. Zandt (1981) interpreted the low-velocity zone to represent thinned lithosphere at the Pacific–North American plate boundary, which suggests some decoupling between the motions of the crust and the upper mantle to account for the oblique trends.

In southern California, the study of Humphreys et al (1984) revealed a narrow, east-west trending high-velocity zone dipping steeply into the upper mantle beneath the Transverse Ranges at the San Andreas' Big Bend, reaching to a depth of as much as 250 km. Two alternative explanations were proposed for this slab-like anomaly: subduction of the Pacific lithosphere due to plate convergence at the Big Bend, or convective downwelling (delamination) of the lower lithosphere actually causing the convergence. Either scenario requires some decoupling between the crust and upper mantle.

High-resolution studies of crustal structure along the central San Andreas have been accomplished using P-wave data from local earthquakes. In a pioneering study, Aki & Lee (1976) investigated the Bear Valley area, just north of Parkfield, where the creeping San Andreas and subparallel Calaveras faults are separated by about 5 km. The principal feature of their model is a wedge of low seismic velocity (15%) located between the two faults, extending to a depth of at least 5 km. In a more recent study for the area just to the northwest of Bear Valley, Thurber (1983) detected the continuation of this wedge along the San Andreas to a point a few kilometers beyond San Juan Bautista, at the transition from the creeping to locked sections of the fault. Thus the presence of the low-velocity, presumably weak, wedge material may play a role in facilitating creep along the central San Andreas fault.

These two studies arrived at different conclusions regarding the puzzling southwestward offset of routinely located epicenters from the trace of the San Andreas in this region (Brown & Lee 1971). Thurber concluded that, within the location uncertainty, epicenters in the vicinity of San Juan Bautista did lie along the San Andreas fault trace, whereas Aki & Lee's results suggested that the Bear Valley epicenters were indeed offset 3 km to the west of the fault trace. However, the latter authors recognized that the results could be biased due to the single-step nature of their inversion.

Volcanic and Geothermal Areas

Several areas of recent or active volcanism have been studied by both the teleseismic and local earthquake methods, including the Island of Hawaii and Kilauea volcano (Ellsworth 1977, Ellsworth & Koyanagi 1977, Thurber 1984), Yellowstone, Wyoming (Zandt 1978, Iyer 1979, Benz & Smith 1984), and The Geysers–Clear Lake region, California (Oppenheimer & Herkenhoff 1981, Eberhart-Phillips 1986). With such parallel studies, we are able to view the detailed shallow structure of the region as well as trace features deep into the Earth. Goals of these studies include mapping the location and size of partial-melt zones and crustal magma reservoirs and providing clues to the origin of the magma itself.

Hawaii, Yellowstone, and The Geysers–Clear Lake area represent somewhat different classes of volcanic/geothermal activity: Hawaii—intraplate oceanic; Yellowstone—intraplate continental; and The Geysers—transform boundary. The crust and upper-mantle velocity anomalies associated with each reflect these differences. All three have underlying low-velocity zones in the upper mantle, but their magnitudes and lateral and depth extents vary. Beneath Hawaii, Ellsworth (1977) found two vertical zones penetrating into the asthenosphere, one beneath the center of the island and one off the southeastern coast, with lateral dimensions increasing from 50 to 100 km. The amplitude of the velocity contrast increases from 4–5% in the lithosphere (15–105 km depth) to about 10% in the depth range 105–165 km, the depth limit of the model; how deep the anomalies continue could not be determined. Zandt's (1978) results indicate the presence of two anomalies of similar dimensions, one beneath Yellowstone Caldera and another to the northeast, but in this case the low-velocity anomalies diminish in amplitude from 20% around 40–90 km depth to only 4% below that, disappearing completely by about 250 km. From the depth trends of the anomalies, Ellsworth concluded that Hawaii represents a mantle plume or hotspot, while Zandt preferred for Yellowstone the hypothesis of a propagating lithosphere fracture. In the upper mantle beneath The Geysers–Clear Lake region, Oppenheimer & Herkenhoff (1981) found a low-velocity anomaly with a contrast of 4% and lateral dimensions of about 30 km by 60 km. They report that this anomaly does not extend below 60 km, supporting the hypothesis of Zandt (1981) that its origin is due to a former "window" in the lithosphere related to the passing of the Mendocino triple junction.

The crustal signatures of these different anomalous regions are also somewhat distinctive. On a large scale, both Yellowstone and The Geysers have major velocity lows in both the upper crust (20% contrast) and lower crust (10% contrast) (Zandt 1978, Oppenheimer & Herkenhoff 1981), while Kilauea displays a crustal velocity high (about 10%) along its summit and rift zones (Ellsworth & Koyanagi 1977). The former are thought to be partial-melt zones, while the latter is likely due to competent intrusive rocks. The coarse-scale local earthquake inversion of Benz & Smith (1984) for Yellowstone confirmed the upper crustal low-velocity anomaly there, resolving it into two separate features in the northeast and southwest portions of the caldera. In contrast, Eberhart-Phillips' (1986) study of the shallow crust (less than 10 km deep) in The Geysers region found only a tiny velocity low associated with the geothermal area, with no evidence for Oppenheimer & Herkenhoff's major crustal low-velocity anomaly. Instead, Eberhart-Phillips discovered a linear high-velocity body southeast of The Geysers, presumably either an intrusive or metamorphosed zone related to the underlying partial-melt zone. Finally, a detailed study of

Kilauea's crustal structure by Thurber (1984) was able to resolve the small low-velocity zone of the volcano's summit magma reservoir within the otherwise high-velocity core of the volcano.

Subduction Zones

Subduction zones have long been recognized as regions of heterogeneous seismic structure from observations of arrival times and amplitudes of body waves (Utsu 1967, Davies & McKenzie 1969) and from theoretical calculations of the structure of descending lithospheric slabs (Toksöz et al 1971, Sleep 1973). The motivations for detailed seismic-imaging studies of subduction zones are numerous. The dense subducting slabs play a major role in driving lithospheric plate motions (Forsyth & Uyeda 1975, Richardson et al 1979); seismic imaging of structure can provide vital information on the density within the downgoing plate (Roecker 1985). Clues to the source and generation of the magma that is feeding arc volcanoes may be obtained from high-resolution imaging studies (Hirahara et al 1986). The location (and mechanism) of earthquakes within the downgoing slab, which require realistic seismic structure models for their accurate determination (Spencer & Engdahl 1983), can constrain the stress state of slabs and hence the viscosity contrast between the upper and lower mantle (Vassiliou et al 1984).

The subduction zones beneath the Japan Islands (Figure 5) have been the subject of extensive body-wave imaging investigations. Large-scale, region-wide studies of P and S velocity structure using a hybrid local-teleseismic method (Hirahara 1977, 1980, Hirahara & Mikumo 1980) have successfully characterized the location and approximate thickness (100 km) of the descending Pacific plate along with the amplitudes of the velocity anomalies (up to 8%). A representative cross section is illustrated in Figure 6.

Subsequent studies have focused on finer-scale investigations. Hirahara (1981) examined the seismic structure beneath southwest Japan in order to determine the nature and extent of subduction of the Phillipine Sea plate. His results indicate the existence of major variations along the Nankai Trough, ranging from collision and extremely shallow subduction in the north (Izu to Tokai) to a more normal 40° dipping slab beneath Kyushu. Hirahara (1981) also observed two lateral gaps within the high-velocity plate corresponding to the extensions of the Kyushu-Palau and Shichito–Iwo Jima seismic ridges, and another down-dip gap at 100 km depth beneath Kii peninsula interpreted to be a detachment in the downgoing slab.

The structure beneath the northeastern part of the Japan arc has been investigated in detail by Hasemi et al (1984) using the local earthquake

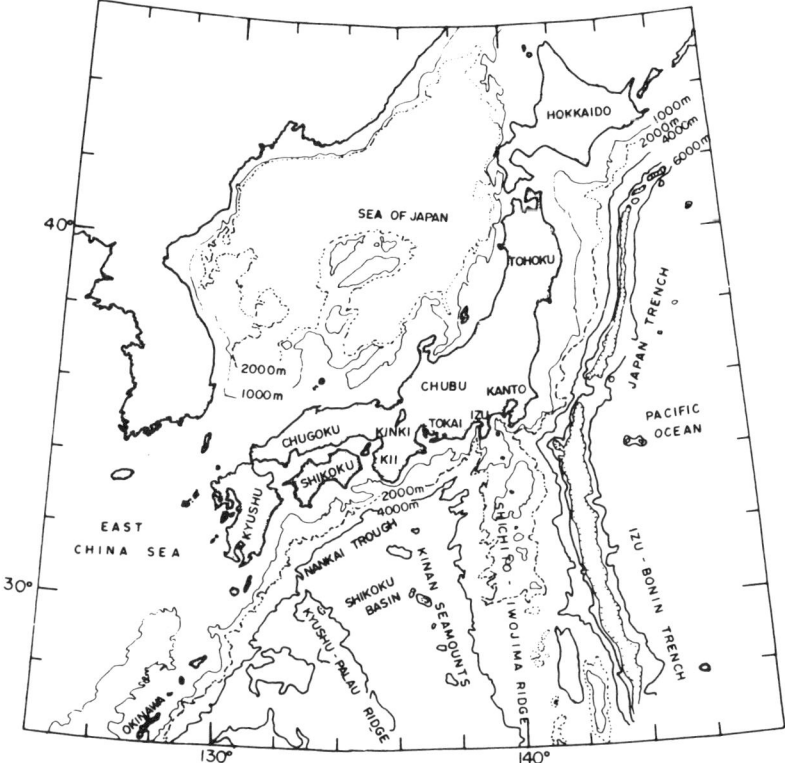

Figure 5 Index map of the Japan Islands region. From Hirahara (1981), with permission.

method. In addition to detecting the high-velocity dipping slab of the Pacific plate, the authors observed two strong low-velocity anomalies (about -4%) in the wedge above the plate, extending from the surface down to about 65 km depth, which coincide with the location of young volcanoes and geothermal areas. A similar observation was reported by Hirahara et al (1986) for central Japan. Dome-shaped low-velocity bodies reaching to depths of 100 to 150 km were found in areas beneath active volcanoes. In both cases, the authors interpreted these low-velocity zones to represent upwelling mantle diapirs with small fractions of partial melt.

Subduction-zone studies in other regions generally have not achieved such fine-scale resolution as in Japan. Michaelson & Weaver (1986) investigated the geometry of the subduction zone beneath Washington and northern Oregon using the ACH method. Their results indicate that the

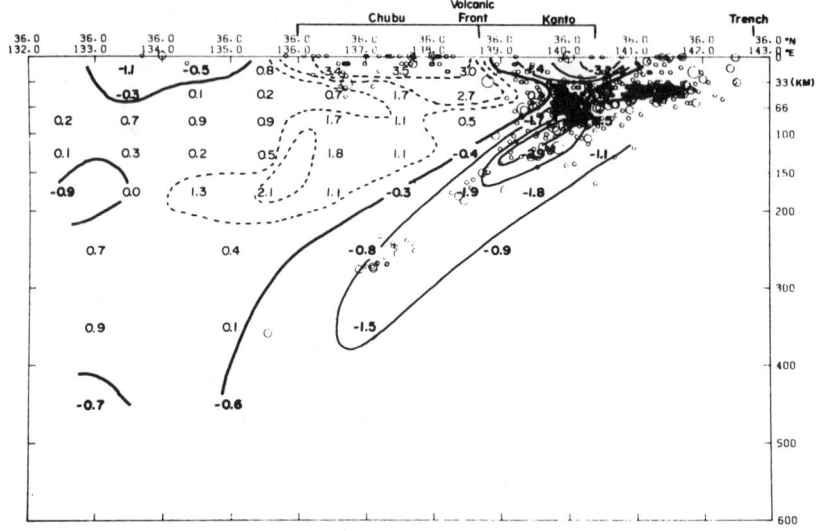

Figure 6 Cross section through the Japan subduction zone model, indicating slowness perturbations (in percent). From Hirahara (1981), with permission.

down-going slab, with a high-velocity anomaly of about 5%, is segmented laterally into three sections with greatly differing dips separated perhaps by tears or warps in the lithosphere. A pair of studies in New Zealand (Spencer & Gubbins 1980) and the central Aleutians (Spencer & Engdahl 1983) used a simple parameterized function to represent the slab structure, obtaining velocity amplitudes of 8% for the slab anomaly in each case. The latter study also illustrated the importance of accounting for the slab structure in locating subduction zone earthquakes. They found that an apparent sharp increase in slab dip below 100 km depth disappeared when locations were calculated using their final slab model.

Exciting results have also been obtained for sites associated with present or former subduction zones that have undergone continental collision. Roecker (1982) detected a dipping low-velocity zone in the Pamir–Hindu Kush region of central Asia, extending continuously from the lower crust to nearly 200 km depth. Roecker interpreted this zone to be subducted continental crust. Similarly inclined low-velocity zones, with anomalies of as much as 10%, have been reported for the Hidaka Mountains, Japan (Takanami 1982, Miyamachi & Moriya 1984), and Taiwan (Roecker et al 1986). All these studies report significant seismic activity concentrated within the low-velocity zone, although Roecker et al (1986) suggest that these events are small, in general.

Regional and Global Studies

Seismologists have begun the process of characterizing the Earth's three-dimensional structure on the scale of hundreds to thousands of kilometers. Body-wave and surface-wave studies have the potential to provide complementary views of the upper mantle at these length scales, while information on the lower mantle is readily obtained only from body waves. Of course, free-oscillation data are also valuable for exploring heterogeneous deep-Earth structure (Cormier 1983).

Regional studies of subcontinental structure have revealed significant lateral variations of seismic velocity, both in the crust and upper mantle. Beneath stable shield areas, like the eastern United States, Scandinavia, and Siberia, seismic velocities in the upper 250 km or so are generally high by roughly 5% (Romanowicz 1979, 1980a, Husebye & Hovland 1982, Feng & Teng 1983). One can interpret this as a temperature effect by which regions not exposed to recent tectonic activity are colder, or conversely the geologic stability can be attributed to the strength of the shield lithosphere (Feng & Teng 1983). Anomalies with smaller amplitudes are present at greater depths as well, perhaps due to undulations of the 400-km (Calcagnile & Scarpa 1985) and 650-km discontinuities (Romanowicz 1979, Hovland et al 1981), the existence of deep continental roots (Jordan 1978), or remnant subducted slabs (Romanowicz 1980b).

Regional surface-wave inversion studies in the Pacific and Atlantic oceans have documented a positive correlation between seafloor age and surface-wave phase and group velocities (Yoshii 1975, Forsyth 1977), reflecting the general thickening of the oceanic lithosphere with age. These studies adopted a priori age regionalizations, however, so that anomalous features unrelated to lithosphere age are masked.

Some recent Rayleigh-wave investigations have employed block regionalization (Suetsugu & Nakanishi 1985) and Tarantola & Valette's (1982) method (Montagner 1986, Suetsugu & Nakanishi 1986) in order to avoid this bias. The age-velocity correlation is generally confirmed except at the the longest periods (around 200 s), where the group velocity of Montagner (1986) is fast in the central Pacific and near the Galapagos triple junction and slow elsewhere, while the phase velocity pattern of Suetsugu & Nakanishi (1985, 1986) is just the opposite. Suetsugu & Nakanishi (1986) inverted the phase velocity results for the three-dimensional SV structure, finding that the 200-s pattern can be attributed simply to the deepening of the low-velocity zone with age. They also seemed to detect the high-velocity subducting slabs at depths of about 150 to 350 km along the marginal basins of the western Pacific.

Global-imaging studies hold great promise for providing clues to the

Earth's geodynamic processes. Some insight into the structure beneath mid-ocean ridges, the nature of hotspots, and the depth of continental lithospheric roots has already been obtained from global surface-wave inversions [see Dziewonski & Anderson (1984) for a recent review]. Above the transition zone, lateral heterogeneities generally can be correlated with the surface tectonic features, including fast shield and old ocean basin areas and slow mid-ocean ridges. Within the transition zone (450–670 km depth), these correlations are greatly diminished, but the resolution of structure is also weakened (Tanimoto 1985). Nevertheless, Woodhouse & Dziewonski (1984) find a degree 2 anomaly pattern at these depths that agrees rather well with the corresponding free-oscillation results of Masters et al (1982). However, Nakanishi & Anderson (1984) find a similar pattern continuing to shallower depths, and they suggest it is simply due to the global configuration of oceans and continents.

Perhaps the most tantalizing results are coming from surface-wave inversions incorporating anisotropy and their interpretation in terms of mantle flow. Tanimoto & Anderson (1985) explored the azimuthal (horizontal) anisotropy of the upper mantle, while Nataf et al (1986) investigated the polarization anisotropy (SV versus SH). In each case, the results must be considered preliminary as a result of inadequate resolution as well as the simplicity of the anisotropy parameterizations. Tanimoto & Anderson find a reasonable correlation between fast phase velocity directions and lithospheric plate motions, for example NW–SE alignment in much of the Pacific Ocean and NNE–SSW alignment in western Australia. The polarization anisotropy results of Nataf et al are interpreted to represent the relative importance of horizontal versus vertical flow. Their principal finding is a dominant horizontal flow related to plate motion beneath ocean basins but vertical flow around mid-ocean ridges (rising warm material) and beneath the oldest ocean basins (sinking cold material).

Coarse images of lower-mantle structure have been obtained by the body-wave studies of Dziewonski (1984) and Sengupta et al (1981). Although these authors represent the Earth's structure differently (spherical harmonic versus block, respectively) and use quite different data sets, some common features are apparent. Both report large lateral variations near the core-mantle boundary, as suggested by Julian & Sengupta (1973). The uppermost lower mantle beneath southern Africa is fast but then becomes slow from about 2000 km depth to the core-mantle boundary; the Indian subcontinent shows the reverse pattern. Siberia and eastern Asia are generally fast throughout the lower mantle, while the Azores triple junction is rather uniformly slow. The resolution of structure is not yet adequate to address the issue of convective flow across the 670-km discontinuity (Dziewonski & Anderson 1984).

THE FUTURE

Although great strides have been made in three-dimensional seismic imaging over the past decade, it is apparent that a great deal remains to be learned regarding both imaging techniques and the heterogeneous structure of the Earth's interior. Advances in several areas should facilitate continuing progress. The IRIS consortium is leading US efforts to improve the instrumentation and data management for both global and portable seismic networks (Smith 1986). The ORFEUS project is coordinating similar efforts in Europe (Romanowicz & Dziewonski 1986). Further developments in seismological theory and analysis techniques will be essential to keep pace with the wealth of broadband digital data. The increasing power of computers must be exploited fully, in terms of raw computational speed, parallel processing, networking, and graphical display. The challenge is great, but with ingenuity and cooperation we can anticipate exciting progress in seismic imaging through the next decade.

ACKNOWLEDGMENTS

We are grateful to Robert Comer, Adam Dziewonski, Robert Nowack, and Steven Roecker for helpful suggestions; in addition we acknowledge the US National Science Foundation for support of our seismic-imaging research programs through grants EAR-8206266 (CHT) and EAR-8407814 (KA).

Literature Cited

Aki, K., Lee, W. H. K. 1976. Determination of three-dimensional velocity anomalies under a seismic array using first P-arrival times from local earthquakes, 1, A homogeneous initial model. *J. Geophys. Res.* 81: 4381–99

Aki, K., Richards, P. 1980. *Quantitative Seismology: Theory and Methods*. San Francisco: Freeman. 932 pp.

Aki, K., Christofferson, A., Husebye, E. S. 1977. Determination of the three-dimensional seismic structure of the lithosphere. *J. Geophys. Res.* 82: 277–96

Backus, G. E. 1964. Geographical interpretation of measurements of average phase velocities of surface-waves over great circular and great semi-circular paths. *Bull Seismol. Soc. Am.* 54: 571–610

Backus, G. E., Gilbert, J. F. 1967. Numerical application of a formalism for geophysical inverse problems. *Geophys. J. R. Astron. Soc.* 13: 247–76

Benz, H. M., Smith, R. B. 1984. Simultaneous inversion for lateral velocity variations and hypocenters in the Yellowstone region using earthquake and refraction data. *J. Geophys. Res.* 89: 1208–20

Brown, R. D. Jr., Lee, W. H. K. 1971. Active faults and preliminary earthquake epicenters (1969–1970) in the southern part of the San Francisco Bay region. *US Geol. Surv. Misc. Field Stud. Map MF-307*

Brune, J., Dorman, J. 1963. Seismic waves and earth structure in the Canadian shield. *Bull. Seismol. Soc. Am.* 53: 167–210

Brune, J. N., Nafe, J. E., Oliver, J. E. 1960. A simplified method for the analysis and synthesis of dispersed wave trains. *J. Geophys. Res.* 65: 287–304

Buland, R. 1982. Towards locating earthquakes in a laterally heterogeneous medium. *Phys. Earth Planet Inter.* 30: 157–60

Calcagnile, G., Scarpa, R. 1985. Deep structure of the European-Mediterranean area from seismological data. *Tectonophysics* 118: 93–111

Chou, C. W., Booker, J. R. 1979. A Backus-Gilbert approach to the inversion of travel time data for three-dimensional velocity structure. *Geophys. J. R. Astron. Soc.* 59: 325–44

Christofferson, A., Husebye, E. S. 1979, On three-dimensional inversion of P wave time residulas: options for geological modeling. *J. Geophys. Res.* 84: 6168–76

Comer, R. P., Clayton, R. W. 1984. Tomographic reconstruction of lateral velocity heterogeneity in the Earth's mantle. *Eos, Trans. Am. Geophys. Union* 65: 236 (Abstr.)

Cormier, V. F. 1983. Deep earth structure. *Rev. Geophys. Space Phys.* 21: 1277–84

Davies, D., McKenzie, D. P. 1969. Seismic travel-time residuals and plates. *Geophys. J. R. Astron. Soc.* 18: 51–63

Der, Z. A., Landisman, M. 1972. Theory for errors, resolution, and separation of unknown variables in inverse problems, with application to the mantle and the crust in southern Africa and Scandinavia. *Geophys. J. R. Astron. Soc.* 27: 137–78

Dorman, J., Ewing, M. 1962. Numerical inversion of surface wave dispersion data and crust-mantle structure in the New York–Pennsylvania area. *J. Geophys. Res.* 67: 5227–44

Dziewonski, A. M. 1984. Mapping the lower mantle: determination of lateral heterogeneity in P velocity up to degree and order 6. *J. Geophys. Res.* 89: 5929–52

Dziewonski, A. M., Anderson, D. L. 1984. Seismic tomography of the earth's interior. *Am. Sci.* 72: 483–94

Dziewonski, A. M., Steim, J. M. 1982. Dispersion and attenuation of mantle waves through waveform inversion. *Geophys. J. R. Astron. Soc.* 70: 503–27

Dziewonski, A. M., Hager, B. H., O'Connell, R. J. 1977. Large-scale heterogeneities in the lower mantle. *J. Geophys. Res.* 82: 239–55

Eberhart-Phillips, D. 1986. Three-dimensional velocity structure in northern California Coast Ranges from inversion of local earthquakes arrival times. *Bull. Seismol. Soc. Am.* 76: 1025–32

Eickemeyer, J., Prothero, W. A. Jr. 1985. Ray tracing in a heterogeneous medium using the simplex method. *Eos, Trans. Am. Geophys. Union* 66: 980 (Abstr.)

Elbring, G. J., Braile, L. W. 1985. Crustal structure of the Yellowstone–Snake River plain region from two-dimensional inversion of seismic refraction data. *Eos, Trans. Am. Geophys. Union* 66: 302 (Abstr.)

Ellsworth, W. L. 1977. *Three-dimensional structure of the crust and upper mantle beneath the island of Hawaii*. PhD thesis. Mass. Inst. Technol., Cambridge. 327 pp.

Ellsworth, W. L., Koyanagi, R. Y. 1977. Three-dimensional crust and mantle structure of Kilauea Volcano, Hawai. *J. Geophys. Res.* 82: 5379–94

Engdahl, E. R., Gubbins, D. 1985. Simultaneous travel-time inversion for earthquake location and subduction zone structure. *Eos, Trans. Am. Geophys. Union* 66: 1086 (Abstr.)

Feng, C., Teng, T. 1983. Three-dimensional crust and upper mantle structure of the Eurasian continent. *J. Geophys. Res.* 88: 2261–72

Firbas, P. 1981. Inversion of travel-time data for laterally heterogeneous velocity structure–linearization approach. *Geophys. J. R. Astron. Soc.* 67: 189–98

Forsyth, D. W. 1977. The evolution of the upper mantle beneath mid-ocean ridges. *Tectonophysics* 38: 89–118

Forsyth, D., Uyeda, S. 1975. On the relative importance of the driving forces of plate motion. *Geophys. J. R. Astron. Soc.* 43: 163–200

Gubbins, D. 1981. Source location in laterally varying media. In *Identification of Seismic Sources—Earthquake or Underground Explosion*, ed. E. S. Husebye, S. Mykkeltviet, pp. 543–73. Dordrecht: Reidel

Haddon, R. A. W., Husebye, E. S. 1978. Joint interpretation of P-wave time and amplitude anomalies in terms of lithospheric heterogeneities. *Geophys. J. R. Astron. Soc.* 55: 19–43

Hager, B. H., Clayton, R. W., Richards, M. A., Comer, R. P., Dziewonski, A. M. 1985. Lower mantle heterogeneity, dynamic topography and the geoid. *Nature* 313: 541–45

Hasemi, A. H., Ishii, H., Takagi, A. 1984. Fine structure beneath the Tohoku district, northeastern Japan arc, as derived by an inversion of P-wave arrival times from local earthquakes. *Tectonophysics* 101: 245–65

Hawley, B. W., Zandt, G., Smith, R. B. 1981. Simultaneous inversion for hypocenters and lateral velocity variations: an iterative solution with a layered model. *J. Geophys. Res.* 86: 7073–76

Hearn, T. M. 1984. P_n travel times in southern California. *J. Geophys. Res.* 89: 1843–55

Hirahara, K. 1977. A large-scale three-dimensional seismic structure under the Japan islands and the Sea of Japan. *J. Phys. Earth* 25: 393–417

Hirahara, K. 1980. Three-dimensional shear

velocity structure beneath the Japan islands and its tectonic implications. *J. Phys. Earth* 28 : 221–41

Hirahara, K. 1981. Three-dimensional seismic structure beneath southwest Japan: the subducting Philippine Sea plate. *Tectonophysics* 79 : 1–44

Hirahara, K., Mikumo, T. 1980. Three-dimensional seismic structure of subducting lithospheric plates under the Japan Islands. *Phys. Earth Planet. Inter.* 21 : 109–19

Hirahara, K., Ikami, A., Ishida, M., Mikumo, T. 1986. Three-dimensional P-wave velocity structure beneath central Japan—low-velocity bodies in the wedge portion of the upper mantle above high-velocity subducting plates. *Tectonophysics*. In press

Hovland, J., Gubbins, D., Husebye, E. S. 1981. Upper mantle heterogeneities beneath Central Europe. *Geophys. J. R. Astron. Soc.* 66 : 261–84

Huang, H., Spencer, C., Green, A. 1986. A method for the inversion of refraction and reflection travel times for laterally varying velocity structures. *Bull. Seismol. Soc. Am.* 76 : 837–46

Humphreys, E., Clayton, R. W., Hager, B. H. 1984. A tomographic image of mantle structure beneath southern California. *Geophy. Res. Lett.* 11 : 625–27

Husebye, E. S., Hovland, J. 1982. On upper mantle seismic heterogeneities beneath Fennoscandia. *Tectonophysics* 90 : 1–17

Iyer, H. M. 1979. Deep structure under Yellowstone National Park, U.S.A.: a continental "hot spot." *Tectonophysics* 56 : 165–97

Jackson, D. D. 1972. Interpretation of inaccurate, insufficient and inconsistent data. *Geophys. J. R. Astron. Soc.* 28 : 97–110

Jordan, T. H. 1978. Composition and development of the continental tectosphere. *Nature* 274 : 544–48

Julian, B. R., Gubbins, D. 1977. Three-dimensional seismic ray tracing. *J. Geophys.* 43 : 95–113

Julian, B. R., Sengupta, M. K. 1973. Seismic travel time evidence for lateral inhomogeneity in the deep mantle. *Nature* 242 : 443–47

Kanasewich, E. R., Chiu, S. K. L. 1985. Least-squares inversion of spatial seismic refraction data. *Bull. Seismol. Soc. Am.* 75 : 865–80

Knopoff, L., Schwab, F. A. 1968. Apparent initial phase of a source of Rayleigh waves. *J. Geophys. Res.* 73 : 755–60

Koch, M. 1985a. A numerical study on the determination of the 3-D structure of the lithosphere by linear and non-linear inversion of teleseismic travel times. *Geophys. J. R. Astron. Soc.* 80 : 73–93

Koch, M. 1985b. Nonlinear inversion of local seismic travel times for the simultaneous determination of the 3D-velocity structure and hypocentres—application to the seismic zone Vrancea. *J. Geophys.* 56 : 160–73

Lay, T., Kanamori, H. 1985. Geometric effects of global lateral heterogeneity on long-period surface wave propagation. *J. Geophys. Res.* 90 : 605–21

Masters, G., Jordan, T. H., Silver, P. G., Gilbert, T. 1982. Aspherical earth structure from fundamental spheroidal mode data. *Nature* 298 : 609–13

Menke, W. 1984. *Geophysical Data Analysis: Discrete Inverse Theory*. Orlando, Fla: Academic. 260 pp.

Michaelson, C. A., Weaver, C. S. 1986. Upper mantle structure from teleseismic P wave arrivals in Washington and northern Oregon. *J. Geophys. Res.* 91 : 2077–94

Miyamachi, H., Moriya, T. 1984. Velocity structure beneath the Hidaka mountains in Hokkaido, Japan. *J. Phys. Earth* 32 : 13–42

Montagner, J. 1986. Regional three-dimensional structures using long-period surface waves. *Ann. Geophys.* 4 : 283–94

Nakanishi, I., Anderson, D. L. 1983. Measurements of mantle wave velocities and inversion for lateral heterogeneity and anisotropy. 1. Analysis of great circle phase velocities. *J. Geophys. Res.* 88 : 267–83

Nakanishi, I., Anderson, D. L. 1984. Measurements of mantle wave velocities and inversion for lateral heterogeneity and anisotropy—II. Analysis by the single-station method. *Geophys. J. R. Astron. Soc.* 78 : 573–617

Nataf, H. C., Nakanishi, I., Anderson, D. L. 1986. Measurements of mantle wave velocities and inversion for lateral heterogeneities and anisotropy. Pt. III: inversion. *J. Geophys. Res.* 91 : 7261–7307

Oppenheimer, D. H., Herkenhoff, K. E. 1981. Velocity-density properties of the lithosphere from three-dimensional modeling at The Geysers–Clear Lake region, California. *J. Geophys. Res.* 86 : 6057–65

Patton, H. 1980. Crust and upper mantle structure of the Eurasian continent from the phase velocity and Q of surface waves. *Rev. Geophys. Space Phys.* 18 : 605–25

Pavlis, G. L. 1986. Geotomography using refraction fan shots. *J. Geophys. Res.* 91 : 6522–34

Pavlis, G. L., Booker, J. R. 1980. The mixed discrete continuous inverse problem: application to the simultaneous determination of earthquake hypocenters and velocity structure. *J. Geophys. Res.* 85: 4801–10

Pavlis, G. L., Booker, J. R. 1983. A sudy of the importance of nonlinearity in the inversion of earthquake arrival time data for velocity structure. *J. Geophys. Res.* 88: 5047–55

Pereyra, V., Lee, W. H. K., Keller, H. B. 1980. Solving two-point seismic ray-tracing problems in a heterogeneous medium, Pt 1. A general adaptive finite difference method. *Bull. Seismol. Soc. Am.* 70: 79–99

Phinney, R. A., Odom, R. I. 1983. Seismic studies of crustal structure. *Rev. Geophys. Space Phys.* 21: 1318–32

Richardson, R. M., Solomon, S. C., Sleep, N. H. 1979. Tectonic stress in the plates. *Rev. Geophys. Space Phys.* 17: 981–1020

Rodi, W. L., Jordan, T. H., Masso, J. F., Savino, J. M. 1980. Determination of the three-dimensional structure of eastern Washington from the joint inversion of gravity and earthquake data. *Rep. SSS-R-80-4516*, Systems Science and Software, La Jolla, Calif. 143 pp.

Roecker, S. W. 1982. The velocity structure of the Pamir–Hindu Kush region: possible evidence of subducted crust. *J. Geophys. Res.* 87: 945–59

Roecker, S. W. 1985. Velocity structure in the Izu-Bonin seismic zone and the depth of the olivine-spinel phase transition in the slab. *J. Geophys. Res.* 90: 7771–94

Roecker, S. W., Yeh, Y. H., Tsai, Y. B. 1986. Three-dimensional P and S wave velocity structures beneath Taiwan: deep structure of an arc-continent collision. Submitted for publication

Romanowicz, B. A. 1979. Seismic structure of the upper mantle beneath the United States by three-dimensional inversion of body wave arrival times. *Geophys. J. R. Astron. Soc.* 57: 479–506

Romanowicz, B. A. 1980a. A study of large-scale lateral variations of P velocity in the upper mantle beneath western Europe. *Geophys. J. R. Astron. Soc.* 63: 217–32

Romanowicz, B. A. 1980b. Large scale three dimensional P velocity structure beneath the western U.S. and the lost Farallon plate. *Geophys. Res. Lett.* 7: 345–48

Romanowicz, B. A., Dziewonski, A. M. 1986. Towards a federation of broad band seismic networks. *Eos, Trans. Am. Geophys. Union* 67: 541–42

Roult, G., Romanowicz, B. 1984. Very long-period data from the Geoscope network: preliminary results on great circle averages of fundamental and higher Rayleigh and Love modes. *Bull. Seismol. Soc. Am.* 74: 2221–43

Santo, T. A. 1961. Division of the southwestern Pacific area into several regions in each of which Rayleigh waves have the same dispersion characters. *Bull. Earthquake Res. Inst. Univ. Tokyo* 39: 603–30

Sato, Y. 1958. Attenuation, dispersion, and the wave guide of the G wave. *Bull. Seismol. Soc. Am.* 48: 231–51

Schwartz, S. Y., Lay, T. 1985. Comparison of long-period surface wave amplitude and phase anomalies for two models of global lateral heterogeneity. *Geophys. Res. Lett.* 12: 231–34

Sengupta, M. K., Hassel, R. E., Ward, R. W. 1981. Three-dimensional seismic velocity structure of the earth's mantle using body wave travel times from intra-plate and deep-focus earthquakes. *J. Geophys. Res.* 86: 3913–34

Shedlock, K. M., Roecker, S. W. 1985. Determination of elastic wave velocity and relative hypocenter location using refracted waves. I. Methodology. *Bull. Seismol. Soc. Am.* 75: 415–26

Sleep, N. H. 1973. Teleseismic P-wave transmission through slabs. *Bull. Seismol. Soc. Am.* 63: 1349–73

Smith, S. W. 1986. IRIS: a program for the next decade. *Eos, Trans. Am. Geophys. Union* 67: 213–19

Spencer, C. P., Engdahl, E. R. 1983. A joint hypocentre location and velocity inversion technique applied to the Central Aleutians. *Geophys. J. R. Astron. Soc.* 72: 399–415

Spencer, C., Gubbins, D. 1980. Travel-time inversion for simultaneous earthquake location and velocity structure determination in laterally varying media. *Geophys. J. R. Astron. Soc.* 63: 95–116

Suetsugu, D., Nakanishi, I. 1985. Tomographic inversion and resolution for Rayleigh wave phase velocities in the Pacific Ocean. *J. Phys. Earth* 33: 345–68

Suetsugu, D., Nakanishi, I. 1986. Three-dimensional velocity map of the upper mantle beneath the Pacific Ocean as determined from Rayleigh wave dispersion. *Phys. Earth Planet. Inter.* In press

Takanami, T. 1982. Three-dimensional seismic structure of the crust and upper mantle beneath the orogenic belts in southern Hokkaido, Japan. *J. Phys. Earth* 30: 87–104

Tanimoto, T. 1985. The Backus-Gilbert approach to the three-dimensional structure in the upper mantle—I. Lateral variation of surface wave phase velocity with its error and resolution. *Geophys. J. R. Astron. Soc.* 82: 105–23

Tanimoto, T., Anderson, D. L. 1985. Lateral heterogeneity and azimuthal anisotropy of the upper mantle: Love and Rayleigh waves 100–250s. *J. Geophys. Res.* 90: 1842–58

Tarantola, A., Nercessian, A. 1984. Three-dimensional inversion without blocks. *Geophys. J. R. Astron. Soc.* 76: 299–306

Tarantola, A., Valette, B. 1982. Inverse problems = quest for information. *J. Geophys.* 50: 159–70

Thurber, C. H. 1981. *Earth structure and earthquake locations in the Coyote Lake area, central California.* PhD thesis. Mass. Inst. Technol., Cambridge. 332 pp.

Thurber, C. H. 1983. Earthquake locations and three-dimensional crustal structure in the Coyote Lake area, central California. *J. Geophys. Res.* 88: 8226–36

Thurber, C. H. 1984. Seismic detection of the summit magma complex of Kilauea volcano, Hawaii. *Science* 223: 165–67

Thurber, C. H., Ellsworth, W. L. 1980. Rapid solution of ray tracing problems in heterogeneous media. *Bull. Seismol. Soc. Am.* 70: 1137–48

Toksöz, M. N., Minear, J. W., Julian B. R. 1971. Temperature field and geophysical effects of a downgoing slab. *J. Geophys. Res.* 76: 1113–38

Um, J., Thurber, C. 1986. A fast algorithm for three-dimensional seismic ray tracing. *Eos, Trans. Am. Geophys. Union* 67: 304 (Abstr.)

Utsu, T. 1967. Anomalies in seismic wave velocity and attenuation associated with a deep earthquake zone, 1. *J. Fac. Sci. Hokkaido Univ. Ser. 7* 3: 1–25

Vassiliou, M. S., Hager, B. H., Raefsky, A. 1984. The distribution of earthquakes with depth and stress in subducting slabs. *J. Geodyn.* 1: 11–28

Weidner, D. J. 1974. Rayleigh wave phase velocities in the Atlantic Ocean. *Geophys. J. R. Astron. Soc.* 36: 105–39

Wesson, R. L. 1971. Travel-time inversion for laterally inhomogeneous crustal velocity models. *Bull. Seismol. Soc. Am.* 61: 729–46

Whitcombe, D. N. 1982. Three-dimensional seismic raytracing for the forward modelling and direct inversion of teleseismic delay times. *Geophys. J. R. Astron. Soc.* 69: 635–48

Wiggins, R. A. 1972. The general linear inverse problem: Implication of surface waves and free oscillations for earth structure. *Rev. Geophys. Space Phys.* 10: 251–85

Woodhouse, J. H., Dziewonski, A. M. 1984. Mapping the upper mantle: three-dimensional modeling of earth structure by inversion of seismic waveforms. *J. Geophys. Res.* 89: 5953–86

Yomogida, K. 1985. Gaussian beams for surface waves in laterally slowly-varying media. *Geophys. J. R. Astron. Soc.* 82: 511–33

Yomogida, L., Aki, K. 1985. Waveform synthesis of surface waves in a laterally heterogeneous earth by the Gaussian beam method. *J. Geophys. Res.* 90: 7665–88

Yoshii, T. 1975. Regionality of group velocities of Rayleigh waves in the Pacific and thickening of the plate. *Earth Planet. Sci. Lett.* 25: 305–12

Zandt, G. 1978. *Study of three-dimensional heterogeneity beneath seismic arrays in central California and Yellowstone, Wyoming.* PhD thesis. Mass. Inst. Technol., Cambridge. 490 pp.

Zandt, G. 1981. Seismic images of the deep structure of the San Andreas fault system, Central Coast Ranges, California. *J. Geophys. Res.* 86: 5039–52

INTEGRATED DIAGENETIC MODELING: A Process-Oriented Approach for Clastic Systems

Ronald C. Surdam

Department of Geology/Geophysics, University of Wyoming, Laramie, Wyoming 82071

Laura J. Crossey

Department of Geology, University of New Mexico, Albuquerque, New Mexico 87131

INTRODUCTION

The decade of the 1970s produced a spectacular revolution in clastic diagenesis. Prior to this time, there was general acceptance of the concept that during progressive burial, primary porosity in sandstones steadily decreased in a constant and regular way until mechanical and/or chemical processes eliminated all effective porosity at a particular and predictable depth. This loss of porosity with depth was thought to be general and inevitable. During the same period, even less was known about the adjacent mudrocks, with the exception of certain mineralogical aspects. With the onset of the 1970s the old bias of viewing the progressive burial of sand-shale sequences as a relatively simple static system was replaced by new ideas suggesting that these systems were anything but static and readily predictable, but instead were highly dynamic diagenetic systems. Each sand-shale package could be considered to be a complicated diagenetic system reacting differently to progressive burial depending on the detailed characteristics of the original depositional setting and the subsequent burial and thermal history. Thus it was recognized that significant differences in the porosity-depth profile were to be expected, both for inter- and intrabasinal comparisons. What caused this dramatic change in the way the diagenetic systems were expected to operate during burial? It started with the work of Heald & Larese (1973), who documented that aluminosilicate framework grains in sandstone could become unstable during burial and undergo dissolution; in addition, they showed that the reaction products could be mobile, resulting in mass transfer and significant increases in porosity in the subsurface during burial.

Subsequent work by Schmidt & McDonald (1979a,b), Hayes (1979), and McBride (1977) documented extensive dissolution of both aluminosilicate and carbonate framework grains and carbonate cement. Pittman (1979) made a very important contribution by showing that dissolution not only affected porosity but also had an even more pronounced effect on permeability of the reservoir facies. Loucks et al (1979), working on Tertiary sandstones of the Gulf Coast, generated a large volume of data relating to the stratigraphic distribution and volumetric significance of diagenetically enhanced porosity ("secondary porosity"). Thus, by the end of the 1970s the concept of dynamic diagenetic systems was firmly in place and widely accepted, even though the processes responsible for the enhanced porosity were not understood.

At the same time, Hower and colleagues (Perry & Hower 1970, Hower et al 1976, Reynolds & Hower 1970, Aronson & Hower 1976, and Yeh & Savin 1977, among others) were delineating the mechanisms controlling the major diagenetic changes in mudrocks. They documented that mixed-layer smectite/illite dominates the mineralogy of many mudrocks and undergoes a continuous decrease in expandability from about 80% to a limit of 20% montmorillonite layers with increasing depth. In addition, they showed that the interstratification changes from random to ordered at about 35% expanded layers. Later Boles & Franks (1979), investigating Tertiary sand-shale systems of the Gulf Coast, explored the implications of smectite diagenesis on sandstone cementation from a chemical standpoint.

Concurrent research on the organic geochemistry of petroleum source rocks has been summarized by Tissot & Welte (1978) and Hunt (1979). Considerable progress had been made in describing kerogen maturation and petroleum-forming reactions during increased burial.

In summary, by the end of the 1970s the major aspects of sandstone diagenesis, shale diagenesis, and organic diagenesis had been described; however, the construction of an integrated model explaining clastic diagenesis in a process-oriented fashion remained for the 1980s.

PROCESS ORIENTATION

Mass Balance

Mass-balance considerations have been applied to diagenetic problems in numerous case studies. This method consists of budgeting the diagenetic changes in detrital mineralogy with authigenic minerals and changes in pore fluid chemistry. Techniques have included chemical and isotopic analysis of rock units, individual phases, and pore fluids; quantitative petrology to determine phase distributions; and theoretical modeling. The scale of the studies varies from the detailed examination of individual reactions and phase relationships to basin-wide evaluations of mass transfer.

REACTION STUDIES   Individual reactions occurring within clastic sediments may be evaluated using several techniques: bulk chemistry and electron microprobe for chemical changes, and petrology, scanning electron microscopy, and X-ray diffraction (essential for studies of fine-grained sediments) for mineralogical changes and phase relationships.

In studies of clay diagenesis of the Gulf Coast Tertiary section, numerous workers (including Perry & Hower 1970, Hower et al 1976, and Boles & Franks 1979) have documented progressive diagenetic changes occurring during the transformation of the expandable smectite component in mixed-layer clays to illite. This mineralogical change is accompanied by chemical changes, including the uptake of potassium by the clay and the release of silica, sodium, calcium, magnesium, iron, and possibly aluminum. The released constituents are then incorporated into authigenic phases such as kaolinite, ankerite, chlorite, calcite, and quartz. The details of the clay reaction continue to be a subject of study (Nadeau et al 1984a,b). Models for the diagenesis of Tertiary sandstones within the Gulf Coast section incorporate the reactions occurring in associated fine-graned rocks to explain cementation patterns and alteration of detrital grains (Boles & Franks 1979).

Albitization of detrital feldspars is a commonly observed process in sandstones that have been deeply buried, and examples have been found in New Zealand (Coombs 1954), the Gulf Coast (Land & Milliken 1981, Boles 1982), and the Great Valley Sequence of California (Dickinson & Rich 1972). Mass transfer associated with this reaction is immense and affects other diagenetic processes involving feldspar (and feldspar constituents) such as calcite, kaolinite, and quartz cementation; the chemical evolution of formation waters; and porosity enhancement. Boles' (1982) study of albitization in the Frio and Wilcox sandstones from the Texas Gulf Coast documented the trend of increasing albitic content of detrital feldspars with increasing burial depth, and he delineated a zone of active albitization (where partial albitization has occurred) at present-day depths of 2.3–2.8 km (100–110°C).

Pervasive laumontite cementation is a commonly observed diagenetic phenomenon associated with the volcanogenic sandstones. These sediments are characterized by high proportions of altered feldspars (Coombs 1954, Boles & Coombs 1977). Boles & Coombs demonstrated that with the addition of silica, calcic plagioclase could be altered to form albite and laumontite. In a detailed study of laumontite occurrence in Paleogene lithic arkoses from California, Helmhold & van de Camp (1984) also related laumontite formation and albitization. In cases where albitization has already occurred, a source of sodium is required for the albitization to proceed. The clay transformation of smectite to illite could supply both sodium and silica. Crossey et al (1984) have examined conditions required for the destabilization of laumontite, which is an important consideration in the prediction of regions of enhanced porosity in these sandstones.

The reaction studies demonstrate the interdependence of such ubiquitous processes as clay mineral diagenesis, albitization, laumontite formation, carbonate and quartz cementation, and porosity enhancement. Taken together, these studies indicate a highly reactive diagenetic zone occurring at temperatures ranging from 80–110°C in sand/shale sequences from a variety of locations. The reaction studies described above have dealt primarily with inorganic constituents of the sequence. Discussed below (see section on "Organic/Inorganic Interactions") is a diagenetic model relating organic reactions to this highly reactive zone of "intense" diagenesis.

ISOTOPIC STUDIES Stable isotopes have provided information pertaining to the temperature, timing, and source of components for many diagenetic reactions. Oxygen and hydrogen isotopes have been used in determining the origin of formation waters in numerous basins (summarized by Kharaka et al 1985). Oxygen and carbon isotopes have been used to estimate the temperature of formation and the extent of organic contribution of diagenetic phases. Less commonly employed are sulfur isotopes of oil and H_2S in gas and waters produced with oils to determine liquid hydrocarbon maturation processes (Orr 1974).

Results of isotopic studies of carbonates are in agreement with the reaction studies outlined above: A zone of "intense" diagenesis is indicated at temperatures between 90 and 140°C (Milliken et al 1981, Land 1984, Franks & Forester 1984, and Suchecki & Land 1983, among others). Studies from a variety of locations (including the Gulf Coast, Texas, and Great Valley, California, areas) indicate that carbon depleted in ^{13}C is incorporated into authigenic carbonates formed in this zone. Based on correlations with organic-acid anions, Carothers & Kharaka (1980) interpreted these isotopically light values to result from microbial degradation of organic matter (at temperatures below 80°C) and from thermal degradation (decarboxylation) at higher temperatures. Methane formed at depth preferentially incorporates ^{12}C, which causes late-forming carbonates to be isotopically heavy as a result of organic processes.

Sulfur isotopic studies have indicated that sulfur in organic matter and in sulfate minerals may participate in redox reactions in the subsurface. The implications of these redox reactions are described below (in the section on "Organic/Inorganic Interactions"). The key observation leading to these conclusions is the observation of isotopically heavy H_2S associated with oils (Orr 1982). H_2S formed from thermal cleavage of organic sulfur would be isotopically light. The incorporation of isotopically heavy sulfur from sulfate mineral phases appears to be the most reasonable explanation. Toland (1960) demonstrated that the reduction of aqueous sulfate in the presence of H_2S oxidized a variety of organic compounds to CO, CO_2, carboxylic acids, and phenols. The reduced sulfur forms H_2S.

In summary, the isotopic studies indicate a zone of "intense" diagenesis at intermediate burial depths (corresponding to temperatures of approximately 90–140°C). These studies also demonstrate that organic-matter diagenesis plays a very active role in the formation of authigenic minerals. This interaction of organic and inorganic matter is reflected in carbon isotopic values of carbonate cements (light where microbial degradation is active, and heavy in zones where methane generation predominates) and in sulfur isotopic values of H_2S.

CHEMICAL MODELING   Theoretical geochemical modeling of water-rock interactions is a completely different approach to the understanding of diagenetic processes. Several computer models are available for predicting the speciation of dissolved constituents and the saturation state of a water, given the total analytical concentration of dissolved species, temperature, pH, and redox potential. The various programs have been summarized and compared by Nordstrom et al (1979). An extension of this type of model predicts successive solution compositions as water-rock reactions progress (e.g. Wolery 1978). Difficulties with the theoretical approach include (*a*) the assumption of complete equilibrium between water and reaction products, (*b*) unavailability of free energy data for many species of interest, (*c*) difficulty of defining the redox potential (particularly in the presence of organic matter), (*d*) compositional variations of many phases (clay and carbonate minerals), and (*e*) uncertainty in complexation behavior of organic compounds. In the context of diagenetic modeling, the endpoint of the theoretical approach is to be able to synthesize a complete paragenetic sequence (Bruton 1985).

Organic/Inorganic Interactions

OIL-FIELD WATERS   Carothers & Kharaka (1978), working primarily with waters from the San Joaquin Valley of California and the Texas Gulf Coast, showed that over the temperature range of 80–120°C there can be significant quantities of carboxylic acids present in oil-field waters (Figure 1). Values as high as 10,000 ppm (as acetate) have been reported (Surdam et al 1984). For 95 formation waters studied by Carothers & Kharaka from relatively young (Eocene-Miocene) oil and gas fields, it was shown that the carboxylic-acid anions generally contribute 50–100% of the measured alkalinity over the 80–200°C temperature range (predominantly in the 80–120°C range). Carothers & Kharaka attributed this distribution to two factors: bacterial consumption at low (<80°C) temperatures, and thermal decarboxylation at higher (>120°C) temperatures. Other workers (Surdam & Crossey 1985a,b, Hanor & Workman 1986, D. MacGowan & R. C. Surdam, in preparation) have supported these observations by analyzing oil-field waters from a variety of locations all over the world. The relative abundances of various organic species vary from location to location, with monofunctional forms (acetic acid) most abundant and the more reactive difunctional forms (maleic, malonic, and oxalic acids) usu-

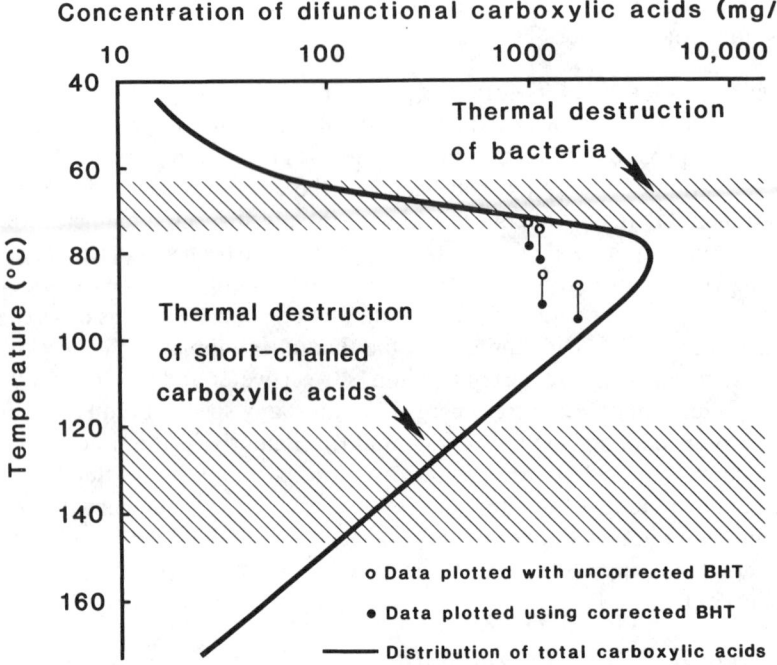

Figure 1 Concentration of difunctional organic acids found in oil-field brines are shown relative to the distribution of total carboxylic acids. Note the abundance of the difunctional forms in the zone of "intense" diagenesis (BHT = bottom hole temperature). From D. MacGowan & R. C. Surdam (in preparation).

ally less abundant (Figure 1). The pH (and Eh; Kharaka et al 1985) of the subsurface waters associated with oil fields is largely controlled by the organic species. An excellent summary of oil-field water sampling and analytical procedures, as well as observed concentrations of both organic and inorganic species, is presented by Kharaka et al (1985).

KEROGEN MATURATION The structure of kerogen is a key element in the evaluation of potential organic-inorganic interactions during progressive burial of sediments. Early diagenetic modification of sedimentary organic material has been described (summarized in Berner 1980). Similarly, much is known about the maturation of kerogen to form liquid hydrocarbons and gases [summarized in Tissot & Welte (1978) and Hunt (1979)]. In contrast, little is known about the processes characterizing the reactive history of kerogen between the stages of early diagenesis (i.e. within 300 m of the surface) and the entrance to the liquid hydrocarbon window. (This depth varies as a function of the time and temperature history of the sediments, but it can generally be considered to be commensurate with a subsurface temperature in excess of 110°C.) In order to fill this gap, a general structural model for the kerogen molecule is essential.

Kerogen composition varies widely. Typically, kerogens are classified on the basis of atomic hydrogen:carbon and oxygen:carbon ratios (H/C and O/C, respectively; see Figure 2) and grouped into three general types (Tissot & Welte 1978, Hunt 1979). Although these types vary compositionally and structurally, several gross similarities can yield insights into the structural changes occurring in the kerogen as temperature increases. Chemical studies (Vitorovic 1980, Crossey et al 1986) have shown that the pervasive oxidation of kerogen results in the generation of carboxylic acids (both mono- and difunctional forms). The model of kerogen that emerges from these studies is a complex core surrounded by unsaturated fatty acid, unbranched monocarboxylic acid, isoprenoid hydrocarbon, aliphatic hydrocarbon, and dicarboxylic acid (Vitorovic 1980). Thus, the carboxylic-acid components are bonded peripherally to the kerogen core by only one bond and are subject to removal by thermal cracking and oxidative degradation.

Carbon-13 nuclear magnetic resonance (NMR) and infrared spectroscopy studies of kerogen support this hypothesis (Miknis et al 1982, Boudou et al 1984). These techniques examine the bonding relationships of the carbon contained in the kerogen and distinguish proportions of carbon contained in aliphatic, aromatic, phenolic, and carbonyl forms (Figure 3). In progressively more deeply buried samples of relatively homogeneous kerogen type, several studies have noted the progressive loss of the phenolic and carbonyl fractions from the kerogen, followed by a

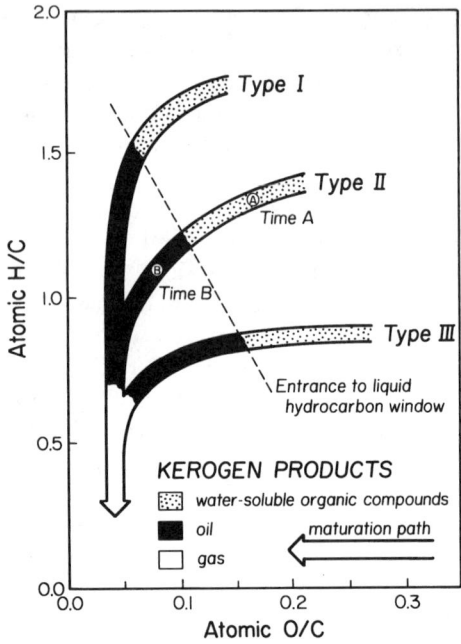

Figure 2 Van Krevelen diagram outlining maturation pathways of typical kerogens of types I, II, and III. Dashed line represents onset of liquid hydrocarbon generation. Stippled areas indicate release of water-soluble organic compounds. Time A, for example, represents kerogen composition prior to hydrocarbon generation; time B is composition after generation. From Crossey (1986).

relative increase in the aromatic fraction as liquid hydrocarbons are produced from the aliphatic fraction (Boudou et al 1984, Surdam & Crossey 1985a). From these changes in the solid kerogen material, it is inferred that oxygen-bearing carboxylic acids as well as phenolic forms are released first, followed by liquid hydrocarbons (Figure 4). The much larger data base consisting of kerogen elemental analyses also supports this hypothesis. The carbon, hydrogen, and oxygen redistribution within kerogen as continued burial occurs may be represented on a van Krevelen diagram (H/C versus O/C; Figure 2). Maturation pathways for the three kerogen types described above are segregated on this diagram, but all three kerogen types show first a decrease in O/C (as oxygen-bearing forms are released), followed by a decrease in H/C (as liquid hydrocarbons are generated; Durand 1985).

In summary, studies of kerogen suggest that oxygen-bearing functional groups such as carbonyls and phenols are released from buried organic

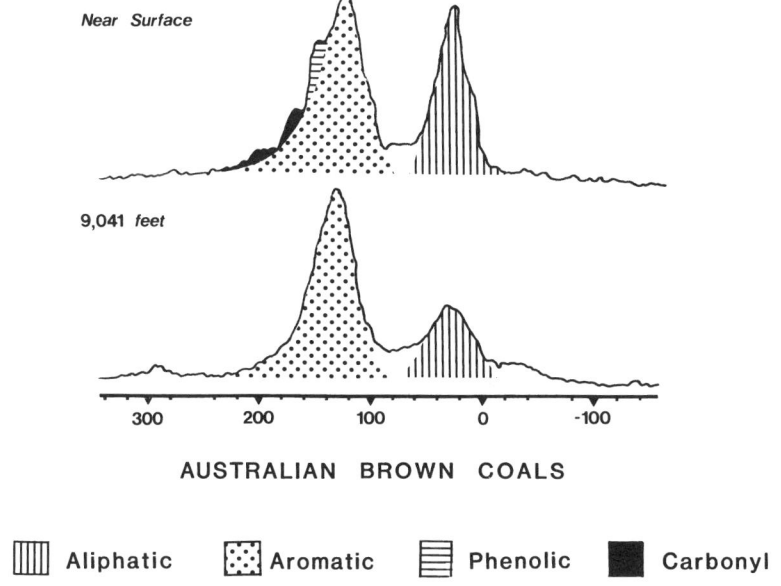

Figure 3 Carbon-13 NMR spectra of Tertiary brown coals of the Latrobe Group (Gippsland Basin, Australia). Phenolic and carbonylic fractions are removed from the kerogen as depth increases. From Surdam & Crossey (1985a).

matter prior to the generation of liquid hydrocarbons. Studies of oil-field waters are in agreement, as significant quantities of these water-soluble organic compounds are observed in these waters.

REDOX REACTIONS An understanding of the mechanisms for the generation of the water-soluble organic compounds from kerogen is the next step in the development of a general model for organic/inorganic interactions during diagenesis. As discussed above, oxidation reactions can result in the generation of water-soluble organic compounds from kerogen. The potential for redox reactions in fine-grained sediments has been discussed by Curtis (1978), Hoffman & Hower (1979), Eslinger et al (1979), and Crossey et al (1986). These workers have focused on the potential of amorphous iron (as grain coatings) and clay diagenesis for reduction accompanied by the oxidation of organic matter. In addition, the sulfur contained in polysulfides (such as pyrite) and sulfate can be reduced in the presence of organic matter (Orr 1982, Berner 1985). These various mechanisms are operable over different stages of diagenesis.

Berner (1980) has summarized many of the important processes characterizing the early stage of diagenesis (burial depth < 300 m). In this zone,

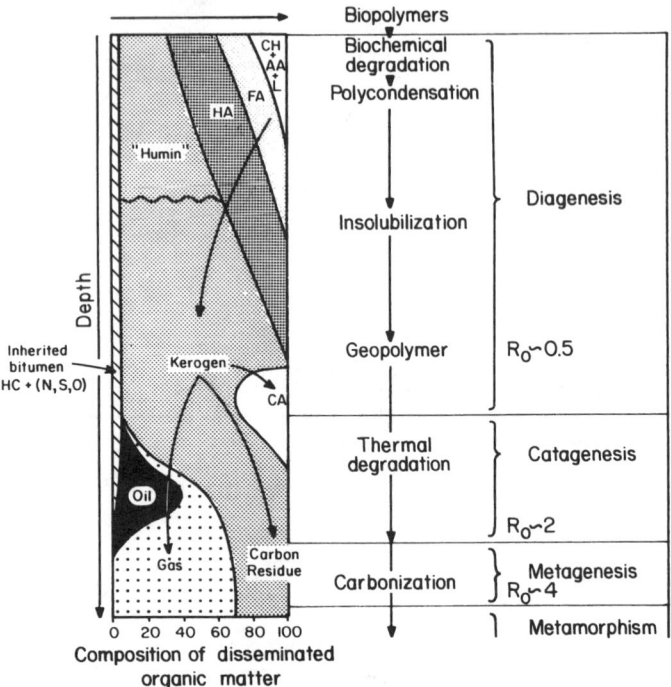

Figure 4 Redistribution of organic matter during diagenesis. (CA) represents carboxylic acids, (CH) carbohydrates, (AA) amino acid, (FA) fulvic acids, (HA) humic acids, (L) lipids, (HC) hydrocarbons, (N, S, O) nitrogen, sulfur, and oxygen compounds. R_0 is vitrinite reflectance in oil. Modified from Tissot & Welte (1978).

microbial processes and the diffusion of constituents from the overlying water column play dominant roles. Redox reactions such as sulfate reduction, ammonia formation, and methanogenesis all involve the oxidation of buried organic matter. These processes result in the precipitation of early-forming authigenic minerals such as pyrite and marcasite, as well as affect clay minerals by adsorption (ammonia). Few data are available on the actual concentrations of dissolved organic acids in this environment, but their presence would certainly affect the stability of the early-forming authigenic minerals. In recent studies of weathering and soil-forming processes, mobilization of cations has been linked to dissolved organic carbon in the ground water in the form of humic acids and short-chain aliphatic acids (Antweiler & Drever 1983, Drever 1980, p. 303).

As burial depth increases and compaction of the sediment restricts fluid communication with the overlying waters, water-rock interactions within the sediments dominate the diagenetic processes. As the burial depth approaches 3000 m (or as temperature exceeds 80°C), a zone of "intense" diagenesis is reached. In this zone a myriad of diagenetic reactions occur, depending on the pore fluid chemistry, detrital mineralogy, and organic matter concentration and composition. The abundance of mixed-layer clays in clastic systems may provide a mineral oxidant in this zone. Johns & Shimoyama (1972) demonstrated that the presence of these clays can catalyze liquid hydrocarbon cracking reactions. Almon (1974) proposed a mechanism for this catalysis: an electron transfer from the kerogen yielding two free radicals (alkane and carboxyl groups), and the acceptance of these electrons by either octahedrally coordinated aluminum at crystal edges or ferric iron within the clay structure. Bulk chemical analyses of Tertiary shales from the Gulf Coast demonstrate a progressive loss of iron from the mixed-layer clay fraction (Hower et al 1976). These results have been used to estimate the potential for the production of organic acids from the shales (Crossey et al 1986) via reduction of the released iron and concomitant oxidation of kerogen. The organic acids produced by this mechanism are sufficient to explain observed framework grain dissolution in associated Tertiary sandstones. Ferric iron contained in amorphous iron oxides may also act as a mineral oxidant in this diagenetic zone, although much of this easily available iron may be consumed in pyrite-forming reactions during early diagenesis (Berner 1980, p. 194).

At higher temperatures ($>120°C$), sulfur species may act as oxidizing agents for the kerogen. In particular, the sulfur contained in sulfate minerals may behave in the manner described by Toland (1960) by providing sulfate to the water. As described above (in the section on "Isotopic Studies"), Toland demonstrated that in the presence of H_2S, aqueous sulfate can oxidize organic material to form CO, CO_2, carboxylic acids, and phenols. The reduced sulfur forms H_2S. This process can account for the sulfur isotope distribution in mature oils (Orr 1974). In this temperature range, bacterial reduction is eliminated and direct chemical reduction is dominant.

In summary, redox processes operating at a given depth in a sand/shale sequence undergoing progressive burial are critically dependent on the ambient temperature, the maturation stage of the enclosed organic matter, the presence of mineral oxidants, and possibly the availability of sulfate minerals. The nature of these redox processes can affect the concentration and type of dissolved organic compounds in the diagenetic fluids.

SYNTHESIS The organic acids released during kerogen maturation, as described above, will interact with the mineral constituents in the sedimentary rocks. The implications for carbonate and aluminosilicate minerals are discussed here, followed by a description of a general diagenetic model for clastic diagenesis in a sand/shale sequence.

Carbonate mineral stability Conventionally the stability of carbonate minerals during progressive burial is evaluated in terms of pH and the aqueous species Ca^{2+}, H_2CO_3, HCO_3^-, and CO_3^{2-}. When only the carbonate system is considered, an increase in P_{CO_2} will result in an increase in carbonate mineral solubility (Holland & Borcsik 1965). Thus, the destabilization of carbonate grains and cements has been attributed to an elevated P_{CO_2} as a result of decarboxylation reactions. This is undoubtedly an important process, both in surface waters and groundwaters as well as at surface temperatures greater than 120–140°C when the organic acids react to form CO_2, CH_4, and H_2O. However, in the temperature range 80–120°C, the presence of organic-acid anions cannot be overlooked because they dominate the alkalinity and because the acetate buffer is orders of magnitude more effective than the carbonate species. If the system is externally buffered by an organic buffer such as acetate, an increase in P_{CO_2} will actually increase carbonate mineral stability (Figure 5; Surdam & Crossey 1985a).

Aluminosilicate mineral stability At 100°C in an aqueous environment the solubility of many aluminosilicate minerals can be elevated by the presence of carboxylic acids and phenols (Figure 6; Surdam et al 1984, Surdam & Crossey 1985a). Under these conditions, aluminosilicate dissolution can be viewed as limited by aluminum solubility (as aluminum is the least-soluble major component). In the absence of a complexing agent, the destabilization of detrital aluminosilicates will result in an in situ alteration to form kaolinite or other secondary mineral phases. Laboratory dissolution experiments have demonstrated the effectiveness of carboxylic acids and phenols in increasing aluminosilicate solubility (Surdam et al 1984, Surdam & Crossey 1985a). In the 80–120°C temperature range, this mechanism can mobilize significant quantities of material in the subsurface.

In order to effectively increase porosity in a sandstone, the dissolved components must be moved beyond the region of interest. In cases where the posority is enhanced at the expense of dissolved aluminosilicate framework grains, the transported material will most likely form kaolinite. The kaolinite forms as the organic complexes are destabilized, probably as a result of a pH change in the fluid (due to mixing of waters in the subsurface or significant mineralogical changes encountered along the migration pathways). Dissolved carbonate material may be transported farther; in any case, carbonate cements are so common that at this point no definitive

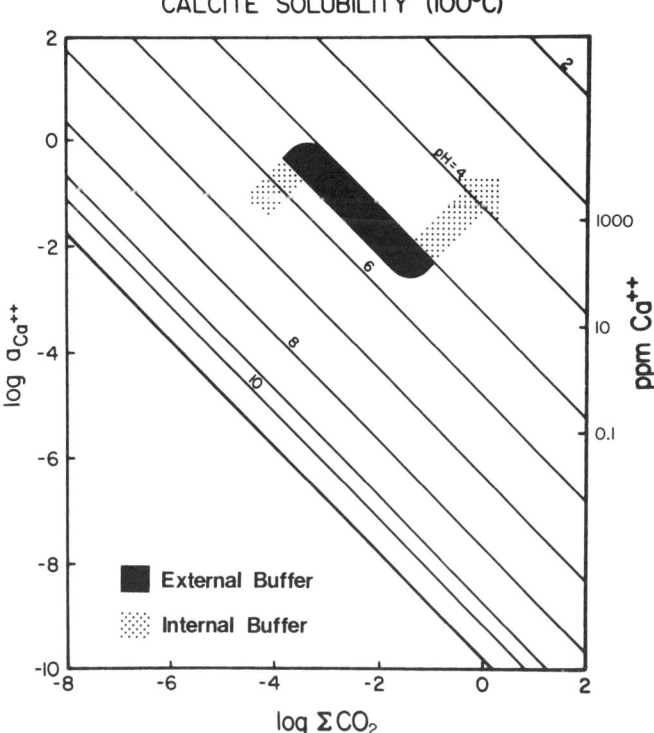

Figure 5 Calculated equilibrium surface for calcite at 100°C projected onto the log $a_{Ca^{2+}}$–log ΣCO_2 plane. Solid lines represent contours of constant pH. ΣCO_2 increases as P_{CO_2} is increased. Arrow represents possible pathways as P_{CO_2} is increased. Modified from Surdam et al (1984).

mass-balance studies have demonstrated a carbonate balance on a basin-wide scale.

Diagenetic model The next step in developing a predictive model for sandstone diagenesis is to integrate the organic and inorganic diagenetic parameters already discussed. Surdam & Crossey (1985a) have attempted this by considering the evolution of a simple diagenetic system consisting of aluminosilicate minerals, carbonate minerals, organic acids, and CO_2 as it passed through the 80–120°C and 120–200°C temperature windows (Figures 7, 8). If we use Figures 7 and 8 as a starting point, a much more complex diagenetic system can be evaluated. For example, Figure 9 represents the integration of most of the significant diagenetic reactions observed in sandstones with the distribution of carboxylic acids in sub-

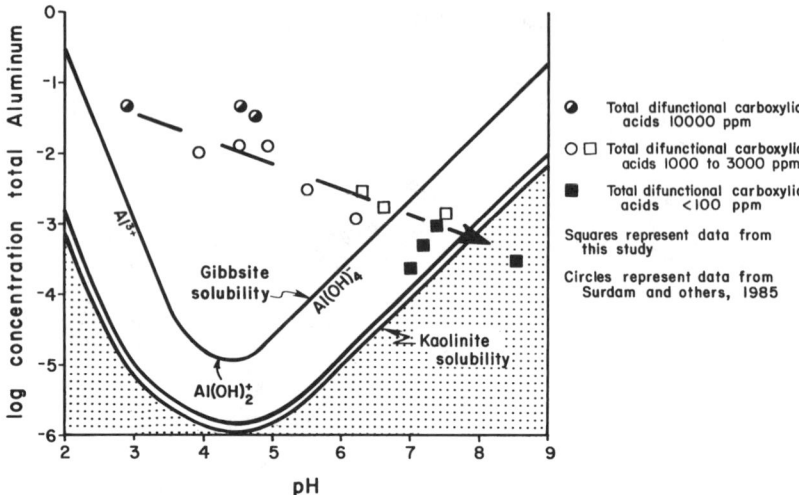

Figure 6 Aluminum concentrations measured in oil-field waters are plotted with reference to gibbsite and kaolinite solubility curves. Upper kaolinite curve represents equilibrium with amorphous silica; the lower curve assumes equilibrium with quartz. Curves were calculated for 100°C, and only three species [Al^{3+}, $Al(OH)_2^+$, $Al(OH)_4^-$] were considered (see labels on gibbsite curve). Thermodynamic data are from Robie et al (1978) and Iler (1979). Also plotted (circles) are experimental feldspar distribution results (Surdam et al 1984). From D. MacGowan & R. Surdam (in preparation).

surface fluids and the reaction-pathway concept of Surdam & Crossey (1985a). Another way to view this more complex diagenetic system is in terms of a general paragenetic sequence (see Figure 10). As is seen in Figures 9 and 10, there is one potential carbonate dissolution episode and one feldspar dissolution episode in the 80–120°C temperature interval. As outlined in Figure 5, the potential for and the degree of carbonate dissolution will depend on the relationship between organic acids and P_{CO_2}, whereas the potential for aluminosilicate dissolution is dependent on the presence of difunctional carboxylic acids or phenols. It should be noted that mineral dissolution (porosity enhancement) and/or the prevention of mineral precipitation (porosity preservation) may characterize these "dissolution" episodes; either possibility will be beneficial to maintaining or developing quality porosity in the sandstone. A second episode of carbonate dissolution is possible in the 120–200°C temperature interval (Figure 10).

Siebert (1985) has suggested that in deeply buried sections characterized by close proximity of organic-rich rocks and evaporites, redox reactions involving sulfate, iron, and methane may precipitate pyrite and lower the pH, resulting in a high-temperature episode of porosity preservation

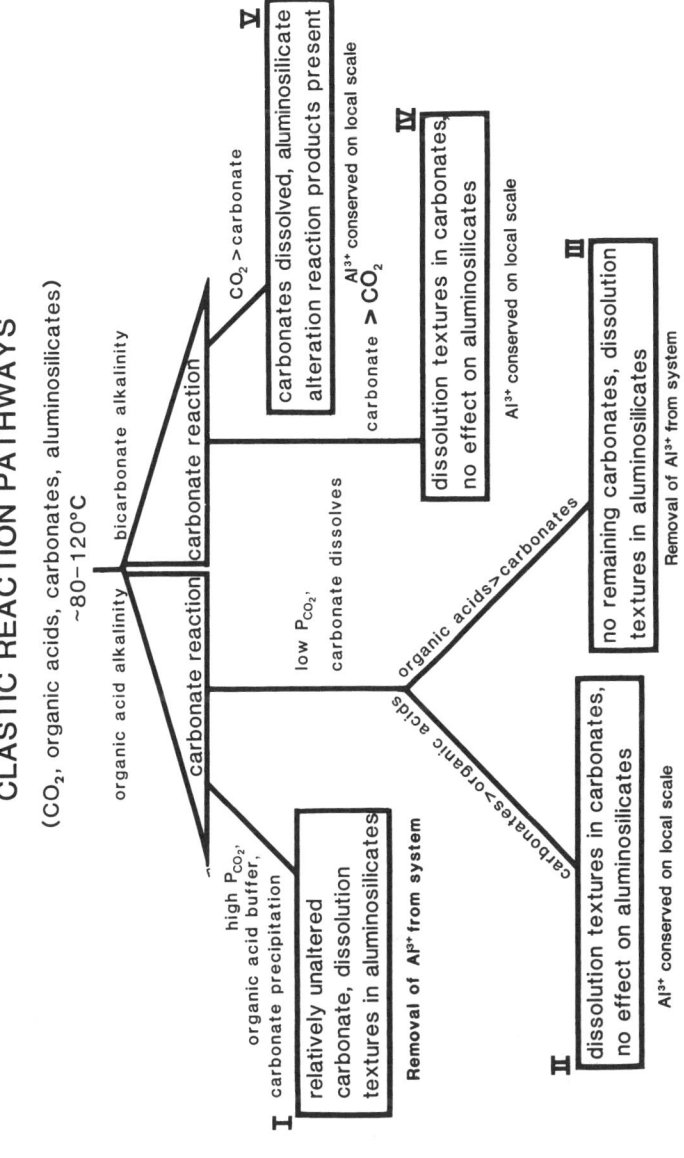

Figure 7 Diagenetic reaction pathways in the temperature zone 80–120°C. Flow diagram combines aspects of organic and inorganic diagenesis. This system represents simplified reservoir conditions. From Surdam et al (1984).

and/or enhancement (due to increased mineral solubility at low pH). This mechanism may be responsible for the occurrence of porosities in excess of 20% at depths greater than 20,000 ft (for example, the Norphlet Formation in the Gulf Coast, southeastern US).

Mass Transfer

All diagenetic reactions require some amount of mass transfer as well as chemical interactions. In some systems, diffusion of components may be sufficient to describe diagenetic modifications, whereas in many cases advection of components into or out of the region of interest is indicated by the observations (cement that could not have been derived locally, or porosity enhanced beyond what can be determined as primary porosity). The analysis of the hydrologic regimes operating in the subsurface can provide constraints on the composition of the subsurface fluids and the time frame over which the reactions occurred, and it can also aid in the prediction of locations and types of diagenetic modification (cementation versus porosity enhancement).

DIFFUSION Diffusion rates in porous media are always slower than the rate in the fluid alone, because the molecules must move around solid particles (tortuosity). In general, diffusion rates are much slower than fluid flow rates in the subsurface; however, diffusion may play a major role in controlling precipitation reactions under certain conditions. For example, Prezbindowski & Pittman (1986) document that calcite cements along the

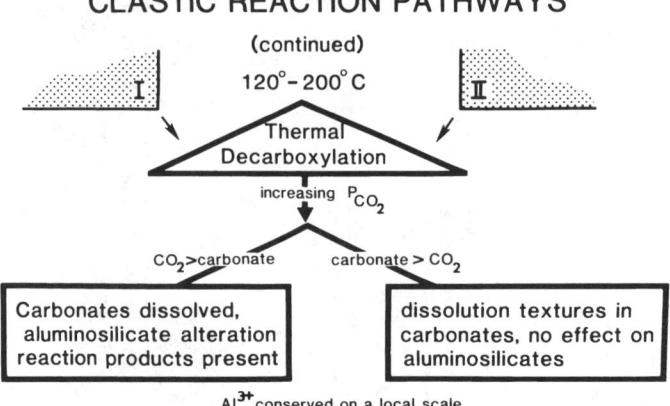

Figure 8 Diagenetic reaction pathways in the temperature zone 120–200°C. Flow diagram combines aspects of organic and inorganic diagenesis. This system represents simplified reservoir conditions. From Surdam et al (1984).

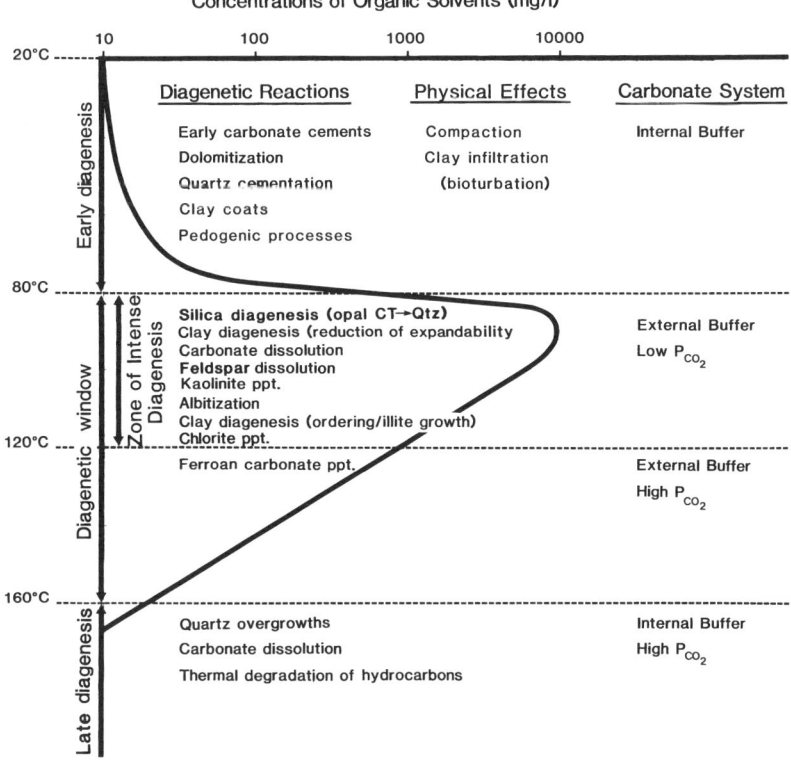

Figure 9 Diagenetic reactions are superimposed on the concentration versus temperature curve for organic acids in oil-field waters (see Figure 1). Modified from Crossey (1985).

upper and lower margins of some Lower Cretaceous turbidite sandstones are related to the proximity of pelagic limestones serving as the calcium source. Although channeled flow occurs in the interior of the massive sands, diffusion of calcium from the margins caused precipitation of calcite. Thin sandstones are entirely cemented by calcite. These workers feel that diffusion can be an important interformational transport mechanism (for distances on the order of a meter). The importance of diffusion in early diagenesis (near the sediment-water interface) has been unequivocally demonstrated [see Berner (1980) for summary].

ADVECTION   Several types of advectional flow are important in sedimentary basins. Convection, compaction, and meteoric water incursion are discussed here.

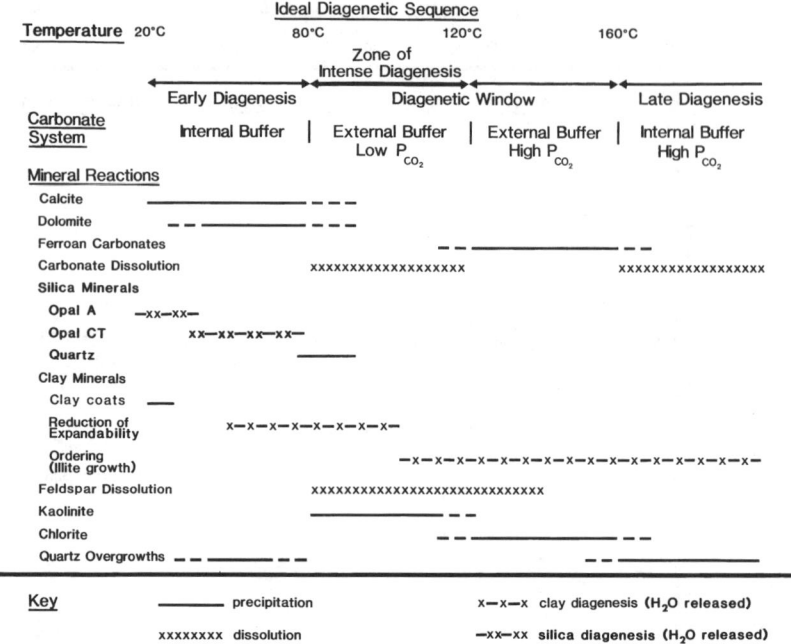

Figure 10 Schematic paragenetic sequence. Commonly observed diagenetic reactions in the context of diagenetic zones defined in Figure 9. Modified from Crossey (1985).

Convection Convectional flow has been postulated by many workers to be a major transport mechanism in the subsurface (e.g. Wood & Surdam 1979, Wood & Hewett 1982, 1984, Blanchard & Sharp 1985). Critics of convection in sedimentary basins question the feasibility of setting up a convective system within a heterogeneous sand/shale sequence. Based on mass-balance considerations, however, Land & Dutton (1979) indicate that the constituents required to form observed cements in the Gulf Coast section require more fluid than can be attained by shale compaction waters. The recycling of basinal fluids provided by convection can explain some of these observations (Wood & Surdam 1979). Aside from chemical considerations, convection may also account for thermal variations observed in many basins. Perhaps the most convincing evidence for convection is found in thermal data from producing oil fields (Blanchard & Sharp 1985), where the distribution of "hot" and "cold" wells can best be described in terms of convection cells.

Compaction Compaction processes have been summarized by Berner (1980) with regard to the volume and upward flow rate of the compaction fluids. Bethke (1985) has constructed a mathematical model of compaction-driven flow rates for intracratonic basins. This model places limits on the amount of transport allowed by the compactional fluids, both chemical and thermal. His model indicates that flow rates are sufficiently slow (millimeters per year) that advective heat transfer cannot occur as a result of compaction alone in intracratonic basins. Another result of this model is the absence of overpressuring. Rapidly subsiding shale-rich basins such as the Ouachita Basin and Gulf Coast behave differently (Sharp 1978); as overpressuring is commonly observed in these settings, compactional flow may be a primary transport medium.

Meteoric water The incursion of meteoric water has been documented to affect diagenesis in sedimentary basins (e.g. Dutton & Land 1985, Longstaffe 1984). Evidence for this influence has been obtained through isotopic and trace-element study of cements (primarily carbonate, but clays have also been examined). Basins exhibiting meteoric effects are characterized by adjacent uplifted recharge areas (Dutton & Land 1985). In general, the diagenetic changes associated with meteoric water occur at relatively shallow depths, and cements are formed at the beginning of the paragenetic sequence. Organic influences have not been documented in this setting, with the exception of possible reactions associated with "water washing" of hydrocarbons in reservoirs. The oxidative degradation of the aliphatic hydrocarbons by meteoric water can potentially generate short-chain aliphatic acids, as described above. Shallow occurrences of extensive dissolution of aluminosilicate framework grains may be due to this effect (Fischer 1986), as meteoric water does not usually carry complexing agents suitable for transporting aluminum (an exception being humic-acid compounds found in surface waters with high dissolved organic carbon contents, usually associated with tropical weathering environments). A major result of meteoric-water diagenesis may be that it affects porosity early in a sedimentary unit's history, either increasing or limiting access to the unit by later fluids associated with hydrocarbon generation.

SUMMARY   Mass-transport phenomena remain some of the most poorly understood aspects of clastic diagenesis. The ability to successfully predict when and where diagenetic reactions have occurred in the subsurface depends to a large extent on an understanding of the fluid migration patterns in a basin throughout its development. Evidence for large-scale mass transfer in sedimentary basins is shown, among other observations, in patterns of silica cementation of sandstones. Houseknecht (1984) petrographically determined the volume of both silica removal (as intergranular pressure solution) and silica addition (quartz cement; see Figure

11). His results indicate mass balance for silica on a regional scale, while individual samples behave as either exporters (predominantly those from areas of greatest thermal maturity) or importers (samples from less thermally mature areas) of silica. A mechanism for large-scale silica transport not requiring exceptionally large volumes of fluid is not clear at this time. Pervasive kaolinite, zeolite (especially laumontite), and carbonate cements document mass transport of other components on a regional scale. Mechanisms for this transport are discussed elsewhere (see the section on "Organic/Inorganic Interactions: Synthesis"). In all cases, some change in fluid composition appears to initiate the precipitation of constituents mobilized from another part of the sedimentary sequence. One likely cause would be a zone where separate fluids come into contact, causing oversaturation with respect to some mineral phase. The ability to track the movement of such diverse fluids as compaction waters, waters derived from dehydration reactions (i.e. opal CT–quartz transformations and clay ordering deeper in the section), and meteoric water through the subsurface at different times in a basin's history will aid in locating positions of these "mixing zones," where numerous reactions are expected to occur.

BASIN-SPECIFIC INTEGRATED DIAGENETIC MODELS

Recent advances in source-rock maturation modeling (summarized in Demaison & Murris 1984) may be applied to diagenetic modeling. As outlined above, the water-soluble organic compounds are produced prior to the generation of liquid hydrocarbons (Figure 4) and have a significant effect on mineral solubilities. Empirical observations suggest that these compounds are most abundant in the 80–120°C temperature range in the

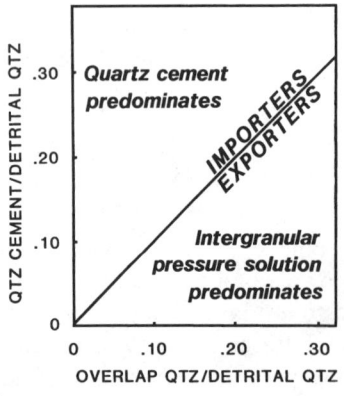

Figure 11 Silica budget diagram, showing volume of quartz cement versus volume of quartz dissolved by intergranular pressure solution (determined petrographically). The 1:1 line separates samples importing silica from those exporting silica. Modified from Houseknecht (1984).

subsurface. By combining these observations with source-rock maturation modeling techniques, a time-temperature integrated diagenetic model can be constructed for diagenetic reactions. The steps to this predictive methodology are as follows: (a) determine the heat flow versus time relationship for the basin of interest, (b) reconstruct the burial history, (c) generate a time-temperature profile for both source and reservoir sequences of interest, (d) locate the position of the 80–120°C temperature window through time, and (e) evaluate the diagenesis in the context of the reaction pathways described above (see the section on "Organic/Inorganic Interactions: Synthesis").

A detailed discussion of steps (a) through (d) is beyond the scope of this paper. Excellent summaries (from a source-rock evaluation standpoint) are presented in Demaison & Murris (1984). Diagenetic applications are discussed by Surdam et al (1986). A simple example is used here to demonstrate how organic and inorganic diagenesis may be incorporated into basin models.

Figure 12 shows a burial history reconstruction for a basin depocenter.

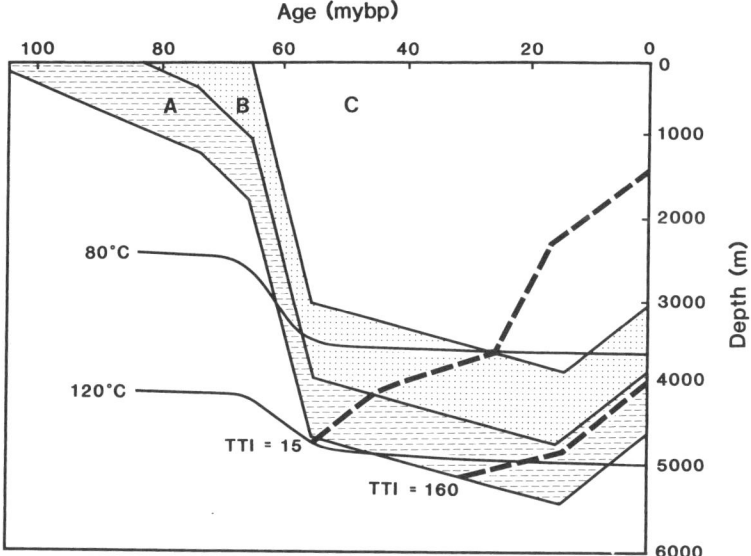

Figure 12 Schematic burial history diagram. Solid lines represent the 80 and 120°C isotherms. Dashed lines represent Time-Temperature Index (TTI) values of 15 and 160 (approximate entrance and exit, respectively, to the liquid hydrocarbon generation window). A gradient of 18°C km^{-1} was used for 60 Myr ago to the present, and 23°C km^{-1} prior to 60 Myr ago. Dashed layer (A) represents a hypothetical source rock, stippled layer (B) reservoir sand, and blank layer (C) undifferentiated overburden. From Crossey (1985).

Layer A represents an organic-rich source rock, layer B a sandstone unit, and layer C undifferentiated overburden. The source rock is presently at a depth of 4000 m. The section is not at its maximum burial, and it is assumed that 800 m were removed between 20 Myr ago and the present. Prior to that, the section experienced continuous burial. Compaction effects are omitted, as are sea-level changes. The thermal history is also assumed to be simple: A constant gradient of 23°C km^{-1} is assumed prior to 60 Myr ago, and a gradient of 18°C km^{-1} is assumed from 60 Myr ago to the present. Thermal conductivities are assumed to be equal throughout the section. The resulting 80 and 120°C isotherms are superimposed on the diagram. As discussed above, this is the temperature range where the organic compounds are observed in the highest quantities in oil-field waters. Thus, the intersection of the reservoir rock (layer B) with this temperature zone represents the time/temperature "window" where reactions due to the organic compounds are most likely to occur. The source of the water-soluble organic compounds may also be evaluated on such a diagram.

A common technique for estimating the maturation level of a hydrocarbon source rock is the use of a Time-Temperature Index (TTI) parameter, based on the time-temperature history of the sediment. Kinetic kerogen maturation models are summarized by Tissot & Welte (1978). Where specific kinetic parameters are unavailable, another maturation estimation, based on the average behavior of kerogen, is obtained via the Lopatin technique (summarized in Waples 1981). This model is based on the observation that kerogen reaction rates double with each 10°C increase in temperature. Once the thermal history of a layer is specified, the maturity index (TTI) is calculated as follows: $TTI_L = \Sigma_i t_{iL} 2^{ni}$, where TTI_L is the maturity factor for layer L, t_{iL} is the time (in million years) spent by layer L in time interval i, and ni is a factor for the ith temperature interval. Temperature factors and a more detailed explanation of the Lopatin method of maturity estimation are available in Waples (1981). Conventionally, a TTI value of 15 is regarded as the point where a source rock begins to generate liquid hydrocarbons, 75 the value at maximum hydrocarbon production, and 160 the exit to the liquid "window." As discussed above, the water-soluble organic compounds are primarily generated *prior* to the generation of liquid hydrocarbons. It seems reasonable, therefore, to assume that most of this generation would occur prior to a TTI value of 75, and possibly prior to a TTI value of 15. At temperatures below 80°C, microbial degradation of water-soluble organic compounds limits their concentration. The configuration of source and reservoir rocks on a burial history diagram can indicate the presence of optimum conditions for the release of substantial amounts of the water-soluble organic compounds into the reservoir. For example, in Figure 12, the source rocks attain a TTI value of 15 at 60 Myr ago. The overlying reservoir rocks at that time and continuing on toward the present are in the critical 80–120°C temperature window. This is an optimum configuration for a reservoir, as liquid hydrocarbons can migrate into any rocks with enhanced porosity before a porosity-reducing reaction occurs. Because of the time and temperature dependence of the kerogen maturation reactions, this configuration is not always attained. As kerogen maturation experiments continue, kinetic parameters for water-soluble organic compounds as well as for liquid hydrocarbons (Lewan 1985) will enable better modeling of the source rock.

Other diagenetic reactions have been calibrated to source-rock maturation levels. Hoffman & Hower (1979) showed how clay diagenesis (conversion of mixed layer illite/smectite to a more illite-rich clay structure) can be correlated to specific temperatures in shales. Isotopic data on carbonate minerals may indicate temperatures of specific mineral phase formation. Fluid inclusion data can yield similar information and even link these mineral precipitation events to the hydrocarbon maturation level if liquid hydrocarbons are present in the inclusions. To date, no studies of water-soluble organic compounds in fluid inclusions have been described. All of these types of temperature data permit the diagenetic pathways to be included in a burial history reconstruction and contoured according to isotherms or time/temperature parameters (such as liquid hydrocarbon production). These conceptual models enable the delineation of regions where optimum conditions for diagenetic modification are present.

The position of the organic-acid maximum in Figure 9 is significant, particularly when viewed in the context of hydrocarbon generation (see especially Surdam et al 1984). The maximum occurs just before liquid hydrocarbon generation and generally overlaps a wide range of diagenetic phenomena. Some of the diagenetic events generally initiated in or associated with the 80–120°C temperature interval are smectite-illite ordering, albitization, kaolinite formation, carbonate formation or dissolution, plagioclase dissolution, chlorite formation, and overpressuring (for more details, see Boles & Franks 1979). Organic acids may be affected by any of these diagenetic factors. Therefore, to maximize the effectiveness of the organic acids in preserving or enhancing porosity, the generation of these solvents must be spatially compressed as close as possible to the generation of liquid hydrocarbons. Ideally, the organic acids would sweep through the sandstone just ahead of hydrocarbon migration. If too much time is available between the organic-acid maximum and the zone of intense hydrocarbon generation, then other diagenetic mineral phases (such as quartz and chlorite) may precipitate into the pores and pore throats, thus eliminating any gains in porosity made by the organic acids.

The experimental data and conceptual diagenetic reaction pathway framework incorporating organic/inorganic reactions allow much more refined estimates than have been available in the past of both the timing of and potential for diagenetic modification in sedimentary basins undergoing progressive burial. The present integrated models can track reaction progress within a sedimentary unit throughout its burial history, provided that kinetic data are available. Empirical observations (for example, linking a diagenetic reaction to a given temperature) may also be used in the models to predict diagenetic reaction pathways in an area of interest.

CONCLUSIONS

Predictive Capabilities

The ability to predict the occurrence of a given diagenetic reaction in specific locations at various points of time in a basin's history requires both spatial and temporal views of the basin's development. As discussed above, basin-modeling techniques have made vast strides in the recent past. These models can be applied to a series of control points throughout

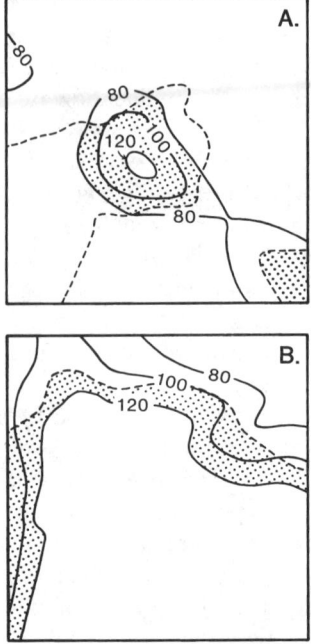

Figure 13 Model results for a hypothetical Cretaceous basin. Solid lines represent temperature contours for a reservoir sequence based on a detailed burial history analysis (including decompaction of sediments, and conductivity and heat-flow variations through time). Dashed lines represent the liquid hydrocarbon expulsion point for the source-rock interval (see text for discussion). Frame *A* represents conditions 40 Myr ago, and *B* represents conditions 20 Myr ago. The stippled region displays the vertical overlap of conditions where the reservoir sequence is within the zone of "intense" diagenesis and the source interval is actively expulsing liquid hydrocarbons.

a basin. The control points may be actual wells for which the stratigraphy has been described, points along a seismic section, or theoretical points from a general model (such as points taken along a cross section of a thermally subsiding rift basin). After each point is modeled, the results can be contoured to allow interpolation to other points in the basin. Figure 13 shows contour plots of this type for a hypothetical Cretaceous basin at two times in the past (40 and 20 Myr ago).Two different types of information can be used to predict diagenetic reaction occurrences: positions of isotherms in the basin at various times, and a measure of reaction progress for reactions described kinetically. Figure 13 shows the outward migration of the zone of "intense" diagenesis (80–120°C temperature window) in the reservoir sequence. Any reaction that is linked to a specific temperature threshold (for example, the albitization window in the Gulf Coast, clay mineral transformations, or opal CT–quartz transformations) can be traced with this technique. The dashed line represents the expulsion point in the source sequence (stratigraphically below the reservoir sequence in this case). The expulsion point is determined in this case by comparing a theoretical measure of kerogen maturation (transformation ratio) with a measured maturation parameter (production index). The depth at which these values diverge is considered to be the expulsion point (see Hagen 1986). Any kinetically controlled reaction, such as liquid hydrocarbon production (Lewan 1985) or sulfate redox reactions (Orr 1982), may be described with this technique. An additional result of this type of model is the ability to determine the spatial relationships of the various reactions occurring in the basin. The stippled region in Figure 13 represents the area of vertical overlap of the zone of intense diagenesis in the reservoir sequence and the expulsion of liquid hydrocarbons in the source interval. In terms of hydrocarbon exploration, this relationship may help pinpoint regions of interest if vertical migration of hydrocarbons is anticipated (i.e. if faults or fractures are common in the section). If expulsed fluids from the source interval are suspected as the primary migration medium, then an updip configuration would be of greater interest.

The techniques described above provide a framework for a predictive approach to diagenetic reactions, both organic and inorganic. As better information becomes available, the model results will become increasingly accurate.

Economic Significance

The diagenetic modeling techniques described above have significant economic implications with regard to the search for hydrocarbons. Armed with grossly improved seismic technology and new organic geochemical techniques, explorationists have significantly reduced risk by improving

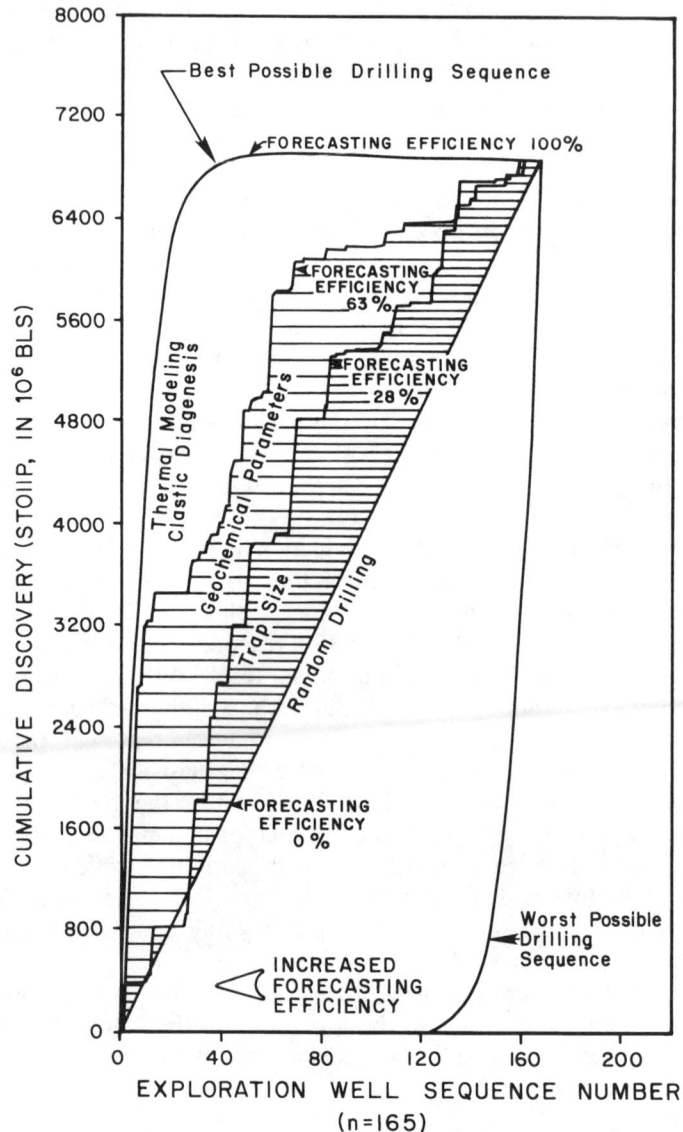

Figure 14 Forecasting efficiency based on different levels of input data. Theoretical forecasting efficiency is improved to 63–100% by viewing organic/inorganic diagenesis in a time-temperature framework. Modified from Demaison & Murris (1984).

forecasting efficiencies (Figure 14). In brief, these improved seismic methods have greatly facilitated the delineations of structural and stratigraphic traps, while the new applied organic geochemistry now allows a careful determination of the extent of generation and timing of migration of the liquid hydrocarbons. The largest remaining uncertainty in most hydrocarbon prospects is the evaluation of the reservoir properties (porosity and permeability) of the targeted sandstone interval prior to drilling. Applying the process-oriented integrated diagenetic modeling techniques developed by Surdam & Crossey (1985a) and Surdam et al (1986), one can also reduce this risk. Using these techniques, the potential for, the quality of, and the timing of the enhanced and/or preserved porosity can be predicted in a forward fashion. R. Murris (in Demaison & Murris 1984) summarizes the forecasting efficiency for hydrocarbon prospects (Figure 14). The integration of structural and organic geochemical (source-rock) parameters can significantly improve forecasting efficiency. This forecasting efficiency can be further significantly increased by the additional consideration of diagenetic modeling of source/reservoir sequences.

Literature Cited

Almon, W. R. 1974. *Petroleum-forming reactions: clay catalyzed fatty acid decarboxylation.* PhD thesis. Univ. Mo., Columbia. 117 pp.

Antweiler, R. C., Drever, J. I. 1983. The weathering of a late Tertiary volcanic ash: importance of organic solutes. *Geochim. Cosmochim. Acta* 47: 623–29

Aronson, J. L., Hower, J. 1976. Mechanism of burial metamorphism of argillaceous sediment, 2, radiogenic argon evidence. *Geol. Soc. Am. Bull.* 87: 738–44

Berner, R. A. 1980. *Early Diagenesis*. Princeton, NJ: Princeton Univ. Press. 241 pp.

Berner, R. A. 1985. Sulphate reduction, organic matter decomposition and pyrite formation. *Philos. Trans. R. Soc. London Ser. A* 315: 25–38

Bethke, C. M. 1985. A numerical model of compaction-driven groundwater flow and heat transfer and its application to the paleohydrology of intracratonic sedimentary basins. *J. Geophys. Res.* 90: 6817–28

Blanchard, P. E., Sharp, J. M. Jr. 1985. Possible free convection in thick Gulf Coast sandstone sequences. *Trans. Southwest Sect. Am. Assoc. Pet. Geol.* 1985: 6–12

Boles, J. R. 1982. Active albitization of plagioclase, Gulf Coast Tertiary. *Am. J. Sci.* 282: 165–80

Boles, J. R., Coombs, D. S. 1977. Zeolite facies alteration of sandstones in the Southland syncline, New Zealand. *Am. J. Sci.* 277: 982–1012

Boles, J. R., Franks, S. J. 1979. Clay diagenesis in Wilcox sandstones of southwest Texas: implications of smectite diagenesis on sandstone cementation. *J. Sediment. Petrol.* 49: 55–70

Boudou, J. P., Durand, B., Oudin, J. L. 1984. Diagenetic trends of a Tertiary low-rank coal series. *Geochim. Cosmochim. Acta* 48: 2005–10

Bruton, C. J. 1985. Predicting mineral dissolution and cementation during burial; synthetic diagenetic sequences. In *SEPM, Gulf Coast Sect. Program and Abstr.* 6: 2–3 (Abstr.)

Carothers, W. W., Kharaka, Y. K. 1978. Aliphatic acid anions in oil-field waters—implications for origin of natural gas. *Am. Assoc. Pet. Geol. Bull.* 62: 2441–53

Carothers, W. W., Kharaka, Y. K. 1980. Stable carbon isotopes of HCO_3 in oil field waters—implications for the origin of CO_2. *Geochim. Cosmochim. Acta* 44: 323–32

Coombs, D. S. 1954. The nature and alteration of some Triassic sediments from Southland, New Zealand. *R. Soc. N.Z. Trans.* 82: 65–109

Crossey, L. J. 1985. *The origin and role of water soluble organic compounds in clastic diagenetic systems*. PhD thesis. Univ. Wyo., Laramie. 134 pp.

Crossey, L. J. 1986. Diagenesis. In *Yearbook of Science and Technology 1987*, pp. 156–61. New York: McGraw-Hill.

Crossey, L. J., Frost, B. R., Surdam, R. C. 1984. Secondary porosity of laumontite-bearing sandstones. See McDonald & Surdam 1984, pp. 225–37

Crossey, L. J., Surdam, R. C., Lahann, R. W. 1986. Application of organic/inorganic diagenesis to porosity prediction. In *Roles of Organic Matter in Sediment Diagenesis*, ed. D. Gautier, pp. 147–55. Tulsa: SEPM

Curtis, C. D. 1978. Possible links between sandstone diagenesis and depth related geochemical reactions occurring in enclosing mudstones. *J. Geol. Soc. London* 135: 107–17

Demaison, G., Murris, R. J., eds. 1984. *Petroleum Geochemistry and Basin Evaluation*, Am. Assoc. Pet. Geol. Mem. No. 35. 426 pp.

Dickinson, W. R., Rich, E. I. 1972. Petrologic intervals and petrofacies in the Great Valley Sequence, Sacramento Valley, California. *Geol. Soc. Am. Bull.* 83: 3007–24

Drever, J. I. 1980. *The Geochemistry of Natural Waters*. Englewood Cliffs, NJ: Prentice-Hall. 388 pp.

Durand, B. 1985. Diagenetic modification of kerogens. *Philos. Trans. R. Soc. London Ser. A* 315: 77–90

Dutton, S. P., Land, L. S. 1985. Meteoric burial diagenesis of Pennsylvanian arkosic sandstones, southwestern Anadarko Basin, Texas. *Am. Assoc. Pet. Geol. Bull.* 69: 22–38

Eslinger, E., Highsmith, P., Albers, D., Demayo, B. 1979. Role of iron reduction in the conversion of smectite to illite in bentonites in the disturbed belt, Montana. *Clays Clay Miner.* 27: 327–38

Fischer, K. 1986. *Diagenesis and mass transfer in an upper Miocene source/reservoir system, southern San Joaquin Basin, California*. MSc. thesis. Univ. Wyo., Laramie. 237 pp.

Franks, S., Forester, R. 1984. Relationships among secondary porosity, pore fluid chemistry and carbon dioxide, Texas Gulf Coast. See McDonald & Surdam 1984, pp. 63–80

Hagen, E. S. 1986. *Hydrocarbon maturation in Laramide-style basins: constraints from the northern Bighorn Basin, Wyoming and Montana*. PhD thesis. Univ. Wyo., Laramie. 215 pp.

Hanor, J. S., Workman, A. L. 1986. Distribution of dissolved volatile fatty acids in some Louisiana oil field brines. *Appl. Geochem.* 1: 37–46

Hayes, J. B. 1979. Sandstone diagenesis—the hole truth. See Scholle & Schluger 1979, pp. 127–39

Heald, M. T., Larese, R. E. 1973. The significance of solution of feldspar in porosity development. *J. Sediment. Petrol.* 43: 458–60

Helmhold, K. P., van de Camp, P. C. 1984. Diagenetic mineralogy and controls on albitization and laumontite formation in Paleogene arkoses, Santa Ynez Mountains, California. See McDonald & Surdam 1984, pp. 239–76

Hoffman, J., Hower, J. 1979. Clay mineral assemblages as low grade metamorphic geothermometers: applications to the thrust faulted disturbed belt of Montana, U.S.A. See Scholle & Schluger 1979, pp. 55–79

Holland, H. D., Borcsik, M. 1965. On the solution and deposition of calcite in hydrothermal systems. *Symp. Probl. Postmagmat. Ore Deposition, Prague*, 2: 364–74

Housenecht, D. W. 1984. Influence of grain size and temperature on intergranular pressure solution, quartz cementation, and porosity in a quartzose sandstone. *J. Sediment. Petrol.* 54: 348–61

Hower, J., Eslinger, E. V., Hower, M. E., Perry, E. A. 1976. Mechanism of burial metamorphism of argillaceous sediment, 1, mineralogical and chemical evidence. *Geol. Soc. Am. Bull.* 87: 725–37

Hunt, J. M. 1979. *Petroleum Geochemistry and Geology*. San Francisco: Freeman. 617 pp.

Iler, R. K. 1979. *The Chemistry of Silica*. New York: Wiley-Interscience. 866 pp.

Johns, W. D., Shimoyama, A. 1972. Clay minerals and petroleum-forming reactions during burial and diagenesis. *Am. Assoc. Pet. Geol. Bull.* 56: 2160–67

Kharaka, Y. K., Hull, R. W., Carothers, W. W. 1985. Water-rock interactions in sedimentary basins. In *Relationship of Organic Matter and Mineral Diagenesis, SEPM Short Course No. 17*, pp. 79–176

Land, L. S. 1984. Frio sandstone diagenesis, Texas Gulf Coast: a regional study. See McDonald & Surdam 1984, pp. 47–62

Land, L. S., Dutton, S. P. 1979. Cementation of sandstone, reply. *J. Sediment. Petrol.* 49: 1359–61

Land, L. S., Milliken, K. L. 1981. Feldspar diagenesis in the Frio Formation, Brazoria County, Texas Gulf Coast. *Geology* 9: 314–18

Lewan, M. D. 1985. Evaluation of petroleum generation by hydrous pyrolysis experimentation. *Philos. Trans. R. Soc. London Ser. A* 315: 123–34

Longstaffe, F. J. 1984. The role of meteoric water in diagenesis of shallow sandstones: stable isotope studies of the Milk River aquifer and gas pool, southeastern Alberta. See McDonald & Surdam 1984, pp. 81–98

Loucks, R. G., Dodge, M. M., Galloway, W. E. 1979. Importance of secondary leached porosity in Lower Tertiary sandstone reservoirs along the Texas Gulf Coast. *Trans. Tex. Gulf Coast Assoc. Geol. Soc.* 29: 164–71

McBride, E. F. 1977. Secondary porosity—importance in sandstone reservoirs in Texas. *Trans. Tex. Gulf Coast Assoc. Geol. Soc.* 27: 121–22

McDonald, D. A., Surdam, R. C., eds. 1984. *Clastic Diagenesis*, Am. Assoc. Pet. Geol. Mem. No. 37. Tulsa: SEPM. 434 pp.

Miknis, F. P., Smith, J. W., Maughan, E. K., Maciel, G. E. 1982. Nuclear magnetic resonance: a technique for direct nondestructive evaluation of source-rock potential. *Am. Assoc. Pet. Geol. Bull.* 66: 1396–1401

Milliken, K. L., Land, L. S., Loucks, R. G. 1981. History of burial diagenesis determined from isotopic geochemistry, Frio Formation, Brazoria County, Texas. *Am. Assoc. Pet. Geol. Bull.* 65: 1397–1413

Nadeau, P. H., Tait, J. M., McHardy, W. J., Wilson, M. J. 1984a. Interstratified XRD characteristics of physical mixtures of elementary clay particles. *Clay Miner.* 19: 67–76

Nadeau, P. H., Wilson, M. J., McHardy, W. J., Tait, J. M. 1984b. Interparticle diffraction: a new concept for interstratified clays. *Clay Miner.* 19: 757–69

Nordstrom, D. K., Plummer, L. N., Wigley, T. M. L., Wolery, T. J., Ball, J. W., et al 1979. A comparison of computerized chemical models for equilibrium calculations in aqueous systems. In *Chemical Modeling in Aqueous Systems*, Am. Chem. Soc. Symp. Ser. No. 93, ed. E. A. Jenne, pp. 857–92

Orr, W. L. 1974. Changes in sulfur content and isotopic ratios of sulfur during petroleum maturation—study of Big Horn Basin Paleozoic oils. *Am. Assoc. Pet. Geol. Bull.* 50: 2295–2318

Orr, W. L. 1982. Rate and mechanism of non-microbial sulfate reduction. *Geol. Soc. Am. Abstr. With Programs* 14: 580 (Abstr.)

Perry, E. A., Hower, J. 1970. Burial diagenesis in Gulf Coast pelitic sediments. *Clays Clay Miner.* 18: 165–77

Pittman, E. D. 1979. Porosity, diagenesis and productive capability of sandstone reservoirs. See Scholle & Schluger 1979, pp. 159–73

Prezbindowski, D. R., Pittman, E. D. 1986. Petrology and geochemistry of sandstones in the Lower Cretaceous submarine fan complex, DSDP Site 603B. In *Initial Reports of the Deep Sea Drilling Project*. Washington, DC: Govt. Print. Off. In press

Reynolds, R. C., Hower, J. 1970. The nature of interlayering in mixed-layer illite-montmorillonites. *Clays Clay Miner.* 18: 25–36

Robie, R. A., Hemingway, B. S., Fisher, J. R. 1978. Thermodynamic properties of minerals and related substances at 298.15 K and 1 bar (10^5 pascals) pressure and at higher temperatures. *US Geol. Surv. Bull.* 1452. 456 pp.

Schmidt, V., McDonald, D. A. 1979a. The role of secondary porosity in the course of sandstone diagenesis. See Scholle & Schulger 1979, pp. 175–207

Schmidt, V., McDonald, D. A. 1979b. Texture and recognition of secondary porosity in sandstones. See Scholle & Schluger 1979, pp. 209–25

Scholle, P. A., Schluger, P. R., eds. 1979. *Aspects of Diagenesis*, SEPM Spec. Publ. No. 26. Tulsa: SEPM. 443 pp.

Sharp, J. M. Jr. 1978. Energy and momentum transport model of the Ouachita Basin and its possible impact on formation of economic mineral deposits. *Econ. Geol.* 73: 1057–68

Siebert, R. M. 1985. The origin of hydrogen sulfide, elemental sulfur, carbon dioxide, and nitrogen in reservoirs. In *SEPM, Gulf Coast Sect. Program and Abstr.* 6: 30–31 (Abstr.)

Suchecki, R. K., Land, L. S. 1983. Isotopic geochemistry of burial-metamorphosed volcanogenic sediments, Great Valley Sequence, northern California. *Geochim. Cosmochim. Acta* 47: 1487–99

Surdam, R. C., Crossey, L. J. 1985a. Organic-inorganic reactions during progressive burial: key to porosity/permeability enhancement and/or preservation. *Philos. Trans. R. Soc. London Ser. A* 315: 135–56

Surdam, R. C., Crossey, L. J. 1985b. Mechanisms of organic/inorganic interactions in sandstone/shale sequences. In *Relationship of Organic Matter and Mineral Diagenesis*, SEPM Short Course No. 17, pp. 177–232

Surdam, R. C., Boese, S. W., Crossey, L. J. 1984. The chemistry of secondary porosity. See McDonald & Surdam 1984, pp. 127–51

Surdam, R. C., Crossey, L. J., Hagen, E. S., Heasler, H. P. 1986. Predictive model for sandstone diagenesis. *Am. Assoc. Pet. Geol. Bull.* In press

Tissot, B. P., Welte, D. H. 1978. *Petroleum Formation and Occurrence.* New York: Springer-Verlag. 638 pp.

Toland, W. G. 1960. Oxidation of organic compounds with aqueous sulfate. *J. Am. Chem. Soc.* 82: 1911–16

Vitorovic, D. 1980. Structure elucidation of kerogen by chemical methods. In *Kerogen*, ed. B. Durand, pp. 301–38. Paris: Technip. 519 pp.

Waples, D. W. 1981. *Organic Geochemistry for Exploration Geologists.* Minneapolis: Burgess. 151 pp.

Wolery, T. J. 1978. *Some chemical aspects of hydrothermal processes at mid-ocean ridges—a theoretical study.* PhD thesis. Northwestern Univ., Evanston, Ill. 275 pp.

Wood, J. R., Hewett, T. A. 1982. Fluid convection and mass transfer in porous sandstones—a theoretical model. *Geochim. Cosmochim. Acta* 46: 1707–13

Wood, J. R., Hewett, T. A. 1984. Reservoir diagenesis and convective fluid flow. See McDonald & Surdam 1984, pp. 99–110

Wood, J. R., Surdam, R. C. 1979. Application of convective-diffusion models to diagenetic processes. See Scholle & Schluger 1979. pp. 243–50

Yeh, H. W., Savin, S. M. 1977. Mechanism of burial metamorphism of argillaceous sediment, 3, O-isotope evidence. *Geol. Soc. Am. Bull.* 88: 1321–30

THE ATMOSPHERES OF VENUS, EARTH, AND MARS: A CRITICAL COMPARISON

Ronald G. Prinn and Bruce Fegley, Jr.

Department of Earth, Atmospheric, and Planetary Sciences, Massachusetts Institute of Technology, Cambridge, Massachusetts 02139

INTRODUCTION

Venus, Earth, and Mars are sibling planets. They all have atmospheres, weathered surfaces, massive volcanoes, and chemically and thermally evolved interiors. Their atmospheres all possess clouds and circulate in response to the thermal forcing by the Sun, modulated by the effects of surface friction and planetary spin. Yet despite these familial characteristics, these three planets show remarkable diversity evident most dramatically in their volatile outer envelopes. We know that the Earth's nitrogen- and oxygen-rich atmosphere is a direct product of its complex biology and that its oceans and freshwater play crucial roles in this biology and in the Earth's climate. Paradoxically, the Earth's proximal neighbor Venus has a comparatively miniscule endowment of water, a carbon dioxide–rich atmosphere about 100 times more massive and with surface temperatures some 450 K warmer than the Earth's atmosphere, and no signs of life. Its next-proximal neighbor Mars shows equally few signs of atmospheric kinship; its predominantly CO_2 atmosphere has a surface pressure about 100 times less than that of Earth, its surface temperature is 60 K cooler than Earth, and it is episodically buffeted by great global storms that shroud the planet in dust.

What can we learn from these similarities and differences? First, they provide invaluable clues concerning atmospheric evolution. Our knowledge of Earth's past atmosphere is gleaned from relics in sedimentary rocks and glacial ice, and the relics usually become more and more difficult to interpret as we go further back in time. In many cases the relic evidence is insufficient to constrain usefully our models of atmospheric evolution. Our planetary neighbors can provide further constraints, since theories developed to explain the evolution of the Earth's atmosphere also ought to yield conclusions about Venus and Mars that are in accord with observations of these planets. For example, proposals concerning the roles of cometary and asteroidal bombardment and atmospheric escape in the evolution of Earth's atmosphere should recognize that these phenomena also occur on Venus and Mars.

The atmospheric compositional extremes on these three terrestrial planets are also very informative. There is evidence that the Earth's atmosphere is subject to change on regional and global scales: first, the observed increase in atmospheric CO_2 is expected to cause a global warming; second, the observed increase in the chlorine species $CFCl_3$ and CF_2Cl_2 in the atmosphere may cause a depletion of the ozone layer; and third, the photooxidation of combustion-derived sulfur gases is causing acid rain in sensitive environments. The processes involved in these present or impending changes are not unique to Earth. The greenhouse effect due to CO_2 is important also on both Venus and Mars; on Venus the amount of CO_2 is so great that its CO_2 greenhouse is the major reason for its very highly elevated surface temperatures. Levels of HCl on Venus are about 1000 times greater than the total mixing ratio of chlorine (as $CFCl_3$, CF_2Cl_2, etc) in the Earth's ozone layer; in addition, levels of sulfur gases on Venus are some five orders of magnitude greater than those on Earth, and photooxidation of these sulfur species produces massive concentrated sulfuric acid clouds that totally shroud the planet.

The atmospheres of Venus, Earth, and Mars provide three natural experiments in geophysical fluid dynamics. Both Earth and Mars rotate rapidly on their axes, possess massive topography and surface thermal contrasts that can force atmospheric motions, and have spin equators tilted relative to the ecliptic, which produces pronounced seasons. On the other hand, Venus rotates extremely slowly on its axis, has a massive thermal inertia, has no tilt-related seasons, and receives most of its solar heating within the atmosphere rather than at the surface. Fluid-dynamical theories and numerical climate models developed for the Earth can be tested critically by application to these other atmospheres.

In this paper we discuss comparatively the physics, dynamics, chemistry, and evolution of the atmospheres of these three terrestrial planets, emphasizing the importance of comparative planetary studies for elucidating and constraining better the fundamental processes at work in planetary atmospheres.

ATMOSPHERIC STRUCTURE AND CIRCULATION

A comparison of basic meteorological parameters on Venus, Earth, and Mars is given in Table 1. Venus possesses the highest albedo of any planet in the solar system, and as a result it absorbs much less solar energy than the Earth (150 vs 240 W m^{-2}), despite its greater proximity to the Sun. Clouds are present on all three planets and play an important role in determining the albedos of the planets. (For Mars, the major clouds are composed of silicate particles, and their thickness is highly variable.) Forcing of motions by release of latent heat during cloud condensation is important only on Earth. The spin angular momentum vectors of Earth and Mars are both inclined significantly to their orbital angular momentum vectors, leading to the variations in hemispheric solar inputs over their orbital periods that produce the distinct seasons seen on both planets. At the present time the southern hemispheres of both Earth and Mars have

Table 1 Comparative meteorological parameters for the tropospheres on Venus, Earth, and Mars

	Venus	Earth	Mars
Solar constant (kW m^{-2})	2.62	1.38	0.59
Fraction of incident solar energy absorbed	0.23	0.70	0.86
Percent cloud cover	100	50	variable (dust storms)
Cloud composition	H_2SO_4	H_2O	dust, H_2O, CO_2
Orbital eccentricity	0.007	0.017	0.093
Spin inclination (obliquity)	2°36'	23°27'	25°12'
Orbital period (days)	225	365	687
Spin period (days)	243 (retrograde)	1 (prograde)	1.03 (prograde)
Solar day (days)	117	1	1.03
Gravitational acceleration (cm s^{-2})	887	980	372
Radius (km)	6051	6378	3394
Surface pressure (bars)	95	1	0.007–0.01
Surface temperature (K)	737	288	220
$d\theta/dz$ (K km^{-1})	−1 to +5	+0.5 (wet), +3.5 (dry)	+2
ΔT (equator−pole) (K)	5 to 15	45	90
Δh (topography) (km)	13	9	25
45° Coriolis parameter (day^{-1})	0.037	8.9	8.6
45° Rossby number (Ro)	23	0.1	0.1

summers near perihelion and thus receive more total energy over an orbital period than their northern counterparts. However, as a result of a slow precession of the longitude of the perihelia of their orbits, this situation reverses on time scales of 10,500 yr (Earth) and 25,500 yr (Mars). On both of these planets the temporal variations in the eccentricity of their orbits and in the inclination of their spin and orbital vectors (obliquity) on time scales of 10^5–10^6 yr produce oscillations in global solar energy inputs that are believed to have caused the ice ages on Earth and the successive layers of polar sedimentary deposits on Mars (Hays et al 1976, Ward 1974). These seasonal and longer-term variations in solar input are, in contrast, muted or negligible on Venus.

Venus spins very slowly on its axis compared with the rotation rates of Earth and Mars, and this has two very important effects on its atmospheric circulation. The Venus solar day (whose inverse equals the sum of the inverses of its orbital and spin periods) is 117 days compared with 1 day on Earth and Mars, and the Venus Coriolis "force" is two orders of magnitude smaller than on Earth and Mars at the same latitude. Venus also has a far greater surface temperature than Earth or Mars due to the greenhouse effect of its massive CO_2 atmosphere. As a rough approximation the surface temperature of a planet with an atmosphere in radiative equilibrium with the Sun is given by

$$T_s = \left[\frac{\alpha S}{4\sigma}\left(1 + \frac{3}{4}\tau_s\right)\right]^{1/4}, \tag{1}$$

where S is the solar constant, α is the fraction of incident solar energy absorbed by the planet, σ is Stefan's constant, and τ_s is the average (gray) infrared optical depth of the atmosphere. Evidently, to explain the T_s values in Table 1 using (1) we require a τ_s value of about 150 for Venus compared with τ_s values of only 0.8 and 0.08 for Earth and Mars. (The actual τ_s values are larger than these because the surfaces are not in radiative equilibrium.)

The convective stabilities of the lower atmospheres (troposphere) of these three planets are quite different. Convective stability is conveniently discussed in terms of potential temperature θ, defined as the temperature that an air parcel would have if it were moved downward adiabatically to the surface. If $d\theta/dz < 0$, then the atmosphere is convectively unstable and an air parcel displaced a small vertical distance δz will have an upward acceleration $d^2z/dt^2 = -g\,\delta z\,d\ln\theta/dz > 0$. Conversely, if $d\theta/dz > 0$ then the atmosphere is convectively stable and a vertically displaced air parcel is accelerated back to its original position, subsequently oscillating about that position with a frequency equal to $(g\,d\ln\theta/dz)^{1/2}$ (the Brunt-Väisälä frequency). On Venus $d\theta/dz$ is often very close to zero and occasionally even negative in the cloudy region between 50 and 60 km altitude (Schubert 1983). In contrast, the Earth's troposphere is generally stable, even when the additional destabilizing influence of latent heat release in a wet atmosphere is taken into account (see Table 1). Convection on Earth and Mars is a transient and patchy phenomenon, whereas it is apparently a ubiquitous phenomenon in the clouds and near the surface of Venus.

Another measure of atmospheric stability and the efficiency of motions in transporting heat is the equator to pole temperature difference ΔT. From Table 1 it is evident that the Venusian atmosphere is apparently the most efficient (and the Martian atmosphere the least efficient) at redistributing the solar energy deposited at low latitudes over the planet. This trend in efficiency is undoubtedly due in part to the very long time needed to radiatively cool the massive lower Venusian atmosphere in contrast to the time needed to move an air parcel from equator to pole (Stone 1975).

Mountains with heights Δh of several kilometers (Table 1) are present on all three of these planets, and horizontal motions must be accompanied by vertical motions forced by flow over these mountains. If the Brunt-Väisälä frequency is sufficiently large (e.g. compared with the radiative damping rate), then these forced vertical motions can produce atmospheric waves. Such topographically forced waves are evident particularly on Earth and Mars.

If a planet is rotating, an air parcel moving with horizontal velocity V will have its motion affected significantly by the rotation if its acceleration dV/dt is less than the Coriolis acceleration $4\pi V \sin \lambda/\tau$, where t is time, λ is latitude, and τ is the spin period. This comparison of scales is usually expressed in terms of the Rossby number Ro:

$$\text{Ro} = \frac{d \ln V/dt}{4\pi \sin \lambda/\tau}$$

$$= \frac{d \ln V/dt}{f}, \qquad (2)$$

where $f = 4\pi \sin \lambda/\tau$ is the Coriolis parameter. Thus Ro is simply the ratio of $1/4\pi$ times the Foucault Pendulum Day to the e-folding time for velocity. If Ro < 1, then the Coriolis force is very important; evidently this is so on Earth and Mars but not on Venus (Table 1).

The latitude-height variations in tropospheric and stratospheric temperature and zonal winds for Earth (Northern Hemisphere winter–Southern Hemisphere summer) and Mars (Northern Hemisphere winter) are shown in Figures 1 and 2. There are important similarities in the temperature structures. At the surface (tropospheric) temperatures decrease

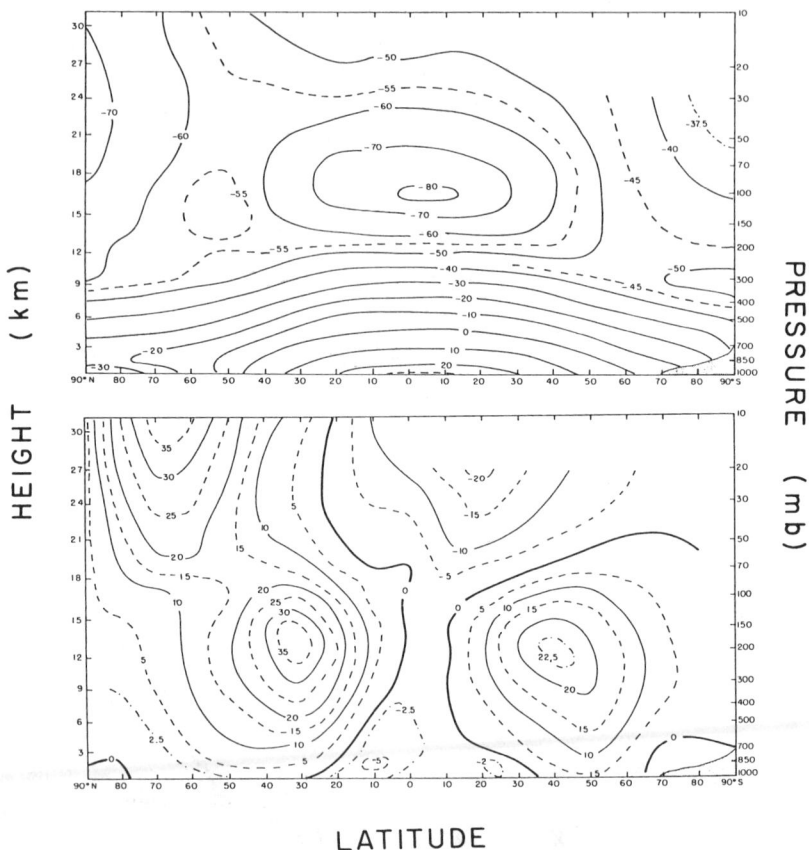

Figure 1 Observed seasonally and zonally averaged temperatures (upper graph) and zonal wind velocities (lower graph) for the atmosphere of Earth (after Newell et al 1972). Velocities are in m s^{-1} (with westerly winds positive), and temperatures are in °C. The Northern Hemisphere is in winter and the Southern Hemisphere in summer. The tropopause separating the troposphere and stratosphere lies at an altitude around 16 km at the equator and 9 km at the poles.

rapidly from equator to pole, but at lower (stratospheric) pressures (100 mbar on Earth, 0.5 mbar on Mars) the winter temperatures increase from the equator to midlatitudes. In the simplest terms, these temperature variations on both planets can be understood in the following way. Near the equatorial surface, air is heated directly or indirectly by the Sun and thus rises on the average and moves toward higher latitudes, where as a result of radiative cooling it subsides again toward the surface. The circulation is thermodynamically direct, since warm buoyant air rises and

ATMOSPHERES OF VENUS, EARTH, & MARS   177

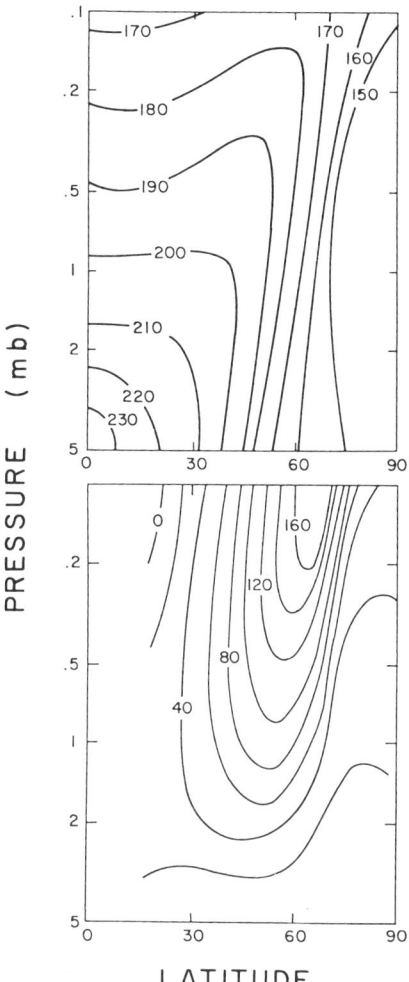

Figure 2 Observed temporally and zonally averaged temperatures (upper graph) and the zonally averaged winds deduced from them using the thermal wind equation (lower graph) as functions of latitude and altitude for a 40-day period in the late Northern Hemispheric winter on Mars (after Conrath 1981). Temperatures are in K and velocities in m s^{-1} (with westerly winds positive). The tropopause lies at about the 0.5-mbar level.

cool dense air sinks. However, in the equatorial lower stratosphere, cool air is being mechanically forced (from below) to rise and thus cool adiabatically, whereas in the midlatitude lower stratosphere warm air is being mechanically forced to sink and warm adiabatically. Thus the circulation in the lower stratosphere is thermodynamically indirect (somewhat like a refrigerator), producing the observed paradoxical equator to midlatitude temperature increase.

Motions on Mars and Earth are both strongly affected by the Coriolis force, outside of the tropics where Ro < 1. Away from the surface, where friction is not important, steady large-scale motions are approximately geostrophic, a state in which the Coriolis force (which is perpendicular to the motion) is balanced by the horizontal pressure gradient force. Also, since vertical wind accelerations are generally small compared with the gravitational acceleration, the upward pressure gradient force is balanced approximately by the downward gravitational force (hydrostatic balance). Under these two combined states of balance, the mean zonal wind ($\bar{u}$) and temperature ($\bar{T}$) outside the tropics are related by the thermal wind equation

$$\frac{\partial \bar{u}}{\partial \ln P} = \frac{R}{f}\left(\frac{\partial \bar{T}}{\partial y}\right)_P, \qquad (3)$$

where P is pressure, R is the gas constant, and y is the meridional (equator-pole) coordinate. Inspection of $\bar{T}$ and $\bar{u}$ for the Earth shown in Figure 1 demonstrates the reality of Equation (3). In particular, $\partial \bar{T}/\partial y < 0$ in the troposphere so $\bar{u}$ increases with decreasing pressure, while $\partial \bar{T}/\partial y > 0$ in the lower stratosphere so $\bar{u}$ decreases with decreasing pressure; the result is the familiar westerly jet stream in the upper troposphere. A similar change in the sign of $\partial \bar{T}/\partial y$ between the middle stratosphere and mesosphere is associated with the presence of the stratospheric westerly polar night jet and easterly summer jet (both evident in Figure 1). For Mars we expect a similar adherence to the thermal wind equation, and the $\bar{u}$ values shown in Figure 2 are in fact derived from the observed temperatures using this equation.

Wave motions in the form of traveling cyclones and anticyclones are very familiar on Earth (Palmen & Newton 1969) and are also evident on Mars (Conrath 1981). On Earth, maximum wave activity occurs during the winter months, and these waves are generated as a result of instabilities of zonal flows in states where isobaric surfaces are inclined to isothermal surfaces [the so-called baroclinic instability; see Holton (1972) for a review]. These traveling waves, which are approximately geostrophic (i.e. quasi-geostrophic) are largely responsible for the observed atmospheric transport of heat from tropical to polar regions. Quasi-geostrophic waves are also forced by topography and by ocean-land temperature contrasts, and in this case these waves are usually stationary and of planetary scale. Both strong high-altitude westerly winds and high-altitude easterly winds on Earth and Mars serve to effectively reflect downward all but the largest scale quasi-geostrophic waves, thus trapping them in the lower atmosphere (Charney & Drazin 1961, Conrath 1981).

While there are certain similarities between the circulations of Earth and Mars, there are also a number of differences that should be emphasized. Mars has episodic global-scale dust storms forced by absorption of solar radiation by wind-raised dust (Gierasch 1974, Zurek 1982), and there are no analogous events to these on Earth. Conversely, the stability and thermodynamics of the Earth's atmosphere are affected very strongly by latent heat release evident most dramatically in hurricanes, typhoons, and intense tropical convective storms (see Holton 1972), whereas latent heating is unimportant on Mars (and on Venus). Temperatures increase upward in the stratosphere on Earth as a result of the absorption of near-ultraviolet and visible solar radiation by the ozone layer, but there is no analogous Martian (or Venusian) stratospheric near-ultraviolet absorber. The oceans on Earth (and the lower atmosphere on Venus) have a large thermal inertia that serves to modulate diurnal and seasonal surface temperature variations, while such a modulation mechanism is not present on Mars.

Temperature and zonal wind structures in the Venusian atmosphere are given in Figures 3 and 4. The large Rossby number on Venus means that the geostrophic balance and thermal wind relation are not valid on this planet. Temperature drops steadily with altitude in the troposphere and strato-mesosphere up to 100 km. Above this level, absorption of solar ultraviolet (UV) photons by CO_2 during the long Venusian day produces an increase in temperature with altitude in the thermosphere. The hot thermosphere totally disappears at night, because cooling by radiation and downward transport of heat apparently occurs on time scales short compared with the Venusian night. In contrast, on Earth and Mars the hot thermosphere, although cooled, is still maintained over the relatively short night. (On Earth, O_2 is the major thermospheric UV absorber, and on Mars it is CO_2.)

The zonal-average winds $\bar{u}$ on Venus are remarkable in that the entire atmosphere superrotates in the same (retrograde) direction as the planet. At the cloud-top altitude of 70 km the zonal velocity is 120 m s^{-1}, so cloud-top features rotate around the planet in only 4 days compared with the planet's spin period of 243 days. For comparison, the Earth (Figure 1) has both westerly winds rotating in the same (prograde) direction as the planet and easterly winds rotating in the opposite (retrograde) direction. Westerly angular momentum is canceled to a large extent by easterly angular momentum, so that total angular momentum in the Earth's atmosphere is over two orders of magnitude less than on Venus.

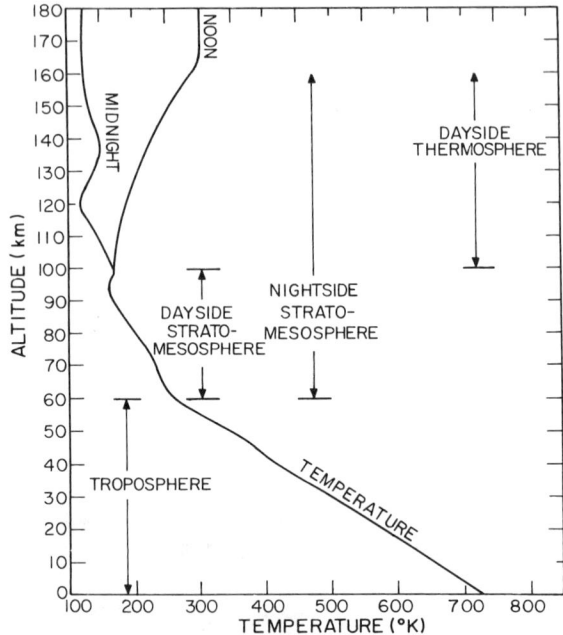

Figure 3 Variation of temperature with altitude at 30° latitude on Venus in the standard atmosphere of Seiff (1983). Also shown are various regions of the atmosphere as discussed in the text. The Venus sulfuric acid clouds lie between 50 and 70 km altitude.

Maintenance of the superrotation of the Venusian atmosphere against down-gradient vertical diffusion of momentum is not fully understood, although there are a number of hypotheses concerning ways in which forced eddy or wave motions can transport zonal momentum in the required upward direction (see Schubert 1983). Waves on Venus (i.e. motions not having zonal symmetry) are observed in UV cloud images, and their dynamic properties have been analyzed (e.g. Rossow et al 1980). As on Earth and Mars, waves on a wide variety of space and time scales are expected on Venus as a result of either thermal forcing by the Sun (producing tides, etc) or mechanical forcing caused by convection or flow over mountains.

The strong zonal winds $\bar{u}$ in the Venusian atmosphere produce a significant equatorward centrifugal force, while higher temperatures at the equator than at the pole produce a significant poleward pressure-gradient force. (This is best visualized by taking a polar view of the zonal wind flow.) Leovy (1973) suggested correctly that these two forces are approximately in

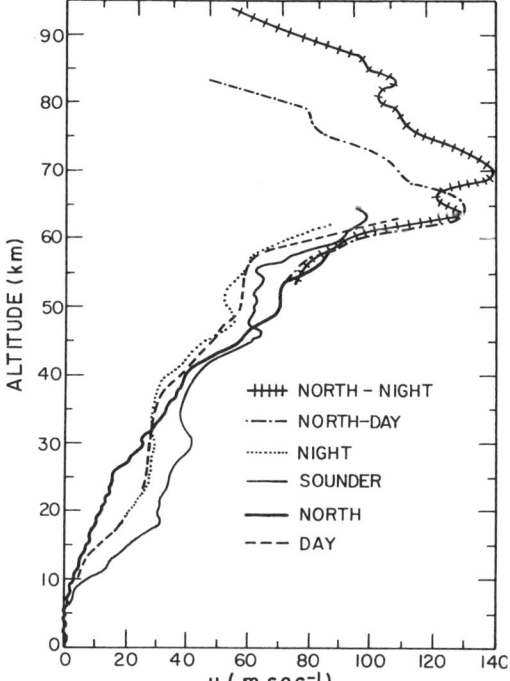

Figure 4 Retrograde zonal wind velocity u as a function of altitude on Venus. The values were obtained by tracking the four Pioneer Venus entry probes (designated night, sounder, north, and day) and by using the temperatures from the north plus night or north plus day entry probes and then deducing the wind velocity from them by using the cyclostrophic thermal wind equation (Counselman et al 1980, Seiff 1983).

balance on Venus. This so-called cyclostrophic balance is familiar on Earth in intense tropical cyclones, where Ro > 1. Combining the cyclostrophic balance with the hydrostatic balance yields an equation analogous to the thermal wind equation (3):

$$\frac{\partial \bar{u}}{\partial \ln P} = \frac{R}{2\bar{u} \tan \lambda}\left(\frac{\partial \bar{T}}{\partial \lambda}\right)_P, \qquad (4)$$

where λ is latitude (Schubert 1983). A comparison of the directly measured winds in the 55–65 km region with those deduced using (4) and the observed latitudinal temperature gradients indicates good agreement (see Figure 4). In particular, as required by (4), $\bar{u}$ increases with decreasing pressure, while temperature decreases from equator to pole at these altitudes on Venus.

ATMOSPHERIC COMPOSITION AND CHEMISTRY

The abundances of the known gases in the atmospheres of Earth, Venus, and Mars are summarized in Tables 2–4, along with information about their sources and sinks where available and relevant. In view of the major objective of this review, we do not attempt a detailed discussion of the chemistry of all the listed species. Appropriate comprehensive discussions of atmospheric chemistry on these three planets are already available (Lewis & Prinn 1984, Levine 1985). Instead, we focus on certain topics where the value of a comparative study of these three atmospheres is particularly fruitful. First, we address the role of living organisms in atmospheric chemical cycles by comparing and contrasting the Earth's

Table 2 Composition of the Earth's nonurban troposphere and the major processes that currently control it (Graedel 1978, Lewis & Prinn 1984)

Gas	Volume mixing ratio	Major source	Major sink
N_2	7.81×10^{-1} [a]	Biology	Biology
O_2	2.09×10^{-1} [a]	Biology	Biology
^{40}Ar	9.3×10^{-3} [a]	Outgassing (^{40}K)	—
H_2O	$\leq 4 \times 10^{-2}$	Evaporation	Condensation
CO_2	3.4×10^{-4}	Combustion, biology	Biology
$^{36,38}Ar$	3.7×10^{-5}	Outgassing (primordial)	—
$^{20,22}Ne$	1.82×10^{-5}	Outgassing (primordial)	—
^4He	5.24×10^{-6}	Outgassing (U, Th)	Escape
CH_4	$1.7–3 \times 10^{-6}$	Biology	Photooxidation
$^{80,82-84,86}Kr$	1.14×10^{-6}	Outgassing (^{235}U)	—
H_2	5×10^{-7}	Photochemistry (H_2O)	Escape (as H)
N_2O	3.1×10^{-7}	Biology	Photodissociation
C_2H_4, etc	$\leq 7 \times 10^{-7}$	Incomplete combustion	Photooxidation
C_2H_2, etc	$\leq 2 \times 10^{-7}$	Incomplete combustion	Photooxidation
C_4H_{10}, etc	$\leq 2 \times 10^{-7}$	Incomplete combustion	Photooxidation
Toluene, etc	$\leq 1 \times 10^{-7}$	Incomplete combustion	Photooxidation
CO	$(0.4–2) \times 10^{-7}$	Photochemistry	Photochemistry
$^{128-132,134,136}Xe$	8.7×10^{-8}	Outgassing (U, I)	—
O_3	$(0.1–1) \times 10^{-7}$	Photochemistry (NO_2)	Photochemistry
CH_3O_2H, etc	$\simeq 10^{-9}$	Photochemistry	Photochemistry
HCl	$\simeq 1 \times 10^{-9}$	Acidification (sea salt)	Rainout
NH_3	$(0.1–1) \times 10^{-9}$	Biology	Photooxidation
HNO_3	$(0.05–1) \times 10^{-9}$	Photochemistry (NO_2)	Rainout
COS	5×10^{-10}	Biology	Photodissociation
CH_3Cl	5×10^{-10}	Biology	Photooxidation
NO, NO_2	$(0.2–5) \times 10^{-10}$	Combustion, biology	Photooxidation
$(CH_3)_2S$	$\simeq 4 \times 10^{-10}$	Biology	Photooxidation
CF_2Cl_2	3.7×10^{-10}	Industry	Photodissociation
SO_2	$\simeq 3 \times 10^{-10}$	Combustion, photochemistry	Photooxidation
$CFCl_3$	2.2×10^{-10}	Industry	Photodissociation
H_2S	$\simeq 2 \times 10^{-10}$	Biology	Photooxidation

[a] Values quoted are for dry air.

Table 3 Composition of the Venusian troposphere and the (probable) major processes that control it (von Zahn et al 1983, Moroz 1983, Prinn 1985)

Gas	Volume mixing ratio	Major source	Major sink
CO_2	9.65×10^{-1}	Outgassing	$CaCO_3$ formation?
N_2	3.5×10^{-2}	Outgassing	—
CO	2×10^{-5} (22 km), 10^{-3} (100 km)	Photochemistry (CO_2)	Photooxidation
SO_2	1.5×10^{-4} (22 km), 5×10^{-8} (70 km)	Photochemistry	$CaSO_4$ formation
$^{36,38}Ar$	3.7×10^{-5}	Outgassing (primordial)	—
^{40}Ar	3.3×10^{-5}	Outgassing (^{40}K)	—
H_2O	10^{-4} (22 km), $(1-40) \times 10^{-6}$ (70 km)	Outgassing, impacts	Silicate hydration, Fe^{++} oxidation plus H escape
H_2	$\leq 2.5 \times 10^{-5}$[a]	Photochemistry	Escape (as H)
^4He	1.2×10^{-5}	Outgassing (U, Th)	Slow escape
H_2S	$(3-40) \times 10^{-6}$[a]	Outgassing (FeS_2)	Photooxidation
COS	$\leq 4 \times 10^{-5}$[a]	Outgassing (FeS_2)	Photooxidation
$^{20,22}Ne$	7×10^{-6}	Outgassing (primordial)	—
$^{80,82-84,86}Kr$	7×10^{-7}, 5×10^{-8}[a]	Outgassing (primordial, ^{235}U)	—
HCl	4×10^{-7}	Outgassing (NaCl)	NaCl formation
HF	5×10^{-9}	Outgassing (CaF_2)	CaF_2 formation

[a] Important disagreements exist between the different instruments that have measured these species.

atmospheric composition to the atmospheric compositions on Venus and Mars, which are determined by strictly abiotic processes. We then discuss three chemical cycles: the sulfur cycles on Venus and Earth, the chlorine-catalyzed cycles on Venus and Earth, and the major photooxidation (hydroxyl radical) cycles on Earth, Mars, and Venus. Finally, we address recent evidence for contemporary global-scale compositional changes on all three planets.

Biology and Atmospheric Composition

As summarized in Tables 2–4, the major components of the Earth's atmosphere are currently controlled by biological processes, whereas the Venusian and Martian atmospheres are controlled by strictly abiotic processes. On Earth, the major N_2 source is the denitrifying bacteria in soils and oceans that convert ammonium and nitrate compounds into N_2. The rates are such that the present atmospheric N_2 amount is produced in about 17 Myr. At the same time, a combination of the activities of nitrogen-fixing bacteria such as those in legume root nodules (which convert atmospheric N_2 into NH_4^+, NO_3^-, and organic nitrogen), together with the abiological processes of lightning and combustion (which convert atmospheric N_2 and O_2 into NO and NO_2, which rain out ultimately as HNO_3), serves to

Table 4 Composition of the Martian troposphere and the (probable) major processes that control it (Barth 1985, Lewis & Prinn 1984)

Gas	Volume mixing ratio	Major source	Major sink
CO_2	9.53×10^{-1}	Evaporation, outgassing	Condensation
N_2	2.7×10^{-2}	Outgassing	Escape (as N)
^{40}Ar	1.6×10^{-2}	Outgassing (^{40}K)	—
O_2	1.3×10^{-3}	Photochemistry (CO_2)	Photoreduction
CO	7×10^{-4}	Photochemistry (CO_2)	Photooxidation
H_2O	$\simeq 3 \times 10^{-4}$	Evaporation, desorption	Condensation, adsorption
$^{20,22}Ne$	2.5×10^{-6}	Outgassing (primordial)	—
^{36}Ar	5×10^{-6}	Outgassing (primordial)	—
Kr	3×10^{-7}	Outgassing (primordial, ^{235}U)	—
Xe	8×10^{-8}	Outgassing (primordial, U, I)	—
O_3	$(0.1–20) \times 10^{-8}$	Photochemistry (CO_2)	Photochemistry
NO	7×10^{-5} (120 km)	Photochemistry (N_2, CO_2)	Photochemistry

remove the current N_2 amount on about the same time scale. If we totally remove the biological processes, then lightning and combustion continued at *current* rates would remove all the atmospheric N_2 in about 80 Myr.

Would the Earth without life therefore lose its N_2 atmosphere? To help answer this question, we could look at our two planetary neighbors and observe that the surface pressure of N_2 on Venus is 3 bars and that N_2 is 2.7% of the atmosphere. Evidently, N_2 can exist without biological activity.

The abiotic removal rates of N_2 by combustion and lightning in the current atmosphere bear little resemblance to those that would occur in the absence of biology. First, there would be no organic material to fuel combustion. Second, the yield of NO in lightning strokes depends sensitively on the total oxygen mixing ratio (the same is true for two other N_2 sinks, namely production of NO from N_2 by cosmic ray and extreme UV bombardment of the upper atmosphere). The major atmospheric oxygen-bearing species are O_2, H_2O, and CO_2. Water vapor is controlled by simple evaporation-condensation. Atmospheric O_2 and CO_2 are currently under biological control, with the opposing biological processes of photosynthesis and respiration/decay producing and removing the Earth's O_2 and CO_2 on time scales of about 3000 and 5 yr, respectively. In the absence of biology, CO_2 levels are not likely to be very much different than those currently observed as long as the oceans exist to buffer CO_2 through carbonate formation. However, only very small amounts (parts per million or less) of O_2 are expected on an abiotic Earth due to UV photodissociation of CO_2 and H_2O (Levine 1985). Such abiotic O_2 production is in fact the major source of O_2 on Mars and Venus today [see reviews by Barth (1985) and Prinn (1985)], but O_2 is only 0.13% of the Martian atmosphere and has never been definitively observed on Venus. Evidently, on a lifeless Earth the major oxygen sources would be H_2O not O_2 in lightning strokes, and CO_2 not O_2 in upper atmospheric cosmic ray and extreme UV processes. Thus, the abiotic N_2 removal rates will be at least one or two orders of magnitude less than today, and the lifetimes of N_2 therefore longer than 1 Gyr rather than as short as 80 Myr. In addition, whatever N_2 is removed through HNO_3 rainout and sedimentary nitrate formation can be recycled through plate subduction and associated volcanism, processes that operate on time scales of 100 Myr.

Apparently, biological processes are therefore essential for maintaining high levels of O_2 in the Earth's atmosphere but not high levels of N_2, CO_2, and H_2O. For the many reactive trace species at levels of parts per million or less, it is evident from Table 2 that the major sources are almost exclusively biological or industrial, and the major sinks almost exclusively are photooxidation or photodissociation. As we discuss shortly, photooxidation of reduced species depends strongly on ambient O_3 (and thus O_2 levels). Thus (as we have emphasized for N_2) removal of all the biological sources is accompanied by a short reduction in the efficiency of the photooxidation sink, making predictions of the levels of these trace species on a lifeless Earth much more difficult. Nevertheless, for species such as the hydrocarbons for which the only known significant sources are biological in origin it is safe to predict that their levels will be much lower despite the less oxidizing nature of the lifeless environment. The same is true for biogenic and anthropogenic species like N_2O, COS, $CFCl_3$, and CF_2Cl_2, whose principal sinks are photodissociation reactions whose rates would actually increase at low O_2 (and thus low O_3) levels.

Chemical Cycles and Atmospheric Composition

SULFUR CYCLES ON EARTH AND VENUS   On Earth we are becoming increasingly aware of the environmental importance of the sulfur cycle (see Duce et al 1984). As depicted in Figure 5, sulfur is injected directly into the atmosphere as SO_2 from fossil-fuel combustion. (Volcanoes are minor in comparison.) Sulfur is also injected in the form of the chemically reduced species $(CH_3)_2S$, CS_2, H_2S, and COS as a result of microbial activity in the oceans and in marshlands (Andreae & Raemdonck 1983, Carroll et al 1986); putatively, the emissions of these reduced species are comparable to SO_2 emissions on a global scale. Successive oxidation reactions involving the OH and NO_3 radicals and UV photons (hν) serve to convert the reduced species (with the exception of COS) to SO_2 on time scales of hours to days. Further oxidation reactions in the gas phase and in droplets then convert the photochemically produced and combustion-derived SO_2 to H_2SO_4 on a time scale of a few days. The H_2SO_4 rains out as a dilute acid, and

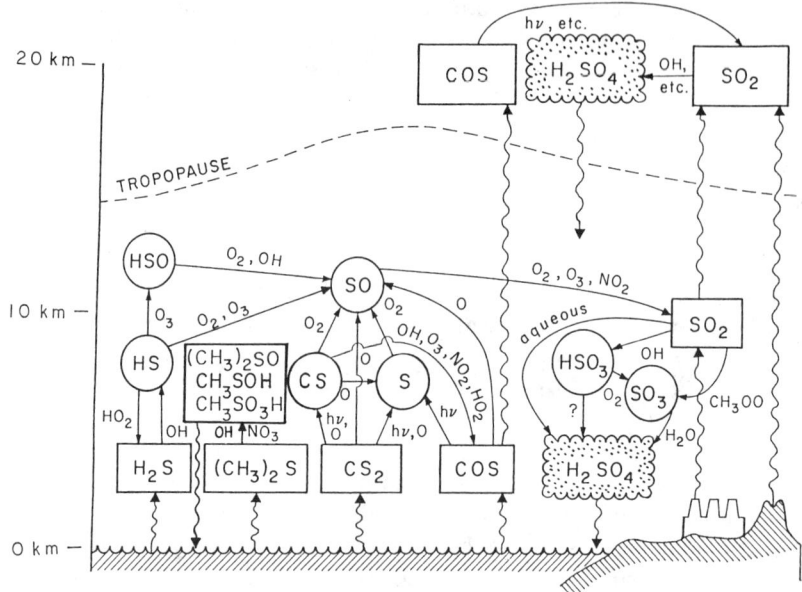

Figure 5 The cycle of sulfur compounds in the Earth's atmosphere. The ocean and marshlands are important sources of H_2S, $(CH_3)_2S$, CS_2, and COS, while fossil-fuel combustion and volcanoes produce SO_2. These source species are converted by a variety of reactions into short-lived species (enclosed in circles) and thence either to SO_2 or to various organic sulfoxyl compounds. The SO_2 is oxidized to yield concentrated sulfuric acid particles in the stratosphere and dilute sulfuric acid particles in the trophosphere. Eventually all these product species are rained out or absorbed at the surface.

because the sources of SO_2 are geographically very patchy and the SO_2 lifetime is short, this acid rain is also geographically patchy and has greatest fluxes within 1000 km or so of SO_2 source regions.

The lifetime of COS is significantly longer than those of the other sulfur species, and it can therefore be transported horizontally over global scales and vertically into the stratosphere before being destroyed. In the stratosphere, COS photodissociates to yield SO_2 and thence concentrated sulfuric acid particles (Crutzen 1976); these particles form a very thin global-scale layer in the stratosphere called the Junge layer. Episodically, volcanoes inject SO_2 rapidly and directly into the stratosphere, providing a transient additional source of sulfuric acid.

Since the presence of sulfur species in our atmosphere is associated with microbial and human activity, we would not expect to see sulfur gases on Venus and Mars. In fact, sulfur gases on Venus have mixing ratios about 10^6 times greater and total amounts about 10^8 times greater than on Earth!

As depicted in Figure 6, the source of Venusian sulfur gases putatively involves reactions between atmospheric CO_2, H_2O, and sulfur-containing minerals such as pyrite (FeS_2) contained in volcanically derived surface lavas or within volcanoes. The emitted gases are COS, H_2S, and to a lesser extent SO_2. The existence of this source depends on the combination of (a) a very hot and chemically reduced surface relative to the Earth and (b) extant volcanism. Photochemical reactions driven by UV photons (hv) convert COS and H_2S to elemental sulfur and (using oxygen derived from CO_2 photodissociation) to SO_2 (Prinn 1973). The SO_2, which is the dominant atmospheric sulfur-bearing gas, is oxidized to sulfuric acid particles (see Yung & DeMore 1982), which form a low-density haze in the 50–70 km altitude region that totally shrouds the planet. The formation of UV-absorbing elemental sulfur particles along with the sulfuric acid provides a plausible cause of the distinct absorption patterns observed in UV images of the planet. The net oxidation of COS and H_2S leads to a SO_2 concentration that exceeds its value in equilibrium with calcium-bearing surface minerals. Various weathering reactions therefore proceed,

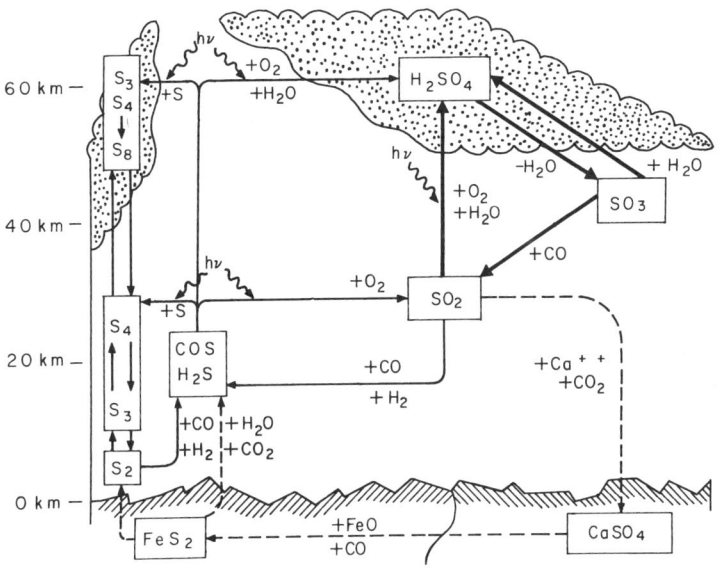

Figure 6 The cycle of sulfur compounds in the Venusian atmosphere (Prinn 1985). Volcanic eruptions or reactions of H_2O and CO_2 with volcanic surface rocks yields COS, H_2S, S_2, and SO_2. Various photochemical reactions convert these species to concentrated H_2SO_4 or elemental sulfur particles in the Venusian clouds. The H_2SO_4 evaporates at the cloud base, producing SO_3, which can then either recondense or be reduced to SO_2. Reactions of SO_2 with Ca^{2+} in rocks provides a sink that must be balanced by the volcanic and surface sources.

involving the leaching of calcium from surface minerals to form anhydrite ($CaSO_4$). X-ray fluorescence measurements indicate that Ca^{2+} is several times more abundant than SO_4^{2-} in Venusian surface rocks, so there is ample CaO available for removing SO_2.

There are obviously many informative comparisons to be made between these two sulfur cycles. The Venusian clouds, in both composition and mode of formation, are not unlike an enormously exaggerated version of the Earth's Junge layer. The role of human and microbial processes on Earth in recycling sulfates and organic sulfur into the atmosphere can be compared to the role of geological processes on Venus in recycling sulfur in buried calcium sulfate into pyrite and thence back into gaseous sulfur compounds. Finally, while the photooxidation processes are similar, the very large amounts of sulfur gases on Venus lead to long time scales in the Venusian cycle relative to Earth; in particular, the lifetime for SO_2 oxidation to H_2SO_4 is about 200 yr on Venus compared with a few days on Earth.

CHLORINE CYCLES ON VENUS AND EARTH   Chlorine as HCl is remarkably abundant on Venus (Table 3). Unlike SO_2 on Venus, the observed HCl abundance is rather close to that expected for equilibrium at the hot surface between HCl, aluminosilicates, H_2O, and halite (Lewis 1970). Unlike the sulfur gases, extant volcanism is therefore not required to sustain the HCl. The Earth's atmosphere also contains chlorine compounds, most notably in the form of CH_3Cl (produced naturally by marine microorganisms) and the synthetic (freon) gases $CFCl_3$ and CF_2Cl_2 (Table 2). The levels of chlorine on Venus are about 1000 times greater than those on Earth, but on both planets the chlorine species play important roles as sources of catalysts for photochemical reactions [see Prinn (1971) for Venus; Stolarski & Cicerone (1974) and Molina & Rowland (1974) for Earth].

On Earth the compounds CH_3Cl, CF_2Cl_2, $CFCl_3$, and other chlorofluorocarbons and hydrochlorofluorocarbons are emitted at the surface, and since their lifetimes range from decades to a century or more, they are transported up to the stratospheric ozone layer (altitudes of 20–40 km) without being significantly depleted. There, these species are destroyed by photodissociation and reaction with $O(^1D)$ and (for the hydrogen-containing species such as CH_3Cl) by reaction with OH (see Figure 7). This leads to production of Cl atoms. The following chlorine atom-conserving catalytic reactions

$$Cl + O_3 \rightarrow ClO + O_2,$$

$$h\nu + O_3 \rightarrow O + O_2,$$

$$ClO + O \rightarrow Cl + O_2,$$

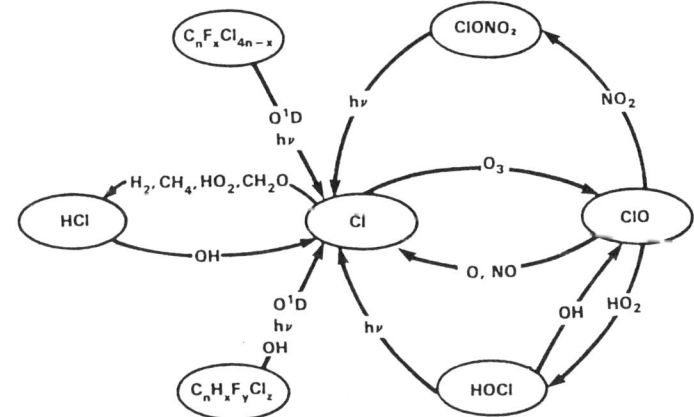

Figure 7 The roles of chlorine compounds in the Earth's stratospheric ozone layer. Chlorofluorocarbons ($C_nF_xCl_{4n-x}$) and hydrochlorofluorocarbons ($C_nH_xF_yCl_z$) produce Cl atoms and ClO radicals that catalytically destroy ozone. Formation of HCl moderates this catalytic destruction (after Watson et al 1986).

then convert two O_3 molecules into three O_2 molecules. The catalytic destruction is partially short-circuited by the reaction of ClO with NO instead of O, and the concentrations of destructive Cl and ClO are decreased by formation of temporary reservoir species HCl, $ClONO_2$, and HOCl.

On Venus, HCl like the freon gases photodissociates in the stratosphere above the clouds to produce Cl atoms. Current ideas are summarized in Figure 8. The Cl atoms form ClO by reaction with O_3, as on Earth, but since O_3 levels are much less on Venus than Earth the ratio of Cl to ClO is $\gg 1$ on Venus while $\ll 1$ on Earth. Large CO and Cl abundances on Venus lead to ClCO and $COCl_2$ formation. Yung & DeMore (1982) have proposed that the formation of $ClCO_3$ from CO, Cl, and O_2 followed by its decomposition to CO_2 and ClO is an important pathway for recombining the CO and O produced by CO_2 photodissociation on this planet.

OXIDATION CYCLES ON EARTH, MARS, AND VENUS   In the Earth's troposphere the major oxidizer is the OH radical (Levy 1971, Duce et al 1984). As depicted in Figure 9, OH is produced primarily by reaction of excited oxygen atoms $O(^1D)$ with water vapor H_2O and removed primarily by reactions with CO. The $O(^1D)$ is produced by photodissociation of O_3 by UV light with wavelengths < 310 nm. The ozone is produced in turn by photodissociation of NO_2 when NO_2 levels are sufficiently high (Chameides & Walker 1973) or by transport down from the stratosphere (where it was produced by O_2 photodissociation). As we have already noted, the

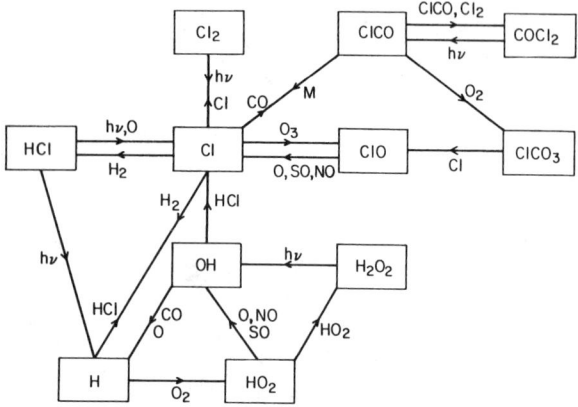

Figure 8 The roles of chlorine and hydrogen compounds in the Venus stratomesosphere. HCl photodissociation produces Cl and H, which then initiate a variety of reactions that recombine CO, O, and O_2 (after Yung & DeMore 1982).

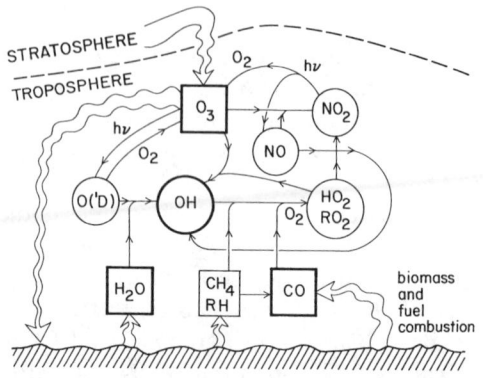

Figure 9 The major oxidation cycle in the Earth's troposphere. The cycle is initiated by the combination of sunlight, H_2O, and ozone that is produced locally from NO_2 or transported down from the ozone layer. The major oxidizer is the OH radical, which is responsible for oxidation of most atmospheric trace species emitted from the surface, including hydrocarbons, CO, and sulfur gases.

composition of the Earth's atmosphere is determined to an important degree by the rate of tropospheric oxidation of biogenic gases by OH.

A remarkably similar oxidation mechanism operates on Mars, where it is responsible for reformation of the CO_2 destroyed by photodissociation (McElroy & Donahue 1972, Parkinson & Hunten 1972). As illustrated in

Figure 10, the major oxidizer is again OH, formed initially by dissociation of H_2O. Once formed, however, it enters a catalytic cycle

$$CO + OH \rightarrow CO_2 + H,$$
$$H + O_2 + M \rightarrow HO_2 + M,$$
$$O + HO_2 \rightarrow OH + O_2,$$

(where M is any molecule) which serves to combine CO and O to form CO_2 with conservation of the OH, H, and HO_2. The same catalytic cycle also occurs on Venus, where it is initiated by the H atoms from HCl photodissociation (see Figure 8). The reaction of O with HO_2 is not important in Earth's trophosphere, and the above catalytic cycle is replaced on Earth by the more complex and less efficient cycle depicted in Figure 9 requiring NO_2 photodissociation and O_3 formation.

Global Changes in Atmospheric Composition

Observations made principally over the last decade have challenged the traditional viewpoint that global atmospheric composition is stable on time scales less than a million years or so. On Earth, the long-lived atmospheric gases CO_2, N_2O, CH_4, $CFCl_3$, CF_2Cl_2, CH_3CCl_3, and CCl_4 are observed today to be increasing over the globe at average rates of about 0.34, 0.26, 1.0, 5.1, 5.0, 6.7, and 1.3% per year, respectively (Komhyr et al 1985, Blake & Rowland 1986, Prinn et al 1983, Cunnold et al 1986). These increases represent a current imbalance between the sources and sinks for these gases (as summarized in Table 2). Some of the imbalances are well understood (e.g. the known industrial source of $CFCl_3$ exceeds its known photodissociation sink), while others are very poorly understood (e.g. is the CH_4 increase due to an increase in microbial production of CH_4

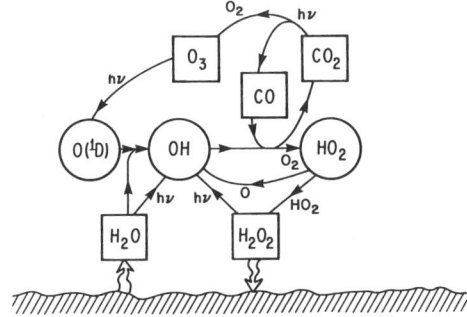

Figure 10 The major oxidation cycle in the Martian atmosphere. The cycle is initiated by photodissociation of H_2O, producing OH that oxidizes CO back to CO_2.

or to a decrease in photochemical destruction of CH_4 resulting from decreasing atmospheric OH levels?) All of the above long-lived gases are important in the Earth's greenhouse effect, and if today's rates of increase are extrapolated into the future a global surface warming of a few degrees Kelvin is predicted early in the next century (Ramanathan et al 1985). All of these gases are also important in the chemistry of the ozone layer through production or removal of catalysts and/or through radiative effects.

Global changes on these short time frames are not confined to Earth (see Figure 11). The global average abundance of SO_2 in the visible part of the Venusian atmosphere decreased tenfold between 1978 and 1983 putatively as a result of the recovery of the atmosphere from a massive (circa 1978) volcanic eruption with important implications for the Venus sulfur cycle (Esposito 1984). The dust-loading in the Martian atmosphere (as measured by the vertical optical depth of the atmosphere) increased fivefold as a result of a mid-1977 global dust storm with very important effects on global Martian climate (Pollack et al 1979).

ATMOSPHERIC ORIGIN AND EVOLUTION

A central problem presented to us by Venus, Earth, and Mars is to explain how these three very different atmospheres can be derived from the same parent material—the cloud of gas and dust making up the solar nebula. The magnitude of this problem is only fully realized when one considers that about 98% (by mass) of the solar nebula was $H_2 + He$, and that only 2% was made up of the volatiles and grains that eventually formed these three planets and their atmospheres. Indeed, no completely satisfactory solution has yet been found after years of work by many investigators. Recent reviews by Lewis & Prinn (1984) and Black & Matthews (1985) describe this work from a variety of perspectives. Instead of repeating these discussions, we review the chemical and physical processes influencing atmospheric origin and evolution, emphasizing in particular the roles of common evolutionary processes.

Figure 11 Examples of global changes on Venus, Earth, and Mars. (*a*) Global-mean concentration of SO_2 at the 40-mbar level on Venus as a function of time (measured in days) beginning at the arrival time of the Pioneer Venus Orbiter on 4 December 1978 (after Esposito 1984). (*b*) Concentration of the freon gas $CFCl_3$ at two remote locations on Earth as measured by the ALE/GAGE trace-gas network (Prinn et al 1983, Cunnold et al 1986). (*c*) Optical depth of the atmosphere at the Viking I Lander site on Mars as a function of time measured in Martian days (1.03 Earth days) beginning on the Viking landing date (20 July 1976). A great dust storm began on about Martian day 310 (after Pollack et al 1979).

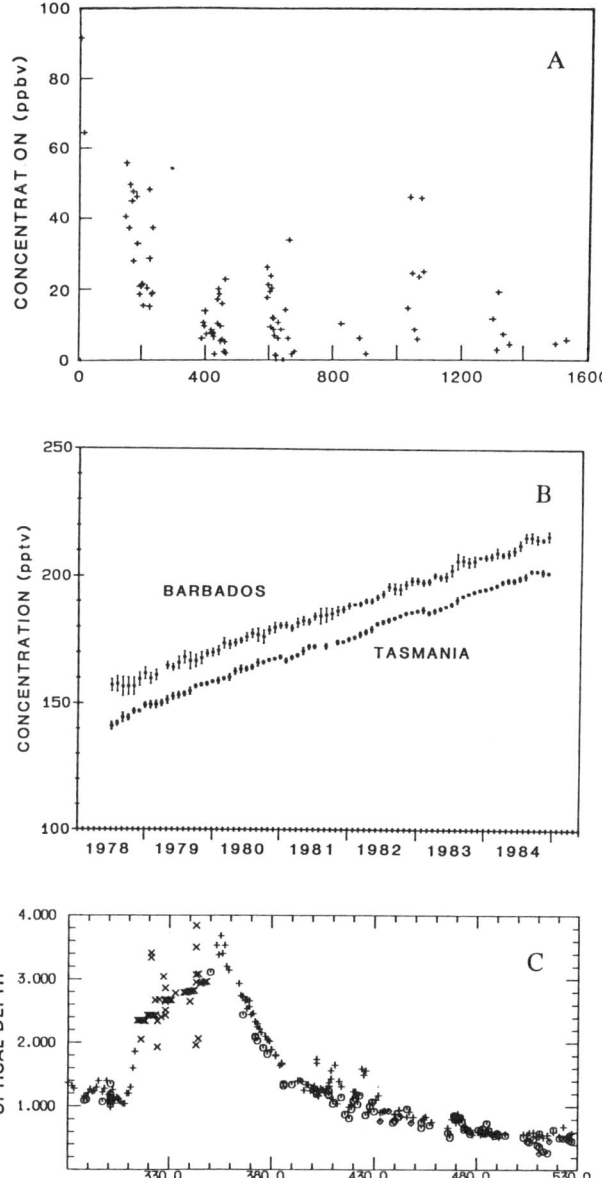

Secondary Origin of the Atmospheres of Venus, Earth, and Mars

One shared characteristic of these three atmospheres is their secondary origin. This was first pointed out for the Earth by several authors (e.g. Brown 1949, Russell & Menzel 1933, Aston 1924), who noted that the rare gases (Ne, Ar, Kr, Xe) are much less abundant than other volatiles on the surface of the Earth. These large depletions are shown in Table 5. The large depletion of Ne to N_2 is particularly striking because Ne and N have very similar solar abundances and similar atomic weights.

Table 5 also shows that the nonradiogenic rare gases are generally more depleted than the chemically reactive volatiles in the atmospheres of Venus and Mars. Several exceptions to this trend are also evident and are discussed later. Nevertheless, the much larger depletions of nonradiogenic ^{20,22}Ne and ^{36,38}Ar compared with the CO_2 and N_2 depletions on all three planets show that the planetary atmospheres did *not* originate principally by capture of gas from the solar nebula. Instead, these atmospheres are almost entirely secondary and originated as the result of chemical processes contemporaneous with or subsequent to the formation of the planets.

One exception to this conclusion may be ^{36}Ar on Venus, which Wetherill (1981) has proposed is due to solar wind implantation in the solid grains accreted by Venus. This hypothesis is attractive because it explains the

Table 5 Depletions of important volatiles in Venus, Earth, and Mars relative to solar abundances $[(g/gSi)/(g/gSi)]$[a]

Volatile	Venus[b]	Earth	Mars
CO_2	3×10^{-5}	3×10^{-5}	2×10^{-8}
H_2O	2×10^{-9}	2×10^{-4}	4×10^{-12}
F	3×10^{-9}	2×10^{-2}	—
^{20,22}Ne	1×10^{-9}	4×10^{-11}	2×10^{-13}
N_2	1×10^{-5}	2×10^{-5}	5×10^{-9}
S	1×10^{-7}	7×10^{-5}	—
Cl	3×10^{-8}	7×10^{-3}	—
^{36,38}Ar	1×10^{-7}	2×10^{-9}	1×10^{-11}
^{84}Kr[c]	$(0.04-1) \times 10^{-5}$	1×10^{-7}	2×10^{-9}
^{132}Xe	—	9×10^{-8}	3×10^{-9}

[a] Solar abundances from Cameron (1982). Atmospheric inventories only were considered for Venus and Mars and were obtained from Tables 3 and 4. Atmospheric plus oceanic plus crustal inventories were considered for Earth and were obtained from Ronov & Yaroshevsky (1976), Turekian (1969), and Ozima & Podosek (1983).
[b] Bulk composition models V2, E5, and Ma2 from the Basaltic Volcanism Study Project (1981, p. 641) were used to determine Si contents for Venus, Earth, and Mars, respectively.
[c] Values for Venus reflect differences between various instrumental measurements. The value for Mars is total Kr; its depletion factor was calculated assuming terrestrial isotopic abundances.

remarkable decrease in ^{36}Ar from Venus to Earth to Mars (see Table 5). However, Wetherill's model and a similar hypothesis advanced by McElroy & Prather (1981) require that radial transport in the inner regions of the solar nebula be sufficiently weak to prevent the ^{36}Ar-rich grains from being mixed in any important amounts out to the accretion zones of Earth and Mars. This requirement argues against a common origin for the grains accreted by Venus and Earth during their formation. We return to this point in connection with the dramatically different H$_2$O inventories of Venus and Earth.

Finally, we note that models for accretion of the Earth (and by implication Venus and Mars) in the gaseous solar nebula are also inconsistent with the observed inventories of nonradiogenic rare gases on these three planets. In these models the planetary inventories of the nonradiogenic rare gases are established by solubility equilibria between the solar nebula gas and magma oceans on the surfaces of the protoplanets (Mizuno et al 1982). However, major predictions of these models (e.g. a terrestrial Ne inventory 3 to 200 times larger than the atmospheric inventory and with a solar ^{20}Ne/^{22}Ne ratio of 14.3 vs the terrestrial value of 9.8) are very difficult to reconcile with the observed rare-gas inventories and isotopic ratios. Comprehensive reviews of constraints imposed upon these models by current knowledge of planetary rare-gas systematics are given by Ozima & Podosek (1983).

Volatile Retention by Solid Grains in the Solar Nebula

Evidently, in order to understand the origin of the atmospheres of Venus, Earth, and Mars we must first understand the chemical processes responsible for retention of volatiles (e.g. H$_2$O, C, N, F, Cl, S) by the solid grains that accreted to form the planets. We will then be able to address some of the questions posed by Table 5. For example, is the difference between the H$_2$O inventories of Venus and Earth genetic or evolutionary? If it is genetic, why do the two planets have very similar inventories of CO$_2$ and N$_2$? If it is evolutionary, what does the loss of a terrestrial H$_2$O inventory imply for the chemistry of Venus? Similarly, we would like to know if Mars is really more volatile poor than Earth or only appears to be so. We address these (and related) questions within the framework of currently accepted solar nebula models in which radial temperature gradients are presumed to be a major influence on the composition of the gas and dust grains.

As the interstellar gas and dust were accreted by the solar nebula, they were thermally and chemically equilibrated to varying degrees. Accreting gases may have been only partially equilibrated (or not at all) as they were warmed and compressed. The extent to which this may have occurred depended on the distance of the gas parcel from the proto-Sun and the rate of radial transport in the nebula relative to the rate of the equilibrating reactions in the gas parcel.

Similar considerations apply to the accreted interstellar dust grains. Recent work (Cameron & Fegley 1982, Morfill & Völk 1984) suggests that the accreting dust grains may have evaporated only partially or not at all, depending on the type of grain, the strength of radial mixing in the nebula, and the distance from the proto-Sun. Indeed, isotopic data for the ancient Ca, Al-rich inclusions in the Allende meteorite imply that these inclusions formed by a complex sequence of condensation, evaporation, and recondensation (Niederer & Papanastassiou 1984). Such observations and theoretical models suggest that evaporation and recondensation leading to thermal and chemical equilibration were very probable in the inner regions of the solar nebula.

However, these arguments become less and less convincing with increasing radial distances (and thus lower temperatures) in the solar nebula. Again, inferences from meteorites are instructive. Observed isotopic anomalies in several light elements in the volatile-rich carbonaceous chondrites (e.g. see the review by Pillinger 1984) imply interstellar material (or at least its chemical and physical signature) is preserved in these meteorites. These caveats should be kept in mind as we discuss the implications of a chemical model of the solar nebula, which assumes complete evaporation and recondensation of the grains and complete chemical equilibration of the gas and dust, for the volatile inventories of Venus, Earth, and Mars.

Table 6 lists the predicted volatile-bearing phases that are stable in this model of the inner region of the solar nebula. Several of these phases are also included in Figure 12, which displays their abundance as a function of temperature and radial distance in the solar nebula. The most important conclusion from these results is that the solid grains equilibrated at lower temperatures (i.e. farther from the proto-Sun) are predicted to be more volatile-rich than the solid grains equilibrated at higher temperatures (i.e. closer to the proto-Sun). However, important exceptions to this behavior are the amounts of elemental carbon and nitrogen, dissolved in Fe-Ni alloy, that go through maxima in the inner region of the nebula (Lewis & Prinn 1984, Fegley 1983).

Table 7 presents a synthesis of the observed volatile inventories for Venus, Earth, and Mars and the predicted volatile inventories of the complete equilibrium model of Table 6 and Figure 12. This comparison provides instructive answers to the questions posed at the beginning of this section. The difference between the H_2O inventories of Venus and Earth is predicted to be genetic. The Earth accreted significantly more hydrated phases (e.g. tremolite, serpentine, hydroxyapatite, talc) than did Venus because these phases only became thermodynamically stable in the

Table 6 Predicted volatile-bearing phases stable at thermochemical equilibrium in solar nebula gas and potential outgassed volatiles[a]

Equilibration temperature (K)[b]	Volatile-bearing phase	Potential volatile(s)[c]
1825	U, Th in Ca-bearing refractories	^{4}He
1520	C,N in Fe-Ni alloy	CO, CO_2, CH_4 N_2, NH_3
1225	Schreibersite $(Fe, Ni)_3P$	P_x, PO_x, PH_3
950–1050	Feldspars and feldspathoids $(Na,K)AlSi_3O_8$, $(Na,K)AlSiO_4$	^{40}Ar
895	Sodalite $(3NaAlSiO_4 \cdot NaCl)$	Cl_2, HCl
766	Fluorapatite $(Ca_5(PO_4)_3F)$	HF
714	Whitlockite $(Ca_3(PO_4)_2)$	P_x, PO_x, PH_3
687	Troilite (FeS)	S_x, H_2S, COS, SO_2
480	Tremolite $(Ca_2Mg_5Si_8O_{22}(OH)_2)$	H_2O, H_2, O_2
460	Hydroxyapatite $(Ca_5(PO_4)_3OH)$	H_2O, H_2, O_2
≃400	Talc $((Mg,Fe)_3Si_4O_{10}(OH)_2)$	H_2O, H_2, O_2
≃400	Serpentine $((Mg,Fe)_3Si_2O_5(OH)_4)$	H_2O, H_2, O_2
350	Bromapatite $(Ca_5(PO_4)_3Br)$	Br_2, HBr

[a] Complete gas phase, gas-solid, and solid-solid equilibrium assumed. See Barshay (1981), Fegley (1983), and Lewis & Prinn (1984) for details.

[b] Highest temperature at which the phase is stable along the model solar nebula adiabat. The relative sequence does not change with total pressure over very wide ranges.

[c] The exact nature of the potential volatiles depends on several factors including the pressure, temperature, and oxygen fugacity during outgassing. F_2 is so reactive it probably never forms; H_2 and O_2 are generated by equilibria with a H_2O-bearing vapor and with other phases such as Fe metal.

cooler nebular region outside of the Earth's orbit. Furthermore, Mars accreted even more of the hydrated phases than the Earth did and is predicted to be even more H_2O-rich. The (inferred) presence of hydrated silicates on the surface of the asteroid 1 Ceres is in qualitative agreement with this trend (Lebofsky 1978). An alternative model requiring massive H_2O loss from an initially "wet" Venus is discussed in the section on evolutionary processes.

Another prediction is that Venus and Earth formed with initially similar inventories of CO_2, S, Cl, and F, while Mars was initially richer in S, Cl, and F than the Earth but formed with less CO_2. However, Mars may have formed with an Earth-like CO_2 inventory if nonequilibrium effects prevailed, as we show below. Comparison of these predictions with the observed volatile inventories is complicated by atmosphere-lithosphere interactions on both Venus and Mars that may reduce the atmospheric inventories of CO_2, S, Cl, and F. These effects may not be significant for CO_2 on Venus because the atmospheric CO_2 inventory is very similar to the terrestrial bulk inventory (see Tables 5 and 7). However, as Table 3

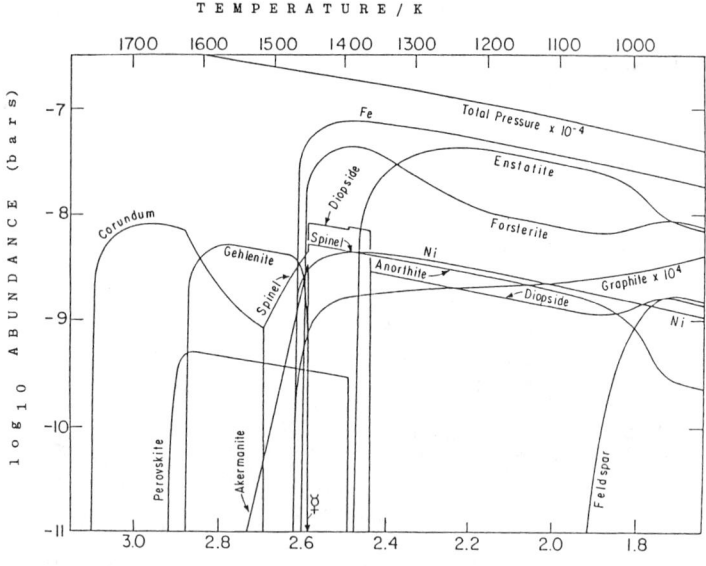

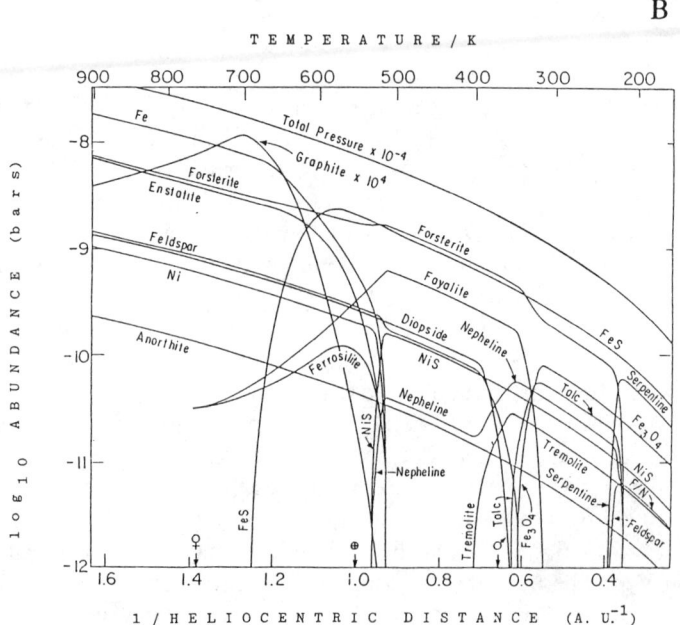

Table 7 Comparison (g/g) of observed volatile inventories of Venus, Earth, and Mars with predictions of the complete equilibrium model shown in Figure 12[a]

		Venus		Earth		Mars	
Volatile		Atmosphere	Bulk	Atmosphere	Bulk	Atmosphere	Bulk
CO_2	(observed)	10^{-4}		4×10^{-10}	8×10^{-5}	4×10^{-8}	
	(predicted)[b]		2×10^{-5}	7×10^{-6} to 3×10^{-4}		10^{-9} to 8×10^{-5}	
N_2	(observed)	2×10^{-6}		6×10^{-7}	3×10^{-6}	7×10^{-10}	
	(predicted)		4×10^{-9}		3×10^{-11}		10^{-13}
H_2O	(observed)	4×10^{-9}		5×10^{-9}	4×10^{-4}	5×10^{-12}	
	(predicted)		0		10^{-5}		8×10^{-3}
S	(observed)	10^{-8}	$(4-6) \times 10^{-3}$	10^{-15}	6×10^{-6}		4×10^{-2}
	(predicted)		10^{-2}		5×10^{-2}		7×10^{-2}
Cl	(observed)	3×10^{-11}	$<3 \times 10^{-3}$	10^{-15}	6×10^{-6}		8×10^{-3}
	(predicted)		10^{-4}		7×10^{-4}		9×10^{-4}
F	(observed)	2×10^{-13}		4×10^{-16}	10^{-6}		
	(predicted)[b]		10^{-5}		6×10^{-5}		8×10^{-5}

[a] Atmospheric inventories based on Tables 2–4. Terrestrial bulk inventory based on Table 5. The S and Cl bulk inventories for Venus and Mars are based on Surkov et al (1984) and Toulmin et al (1977), respectively. In both cases the observed S and Cl contents (or upper limits) are taken as the bulk inventories. Note that this table is not normalized to Si and is simply on a (g/g) basis.
[b] Predictions based on Barshay (1981), Lewis & Prinn (1984), and Fegley (1983). The range of CO_2 values for Earth and Mars includes the effects of kinetic inhibition on dissolved carbon in grains.

indicates, $CaCO_3$ formation is a possible sink for CO_2 in the atmosphere of Venus. By contrast, $CO_2(s)$ in the Martian polar caps is a known CO_2 sink on Mars. Furthermore, S and Cl have been observed on the surface of Mars (Toulmin et al 1977), and S has been observed on the surface of Venus (Surkov et al 1984). The geochemical similarity of F and Cl suggests that F may also be present on the Martian surface, while the (probable) buffering of HCl and HF in the atmosphere of Venus suggests the presence of Cl and F in surface minerals (Lewis 1970).

Finally, two other important trends shown in Table 7 deserve attention. First, both Venus and Earth contain more CO_2 than the complete equilibrium model predicts. Second, Venus, Earth, and Mars all contain substantially more N_2 than predicted. Significantly, a similar situation holds for the chondritic meteorites, especially for the volatile-rich carbonaceous chondrites (Lewis & Prinn 1984, Fegley 1983). The latter meteorites may contain several percent (by mass) of nonequilibrium organic material,

Figure 12 Equilibrium abundances of volatile-bearing phases and major-element condensates along an adiabatic (P, T) profile in the inner regions of the solar nebula. The astrological symbols for Mercury, Venus, Earth, and Mars are shown at the appropriate places on the distance scale, which is in inverse astronomical units (AU^{-1}) (after Barshay 1981).

which is the dominant reservoir of C and N in the meteorites (e.g. see Kung & Clayton 1978). Thus we now explore nonequilibrium effects on volatile retention by solid grains.

Nonequilibrium effects in a cooling parcel of gas and dust in the solar nebula will be favored when the characteristic cooling time (t_{cool}) is less than the characteristic chemical time scales for the gas phase (t_g), gas-solid (t_{gs}), and solid-solid (t_{ss}) reactions that may occur inside this parcel. This is expressed by the inequalities $t_{cool} < t_g$, $t_{cool} < t_{gs}$, and $t_{cool} < t_{ss}$. These inequalities will be favored by low temperatures, fast radial mixing rates, and fast nebular cooling rates; for reactions involving solids, the inequalities will also be favored by large grain sizes and fast accretion rates for these grains. How will nonequilibrium effects influence volatile retention by solid grains?

We can gain some insight into this question by considering three reactions that exemplify volatile retention reactions in a cooling parcel of gas and dust in the solar nebula. First, consider solid-solid reactions, which are likely to be the most sluggish and hence the most susceptible to nonequilibrium effects. The retention of H_2O as the hydrous mineral serpentine (see Table 6) proceeds by the reaction

$$Mg_2SiO_4(s) + MgSiO_3(s) + 2H_2O(g) = Mg_3Si_2O_5(OH)_4(s), \qquad (5)$$

which because it requires the transport and reaction of elements between two minerals, may proceed very slowly at the low temperatures ($\simeq 400$ K) where serpentine is thermodynamically stable in the solar nebula. If this is the case, then $t_{cool} \ll t_{ss}$ may hold, and in the absence of "fast" pathways for forming equal amounts of other hydrated phases, H_2O may not be retained in solid grains until below 200 K, when H_2O (ice) becomes stable. Tremolite and hydroxyapatite, which are stable at higher temperatures, are much less important for H_2O retention because of the significantly lower solar abundances of Ca and P relative to Mg and Si. The implications for H_2O retention by Venus, Earth, and Mars are twofold. First, H_2O must then be delivered to these planets by icy planetesimals gravitationally scattered into the inner solar system during the late stages of planetary accretion (e.g. Wetherill 1975). Then Venus and Earth would be predicted to have similar initial H_2O inventories. Second, we would then require a mechanism to remove the equivalent of the Earth's oceans from Venus in order to explain its present H_2O-depleted state.

An exemplary gas-solid reaction, the formation of troilite, proceeds by H_2S permeation into and reaction with Fe metal grains:

$$Fe(s) + H_2S(g) = FeS(s) + H_2(g). \qquad (6)$$

Given sufficiently rapid radial mixing or nebular cooling, then $t_{cool} \ll t_{gs}$

may hold and this reaction may not proceed significantly before being quenched at some intermediate conversion of H_2S to troilite (FeS). Although a few percent of the H_2S may be retained as an alkali sulfide at similar temperatures (Lewis & Prinn 1984), the bulk of the available S cannot be retained in solid grains until below 200 K, when $NH_4HS(s)$ or $H_2S(s)$ forms (Lewis & Prinn 1984). Again, the major implication is that planetesimals formed in the outer solar system must impact Venus, Earth, and Mars to provide these planets' observed sulfur inventories.

Finally, let us consider a gas phase reaction such as the conversion of CO to CH_4:

$$CO(g) + 3H_2(g) = CH_4(g) + H_2O(g). \qquad (7)$$

The kinetic inhibition of the CO to CH_4 conversion has in fact been studied quantitatively (reviewed in Lewis & Prinn 1984). The presence of metastable CO(g) inside the $CH_4(g)$ stability field leads to supersaturation of elemental carbon in the gas phase. Figure 13, which compares the equilibrium and nonequilibrium cases for reaction (7), illustrates that one consequence of this situation may be greatly increased amounts of carbon dissolved in Fe-Ni alloy equilibrated with this gas. The range of predicted CO_2 inventories given in Table 7 for Earth and Mars incorporates this enhancement. In this instance, nonequilibrium effects increase the ease of volatile retention by solid grains in the inner regions of the nebula.

Another possible consequence of the kinetic inhibition of the CO to CH_4 conversion is that reactions in the supersaturated carbon-bearing gas will proceed to the point of making organic material, as in the Fischer-Tropsch reactions studied by Anders and coworkers (Hayatsu & Anders 1981). Large Martian CO_2 inventories are also plausible in this case. In fact, such nonequilibrium organic matter may have been an important source of the CO_2 and N_2 in the atmospheres of Venus, Earth, and Mars.

Volatile Degassing and Atmospheric Formation

We now have some insight into the chemical and physical processes responsible for influencing volatile retention by solid grains in the solar nebula. Once these grains are incorporated into Venus, Earth, and Mars, their volatiles are released (to varying degrees) by degassing and atmospheric formation occurs. How can our knowledge of the atmospheres of Venus, Earth, and Mars help us to develop models of this complex process? What factors are responsible for influencing the amount and initial composition of the degassed volatiles? Again, we address these questions by emphasizing the roles of common chemical and physical processes acting on all three planets. More detailed models that are specific for the Earth have

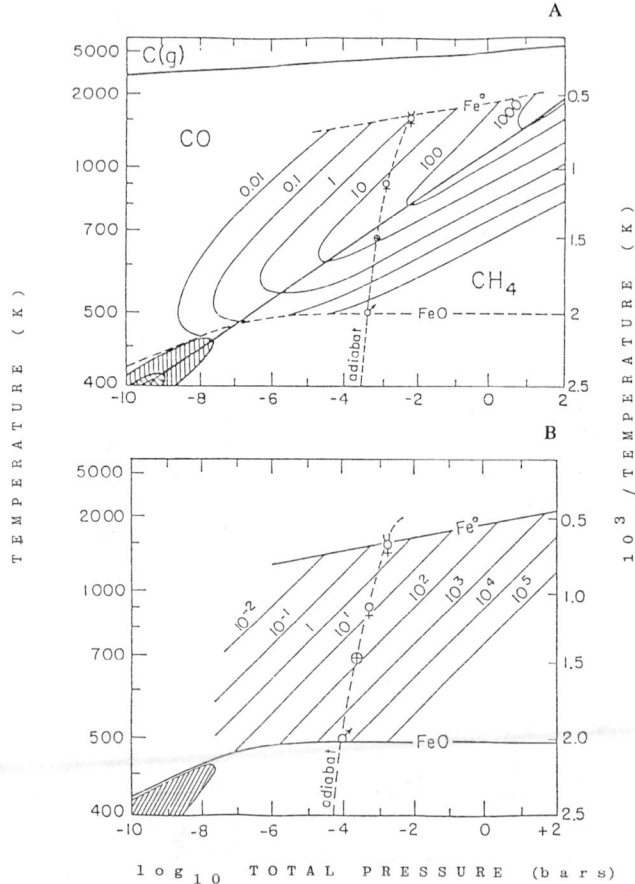

Figure 13 Calculated amounts (parts per million by mass) of elemental carbon dissolved in Fe-Ni alloy equilibrated with nebular gas. The (P, T) adiabat from Figure 12 is also shown. (*A*) Complete equilibrium case; (*B*) kinetic inhibition of the CO to CH_4 conversion (after Lewis & Prinn 1984).

been reviewed by Ozima & Podosek (1983), Holland (1984), and Lewis & Prinn (1984).

Volatile degassing will occur when the equilibrium vapor pressure of a volatile species over its condensed phase exceeds the pressure due to the overlying materials. This situation is favored by high temperatures and low pressures. Several potential heat sources are available during the formation and early history of Venus, Earth, and Mars. These include (*a*) heating during planetary accretion, (*b*) heating by strong coupling of a planetary body with the T-Tauri solar wind, (c) heating by short-lived radionuclides such as ^{26}Al, (d) heating by long-lived radionuclides such as ^{40}K, ^{232}Th, ^{235}U, and ^{238}U, and (e) heating during planetary differentiation and core formation. Several of these processes, such as accretion heating, heating from long-lived radionuclides, and heating from planetary differentiation, are currently thought to have been the most important processes for Venus, Earth, and Mars. Thus, since the accretion rates, radionuclide abundances, and degrees of differentiation may have been very different for these three planets, the times and rates of volatile degassing may also have varied significantly. In particular, the smaller size of Mars relative to Venus and Earth suggests that accretion heating may not have been as great for Mars. Thus, Mars may be less efficiently degassed than Venus or Earth.

Furthermore, the differing compositions of these three planets may have also influenced the composition and oxidation state of degassed volatiles. Thus, if Venus was initially H_2O poor, the degassed volatiles would reflect this and would be dominated by C-, N-, and S-bearing gases rather than by H_2O. In fact, calculations by Gerlach & Nordlie (1975) clearly show an increase in C- and S-bearing gases (N was excluded from their calculations) as H_2O drops off. The latter volatile is generally the dominant constituent of present-day volcanic gases on the Earth (e.g. see Gerlach & Nordlie 1975).

Likewise, the equilibrium models suggest that degassed volatiles on Mars may also have been initially H_2O rich. The (inferred) presence of abundant H_2O on Mars (see the review in Lewis & Prinn 1984) is in qualitative agreement with this prediction.

The oxidation state of the degassed volatiles on Venus, Earth, and Mars may also have been initially different as a consequence of the solid grains accreted by each planet equilibrating with the nebular gas at different temperatures (see Figure 12). The shifting equilibrium with temperature between H_2 and H_2O in the gas phase controls the oxygen fugacity (fO_2) of the gas and dust grains in equilibrium with it. In this case the solid grains become more oxidized (e.g. more FeO rich) with decreasing temperature (i.e. increasing radial distance) in the nebula. The (initially) degassed volatiles on Venus, Earth, and Mars would then follow the same trend. However, if the time scale (t_{gs}) for the relevant gas-solid reactions controlling the fO_2 of the solid grains is sufficiently large relative to the characteristic cooling time scale (t_{cool}) defined earlier, then the fO_2 of the solid grains may be frozen in (or quenched) at a sufficiently high temperature $T = T_Q$, which may be the same for all three planets. In this case, the initially degassed volatiles may have had the same oxidation state on all three planets. At present we are unable to distinguish between these two (extreme) possibilities. This situation is partly due to our lack of information on the composition of volatiles initially outgassed on any of these three planets and partly due to our present ignorance of the quench temperatures of the relevant reactions in the solar nebula. However, while information on the former topic is unlikely to become available anytime soon, information on the latter could be provided by suitably designed theoretical models of nebular chemistry.

Similar considerations are also relevant for volatile degassing on the three planets. Figure 14 illustrates several possible paths (e.g. volcanism, evaporation, sublimation, slow upward permeation, effusion) that may be involved to different degrees (and at different times) in volatile degassing. Some degassing mechanisms, such as mineral devolatilization at depth, may be coupled with subsequent reactions with gaseous, liquid, or solid phases that will change the composition and oxidation state of the volatile phase. By analogy with our treatment of nebular reactions, it is convenient to define characteristic chemical lifetimes for the gas phase (t_g), gas-magma (t_{gm}), and gas-rock (t_{gr}) reactions involved in degassing. It is also convenient to define a characteristic time scale (t_{trans}) for the upward transport of the volatile phase and a critical depth D^* where $t_{chem} = t_{trans}$. Here we take $t_{chem} = t_g, t_{gm}$, or t_{gr} as appropriate for the type of reaction being considered. Thus we see that the relevant reactions will only be effective for changing the composition and oxidation state of the volatile phase at depths $D > D^*$, where $t_{chem} < t_{trans}$. Conversely, the gas phase, gas-magma, and gas-rock reactions will be ineffective for altering the volatile phase at depths $D < D^*$, where $t_{chem} > t_{trans}$. At depths $D = D^*$ ($t_{chem} = t_{trans}$) quenching of the relevant reaction will occur, and the composition and fO_2 of the ascending volatile phase will also be frozen in.

In general, the chemical time scales will decrease rapidly with increasing depth in the planet. The critical depth D^* will be different for each reaction considered because of the different chemical time scales. Just as importantly, D^* will vary with the mode of degassing because of the inherent variations in the time scales for upward transport of the volatile phase. For example, reactions occurring in a volatile phase slowly permeating through a layer of rock will quench closer to the surface than reactions in a volatile phase carried to the planetary surface by a volcanic eruption.

Such constraints are potentially very important for defining the composition and fO_2 of degassed volatiles. A simple example based on the calculations of Heald et al (1963) for the equilibrium abundances in a terrestrial volcanic gas illustrates this point. If the gas phase were quenched at 1400 K, the resulting CO/CH_4 molecular ratio would be $10^{9.9}$, while the same gas quenched at 400 K would have a CO/CH_4 molecular ratio of only $10^{-4.4}$. We also note that Fe metal grains will directly influence the

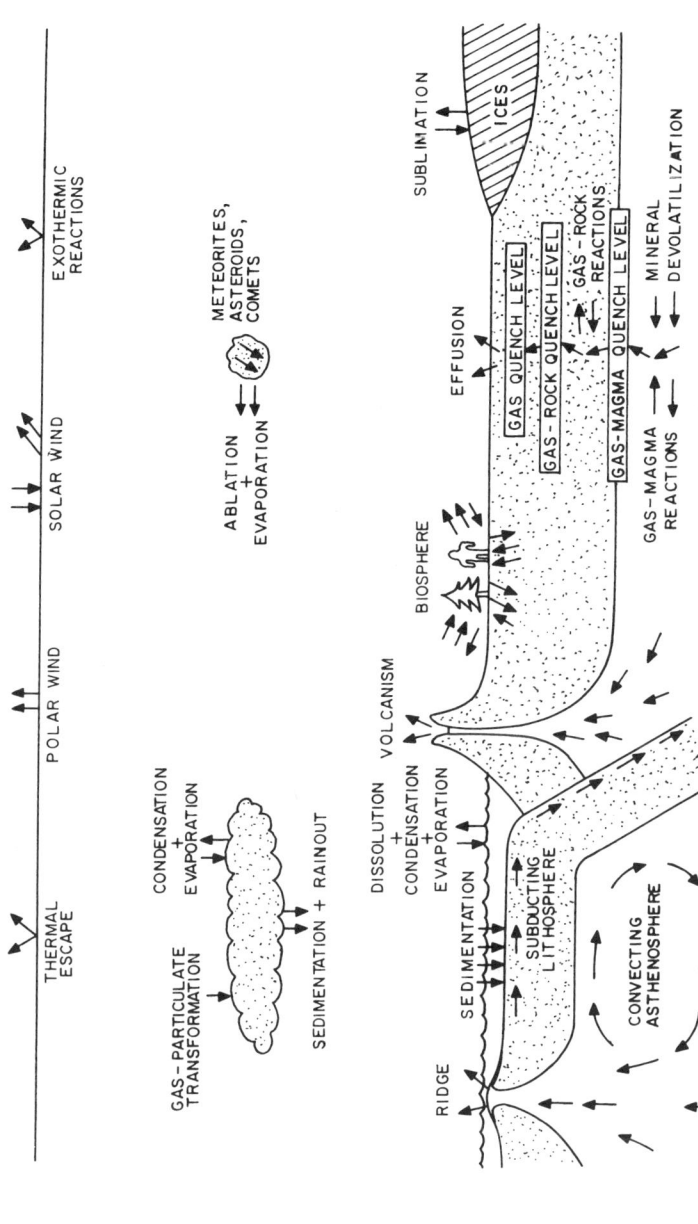

Figure 14 Schematic diagram illustrating the processes affecting volatiles in planetary atmospheres. Net fluxes are illustrated by arrows (after Prinn 1982).

fO_2 of the volatile phase (e.g. see Holland 1984) and will also act as a catalyst for conversions of $CO+CO_2$ to CH_4, N_2 to NH_3, etc, and thus influence the quench levels of gas phase reactions. More work is clearly required in this area of kinetic constraints on the degassing process.

Finally, we note that relative to the total planetary mass the present atmospheric inventories of ^{40}Ar on Venus, Earth, and Mars are 3×10^{-9}, 11×10^{-9}, and 0.6×10^{-9} (g/g), respectively. These inventories imply similar degassing efficiencies for Venus and Earth and an order of magnitude lower degassing efficiency for Mars. This trend is consistent with (but does not prove) the hypothesis that Mars is volatile rich and only appears volatile poor because the volatiles are sequestered in the interior.

Evolutionary Processes, Sources, Cycles, and Sinks

After volatiles are present in a planetary atmosphere, they are subject to a variety of competing influences that may affect them in several ways (see Figure 14). Several of these influences, such as thermal escape, interactions with electromagnetic fields in polar regions (i.e. polar wind), interactions with the solar wind, and exothermic reactions, cause loss of the volatiles to space. Other influences, such as solar ultraviolet radiation and high temperatures (e.g. possibly generated by lightning or by shock heating from impacting bolides), lead to chemical conversions and reprocessing. The synthesis and destruction of volatiles by living organisms also lead to chemical transformations and are unique to the Earth. A variety of exchange and weathering reactions between the atmosphere and lithosphere (and hydrosphere on Earth) further serve to cycle and modulate the atmosphere's composition and overall inventory of various degassed species. Instead of reviewing all these processes, we discuss here an episodic phenomenon (large cometary and asteroidal impacts into planetary atmospheres) and a chronic phenomenon [the (inferred) loss of H_2O from an initially "wet" Venus] that exemplify the differing influences of these two classes of processes on atmospheric evolution.

LARGE ASTEROIDAL AND COMETARY IMPACTS   Several recent studies have shown that large bolide impacts may have affected the Earth throughout its history. Planetesimal impacts during and shortly after the accretion of the Earth may have formed the Moon (Cameron & Ward 1976) and have released atmospheric gases by impact devolatilization (Lange & Ahrens 1982). Large impacts, especially on Mars, may also have been a mechanism for atmospheric erosion and blowoff (Cameron 1983). A late heavy bombardment (Wetherill 1975) may also have caused chemical reprocessing of the Earth's early atmosphere (Fegley et al 1986). Significantly later in the Earth's history, the impact of an $\simeq 10$-km bolide may have terminated the

Cretaceous era (Alvarez et al 1980), leading to severe environmental effects and biospheric disruptions (Lewis et al 1982, Prinn & Fegley 1986). Finally, in recent times the 1908 impact event at Tunguska, Siberia, led to global perturbations in atmospheric ozone (Turco et al 1981), severe environmental effects in the immediate vicinity of the endpoint of the trajectory, and atmospheric pressure waves recorded across Russia and Europe (Shoemaker 1983).

Shoemaker (1983) reviews estimates of the energy released by the 1908 Tunguska impact. These estimates are in the range of 12–30 megatons TNT equivalent. He further estimates the frequency of such encounters with the Earth as once every $\simeq 10^{2.5 \pm 0.3}$ yr (12 megatons) and once every $\simeq 10^{2.8 \pm 0.3}$ yr (30 megatons). Shoemaker's (1983) estimated mean frequency of encounters with $\simeq$ 10-km impactors, such as the putative Cretaceous-Tertiary impactor 65 Myr ago, is about once every 10^8 yr. This frequency is small but still impressive on a geological time scale. It is therefore instructive to review the atmospheric effects of large asteroidal and cometary impacts.

Large bolides ($\simeq$ 10 km size) impacting a planet will cause severe shock heating and chemical reprocessing of the planetary atmosphere during atmospheric entry and as a result of the supersonic plume ejected on impact (Prinn & Fegley 1986). The nature of the products of this shock heating and chemical reprocessing depend on the oxidation state of the planetary atmosphere. Nitrogen-bearing shock products are particularly interesting because of their potential environmental effects. NO is produced in oxidizing atmospheres (atomic C/O < 1), while HCN is produced in reducing atmospheres (atomic C/O ≥ 1) (Fegley et al 1986). The former case is relevant to the present atmospheres of the Earth, Venus, and Mars, while the latter case is relevant to (postulated) primitive atmospheric compositions on the Earth. The scope of the chemical and physical consequences of the impacts also depends on the nature of the impactor, although these consequences are likely to be very severe in all cases.

Detailed studies (Lewis et al 1982, Prinn & Fegley 1986, Fegley et al 1986) illustrate some potential environmental effects of such large impacts. These may extend to the atmosphere, lithosphere, hydrosphere, and biosphere. For example, if the putative Cretaceous-Tertiary impactor were cometary, it would produce semiglobal atmospheric NO_2 volume mixing ratios of 0.1% and concentrated nitrous and nitric acid rain with a pH $\simeq$ 0–1.5 globally (Prinn & Fegley 1986). The resulting consequences may include (*a*) inhibition of photosynthesis due to extinction of solar radiation by NO_2, (*b*) foliage damage due to exposure to NO_2 and HNO_3, (*c*) toxicosis resulting from massive mobilization of soil trace metals, (*d*) respiratory damage due to exposure to NO_2, (*e*) decreases in the pH of the oceanic mixed layer, (*f*) global warming due to this sudden CO_2 injection into the atmosphere, (*g*) extinction of many calcareous-shelled organisms, and so on (Prinn & Fegley 1986). Similarly, impacts into reducing atmospheres postulated for the primitive Earth may lead to production of large amounts of HCN, an important precursor for the abiotic synthesis of complex organic molecules (Fegley et al 1986). Thus, although episodic in character, asteroidal and cometary impacts may have had a significance for atmospheric evolution far beyond their potential role as suppliers of volatiles.

LOSS OF WATER FROM AN INITIALLY "WET" VENUS The proposed loss of water from an initially "wet" Venus exemplifies the type of chronic phenomena that have influenced atmospheric evolution. If the initial volatile endowments of the Earth and Venus were not controlled by the radial temperature gradient in the solar nebula, then both Venus and Earth may have formed with similar H_2O inventories. Thus the greater proximity of Venus to the Sun may have led to a H_2O-rich atmosphere with a steadily increasing atmospheric opacity (the so-called runaway greenhouse). The H_2O is then irreversibly dissociated to H_2 and O, with loss of these species. The H_2 is lost by escape to space, and the O is lost either by escape or by reactions with the Venusian surface. The end result of these processes is the present "dry" Venus.

This hypothesis has recently received support from the D/H ratio of $\simeq 1.6 \times 10^{-2}$ measured by the Pioneer Venus mass spectrometer (Donahue et al 1982). This value is significantly higher than the terrestrial value of 5×10^{-5} and is consistent with the depletion of an 8–9-m-thick layer of H_2O from Venus over geologic time (McElroy et al 1982, Donahue et al 1982). This amount of H_2O is $\simeq 0.3\%$ of the amount of H_2O on Earth and implies an initially "damp" Venus that had managed to accrete some hydrated minerals, but still not nearly as much as the Earth. However, Donahue et al (1982) also note that for H_2O volume mixing ratios $\geq 2 \times 10^{-2}$, the loss of hydrogen to space involves an upward fluid hydrodynamic flow in which H and D are equally depleted. Thus, enhancement of the D/H ratio can only begin once the H_2O volume mixing ratio drops below $\simeq 2 \times 10^{-2}$, which is equivalent to the 8–9-m-thick layer of H_2O mentioned above.

However, other constraints may help us to distinguish between the "dry," "damp," and "wet" Venus scenarios. For example, what is the fate of oxygen in the latter two cases? This depends to some extent on the oxidation state of the volatile-bearing phases responsible for bringing carbon to Venus. If carbon arrived in a reduced form, either as elemental carbon dissolved in Fe-Ni alloy (as equilibrium considerations predict) or as nonequilibrium organic matter, then it must react with O_2 to produce the present CO_2-rich atmosphere. If this O_2 was produced from H_2O dissociation, then the present CO_2 inventory on Venus implies a past sink for $\simeq 900$ m of H_2O on a "wet" primordial Venus.

On the other hand, reaction with Fe-bearing phases in the interior of Venus may have provided an O_2 sink. Again, this mechanism depends to some extent on the oxidation state of the solid grains accreted by Venus [e.g. their Fe(metal), Fe^{2+}, and Fe^{3+} ratios] and on the thermal history and differentiation of Venus. However, the reaction of the oxygen contained in a "wet" (Earth-like) Venus requires oxidation of $\simeq 5 \times 10^{24}$ g of Fe to FeO or of $\simeq 2 \times 10^{25}$ g of FeO to Fe_3O_4. This sink then requires reaction of 1–4% of the mass of Venus [for a planet with 10% (by mass) of Fe or FeO] in order to remove the required amount of oxygen. This appears to require a very efficient mechanism for exposing the interior of Venus to the atmosphere.

Finally, we note that the solar wind irradiation models (Wetherill 1981, McElroy & Prather 1981) discussed earlier can provide constraints relevant to H_2O retention by Venus. These models require weak radial transport in the accretion regions of Venus and Earth to preserve the segregation of ^{36}Ar-rich grains presumably accreted by Venus. Suitably designed theoretical models may therefore be useful for exploring mutually compatible ^{36}Ar and H_2O inventories (assuming some radial temperature gradient influence on solid grain composition) on Earth and Venus.

CONCLUDING REMARKS

Intensive investigations of our own atmosphere in recent years coupled with Earth-based, satellite, probe, and lander observations of our two nearest planetary neighbors have allowed us to critically compare and contrast the remarkable atmospheres on these three terrestrial planets. This exercise in comparative planetology has provided valuable insight into atmospheric origin and circulation and the biogeochemical processes that control atmospheric composition. We have been able to investigate the common aspects of the meteorology, chemistry, and evolution of the atmospheres of Venus, Earth, and Mars and puzzle over the unique aspects of each atmosphere, exemplified by the remarkable role of the biota in determining and altering the Earth's atmospheric chemical state. As more data are gathered on these three planets, there seems little doubt that such comparisons and contrasts will continue to be stimulating, provocative, and fruitful.

ACKNOWLEDGMENTS

We thank Gail Rodriguez for help in preparing the manuscript. This research was supported by the National Science Foundation (Grant ATM-84-01232) and NASA (Grants NAG9-108 and NAGW-821).

Literature Cited

Alvarez, L., Alvarez, W., Asaro, F., Michel, H. 1980. Extraterrestrial cause for the Cretaceous-Tertiary extinction. *Science* 208: 1095–1108

Andreae, M., Raemdonck, H. 1983. Dimethyl sulfide in the surface ocean and the marine atmosphere: a global view. *Science* 221: 744–47

Aston, F. W. 1924. The rarity of the inert gases on the Earth. *Nature* 114: 786

Barshay, S. S. 1981. *Combined condensation-accretion models of the terrestrial planets.* PhD thesis. Mass. Inst. Technol., Cambridge. 67 pp.

Barth, C. 1985. The photochemistry of the atmosphere of Mars. In *The Photochemistry of Atmospheres*, ed. J. Levine, pp. 337–92. New York: Academic. 518 pp.

Basaltic Volcanism Study Project. 1981. *Basaltic Volcanism on the Terrestrial Planets.* New York: Pergamon. 1286 pp.

Black, D. C., Matthews, M. S., eds. 1985. *Protostars and Planets II.* Tucson: Univ. Ariz. Press. 1293 pp.

Blake, D., Rowland, R. 1986. Worldwide increase in tropospheric methane, 1978–1983. *J. Atmos. Chem.* In press

Brown, H. 1949. Rare gases and the formation of the Earth's atmosphere. In *The Atmospheres of the Earth and Planets*, ed. G. P. Kuiper, pp. 260–68. Chicago: Univ. Chicago Press. 365 pp.

Cameron, A. G. W. 1982. Elementary and nuclidic abundances in the solar system. In *Essays in Nuclear Astrophysics*, ed. C. A. Barnes, D. D. Clayton, D. N. Schramm, pp. 23–43. Cambridge: Cambridge Univ. Press. 562 pp.

Cameron, A. G. W. 1983. Origin of the atmospheres of the terrestrial planets. *Icarus* 56: 195–201

Cameron, A. G. W., Fegley, B. Jr. 1982. Nucleation and condensation in the primitive solar nebula. *Icarus* 52: 1–13

Cameron, A. G. W., Ward, W. R. 1976. The origin of the Moon. *Lunar Sci. VII*, pp. 120–22

Carroll, M., Heidt, L., Cicerone, R., Prinn, R. 1986. OCS, H_2S, and CS_2 fluxes from a salt water marsh. *J. Atmos. Chem.* In press

Chameides, W., Walker, J. C. G. 1973. A photochemical theory of tropospheric ozone. *J. Geophys. Res.* 78: 8751–60

Charney, J., Drazin, P. 1961. Propagation of planetary scale disturbances from the lower into the upper atmosphere. *J. Geophys Res.* 66: 83–109

Conrath, B. 1981. Planetary-scale wave structure in the Martian atmosphere. *Icarus* 48: 246–55

Counselman, C., Gourevitch, S., King, R., Loriot, G. 1980. Zonal and meridional circulation of the lower atmosphere of Venus determined by radio interferometry. *J. Geophys. Res.* 85: 8026–30

Crutzen, P. J. 1976. The possible importance of COS for the sulfate layer of the stratosphere. *Geophys. Res. Lett.* 3: 73–76

Cunnold, D., Prinn, R., Rasmussen, R., Simmonds, P., Alyea, F., et al. 1986. Atmospheric Lifetime and Annual Release estimates for $CFCl_3$ and CF_2Cl_2 for 5 years of ALE data. *J. Geophys. Res.* 91: 10,797–10,817

Donahue, T. M., Hoffman, J. H., Hodges, R. R. Jr., Watson, A. J. 1982. Venus was wet: a measurement of the ratio of deuterium to hydrogen. *Science* 216: 630–33

Duce, R., Cicerone, R., Davis, D., Delwiche, C., Dickinson, R., et al. 1984. *Global Tropospheric Chemistry: A Plan for Action.* Washington, DC: Natl. Acad. Press. 194 pp.

Esposito, L. 1984. Sulfur dioxide: episodic injection shows evidence for active Venus volcanism. *Science* 223: 1072–74

Fegley, B. Jr. 1983. Primordial retention of nitrogen by terrestrial planets and meteorites. *Proc. Lunar Planet. Sci. Conf., 13th, J. Geophys. Res.* 88: A853–68

Fegley, B. Jr., Prinn, R. G., Hartman, H., Watkins, G. H. 1986. Chemical effects of large impacts on the Earth's primitive atmosphere. *Nature* 319: 305–8

Gerlach, T. M., Nordlie, B. E. 1975. The C-O-H-S gaseous system. *Am. J. Sci.* 275: 353–410

Gierasch, P. 1974. Martian dust storms. *Rev. Geophys. Space Phys.* 12: 730–35

Graedel, T. 1978. *Chemical Compounds in the Atmosphere.* New York: Academic. 440 pp.

Hayatsu, R., Anders, E. 1981. Organic compounds in meteorites and their origins. *Top. Curr. Chem.* 99: 1–37

Hays, J. D., Imbrie, J., Shackleton, N. J.

1976. Variations in the Earth's orbit: pacemaker of the ice ages. *Science* 194: 1121–32

Heald, E. F., Naughton, J. J., Barnes, I. L. Jr. 1963. The chemistry of volcanic gases. 2. Use of equilibrium calculations in the interpretation of volcanic gas samples. *J. Geophys. Res.* 68: 545–57

Holland, H. D. 1984. *The Chemical Evolution of the Atmosphere and Oceans.* Princeton, NJ: Princeton Univ. Press. 582 pp.

Holton, J. 1972. *An Introduction to Dynamic Meteorology.* New York: Academic. 319 pp.

Komhyr, W., Gammon, R., Harris, T., Waterman, L., Conway, T., et al. 1985. Global atmospheric CO_2 distribution and variations from 1968–1982. *J. Geophys. Res.* 90: 5567–96

Kung, C. C., Clayton, R. N. 1978. Nitrogen abundance and isotopic compositions in stony meteorites. *Earth Planet. Sci. Lett.* 38: 421–35

Lange, M. A., Ahrens, T. J. 1982. The evolution of an impact-generated atmosphere. *Icarus* 51: 96–120

Lebofsky, L. A. 1978. Asteroid 1 Ceres: evidence for water of hydration. *Mon. Not. R. Astron. Soc.* 182: 17P–21P

Leovy, C. 1973. Rotation of the upper atmosphere of Venus. *J. Atmos. Sci.* 30: 1218–20

Levine, J. S. 1985. The photochemistry of the early atmosphere. In *The Photochemistry of Atmospheres*, ed. J. S. Levine, pp. 3–33. New York: Academic. 518 pp.

Levy, H. 1971. Normal atmosphere: large radical and formaldehyde concentrations predicted. *Science* 173: 141–43

Lewis, J. S. 1970. Venus: atmospheric and lithospheric composition. *Earth Planet. Sci. Lett.* 10: 73–80

Lewis, J. S., Prinn, R. G. 1984. *Planets and Their Atmospheres: Origin and Evolution.* New York: Academic. 470 pp.

Lewis, J. S., Watkins, G. H., Hartman, H., Prinn, R. G. 1982. Chemical consequences of major impacts on Earth. In *Geological Implications of Impacts of Large Asteroids and Comets on the Earth, Geol. Soc. Am. Spec. Pap. No. 190*, ed. L. T. Silver, P. H. Schultz, pp. 215–21. Boulder, Colo: Geol. Soc. Am. 578 pp.

McElroy, M. B., Donahue, T. 1972. Stability of the Martian atmosphere. *Science* 177: 986–88

McElroy, M. B., Prather, M. J. 1981. Noble gases in the terrestrial planets. *Nature* 293: 535–39

McElroy, M. B., Prather, M. J., Rodriguez, J. 1982. Escape of hydrogen from Venus. *Science* 215: 1614–15

Mizuno, H., Nakazawa, K., Hayashi, C. 1982. Gas capture and rare gas retention by accreting planets in the solar nebula. *Planet. Space Sci.* 30: 765–72

Molina, M., Rowland, F. 1974. Stratospheric sink for chlorofluoromethanes: chlorine-catalysed destruction of ozone. *Nature* 249: 810–12

Morfill, G. E., Völk, H. J. 1984. Transport of dust and vapor and chemical fractionation in the early protosolar cloud. *Astrophys. J.* 287: 371–95

Moroz, V. 1983. Summary of preliminary results of the Venera 13 and Venera 14 missions. In *Venus*, ed. D. Hunten, L. Colin, T. Donahue, V. Moroz, pp. 45–68. Tucson: Univ. Ariz. Press. 1143 pp.

Newell, R., Kidson, J., Vincent, D., Boer, G. 1972. *The General Circulation of the Tropical Atmosphere and Interactions With Extratropical Latitudes*, Vol. 1. Cambridge, Mass: MIT Press. 258 pp.

Niederer, F. R., Papanastassiou, D. A. 1984. Ca isotopes in refractory inclusions. *Geochim. Cosmochim. Acta* 48: 1279–93

Ozima, M., Podosek, F. A. 1983. *Noble Gas Geochemistry.* Cambridge: Cambridge Univ. Press. 367 pp.

Palmen, E., Newton, C. 1969. *Atmospheric Circulation Systems.* New York: Academic. 420 pp.

Parkinson, T., Hunten, D. 1972. Spectroscopy and aeronomy of O_2 on Mars. *J. Atmos. Sci.* 29: 1380–90

Pillinger, C. T. 1984. Light element stable isotopes in meteorites—from grams to picograms. *Geochim. Cosmochim. Acta* 48: 2739–66

Pollack, J., Colburn, F., Flasar, F., Kahan, R., Carlston, C., Pidek, D. 1979. Properties and effects of dust particles suspended in the Martian atmosphere. *J. Geophys. Res.* 84: 2929–45

Prinn, R. G. 1971. Photochemistry of HCl and other minor constituents in the atmosphere of Venus. *J. Atmos. Sci.* 28: 1058–67

Prinn, R. G. 1973. Venus: composition and structure of the visible clouds. *Science* 182: 1132–34

Prinn, R. G. 1982. Origin and evolution of planetary atmospheres: an introduction to the problem. *Planet. Space Sci.* 30: 741–53

Prinn, R. G. 1985. The photochemistry of the atmosphere of Venus. In *The Photochemistry of Atmospheres*, ed. J. Levine, pp. 281–336. New York: Academic. 518 pp.

Prinn, R. G., Fegley, B. Jr. 1986. Bolide impacts, acid rain, and biospheric traumas at the Cretaceous-Tertiary boundary. Submitted for publication

Prinn, R. G., Simmonds, P. G., Rasmussen,

R. A., Rosen, R. D., Alyea, F. N., et al. 1983. The Atmospheric Lifetime Experiment 1. Introduction, instrumentation, and overview. *J. Geophys. Res.* 88: 8353–67

Ramanathan, V., Cicerone, R., Singh, H., Kiehl, J. 1985. Trace gas trends and their potential role in climate. *J. Geophys. Res.* 90: 5547–66

Ronov, A. B., Yaroshevsky, A. A. 1976. A new model for the chemical structure of the Earth's crust. *Geochem. Int.* 12: 89–121

Rossow, W., Del Genio, A., Limaye, S., Travis, L., Stone, P. 1980. Cloud morphology and motions from Pioneer Venus images. *J. Geophys. Res.* 85: 8107–28

Russell, H. N., Menzel, D. H. 1933. The terrestrial abundance of the permanent gases. *Proc. Natl. Acad. Sci. USA* 19: 997–1001

Schubert, G. 1983. General circulation and dynamical state of the Venus atmosphere. In *Venus*, ed. D. Hunten, L. Colin, T. Donahue, V. Moroz, pp. 681–765. Tucson: Univ. Ariz. Press. 1143 pp.

Seiff, A. 1983. Thermal structure of the atmosphere of Venus. In *Venus*, ed. D. Hunten, L. Colin, T. Donahue, V. Moroz, pp. 215–79. Tucson: Univ. Ariz. Press. 1143 pp.

Shoemaker, E. M. 1983. Asteroid and comet bombardment of the Earth. *Ann. Rev. Earth Planet Sci.* 11: 461–94

Stolarski, R., Cicerone, R. 1974. Stratospheric chlorine: a possible sink for ozone. *Can. J. Chem.* 52: 1610–15

Stone, P. 1975. The dynamics of the atmosphere of Venus. *J. Atmos. Sci.* 32: 1005–16

Surkov, Yu. A., Barukov, V. L., Moskalyeva, L. P., Kharyukova, V. P., Kemurdzhian, A. L. 1984. New data on the composition, structure, and properties of Venus rock obtained by Venera 13 and Venera 14. *Proc. Lunar Planet. Sci. Conf., 14th, J. Geophys. Res.* 89: B393–402

Toulmin, P. III, Baird, A. K., Clark, B. C., Keil, K., Rose, H. J. Jr., et al. 1977. Geochemical and mineralogical interpretation of the Viking inorganic chemical results. *J. Geophys. Res.* 82: 4625–34

Turco, R. P., Toon, O. B., Park, C., Whitten, R. C., Pollack, J. B., Noerdlinger, P. 1981. Tunguska meteor fall of 1908: effects on stratospheric ozone. *Science* 214: 19–23

Turekian, K. K. 1969. The oceans, streams, and atmosphere. In *Handbook of Geochemistry*, ed. K. H. Wedepohl, 1: 297–323. Berlin: Springer-Verlag

von Zahn, U., Kumar, S., Niemann, H., Prinn, R. G. 1983. Composition of the Venus atmosphere. In *Venus*, ed. D. Hunten, L. Colin, T. Donahue, V. Moroz, pp. 299–430. Tucson: Univ. Ariz. Press. 1143 pp.

Ward, W. 1974. Climatic variations on Mars. 1. Astronomical theory of insolation. *J. Geophys. Res.* 79: 3375–86

Watson, R., Geller, M., Stolarski, R., Hampson, R. 1986. Present state of knowledge of the upper atmosphere: An assessment report. *NASA Ref. Publ. No. 1162.* 134 pp.

Wetherill, G. W. 1975. Late heavy bombardment of the Moon and the terrestrial planets. *Proc. Lunar Sci. Conf., 6th,* pp. 1539–61

Wetherill, G. W. 1981. Solar wind origin of ^{36}Ar on Venus. *Icarus* 46: 70–80

Yung, Y., DeMore, W. 1982. Photochemistry of the stratosphere of Venus: implications for atmospheric evolution. *Icarus* 51: 199–247

Zurek, R. 1982. Martian great dust storms: an update. *Icarus* 50: 288–310

TECTONICS OF THE TETHYSIDES: Orogenic Collage Development in a Collisional Setting

A. M. Celâl Şengör

İstanbul Teknik Üniversitesi, Maden Fakültesi, Jeoloji Bölümü, 80394 Teşvikiye, İstanbul, Turkey

INTRODUCTION AND HISTORICAL REVIEW

The concept of Tethys, a former marine realm lying along the Alpine-Himalayan-Indonesian mountain ranges, was a product of Suess' global geology school in Vienna. It has occupied a central place in thinking on the evolution of Eurasia since its invention in 1885 by Suess' son-in-law, the famous stratigrapher Melchior Neumayr, on the basis of the distribution of Jurassic marine strata. Neumayr (1885) called this Jurassic seaway stretching from the Caribbean to Southeast Asia simply the "Central Mediterranean" and viewed it only as a paleogeographic feature. In 1888 Suess noted that it had existed since at least the Triassic and had episodically flooded wide tracts of Eurasia on both sides of the Alpine-Himalayan ranges. In 1893 Suess attached to it a tectonic connotation when he pointed out that the Alpine-Himalayan system had formed by its contractional obliteration and renamed it Tethys, after the sister and consort of Okeanos, the Greek god of the ocean (Şengör 1984). In *The Face of the Earth*, Suess (1901) further elaborated on the fundamental role of the Tethys in Mesozoic paleogeography and subsequent tectonic evolution of Eurasia, and he hinted at its presence in some locations as early as the late Paleozoic (cf. Tollmann 1984).

In the twentieth century, Tethys has been regarded as both a paleobiogeographic concept and a tectonic concept. Paleobiogeographers viewed it as a body of water characterized by a distinct fauna (e.g.

Tollmann & Kristan-Tollmann 1985), whereas tectonicians saw in it the "mother basin" of the Alpine-Himalayan ranges (e.g. Stille 1948). The confusion of these two entirely distinct connotations has led to frequent errors in the past, such as the presumed Mesozoic and Cenozoic tectonic connections between the Caribbean and the Mediterranean (e.g. Bucher 1924). In this review, "Tethys" refers exclusively to the tectonic concept. For good overviews of the paleobiogeographic Tethys, see Buffetaut et al (1984), Fourcade (1985), and Nakazawa & Dickins (1985).

Pre–plate tectonic views on the tectonic nature of the Tethys were polarized into two camps: The fixists (e.g. Stille 1948) considered it a "geosyncline" that had existed since the later Proterozoic and that had diminished in size progressively throughout the Phanerozoic, while the mobilists (e.g. Argand 1924) saw it as a narrow mobile marine realm caught up between the drifting continental rafts of Laurasia and Gondwana-Land. All these diverse, even incompatible, views had the common attribute of regarding the Tethys as a single feature, a relatively narrow seaway bordered by two major continental masses.

When Wilson (1963) reconstructed the Permo-Triassic Pangea, he realized that an eastward-gaping triangular oceanic embayment had to open along what are now the Alpine-Himalayan ranges, confirming the existence and the singularity of the Tethys. Its later obliteration during the dispersal of Pangea, coeval with the growth of the Alpine-Himalayan orogenic belts, was hailed as one of the great predictive successes of plate tectonics (e.g. Dietz & Holden 1970).

In the 1970s it became obvious, however, that the then-known suture zones along the Alpine-Himalayan system represented oceans that began opening in the Triassic and/or later, for neither older ophiolites nor older rifted margins existed along them (e.g. Smith 1973, Dewey et al 1973, Crawford 1974). This led to a conflict, called the "Tethyan paradox" (Şengör 1984, Stöcklin 1984), between the continental reconstructions demanding a vast ocean along the Alpine-Himalayan belt since the late Carboniferous and the field data showing that no such ocean could have possibly existed until after the middle Triassic.

There seemed to be three ways in which the Tethyan paradox could be resolved: The first was to question the validity of plate tectonic mobilism as a whole or to resort to a mobilism based on expanding-Earth models (e.g. Stöcklin 1983), but the weight of the corroborative evidence in favor of plate tectonics on a roughly constant-size Earth has made this an unfruitful line of attack. The second way was to try Pangea reconstructions different from Wilson's (1963) so as to diminish the size of the Tethyan embayment. Morel & Irving's (1981) Pangea B and Smith et al's (1981) end-Paleozoic Pangea (Pangea C) based exclusively on paleomagnetic data represent such attempts. However, Van der Voo et al (1984) showed that both Pangea B and the Pangea reconstructed by Van der Voo & French (1974)—"Pangea A2," which is similar to Wilson's (1963) reconstruction in requiring a large late Paleozoic Tethys—are compatible with paleomagnetic data, with the Pangea A2 configuration having superior statistical precision. Moreover, a reconsideration of the stratigraphic basis of the pole ages now suggests that the Pangea B and C reconstructions may have resulted from averaging poles of dissimilar age along common polar wander paths, creating unreal displacements (Livermore et al 1986). Because the geological evidence is also in favor of the Pangea A2 geometry (e.g. Pindell & Dewey 1982, Hallam 1983, Burke et al 1984, Pindell 1985), hopes of resolving the Tethyan paradox by alternative Pangea reconstructions have not come to fruition.

The third possible way to resolve the Tethyan paradox was to consider the adduced field evidence from the Alpine-Himalayan ranges insufficient to discuss whether there really was a paradox. Şengör (1985a) pointed out that the paradox was based entirely on negative evidence, i.e. on the absence of indicators of a Permo-Triassic ocean in areas where regional geological work is still at a reconnaissance stage and from where the results of ongoing studies only irregularly reach the international audience. Şengör (1979, 1984, 1985a,b, 1986a) also showed, on the basis of a comprehensive evaluation of all the available information from the Alpine-Himalayan system and following earlier suggestions by Stöcklin (1974) and Hsü & Bernoulli (1978), that two (instead of the previously assumed one) major Tethyan oceans had disappeared along the Alpine-Himalayan system.

The earlier of these two Tethyan oceans (i.e. that which was identical with the Permo-Triassic Pangean Tethyan embayment) Şengör (1979) called, following Stöcklin (1974), Paleo-Tethys [="old Tethys," not an exclusively Paleozoic Tethys (see Şengör 1986a, footnote 2)]. This ocean closed, largely during the early to middle Mesozoic, as a partial consequence of the rifting from northern Gondwana-Land and counterclockwise rotation of a thin continental stripe, called the Cimmerian continent (Şengör 1979). While closing Paleo-Tethys, this rotation opened Neo-Tethys (="new Tethys") from the Triassic onward (Şengör 1979). Neo-Tethys is thus largely equivalent to Suess' (1888, 1893) original Tethys concept.

Since Şengör's (1979) original reconstruction of the Tethyan domain showing both Paleo- and Neo-Tethys with the intervening Cimmerian continent, enormous progress has been made in mapping their suture zones. The resulting picture is one of an anastomosing suture network enclosing numerous continental and otherwise buoyant crustal blocks of various sizes and provenances forming a veritable orogenic collage, which resembles that of the North American Cordillera (Saleeby 1983) and whose structure and tectonic history increase in complexity from west to east (Figure 1). When and how did these blocks accrete to Eurasia and/or to dispersed major pieces of Gondwana-Land? What effects did their accretion have on the host continent and on their previously accreted brethren? Where did they come from? Have they been rearranged following accretion? Much of the present Tethyan research is directed toward providing answers to such questions, and the following review is a brief

―――――――――――――――――――――――――――――――――――――――→

Figure 1 Generalized tectonic map showing the major tectonic subdivisions of Eurasia and the suture distribution within the Tethysides. Cimmeride sutures: I—Paleo-Tethyan suture in the Balkan/Carpathian Cimmerides, II—Karakaya, III—Luncavita-Consul, IV—North Turkish, V—Paleo-Tethyan suture in the Caucasus, V'—Chorchana-Utslevi zone, VI—Talesh-Mashhad, VII—Waser (Farah-Rud), VIII—Paropamisus–Hindu Kush–North Pamir, VIII'—Kopet Dagh, IX—South Ghissar, X—Northern "synclinorium" of western Kuen-Lun, X'—Qiman-Dagh, XI—Altin Dagh, XII—Suelun Hegen Mts., XIII and XIII'—Inner Mongolian, XIV—Suolun-Xilamulun, XV—Da Hingan (G. Khingan), XVI—Tergun Daba Shan–Qinhai Nanshan, XVII—Southern "synclinorium" of western Kuen-Lun, XVIII—Burhan Budai Shan–Anyemaqen Shan, XIX—Lancan Jiang–Litien, XX—Jinsha Jiang, XXI—Litang, XXI'—Luochou "arc-trench belt," XXII—Banggong Co–Nu Jiang, XXII'—Mid-Qangtang, XXIII—Shiquanhe, XXIV—Southwest Karakorum, XXV—Nan-Uttaradit-Sra Kaeo, XXVI—Tamky-Phueson, XXVII—Song Ma (Red River), XXVIII—Song Da (Black River), XXIX—Bentong-Raub, XXX—West Borneo, XXXI—Qin-Ling, XXXII—Longmen Shan–Qionglai Shan, XXXIII—Mid-South China, XXXIV—Korea, XXXV—Helan Shan, XXXVI—Mandalay, XXXVII—Shilka. Alpide sutures: 1—Pyrenean, 2—Betic,* 3—Riff,* 4—High Atlas,* 5—Saharan Atlas,* 6—Kabylian,* 7—Apennine, 8—Alpine, 9—Pieniny Klippen belt, 10—Circum-Moesian, 11—Mures, 12—Srednogorie, 13—Peonias–Intra-Pontide, 14—Almopias-İzmir-Ankara, 15—Pindos-Budva-Bükk, 16—Ilgaz-Erzincan, 17—Inner-Tauride, 18—Antalya, 19—Cyprus, 20—Assyrian, 21—Maden, 22—Sevan-Akera-Qaradagh, 23—Slate-Diabase zone, 24—Zagros, 25—Circum-Central Iranian microcontinent, 26—Oman, 27—Waziristan, 28—Kohistan sutures, 29—Ladakh sutures (northeast: Shyok; southwest: Indus), 30—Indus-Yarlung-Zangbo, 31—Burma, 32—Mid-Sumatra, 33—Meratus. (Asterisks indicate sutures of sialic oceans *sensu* Şengör & Monod 1980.) Tethyside block: a—Moroccan Meseta, b—Oran Meseta, c—Alboran, d—Iberian Meseta, e—African promontory, f—Rhodope-Pontide, g—Sakarya, h—Kırşehir, i—Northwest Iran, j—Central Iranian, k—Aghdarband arc, l—Farah, m—Helmand (*sensu* Şengör 1984), m'—Kohistan arc, n—Western Kuen-Lun Central Meganticlinorium, o—Qaidam, p—Alxa, q—North China (Sino-Korean) platform, r—North China fold belt, s—Qangtang (possibly divided into s'—East Qangtang and s''—West Qangtang: Chang Cheng-fa, oral communication, 1985), t—Lhasa (possibly divided into t'—Bongthol Tangla, t''—Nagqu, and t'''—Lhasa proper: A. Gansser, oral communication, 1985), t''''—Ladakh arc, u—Shaluli Shan arc, v—Chola Shan arc, w—Yangtze, x—Annamia, y—Huanan, z—Songpan Massif. E, M, and S are East Anatolian, Makran, and Songpan-Ganzi accretionary complexes, respectively. Updated from Şengör (1984) using Adamia & Belov (1984), Bakirov & Burtman (1984), Baud (1985a,b), Chang et al (1986a), Gatinsky (1986), Mitchell (1986), Şengör (1986a,b,c), and Yang et al (1986).

summary of where we now stand with regard to these questions, with emphasis on the less well-known earlier history between the Permian and the Cretaceous.

TETHYAN SUTURES, BLOCKS, AND OROGENS

Tethyan Sutures

Şengör (1986c) suggested that "orogenic zones," consisting of orogenic collages (Helwig 1974) plus their associated areas of marginal deformation along the edges of host continents, and the host continental nuclei themselves form the first-order tectonic subdivisions of continents. Figure 1 shows the tectonic subdivisions of Eurasia. The wide and long orogenic zone between the Laurasian fragments in the north and the Gondwanian

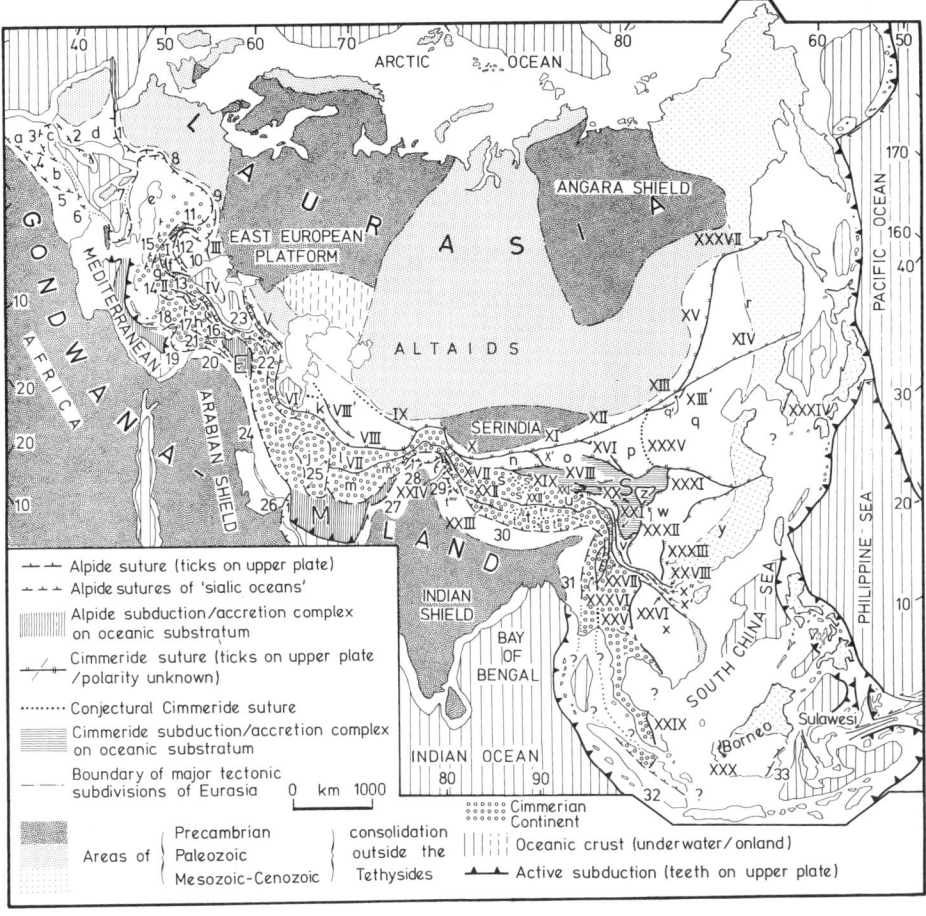

fragments in the south is the product of the obliteration of Tethys *sensu lato*. Şengör (1984, 1985b) called this the Tethyside super orogenic complex (Figure 2). It contains two independent but largely superimposed orogens, of which the older one was formed largely between the late Carboniferous and the early Cretaceous as the product of the closure of Paleo-Tethys. It is called the Cimmerides, after the Cimmerii, the oldest known inhabitants of the northwestern shores of the Black Sea, where evidence for Cimmeride orogenic events was first discovered (Suess 1901, p. 22). The younger one, called the Alpides, has been evolving since the Jurassic. Figure 1 shows the suture zones of the Cimmerides (I–XXXVII)[1] and those of the Alpides (1–33).[1] The numerous continental blocks [a–z[1]; I avoid the misleading and confusing term "terrane" (see below)] enclosed by these sutures constitute the Tethyside orogenic collage (Şengör 1986a).

Tethyan Blocks

Over 90% of the continental and otherwise buoyant objects within the Tethyside collage belong to the Cimmerides and the rest to the Alpides. Sutures segmenting the Cimmerian continent and thus representing intra-Cimmerian continental oceans (VII, XXII, XXII', XXIII, XXXVI, 13, 14, 16, 17, 22, 25) are assigned to the Cimmerides or to the Alpides depending on the timing of their closure. If a closure occurred before the onset of the main Neo-Tethyan subduction during the Aptian-Albian, the suture is considered Cimmeride; if the closure occurred after this time, the suture

[1] These are cited hereinafter without reference to Figure 1.

Figure 2 The Tethyside super orogenic complex. Larger bold letters A, T, P, and Y are the Alpine, Turkish, Pamir, and Yunnan syntaxes. Smaller letters: A—Alps, AG—Akçakale graben, AGr—An Chau graben, Al—Alborz, Ap—Apennines, At—Atlas Mountains (*sensu lato*), B—Betics, BF—Bogdo fault, BG—Bresse graben, C—Carpathians, Ca—Caucasus, CAGS—Central Arabian graben system, CF—Chaman fault, CG—Central graben, D—Dinarides, DA—Dnyepr-Donetz aulacogen, EAB—East Arabian block, EAf—East Anatolian fault, EI—East Ili basin, GKF—Great Kavir fault, GT—Gerze thrust, H—Hellenides, HF—Herat fault, HRF—Harirud fault, H-RR—Hantaj-Rybninsk aulacogen, IG—Issyk Gol basin, IR—Irkineev aulacogen, KDF—Kopet Dagh fault, KF—Karakorum fault, KKU—Kızıl Kum uplift, KTF—Kang Ting fault, MF—Mongolian faults, MR—Main Range of the Greater Caucasus, NAF—North Anatolian fault, NCD—North Caspian depression, P—Pyrenees, PA—Pachelma aulacogen, PNT—Palni–Nilgiri Hills thrust, PT—Polish trough, R—Riff cordillera, RG—Upper Rhine graben, RRF—Red River fault, S—Sichuan basin, SF—Sagain fault, SGS—Shanxi graben system, SMÜR—South Mangyshlak–Üst Yurt ridge, SUF—South Ural faults, T—Turkish ranges, TD—Turfan depression, T-LF—Tan-Lu fault, UR—Ura aulacogen, VG—Viking graben, WSB—West Siberian basin, Z—Zagrides. Modified from Şengör (1984) using Volchanskaya & Korytov (1982), Ziegler (1982), Burke (1983), Şengör (1986c), Tapponnier et al (1986), and Yang et al (1986).

is assigned to the Alpides. A better, more satisfactory criterion would have been to group together those sutures whose closures were parts of one kinematic evolution, but the available data are not now sufficient to apply it to the Tethysides.

The Cimmerides are the products of the convergence and terminal collision with Eurasia of two main groups of continental objects. One of them, between the Balkans and Malaysia, was formed from the Cimmerian continent (Figure 1). Although probably originally consisting of a single piece, it disintegrated into a number of blocks (f, g, h, i, l, m, s, t) as it moved across the Tethyan domain during the early Mesozoic, but without disrupting its essential east-west continuity (Şengör 1979, 1984, Boulin 1981, Audley-Charles 1984; Figures 3A–C). Chang et al (1986a) report

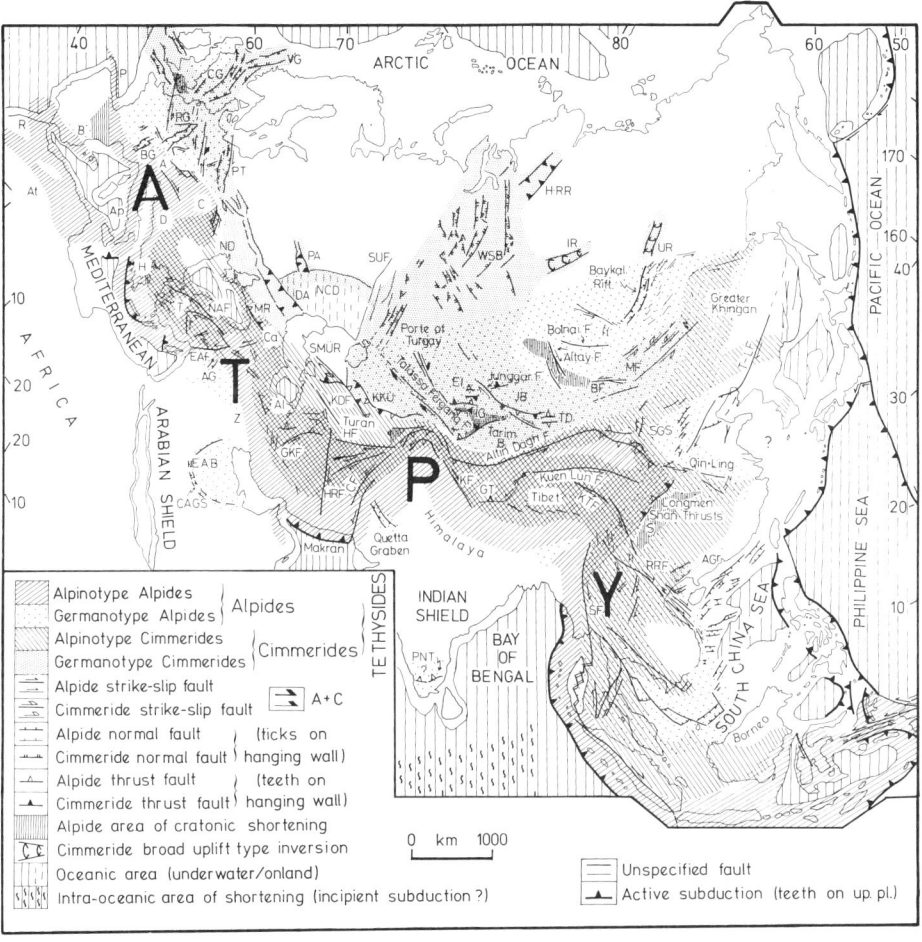

total sediment thicknesses in excess of 15 km from within the Quangtang block (s). This finding, when combined with earlier reports of glaucophane schists from within the same block (Hennig 1915), may indicate the existence of a suture (XXII') within the Quangtang block dividing it into two (s' and s") and thus possibly disrupting the east-west continuity of the Cimmerian continent. A. Gansser (oral communication, 1985) also suggested that the Xainxa ophiolites (Girardeau et al 1985) and similar occurrences (Geological Map of Tibet 1980) within the Lhasa block (t) may represent early Mesozoic sutures dividing it into at least three subblocks (t', t", t'''). The most recent field observations by the 1985 Royal Society–Academia Sinica geotraverse team on the intra-Lhasa block ophiolites suggest, however, that all of them may be remnants of one giant ophiolite nappe rooted originally into the Banggong Co–Nu Jiang suture zone (XXII) (Chang et al 1986b). Until more data are obtained, I adhere to my original suggestion of the east-west continuity of the Cimmerian continent between the Balkans and Malaysia, at least for that part of it lying north and east of sutures VII, XXII, XXXVI, 13, 16, and 24 (Şengör 1979, 1984).

The second group of major continental blocks that collided with Eurasia

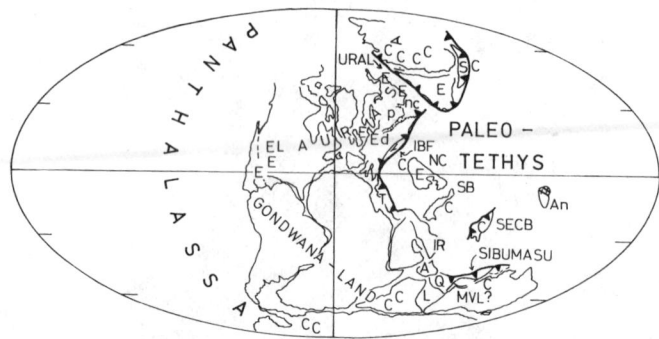

Figure 3A Schematic reconstruction showing the paleotectonics of the Tethyan domain in the late Permian (Kazanian). Base map is from Scotese (1984), modified using Şengör & Hsü (1984), Lin et al (1985a), and Şengör (1986c). Distribution of coals (C) and evaporites (E) were taken from Ziegler et al (1979). Abbreviations for all reconstructions: A—Afghan blocks, An—Annamia, B—Bitlis/Pötürge fragment, BNJ—Banggong Co–Nu Jiang ocean, CI—Central Iranian microcontinent, CS—Chola Shan, d—Dnyepr-Donetz aulacogen, F—Farah block, H—Helmand block (*sensu* Şengör 1984), IBF—Istanbul-Balkan fragment, IR—Iran block, K—Kırşehir block, L—Lhasa block, LB—Luochou arc, MVL—Mount Victoria Land block, NC—North China block, nc—North Caspian depression, No—Northern branch of Neo-Tethys, p—Pachelma aulacogen, Q—Qangtang block, Qu—Quetta graben, RRF—Red River fault, S—Serindia, Sa—Sakarya continent, SB—Yangtze block, SECB—Huanan block, SG—Songpan-Ganzi system, ShS—Shaluli Shan arc, SIBUMASU—China-Burma-Malaya-Sumatra portion of the Cimmerian continent (after Metcalfe 1984), So—Southern branch of Neo-Tethys, T—Turkish blocks.

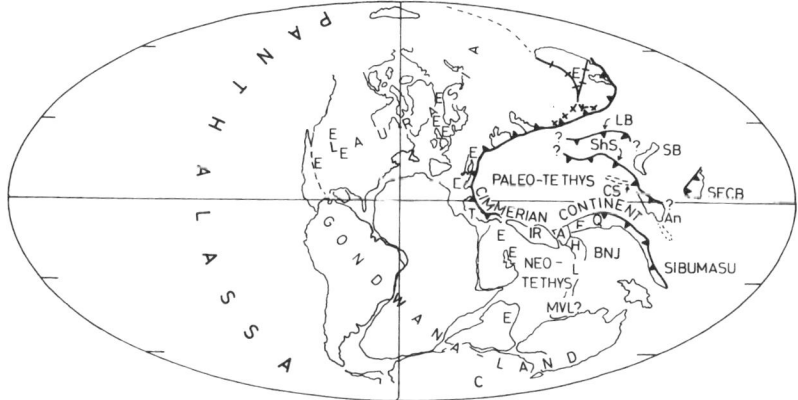

Figure 3B Early Triassic (Induan) paleotectonics of the Tethyan domain. Base map and distribution of coals (C) and evaporites (E) for this and all subsequent reconstructions are from Parrish et al (1982). This map was modified using Şengör & Hsü (1984), Yang et al (1986), Zhang Qinwen & K. J. Hsü (personal communication, 1985). For abbreviations, see Figure 3*A*.

and with the Cimmerian continent to eliminate Paleo-Tethys was located east of the 100°E meridian and consisted of five major pieces, namely the North China fold belt (r; Klimetz 1983), the North China platform (q), the Yangtze block (w), the Huanan block (y) [Yangtze block = Sichuan

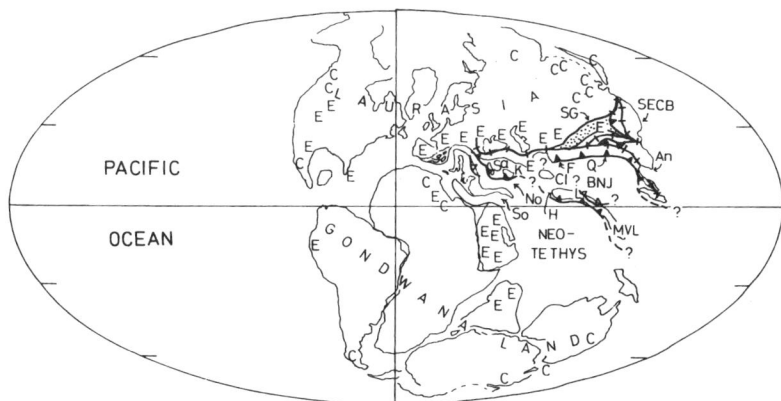

Figure 3C Late Jurassic (Volgian) paleotectonics of the Tethyan domain. Base map modified using Şengör & Hsü (1984) and Zhang Qinwen (personal communication, 1985). For abbreviations, see Figure 3*A*.

block, Huanan block = Southeast China block in Şengör & Hsü (1984)], and nuclear Indochina, or Annamia (x), plus a host of smaller continental and some possibly buoyant ensimatic slivers and fragments (q′, u, v, x′, x″, z; Şengör & Hsü 1984). Some of the latter may represent later separated original pieces of the larger blocks based mainly on their Paleozoic stratigraphic record [e.g. u, v, and z may have been pieces of w before the Permian (Yang et al 1986)]. The Huanan block (y) itself may have consisted of three separate continental blocks, namely the Hunan, the Huanan *sensu stricto*, and the coastal blocks before the Mesozoic, but the data are not now sufficient to be sure of the timing of its unification (K. J. Hsü, personal communication).

Fragment k, the Aghdarband arc (the Kopet Dagh arc of Baud 1985a), was clearly always a part of Turan. However, at least a portion of the southern part of Turan itself (region between VIII and IX) may have been a piece foreign to Eurasia and more related to the Cimmerian continent [and therefore to Gondwana-Land (see below)] during the earlier late Paleozoic (Soffel & Förster 1984). In any case, by latest Carboniferous time at least its northern portion had already docked against nuclear Asia along the South Ghissar suture (IX) (Kravchenko 1979, Bakirov & Burtman 1984).

Fragments n, o, and p are much less well known. Fragment o, probably an assembly of arcs, clearly docked against p in early Paleozoic time along the Qinghai Nanshan suture (XVI; Yang et al 1986). Both of them and n were welded against the Serindian nucleus by late Paleozoic time (Yang et al 1986).

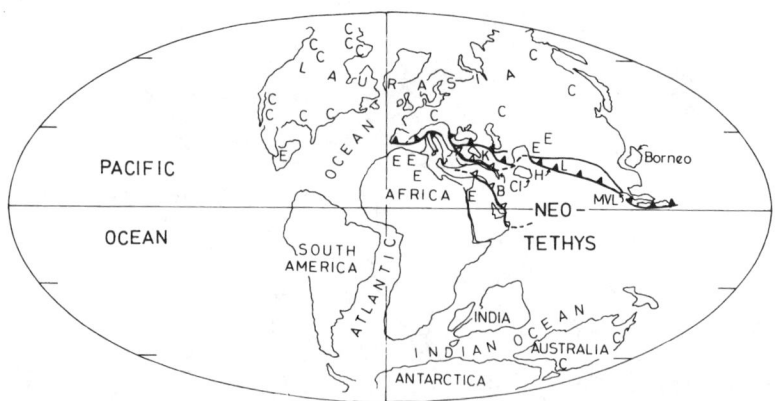

Figure 3D Late Cretaceous (Cenomanian) paleotectonics of the Tethyan domain. Base map modified using Achache (1984) and Soffel & Förster (1984). For abbreviations, see Figure 3*A*.

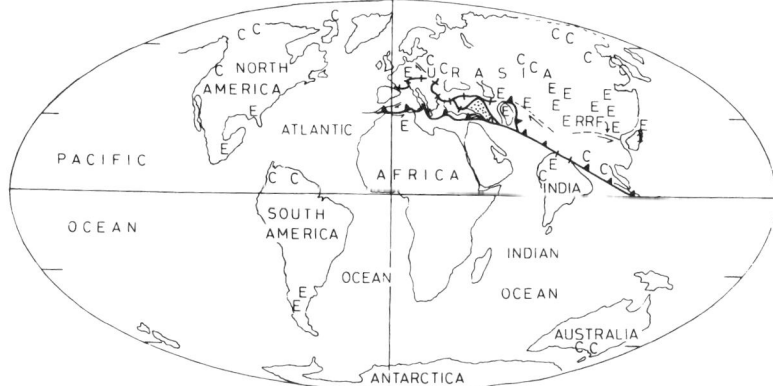

Figure 3E Middle Eocene (Lutetian) paleotectonics of the Tethyan domain. Base map modified using Tirrul et al (1983) and Tapponnier et al (1986). For abbreviations, see Figure 3*A*.

The Alpides are the products of the convergence and collision with Eurasia of dispersed pieces of Gondwana-Land such as Africa, Arabia, India, and Australia (Figure 1), plus smaller blocks in Turkey [such as the Rhodope-Pontide fragment (f), the Sakarya continent (g), and the Kırşehir block (h)] and in the Himalaya [such as the Kohistan (m′) and the Ladakh (t′′′′) arcs] during the Jurassic to the present interval (Sborschikov 1983, Görür et al 1984, Vergely 1984, Fourcade 1985, Aubouin et al 1986, Şengör

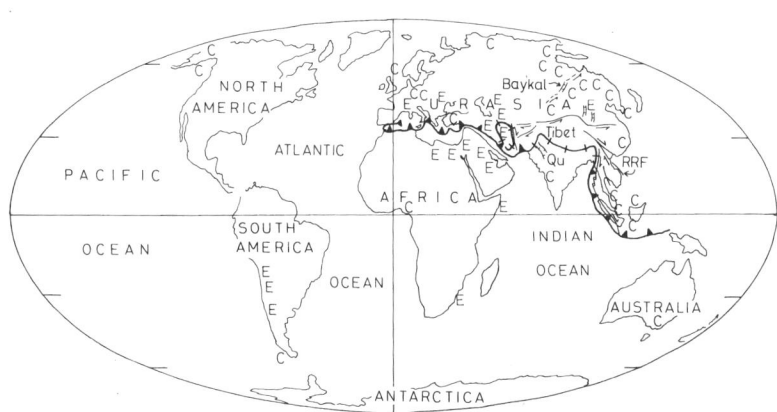

Figure 3F Late Miocene (Vindobonian) paleotectonics of the Tethyan domain. Base map modified using Dercourt et al (1986) and Tapponnier et al (1986). For abbreviations, see Figure 3*A*.

et al 1984a, Windley 1983, Bender 1983, Hamilton 1979). Within these blocks and also along the southern continental margin of Eurasia, back-arc basins opened and closed during the Alpide evolution (Antonijević et al 1974, Aiello et al 1977, Cameron et al 1980, Şengör & Yılmaz 1981, Bard 1983, Berberian 1983, Dietrich et al 1983, Zonenshain & Le Pichon 1986), which increased the number of Alpide sutures and complicated the local structural geometry.

Provenance and Possible Migration Routes of Tethyside Blocks

Four main lines of evidence exist to ascertain the site of origin and the subsequent path of migration of the blocks forming the Tethyside collage: paleomagnetism, paleobiogeography, paleoclimatology, and the geology of the late Precambrian–early Paleozoic basement of the Tethyside blocks.

The pre-1982 paleomagnetic data from the blocks of the Cimmeride part of the Tethyside collage were reviewed by Şengör (1984). Westphal et al (1986) reviewed the recent Mesozoic-Cenozoic data west of the Pamir syntaxis for the entire Tethyside collage, while Achache (1984) and Lee (1984) present new data and syntheses on (a) Tibet and Indochina and (b) Korea, respectively. New Chinese data relevant to the evolution of the Tethyside collage are presented by Lin et al (1985a,b) and Chan (1986). Although newer data were collected from Tibet, both from the Lhasa and the Qangtang blocks, by the 1985 Royal Society–Academia Sinica geotraverse team, their analysis has not yet been published except for one abstract (Lin & Watts 1986). New data and syntheses on Iranian paleomagnetism not included in Westphal et al are given in Soffel & Förster (1984).

Most published new data from Tibet and Indochina are from post-Paleozoic rocks and say little about the original provenance of these regions. Data from both the North China (q) and the Yangtze (w) blocks clearly indicate that during the Cambrian both were in the Southern Hemisphere, adjacent to (a) Tibet, Iran, and northern India and (b) northwestern Australia, respectively. Data from the Central Iranian microcontinent (j) show that it became an independent block after the middle Triassic and rotated about 135° counterclockwise by the late Cretaceous. Data also show that at least part of Turan was considerably farther south (15°N) than the rest of Eurasia during the Permian, thus corroborating the earlier expressed suspicion that it too may have come from the southern continents. In general, the new paleomagnetic data do not impose significant changes to the evolutionary model proposed in Şengör & Hsü (1984).

Much more abundant and decisive for the late Paleozoic are the paleobiogeographic and paleoclimatic data summarized in Figure 4.[2] They

[2] *Note added in proof*: The questionable Gondwana-type flora shown of the Qangtang block is now claimed to be Cathaysia type (Chang Cheng-fa and W. S. F. Kidd, personal communication, 1986), although this does not affect the interpretation of the provenance of the entire Cimmerian continent as presented above.

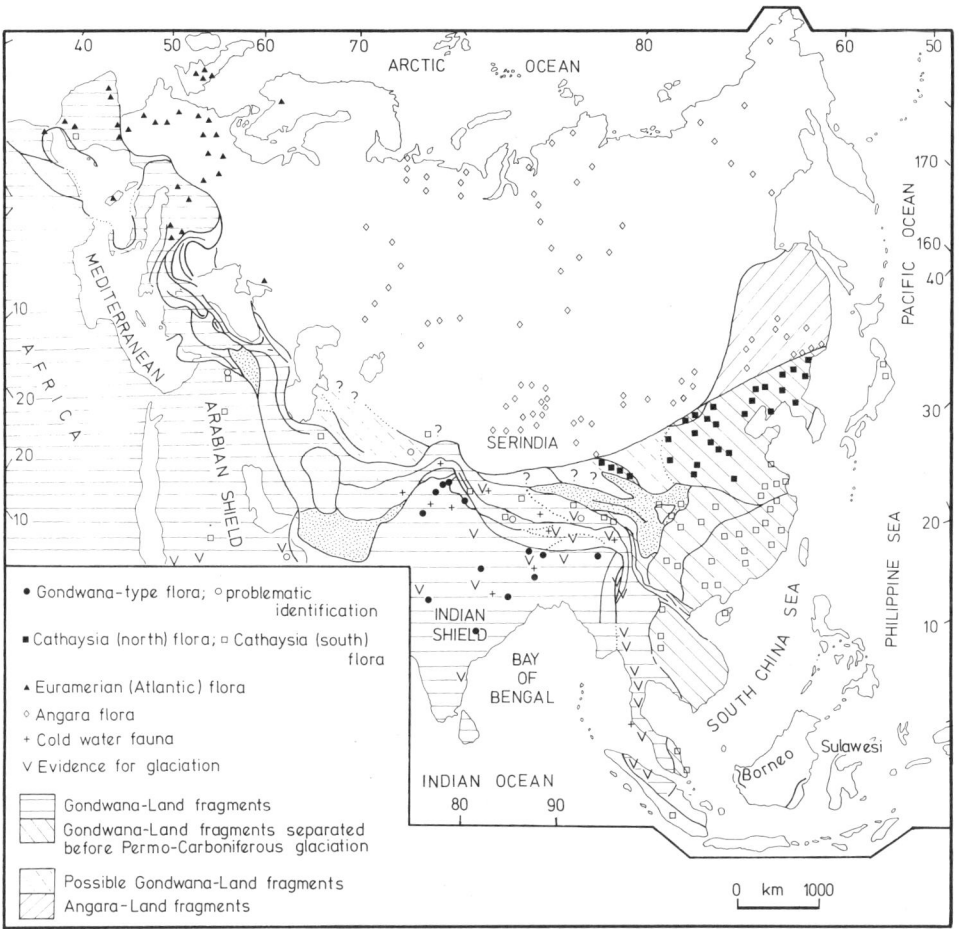

Figure 4 Distribution of late Paleozoic (mainly Late Carboniferous–Early Permian) floras, cold-water faunas, and traces of glaciation in and around the Tethysides. Dotted areas represent the flysch fill of the Songpan-Ganzi accretionary complex. Modified from Şengör (1985a) using Ctyroky (1973), Lemoigne (1981), Thenius (1981/1982), Braakman et al (1982), Waterhouse (1982), Asama (1984), Vozenin-Serra (1984), Dickins (1985), Nakazawa (1985), Chang et al (1986a), and Stauffer & Lee (1986).

collectively show that the entire Cimmerian continent was a part of Gondwana-Land until at least the early Permian. Burrett & Stait (1985) supported this view on the basis of Ordovician marine faunas from western Thailand and Malaysia.

The results of paleomagnetic, paleobiogeographic, and paleoclimatic data are further corroborated by the similarity of the late Proterozoic–early Paleozoic evolution of the basement of the Cimmerian continent, Annamia, North China, Yangtze, and the Huanan blocks. Especially widespread in the Cimmerian continent basement and in Annamia is the "Pan-African" event, which gives dates between 600 and 400 Myr ago [e.g. Şengör et al (1984b) for Turkey, Berberian & King (1981) and Davoudzadeh et al (1986) for Iran, Montenat et al (1981) for the Helmand block in Afghanistan, Chang et al (1986a) for Tibet; for a synthesis, see Şengör (1986a)]. These ages correlate with similar ages and structural sequences from the basements of Africa, Arabia, and India. Lin et al (1985b) and Şengör (1986a) compared the Proterozoic and Cambrian geological evolution of the Chinese blocks and Australia and noted remarkable similarities.

Reconstructions shown in Figures 3*A–F* were drawn largely on the basis of the constraints provided by the data reviewed above. Seafloor magnetic anomalies from the Indian Ocean were not used to infer the possible motion paths of any of the Tethyside blocks, for none of the blocks can be shown to have left their original sites during or after the late Jurassic. Although Rowley et al (1986) speculate that the Lhasa block may have rifted from northwest Australia in the late Jurassic, this contradicts the available geological evidence. Audley-Charles' (1984) model in which the entire Cimmerian continent and Southeast Asia are rifted from Gondwana-Land in the late Jurassic is similarly refuted by the present information. The only possible candidate that might have rifted off northwest Australia in late Jurassic time is the Mount Victoria Land block of Mitchell (1986) in Burma. This is shown as a possibility in Figure 3*B*.

Tethyan Orogens

Şengör (1984, 1986c) defined an orogenic complex as comprising both the orogenic zone proper and the fore- and hinterland areas of deformation that formed as a consequence of the evolution of the orogenic zone. In each orogenic complex, the orogenic zone itself is made up of "alpinotype" structures such as nappes, metamorphic cores, and mélange complexes, whereas the fore- and hinterland areas of deformation contain "germanotype" structures (i.e. dominantly brittle, "blocky" structures such as grabens, block uplifts, and large strike-slip faults).

Figure 2 shows the areal distribution of the Cimmeride and Alpide orogenic complexes that together constitute the Tethyside super orogenic complex. Notice in that figure the extent of superposition of the Cimmerides by the Alpides in both their alpinotype and germanotype areas. Where alpinotype Cimmerides are superimposed by alpinotype Alpides, the recognition and reconstruction of the former are very difficult, in places even impossible, as in the eastern Mediterranean region or in the Indonesian archipelago (Figure 2). Superposition of alpinotype Cimmerides by germanotype Alpides, as in wide regions of China (Figure 2), largely preserves the structures of the former, provided that they are not buried under sediments filling germanotype basins. If germanotype Cimmerides are superimposed by alpinotype Alpides, as, for example, happened along the Sistan suture zone (25; Tirrul et al 1983), the earlier structure generally can be recognized and is often interpreted as the "rifting phase" of the later alpinotype tectonism, although the two events belong to two entirely different orogenic systems. The Pindos-Budva trough is another such example (Şengör 1985b). Superposition of germanotype Cimmeride structures by germanotype Alpide elements frequently produces the effect of "long-lived structures" subject to repeated "posthumous" reactivations. This has led to the formulation of fixist hypotheses in the past, because the same—or similar—sorts of events are seen to recur in the same places and thus give the impression that their cause must be stationary and located immediately beneath the reactivated structures. The inversion of basins during the early and middle Mesozoic both in Europe and Asia, an idea much publicized by Soviet geologists (e.g. Vasiliev et al 1981) and now becoming popular in the western world (e.g. Bally 1984), is also a manifestation of the reactivation of germanotype structures. In this case, however, a previously extensional structure becomes reactivated in a compressional mode. Figure 2 shows many examples of "long-lived" germanotype structures in Eurasia that were either formed or first reactivated by the germanotype Cimmeride tectonism and then later reactivated by the germanotype Alpide events.

Both the germanotype Cimmerides and the germanotype Alpides considerably distorted Eurasia. No paleogeographic reconstruction that does not take these strains into account, therefore, can be taken seriously.

TECTONIC EVOLUTION OF THE TETHYSIDES

Table 1 and Figures 3A–F summarize the essential steps in the tectonic evolution of the Tethyside super orogenic complex. In simplest terms this evolution consisted of the destruction of Gondwana-Land by continuous calving of continental blocks from it and their northerly flight to enlarge

Table 1 Generalized timing of ocean opening and closure along the Tethyside sutures[a]

	Cimmerides			Alpides			
	Closing		Closing	Opening	Closing	Opening	Closing
I	mJ	XXIII	?J	1 eK,	mE	28 Tr,	lK–lE
II	lTr	XXIV	?	2 emJ,	M	29 Tr,	lK–lE
III	emJ	XXV	lTr	3 emJ,	M	30 eTr,	m–lE
IV	mJ	XXVI	ePale	4 emJ,	lK	31 ?lP,	E
V	lTr–eJ	XXVII	eC	5 emJ,	lK	32 eK,	lK
VI	lTr	XXVIII	lTr–eJ	6 mJ,	M	33 ?,	lK
VII	eK	XXIX	lTr	7 mJ,	Ol		
VIII	eJ	XXX	?mJ	8 mJ,	Pal–eE		
VIII'	lC–lTr	XXXI	lTr	9 eJ,	E		
IX	lC	XXXII	lTr–eK	10 lJ,	eK–M		
X	?lC	XXXIII	lTr–mJ	11 eJ,	lK–Pal		
X'	P	XXXIV	lTr	12 lK,	E		
XI	D	XXXV	?P	13 eJ,	lK–E		
XII	P	XXXVI	mJ	14 eJ,	Pal–eE		
XIII	P	XXXVII	eK	15 eTr,	Ol		
XIII'	P			16 eJ,	lE		
XIV	lC			17 eTr,	mE		
XV	P			18 eTr,	Pal		
XVI	D			19 m–lTr,	lK		
XVII	eJ			20 m–lTr,	lK		
XVIII	eJ			21 lK,	mE–mM		
XIX	mTr			22 ?mJ,	Pal		
XX	mTr			23 eJ,	E		
XXI	mJ			24 e-mTr,	M		
XXI'	J			25 lTr,	Ol		
XXII	eK			26 mTr,	lK		
XXII'	eMe			27 Tr,	E		

[a] Numbers identifying individual suture segments correspond with those in Figure 1. Abbreviations: Pale—Paleozoic, Me—Mesozoic, D—Devonian, C—Carboniferous, P—Permian, Tr—Triassic, J—Jurassic, K—Cretaceous, T—Tertiary, Pal—Paleocene, E—Eocene, Ol—Oligocene, M—Miocene, e—early, m—middle, l—late.

Eurasia by the formation of the large Tethyside orogenic collage. The story of the Tethysides properly began when Paleo-Tethys first formed during the late Paleozoic assembly of Pangea and its floor started to be destroyed by both internal and peripheral subduction. Figure 3A shows that even by late Permian time, Pangea was not yet assembled completely (the Uralide collision was not yet completed), although the destruction of Paleo-Tethyan ocean floor had already commenced both along peripheral subduction zones [e.g. along the Hindu Kush–Paropamisus line (VIII), in northern Turkey (IV), in northeastern Tibet (XX), and in Thailand

(XXXVI); Şengör 1984] and along subduction zones located along continental blocks within the Paleo-Tethyan realm [e.g. along the northwestern margin of the Huanan block (XXXIII; Yang et al 1986)].

The oldest Cimmeride sutures are those between the Tethyan continental blocks. The best example of these is the Song Ma (Red River) suture (XXVII) of Tournaisian-Visean age recording the collision between Annamia and a late Palcozoic island arc (x''). The "Central Meganticlinorium of western Kuen-Lun" (n) docked against Eurasia and against the Qaidam block (o) by late Permian time. A similar, probably mid-Carboniferous suture is the Suolun-Xilamulun suture (XIV; Yang et al 1986) that welded the North China platform (q) to the North China fold belt (r). Both Klimetz (1983) and (following him) Şengör (1984) thought that the Helan Shan (XXXV)–Inner Mongolian (XIII and XIII') and the Da Hingan (Greater Khingan mountains: XV) sutures had closed during Jurassic time mainly on the basis of the strong late Jurassic–early Cretaceous (Yenshanian) deformation [sutures XIII and XIII' coincide with the typelocality of the Yenshanian (Wong 1929)] along this entire belt and the strong felsic magmatism mainly along the Da Hingan mountains. A complete lack of marine sediments younger than the Permian along the entire belt, on the other hand, led authors such as Zhang et al (1984) to deny the presence of this suture system. It now seems that although the sutures are there, they are older (Permian) than originally suggested by Klimetz (1983) (K. J. Hsü, personal communication, 1986). Yang et al (1986) recently reemphasized the differences between the Alxa block (p) and the North China platform, suggesting the presence of some sort of discontinuity between them along the Helan Shan (XXXV) suture of Klimetz (1983) and Şengör (1984) and thus supporting the possible existence of at least a Paleozoic suture there.

With the exception of sutures X, XXXV, XIII, XIII', and XV, none of the late Paleozoic sutures mentioned above are Cimmeride in the strict sense of the word because they represent tectonic events in Panthalassa whose connections with a "native" Tethyan tectonics are not clear. They are included in the Tethysides here, as are the early Paleozoic sutures XVI and XXVI, because they are today located within the Cimmeride orogenic zone and any other tectonic connections they may have had are now even less clear.

Rifting associated with the beginning separation of the Cimmerian continent from Gondwana-Land commenced during the Permian, although actual continental separation took place diachronously from east to west in the earliest (in the Himalaya) to middle Triassic time (in Turkey) to open Neo-Tethys. Also, in about the same time interval the Cimmerian continent split longitudinally from central South Afghanistan all the way to Burma to open the Waser-Banggong Co-Nu Jiang ocean (BNJ in Figure 3B; Meso-Tethys of Belov 1981; also see Belov et al 1986). Later, in the late Jurassic-early Cretaceous interval this ocean closed along the Waser (Farah-Rud)/Rushan-Pshart/Banggong Co-Nu Jiang/Mandalay suture zones (VII-XXII-XXXVI).

By the early Triassic most of the Paleo-Tethyan subduction systems had been activated (Figure 3B). The earliest collisional contact between the Cimmerian continent and Laurasia was in northern Iran, along the Talesh-Mashhad suture (VI), in the late Triassic. However, because the position of southern Turan in Triassic time is now thought to have been considerably farther south than the rest of nuclear Asia at the time, it is not certain whether the Talesh-Masshad suture really marks the initial contact of the Cimmerian continent with Laurasia. It is likely that the western end of suture VIII' closed later, sometime during the Jurassic, and that this collision represented the final welding of the Cimmerian continent onto Laurasia. But the presence of an Early Jurassic Central Asiatic flora in Central Iran suggests that some kind of land bridge must have existed between Iran and the rest of Asia by that time (Şengör 1984). The Triassic closure along the Talesh-Mashhad suture probably also resulted in the splitting of the Cimmerian continent in Iran and isolated the Central Iranian microcontinent (j) (Şengör 1984), which later rotated about 135° counterclockwise (Figure 3C) (Soffel & Förster 1984).

Before the final closure of Paleo-Tethys, three back-arc opening events successively disrupted the Cimmerian continent in the eastern Mediterranean region. The first opened in latest Permian-earliest Triassic time along the Pindos-Budva (15) and Karakaya (II) suture traces. This event produced what is thought to be a small ocean basin, whose eastern half closed by latest Triassic time (Şengör et al 1984a), whereas the Pindos-Budva segment survived until the Oligocene. Alternatively, Şengör (1986c) suggested that the Karakaya suture trace may represent the rest of a continental-margin subparallel strike-slip zone along which a Hercynian Istanbul-Balkan fragment (Figure 3A) may have been transported from southern Laurasia to northern Gondwana-Land in the early Permian-middle Triassic interval. This interpretation provides one explanation for the otherwise anomalous late Carboniferous Laurasian flora known from the Istanbul-Balkan fragment (Kerey 1984) now lying to the south of the main Paleo-Tethyan suture in northern Turkey (IV).

The second opening in the Middle Triassic was the continuation of the Neo-Tethyan rifting that produced the present eastern Mediterranean (the southern branch of Neo-Tethys in Figure 3C). The third and latest opening event, during the Lias, generated a multiarmed ocean and linked with the opening Central Atlantic during the middle Jurassic (Vergely 1984,

Dercourt et al 1986), becoming the northern branch of Neo-Tethys (Figure 3C), whose remnants are today seen along the suture zones of the Mediterranean Alpides (2–11, 13, 14, 16, 17).

East of Iran, collision along the Paropamisus–Northern Pamir (VIII)–Southern Kuen-Lun (XVII)–Burhan Budai Shan–Anyemaqen Shan-Litien (XVIII–XIX) suture systems occurred mainly in early Jurassic time. In these regions intracontinental convergence continued long after the collision, as shown by widespread Jurassic compressional structures and persistent topographic highs until early Cretaceous times (Parrish et al 1982). Hallam (1984a) pointed out that the late Jurassic arid belt of western and central Asia (shown by evaporites in Figure 3C) may have developed in the rain shadow of the Cimmeride mountains that flanked the southern periphery of Eurasia at the time.

East of the 100°E meridian, the Cimmeride part of the Tethyside orogenic collage rather suddenly becomes very much enlarged, including the flysch- and mélange-filled Songpan-Ganzi accretionary complex of China (S in Figure 1). The suture zones, XX, XXI, XXVIII, XXXI, XXXIV, and XXXVI had already formed by latest Triassic time and had welded the four major continental blocks of the eastern Tethyside collage onto Laurasia (Şengör 1986b). Suture XXXIII probably closed by the Lias.

Continued tightening of this collage further constricted the Songpan-Ganzi accretionary complex, beneath which Paleo-Tethyan ocean floor subduction was continuing "subcutaneously" ("hidden subduction"; cf. Şengör 1984) under the peripheral arcs of the Songpan-Ganzi system until the early Cretaceous. Paleomagnetic evidence shows that the tightening of the suture zones and the Songpan-Ganzi system resulted in a 1650 ± 750 km northward motion of Annamia and South China in the Jurassic (Achache 1984).

By the early Cretaceous, the formation of the Cimmeride orogenic complex and the elimination of Paleo-Tethys were completed by the closure of the Shilka suture (XXXVII) and the destruction of the Waser–Banggong Co–Nu Jiang ocean. The latter was preceded by the obduction of a vast ophiolite sheet southward onto the Lhasa block (Figure 3C), largely covering it (Chang et al 1986b).

Primarily during the Aptian-Albian interval [locally already during the Jurassic in the Hellenides and in Ladakh (Roddick et al 1979, Dietrich et al 1983)], i.e. shortly after the final closure of the Waser–Banggong Co–Nu Jiang ocean, north-directed subduction of Neo-Tethyan ocean floor commenced all along the southern periphery of the pieces of the former Cimmerian continent, thus initiating the growth of the Alpides on the ruins of the Cimmerides. The story of the demise of Neo-Tethys as a partial compensation for the opening of the Atlantic and Indian oceans has been recently reviewed (Nairn & Stehli 1982, Sborschikov 1983, Mercier & Li 1984, Robertson & Dixon 1984, Vergely 1984, Dercourt et al 1986) and is not repeated here, although Figures 3D–F show its essential outlines.

DISCUSSION AND CONCLUSIONS

In the preceding section I sketched a very simplified evolutionary history of the Tethyside super orogenic complex, concentrating mainly on the agglomeration of its vast orogenic collage, similar to that of the North American Cordillera (Şengör 1984, 1985b); some 90% of this orogenic collage accreted to Eurasia during the Cimmeride evolution. The paleotectonic maps presented in Figures 3A–F were drawn only to facilitate the description and certainly not as final pictures, as emphasized above. Great uncertainties are implicit in all such maps with respect to initial reconstructions, motion paths of most of the collage components, palinspastic restoration of block shapes, and postcollisional distortion of entire collages and the host nuclei around which they accrete. Ziegler et al (1985) recently reviewed the problems associated with such paleotectonic/paleogeographic analyses.

In the tectonic evolution of the Tethysides we saw that the elimination of Paleo-Tethys (accomplished by the collision with Eurasia not only of the Cimmerian continent but also of a large orogenic collage that had accreted mostly around its eastern end) formed the huge, multibranched orogenic zone of the Cimmerides. By contrast, the closure of Neo-Tethys involved a much smaller number of continental blocks, and the Alpide orogenic zone consequently includes a much smaller collage; the Alpides were generated largely on the ruins of the Cimmerides. This overprinting is one of the main reasons why the recognition of the Cimmerides as an entirely independent Tethyan orogenic system has been so difficult.

Another reason for the late recognition of the Cimmeride orogenic system was that the late Paleozoic and Triassic Cimmeride events have so far been regarded as "Hercynian" events, whereas Jurassic and early Cretaceous events have been lumped into the "Alpine orogenic cycle." Thus, the traditional tripartite subdivision of all Phanerozoic orogenic phenomena into "Caledonian," "Hercynian," and "Alpine" cycles has resulted in an artificial splitting of related events in the Tethyan realm into different groups, hindering their correct recognition (cf. Şengör 1984, 1985b, 1986b).

A common problem in deciphering both the Cimmeride and the Alpide evolutionary histories is to ascertain the presence, sense, and amount of motion along latitudes of the constituent blocks of the Tethysides. While latitudinal motion is relatively easy to establish both from paleomagnetic and paleoclimatic data, longitudinal motion requires more circumstantial and diverse sets of evidence to be recognized. Figure 5A and Table 2 summarize our present knowledge on major amounts of sideways motion only along the sutures of the Tethysides, because such motions are the most difficult to recognize and map and thus they most affect our attempts at paleotectonic reconstructions. Strike-slip faults slicing across several units obliquely are relatively easier to catalog.

An inspection of Figure 5A and Table 2 shows that an absolute minimum of 60% (by length) of the Tethyside sutures had considerable strike-slip motion on them, more than 15% having experienced more than one major episode. Of these sutures, 25% had more than 1000 km of motion on them, and over 50% had more than a few hundred kilometers of motion. Table 2 shows that strike-slip motion along the trend of the orogen parallel with the sutures took place at all times with no recognizable temporal preference.

Figure 5B shows some of the complications that may result from both pre- and postcollisional, suture-subparallel strike-slip faults. In these series of diagrams, 3a shows, for example, the creation of ribbon continents attached to the mainland (such as Baja California) or completely isolated microcontinents (such as Madagascar). In 4a this geometry is compressed across the strike, and an anastomosing suture network is obtained. Some sutures in this case pass along the strike into fossil strike-slip systems that may be mistakenly identified as "cryptic sutures" forming "hallucinocryptic sutures." Segments of sutures labeled "fz" may contain ophiolites obducted across former fracture zones.

In the b series of figures in Figure 5B, the resulting geometry not only contains fossil strike-slip systems and abrupt "linkages" (1 in 5b) in the suture geometry but also sudden reversals of suture polarity (S) along a segment of an otherwise continuous suture zone. Both the a and b series of figures in Figure 5B produce geometries common along the Tethyside suture network.

Figure 5C illustrates another complexity associated with suture-parallel strike-slip faulting: It is a geological map of a rather well-mapped paleo-strike-slip zone along the Paleo-Tethyan suture of the Caucasus. Its total outcrop length is less than 20 km, and had it not been fortuitously exposed because of the late uplift of the Dzirula Massif, we would have entirely missed it! Even now, although we know that major strike-slip motion did occur along the Paleo-Tethyan suture in the Caucasus, it is difficult to ascertain exactly how much, simply because there is no sufficient exposure.

Figure 5D (column 1) illustrates how major blocks (e.g. b) may be sliced up by strike-slip faulting during collision, and how syncollisional block pinning and rotation (as in block b_1) may induce reversal of strike-slip

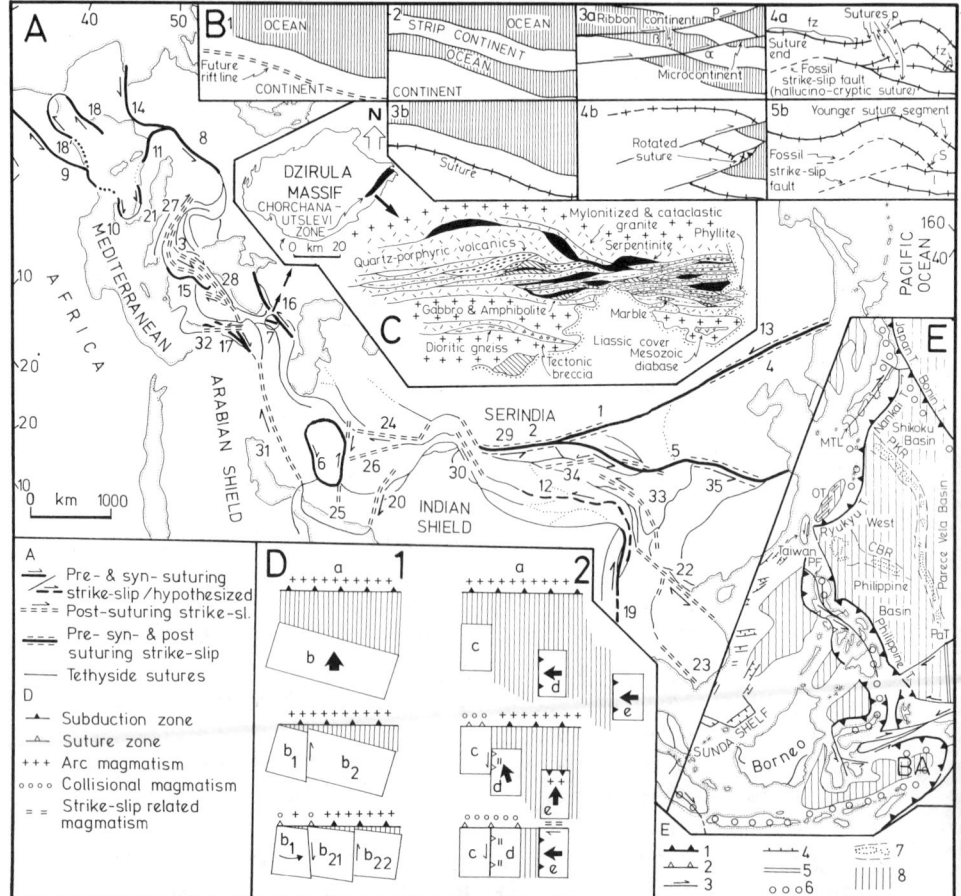

Figure 5 (*A*) Map showing pre-, syn-, and postcollisional strike-slip motion along the Tethyside sutures. For numbers and sources, see Table 2. (*B, D*) Diagrams showing possible complications introduced by strike-slip faulting during and after continental collision. (*C*) Sketch of the geology of the Chorchana-Utslevi sheared suture in the Dzirula Massif (redrawn after Adamia & Belov 1984). (*E*) Map showing the present-day tectonic settings of pre- and syncollisional strike-slip faults in eastern and southeastern Asia. Key to numbers: 1—subduction zone (teeth on upper plate), 2—young zone of collision, 3—strike-slip fault, 4—normal fault (ticks on hanging wall), 5—spreading center, 6—arc magmatism, 7—aseismic ridge, 8—oceanic area. Abbreviations: BA—Banda arc, CBR—Central Basin Ridge, MTL—median tectonic line, OT—Okinawa trough, PaT—Palau "Trench," PF—Philippine fault, PKR—Palau-Kyushu Ridge [modified after Şengör (1986c) using Tapponnier et al (1986)]. Discussion in text.

TECTONICS OF THE TETHYSIDES   235

Table 2 Sense, amount, and timing of strike-slip motion on some Tethyside sutures[a]

Suture[b]	Sense	Amount (km)	Time	Reference
1 (XI)	Left	?	ePale	Yang et al (1986)
2 (X+XI)	Left	?	?C–P	Bally et al (1980)
3 (II)	Right	>1000	C–?eTr	Şengör (1986c)
4 (XIV)	Left	>4000	P	Lin et al (1985a)
5 (XVIII–XXXI)	Left	4300±1200	P–eJ	Lin et al (1985a)
6 (25)	Right	~1000	lTr–lK	Davoudzadeh & Schmidt (1984)
7 (V′)	?	?	P–lTr	Adamia & Belov (1984)
8 (8)	Left	~1500	eJ–eK	Weissert & Bernoulli (1985)
9 (4–5)	Left / Right	10–20 / 10–20	e–mJ / lJ–eK	Mattauer et al (1977)
10 (7 south)	?	?	eJ	Catalano & D'Argenio (1982)
11 (7 north)	?	?	m–lJ	Gianelli & Principi (1977)
12 (XXII)	Left	>500	lJ–eK	This paper
13 (XIV)	Left	?	lJ–eK	Klimetz (1983)
14 (I)	Left	150–200	e–lK	Trümpy (1976)
15 (14)	?	?	lK	Şengör et al (1984b)
16 (23)	Left	350–700	lK–Pal	Şengör et al (1984a)
17 (17, 21)	Right	~1000	lK–eE	Y. Yılmaz & O. Sungurlu (personal communication, 1984)
18 (2) / (3)	Right / Left	~300	lE–lM	Wildi (1983)
19 (XXXVI)	Right	~460	lOl–present	Curray et al (1982)
20 (27)	Left	>200	u–mT	Lawrence & Yeats (1979)
21 (7 south)	(north) Left (south) Right	~50	T	B. D'Argenio (personal communication, 1980)
22 (XXVII)	Left / Right	~500 / A few tens	Ol–M / Q	Tapponnier et al (1986) / Allen et al (1984)
23 (XXV)	Left	~300	Ol–M	Tapponnier et al (1986)
24 (VIII)	Right	?	end E–?M	Tapponnier et al (1981)
25 (25)	Right	?	post eM	Tirrul et al (1983)
26 (VII)	Right	?	end E	Tapponnier et al (1981)
27 (14)	Right	>100	M–Pli	Burchfiel (1980)
28 (13, 16)	Right	>100	lM–present	Şengör (1979), A. M. C. Şengör & N. Görür (unpublished data)
29 (X, XI)	Left	500–600	M–present	Tapponnier et al (1981)
30 (30)	Right	~200	lT	Molnar & Tapponnier (1978)
31 (24)	Right	60	Pli–present	Tchalenko & Braud (1974)
32 (20, 21)	Left	~20	Pli–present	Şengör et al (1985)
33 (XXI, XXI′)	Left	?	lT	Molnar & Tapponnier (1975)
34 (XVIII)	Left	?	lT	Molnar & Tapponnier (1975)
35 (XXXI)	Left	Several hundred	post E	Peltzer et al (1985)

[a] Abbreviations: Pale—Paleozoic, C—Carboniferous, P—Permian, Tr—Triassic, J—Jurassic, K—Cretaceous, T—Tertiary, Pal—Paleocene, E—Eocene, Ol—Oligocene, M—Miocene, Pli—Pliocene, Q—Quaternary, e—early, m—middle, l—late.

[b] In the enumeration of sutures, left-hand numerals refer to those in Figure 5A and right-hand numerals (in parentheses) refer to those in Figure 1.

motion on the faults that segment the original block. Figure 5D (column 2) shows the generation of a complex sequence of suture-parallel strike-slip faulting during the assembly of an orogenic collage. In Figure 5E some of the rather bewildering complexities of pre- and syncollisional strike-slip faulting in orogenic collages are illustrated on the example of the present Southeast Asian subduction systems.

The data and the rather simple-minded models displayed in Figure 5 show how unrealistic, indeed how "fixist," as yet all reconstruction attempts such as those shown in Figures 3A–F must be and how critical it is to be able to recognize and map in detail all major strike-slip faults and precollisional sideways motions in an orogenic complex, especially those parallel with suture zones.

Not only strike-slip faulting but also major thrusting and extensive low-angle normal faulting may result in significant redistribution of crustal material in orogenic zones. This is one reason why the terrane concept, now so popular in the circum-Pacific orogenic belts, defined as a "fault-bounded geologic entity of regional extent that is characterized by a geologic history different from that of neighboring terranes" (Schermer et al 1984, p. 110) and treated as dispersed miniblocks that were later accreted to a host continent, is so dangerously misleading. It has long been realized that there is no necessary one-to-one correspondence between paleogeographic and structural units in orogenic belts, as emphasized more than three decades ago by Trümpy (1955) using the example of the Alps. That is why in mapping orogenic collages instead of "terranes" we should perhaps distinguish (a) primary collage components consisting of original nonsubductable lithospheric or crustal entities embedded in or lying on subductable oceanic lithosphere that are scraped off the underthrusting slab at subduction zones and accreted to the prow of the overlying plate, and (b) secondary collage components that are fault- (or any sort of structural discontinuity such as a slide) bound crustal or lithospheric fragments that form within an already assembled collage (Figure 6). Category (a) commonly includes microcontinents, island arcs, nonsubductable oceanic plateaus, and more rarely large guyots, aseismic ridges, and large deltas, whereas (b) involves compressional and extensional nappes and strike-slip-generated slivers (Şengör 1986c). All Tethyside blocks shown in Figure 1 belong to category (a), whereas blocks in Figure 2 belong to category (b). The "terrane" jargon regrettably has lumped these two entirely different classes of objects into one and thus has caused much unnecessary vexation.

Another interesting aspect of the evolution of the Tethysides is the remarkable correlation of times of major Cimmeride collisions with those of major falls in sea level, as pointed out by Şengör & Hsü (1984) and

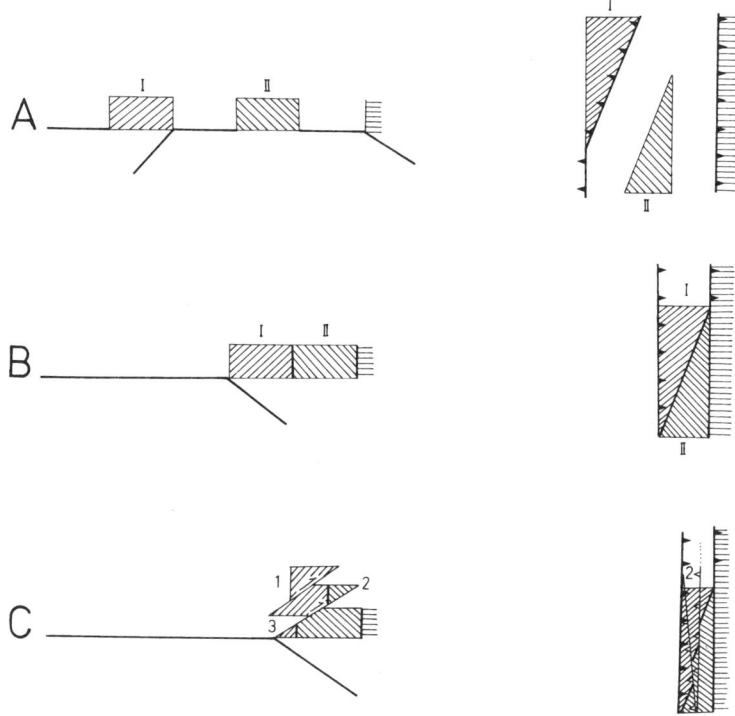

Figure 6 Primary (I and II) and secondary (1, 2, and 3) orogenic collage components.

Şengör (1985a). Although major Alpide collisions in the middle to late Eocene and the subsequent shortening also correlate with a sudden drop in worldwide sea level at the Rupelian/Chattian boundary (middle Oligocene), it is not possible to distinguish the effects of the Alpide collisions from the beginning growth of the Antarctic ice sheet at the same time. It appears that the Cimmeride collisions provide at least one possible factor affecting the early and middle Mesozoic sea-level changes (cf. Hallam 1984b). The creation of the long Alpide subduction zone during the early Cretaceous may have been a factor that led to the late Cretaceous speeding up of plate-motion rates and to the consequent major transgression. The late Jurassic spread of arid conditions into much of Eurasia may have been helped along also by the sudden Kimmeridgian sea-level drop that was possibly associated with the Cimmeride-related crustal shortening in Central Asia (Şengör 1985a).

Finally, the evolution of the Tethyside orogenic collage as a whole has disclosed a common (although not ubiquitous) trend in the evolution of

Pangea since at least the Permian, namely the progressive disintegration of the southern supercontinent, Gondwana-Land, and the northerly flight of its dispersed pieces to unite with Eurasia (Figures 3*A–F*). This tendency "to go north" is also seen in the variety of much smaller blocks that have been accreted to Japan and to the North American Cordillera, also since the Permian (Mascle & Marcoux 1982, Saleeby 1983). This large-scale and persistent northward migration of continental blocks appears significant and somehow must reflect the kinematics of the first-order convective circulation, at least in the upper mantle.

In summary, a review of the available field evidence from the entire Alpine-Himalayan mountain ranges shows that during early and middle Mesozoic time the Tethyan domain consisted of two independent oceanic realms separated by a continental strip, or perhaps an archipelago, called the Cimmerian continent. It began separating from northern Gondwana-Land mainly in Triassic time. Until the middle Jurassic it rotated counter-clockwise around the western apex of the triangular Pangean embayment, the Paleo-Tethys. This rotation, along with the collision both with the eastern part of the Cimmerian continent and with Laurasia of a number of continental blocks now making up much of China and Indochina, progressively eliminated Paleo-Tethys while Neo-Tethys evolved in the wake of the Cimmerian continent. Neo-Tethys, which is identical with Suess' "classical" Tethys, was later obliterated (mainly during the middle Cretaceous to the latest Cenozoic interval) by the collision with Eurasia of the dispersed pieces of Gondwana-Land.

The evolution of the Tethysides not only extensively and repeatedly deformed cratonic Asia, creating vast areas of commonly reactivated germanotype structures, but it also involved significant pre-, syn-, and post-collisional wideways motion of microcontinental blocks making up the Tethyside collage along the trend of the orogenic zone, rendering paleogeographic reconstructions extremely laborious. The analysis of the Tethyside collage also shows that major continental collisions significantly disrupt preexisting collage fabric and multiply fault-bounded structural entities, thus making the recently-developed "terrane analysis" misleading. Major worldwide sea-level lowerings correlate with times of major collisions in the Tethysides, as do widespread arid episodes in Central Eurasia.

An urgent need now exists for more detailed geological field work to map the Tethyside sutures and to establish geological criteria for the place of origin and paths of migration of the blocks they bound. Ages and amounts of major strike-slip displacement along the Tethyside sutures are among the most outstanding problems in the study of this immense orogenic complex.

ACKNOWLEDGMENTS

I thank K. J. Hsü for awakening my interest in the Paleo-Tethys. Fruitful discussions with him and with A. A. Belov, B. C. Burchfiel, K. Burke, Chang Cheng-fa, J. F. Dewey, A. Gansser, D. E. Karig, J. Marcoux, O. Monod, I. Ketin, V. E. Khain, W. S. F. Kidd, O. Sungurlu, R. Trümpy, and Y. Yılmaz contributed much to the maturation of my ideas on the evolution of the Tethysides. I am grateful to my wife Oya for patiently compiling the interminable references, for critically reading the manuscript, and for providing decisive inspiration at times of serious desperation.

Literature Cited

Achache, J. 1984. *Paléomagnetisme des zones actives. Croissance, destruction et formation des marges continentales*. Thèse de Doctorat d'Etat. Univ. Paris VII (unpaginated)

Adamia, S. A., Belov, A. A. 1984. Excursion 008A + C. Pre-Mesozoic complexes of the Caucasus. In *Georgian Soviet Socialist Republic, Excursions: 001, 007, 008, 012, 014, 017 Guidebook. Int. Geol. Congr., 27th, Moscow*, pp. 108–53. Tbilisi: Khelovneba

Aiello, E., Bartolini, C., Boccaletti, M., Gocev, P., Karagjuleva, J., et al. 1977. Sedimentary features of the Srednogorie zone (Bulgaria): an Upper Cretaceous intra-arc basin. *Sediment. Geol.* 19: 36–68

Allen, C. R., Gillespie, A. R., Han Yuan, Sieh, K. E., Zhang, Buchum, et al. 1984. Red River and associated faults, Yunnan Province, China. Quarternary geology, slip rates, and seismic hazard. *Geol. Soc. Am. Bull.* 95: 686–700

Antonijević, I., Grubić, A., Djordjević, M. 1974. The Upper Cretaceous paleorift in eastern Serbia. In *Metallogeny and Concepts of the Geotectonic Development of Yugoslavia*, pp. 315–39. Belgrade: Fac. Min. Geol., Belgrade Univ., Dep. Econ. Geol.

Argand, E. 1924. La Tectonique de l'Asie, *Congr. Géol. Int.*, *13th*, 1: 171–372. Liège: Vaillant-Carmanne

Asama, K. 1984. Gigantopteris flora in China and Southeast Asia. *Geol. Paleontol. SE Asia* 25: 311–23

Aubouin, J., Le Pichon, X., Monin, A. S., eds. 1986. Evolution of the Tethys. *Tectonophysics* 123: 1–315

Audley-Charles, M. G. 1984. Cold Gondwana, warm Tethys and the Tibetan Lhasa block. *Nature* 310: 165–66

Bakirov, A. A., Burtman, V. S. 1984. *Kirghiz Soviet Socialist Republic Excursion 032. Int. Geol. Congr., 27th, Moscow*. Frunze: "Kirgizstan." 74 pp.

Bally, A. W. 1984. Tectogenèse et sismique reflexion. *Bull. Soc. Géol. Fr., Ser. 7* 26: 279–85

Bally, A. W., Allen, C. R., Geyer, R. B., Hamilton, W. B., Hopson, C. A., et al. 1980. Notes on the geology of Tibet and adjacent areas—report of the American plate tectonic delegation to the People's Republic of China. *US Geol. Surv. Open-File Rep. No. 80-501*. 71 pp.

Bard, J. P. 1983. Metamorphism of an obducted island arc: example of the Kohistan sequence (Pakistan) in the Himalayan collided range. *Earth Planet. Sci. Lett.* 65: 133–44

Baud, A. 1985a. Tectogenesis of a segment of the Cimmerides: the volcano—sedimentary Triassic of Aghdarband (Kopeth Dagh, Iran). In *Tectonic Evolution of the Tethyan Regions, Abstracts*, p. 3. Istanbul: Istanbul Tech. Univ., Fac. Mines

Baud, A. 1985b. Geological observations on a geotraverse from the Sutlej to the Yarkand River (W. Tibet). In *Himalayan Workshop, Abstracts*. Leicester: Univ. Leicester, Dep. Geol. (unpaginated)

Belov, A. A. 1981. *Tektonicheskoye Razvitie Alpinskoi Skladchatoi Oblasti v Paleozoe*. Moscow: Nauka. 211 pp.

Belov, A. A., Gatinsky, Y. G., Mossakovsky, A. A. 1986. A precis on pre-Alpine tectonic history of Tethyan oceans. *Tectonophysics* 127: 197–211

Bender, F. 1983. *Geology of Burma*. Berlin: Gebrüder Borntraeger. 293 pp.

Berberian, M. 1983. The southern Caspian: a compressional depression floored by a trapped, modified oceanic crust. *Can. J. Earth Sci.* 20: 163–83

Berberian, M., King, G. C. P. 1981. Towards a paleogeography and tectonic evolution of Iran. *Can. J. Earth Sci.* 18: 210–65

Boulin, J. 1981. Afghanistan structure, greater India concept and eastern Tethys evolution. *Tectonophysics* 72: 261–87

Braakman, J. H., Levell, B. K., Martin, J. H., Potter, T. L., Van Vliet, A. 1982. Late Palaeozoic Gondwana glaciation in Oman. *Nature* 299: 48–50

Bucher, W. 1924. The pattern of the Earth's mobile belts. *J. Geol.* 32: 265–90

Buffetaut, E., Jaeger, J.-J., Rage, J.-C., eds. 1984. Paléogéographie de l'Inde, du Tibet et du Sud-Est asiatique: Confrontation des données paléontologiques avec les modèles géodynamiques. *Mém. Soc. Géol. Fr.* N.S. 147: 1–194

Burchfiel, B. C. 1980. East European Alpine System and the Carpathian orocline as an example of collision tectonics. *Tectonophysics* 63: 31–61

Burke, K. 1983. Is the Kapuskasing structure the site of a cryptic suture? In *Workshop on Cross Section of Archean Crust, LPI Tech. Rep. 83-03*, ed. L. D. Ashwal, K. D. Card, pp. 20–23. Houston: Lunar Planet. Inst.

Burke, K., Cooper, C., Dewey, J. F., Mann, P., Pindell, J. L. 1984. Caribbean tectonics and relative plate motions. *Geol. Soc. Am. Mem.* 162: 31–63

Burrett, C., Stait, B. 1985. South East Asia as a part of an Ordovician Gondwanaland—a paleobiogeographic test of a tectonic hypothesis. *Earth Planet. Sci. Lett.* 75: 184–90

Cameron, N. R., Clarke, M. C. G., Aldiss, D. T., Aspden, J. A., Djunuddin, A. 1980. The geological evolution of northern Sumatra. *Proc. Indones. Pet. Assoc. Congr., 9th*, pp. 149–87

Catalano, R., D'Argenio, B. 1982. Infraliassic strike-slip tectonics in Sicily and southern Apennines. *Rend. Soc. Geol. Ital.* 5: 4–10

Chan, L. S. 1986. Paleomagnetic research in Hong Kong. *Geol. Soc. Hong Kong Newsl.* 4: 17–22

Chang Cheng-fa, Pan Yu-shen, Sun Yi-ying. 1986a. The tectonic evolution of the Qinghai-Tibet plateau: a review. In *Tectonic Evolution of the Tethyan Regions*, ed. A. M. C. Şengör. In press

Chang Cheng-fa, Shen Nanfheng, Coward, M. P., Dewey, J. F., Deng Wanming, et al. 1986b. Royal Society–Academia Sinica 1985 geotraverse of Tibet: preliminary conclusions. *Nature*. In press

Crawford, A. R. 1974. The Indus Suture Line, the Himalaya, Tibet, and Gondwanaland. *Geol. Mag.* 111: 369–83

Ctyroky, P. 1973. Permian flora from the Ga'ra region (Western Iraq). *Neues Jahrb. Geol. Paläontol. Monatsh.* 7: 383–88

Curray, J. R., Emmel, F. J., Moore, D. G., Raitt, R. W. 1982. Structure, tectonics and geological history of the northeastern Indian Ocean. See Nairn & Stehli 1982, pp. 399–450

Davoudzadeh, M., Schmidt, K. 1984. A review of the Mesozoic paleogeography and paleotectonic evolution of Iran. *Neues Jahrb. Geol. Paläontol. Abh.* 168: 182–207

Davoudzadeh, M., Lensch, G., Weber-Diefenbach, K. 1986. Contribution to the paleogeography, stratigraphy and tectonics of the Infracambrian and lower Paleozoic of Iran. *Neues Jahrb. Geol. Paläontol. Abh.* 172: 245–69

Dercourt, J., Zonenshain, L. P., Ricou, L.-E., Kazmin, V. G., Le Pichon, X., et al. 1986. Geological evolution of the Tethys belt from the Atlantic to the Pamirs since the Lias. See Aubouin et al 1986, pp. 241–315

Dewey, J. F., Pitman, W. C. III, Ryan, W. B. F., Bonnin, J. 1973. Plate tectonics and the evolution of the Alpine System. *Geol. Soc. Am. Bull.* 84: 3137–80

Dickins, J. M. 1985. Palaeobiofacies and palaeobiogeography of Gondwanaland from Permian to Triassic. See Nakazawa & Dickins, pp. 83–92

Dietrich, V. J., Frank, W., Honegger, K. 1983. A Jurassic-Cretaceous island arc in the Ladakh-Himalayas. *J. Volcanol. Geotherm. Res.* 18: 405–33

Dietz, R. S., Holden, J. C. 1970. Reconstruction of Pangaea: breakup and dispersion of continents, Permian to present. *J. Geophys. Res.* 75: 4939–55

Fourcade, E., ed. 1985. Paléobiogéographie de la Téthys. *Bull. Soc. Géol. Fr., Ser. 8* 1: 623–790

Gatinsky, Y. G. 1986. *Lateralni Strukturno-Formachionniy Analiz*. Moscow: Nedra. 194 pp.

Geological Map of Tibet. 1980. 1: 1,500,000, 8 sheets. Chengdu, Sichuan, People's Republic of China

Gianelli, G., Principi, G. 1977. Northern Apennine Ophiolite: an ancient transcurrent fault zone. *Bull. Soc. Geol. Ital.* 96: 53–58

Girardeau, J., Marcoux, J., Fourcade, E., Bassoullet, J. P., Tang Youking. 1985. Xainxa ultramafic rocks, central Tibet, China: tectonic environment and geodynamic significance. *Geology* 13: 330–33

Görür, N., Oktay, F. C., Seymen, İ., Şengör, A. M. C. 1984. Paleotectonic evolution of the Tuzgölü basin complex, Central Turkey: sedimentary record of a neo-Tethyan closure. *Geol. Soc. London Spec. Publ.* 17: 467–82

Hallam, A. 1983. Supposed Permo-Triassic megashear between Laurasia and Gondwana. *Nature* 301: 499–502

Hallam, A. 1984a. Continental humid and arid zones during the Jurassic and Cretaceous. *Palaeogeogr. Palaeoclimatol. Palaeoecol.* 47: 195–223

Hallam, A. 1984b. Pre Quaternary sea-level changes. *Ann. Rev. Earth Planet. Sci.* 12: 205–43

Hamilton, W. B. 1979. Tectonics of the Indonesian region. *US Geol. Surv. Prof. Paper 1078.* 345 pp.

Helwig, J. E. 1974. Eugeosynclinal basement and a collage concept of orogenic belts. *Soc. Econ. Paleontol. Mineral. Spec. Publ.* 19: 359–76

Hennig, A. 1915. *Zur Petrographie und Geologie von Südwest Tibet, Southern Tibet*, Vol. 5. Stockholm: Lithogr. Inst. Gen. Staff Swed. Army. 220 pp.

Hsü, K. J., Bernoulli, D. 1978. Genesis of the Tethys and the Mediterranean. In *Initial Reports of the Deep-Sea Drilling Project*, ed. K. J. Hsü, L. Montadert, et al, 42(1): 943–49. Washington, DC: US Govt. Print. Off.

Kerey, I. E. 1984. Facies and tectonic setting of the Upper Carboniferous rocks of northwestern Turkey. *Geol. Soc. London Spec. Publ.* 17: 123–28

Klimetz, M. P. 1983. Speculations on the Mesozoic plate tectonic evolution of eastern China. *Tectonics* 2: 139–66

Kravchenko, K. N. 1979. Tectonic evolution of the Tien-Shan, Pamir and Karakorum. In *Geodynamics of Pakistan*, ed. A. Farah, K. A. De Jong, pp. 25–40. Quetta: Geol. Surv. Pak.

Lawrence, R. D., Yeats, R. S. 1979. Geological reconnaissance of the Chaman fault in Pakistan. In *Geodynamics of Pakistan*, ed. A. Farah, K. A. De Jong, pp. 351–61. Quetta: Geol. Surv. Pak.

Lee, G. 1984. *Paléomagnetisme des formations cretacées de la Corée du sud. Contraintes geodynamiques.* Thèse de Doctorat de 3ᵉ Cycle. Univ. Paris VII. 115 pp.

Lemoigne, Y. 1981. Flore mixte au Permien Supérieur en Arabie Saudite. *Géobios* 14: 611–35

Lin, J., Fuller, M., Zhang, W. 1985a. Preliminary Phanerozoic polar wander paths for the North and South China blocks. *Nature* 313: 444–49

Lin, J., Fuller, M., Zhang, W. 1985b. Paleogeography of the North and South China blocks during the Cambrian. *J. Geodyn.* 2: 91–114

Lin, Jin-Lu, Watts, D. 1986. Paleomagnetic evidence for the displaced terranes in the Tibetan plateau. In *Geol. Soc. London First Lyell Meeting: Gondwana and Tethys, Abstracts* (Unpaginated suppl.)

Livermore, R. A., Smith, A. G., Vine, F. J. 1986. Late Palaeozoic to early Mesozoic evolution of Pangaea. *Nature* 322: 162–65

Mascle, G., Marcoux, J. 1982. Analogies entre la Téthys mésogéen et la Téthys ouest americaine, conséquences. *Atti Accad. Naz. Lincei Cl. Sci. Fis. Mat. Nat. Rend.* 12: 373–79

Mattauer, M., Tapponnier, P., Proust, F. 1977. Sur les mécanismes de formation des chaines intracontinentales. L'exemple des chaines atlasiques du Maroc. *Bull. Soc. Géol. Fr.* (7) 19: 521–26

Mercier, J. L., Li Guangcen, eds. 1984. *Mission Franco-Chinoise au Tibet 1980.* Paris: CNRS. 433 pp.

Metcalfe, I. 1984. Late Palaeozoic palaeogeography of Southeast Asia: some stratigraphical, palaeontological and palaeomagnetic constraints. *GEOSEA V, Abstracts of Papers*, p. 20

Mitchell, A. H. G. 1986. The Shan Plateau and western Burma: Mesozoic-Cenozoic plate boundaries and correlations with Tibet. In *Tectonic Evolution of the Tethyan Regions*, ed. A. M. C. Şengör. In press

Molnar, P., Tapponnier, P. 1975. Cenozoic tectonics of Asia: effects of a continental collision. *Science* 189: 419–26

Molnar, P., Tapponnier, P. 1978. Active tectonics of Tibet. *J. Geophys. Res.* 83: 5361–75

Montenat, C., Blaise, J., Bordet, P., Debon, F., Deutsch, S., et al. 1981. Métamorphisme et plutonisme au Paléozique ancien en domaine gondwan sur la marge nord-ouest des Montagnes Centrales d'Afghanistan. *Bull. Soc. Géol. Fr.*, Ser. 7 23: 101–10

Morel, P., Irving, E. 1981. Paleomagnetism and the evolution of Pangea. *J. Geophys. Res.* 86: 1858–72

Nairn, A. E. M., Stehli, F. G., eds. 1982. *The Ocean Basins and Margins: The Indian Ocean*, Vol. 6. New York: Plenum. 776 pp.

Nakazawa, K. 1985. The Permian and Triassic Systems in the Tethys—their paleogeography. See Nakazawa & Dickins, pp. 93–111

Nakazawa, K., Dickins, J. M., eds. 1985. *The Tethys: Her Paleogeography and Paleobiogeography from Paleozoic to Mesozoic.* Tokyo: Tokai Univ. Press. 317 pp.

Neumayr, M. 1885. Die geographische Verbreitung der Juraformation. *Denkschr. Akad. Wiss. Wien Math.-Naturwiss. Cl.* 15: 57–114

Parrish, J. T., Ziegler, A. M., Scotese, C. R.

1982. Rainfall patterns and the distribution of coals and evaporites in the Mesozoic and Cenozoic. *Palaeogeogr. Palaeoclimatol. Palaeoecol.* 40: 67–101

Peltzer, G., Tapponnier, P., Zhitao, Z., Qin, X.-Z. 1985. Neogene and Quaternary faulting in and along the Qinling Shan. *Nature* 317: 500–5

Pindell, J. L. 1985. Alleghenian reconstruction and subsequent evolution of the Gulf of Mexico, Bahamas, and proto-Caribbean. *Tectonics* 4: 1–39

Pindell, J. L., Dewey, J. F. 1982. Permo-Triassic reconstruction of western Pangea and the evolution of the Gulf of Mexico/Caribbean region. *Tectonics* 1: 179–211

Robertson, A. H. F., Dixon, J. E. 1984. Introduction: aspects of the geological evolution of the eastern Mediterranean. *Geol. Soc. London Spec. Publ.* 17: 1–74

Roddick, J. C., Cameron, W. E., Smith, A. G. 1979. Permo-Triassic and Jurassic ^{40}Ar-^{39}Ar ages from Greek ophiolites and associated rocks. *Nature* 279: 788–90

Rowley, D. B., Lottes, A. L., Nie, S. Y., Yao, J. P., Ziegler, A. M. 1986. Mesozoic-Cenozoic tectonic evolution of China and adjacent Asia. *Geodyn. Symp., Mesozoic and Cenozoic Plate Reconstructions, Texas A & M Univ., Abstr.* (unpaginated)

Saleeby, J. B. 1983. Accretionary tectonics of the North American Cordillera. *Ann. Rev. Earth Planet. Sci.* 11: 45–73

Sborschikov, I. M. 1983. *Tektonicheskaya evolutsia vostochnoi chasti okeana Tetis.* DSc thesis (Avtoreferat). Akad. Nauk SSSR, Inst. Okeanol. im P. P. Shirshova, Moskow. 43 pp.

Schermer, E. R., Howell, D. G., Jones, D. L. 1984. The origin of allochthonous terranes: perspectives on the growth and shaping of continents. *Ann. Rev. Earth Planet. Sci.* 12: 107–31

Scotese, C. 1984. An introduction to this volume: Paleozoic paleomagnetism and the assembly of Pangea. In *Plate Reconstruction from Paleozoic Paleomagnetism, Geodyn. Ser.*, 12: 1–10. Washington, DC: Am. Geophys. Union

Şengör, A. M. C. 1979. Mid-Mesozoic closure of Permo-Triassic Tethys and its implications. *Nature* 279: 590–93

Şengör, A. M. C. 1984. The Cimmeride orogenic system and the tectonics of Eurasia. *Geol. Soc. Am. Spec. Pap. No. 195.* 82 pp.

Şengör, A. M. C. 1985a. The story of Tethys: how many wives did Okeanos have? *Episodes* 8: 3–12

Şengör, A. M. C. 1985b. Die Alpiden und die Kimmeriden: die verdoppelte Geschichte der Tethys. *Geol. Rundsch.* 74: 181–213

Şengör, A. M. C. 1986a. The dual nature of the Alpine-Himalayan System: progress, problems and prospects. *Tectonophysics* 127: 177–95

Şengör, A. M. C. 1986b. Die Alpiden und die Kimmeriden: die verdoppelte Geschichte der Tethys: Reply. *Geol. Rundsch.* 75: 203–12

Şengör, A. M. C. 1986c. Tectonic subdivisions and evolution of Asia. *Bull. Tech. Univ. Istanbul.* In press

Şengör, A. M. C., Hsü, K. J. 1984. The Cimmerides of eastern Asia: history of the eastern end of Paleo-Tethys. See Buffetaut et al 1984, pp. 139–67

Şengör, A. M. C., Monod, O. 1980. Océans sialiques et collisions continentales. *C. R. Acad. Sci. Paris* 290: 1459–62

Şengör, A. M. C., Yılmaz, Y. 1981. Tethyan evolution of Turkey: a plate tectonic approach. *Tectonophysics* 75: 181–241

Şengör, A. M. C., Yılmaz, Y., Sungurlu, O. 1984a. Tectonics of the Mediterranean Cimmerides: nature and evolution of the western termination of Palaeo-Tethys. *Geol. Soc. London Spec. Publ.* 17: 77–112

Şengör, A. M. C., Satır, M., Akkök, R. 1984b. Timing of tectonic events in the Menderes Massif, western Turkey: implications for tectonic evolution and evidence for Pan-African basement in Turkey. *Tectonics* 3: 693–707

Şengör, A. M. C., Görür, N., Şaroğlu, F. 1985. Strike-slip faulting and related basin formation in zones of tectonic escape: Turkey as a case study. In *Strike-Slip Faulting and Basin Formation, Soc. Econ. Paleontol. Mineral. Spec. Publ.*, ed. K. T. Biddle, N. Christie-Blick, 37: 227–64

Smith, A. G. 1973. The so-called Tethyan ophiolites. In *Implications of Continental Drift to the Earth Sciences*, ed. D. H. Tarling, S. K. Runcorn, 2: 977–86. London: Academic

Smith, A. G., Hurley, A. M., Briden, J. C. 1981. *Phanerozoic Paleocontinental World Maps.* Cambridge: Cambridge Univ. Press. 102 pp.

Soffel, H. C., Förster, H. G. 1984. Polar wander path of the Central-East-Iran microplate including new results. *Neues Jahrb. Geol. Paläontol. Abh.* 168: 165–72

Stauffer, P. H., Lee Chai Peng. 1986. Late Paleozoic glacial marine facies in Southeast Asia and its implications. *Geol. Soc. Malays. Bull.* 20: In press

Stille, H. 1948. Ur- und Neuozeane. *Abh. Deutsch. Akad. Wiss. 1945/46, Berlin, Math.-Naturwiss. Kl.*, Vol. 6. 68 pp.

Stöcklin, J. 1974. Possible ancient continental margins in Iran. In *The Geology of Continental Margins*, ed. C. A. Burk, C. L. Drake, pp. 873–87. Berlin: Springer-Verlag

Stöcklin, J. 1983. Himalayan orogeny and Earth expansion. In *Expanding Earth Symposium*, ed. S. W. Carey, pp. 119–30. Hobart: Univ. Tasmania

Stöcklin, J. 1984. The Tethys paradox in plate tectonics. In *Plate Reconstruction from Paleozoic Paleomagnetism, Geodyn. Ser.*, 12: 27–28. Washington, DC: Am. Geophys. Union

Suess, E. 1888. Die Entstehung der Meere. *Deutsch. Zeitung*, 4 Januar 1888, No. 5751. Wien: Beilage

Suess, E. 1893. Are great ocean depths permanent? *Nat. Sci.* 2: 180–87

Suess, E. 1901. *The Face of the Earth*, Vol. 3/I. Oxford: Clarendon. 400 pp. (English edition published in 1908)

Tapponnier, P., Mattauer, M., Proust, F., Cassaigneau, C. 1981. Mesozoic ophiolites, sutures, and large-scale tectonic movements in Afghanistan. *Earth Planet. Sci. Lett.* 52: 355–71

Tapponnier, P., Peltzer, G., Armijo, R. 1986. On the mechanics of the collision between India and Asia. *Geol. Soc. London Spec. Publ.* 19: 115–57

Tchalenko, J. S., Braud, J. 1974. Seismicity and structure of the Zagros (Iran): the main recent fault between 33 and 35°N. *Philos. Trans. R. Soc. London Ser. A* 277: 1–25

Thenius, E. 1981/1982. Das "Gondwana-Land" Eduard Suess 1885: der Gondwanakontinent in erd- und biowissenschaftlicher Sicht. *Mitt. Österr. Geol. Ges.* 74/75: 53–81

Tirrul, R., Bell, I. R., Griffis, R. J., Camp, V. E. 1983. The Sistan suture zone of eastern Iran. *Geol. Soc. Am. Bull.* 94: 134–50

Tollmann, A. 1984. Entshehung und früher Werdegang der Tethys mit besonderer Berücksichtigung des mediterranean Raumes. *Mitt. Österr. Geol. Ges.* 77: 93–113

Tollmann, A., Kristan-Tollmann, E. 1985. Paleogeography of the European Tethys from Paleozoic to Mesozoic and the Triassic relations of the eastern part of Tethys and Panthalassa. See Nakazawa & Dickins 1985, pp. 3–22

Trümpy, R. 1955. Wechselbezichungen zwischen Paläogeographie und Deckenbau. *Vierteljahresschr. Naturforsch. Ges. Zürich* 100: 217–31

Trümpy, R. 1976. Du Pèlerin aux Pyrénées. *Eclogae Geol. Helv.* 69: 249–64

Van der Voo, R., French, R. B. 1974. Apparent polar wandering for the Atlantic-bordering continents: Late Carboniferous to Eocene. *Earth Sci. Rev.* 10: 99–119

Van der Voo, R., Peinado, J., Scotese, C. R. 1984. A paleomagnetic re-evaluation of Pangaea reconstructions. In *Plate Reconstruction from Paleozoic Paleomagnetism, Geodyn. Ser.*, 12: 11–26. Washington, DC: Am. Geophys. Union

Vasiliev, Y. M., Milnichuk, V. S., Arabaji, M. S. 1981. *General and Historical Geology*. Moscow: Mir. 382 pp.

Vergely, P. 1984. Tectonique des ophiolites dans les Hellenides internes (*déformations, métamorphismes et phénomènes sédimentaires*). Consequences sur l'évolution des régions Téthysiennes occidentales. Vols. 1, 2. Thèse de Docteur. Univ. Paris-Sud, Centre d'Orsay. 250 pp., 411 pp.

Volchanskaya, I. K., Korytov, F. Y. 1982. Deep-seated faults of central and eastern Mongolia and the distribution of fluorite mineralization. *Int. Geol. Rev.* 24: 646–54

Vozenin-Serra, C. 1984. État de nos connaissances sur les flores du Paléozoique supérieur et du Mésozoique du Sud-Est asiatique. Interprétations paléogéographiques. See Buffetaut et al 1984, pp. 169–81

Waterhouse, J. B. 1982. An early Permian cool-water fauna from pebbly mudstones in south Thailand. *Geol. Mag.* 119: 337–432

Weissert, H. J., Bernoulli, D. 1985. A transform margin in the Mesozoic Tethys: evidence from the Swiss Alps. *Geol. Rundsch.* 74: 665–79

Westphal, M., Bazhenov, M. L., Lauer, J. P., Pechersky, D. M., Sibuet, J. C. 1986. Paleomagnetic implications on the evolution of the Tethys belt from the Atlantic Ocean to the Pamirs since the Triassic. See Aubouin et al 1986, pp. 37–82

Wildi, W. 1983. La chaîne tello-rifaine (Algérie, Maroc, Tunisie): structure, stratigraphie et évolution du Trias au Miocène. *Rev. Géogr. Phys. Géol. Dyn.* 24: 201–97

Wilson, J. T. 1963. Hypothesis of Earth's behaviour. *Nature* 198: 925–29

Windley, B. F. 1983. Metamorphism and tectonics of the Himalaya. *J. Geol. Soc. London* 140: 849–65

Wong, W. H. 1929. The Mesozoic orogenic movement of eastern China. *Bull. Geol. Soc. China* 8: 1–12

Yang Zunyi, Cheng Yuqi, Wang Hongzhen. 1986. *The Geology of China*. Oxford: Clarendon. 303 pp.

Zhang, Zh. M., Liou, J. G., Coleman, R. G. 1984. An outline of the plate tectonics of China. *Geol. Soc. Am. Bull.* 95: 295–312

Ziegler, A. M., Scotese, C. R., McKerrow, W. S., Johnson, M. E., Bambach, R. K. 1979. Paleozoic paleogeography. *Ann. Rev. Earth Planet. Sci.* 7: 473–502

Ziegler, A. M., Rowley, D. B., Lottes, A. L., Sahagian, D. L., Hulver, M. L., Gierlowski, T. C. 1985. Paleogeographic interpretation: with an example from the Mid-Cretaceous. *Ann. Rev. Earth Planet. Sci.* 13: 385–425

Ziegler, P. A. 1982. *Geological Atlas of Western and Central Europe.* The Hague: Shell Int. Pet. Maatsch. 130 pp.

Zonenshain, L. P., Le Pichon, X. 1986. Deep basins of the Black Sea and Caspian Sea as remnants of Mesozoic back-arc basins. See Aubouin et al 1986, pp. 181–211

TERRESTRIAL IMPACT STRUCTURES

Richard A. F. Grieve

Geophysics Division, Geological Survey of Canada, Ottawa, Ontario K1A 0Y3, Canada

INTRODUCTION

Impact structures are the dominant landform on planets that have retained portions of their earliest crust. The present surface of the Earth, however, has comparatively few recognized impact structures. This is due to its relative youthfulness and the dynamic nature of the terrestrial geosphere, both of which serve to obscure and remove the impact record. Although not generally viewed as an important terrestrial (as opposed to planetary) geologic process, the role of impact in Earth evolution is now receiving mounting consideration. For example, large-scale impact events may have been responsible for such phenomena as the formation of the Earth's moon and certain mass extinctions in the biologic record.

The importance of the terrestrial impact record is greater than the relatively small number of known structures would indicate. Impact is a highly transient, high-energy event. It is inherently difficult to study through experimentation because of the problem of scale. In addition, sophisticated finite-element code calculations of impact cratering are generally limited to relatively early-time phenomena as a result of high computational costs. Terrestrial impact structures provide the only ground truth against which computational and experimental results can be compared. These structures provide information on aspects of the third dimension, the pre- and postimpact distribution of target lithologies, and the nature of the lithologic and mineralogic changes produced by the passage of a shock wave. They also provide data for cratering rate estimates and, in some cases, on the nature of the impacting body.

In this review, emphasis is placed on the nature of terrestrial impact structures, the criteria for their identification, and their contribution to constraining formational processes and cratering rate estimates. The relationship of large-scale impacts to Earth history is also considered. Further details of cratering mechanics, scaling relationships, and temporal aspects of the bombardment of the Earth can be found in reviews by Melosh (1980) and Shoemaker (1983).

BASIC CHARACTERISTICS

Although there are a number of research papers describing aspects of specific impact structures (Roddy et al 1977), there is unfortunately no single English language text that describes the basic geological and geophysical characteristics of terrestrial impact structures (cf. Masaitis et al 1980). These structures are characterized by a circular form and evidence for intense, localized, near-surface structural disturbance and brecciation. Geophysical modeling indicates that they do not have deep-seated roots. The disruption and brecciation lead to associated low seismic velocities and residual negative gravity anomalies (Pohl et al 1977). Their magnetic signature is often in the form of a disruption of the magnetic trends in the target rocks or a magnetic low (Dabizha & Ivanov 1978). Localized magnetic anomalies may also occur and have been attributed to a variety of impact and postimpact causes (Coles & Clark 1978). The principal criterion, however, for the recognition of an impact structure is the occurrence of so-called shock-metamorphic effects (French & Short 1968). These effects are largely related to the progressive destruction of mineral structure and to thermal effects associated with the passage of a shock wave.

Morphology

Terrestrial impact structures are subdivided into two morphological classes: simple and complex. Simple structures have the classic bowl shape exhibited by Meteor (Barringer) Crater, Arizona (Figure 1). They have an uplifted rim area, and in the freshest examples, this is overlain by an overturned flap of near-surface target rocks with inverted stratigraphy, which is in turn overlain by fallout ejecta. Drilling at a number of structures has indicated that the floor of the so-called apparent crater (Figure 1) is underlain by a lens of allochthonous unshocked and shocked target-rock breccia. Bounding the breccia lens are autochthonous brecciated and fractured target rocks. They define the so-called true crater, which is roughly parabolic in cross section (Figure 1). Shock-metamorphic effects in the autochthonous target rocks are limited to the lower wall and beneath the floor of the true crater (Figure 1). Simple structures have the morphometric

TERRESTRIAL IMPACT STRUCTURES 247

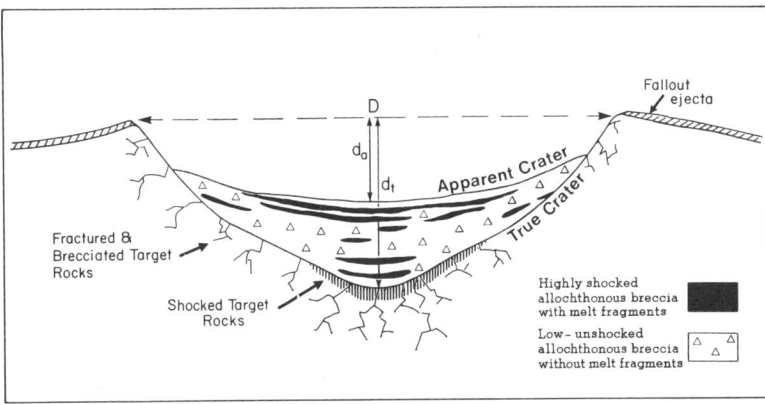

Figure 1 (*top*) Aerial photograph of Meteor (Barringer) Crater, Arizona, an example of a relatively young simple crater. (*bottom*) Schematic cross section of the principal elements of a simple crater in crystalline rocks. Notation defined in text.

relations

$$d_a = 0.14D^{1.02}, \quad (n = 18; \text{Pike 1980})$$

and

$$d_t = 0.29D^{0.93}, \quad (n = 9; \text{this work})$$

where D is the rim diameter, d_a and d_t are the apparent depth and true depth, respectively, n is the number of structures upon which the relation is based, and units are in kilometers (Figure 1). By comparison, lunar simple structures ($d_a = 0.196D^{1.01}$; Pike 1980) are deeper, reflecting the lower gravity of the Moon.

At diameters above ~ 2 km in sedimentary targets and ~ 4 km in crystalline targets, terrestrial impact structures have a complex form (Figure 2). The freshest examples are characterized by a structurally uplifted central area, exposed as a central peak and/or rings, surrounded by a peripheral depression and a faulted rim area. The peripheral depression is partly filled by allochthonous breccia and/or an annular sheet of so-called impact melt rocks. Shock-metamorphic effects in the autochthonous target rocks are present in the uplifted central area.

Various morphologic subtypes of complex structures identified on the other terrestrial planets [e.g. central peak craters, central peak basins, peak ring basins (Wood & Head 1976)] occur on Earth. Unequivocal terrestrial examples of multiring basins have not been identified. As many of the larger complex structures have been eroded to the extent that their principal morphologic elements are a mixture of structural and topographic features, morphometric comparisons with, for example, lunar structures (Pike 1985) should be made with caution. Complex structures are relatively shallow compared with simple structures, with the depth of the apparent crater given by

$$d_a = 0.27D^{0.16}, \quad (n = 11; \text{Pike 1980}).$$

There is considerable uncertainty attached to this formulation (± 0.11 on the exponent; Pike 1980) due to scatter in the data, and other such relationships have been given. There is also some evidence that complex structures in sedimentary targets are systematically shallower than those in crystalline targets (Grieve et al 1981).

The principal difference between complex and simple structures is the occurrence of an uplifted central core of shocked target rocks in the latter. From structures with stratigraphic and structural control, it is possible to define

$$SU = 0.06D^{1.1}, \quad (n = 10; \text{Grieve et al 1981})$$

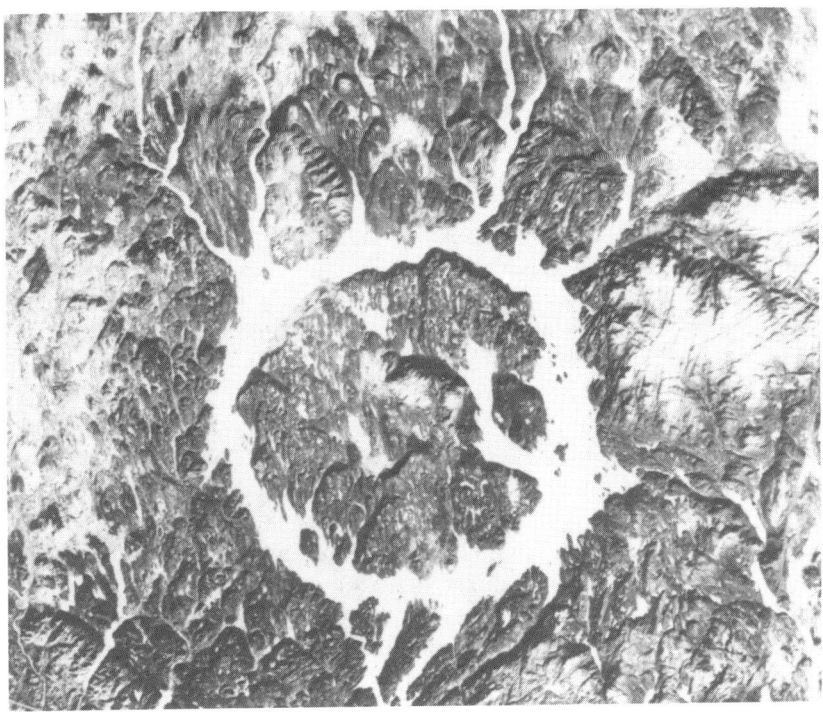
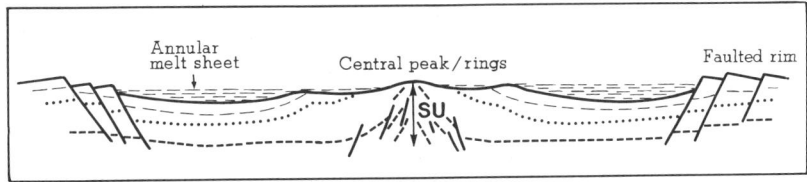

Figure 2 (*top*) LANDSAT image of the Manicouagan impact structure, Quebec, Canada. The annular lake has an outer diameter of ~70 km and surrounds an inner plateau capped by impact rocks and an uplifted central region. (*bottom*) Schematic cross section of the principal elements of a complex structure in crystalline rocks. Note the faulted rim area, the uplifted central area, and the relatively shallow nature compared with simple craters. The distance *SU* represents the net amount of structural uplift undergone by the deepest horizon now exposed in the central core.

where SU is the net amount of structural uplift undergone by the deepest horizon now exposed in the central core (Figure 2). This relationship is a minimum, as the amount of uplift decreases with depth (Brenan et al 1975) and erosion reduces the amount of observed uplift. It indicates, however, that an extensive vertical section of the upper crust is subjected to structural disturbance in the center of the largest structures (e.g. $SU \sim 9.5$ km for $D \sim 100$ km).

Shock Metamorphism

For typical terrestrial impact velocities of 15–25 km s^{-1}, the impacting body penetrates the target rocks to ~ 2–3 times its radius and transfers the bulk of its kinetic energy to the target, where it is partitioned into kinetic and internal energy. Energy transfer is by means of a hemispherically propagating shock wave, which decays exponentially with radial distance until it becomes an elastic wave. The peak pressures occurring close to the point of impact are of the order of several hundred gigapascals or several megabars. Considerable pressure-volume work is done by shock compression. Not all of this is recovered on rarefaction, and some is trapped as waste heat. Close to the point of impact, the latter is sufficient to raise the temperature of the target to several thousand degrees Celsius. These high pressures and temperatures lead to a series of changes known as shock metamorphism in the minerals and rocks of the target. Shock metamorphism differs from traditional endogenic metamorphism in the scales of pressures, temperatures, and time (Figure 3). It is characterized by very high strain rates and disequilibrium. With increasing pressure, shock effects are manifested as fracturing and cataclasis, plastic deformation, phase transitions, thermal decomposition, melting, and finally vaporization. The subject of shock metamorphism is dealt with in detail in the volume edited by French & Short (1968) and in work by Stöffler (1972, 1974).

Above the Hugoniot elastic limit, which is in the range of 2–12 GPa for silicates (Stöffler 1972), material compressed by the shock wave is returned to ambient pressure with reduced density upon rarefaction and has permanent changes with respect to its initial preshock state. The reduction in postshock density is due to the progressive breakdown of crystallographic order with increasing pressure. The most detailed observational and experimental studies in shock metamorphism have been on the common rock-forming tectosilicates—quartz and feldspars—where the most obvious and best-documented characteristic is the occurrence of planar features (Figure 4b). These microscopic sets of planes, a few microns in width, correspond to glide planes now generally filled by solid-state glass (von Engelhardt & Bertsch 1969). Planar features develop along a few specific crystallographic orientations, and experimental studies indi-

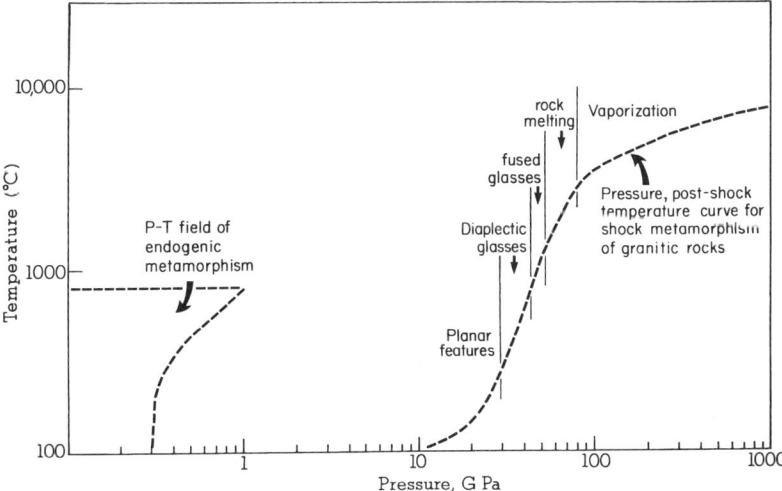

Figure 3 Pressure-temperature regime of shock metamorphism compared with that of endogenic metamorphism. Pressures and temperatures required for various shock-metamorphic features in granitic rocks are indicated.

cate that certain orientations develop in particular pressure ranges (Hörz 1968, Müller & Défourneaux 1968, Robertson 1975). Planar features in tectosilicates are developed over the range ~7.5–30 GPa, with the exact threshold pressure depending on mineral composition, structural state, exsolution lamellae, and alteration (Ostertag 1983). Accompanying planar feature development and with increasing shock pressure, there is increasing mosaicism, asterism, loss of birefringence, and glass formation. The only diagnostic megascopic shock feature in this general pressure range is a conical form of rock fracture known as shatter cones (Figure 4a). Although duplicated in experiments (Schneider & Wagner 1976, Roddy & Davis 1977), shatter-cone formation as a physical process has received little attention. These cones occur over a pressure range of 2–25 GPa, are developed best in fine-grained, structurally isotropic rocks, and are oriented with their apexes pointing in the direction of the origin of the shock wave (Milton 1977).

In the tectosilicates, there is a continuous structural change from crystals to glass, so that by ~30–45 GPa they are converted completely to so-called diaplectic glass (Stöffler & Hornemann 1972; Figure 4c). These solid-state glasses exhibit the original habit of the unshocked mineral and thus are also known as thetomorphic glasses. Metastable high-pressure polymorphs, such as stishovite, diamond, and other phases found on inversion

Figure 4 Shock-metamorphic features. (*a*) Shatter cones in quartzite, Sudbury, Ontario, Canada. (*b*) Microscopic planar features in quartz, Charlevoix, Quebec, Canada. Crossed polars. (*c*) Feldspar grain (grey) partially converted to diaplectic glass (black), Saint Martin, Manitoba, Canada. Crossed polars. (*d*) Erosional remnant of impact melt sheet at Mistastin, Newfoundland and Labrador, Canada. Cliff is ~80 m high.

(such as coesite) form in this pressure range. With increasing pressure, these glasses show signs of melting, so that by ~45–55 GPa many shocked rocks show evidence of flow, with inhomogeneous glasses corresponding to mixed mineral compositions.

Other silicates also show a progressive series of subsolidus shock-metamorphic effects involving the formation of planar features and/or mechanical twins. Onset pressures are not as well documented by experimentation as they are for the tectosilicates, although pressure calibration is possible by comparison with shock features in coexisting tectosilicates. Observations of shock features in the autochthonous rocks at terrestrial impact structures indicate that the recorded shock pressures beneath simple structures are a maximum of ~25 GPa and decrease with depth (Robertson & Grieve 1977). At complex structures, maximum pressures of ~35 GPa are recorded in the centrally uplifted autochthonous rocks and decay radially and with depth (Dence 1968).

Above ~60 GPa, postshock temperatures are sufficient to cause wholerock melting in the target, which leads to the formation of impact melt rocks (Figure 4d). Melt rocks occur as pockets up to a few tens of meters thick in the breccia lens of simple structures, as coherent annular melt sheets up to several hundred meters thick surrounding central peaks in complex structures, as glassy bombs in ejecta deposits, and as glassy clasts in breccias and veins beneath the floors of impact structures. Impact melt rocks are characterized by a high degree of compositional homogeneity corresponding to a mixture of the target rocks, even when occurring in volumes of hundreds of cubic kilometers (Grieve & Floran 1978, Masaitis et al 1980). They are also characterized by considerable textural inhomogeneity, particularly when they occur as coherent sheets. Melt sheets are heavily charged with shocked and unshocked lithic and mineral clasts ($>50\%$) at the top and bottom (Phinney et al 1978). Clast content decreases and matrix grain size increases toward the middle of impact melt sheets. Clasts are removed by resorption and reaction with the matrix melt, and their population is biased toward the more refractory lithologies and minerals.

The chemical homogeneity of melt sheets can be explained by their origin as a mixture of melted and vaporized target rocks, which are driven down into the expanding transient cavity under high-velocity turbulent flow. Textural inhomogeneity is due to the incorporation and later selective destruction of clasts and to variations in postimpact cooling history with vertical position in the melt sheet (Grieve et al 1977). Unlike igneous melts, impact melts are superheated. This is evidenced by their ability to remain liquid and resorb high contents of cold clasts (Onorato et al 1978) and by the occurrence of ultra-high-temperature breakdown minerals, such as baddelyite (El Goresy 1968).

In some cases, the melt rocks have higher K_2O/NaO ratios than the target rocks. This has led some to postulate that these rocks and their associated structures are the result of alkali igneous activity (Currie 1971). One possible explanation for this anomaly is selective elemental vaporization and condensation during melt and vapor formation (Basilevsky et al 1982). An alternative explanation is hydrothermal alteration (Dence 1971), particularly of highly shocked felsic clasts (Grieve 1978). This hypothesis is supported by experimentation, which indicates the rapid and total replacement of Na, and some Ca, by K in labradorite shocked to 26 GPa and exposed to KCl melt (Ostertag & Stöffler 1982).

Other excesses from average target-rock compositions have been noted for trace elements (such as Cr, Ni, and Co) at the parts per million level and other siderophiles (such as Ir) at the parts per billion level (Palme 1982, Basilevsky et al 1984). These enrichments represent the admixture of material from the impacting body. In some cases, the relative abundances of these elements in impact melt rocks has been used to identify the composition of the impacting body to the meteoritic group level, e.g. LL chondrite at the Brent crater (Palme 1982). In other cases, however, specific identification of the type of impacting body is equivocal. Lack of siderophile element enrichments in melt rocks does not necessarily imply the absence of a component of the impacting body. The compositions of basaltic achondrites, some irons, and stony irons are such that a small component of meteoritic material cannot be detected above background target values. In addition, as the total amount of vapor and melt increases with the square of the impact velocity (O'Keefe & Ahrens 1977), particularly high-velocity events will effectively dilute the fraction of the impacting body in the melt, which is seldom greater than a few percent at best.

FORMATIONAL PROCESSES

From experiments (Gault et al 1968, Stöffler et al 1975), calculations (Maxwell 1977, Orphal 1977), and observations (Dence et al 1977, Masaitis et al 1980), the process of forming simple structures is fairly well understood. Details and quantified aspects of cratering mechanics can be found in Croft (1980), Melosh (1980), Basilevsky et al (1983), and Grieve & Garvin (1984). Target material is compressed and accelerated downward and outward at initial particle velocities of several kilometers per second by the shock wave. Rarefaction waves, generated at free surfaces (such as the rear of the impacting body and the ground surface), follow the compressional wave and restore the compressed target material to ambient pressure. The rarefaction wave fronts are not parallel to the compressional wave, except in the volume directly below the impacting body. Consequently, the particle accelerations associated with rarefaction lead to the deflection of the initial downward and outward particle motions to upward and outward motions for that portion of the target near the surface. This results in the excavation of target material. Thus a cratering flow field is established, and a cavity known as the transient cavity is formed, partly by upward and outward excavation and partly by downward and outward displacement (Figure 5).

Cavity reconstructions suggest that this transient cavity has a depth to diameter ratio of $\sim 1/3$ (Dence et al 1977). As its name suggests, the transient cavity is short lived. In fact, it may never truly exist as a physical entity at the moment in time when there is both maximum downward displacement and maximum radial excavation (Schultz et al 1981). Whatever the case, the cavity walls very rapidly collapse inward to form the breccia lens observed at simple structures (Figure 5). In the freshest

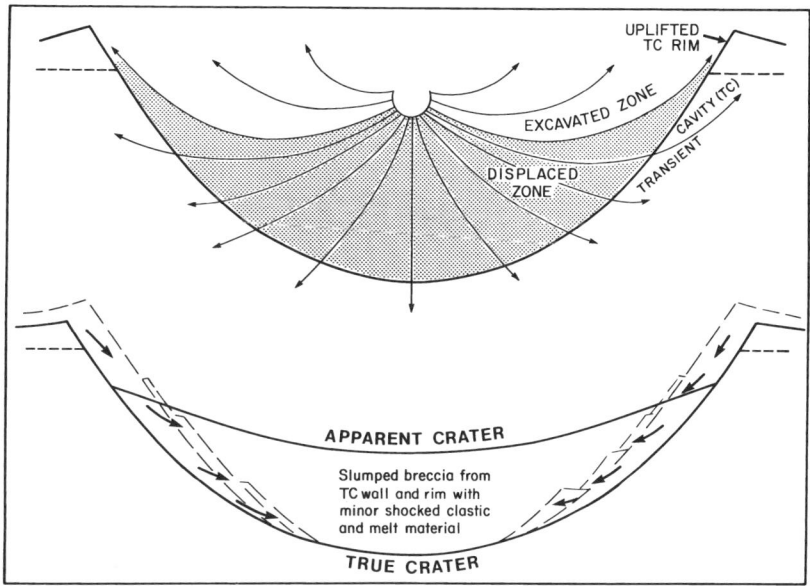

Figure 5 Formation of a simple crater. (*top*) Formation of transient cavity by a combination of excavation and displacement by the cratering flow field. Dashed line is original ground surface. (*bottom*) Subsequent collapse of transient cavity walls to form the final crater with allochthonous breccia lens. See text for details.

examples, this allochthonous breccia lens is overlain by a thin deposit of fallback breccia, which settles out of the ejecta cloud (Shoemaker 1960).

There is less consensus on the details of cratering mechanics at complex structures (Schultz & Merrill 1981). Difficulties in scaling small-scale experiments and extrapolating trends in computational models downgrade the constraints available. There is considerable observational evidence that morphologic features, such as uplifted central structures, indicate extensive structural movement. The majority of the models center on uplift and collapse hypotheses in which the original cavity produced by the cratering flow field undergoes considerable modification under plastic conditions (Melosh 1977; Figure 6). The driving force for modification has been considered to be gravity (Dence 1968, Melosh 1977) or a combination of gravity and elastic rebound (Ullrich et al 1977, Grieve & Robertson 1979).

The debate largely revolves around the validity of the concept of an initial deep transient cavity similar to that for simple structures (Roddy 1977, Pike 1980, Croft 1981). Data from terrestrial complex structures have been cited in favor of similar initial geometries (Dence et al 1977). Regardless of the details of transient cavity geometry, it is apparent that

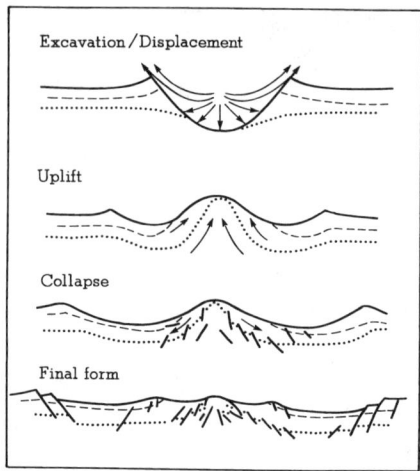

Figure 6 Model of the formation of a complex structure. Top to bottom represents increasing time following impact. Initial excavation and displacement is assumed to be similar to simple craters and is followed by considerable modification through uplift and collapse to produce the final form. See text for details.

deep excavation in complex structures is confined to the central region. The diameter of significant excavation is limited to 0.5–0.65 times the final diameter, which is determined by downfaulting during modification. Where there is sufficient control, estimates of the maximum depth of excavation indicate depths comparable to those of simple structures, provided they are normalized to the excavated rather than the final diameter (Grieve et al 1981). Observational data indicate the depth of excavation is $\sim 0.05D$. The question of the depth of excavation is particularly important in interpreting the provenance of lunar highland samples, which may be related to the formation of multiring impact basins, and in determining the effects of basin-sized events on lunar crustal stratigraphy (Spudis 1984).

THE KNOWN IMPACT RECORD

Small, young impact structures may have associated fragments of the impacting body, since relatively small bodies impact with reduced velocities due to atmospheric retardation or breakup (Melosh 1981). At larger structures, fragments of the impacting body are not expected, since the pressures and temperatures associated with undiminished impact velocities are sufficient to vaporize and melt the impacting body. The absence of physical evidence of the impacting body coupled with complexities in crater form at larger diameters and the modifying effects of erosion have resulted in considerable historical controversy over the origin of many terrestrial impact structures. A variety of internal processes, such as cryptoexplosions of gas, have been cited (Bucher 1963). A considerable literature now exists on impact phenomena, however, such that only a few geologists question the existence of large terrestrial impact structures without associated meteorite fragments (McCall 1979). Recognition of new structures is no longer solely the province of experts in the field of impact processes, and new discoveries are now being made of the order of five per year.

On the basis of the recognition of meteorite fragments and/or shock-metamorphic features, the current listing of known terrestrial impact structures indicates ages from Precambrian to Recent and diameters from $D \sim 140$ km to a few tens of meters (Table 1). Known structures are restricted to the exposed land. Oceanic impact structures or their derivatives must exist, but the current level of detailed knowledge of ocean-floor topography and geology is insufficient to lead to the straightforward discovery of submarine impact structures. Although impact is a spatially random process, the location of known impact structures is not. The concentrations on the cratons of North America and Europe (Figure 7) are the result of the relatively long-term geologic stability and active study and search programs in these areas.

There is also a bias toward younger structures; over 50% of the known structures with $D > 10$ km have ages < 200 Myr (Table 1). This is the result of erosion, which can rapidly render the crater form and the underlying structurally disturbed target unrecognizable. The effect of erosion can be seen in the cumulative size-frequency distribution of terrestrial structures. For large diameters the distribution approximates the relation $N \propto D^{-1.8}$, where N is cumulative number (Figure 8). This relation is considered to be the crater production rate on the other terrestrial planets (Neukum & Wise 1976). At smaller diameters, however, the distribution falls off (Figure 8), which indicates the removal of craters by erosion and difficulties in recognizing small structures.

If crater preservation is considered in terms of the depth to which the effects of impact are visible, the geologic signature of structures with $D > 20$ km that have not been protected from erosion by postimpact sedimentation and that occur in glaciated areas can be removed in as little as ~ 100 Myr (Grieve 1984). It is important, therefore, when commenting on terrestrial cratering rates and their variability through time to remember that the known sample is not necessarily representative, and it may be necessary to consider subsamples with restricted sizes, ages, and erosional histories.

Table 1 Terrestrial impact structures

Name	Latitude	Longitude	Diameter (km)	Age (Myr)
Amguid, Algeria	26°05'N	004°23'E	0.45	<0.1
Aouelloul, Mauritania[a]	20°15'N	012°41'W	0.37	3.1±0.3
Araguainha Dome, Brazil	16°46'S	052°59'W	40	<250
Azuara, Spain	41°01'N	000°55'W	30	<130
Barringer, Arizona, USA[a]	35°02'N	111°01'W	1.2	0.025
Bee Bluff, Texas, USA	29°02'N	099°51'W	2.4	<40
Beyenchime-Salaatin, Russian SFSR, USSR	71°50'N	123°30'E	8	<65
Bigatch, Kazakh SSR, USSR	48°30'N	082°00'E	7	6±3
Boltysh, Ukraine, USSR	48°45'N	032°10'E	25	100±5
Bosumtwi, Ghana	06°32'N	001°25'W	10.5	1.3±0.2
Boxhole, Northern Territory, Australia[a]	22°37'S	135°12'E	0.18	—
B.P. Structure, Libya	25°19'N	024°20'E	2.8	<120
Brent, Ontario, Canada[a]	46°05'N	078°29'W	3.8	450±30
Campo del Cielo, Argentina (20)[a,b]	27°38'S	061°42'W	0.09	—
Carswell, Saskatchewan, Canada	58°27'N	109°30'W	37	117±8
Charlevoix, Quebec, Canada	47°32'N	070°18'W	46	360±25
Clearwater Lake East, Quebec, Canada	56°05'N	074°07'W	22	290±20
Clearwater Lake West, Quebec, Canada	56°13'N	074°30'W	32	290±20
Connolly Basin, Western Australia, Australia	23°32'S	124°45'E	9	<60
Crooked Creek, Missouri, USA	37°50'N	091°23'W	5.6	320±80
Dalgaranga, Western Australia, Australia[a]	27°43'S	117°05'E	0.21	—
Decaturville, Missouri, USA	37°54'N	092°43'W	6	<300
Deep Bay, Saskatchewan, Canada	56°24'N	102°59'W	12	100±50
Dellen, Sweden	61°55'N	016°32'E	15	109.6±1
Eagle Butte, Alberta, Canada	49°42'N	110°30'W	10	<65
El'gygytgyn, Russian SFSR, USSR	67°30'N	172°05'E	23	3.5±0.5
Flynn Creek, Tennessee, USA	36°17'N	085°40'W	3.8	360±20
Glover Bluff, Wisconsin, USA	43°58'N	089°32'W	6	<500
Goat Paddock, Western Australia, Australia	18°20'S	126°40'E	5	<50
Gosses Bluff, Northern Territory, Australia	23°50'S	132°19'E	22	142.5±0.5
Gow Lake, Saskatchewan, Canada[a]	56°27'N	104°29'W	5	<250
Gusev, Russian SFSR, USSR	48°20'N	040°15'E	3	65

Name	Latitude	Longitude	Diameter (km)	Age (Myr)
Haughton, Northwest Territories, Canada	75°22′N	089°40′W	20	21.5 ± 1.2
Haviland, Kansas, USA[a]	37°35′N	099°10′W	0.011	—
Henbury, Northern Territory, Australia (14)[a,b]	24°34′S	133°10′E	0.15	—
Holleford, Ontario, Canada	44°28′N	076°38′W	2	550 ± 100
Ile Rouleau, Quebec, Canada	50°41′N	073°53′W	4	<300
Ilintsy, Ukraine, USSR	49°06′N	029°12′E	4.5	395 ± 5
Ilumetsy, Estonia, USSR	57°58′N	025°25′E	0.08	0.002
Janisjärvi, Russian SFSR, USSR	61°58′N	030°55′E	14	698 ± 22
Kaalijärvi, Estonia, USSR (7)[a,b]	58°24′N	022°40′E	0.11	0.004
Kaluga, Russian SFSR, USSR	54°30′N	036°15′E	15	380 ± 10
Kamensk, Russian SFSR, USSR	48°20′N	040°15′E	25	65
Kara, Russian SFSR, USSR[a]	69°10′N	065°00′E	60	57 ± 9
Karla, Russian SFSR, USSR	57°54′N	048°00′E	10	10
Kelly West, Northern Territory, Australia	19°30′S	132°50′E	2.5	<550
Kentland, Indiana, USA	40°45′N	087°24′W	13	<300
Kjardla, Estonia, USSR	57°00′N	022°42′E	4	510 ± 30
Kursk, Russian SFSR, USSR	51°40′N	036°00′E	5	250 ± 80
Lac Couture, Quebec, Canada	60°08′N	075°20′W	8	425 ± 25
Lac La Moinerie, Quebec, Canada	57°26′N	066°36′W	8	400 ± 50
Lappajärvi, Finland[a]	63°09′N	023°42′E	14	77 ± 4
Liverpool, Northern Territory, Australia	12°24′S	134°03′E	1.6	150 ± 70
Logancha, Russian SFSR, USSR	65°30′N	095°50′E	20	50 ± 20
Logoisk, Byelorussia, USSR	54°12′N	027°48′E	17	40 ± 5
Lonar, India	19°58′N	076°31′E	1.83	0.05
Machi, Russian SFSR, USSR (5)[b]	57°30′N	116°00′E	0.3	<1
Manicouagan, Quebec, Canada	51°23′N	068°42′W	100	210 ± 4
Manson, Iowa, USA	42°35′N	094°31′W	32	61 ± 9
Middlesboro, Kentucky, USA	36°37′N	083°44′W	6	<300
Mien, Sweden[a]	56°25′N	014°52′E	5	118 ± 3
Misarai, Lithuania, USSR	54°00′N	023°54′E	5	395 ± 145
Mishina Gora, Russian SFSR, USSR	58°40′N	028°00′E	2.5	<360
Mistastin, Newfoundland, and Labrador, Canada	55°53′N	063°18′W	28	38 ± 4
Monturaqui, Chile[a]	23°56′S	068°17′W	0.46	1
Morasko, Poland (7)[a,b]	52°29′N	016°54′E	0.1	0.01

Table 1 (continued)

Name	Latitude	Longitude	Diameter (km)	Age (Myr)
New Quebec, Quebec, Canada	61°17'N	073°40'W	3.2	<5
Nicholson Lake, Northwest Territories, Canada[a]	62°40'N	102°41'W	12.5	<400
Oasis, Libya	24°35'N	024°24'E	11.5	—
Obolon', Ukraine, USSR[a]	49°30'N	032°55'E	15	215±25
Odessa, Texas, USA (3)[a,b]	31°45'N	102°29'W	0.168	—
Ouarkziz, Algeria	29°00'N	007°33'W	3.5	<70
Piccaninny, Western Australia, Australia	17°32'S	128°25'E	7	<360
Pilot Lake, Northwest Territories, Canada	60°17'N	111°01'W	6	440±2
Popigai, Russian SFSR, USSR[a]	71°30'N	111°00'E	100	39±9
Puchezh-Katunki, Russian SFSR, USSR	57°06'N	043°35'E	80	183±5
Red Wing Creek, North Dakota, USA	47°36'N	103°33'W	9	200
Riacho Ring, Brazil	07°43'S	046°39'W	4	—
Ries, Fed. Rep. Germany[a]	48°53'N	010°37'E	24	14.8±0.7
Rochechouart, France[a]	45°30'N	000°56'E	23	160±5
Rogozinskaja, Russian SFSR, USSR	58°18'N	062°00'E	8	55±5
Rotmistrovka, Ukraine, USSR	49°00'N	032°00'E	2.5	140±20
Sääksjärvi, Finland[a]	61°23'N	022°25'E	5	<330
Saint Martin, Manitoba, Canada	51°47'N	098°32'W	23	225±40
Serpent Mound, Ohio, USA	39°02'N	083°24'W	6.4	<320
Serra da Canghala, Brazil	08°05'S	046°52'W	12	<300
Shunak, Kazakh SSR, USSR	42°42'N	072°42'E	2.5	12
Sierra Madera, Texas, USA	30°36'N	102°55'W	13	100
Sikhote Alin, Russian SFSR, USSR (122)[a,b]	46°07'N	134°40'E	0.0265	—
Siljan, Sweden	61°02'N	014°52'E	52	368±1
Slate Island, Ontario, Canada	48°40'N	087°00'W	30	<350
Sobolev, Russian SFSR, USSR[a]	46°18'N	138°52'E	0.05	—
Söderfjärden, Finland	63°02'N	021°35'E	5.5	<600
Spider, Western Australia, Australia	16°30'S	126°00'E	5	—
Steen River, Alberta, Canada	59°31'N	117°38'W	25	95±7
Steinheim, Fed. Rep. Germany	48°41'N	010°04'E	3.4	14.8±0.7
Strangways, Northern Territory, Australia[a]	15°12'S	133°35'E	24	<472
Sudbury, Ontario, Canada	46°36'N	081°11'W	140	1850±150
Tabun-Khara-Obo, Mongolia[a]	44°06'N	109°36'E	1.3	<30
Talemzane, Algeria	33°19'N	004°02'E	1.75	<3

Name	Latitude	Longitude	Diameter (km)	Age (Myr)
Teague, Western Australia, Australia	25°50'S	120°55'E	28	1685 ± 5
Tenoumer, Mauritania	22°55'N	010°24'W	1.9	2.5 ± 0.5
Ternovka, Ukraine, USSR	48°01'N	033°05'E	8	330 ± 30
Tin Bider, Algeria	27°36'N	005°07'E	6	<70
Ust-Kara, Russian SFSR, USSR	69°18'N	065°18'E	25	57 ± 9
Upheaval Dome, Utah, USA	38°26'N	109°54'W	5	—
Veevers, Western Australia, Australia[a]	22°58'S	125°22'E	0.08	<450
Vepriaj, Lithuania, USSR	55°06'N	024°36'E	8	160 ± 30
Vredefort, South Africa	27°00'S	027°30'E	140	1970 ± 100
Wabar, Saudi Arabia (2)[a,b]	21°30'N	050°28'E	0.097	—
Wanapitei Lake, Ontaria, Canada[a]	46°44'N	080°44'W	8.5	37 ± 2
Wells Creek, Tennessee, USA	36°23'N	087°40'W	14	200 ± 100
West Hawk Lake, Manitoba, Canada	49°46'N	095°11'W	2.7	100 ± 50
Wolf Creek, Western Australia, Australia[a]	19°10'S	127°47'E	0.85	—
Zeleny Gai, Ukraine, USSR	48°42'N	035°54'E	1.4	120 ± 20
Zhamanshin, Kazakh SSR, USSR[a]	48°24'N	060°48'E	10	0.75 ± 0.06

[a] Structures with meteoritic fragments or geochemical anomalies considered to have a meteoritic source.
[b] Sites with multiple craters, with (n) indicating number of craters. Diameter given corresponds to largest crater.

Cratering Rate Estimates

The ubiquitous nature of cratering in the solar system has resulted in the use of crater counts to estimate the age of unsampled planetary surfaces (Basaltic Volcanism Study Project 1981, pp. 1050–1129). Central to estimating absolute ages is knowledge of the crater production rate and its variation over geologic time for the planet in question. Estimates of the crater production rate with time depend on data from the Earth and Moon, the only bodies for which samples and thus ages are currently available. The lunar data are for relatively large numbers of structures on isotopically dated surfaces >3.0 Gyr. The terrestrial data are for relatively small numbers of structures with isotopic or biostratigraphic ages on relatively young surfaces.

It is insufficient, however, to simply count the number of craters greater than some size and older than some age to derive a terrestrial crater

Figure 7 Location of known terrestrial impact structures.

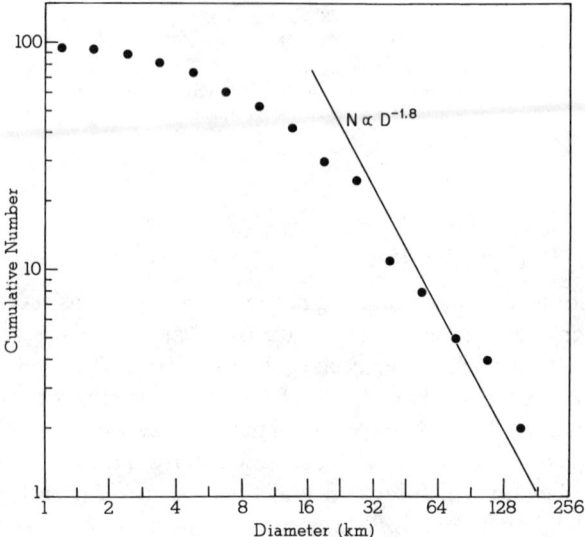

Figure 8 Cumulative size-frequency distribution of known terrestrial impact structures in Table 1. Note departure from $N \propto D^{-1.8}$ at smaller diameters, which indicates a shortage of known craters due to crater recognition and retention problems.

production rate. There are considerable uncertainties in many biostratigraphic ages, and there are also potential difficulties with K/Ar ages of impact melt rocks, which contain only partially degassed lithic and mineral clasts (Mak et al 1976). The major problems, however, are with crater recognition and retention. Average terrestrial cratering rate estimates have been calculated from well-dated, relatively young (< 120 Myr) large craters on the North American and European cratons ($5.4 \pm 2.7 \times 10^{-15}$ km^{-2} yr^{-1} for $D \geq 20$ km, $n = 7$; Grieve 1984) and from smaller craters on a nonglaciated area in the United States with a well-established geologic history ($2.2 \pm 1.1 \times 10^{-14}$ yr^{-1} km^{-2} for $D \geq 10$ km, $n = 4$; Shoemaker 1977). For $N \propto D^{-1.8}$, these independent rate estimates are comparable and similar to that derived from observations on bodies with Earth-crossing orbits ($2.4 \pm 2.2 \times 10^{-14}$ km^{-2} yr^{-1} for $D \geq 10$ km; Shoemaker 1983). All estimates have large uncertainties, reflecting concerns over the small number of observations and the completeness of the searches.

LARGE-SCALE IMPACT AND TERRESTRIAL EVOLUTION

Planetary exploration missions have provided evidence that the terrestrial planets were subjected to an episode of heavy bombardment during their early history. Although clear evidence of this episode is lacking on Earth and possibly Venus (Ivanov et al 1986), the question remains whether it had any lasting effect on terrestrial evolution. If a terrestrial crust and lithosphere did exist during the heavy bombardment, the potential numbers of impact structures can be estimated from the cratering record in the lunar highlands. After corrections for the larger gravitational cross section, higher impact velocity, and stronger gravity of the Earth, estimates for the number of potential impact basins with $D \geq 1000$ km range from 30–60 to 200 (Frey 1980, Grieve & Parmentier 1984). Estimated parameters such as the addition of exogenic energy and impact melt production on the early Earth, when averaged over the several hundred million years of bombardment, are of the same order of magnitude as present-day internal energy losses and island-arc volcanism (Grieve 1980). There is, however, a fundamental difference in the mode of energy release. Each major impact deposited orders of magnitude more energy than present-day internal energy losses at a specific location on time scales that are essentially instantaneous.

The immediate consequences of a basin-sized event would have been the establishment of topography on the order of a few kilometers, disturbance of the subimpact thermal gradient by the addition of postshock waste heat, and the uplift of originally deep-seated materials and considerable crustal fracturing. The combination of excavation, uplift, and a relatively thin lithosphere raises the possibility of impact-induced volcanism resulting from lithospheric thinning or removal and adiabatic decompression. Such a sequence has been suggested to account for the occurrence of the 1.85-Gyr-old Sudbury Igneous Complex (French 1970), and consideration of these effects led Frey (1980) to postulate that basin-sized impacts could produce proto-ocean basins and thus the basic crustal dichotomy of the Earth.

A number of processes would act to modify these impact basins (e.g. thermal subsidence, topographic degradation by erosion, and relaxation and basin loading). Some of these effects have been modeled for lunar basins with the result that lunar basins appear to have appreciable effects for 10^6–10^8 yr on the thermal, tectonic, and volcanic history in and around the basin (Bratt et al 1985, Solomon et al 1982). Calculations for a model 1000-km basin on an early Earth indicates conductive heat losses for 10^7 yr and postimpact thermal and isostatic subsidence of ~ 5–6 km (Grieve & Parmentier 1984). This time scale is similar to that for the Moon but does not account for the more vigorous heat losses associated with convection, which are likely in the terrestrial case. While it is clear that such basin-sized events would have appreciable direct and indirect effects on the early crust and lithosphere, it is not yet clear that they would be sufficiently long lived to significantly affect early crustal evolution.

Some authors have suggested relationships between more recent large impacts and such phenomena as magnetic reversals and plate movements (Glass et al 1979, Clube & Napier 1982). These suggestions remain unproven. It has also been suggested that large-scale impact may have been responsible for certain mass extinctions in the geologic record (de Laubenfels 1956, McClaren 1970). Evidence in favor of this hypothesis was first presented by Alvarez et al (1980) and Ganapathy (1980) in the form of relatively high abundances of Ir and other siderophile elements in the Cretaceous-Tertiary boundary clay layer at Gubbio, Italy, Stevns Klint, Denmark, and Woodside, New Zealand. Since these initial discoveries, numerous workers have confirmed a global enrichment in siderophile elements at Cretaceous-Tertiary boundary sites. Other evidence in favor of a major impact is the roughly chondritic abundances for the siderophiles (Kyte et al 1985), with isotopic data indicating that the boundary layer is distinct from normal detrital sediments (DePaolo et al 1983) and suggesting a meteoritic source (Luck & Turekian 1983). Physical evidence for an impact is given by the occurrence of shocked quartz and feldspar (Bohor et al 1984, Izett & Pillmore 1985), sanidine and microtektite spherules (Smit & Klaver 1981, Graup et al 1986), and high-temperature spinels

(Smit & Kyte 1984) at boundary-layer sites. The evidence is consistent with the global dispersion of projectile-contaminated, early-time ejecta from a major impact event (O'Keefe & Ahrens 1982). Mass-balance calculations suggest a projectile mass of 10^{17}–10^{18} g with associated kinetic energy of $\sim 10^{23}$ J. There is no known structure commensurate in both size (~ 100–200 km) and age (~ 65 Myr) with the values required for such an impact. This is not necessarily a fatal flaw in the hypothesis, as isotopic data suggest the presence of basaltic crustal materials within the boundary layer, implying an oceanic impact (DePaolo et al 1983).

A number of impact-induced extinction mechanisms have been suggested. The first is high dust loading of the stratosphere leading to the cessation of photosynthesis and breakdown of the food chain (Alvarez et al 1980). Model calculations suggest that light levels would be reduced below that required for photosynthesis for several months, and that there would be protracted cooling of the land surface (Pollack et al 1983). Another possibility is short-term heating of the atmosphere due to a greenhouse effect induced by water vapor from an oceanic impact (Emiliani et al 1981). Finally, the extinctions could have been induced through poisoning by noxious chemicals (Hsü 1980) or by the creation of large quantities of oxides of nitrogen by shock heating of the atmosphere, resulting in highly acidic rains (Lewis et al 1982). The hypothesis of impact-induced extinctions at the Cretaceous-Tertiary boundary has not gone unchallenged. Alternate hypotheses centering mainly on large-scale volcanism have been offered (McLean 1985, Officer & Drake 1985). Although there are still considerable uncertainties with respect to the details of the Cretaceous-Tertiary event, the observational evidence is consistent with the occurrence of a major impact, and thus an impact-related mass extinction remains a good working hypothesis.

Searches are currently underway for other siderophile anomalies in the geologic record. A correlation between microtektites, anomalous Ir, and the extinction of some late Eocene radiolaria has been reported (Sanfilippo et al 1985). An Ir anomaly has been detected in the Upper Devonian near the Frasnian-Famennian boundary, but it is possible that this anomaly has a terrestrial source (Playford et al 1984). There are also claims of Ir anomalies at the Permo-Triassic and Cambrian-Precambrain boundaries, but these results are preliminary.

It is tempting to speculate that large-scale impact was the forcing function for a number of, or possibly all, mass extinctions. Raup & Sepkoski (1984) statistically analyzed the marine extinction record over the last 250 Myr and determined a periodicity of 26 Myr with a phase of 13 Myr. A number of authors have suggested that periodic cometary showers, resulting from perturbations of the Oort cloud (the outer solar system source of comets) by an unseen solar companion or passage of the solar system through the galactic plane, could account for the periodic mass extinctions. Evidence in favor of periodic impacts has been presented through analyses of the ages of known craters by Alvarez & Muller (1984) and Rampino & Stothers (1984). They determined periodicities of 28.4 ± 1 Myr with a phase of 13 ± 2 Myr and 31 ± 1 Myr with a phase of 5 ± 6 Myr, respectively. From a review of the record, however, Grieve et al (1985) concluded that, because of its inherent nature, there were considerable uncertainties in using time-series analyses on the cratering record, and that a variety of periods could be defined depending on which sample of the record was considered most representative. They also argued that siderophile data suggested a variety of projectile types were responsible for the analyzed craters, and that the similarity in the cratering rate as determined from known craters and from observations of Earth-crossing Apollo bodies argued against periodic cometary showers.

SUMMARY

Approximately 120 terrestrial impact structures are currently known and several new structures are recognized each year. They are characterized by a circular form, severe surface and near-surface structural disturbance, and the occurrence of shock-metamorphic features. Projectile fragments are restricted to the smaller young structures, but larger structures may contain meteoritic material in the form of enriched siderophile abundances in impact melt rocks. The cratering process is fairly well understood for simple structures, but details for complex structures ($D > 4$ km) are less clear and additional work is required. Impact is a ubiquitous process in the solar system, and terrestrial impact structures provide the only ground-truth to understand large-scale impact as a geologic process. Consequently, recent research has centered less on the identification of new structures and more on the observational constraints terrestrial impact structures can provide to an understanding of impact phenomena. Unlike bodies such as the Moon, there is no clear evidence that large-scale impact had a lasting effect on the early crustal evolution of the Earth during the period of heavy bombardment. Modeling to date suggests fairly long-lived thermal and tectonic effects, but these models are preliminary.

There is good circumstantial evidence that a more recent large-scale impact affected severely the biosphere at the Cretaceous-Tertiary boundary. Details, however, of the killing mechanism(s) and the exact nature of the impact event still require research. There are some suggestions that large-scale impact was related to other global extinction events in the geologic record, but at present these are poorly documented; thus, categorical statements regarding terrestrial impact and other mass extinctions are premature. Similarly, suggestions that the Earth is subjected to periodic cometary showers require evidence beyond statistical arguments based on an incomplete cratering record. The potential for a relationship between large-scale impact events and climatological and biological changes is exciting and may ultimately result in a reevaluation of current thinking. Much work, however, remains to be done before large-scale impact can be considered the panacea for questions related to rapid changes in terrestrial biological evolution.

ACKNOWLEDGMENTS

This article is Geological Survey of Canada Contribution No. 21586.

Literature Cited

Alvarez, L. W., Alvarez, W., Asaro, F., Michel, H. V. 1980. Extraterrestrial cause for the Cretaceous-Tertiary extinction. *Science* 208: 1095–1108

Alvarez, W., Muller, R. A. 1984. Evidence from crater ages for periodic impacts on the Earth. *Nature* 308: 718–20

Basaltic Volcanism Study Project. 1981. *Basaltic Volcanism on the Terrestrial Planets.* New York: Pergamon. 1286 pp.

Basilevsky, A. T., Florensky, K. P., Yakolev, O. I., Ivanov, B. A., Fel'dman, V. I., Granovsky, L. B. 1982. Transformation of planetary material in high-speed collisions. *Geokhimiya* 1982(7): 946–60 (In Russian)

Basilevsky, A. T., Ivanov, B. A., Florensky, K. P., Yakolev, O. I., Fel'dman, V. I., et al. 1983. *Impact Craters on the Moon and Planets.* Moscow: Nauka. 200 pp. (In Russian)

Basilevsky, A. T., Fel'dman, V. I., Kapustkina, I. G., Kolesov, G. M. 1984. On the distribution of iridium in the rocks of terrestrial impact craters. *Geokhimiya* 1984(6): 781–90 (In Russian)

Bohor, B. F., Frood, E. E., Modreski, P. J., Triplehorn, D. M. 1984. Mineralogic evidence for an impact event at the Cretaceous-Tertiary boundary. *Science* 224: 867–69

Bratt, S. R., Solomon, S. C., Head, J. W. 1985. The evolution of impact basins: cooling, subsidence and thermal stress. *J. Geophys. Res.* 90: 12,415–33

Brenan, R. L., Peterson, B. L., Smith, H. J. 1975. The origin of Red Wing Creek structure, McKenzie County, North Dakota. *Wyo. Geol. Assoc. Earth Sci. Bull.* 8(3): 1–41

Bucher, W. H. 1963. Cryptoexplosion structures caused from without or within the Earth? ("Astroblemes" or "Geoblemes"). *Am. J. Sci.* 261: 596–649

Clube, S. V. M., Napier, W. M. 1982. The role of episodic bombardment in geophysics. *Earth Planet. Sci. Lett.* 57: 251–62

Coles, R. L., Clark, J. F. 1978. The central magnetic anomaly, Manicouagan structure, Quebec. *J. Geophys. Res.* 83: 2805–8

Croft, S. K. 1980. Cratering flow fields: implications for the excavation and transient expansion stages of crater formation. *Proc. Lunar Planet. Sci. Conf., 11th*, pp. 2347–78

Croft, S. K. 1981. The excavation stage of basin formation: a qualitative model. See Schultz & Merrill 1981, pp. 207–26

Currie, K. L. 1971. A study of potash fenitization around the Brent crater, Ontario, a Paleozoic alkaline complex. *Can. J. Earth Sci.* 8: 481–97

Dabizha, A. I., Ivanov, B. A. 1978. A geophysical model of the structure of meteorite craters and problems of the mechanics of crater formation. *Meteoritika* 37: 160–67 (In Russian)

de Laubenfels, M. W. 1956. Dinosaur extinction: one more hypothesis. *J. Paleontol.* 30: 207–18

Dence, M. R. 1968. Shock zoning at Canadian craters: petrography and structural implications. See French & Short 1968, pp. 169–84

Dence, M. R. 1971. Impact melts. *J. Geophys. Res.* 76: 5525–65

Dence, M. R., Grieve, R. A. F., Robertson, P. B. 1977. Terrestrial impact structures:

principal characteristics and energy considerations. See Roddy et al 1977, pp. 247–76

DePaolo, D. J., Kyte, F. T., Marshall, B. D., O'Neil, J. R., Smit, J. 1983. Rb-Sr, Sm-Nd, K-Ca, O and H isotopic study of Cretaceous-Tertiary boundary sediments, Caravaca, Spain: evidence for an oceanic impact. *Earth Planet. Sci. Lett.* 64: 356–73

El Goresy, A. 1968. The opaque minerals in impactite glasses. See French & Short 1968, pp. 531–53

Emiliani, C., Kraus, E. B., Shoemaker, E. M. 1981. Sudden death at the end of the Mesozoic. *Earth Planet. Sci. Lett.* 55: 317–34

French, B. M. 1970. Possible relations between meteorite impact and igneous petrogenesis, as indicated by the Sudbury structure, Ontario, Canada. *Bull. Volcanol.* 34: 466–517

French, B. M., Short, N. M., eds. 1968. *Shock Metamorphism of Natural Materials*. Baltimore: Mono. 644 pp.

Frey, H. 1980. Crustal evolution of the early Earth: the role of major impacts. *Precambrian Res.* 10: 195–216

Ganapathy, R. 1980. A major meteorite impact on the Earth 65 million years ago: evidence from the Cretaceous-Tertiary boundary clay. *Science* 209: 921–23

Gault, D. E., Quaide, W. L., Oberbeck, V. R. 1968. Impact cratering mechanics and structure. See French & Short 1968, pp. 87–99

Glass, B. P., Swincki, M. B., Zwart, P. A. 1979. Australasian, Ivory Coast and North American tektite strewnfields: size, mass and correlation with geomagnetic reversals and other Earth events. *Proc. Lunar Planet. Sci. Conf., 10th*, pp. 2535–45

Graup, G., Huth, J., Rast, U., Spettel, B. 1986. Microtektites at the Cretaceous-Tertiary boundary. *Lunar Planet Sci. XVII*, pp. 283–84 (Abstr.)

Grieve, R. A. F. 1978. The melt rocks at Brent crater, Ontario. *Proc. Lunar Planet Sci. Conf., 9th*, pp. 2579–2608

Grieve, R. A. F. 1980. Impact bombardment and its role in proto-continental growth on the early Earth. *Precambrian Res.* 10: 217–48

Grieve, R. A. F. 1984. The impact cratering rate in recent time. *J. Geophys. Res.* 89: B403–8 (Suppl.)

Grieve, R. A. F., Floran, R. J. 1978. Manicouagan impact melt, Quebec. 2. Chemical interrelations with basement and formational processes. *J. Geophys. Res.* 83: 2761–71

Grieve, R. A. F., Garvin, J. B. 1984. A geometric model for excavation and modification at terrestrial simple impact craters. *J. Geophys. Res.* 89: 11,561–72

Grieve, R. A. F., Parmentier, E. M. 1984. Impact phenomena as factors in the evolution of the Earth. *Proc. Int. Geol. Congr., 27th, Moscow*, 19: 99–114. Utrecht: VNU Science

Grieve, R. A. F., Robertson, P. B. 1979. The terrestrial cratering record: I. Current status of observations. *Icarus* 38: 211–29

Grieve, R. A. F., Dence, M. R., Robertson, P. B. 1977. Cratering processes: as interpreted from the occurrence of impact melts. See Roddy et al 1977, pp. 791–814

Grieve, R. A. F., Robertson, P. B., Dence, M. R. 1981. Constraints on the formation of ring impact structures based on terrestrial data. See Schultz & Merrill 1981, pp. 37–57

Grieve, R. A. F., Sharpton, V. L., Goodacre, A. K., Garvin, J. B. 1985. A perspective on the evidence for periodic cometary impacts on Earth. *Earth Planet. Sci. Lett.* 76: 1–9

Hörz, F. 1968. Statistical measurements of deformation structures and refractive indices in experimentally shock loaded quartz. See French & Short 1968, pp. 243–54

Hsü, K. 1980. Terrestrial catastrophe caused by cometary impact at the end of the Cretaceous. *Nature* 285: 201–3

Ivanov, B. A., Basilevsky, A. T., Kryuchkov, V. P., Chernaya, I. M. 1986. Impact craters on Venus: analysis of Venera 15 and 16 data. *J. Geophys. Res.* 91: D413–30 (Suppl.)

Izett, G. A., Pillmore, C. L. 1985. Shock-metamorphic minerals at the Cretaceous-Tertiary boundary, Raton Basin, Colorado and New Mexico provide evidence for an asteroidal impact in continental crust. *Eos, Trans. Am. Geophys. Union* 66: 1149–50 (Abstr.)

Kyte, F. T., Smit, J., Wasson, J. T. 1985. Siderophile interelement variations in the Cretaceous-Tertiary boundary sediments from Caravaca, Spain. *Earth Planet Sci. Lett.* 73: 183–95

Lewis, J. S., Watkins, G. H., Hartman, H., Prinn, R. G. 1982. Chemical consequences of major impact events on Earth. *Geol. Soc. Am. Spec. Pap. No. 190*, pp. 215–22

Luck, J. M., Turekian, K. K. 1983. Osmium-187/Osmium-186 in manganese nodules and the Cretaceous-Tertiary boundary. *Science* 222: 613–15

Mak, E. K., York, D. E., Grieve, R. A. F., Dence, M. R. 1976. The age of the Mistastin Lake crater, Labrador, Canada. *Earth Planet. Sci. Lett.* 31: 345–57

Masaitis, V. L., Danilin, A. N., Mashchak,

M. S., Raykhlin, A. I., Selivanoskaya, T. V., Shadenkov, Y. E. M. 1980. *The Geology of Astroblemes*. Leningrad: Nedra. 231 pp. (In Russian)

Maxwell, D. E. 1977. Simple Z model of cratering, ejection and overturned flap. See Roddy et al 1977, pp. 1003–8

McCall, G. J. H., ed. 1979. *Astroblemes— Cryptoexplosion Structures. Benchmark Papers in Geology*, Vol. 50. Stroudsburg, Pa: Dowden, Hutchinson & Ross. 437 pp.

McClaren, D. J. 1970. Time, life and boundaries. *J. Paleontol.* 44: 801–15

McLean, D. M. 1985. Deccan traps mantle degassing in the terminal Cretaceous marine extinctions. *Cretaceous Res.* 6: 235–59

Melosh, H. J. 1977. Crater modification by gravity: a mechanical analysis of slumping. See Roddy et al 1977, pp. 1245–60

Melosh, H. J. 1980. Cratering mechanics— observational, experimental, and theoretical. *Ann. Rev. Earth Planet. Sci.* 8: 65–93

Melosh, H. J. 1981. Atmospheric breakup of terrestrial impactors. See Schultz & Merrill 1981, pp. 29–36

Milton, D. J. 1977. Shatter cones—an outstanding problem in shock mechanics. See Roddy et al 1977, pp. 703–14

Müller, W. F., Défourneaux, W. 1968. Deformation structures in quartz as an indicator for shock: an experimental study on crystalline quartz. *Z. Geophys.* 34: 483–504 (In German)

Neukum, G., Wise, D. U. 1976. Mars: a standard crater curve and possible new time scale. *Science* 194: 1381–87

Officer, C. B., Drake, C. L. 1985. Terminal Cretaceous environmental effects. *Science* 227: 1161–67

O'Keefe, J. D., Ahrens, T. J. 1977. Impact-induced energy partitioning, melting, and vaporization on terrestrial planets. *Proc. Lunar Sci. Conf., 8th*, pp. 3357–71

O'Keefe, J. D., Ahrens, T. J. 1982. The interaction of the Cretaceous-Tertiary bolide with the atmosphere, ocean and solid Earth. *Geol. Soc. Am. Spec. Pap. No. 190*, pp. 103–20

Onorato, P. I. K., Uhlmann, D. R., Simonds, C. H. 1978. The thermal history of the Manicouagan impact melt sheet, Quebec. *J. Geophys. Res.* 83: 2789–98

Orphal, D. L. 1977. Calculations of explosion cratering—II. Cratering mechanics and phenomenology. See Roddy et al 1977, pp. 907–17

Ostertag, R. 1983. Shock experiments on feldspar crystals. *J. Geophys. Res.* 88: B364–76 (Suppl.)

Ostertag, R., Stöffler, D. 1982. Thermal annealing of experimentally shocked feldspar crystals. *J. Geophys. Res.* 87: A457–64 (Suppl.)

Palme, H. 1982. Identification of projectiles at large terrestrial impact craters and some implications of Ir-rich Cretaceous-Tertiary boundary layers. *Geol. Soc. Am. Spec. Pap. No. 190*, pp. 223–33

Phinney, W. C., Simonds, C. H., Cochran, A., McGee, P. E. 1978. West Clearwater, Quebec, impact structure, part II: petrology. *Proc. Lunar Planet Sci. Conf., 9th*, pp. 2659–93

Pike, R. J. 1980. Formation of complex impact craters: evidence from Mars and other planets. *Icarus* 43: 1–19

Pike, R. J. 1985. Some morphologic systematics of complex impact structures. *Meteoritics* 20: 49–68

Playford, P. E., McClaren, D. J., Orth, C. J., Gilmore, J. S., Goodfellow, W. D. 1984. Iridium anomaly in the Upper Devonian of the Canning Basin, Western Australia. *Science* 226: 437–39

Pohl, J., Stöffler, D., Gall, H., Ernstson, K. 1977. The Ries impact crater. See Roddy et al 1977, pp. 343–404

Pollack, J. B., Toon, O. B., Ackerman, T. P., McKay, C. P., Turco, R. P. 1983. Environmental effects of an impact-generated dust cloud: implications for the Cretaceous-Tertiary extinctions. *Science* 219: 287–89

Rampino, M. R., Stothers, R. B. 1984. Terrestrial mass extinction, cometary impact and the Sun's motion perpendicular to the galactic plane. *Nature* 308: 709–12

Raup, D. M., Sepkoski, J. J. 1984. Periodicity of extinctions in the geologic past. *Proc. Natl. Acad. Sci. USA* 81: 801–5

Robertson, P. B. 1975. Experimental shock metamorphism of maximum microcline. *J. Geophys. Res.* 80: 1903–10

Robertson, P. B., Grieve, R. A. F. 1977. Shock attenuation at terrestrial impact structures. See Roddy et al 1977, pp. 687–702

Roddy, D. J. 1977. Pre-impact conditions and cratering processes at the Flynn Creek crater, Tennessee. See Roddy et al 1977, pp. 277–308

Roddy, D. J., Davis, L. K. 1977. Shatter cones formed in large-scale experimental explosion craters. See Roddy et al 1977, pp. 715–50

Roddy, D. J., Pepin, R. O., Merrill, R. B., eds. 1977. *Impact and Explosion Cratering*. New York: Pergamon. 1301 pp.

Sanfilippo, A., Riedel, W. R., Glass, B. P., Kyte, F. T. 1985. Late Eocene microtektites and radiolarian extinctions on Barbados. *Nature* 314: 613–15

Schneider, E., Wagner, G. W. 1976. Shatter cones produced experimentally by impacts in limestone targets. *Earth Planet. Sci. Lett.* 32: 40–44

Schultz, P. H., Merrill, R. B., eds. 1981. *Multi-Ring Basins*. New York: Pergamon. 295 pp.

Schultz, P. H., Orphal, D. L., Miller, B., Borden, W. F., Larson, S. A. 1981. See Schultz & Merrill 1981, pp. 181–95

Shoemaker, E. M. 1960. Penetration of high velocity meteorites, illustrated by Meteor Crater, Arizona. *Int. Geol. Congr., 21st, Copenhagen*, 18: 418–34

Shoemaker, E. M. 1977. Astronomically observable crater-forming projectiles. See Roddy et al 1977, pp. 617–28

Shoemaker, E. M. 1983. Asteroid and comet bombardment of the Earth. *Ann. Rev. Earth Planet. Sci.* 11: 461–94

Smit, J., Klaver, G. 1981. Sanidine spherules at the Cretaceous-Tertiary boundary indicate a large impact event. *Nature* 292: 47–49

Smit, J., Kyte, F. T. 1984. Siderophile-rich magnetic spheroids from the Cretaceous-Tertiary boundary in Umbria, Italy. *Nature* 310: 403–5

Solomon, S. C., Comer, R. P., Head, J. W. 1982. The evolution of impact basins: viscous relaxation of topographic relief. *J. Geophys. Res.* 87: 3975–92

Spudis, P. 1984. Apollo 16 site geology and impact melts: implications for the geologic history of the lunar highlands. *J. Geophys. Res.* 89: C95–107 (Suppl.)

Stöffler, D. 1972. Deformation and transformation of rock-forming minerals by natural and experimental shock processes. I. Behaviour of minerals under shock compression. *Fortschr. Mineral.* 49: 50–113

Stöffler, D. 1974. Deformation and transformation of rock-forming minerals by natural and experimental shock processes. II. Physical properties of shocked minerals. *Fortschr. Mineral.* 51: 256–89

Stöffler, D., Hornemann, U. 1972. Quartz and feldspar glass produced by natural and experimental shock. *Meteoritics* 7: 371–94

Stöffler, D., Gault, D. E., Wedekind, J., Polkowski, G. 1975. Experimental hypervelocity impact into quartz sand: distribution and shock metamorphism of ejecta. *J. Geophys. Res.* 80: 4062–77

Ullrich, G. W., Roddy, D. J., Simmons, G. 1977. Numerical simulations of a 20-ton TNT detonation on the Earth's surface and implications concerning the mechanics of central uplift formation. See Roddy et al 1977, pp. 959–83

von Engelhardt, W., Bertsch, W. 1969. Shock induced planar deformation structures in quartz from the Ries crater, Germany. *Contrib. Mineral. Petrol.* 20: 203–34

Wood, C. A., Head, J. W. 1976. Comparison of impact basins on Mercury, Mars, and the Moon. *Proc. Lunar Sci. Conf., 7th*, pp. 3629–51

ORIGIN OF THE MOON—THE COLLISION HYPOTHESIS

D. J. Stevenson

Division of Geological and Planetary Sciences, California Institute of Technology, Pasadena, California 91125

INTRODUCTION

In 1871, during his presidential address to the British Association in Edinburgh, Sir William Thompson (later Lord Kelvin) discussed the impact of two Earth-like bodies, asserting that "when two great masses come into collision in space, it is certain that a large part of each is melted" [see Arrhenius (1908, p. 218) for the complete quotation]. Although he did not go on to speculate about lunar origin, it must have been remarkable to see one of the creators of the bastions of nineteenth century conservative science discuss such an apocalyptic event and the debris issuing from it. It is equally remarkable that until recently, lunar origin myths have usually centered around three possibilities (fission, capture, and binary accretion) that exclude any important role for giant impacts. The Origin of the Moon Conference held in Kona, Hawaii, on October 13–16, 1984, saw a megaimpact hypothesis of lunar origin emerge as a strong contender, not because of any dramatic new development or infusion of data, but because the hypothesis was given serious and sustained attention for the first time. The resulting bandwagon has picked up speed (and some have hastened to jump aboard). Most significantly, efforts have been made to simulate giant impacts using three-dimensional hydrodynamic codes on supercomputers. Although all this effort is promising, a sober reflection on the problem after two years suggests that a lot more work is needed. It is not yet clear whether the collision hypothesis satisfies the observational facts.

A definition is in order. By the impact or collision hypothesis, I mean any theory that seeks to derive the Moon-forming material from the outcome of one or more collisions between the Earth and other Sun-orbiting bodies. For reasons that will become apparent, the impacting body or bodies must be large—larger than the Moon and perhaps even larger than Mars. Notice that this definition does *not* assume that the formation of the Moon was necessarily a singular event. Among proponents of the collision hypothesis, there are those who think that a single event overwhelmingly dominated and those who think that a few (or even many) impact events were needed. There are even versions of the collision hypothesis that are not very different from extreme versions of one of the alternative origin scenarios of capture, fission, and binary accretion!

The mainstream view (if one can be said to exist) of one, or at most a few, oblique impacts ejecting material into Earth orbit owes its origin largely to the ideas of Hartmann & Davis (1975) and Cameron & Ward (1976). Hartmann & Davis were among the first to emphasize the possibility of very large "planetesimals" as part of the population of impacting bodies during Earth accretion, a possibility that is consistent with computer simulations by Wetherill (1980, 1985). Cameron & Ward were the first to assess the physical outcome of very large impacts and the important issues posed by the angular momentum budget; this has led to recent numerical simulations (Cameron 1985b, Benz et al 1986a,b). As often happens in this kind of interdisciplinary science, many other characters had (and have) roles to play and are introduced in due course. The biggest danger in this review, however, lies not in the possibility of inadequate attribution but in attempting to review an area where many of the important calculations are in progress or have not yet been done (perhaps cannot be done). This review proceeds by advancing 10 propositions that I believe embody the most important issues confronting the theory. These propositions may or may not be true, but they form a framework for asking the right questions and for organizing the presentation. Figure 1 summarizes the main features of the impact hypothesis embodied in these propositions.

TEN PROPOSITIONS

1. *The other theories of lunar origin are inadequate.* Fission is dynamically implausible; capture and binary accretion have both dynamical and cosmochemical problems.

Figure 1 Possible sequence of events leading to lunar formation. (*A*) A giant impact causes an expanding flow of liquid and vapor away from the impact site, carrying part of the angular momentum of the projectile. (*B*) A disk has formed, consisting of a liquid sublayer and a gas "atmosphere." (*C*) The disk has spread, and protomoons form at its extremity.

ORIGIN OF THE MOON 273

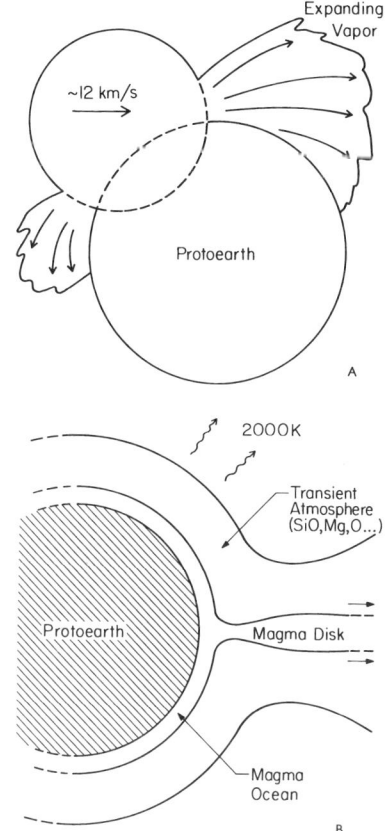

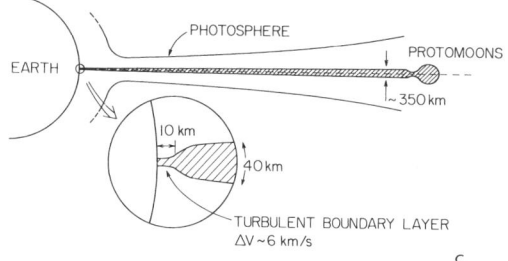

2. *Large impacts occurred during planetary accumulation.* There is no good reason to suppose that the masses of the impacting bodies were always much less than the masses of the resulting planets.
3. *Large impacts lead to qualitatively different outcomes than small impacts at the same velocity. In particular, orbital injection of material may occur.* The failure of scaling arises mainly because of the essential three dimensionality of large (oblique) impacts. Gradients in the gravitational field become important and hydrodynamic effects (especially pressure gradients) can operate over distances comparable to the radius of the Earth. These factors may be essential to the issue of orbital injection efficiency.
4. *The Earth's escape velocity is neither much less than nor much greater than the impact velocity needed for substantial vaporization of rock.* This is important because impact velocities are comparable to escape velocity, and vaporization is needed if pressure-gradient acceleration plays an essential role in orbital injection.
5. *The postimpact Earth may be like a brown dwarf star for about 100 yr.* An immense amount of energy may be dumped into the Earth, causing a transient global magma ocean and a transient atmosphere of silicate vapor. The Earth may radiate from an extended photosphere at $T \sim 2000$ K; such radiation would have been detectable by infrared astronomers orbiting nearby stars! The consequences for Earth evolution may provide a test for the theory.
6. *The material injected into orbit is very hot and probably consists of two phases (liquid with gas bubbles or "foam"). It may form a disk and spread rapidly.* Evolution times of a two-phase disk are very short; material may spread out to and beyond the Roche limit in 10^2–10^3 yr, before it has a chance to cool and solidify.
7. *The Moon-forming material may be derived primarily from the mantles of the projectile and the Earth and therefore iron poor.* Both projectile and target should be hot and well-differentiated bodies. Although not yet convincingly demonstrated by computer simulations, the outer layers of each may contribute most to the protomoon(s). The relative contributions of projectile and target are very uncertain.
8. *The newborn Moon or protomoon(s) are hot because of impact energy rather than because of their gravitational self-energy.* This is because everything happens so quickly that radiative and convective cooling are inadequate to allow solidification prior to Moon formation. The newborn Moon is then at least partially molten; this is relevant to lunar thermal history.
9. *Despite incongruent vaporization and localized differentiation, the system may be almost "closed."* The net hydrodynamic outflow to

infinity may be small, although important for devolatilization. However, major-element fractionation may be unimportant, except to the extent that *physical* separation (e.g. core formation or separation of liquid iron from liquid silicate) dictates the material available for the Moon.

10. *One Moon arises because the largest impact was the last important impact for supplying lunar material.* Accretion is hierarchical, and the largest impact may occur late. The resulting protomoon undergoes more rapid tidal evolution than smaller, earlier protomoons and sweeps up these smaller bodies.

Although not included in this list, an additional proposition is that the giant-impact hypothesis may have implications for comparative planetology, including Earth-Venus dissimilarities, the absence of a substantial moon around Mars, and the obliquities of Saturn, Uranus, and Neptune. These are assessed briefly.

The next section deals with observational data: What properties of the Moon do we seek to explain? This is followed by sections on planetary accretion and on the various lunar origin scenarios (defense of Proposition 1). The section following these (Physics of Large Impacts) motivates Propositions 3–6, and subsequent sections discuss recent and ongoing numerical simulations, efforts to understand the postimpact evolution, and the chemical aspects of the hypothesis.

OBSERVATIONAL DATA

A nice review of these data has been recently provided by Wood (1986). This section, although slanted differently, is accordingly kept to the bare essentials.

Mass and Angular Momentum

The lunar mass, one eightieth of the Earth's mass, seems an anomalously large fraction of the total Earth-Moon mass compared with other planet-satellite systems. However, this is of questionable significance. There are few terrestrial planets with which to compare, and two of these (Mercury, Venus) have been greatly affected by solar tides. It is probably inappropriate to compare the Earth-Moon system with outer solar system planets because the latter may have different satellite origins (e.g. Stevenson et al 1986) and certainly have a large gas component. Actually, the ratio of the total Jovian satellite system mass to the mass of the *core* of Jupiter (or the rock and ice component of Jupiter) is probably not much less than the Moon to Earth mass ratio. Consequently, there is no strong reason to suppose that the Earth-Moon system is "special."

For similar reasons, the angular momentum budget of the Earth-Moon system may not be particularly special or anomalous. We have simply too few other systems with which to compare. Nevertheless, this angular momentum (equivalent to that of an Earth rotating with a $\sim$4-hr period) is a very important constraint on origin models that is not always readily satisfied.

Bulk Chemistry

The Moon's mean density is 3.344 ± 0.002 g cm^{-3} (Bills & Ferrari 1977). If one constructs a body of lunar mass assuming cosmic Mg/Fe and Mg/Si ratios and appropriate combined oxygen, then the resulting mean density depends on the form of the iron (metal or oxide or substituting for Mg in silicates), but it is always at least 10% greater than observed. The only reasonable way to explain this discrepancy is by reducing the iron content by at least a factor of three relative to the cosmic abundance. This argument is independent of, but supported by, evidence that the Moon either has no iron-rich core or has a core that is at most $\sim$400 km in radius (corresponding to $\lesssim$2% of the mass). Constraints on the lunar core arise from a variety of arguments, many of which are geophysical (see Newsom 1984).

The depletion of iron is not in dispute, but its interpretation is still unclear because there is no consensus on the similarity of lunar bulk chemistry and Earth mantle composition. A refractory-rich Moon has had many advocates (Anderson 1972, Cameron 1972, Taylor & Jakes 1974, Ganapathy & Anders 1974), but the idea that the Moon is iron poor because of noncondensation of metallic iron in the region of lunar formation seems implausible, based on both cosmochemical and petrological considerations (Ringwood 1979). Trace-element abundances may be more diagnostic (see below), but they are controversial. Comparisons of the Moon with the *whole* mantle of the Earth are even more uncertain because the lower mantle is not yet well characterized (but see, for example, Jeanloz & Thompson 1983).

Volatile Depletion

Lunar soils and rocks are strongly depleted in volatiles, even more so than the Earth's mantle. It is widely believed that this depletion results from some highly energetic process accompanying lunar formation. Certainly, it cannot be attributed solely to the small size of the Moon, since at least one solar system body of comparable mass and size (Io) has a substantial volatile component. It is likely, however, that the volatile depletion of the Moon is not due to a single event and at least partially predates lunar formation (Taylor 1986). It is also possible that the degree of volatile depletion has been overestimated: There may be a significant component of volatile material deep within the Moon. We should be wary of a lunar origin scenario that extracts volatiles *too* efficiently. We should also be careful about terminology; molecules or atoms that are volatile in one thermodynamic or chemical environment may be involatile in another. Accordingly, each physical scenario has to be modeled directly and not loosely categorized merely by the degree of volatile depletion.

Trace Elements

Most attention has been given to siderophile elements (those that preferentially partition into a metallic iron phase and are therefore believed to be concentrated in planetary cores). Most, but by no means all, siderophiles are also volatile. Ringwood (1979) and Wänke and coworkers (Wänke & Dreibus 1986) have advanced the view that the similarity of siderophile *patterns* in the Earth and Moon argues for cogenesis (e.g. the derivation of the Moon from the Earth's mantle). However, there are differences in the patterns (Drake 1983, Kreutzberger et al 1986), so the inferences are unclear. In fact, it would be surprising if the patterns were extremely similar, since there may have been some further differentiation (e.g. lunar core formation) and further accretion of Sun-orbiting debris after the main lunar formation event(s).

The trace-element questions are both complex and important. It is probably a fair assessment at present to say that the data argue neither conclusively for nor conclusively against deriving lunar matter from the Earth's mantle or from the mantle of a body that has undergone geochemical differentiation similar to that of the Earth. Much more discussion of these issues can be found in several chapters of Hartmann et al (1986).

Primordial High Temperatures

The anorthositic highlands of the Moon have been frequently attributed to fractional crystallization from a primordial magma ocean (reviewed by Warren 1985). In fact, the magma ocean concept arose more out of geochemical convenience than from compelling physical or chemical arguments. Nevertheless, essentially global melting or extensive partial melting to a depth $\gtrsim 100$ km seems to be needed, although this melting need not have been uniform in space and time. It is questionable whether this could have been achieved from the gravitational energy of lunar formation (Wetherill 1975, Kaula 1979). The "magma ocean" is therefore a significant constraint on lunar formation models.

Orbital Evolution

The gradual increase of the Earth-Moon distance has long been known and has been directly measured by astronomical methods (Lambeck 1980,

Ch. 10). However, the backward extrapolation in time of tidal theory is highly uncertain, even aside from the well-known "problem" that the current specific tidal dissipation (or reciprocal of the quality factor Q) is higher than the average over geologic time. It is relatively easy to construct a model that brings the Moon back to the Roche limit at $\sim 4.5 \times 10^9$ yr before present (e.g. Walker et al 1983; see also Lambeck 1986, Walker & Zahnle 1986), but it is not possible to predict the configuration of this primordial orbit. In particular, a near-equatorial orbit cannot be excluded, even though specific calculations (e.g. Goldreich 1966) suggest an inclined orbit. This lack of prediction arises because of incomplete knowledge of tidal dissipation in both the Earth and the Moon, and because of the possibility that one or both bodies were significantly affected by later impacts.

PLANETARY ACCUMULATION

Modern ideas of solar system formation are guided by astrophysical observations and our improved understanding of planetary properties, and they are aided by the rapid recent developments in computing facilities. There are two very distinct views of planetary formation that currently receive the most attention. The less popular view, strongly advocated by Cameron (1985a), involves "giant, gaseous protoplanets" that arise through gravitational instability of the solar nebula. In the terrestrial zone, these bodies are believed to lose their gaseous component, leaving the rock and iron nuclei that are the building blocks of the terrestrial planets. Cameron asserts that there would have been many more nuclei than the current number of terrestrial planets, so that the subsequent evolution must have involved giant impacts between these (Mars-sized?) building blocks. Cameron's theory has received insufficient quantitative development to be assessed fully, but it appears to have some potential problems for explaining both the terrestrial and the giant planets (see, for example, Stevenson et al 1986). For our present purposes, it is sufficient to note that this theory probably provides the kind of impact that Cameron wishes to invoke for lunar origin (Cameron 1985b, 1986).

The more popular view of planetary formation assumes that the growth of solid bodies is by a sequential hierarchical process: condensation of dust grains, aggregation into larger clumps, formation of $\sim$ kilometer-sized planetesimals (possibly by gravitational instability), and progressive growth of larger planetesimals leading eventually to the planets. There are two main versions of this scenario: gas free and gas rich (see review by Wetherill 1980). In the gas-free scenario, which is the focus of the work done by Safronov (1966, 1969) and carried on by Wetherill, most of at least the later stages of terrestrial planetary accumulation occur in the absence of any primordial, hydrogen-rich nebula. The gas-rich scenario is mainly the work of Hayashi and collaborators (Hayashi et al 1985) and assumes that there is still sufficient gas, even at the later stages, to affect dynamical and thermal conditions of accumulation. Although the Hayashi group favor a nonimpact lunar origin, their theory is not necessarily inimical to an impact origin.

In any event, the central issue is the spectrum of planetesimal masses. Can we model the formation of the Earth by runaway growth of a single large embryo that sweeps up much smaller bodies, or are the impacting bodies not much smaller than the protoplanet? There is no unequivocal answer to this question, but current understanding tentatively favors the latter possibility. There are several steps and issues involved in reaching this assessment. First, one must understand the early evolution of a "gas" of small planetesimals, which may be initially monodisperse (i.e. about equal mass). This is best treated by kinetic theory, or "particles in a box," simulations. Although there are uncertainties in the collision physics, most controversy has centered around the correct treatment of gravitational stirring and scattering. Greenberg and coworkers have proposed "runaways," in which certain embryos grow much faster than neighboring bodies because the gravitational cross section can be much larger than the physical cross section when the encounter velocities are small compared with the escape velocity from the embryo (Greenberg 1982). A series of calculations by Wetherill and coworkers (Wetherill & Cox 1984, 1985, Stewart & Wetherill 1986, Wetherill & Stewart 1986) indicates that although runaway is conceivable, a substantial embryo is needed to initiate the process. The presence or absence of gas is not important, and the uncertainties in modeling the collisions do not affect the conclusion. It is important to realize that even if runaway occurs, one is left with a very large number of bodies, probably of lunar size, and not with just four planets. Impacts between large bodies would still take place.

In the absence of runaway, it is usually possible to approximate the mass spectrum as a power law, $N(m) \propto m^{-\alpha}$, where $N(m)\,dm$ is defined as the number of planetesimals with masses between m and $m+dm$. In many models, as discussed by Wetherill (1980), α lies between 5/3 (for which each mass decade contributes an equal amount of surface area) and 2 (for which each mass decade contributes an equal amount of mass). The value of α may be affected by the presence of gas but still be in this range. Suppose we were standing on the protoearth, maintaining an inventory of incoming bodies as they impacted. Table 1 shows what the accumulated inventory might look like. Each line in the table refers to roughly a decade in mass (so that "10^2 Iapetus" means "roughly 100 bodies each within a

Table 1 Illustrative inventories of planetesimals required to form the Earth

$\alpha = 2^a$	$\alpha = 5/3^b$
1 Mars	5 Mars
10 Moons	3 Moons
10^2 Iapetus	12 Iapetus
.	.
.	.
.	.
10^9 1-km planetesimals	10^5 1-km planetesimals
Total = 1 Earth mass	Total = 1 Earth mass

^a Equal amount of mass in each logarithmic mass interval down to 10^{-10} Earth masses.
^b Equal amount of cross-sectional area contributed by each logarithmic mass interval (lower mass cutoff unimportant for total mass).

factor of three of the mass of Iapetus"). Strictly speaking, this table contradicts Safronov's model (Safronov 1966, 1969), which has formed the basis for much of the work on planetary accumulation for more than a decade. Safronov argued that the second biggest body in a given accumulation zone is only $\sim 10^{-2}$ or 10^{-3} of the largest mass because of the enhanced (gravitational) cross section of the largest body. However, this is an overinterpretation of the Safronov model, which artificially isolates a preferred embryo planet in a specified zone of accretion and does not, therefore, make any pretense of explaining why there are 4 (rather than, say, 100) terrestrial planets.

There has never been a computer simulation that goes all the way from kilometer-sized planetesimals to fully grown planets. Ideally, one should use the output from kinetic ("particles in a box") calculations, in which the Keplerian character of the orbits is not important, as the input for an orbital simulation of later stages. Currently, this is computationally prohibitive. However, numerical simulations support many of the most important features of Safronov's (1969) model, especially the notion of a "steady-state" velocity. Safronov showed that the relative velocity of two planetesimals is of the order of the escape velocity from the largest planetesimal in the swarm. Guided by this theory and tests of its validity, Wetherill has performed simulations of planetary accumulation that come closest to satisfying the stated goals. Although other simulations exist (Lecar & Aarseth 1986) and analytical work continues (Horedt 1985, Hayashi et al 1965), the Wetherill simulations (Wetherill 1980, 1985) are closest to a realistic description of terrestrial planetary formation because they are fully three dimensional and involve only minor approximations in the gravitational physics, except for neglecting the effect of possible resonances.

One interesting feature of Wetherill's results is that they indicate that many aspects of the final outcomes are insensitive to most of the initial conditions, so the precise dovetailing of early to later stages may not be needed. In particular, his results are *not* strongly affected by whether he begins with a monodisperse system or a mass spectrum with $\alpha = 1.83$. Wetherill's most recent simulations, largely motivated by the giant-impact hypothesis of lunar origin, yield terrestrial planetary "systems" with characteristics rather similar to those observed (Wetherill 1985). He typically begins with an initial swarm of 500 bodies distributed between 0.7 and 1.1 AU, with initial eccentricity randomly distributed between 0 and 0.05 and inclinations between 0 and 0.025 radians, as expected from Safronov's theory (Safronov 1969). As the Monte Carlo simulation proceeds, bodies grow and eccentricities and inclinations increase. Figure 2 shows the most important aspect of these simulations for the question of lunar origin—namely, that many large impacts occur on the body that eventually becomes Earth. These bodies may be a few times the mass of Mars in some cases. Typical impact velocities are ~ 9 km s^{-1} (slightly less than Earth escape velocity because of the finite size of the projectile and the fact that the Earth has not yet reached its final mass). These impacts

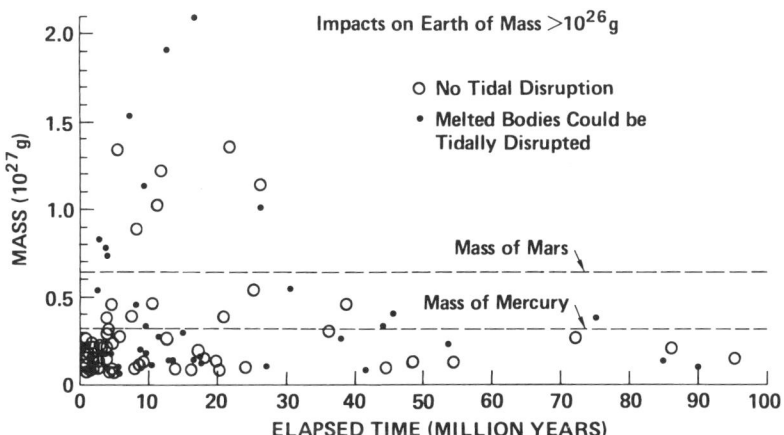

Figure 2 Combined results of 10 simulations of terrestrial planetary accumulation (Wetherill 1985), showing the time and size of giant impacts on Earth. In 5 of the simulations (labeled by open circles), tidal disruption of bodies previously impact melted is allowed for; in the other 5 simulations (closed circles), there is no tidal disruption.

are sufficiently common that they cannot be a special occurrence for Earth.

The final orientation of planetary spin axes (obliquities) and the orbital characteristics provide a measure of the role of large impactors. If all of the impacting mass is in the form of small bodies, then planetary orbital eccentricities, inclinations, and obliquities should all be small (leaving aside the subsequent, nonsecular perturbations among the planets). Wetherill's simulations provide reasonable eccentricities and inclinations because the large excursions excited by the largest impactors tend to be damped by the incoherent stream of smaller impactors. Analytical calculations (Harris & Ward 1982) suggest that the mass spectrum must be fairly "soft" (meaning much mass in smaller bodies, or $\alpha = 2$ in Table 1). Even in this case, Mars-sized impactors occur. The situation for obliquities is not so clear: Giant impacts would tend to randomize their values, yet the terrestrial planets tend to have a preferred (prograde) sense of spin. However, we are dealing with the statistics of small numbers, and two planets (Venus, Uranus) do have reversed spin. The pattern of obliquities generally supports the existence of large impacts, but the quantification of this is uncertain.

ORIGIN SCENARIOS

There are many reviews of lunar origin (most recently Boss 1986, Wood 1986). The main emphasis here is on the inadequacies of alternative ("conventional") origins. First, a philosophical point: It often seems that proponents of one or another theory adopt the attitude that their favorite theory is perfectly acceptable because it cannot be conclusively disproved. In this area of science, this is an invalid criterion. The origin of the Moon is a problem that involves many aspects and complications; it cannot be addressed by "sterilizing" it into an abstract dynamics problem in fission or capture or whatever. It is only by assessing the broad ramifications that a meaningful probability can be assigned. It is in this context that fission, intact capture, and possibly coformation (binary accretion) fail as satisfying explanations of lunar origin. I briefly discuss each of these alternatives in turn, and I also consider less clearly defined intermediate cases or modifications.

Fission

In the original Darwinian version (Darwin 1880), the protoearth rotates so rapidly that it is dynamically unstable to fission. There are two main problems with this: How do you create this state, and how do you explain the fact that the current angular momentum of the Earth-Moon system is lower than that needed for fission by a factor of at least three? There is nothing in astrophysical or solar system experience to suggest that the requisite high angular momentum can be supplied gradually during the planetary accumulation process. On the contrary, the net angular momentum influx during accretion is a subtle and small effect (as mentioned at the end of the last section; see also Harris 1977). On the second issue concerning the total angular momentum budget, this problem could in principle be avoided by initiating fission at an earlier stage (well before the protoearth approaches its final mass) or by guaranteeing that a large amount of mass escapes to infinity, carrying away excess angular momentum (much as satellites outside the Saturnian ring system can "soak up" the outward angular momentum flux present in the rings by gravitational torques). These possibilities cannot be disproved, but neither do they arise naturally in any self-consistent, fully developed theory of planetary accumulation. More recent work on fission has focused on the dynamics and the possibility of disk formation (Durisen & Scott 1984), but the fundamental objections remain unanswered.

"Fission" has also been invoked in a highly modified form by Ringwood (1966, 1979), and the word has been used, perhaps inappropriately, by Stevenson (1984a) to describe the spin-out of a superrotating atmosphere immediately after a giant impact. Although Ringwood's scenario has some dynamical difficulties, many aspects of his idea are remarkably similar to a possible aftermath of a giant impact. Since these "fission" proposals embody the physics of impact, they are more properly discussed later, when large impacts are described.

Capture

If collisions occur between the protoearth and bodies at least as large as the Moon, then close encounters are even more common. However, both Safronov's theory and the numerical simulations indicate that the encounter velocity (i.e. the velocity at infinity) is significant (typically up to a few kilometers per second), so that a substantial amount of kinetic energy must be dissipated. This proves to be not possible by tidal dissipation unless the encounter distance is so close that the body is disrupted. (This case is discussed further below.) One could envisage a scenario involving several, successive nondisruptive encounters and damping of the excess energy, but such a model would be contrived, since the problem is not just one of three bodies (Sun, Earth, Moon) but involves other scattering bodies that will guarantee incoherence and hence a predominance of scattering (i.e. an *increase* of encounter velocity rather than decrease). Gas drag could also be invoked (Nakazawa et al 1983), but this model is also contrived because it requires a very small encounter velocity and because the gas responsible for capturing the Moon must subsequently be removed rather quickly if the Moon is not to spiral inward and accrete onto the Earth. It is possible, however, that disruption does not occur if the viscosity of the planetesimal is too high (Mizuno & Boss 1985).

It seems almost superfluous to point out that no satisfactory explanation of lunar composition has arisen in models where the Moon is made elsewhere in the solar system. However, *disruptive* captures involving encounters within ~ 2 Earth radii and orbital injection of some debris might have happened, and this material may be preferentially from the mantle. Even though disruptive capture would only be important for small encounter velocities (Öpik 1972, Wood & Mitler 1974, Kaula & Beachey 1986, Hayashi et al 1985), it might contribute nonnegligibly to the Moon, provided disruption can occur at all (see Mizuno & Boss 1985).

Coformation (Binary Accretion)

Ruskol (1960), motivated by the ideas of Schmidt (see Ruskol 1982), pointed out that when two planetesimals collide within the Hill sphere of a planet, at least some of the debris may have both low enough energy and high enough angular momentum to end up in orbit. A circumplanetary disk can eventually form, fed by the debris of these collisions and, eventually, from collisions between the disk and later planetesimals. This disk can evolve, with one or more satellites forming at or beyond the Roche limit. This model has been developed further by Harris & Kaula (1975) and considered anew for lunar origin by Weidenschilling et al (1986).

The main virtue of this model is that it invokes a process that almost certainly happens, provided only that one accepts any of the hierarchical accumulation pictures of planetary formation. However, the model has three or four problems. First, it has difficulty explaining the iron depletion of the Moon. The possibility that the disk is a "compositional filter," which selectively excludes iron because of its greater strength or density, has been suggested, but this scenario appears to require very restrictive conditions to work, if it ever works. Second, the disk is cold and particulate; the Moon grows at the same rate as the Earth and is never very hot. Consequently, one cannot easily explain the hot, primordial Moon (or the "magma ocean"). Third, the disk is fed by roughly equal amounts of positive and negative angular momentum; it seems difficult to ensure that the material has the requisite angular momentum to make the Moon. The fourth point, related to the third, concerns the evolution of the disk: The natural time scale for redistributing angular momentum within the disk is short compared with the formation time of Earth. Is it possible to even maintain the disk?

In the final analysis, the significance of coformation may rest mainly on the mass spectrum of the planetesimals. As Stevenson et al (1986) discuss, it is only possible to put a significant amount of material into orbit by the Ruskol mechanism if most of the incoming mass is in small planetesimals. Given the other problems outlined above, it still seems likely that co-accretion is not *the* explanation of lunar origin, but at best a contributor of mass.

PHYSICS OF LARGE IMPACTS

The literature on impact physics is extensive but disappointing. It is extensive because impacts are an important planetary process and because there are obvious connections with the physics of explosions. It is disappointing because much of the work is empirical or "modular" (meaning that the person or persons responsible often do not understand what is going on in all parts of the calculation or interpretation because they are connecting together independently developed algorithms or procedures). Compilations of work in this area include Kinslow (1970), Roddy et al (1977), and the relevant subsection of Silver & Schultz (1982). The problem of a collisional lunar origin is even more challenging because it would have involved an impact far beyond any human experience. This means that any attempt to "scale" known impacts is probably doomed. A better understanding of the fundamentals is needed.

There are three sets of issues confronting this better understanding:

1. *Thermodynamics* (*equation of state*). How does the material respond to high shock pressures? In particular, what is the irreversible entropy production and the extent of melting and vaporization following shock release?
2. *Constitutive law* (*rheology*). What is the relationship between stress and strain during postimpact expansion? What viscosity (turbulent or otherwise) or small-scale instabilities characterize the macroscopic flow?
3. *Dynamics* (*equation of motion*). How does the shocked material flow, subject to nonuniform gravity, pressure gradients, and "viscous" stresses?

I discuss each of these issues in turn, but it is first valuable to motivate the important questions. If we wish to explain lunar origin by one or more giant impacts, then some material must be emplaced in an Earth orbit. In impact events, one normally thinks of the positive acceleration of debris as confined to the immediate vicinity of the impact site. Upward-moving debris is subsequently subject only to the r^{-1} gravitational potential of the (approximately) spherical Earth and suffers one of two fates. If it has a total (gravitational plus kinetic) energy that is positive, then it escapes on a hyperbolic trajectory. If it has a total energy less than zero, then it traverses a closed elliptical orbit that must eventually reimpact the Earth. This is illustrated in Figure 3a. Actually, reimpact of all negative energy debris also occurs if one allows for the higher order field of an oblate (rapidly rotating) Earth. The important point is that one needs a "second burn"—that is, some way of raising the periapse of negative energy material above the surface of the Earth. Each of the three issues listed above is connected to possible ways of achieving this "second burn."

Figure 3b shows one way to do this. Following a large impact in which significant vaporization occurs, the outflowing material is subject to pressure gradients as well as to gravity. This hydrodynamic effect increases the kinetic plus gravitational energy of the material but, more importantly, can also increase its angular momentum and lift the periapse above the surface. In order for this to happen, a significant pressure gradient must act over a substantial fraction of a planetary radius. (This is quantified more fully below.)

Figure 3c shows another way of achieving orbital injection. "Viscous" stresses within the jet of outgoing material redistribute angular momentum, allowing some material to achieve orbit at the expense of other material that loses angular momentum and falls back to Earth. This is a less plausible scenario than pressure-gradient acceleration because it requires a very large effective viscosity (sufficient to redistribute a large amount of angular momentum in just one orbital period).

Figure 3d shows a third way of achieving orbital injection, which involves gravitational torques (or, more generally, a severe time-dependent perturbation of the gravitational potential). Either a bulge on the planet or a separate body could transfer angular momentum to a more distant body, leading to orbital injection. This occurs in some recent numerical simulations (Benz et al 1986b) and is discussed further below and in the next section.

Thermodynamics of Impact

Contrary to what one might suppose from a cursory inspection of the literature, the thermodynamics of impacts are not well understood,

Figure 3 Schematic outcomes for outflow from a giant impact. (*a*) The material follows a purely Keplerian trajectory and must either escape or reimpact. (*b*) Pressure gradients near the impact site increase the angular momentum of (negative energy) material, which allows it to achieve orbital injection. (*c*) Redistribution of angular momentum within a viscous jet allows some material to gain angular momentum and achieve orbital injection at the expense of other material that reimpacts. (*d*) The gravitational torque exerted by the bulge or discrete body on the more distant protomoon allows the latter to gain angular momentum and avoid reimpact.

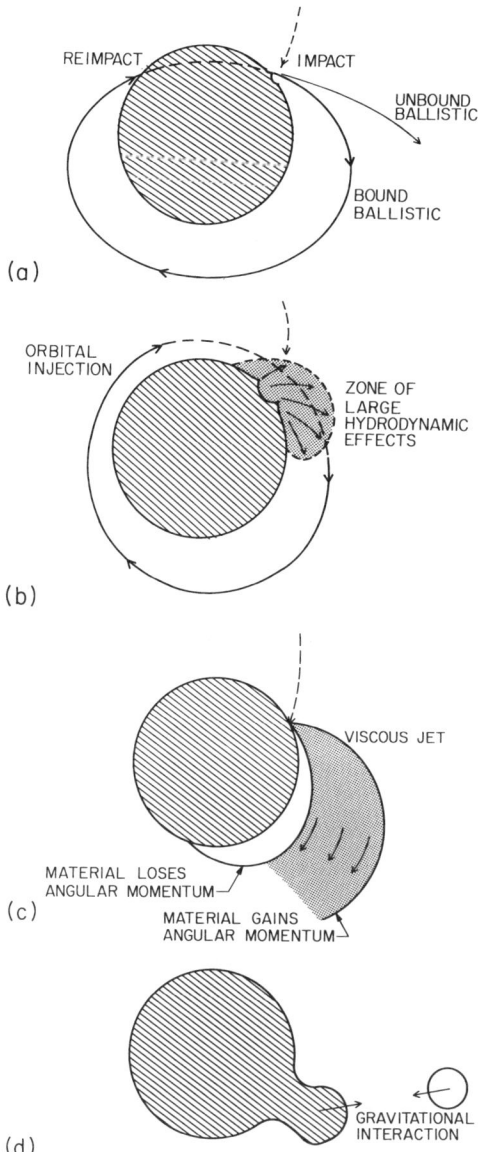

especially when substantial vaporization takes place. The general principles are known, but the detailed quantification is incomplete. During impact, material is very rapidly shocked to high pressure and temperature; irreversible entropy production occurs at this stage. This material then expands approximately isentropically from the peak pressure state. It is, of course, not a thermodynamic question as to whether the expansion phase is well approximated as isentropic; this depends on opacity, turbulence, etc. For the moment, isentropy is assumed. Often the peak pressure state is supercritical (meaning that it cannot be characterized as being either gas or liquid); during expansion it eventually hits a phase boundary, and the extent of vaporization is then well defined. It is popular in the literature to state the degree of vaporization at a nominal 1-bar pressure. This is arbitrary and somewhat misleading, since (as we shall see) there is nothing special about a pressure of 1 bar in terms of understanding giant impacts on the protoearth. More importantly, the degree of vaporization can be substantially different if one chooses (say) a kilobar or a millibar as the nominal state. In the brief historical survey below, the nominal state is 1 bar.

Stanyukovich (1950) was first to estimate vapor production and obtained $\sim (V_{imp}/14 \text{ km s}^{-1})^2$ in units of projectile mass, assuming cold (terrestrial rock) starting material. Notice that the kinetic-energy content of the projectile has to be over an order of magnitude larger than the heat of vaporization of rock ($\sim 10^{11}$ erg g^{-1}). There are two reasons for this low yield: (a) Much of the energy remains kinetic (at least initially), and (b) much of the energy is stored in the internal energy of compression. Subsequent calculations, based on more substantial data, predict even less vaporization (Ahrens & O'Keefe 1972, O'Keefe & Ahrens 1977, 1982), except when the target is a magma ocean (Rigden & Ahrens 1981). All of these calculations are potentially misleading because they assess vaporization in terms of the shock process alone; no contribution due to gradual degradation of kinetic energy is computed. In a giant impact, all of the impact energy must eventually be accounted for in assessing vaporization. Whereas in small impacts it is valid to neglect energy release from the rain-out of debris far from the impact site, there is no such place as "far from the impact site" on Earth if the projectile is the size of Mars!

It is instructive to perform a computation of prompt, irreversible entropy production in a shock event to show where the uncertainties arise. For this purpose I consider SiO_2, where the data base is far more complete than for more appropriate starting materials (e.g. Mg_2SiO_4 or $MgSiO_3$). Approximate calculations for more realistic silicate assemblages indicate that this is not a serious deficiency [mainly because the vapor pressure of

MgO is not much different than that for SiO_2 (Krieger 1967)]. The entropy production upon shock compression is given by

$$\Delta S = \int_{T_s}^{T_h} C_v \, dT/T, \tag{1}$$

where T_h is the Hugoniot temperature (the temperature behind the shock wave), T_s is the temperature that the material would have at the same density if the compression had been isentropic, and C_v is the specific heat at constant volume (usually not constant). To get a rough idea as to how ΔS scales, we can use the empirical fact that at very high (megabar) pressures, we have

$$T_h \simeq T_1 P, \tag{2}$$

where P is the peak pressure and T_1 is some constant. This behavior is very frequently observed [e.g. Lyzenga et al (1983) for SiO_2]. We also have

$$T_s \simeq T_0 (\rho/\rho_0)^\gamma, \tag{3}$$

where T_0 is the initial temperature, ρ is the shock pressure, ρ_0 is the initial density, and γ is the Gruneisen parameter (here assumed constant). Moreover, we have

$$P = \rho_0 U_s U_p \propto \rho_0 V_{imp}^2, \tag{4}$$

where U_s is the shock velocity and U_p is the particle velocity [again assuming very high pressure ($P \gg$ bulk modulus)]. If we let $\beta \equiv d \ln P/d \ln \rho$, then we have

$$\Delta S \simeq (1-\gamma/\beta) C_v \ln (V_{imp}^2) + \text{constant}. \tag{5}$$

To estimate vapor production we then compare this expression with ΔS_v, the entropy difference between liquid and vapor. It is interesting, but perhaps counterintuitive, that ΔS is only a weak (logarithmic) function of impact velocity, since we would have expected the total vapor production to scale as the kinetic energy of the projectile (e.g. O'Keefe & Ahrens 1982). The resolution of this apparent conflict lies in the realization that at very high impact velocities, a volume much larger than the projectile volume is at least partially vaporized, and it is the integral over all this volume (which scales roughly linearly with peak shock pressure) that is relevant. For our *present* purposes, it is likely to be the local vapor content that is important, since we wish to consider the role of pressure-gradient acceleration.

A detailed calculation was carried out using the following data and theory: The equation of state and shock temperatures are from Lyzenga & Ahrens (1980) and Lyzenga et al (1983). These data were also used to estimate γ and C_v. An extended Debye model was used for C_v (Kieffer 1979). The Dulong-Petit limit of $C_v = 6$ cal (mole-atom-K)$^{-1}$ appears to be exceeded at $T_h \gtrsim 6000$ K ($P \gtrsim 1$ Mbar), probably because of electronic excitation. The Gruneisen parameter is also mildly temperature dependent. It was also assumed that the initial condition was a 50% molten Earth at $T \sim 1800$ K. The motivation for this choice is work done on planetary accretion (Kaula 1979, 1980, Stevenson 1981, 1983a, Davies 1985) suggesting that the heat retention due to prior impacts is sufficiently large to guarantee a hot target. The actual surface of the protoearth may be cold (because of radiative cooling), but the thermal boundary-layer thickness would be $\sim$ few kilometers, negligible compared with the excavation depth of the projectile. The protoearth may even have had a magma ocean, as assumed by Rigden & Ahrens (1981) and Hofmeister (1983), but the entropy difference between solid and liquid is small compared with that needed for vaporization, so the prior presence of a magma ocean is not important. It *is* important that the target be hot; the same calculation with an initial temperature of 300 K yields much less varporization. The hot initial state means that the shock data on highly porous silica (aerogel) of Holmes et al (1984) were especially useful.

The results from numerical integrations of Equation (1) are shown in Figure 4, superimposed on the SiO_2 phase diagram. This entropy-pressure representation, popularized in volcanology by Kieffer (1982), seems particularly suitable for understanding postimpact expansion. Each vertical line represents an isentropic pressure release path from the peak pressure achieved for a given impact velocity under the assumptions of normal impact and equal material properties for target and projectile. The phase boundary was computed using JANAF thermochemistry tables (*JANAF Thermochemical Tables* 1971) and allowing for the following species in the gas phase: SiO_2, SiO, Si, O_2, and O. In fact, SiO and O_2 dominate. Pressure corrections for the vapor-phase chemical potentials were made assuming ideal gas partition functions with fully excited internal degrees of freedom. The resulting vapor pressure agrees with the calculation of Krieger (1967) but *not* with some data (Ruff & Schmidt 1921). The source of the discrepancy is not known. The extrapolation to the critical point is highly uncertain, but it can be crudely estimated using corresponding-states arguments (see also Ahrens & O'Keefe 1972).

These results are revealing in several ways. First, the fact that entropies are additive means that a lever rule can be applied to determine the extent of vaporization at a given pressure on the release path. For example, material subject to a 10 km s^{-1} impact is about 20% vapor, 80% liquid *by mass* at $P = 1$ kbar ($T \sim 4000$ K). As expansion proceeds, the liquid boils, producing more vapor. (This is directly analogous to the boiling of

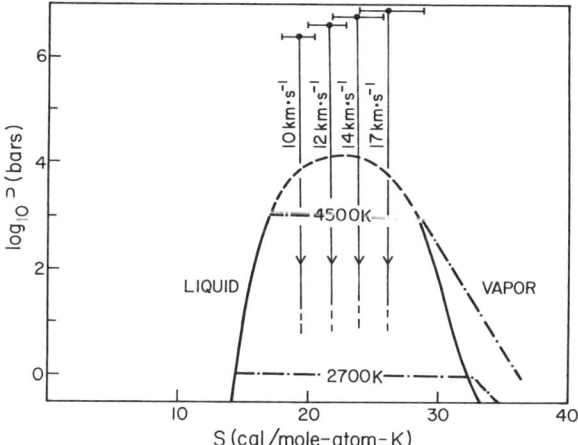

Figure 4 Phase diagram for SiO_2, with superimposed isentropic pressure-release lines for postimpact flows corresponding to the impact velocities indicated. The critical region (where liquid and vapor become indistinguishable) is not well known and is indicated by a dashed line. The error bars at the top of each trajectory indicate the uncertainties in entropy computation. By applying the lever rule, this diagram indicates the degree of vaporization for a given irreversible entropy production.

water during isentropic expansion to lower pressure and temperature.) If the postimpact entropy is larger than the critical value of ~ 22 cal (mole-atom-K)$^{-1}$, then the expansion involves condensation of silica droplets (i.e. the mass fraction of liquid increases as expansion proceeds). This applies for impact velocities $\gtrsim 14$ km s^{-1}. In any event, a substantial fraction of the mass (and almost all the volume) is eventually in vapor form, even for an impact at 10 km s^{-1}, a value that coincidentally is comparable to the escape velocity from Earth or the impact velocity on Earth (cf. Proposition 4). A second interesting feature of Figure 4 is indicated by the error bars associated with each vertical curve. These attempt to incorporate all the shock data and theoretical extrapolation uncertainties entering computation of ΔS. The uncertainties become very large at high impact velocity because large extrapolations of shock data are required and electronic corrections to γ and C_v become important, yet uncertain.

Of course, this calculation does not deal with the actual conditions of an oblique impact because of its highly idealized nature of a plane, normal shock, but it does provide a useful guide to the possible extent of prompt vaporization. The differences between oblique and normal impacts are discussed in the next section.

There is another way of analyzing vaporization, on purely energetic grounds, that relies on the expectation that the impact is so large that the postimpact Earth can be treated as spherically symmetric, with the surface layer heated more than the interior. [Clearly, this would be a ridiculous assumption for small impacts, even for the body that supposedly caused biological trauma at the end of the Cretaceous (see Silver & Schultz 1982).] Calculations were carried out assuming injection of a specified total amount of energy ΔE (some fraction of the kinetic energy of the projectile), which is then partitioned among internal energy (mostly thermal) and increased gravitational energy (because the planet puffs up). As a rough guide for what to expect, a Mars-sized projectile impacting at 10 km s^{-1} has an energy of 3×10^{38} erg, which, if completely converted to heat, would increase the average temperature of an Earth mass by roughly 5000 K (if we assume no latent heat buffering). Figure 5 shows estimated temperature profiles for an arbitrary heat injection of 1.5×10^{38} erg and a range of energy emplacements varying from uniform to that obtained by assuming a pressure drop-off away from the impact site as r^{-2} [a likely upper bound to the rapidity of the drop-off (e.g. Orphal et al 1980)]. These calculations show that the postimpact Earth certainly has a deep magma ocean with a supercritical near-surface layer that merges continuously with a vapor atmosphere. The cooling time of this system is defined by

$$\tau_{\text{cool}} = \frac{\Delta E}{\sigma T_e^4 4\pi R^2}, \tag{6}$$

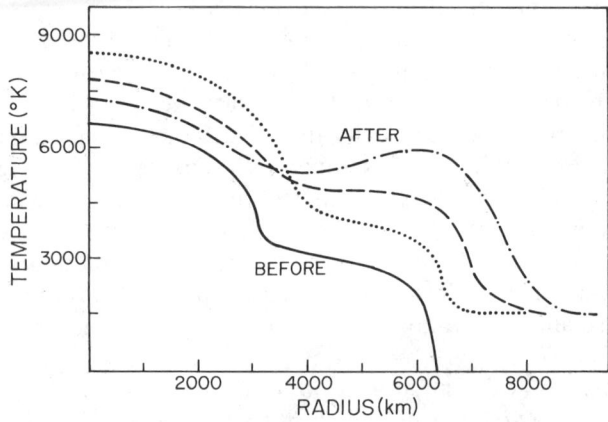

Figure 5 Several possible outcomes arising from injecting 1.5×10^{38} erg into Earth because of a giant impact. The preimpact state (solid line) is assumed to be 50% molten because of previous smaller impacts. Three possible postimpact states vary from uniform energy deposition (dotted line) to a concentration in the mantle (dash-dot line). In each case, the planet "puffs up" and an extended transient atmosphere ($T \sim 2000$ K) develops.

where T_e is the effective temperature, σ is the Stefan-Boltzmann constant, and R is the radius of the photosphere. The radiating level is probably determined by the opacity of dust grains that condense (Thompson & Stevenson 1986) and corresponds to $T_e \sim 2000$ K. For this choice, we obtain $\tau_{cool} \sim 10^3$–10^4 yr. Notice that this is an enormous time compared with dynamical times. This is a fundamental distinction between giant impacts and small impacts (such as the event at the end of the Cretaceous). Perhaps the most interesting point of Figure 5 is that it demonstrates how traumatic this giant impact would have been for Earth. Perhaps the best evidence for the event will be found in the deep mantle in the form of residual evidence for magma ocean differentiation and layering (e.g. Ohtani 1985).

Clearly, a description of the postimpact behavior requires consideration of two-phase media (a liquid with bubbles or a gas with droplets). One property of a two-phase medium is that it is not completely specified by two thermodynamic variables in the usual way. For example, a medium with specified P and T (these two variables being linked by the Clausius-Clapeyron equation for vapor equilibrium) can have any liquid mass fraction between zero and unity. This means it can have an enormous range of average densities and average specific entropies for the same P and T. It is also possible to have an enormous range of pressures for a given internal energy. Unfortunately, this thermodynamic requirement is not satisfied by all equations currently in use. In particular, it is not satisfied by the Tillotson equation of state (Tillotson 1962, Allen 1967), despite its plethora of adjustable constants. This is the equation used initially by Benz et al (1986a,b), and thus it casts doubt on the accuracy of their vaporization estimates and pressure-gradient effects. More realistic equations of state are now being used. Regrettably, it is not possible at the time of this review to assess whether this greatly modifies the results.

Constitutive Properties

The focus here is on "viscosity," due both to microscopic processes and to fluid dynamical processes. This is less well understood than the thermodynamics but possibly also less important. In fact, numerical simulations of impacts do not usually include viscosity except as a numerical artifice to promote stability of the code. We return to this "artificial" viscosity in the next section; here, we pose the question, What is a dynamically interesting viscosity and might it exist? A dynamically interesting viscosity would be one for which an element of material ejected from the impact site is subjected to a viscous couple in one orbital period large enough to significantly increase its angular momentum and to make orbital injection possible. Of course, this would be done at the expense of other fluid elements that lose angular momentum. Clearly, a complete quantitative answer requires a detailed model of jets, but we can get a rough idea by assuming Keplerian differential rotation of a disk of material and by borrowing from the physics of accretion disks (Lynden-Bell & Pringle 1974). In steady state, we have

$$F \frac{dh}{dR} = \frac{dg}{dR}, \tag{7}$$

where F is the total mass flux radially outward at radius R of a disk of material with local specific angular momentum $h = R^2\Omega$ (Ω is the angular velocity). The viscous couple is g, and we assume that the radial velocity is $\sim 0.1 R\Omega$ (to raise the periapse of outer material in the disk in one orbital period). Then, since

$$F = 0.1\pi R^3 \sigma \Omega,$$

$$g = 2\pi R^3 v \sigma \Omega, \tag{8}$$

where σ is the surface density of the disk (mass per unit area), Equation (7) implies that we need a viscosity

$$v \gtrsim 0.01 R^2 \Omega, \tag{9}$$

or $v \gtrsim 10^{14}$–10^{15} cm^2 s^{-1} typically. (This is about the viscosity of glacier ice on Earth.) The microscopic viscosity of a liquid ($\sim 10^{-2}$ cm^2 s^{-1}) or of a gas ($\sim 10^2$ cm^2 s^{-1} typically) are small by comparison. The *bulk* viscosity of a foam (bubbly liquid) can be remarkably high [$\sim 10^{11}$ cm^2 s^{-1}; Stevenson 1983b) because of the irreversible entropy production accompanying the induced phase change between gas and liquid, but even this value is not large enough to be important. However, fluid dynamical instabilities could produce viscosities approaching that required. In the Prandtl picture of turbulent viscosity, we can imagine blobs of fluid with relative velocity u and size l; the resulting viscosity is then $\sim ul$ and could be $\sim 10^{14}$ cm^2 s^{-1} for $u \sim 10$ km s^{-1} and $l \sim 1000$ km. More specifically, Thompson & Stevenson (1983, 1986) point out that a two-phase medium is susceptible to gravitational patch instabilities, even close to the Earth, because the two-phase medium is highly compressible. These instabilities promote turbulence because they cannot evolve all the way to formation of a self-gravitating sphere if they occur within the Roche limit. However, the estimated turbulent viscosity is then $\lesssim l_{\text{crit}}^2 \Omega$, where l_{crit} is the wavelength of the instability, probably only about 10^2 km. The resulting turbulent viscosity is then $\lesssim 10^{12}$ cm^2 s^{-1}, less than the constraint give by Equation (9). A viscosity of 10^{12} cm^2 s^{-1} (or even 1000 times less) is still extremely

ORIGIN OF THE MOON   295

interesting for lunar formation, however, if the Moon forms from a disk. This is discussed later.

Dynamics

Even with a complete understanding of thermodynamics and rheology, there is a remarkable range of possible fluid dynamical outcomes. These can only be fully understood by numerical simulation. In this section the possibilities are described in very simple terms.

Suppose that non-Keplerian effects (primarily pressure-gradient acceleration) act from the impact site out to a height h above the Earth's surface (radius $R_\oplus$), but that the subsequent motion of an element of material is Keplerian. The initial vertical and horizontal components of velocity at height h are taken to be V_r and V_h, respectively. In the limit $h \ll R_\oplus$, the best candidate trajectories for orbital injection are those for which $V_h \gg V_r$, but with negative total energy. We can express the velocity components as

$$V_r = A \left(\frac{GM_\oplus}{R_\oplus} \right)^{1/2},$$

$$V_h = B \left(\frac{GM_\oplus}{R_\oplus} \right)^{1/2}. \qquad (10)$$

The requirement that a bound orbit result is that $A^2 + B^2 < 2/(1+x)$, where $x \equiv h/R_\oplus$. The requirement that periapse lie above the Earth's surface is

$$B^2 > \frac{A^2 + 2x}{x^2 + 2x}. \qquad (11)$$

For each value of total velocity $V_t = (V_r^2 + V_h^2)^{1/2}$, one can define a cone of trajectories for which both criteria are satisfied. The solid angle of this cone, divided by 2π, can be thought of as the "probability" P of orbital injection for each V_t and x. This is shown in Figure 6. Even for quite high starting elevations, represented by x, the value of P is low, since it is truncated at high velocities by escape. In a crude way this calculation gives the requirement that must be satisfied to inject material: Nonballistic processes (the "second burn") must be able to achieve a value of x and V_t so that injection is possible. Notice that other factors being equal, injection into a highly elliptical initial orbit (i.e. total energy only slightly negative) is favored because such an orbit maximizes P.

We turn next to the question of whether pressure-gradient acceleration is capable of acting out to a significant fraction of the Earth's radius, so that the initial conditions in the above calculation could be achieved. For

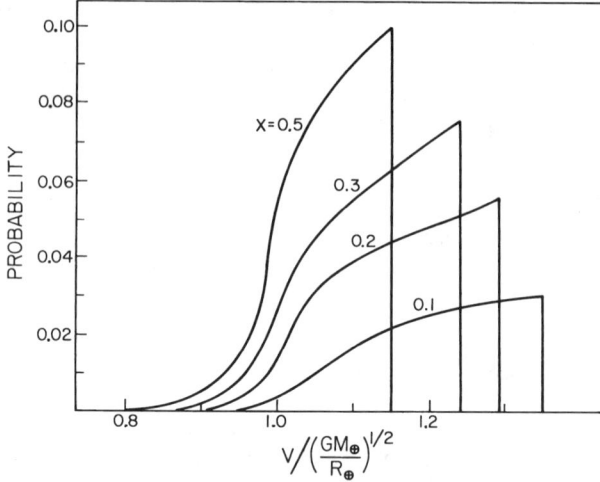

Figure 6 Probability of orbital injection, here defined as the fraction of all possible Keplerian trajectories (equally distributed in solid angle), starting from a nondimensional height $x \equiv h/R_\oplus$ with velocity V, for which orbital injection occurs. The truncation at high V occurs because more energetic trajectories lead to escape. The value of x can be thought of as a measure of the regions within which hydrodynamic effects are very important. Hence, large (small) x applies to large (small) impactors.

simplicity, consider steady-state gas flow expanding hemispherically away from a point source. We estimate the total kinetic energy gain between some radius r_o from the source and infinity from the Euler equations:

$$\int_{r_o}^{\infty} u \frac{\partial u}{\partial r} dr = \frac{1}{2}(u^2|_\infty - u^2|_{r_o})$$

$$= \int_{r_o}^{\infty} -\frac{1}{\rho}\frac{\partial p}{\partial r} dr$$

$$= \frac{\gamma p}{(\gamma-1)\rho}\bigg|_{r_o}, \quad (12)$$

where u is the radial velocity and we assume $p \propto \rho^\gamma$ as the adiabatic equation of state. We wish to define the hemisphere of radius r_o as the region outside of which only small hydrodynamic effects occur. Somewhat arbitrarily, let us require that the change in $u^2/2$ is only 10% of the gravitational energy during the expansion from r_o to ∞. (The flow need not actually go to infinity; the results are similar if the flow is bound.) We then must define r_o as the place where $p \approx 10^3$ bar, $\rho \approx 0.1$ g cm^{-3} ($T \approx 5000$ K, $\gamma \approx 1.2$), where we assume most of the mass is in the form of vapor. This corresponds to a release path at $V_{\text{imp}} \sim 12$ km s^{-1} (Figure 4). For a projectile mass M_{proj}, it follows that

$$\frac{r_o}{R_\oplus} \sim 3(M_{\text{proj}}/M_\oplus)^{1/3}(V_{\text{imp}}/14 \text{ km s}^{-1})^{2/3}. \tag{13}$$

This formula can be loosely translated as the predicted value of x for Figure 6. It suggests that all impacting bodies with mass $\gtrsim 10^{-3}$ $M_\oplus$ are capable *in principle* of injecting a significant fraction of their own mass (a mixture of target and projectile) into orbit. Most importantly, Equation (13) suggests a rather weak dependence of injection efficiency on projectile mass. Of course, the actual efficiency can only be assessed by numerical simulation.

Melosh and Sonett (1986) emphasize a particular aspect of impacts that may be very important for orbital injection. They point out that the highest speed ejecta thrown out in the earliest stage of impact cratering is a *jet* of very fast moving material, usually less than 10% of the projectile mass. They find that the fraction of jetted material is greater during oblique impacts and favor this material for orbital injection because it is the most likely material to be vaporized. The existence of a jet is not in dispute (Gault et al 1968, Kieffer 1975, 1977a), but its role in the impact theory of lunar origin is unclear, since (as argued earlier) substantial vaporization is achievable even at 10 km s^{-1} impact velocity, provided that both the projectile and target are hot. Therefore, it is not clear whether the jet is essential or even desirable—since this material is also the most likely to escape.

The final issue we address here concerns large deviations from an r^{-1} gravitational potential. In appropriate circumstances, this could raise the periapse of ejecta above the Earth's surface and allow orbital injection. Consider, for example, the situation in which a small body is ejected into orbit about two larger bodies (the protoearth and the remainder of the projectile) that are undergoing merger. There are two effects here: The center of mass is moving toward the center of the protoearth, and there are higher-order terms in the gravitational potential. The former means that even a *closed* orbital trajectory need not reimpact the Earth. In fact, the periapse can be "raised" by an order of the distance that the center of mass moves relative to the protoearth (leaving out the small change in Earth radius due to merging). This could easily be ~ 0.1 $R_\oplus$. The deviation from r^{-1} in the gravitational potential is more complicated, but it can be thought of as a gravitational torque, analogous to that responsible for the current recession of the lunar orbit, only much larger. Consider a mass anomaly ΔM exerting a torque on an orbiting mass m at distance R (Figure

3d). This torque is of order $Gm\Delta M/R$ times some numerical factors involving the geometry. In a time τ, the angular momentum transferred to m is of order $Gm\Delta M\tau/R$. This value should be compared with $m(GM_\oplus R)^{1/2}$. To achieve a 10% change in angular momentum would require

$$\frac{\Delta M}{M_\oplus} \sim \frac{0.1}{\Omega\tau}, \tag{14}$$

where Ω is the angular velocity of the orbit of m. Plausibly, we could have $\Omega\tau \sim 0.3$–0.6, and a mass of order two Mars masses would be sufficient. Notice that the result is independent of m, provided $m \ll M_\oplus$. Of course, this *only* works if ΔM is transient, since otherwise the torque varies in sign as this bulge rotates beneath the orbiting protomoon. It is desirable that the bulge relax on a dynamical (free-fall) time scale.

NUMERICAL SIMULATIONS

At the Origin of the Moon Conference in 1984, Cameron presented pioneering simulations of an impact origin of the Moon (Cameron 1985b). These early calculations have already been completely superceded, but even so, simulation work is only in its infancy. I concentrate here on the efforts of Benz et al (1986a,b), which are the only reasonably well-documented computations at the time of this writing. Calculations in progress by Kipp & Melosh (1986) are difficult to assess at this stage because these authors' earliest results omitted self-gravity of the ejecta.

Benz et al studied three-dimensional numerical simulations of oblique impacts of Mars-sized bodies on Earth. They chose to explain the present angular momentum of the Earth-Moon system (3.5×10^{41} g cm^2 s^{-1}) entirely by a single impact; this constrains the relationship between projectile mass M_{proj}, impact parameter d, and impact velocity V_{imp}:

$$0.085 \simeq \left(\frac{M_{proj}}{M_\oplus}\right)\left(\frac{d}{R_\oplus}\right)\left(\frac{V_{imp}}{11 \text{ km s}^{-1}}\right). \tag{15}$$

This equation makes the preferred choice of a Mars-sized body ($M_{proj} \approx 0.1$ $M_\oplus$) self-evident, although one could go to somewhat more massive bodies striking more nearly head-on. The equation also illustrates why at most a small number of projectiles can provide the required angular momentum, since their contributions tend to add incoherently and make the angular momentum constraint harder to satisfy. The constraint assumes that the loss of angular momentum to infinity, carried by fast-moving ejecta, is small. The simulations tend to be consistent with this assumption. The system is assumed energetically closed in the sense that radiation loss is small on the relevant dynamical time scale of hours. This is already suggested by Equation (6) and is more explicitly demonstrated in the next section. Benz et al use a Tillotson equation of state that agrees with shockwave data on granite in the high-density limit (Allen 1967) and with an ideal gas in the low-density limit. As briefly discussed earlier, this equation of state does *not* describe a two-phase medium. Moreover, the quoted degree of vaporization in Benz et al (1986a) is incorrect, since it is based only on the sum of the mass elements that are *completely* vaporized. In fact, most of the mass is partially vaporized. The seriousness of these deficiencies is not known at the time of this writing.

Benz et al used a method known as smoothed particle hydrodynamics (SPH) in which the medium is represented by a finite number of mass points whose trajectories are followed in a Lagrangian sense. In this way, the Navier-Stokes equations are translated into an N-body problem, a desirable feature in a problem that has a complicated three-dimensional geometry. The SPH method is a recent development in astrophysical computational fluid dynamics (Lucy 1977, Gingold & Monaghan 1979; see also other references in Benz et al 1986a). Each particle is assumed to have its mass spread out in space according to a given distribution called the kernel. Details of the method can be found in the references given above. One other feature of interest is that an "artificial viscosity" is introduced in order to avoid unacceptably large postshock oscillations. However, this viscosity corresponds to a Reynolds number of $\gtrsim 200$ in the postshock flow (W. Benz, private communication), which is better than what standard finite-difference techniques achieve and large enough to suggest that (unphysical) angular momentum redistribution in the ejecta flow is tolerably low.

Two sets of calculations have been carried out. In Benz et al (1986a), there was no iron core present. For a Mars-sized impactor with grazing incidence, they found that the impactor is not completely destroyed. Instead, a clump of the projectile most distant from the Earth's center is injected into a highly elliptical orbit. However, this orbit brings the material back to within the Earth's Roche limit, so they conjecture that the material is sheared out by tidal forces and forms a disk. In the more recent calculations (Benz et al 1986b, W. Benz, private communication), a core is included in the projectile and more massive projectiles are considered. The results were somewhat different and perhaps surprising. Most analysis has been for low-velocity impacts (meaning that the velocity at infinity is small, so $V_{\rm imp} \sim 11$ km s^{-1}). Assuming always that Equation (15) is satisfied, Benz et al find that if $M_{\rm proj} < 0.12\ M_\oplus$, then too much iron ends up in orbit. If $M_{\rm proj} \gtrsim 0.16\ M_\oplus$, then the impact is closer to head-on and too little mass ends up in orbit to explain the Moon. If $0.12\ M_\oplus \lesssim M_{\text{proj}} \lesssim 0.16\ M_\oplus$, then *three* substantial bodies are present (see Figure 7 for the simulation depicting $M_{\text{proj}} = M_\oplus/7$). These bodies are Earth, the core of the projectile, and a small ($\sim$ Moon mass) body in a more distant orbit, which may pick up enough angular momentum by gravitational torque (Equation 14) to remain intact. Alternatively, the latter body may break up tidally to form a disk. Meanwhile, the intermediate-mass (iron-rich) body merges with Earth. If, instead, the impact velocity is higher, then some mass leaves the system completely, but it seems likely that either there is insufficient mass placed in orbit or too much iron in the material placed in orbit.

It would be premature to reach firm conclusions on the basis of these results, for at least three reasons. First, it is not clear whether the thermodynamic code in use is adequate. Second, the final outcome is not determined. (The simulations only cover the first ~ 20 hr.) Third, there have been insufficient simulations to understand the entire range of possibilities. In particular, the requirement that all the Earth-Moon angular momentum is explained by a single event is still a questionable simplification or application of Occam's razor. (All applications of Occam's razor in planetary science should be viewed with skepticism.) Nevertheless, the results yield some interesting insights. The following four suggested implications are most striking:

1. Pressure gradients may not always play an essential role in putting material into orbit.
2. "Clumpiness" of ejecta may happen, rather than well-dispersed clouds.
3. The Moon-forming material may come primarily from the projectile.
4. There are often difficulties in *preventing* the incorporation of metallic iron in Moon-forming material.

I think it is wise to be skeptical about all four "conclusions" at our current premature state.

DISK EVOLUTION

One possible outcome of a giant impact is a gravitationally bound clump of material, of order one Moon mass, that retains its integrity and evolves outward by tidal friction. A more likely outcome is the formation of a disk, either directly from a broadly disseminated "cloud" of two-phase (liquid-gas) ejecta or indirectly by the tidal disruption of a clump of material placed in an orbit so elliptical that it undergoes grazing encounters with Earth. Here, we focus on the evolution of this disk and how the Moon might form from it.

Cameron & Ward (1976) and Ward & Cameron (1978) were first to discuss this disk, but they assumed that it would cool and solidify rapidly to form a massive analogue of the rings of Saturn. Gravitational instabilities within the Roche limit would then provide sufficient dissipation (hence "viscosity") to spread the disk out beyond the Roche limit, which would allow the Moon to form essentially cold. Thompson & Stevenson (1983, 1986) reconsidered this problem and found that the disk stays hot (liquid and gas) for 10^2–10^3 yr, and that it can spread in this time, allowing the Moon to form. The main points are these:

1. The cooling time for a disk of ~ 2 Moon masses placed within the Earth's Roche limit is $\gtrsim 10^2$ yr.
2. The disk is a two-phase medium that is highly compressible and unstable with respect to gravitational "patch" instabilities, even when it is very hot.
3. The resulting turbulence and eddy viscosity allow the disk to spread in a time comparable to its cooling time. The disk self-regulates, maintaining its two-phase character because of the gravitational energy dissipated as the spreading proceeds.
4. Material spreads beyond the Roche limit, still maintaining its ability to undergo patch instabilities. Progressive cooling allows the instabilities to proceed all the way to the formation of protomoons. These protomoons are probably nearly fully molten and subsequently may coalesce to form the Moon.

We now elaborate on these points, basing our discussion on that of Thompson & Stevenson (1986). Consider 2 Moon masses ($2M_{\mathrm{☾}}$) of material spread out in a disk between $1.5R_{\oplus}$ and $3R_{\oplus}$, initially in a molten state with coexisting vapor. The characteristic cooling time of this material is at least

$$\frac{2M_{\mathrm{☾}}C_{\mathrm{p}}T}{2\sigma T^4 A} \sim 10 \text{ yr}, \tag{16}$$

where A is the area of the disk ($\sim 7\pi R_{\oplus}^2$), the latent heat of condensation is ignored, and $T \sim 2000$ K is assumed. Actually, cooling turns out to be much slower (10^2–10^3 yr) because of the latent heat and because the disk creates its own heat as it spreads, allowing part of the mass to sink down into the Earth's gravity field while another part climbs farther out in the gravitational potential (so as to conserve angular momentum). For example, suppose all the mass ($2M_{\mathrm{☾}}$) were initially at R_{i}. Now suppose half of this mass moves out to $1.5R_{\mathrm{i}}$ and the other half to $[2-(1.5)^{1/2}]^2 R_{\mathrm{i}}$, thereby conserving angular momentum. The energy release is

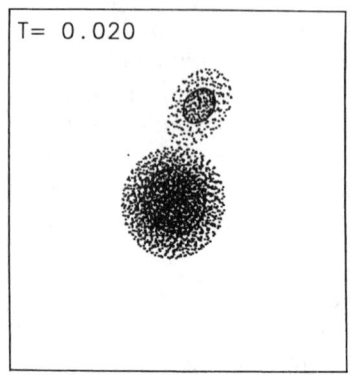
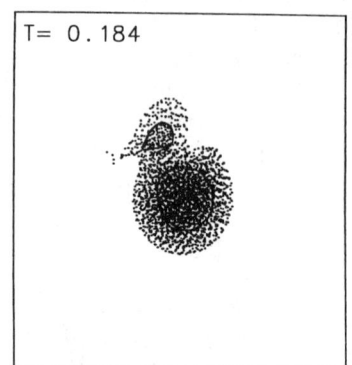
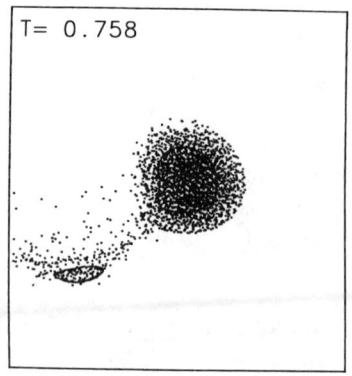
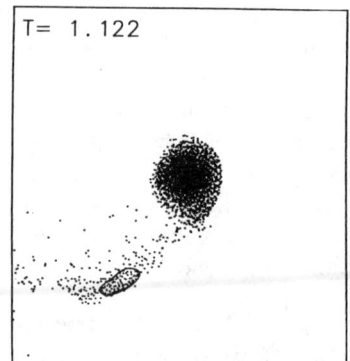
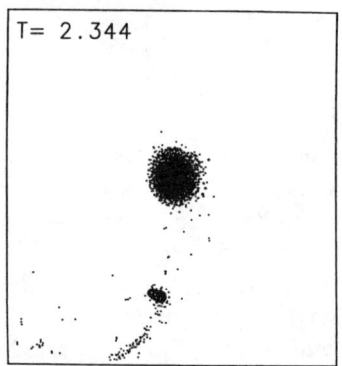
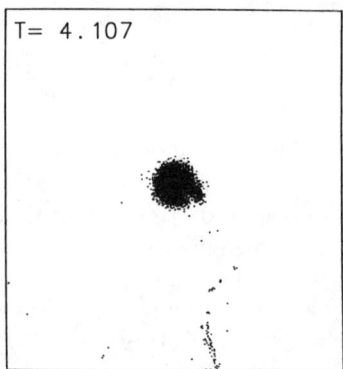

ORIGIN OF THE MOON 303

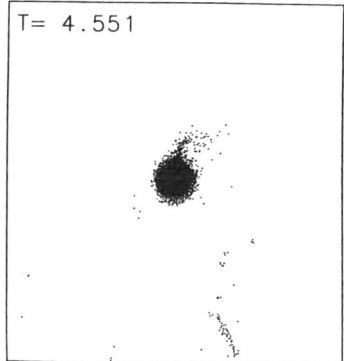
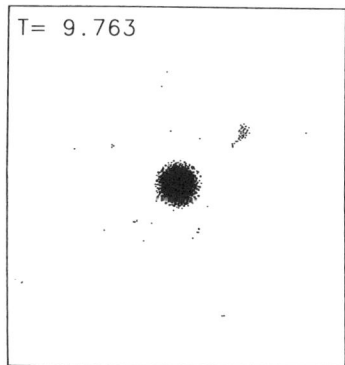
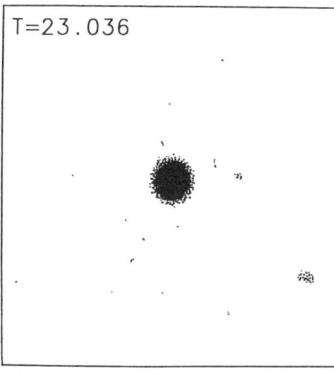

Figure 7 Numerical simulation by Benz et al (1986b) for impact of Earth by a body of mass $M_\oplus/7$. The initial conditions are 0 km s^{-1} relative velocity at infinity but the present Earth-Moon system angular momentum. The iron cores are circled. Time is given in hours from impact. The clump in the lower right corner of the last frame has exactly one Moon mass (35 particles). There is another 1/2–1/3 Moon mass in a disk around Earth. No iron is in orbit. Subsequent evolution after the final frame is not known. Results courtesy of W. Benz (Los Alamos National Laboratory).

$$\frac{GM_\oplus M_{\mathbb{C}}}{R_i}\left(\frac{1}{0.6}+\frac{1}{1.5}-2\right) \simeq \frac{0.33\ GM_\oplus M_{\mathbb{C}}}{R_i}, \tag{17}$$

a value that is larger than $2M_{\mathbb{C}} C_p T$ by a factor of roughly five (assuming $R_i \approx 2R_\oplus$). We show below that something like this happens, so that much energy is available.

Now, the material emplaced in orbit settles into a disk on a dynamical time scale ($\sim$ hours) after having undergone some adiabatic expansion and cooling since leaving the impact site. According to Figure 4, the cooling does *not* imply a large reduction in the mass fraction of vapor. Moreover, the material has great difficulty in cooling below a temperature ~ 2000 K because of the energy generated by gravitational instabilities. These arise because the medium contains bubbles and is therefore highly compressible. Consider, for example, a thin disk of surface density σ, sound speed c, and (local) orbital angular velocity Ω, assumed Keplerian. The dispersion relationship for waves of the form $\exp(i[kr+\omega t])$ is

$$\omega^2 = k^2 c^2 - 2\pi G\sigma |k| + \Omega^2, \tag{18}$$

where r is the radial distance, and $kr \gg 1$ is assumed (e.g. Lin & Shu 1966; see also Goldreich & Ward 1973). Each term on the right-hand side has a simple physical explanation. The first term describes dispersionless sound waves ($\omega = ck$ if k is very large). The second term is negative and describes the possibility of gravitational collapse [analogous to the well-studied Jeans collapse in astrophysics (e.g. Chandrasekhar 1961, Ch. 13)]. The third term is the stabilizing effect of rotation (a combination of the effects of the Coriolis force and Keplerian shear). The important point is that because of the two-phase nature of the medium, the sound speed c is much smaller than the value appropriate to pure liquid or pure gas. This reduces one of the stabilizing terms and makes instability ($\omega^2 < 0$) much more likely. This is best understood graphically (see Figure 8).

The reduction in sound speed is very dramatic and is a well-known effect, for example, in a frothy air-water mixture, where the sound speed can be ~ 20 m s^{-1} compared with 1460 m s^{-1} for pure water (Kieffer 1977b). It is even more dramatic when the liquid and vapor are composed of the same material, since much of the compression is then accommodated by gas molecules within bubbles changing phase, which allows the bubbles to shrink. Thompson & Stevenson (1983, 1986) find, for example, that the sound speed of the two-phase medium $SiO_2(l)$–$SiO(g)$–$1/2\ O_2(g)$ can be three orders of magnitude lower than pure $SiO_2(l)$. This situation is illustrated in Figure 9. Equation (18) predicts that instability is possible for

$$\sigma > \sigma_{crit} \equiv \frac{\Omega c}{\pi G}. \tag{19}$$

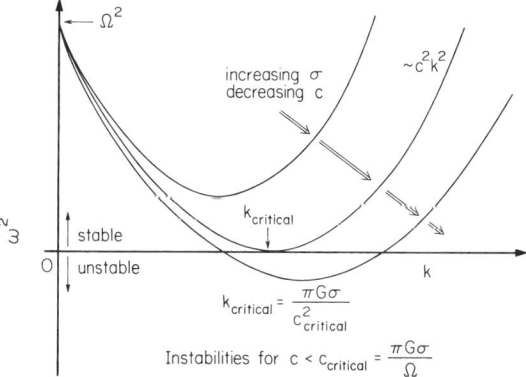

Figure 8 Dispersion relationship [Equation (18)] showing how a reduction in sound speed can cause instability (negative ω^2). This promotes turbulence in the disk.

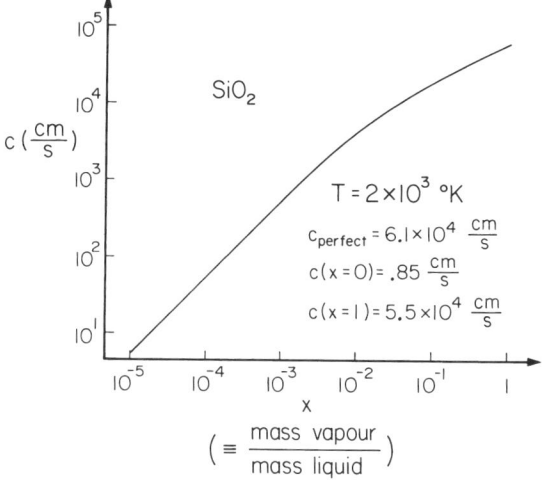

Figure 9 Sound speed of a two-phase medium (SiO_2 liquid and coexisting vapor). The ideal gas value is c_{perfect}. Notice that a medium that is mostly liquid but contains bubbles can have a very low sound speed. The value of c as $x \to 0$ is low and corresponds to the (artificial) case of an infinitesimal population of bubbles.

For a disk of mass $2M_{\mathbb{C}}$ distributed over $\sim 7\pi R_{\oplus}^2$, we have $\sigma \sim 1.5 \times 10^7$ g cm^{-2}. Instability occurs for $c \lesssim 8 \times 10^3$ cm s^{-1}. This is well below the value of 10^5 cm s^{-1} that roughly characterizes pure liquid or gas, but it is easily achievable in a two-phase medium. Notice that if Equation (19) is satisfied by a substantial factor, then the instabilities grow on a time scale $\lesssim 1$ orbital period, because $2\pi G \sigma |k|$ is comparable to Ω^2 and $c^2 k^2$ for the fastest-growing instabilities ($k \sim c/\Omega$). These instabilities are patches of collapsing fluid elements that *cannot* collapse all the way to a protomoon because they are within the Roche limit. Instead, they are tidally sheared. The resulting turbulent dissipation is at the expense of the gravitational energy of the disk [cf. Equation (17)] and can be envisaged as a turbulent viscosity that provides the coupling whereby the disk spreads. This idea of Ward & Cameron (1978) does not require a precise identification of the detailed kinematics of the turbulent viscosity, but rather it follows just from the energy argument stated here.

The disk self-regulates for $\sim 10^2$ yr, dissipating just enough energy to maintain itself at a marginally unstable state. It cannot dissipate less rapidly, since that would require cooling to a still more unstable state (one in which the mass in gas is a smaller fraction and the sound speed is even lower—see Figure 9). It cannot dissipate more rapidly, since that would heat the disk up, which makes the sound speed higher and stabilizes the disk. Consequently, the disk spreading time R^2/v, where v is the effective eddy viscosity, is of order 10^2 yr. This implies that $v \sim 10^9$ cm^2 s^{-1}.

Material spreads beyond the Roche limit, but because the disk is only marginally unstable, instabilities cannot proceed all the way to self-gravitating spheres immediately. The disk must continue to cool until the Roche limit evolves inward and material stranded beyond this limit can coalesce. This occurs at 10^2–10^3 yr after disk formation. The initial protomoons formed have a mass δm comparable to the "patch" instability mass suggested by marginal instability [Equation (18)]:

$$\delta m \sim \pi(\pi/k)^2 \sigma,$$

$$k \sim \Omega^2/\pi G \sigma. \qquad (20)$$

This implies a body of mass 10^{20}–10^{22} g (radius 10–100 km). These bodies are rather closely packed and undergo coagulation on a short time scale ($\sim$ years). However, this process has not been modeled quantitatively.

Once even larger bodies are formed, tidal evolution is also rapid. The time that it takes a body of mass m to double its orbital radius from $3R_\oplus$ to $6R_\oplus$, due to tides excited on Earth, is

$$\tau_{\text{tidal}} \approx (500 \text{ yr}) \left(\frac{M_\mathrm{C}}{m}\right)\left(\frac{Q}{10}\right), \qquad (21)$$

where Q is the tidal quality factor for Earth. This estimate is based on standard tidal theory (e.g. Lambeck 1980). Notice that the time scale is inversely proportional to mass, so smaller bodies can be swept up by larger bodies. Gravitational perturbations (excitation and eccentricity) will play a role also, just as they do in planetary accumulation, but the associated time scale is much shorter than in terrestrial planet formation because the protolunar disk is very compact and has a short dynamical time scale (orbital period $\ll 1$ year).

The time scale to go from the initial instabilities to a mass of lunar magnitude has an important implication for the thermal state of the initial Moon. Smaller bodies can cool more rapidly, forming transient crusts. As these bodies coalesce, rather little extra energy is dissipated, and the resulting body is only partially molten. This has not been quantified.

CHEMISTRY OF THE IMPACT HYPOTHESIS

If the dynamics are uncertain, then the chemistry is more so, in part because it depends on the dynamics and in part because some aspects of the chemistry involve regimes that have received little attention in the laboratory. The important unresolved dynamical issues are these: Where does the lunar material come from? Is it mostly from the projectile (Benz et al 1986a,b) or from both the projectile and the target (Melosh & Sonett 1986, Kipp & Melosh 1986)? How much iron is put into orbit? In some of the simulations by Benz and coworkers, a substantial amount of iron may be put in orbit. How much material is lost, and is the loss differential (i.e. do some elements or molecules escape preferentially)? It is only on this latter point that a partial answer seems possible at present, but we discuss each issue in order.

The provenance of lunar-forming material is especially important to those who advocate a terrestrial origin (most notably Ringwood 1966, 1979, 1986a,b). In Ringwood's view, the Earth's mantle would be a distinctive reservoir that would differ from the mantle of the projectile unless the latter were sufficiently larger to have undergone the same differentiation and processing as the Earth's mantle. This places a lower bound on the projectile mass that is uncertain but surely larger than the mass of Mars, since models of Mars constructed by Ringwood are significantly different from Earth. As discussed earlier, the similarities and differences between the Moon and the Earth's mantle remain controversial. There seems little doubt that the projectile would be differentiated, since core formation is very effective in bodies with mass $\gtrsim$ Mars (e.g. Stevenson 1981, 1983a), and its mantle might therefore be suitable for Moon-forming material. The relative amounts of projectile and target in the Moon have not been well established yet.

The question of iron is a serious one, given the low iron abundance in the Moon. If a disk that includes iron forms, then it is unlikely that the iron is selectively filtered out of the Moon-forming matter. It is comparably volatile to the silicates (in the sense that its vaporization temperature is similar), but, more importantly, it would participate in the same patch instabilities that stir the silicate fluid and promote spreading of the disk. In other words, the liquid iron may form a sublayer at the equatorial plane of the disk but spread at the same rate as the rest of the disk. Wood (1986) discusses the data of Hashimoto (1983) indicating that iron (as FeO) is volatile, and he suggests that the iron could be evaporated away. As discussed below, this is insufficiently efficient to be significant. I conclude that if the Moon formed from a molten disk, then it would have the same iron content that at least the outer half of the disk has. The possibility (Benz et al 1986b) that the core of the projectile maintains its integrity and loses angular momentum, merging with the Earth, is a potential alternative.

After the impact, some material may be lost promptly on a hyperbolic trajectory. This material, possibly a jet of the most highly shocked material, would leave hydrodynamically (i.e. preserving composition). But what about "diffusive" loss from a disk emplaced in orbit? At a distance of $5R_\oplus$ from Earth center, a molecule of atom of mass μ in orbit needs only $E_{\rm esc} = GM_\oplus \mu/10R_\oplus$ energy to escape. (This is only half the gravitational binding energy because of the Keplerian orbital motion.) Comparing this value with the kinetic energy $E_{\rm kin} \simeq 3/2(kT)$, we have

$$\frac{E_{\rm kin}}{E_{\rm esc}} \simeq \left(\frac{4}{\mu}\right)\left(\frac{T}{2000}\right), \tag{22}$$

where μ is expressed in units of the proton mass. Hydrogen, derived from dissociation of H_2O advected to the upper levels of the disk, can clearly escape. The major constituents [MgO ($\mu = 40$), O_2 ($\mu = 32$), and SiO ($\mu = 44$)] cannot escape by a Jeans (thermally activated) process. Actually, the thermal escape would be in a cone of forward velocities, with the axis of the cone defined by the Keplerian orbital motion. In reality, the flow would be more like that of the solar wind because the exobase is at an enormous distance from the disk. It is possible that acoustic waves propagating into the hot-disk atmosphere may drive a wind (Thompson & Stevenson 1986). Since the turbulent velocities in the upper, rarefied layers of the disk are close to the sound speed, a rough estimate of the available energy input is

$$\dot{E} \simeq 2\pi R_{\rm d}^2 \rho_{\rm g} C_{\rm g}^3, \tag{23}$$

where $R_{\rm d}$ is the disk radius and $\rho_{\rm g}$, $C_{\rm g}$ are the gas density and sound speed, respectively. If all this energy goes into mass loss, then we have

$$\dot{M} \simeq 10^{-5} M_\mathrm{C} \left(\frac{\rho_{\rm g}}{10^{-7}\ {\rm g\ cm^{-3}}}\right)\left(\frac{C_{\rm g}}{10^4\ {\rm cm\ s^{-1}}}\right)\ {\rm yr}^{-1}.$$

The disk photosphere is defined by the pressure level at which $p = g/K$ (g = disk self-gravity, K = opacity). As Thompson & Stevenson (1986) discuss, the opacity is very uncertain but is probably grain dominated, which suggests that $K \sim 1$ cm^2 g^{-1}. The corresponding density is $\sim 10^{-7}$ g cm^{-3}, and in view of the turbulent nature of the disk, there may already be enough energy input at this level to excite waves at the rate given in Equation (23). For this upper bound, the mass loss is still only about 0.1% of the mass of the *Moon* and hydrodynamic in nature. It is unlikely that elemental and isotopic fractionation occurs, as suggested by Cameron (1983), except perhaps for a preferential loss of hydrogen, derived from dissociation of water. Water would be present as a molecular species high in the disk atmosphere, advected there by silicate droplets and then exsolved from the silicates in the low-pressure region of grain formation. At $T \sim 2000$ K, $P(H_2O) \lesssim 10^{-6}$ bar, plenty of atomic hydrogen would be available by thermal dissociation.

To summarize, the disk is essentially a *closed system*. With the possible exception of hydrogen (and hence water), the loss of material by outflow from the disk is neither large nor capable of differentiation. This partially validates attempts to compare chemically the Moon with the mantle of the Earth or with a large projectile.

COMMENTS AND QUESTIONS

It should be clear that the collision hypothesis of lunar origin is still in a primitive form. This is readily apparent by returning to the 10 propositions listed early in this review and critically assessing their merits and uncertainties:

1. Inadequacies certainly exist in the other "conventional" lunar origin scenarios (fission, capture, binary accretion). But this is not an argument in favor of megaimpact; it may simply be that we have had insufficient time to recognize all the shortcomings of the current bandwagon.
2. The theoretical evidence certainly points to large impacts during planetary formation. However, no simulations exist that cover the entire sequence of events from the formation of small planetesimals all the way to the final planets. Even so, this remains as one of the better-justified propositions.
3. Large impacts should certainly be very different dynamically than small impacts. Quantification of this proposition remains difficult and imperfect, however, and three-dimensional hydrodynamic simulations are still rather primitive and incomplete. Much more work is needed

to characterize the range of outcomes and the efficiency of orbital injection.
4. Estimates of vaporization accompanying large impacts vary widely because they involve extrapolations of existing equation of states or uncertain theoretical modeling (e.g. to evaluate electronic excitations). Current estimates arising from the three-dimensional computer simulations may be incorrect. As a consequence, the role of pressure gradients in the postimpact flow is not yet understood.
5. The impact trauma is very substantial for Earth and almost certainly causes a hot, silicate vapor atmosphere and an underlying magma ocean. But is there a diagnostic consequence of this in Earth history (e.g. the geochemistry of the deep mantle)? More work is needed on the behavior of magma oceans.
6. The material placed in orbit may be clumpy or dispersed. Current calculations do not demonstrate conclusively that a disk will form. Numerical simulations have to be carried out to longer times.
7. The origin of the orbital material is unclear. It may come mainly from the projectile, or it may come partly or wholly from the target. It is also not clear whether this orbital material is devoid of iron from the core of the projectile. Again, more simulations are needed to understand this better.
8. It seems likely that the dynamics of lunar formation are *very fast* irrespective of the details. This seems to be one of the safer propositions. Nevertheless, detailed calculations do not exist.
9. Despite the high energy of a giant impact, the thermal energy of the resulting material is small compared with the escape energy, and loss during the history of the disk is small. Consequently, the system is almost "closed," and this proposition emerges as one of the better-justified ones.
10. Crude estimates of orbital injection suggest that it should work for lower-mass impactors but with gradually diminishing efficiency. Consequently, we must *expect* that the Moon is a combination of several protomoons. The number cannot be too great, since otherwise the angular momentum problem is very serious. (Roughly speaking, the protomoons should have comparable likelihoods of prograde and retrograde orbits.) Formation of the Moon from a *large* number of impacts (e.g. Ringwood 1986a,b) does not seem tenable. We can rationalize the existence of only one Moon by arguing that the largest impact occurred late in Earth accretion. However, more work is needed on the question of whether smaller protomoons are either swept up or lose angular momentum and collide with Earth.

Another way of assessing lunar origin by impact is to ask how well the scenario explains the data. Such an assessment is presented in Table 2. A quick look confirms that the uncertainties dominate the positive attributes. The best that can be said for the impact hypothesis is that it has no *strongly* negative attributes at present.

Finally, what are the implications of giant impacts for comparative planetology? Here, we see many attractive possibilities. Cameron (1983) has suggested that the differences between Earth and Venus may be partly attributed to the Earth (but not Venus) suffering a very large impact. For example, Cameron speculates that a large impact could "blow away" a massive CO_2 atmosphere. Our discussion here does not provide very strong support for this idea, but it merits further attention. An equally interesting idea concerns Mars. Since the escape velocity from Mars is too low for significant vaporization of incoming impactors with low velocity at infinity, and since fast-moving projectiles would cause ejecta to escape hyperbolically, we conclude tentatively that Mars (or any small planet) should never have a substantial satellite, at least by the impact process. Indeed, Mars has only two small satellites, both plausibly derived by capture (see discussion in Stevenson et al 1986). Last, but certainly not least, the Uranian system is a natural candidate for giant-impact effects. An impact by a body of ~ 2 Earth masses may have stirred the interior (lowering the moment of inertia) and caused the formation of a disk from which the satellites formed. The compositions of these satellites [see Smith et al (1986)

Table 2 Comparison between constraints from lunar data and the impact scenario

Constraints from lunar data	Impact model
1. Mass	Need projectile $\gtrsim$ Mars mass and $\sim 10\%$ efficiency of orbital injection
2. Angular momentum	Implies one or (at most) a few large impacts; many small impacts have canceling angular momenta
3. Low iron content	Iron in projectile must avoid orbital injection; it is not clear whether this is possible
4. Volatile depletion	Uncertain predictions; hydrogen (hence water) probably lost but other volatiles may be partly retained
5. Trace elements and other chemical constraints	Lunar-forming material is an uncertain mix of projectile and Earth mantles
6. High initial temperatures (putative lunar magma ocean)	Readily provided by impact energy release
7. Orbital evolution (tidal theory)	Initial moon may be in equatorial plane. Tidal theory does not clearly preclude this

for Voyager data] are clearly more rock rich than the Saturnian satellites, an observation possibly consistent with an impact origin (Stevenson 1984b, Stevenson & Lunine 1986). The Uranian system is an enticing locale for those who tire with modeling their own backyard.

ACKNOWLEDGMENTS

I thank W. Benz, A. G. W. Cameron, and H. J. Melosh for discussing their unpublished work, A. E. Ringwood and G. W. Wetherill for useful conversations, and the organizers of the Kona Conference (Hartmann et al 1986) for motivating much of this effort. This work was supported by NASA Planetary Geophysics grant NAGW-185 and is contribution number 4348 from the Division of Geological and Planetary Sciences, California Institute of Technology, Pasadena, CA 91125.

Literature Cited

Ahrens, T. J., O'Keefe, J. D. 1972. Shock melting and vaporization of lunar rocks and minerals. *The Moon* 4: 214–49

Allen, R. T. 1967. Equation of state of rocks and minerals. *Rep. GAMD-7834*, General Atomic Div., General Dynamics, San Diego, Calif.

Anderson, D. L. 1972. The origin of the Moon. *Nature* 239: 263–65

Arrhenius, S. 1908. *Worlds in the Making*. New York/London: Harper & Brothers

Benz, W., Slattery, W. L., Cameron, A. G. W. 1986a. The origin of the Moon and the single impact hypothesis, I. *Icarus* 66: 515–35

Benz, W., Slattery, W. L., Cameron, A. G. W. 1986b. The origin of the Moon: 3D numerical simulations of a giant impact. *Lunar Planet. Sci. XVII*, pp. 40–41 (Abstr.)

Bills, B. G., Ferrari, A. J. 1977. A harmonic analysis of lunar topography. *Icarus* 31: 244–59

Boss, A. P. 1986. The origin of the Moon. *Science* 231: 341–45

Cameron, A. G. W. 1972. Orbital eccentricity of Mercury and the origin of the moon. *Nature* 240: 299–300

Cameron, A. G. W. 1983. Origin of the atmospheres of the terrestrial planets. *Icarus* 56: 195–201

Cameron, A. G. W. 1985a. Formation and evolution of the primitive solar nebula. In *Protostars and Planets II*, ed. D. C. Black, M. S. Matthews, pp. 1073–99. Tucson: Univ. Ariz. Press

Cameron, A. G. W. 1985b. Formation of the prelunar accretion disk. *Icarus* 62: 319–27

Cameron, A. G. W. 1986. The impact theory for origin of the Moon. See Hartmann et al 1986, pp. 609–16

Cameron, A. G. W., Ward, W. R. 1976. The origin of the Moon. *Lunar Sci. VII*, pp. 120–22 (Abstr.)

Chandrasekhar, S. 1961. *Hydrodynamic and Hydromagnetic Stability*. Oxford: Oxford Univ. Press

Darwin, G. H. 1880. On the secular change in elements of the orbit of a satellite revolving around a tidally distorted planet. *Philos. Trans. R. Soc. London* 171: 713–891

Davies, G. F. 1985. Heat deposition and retention in a solid planet growing by impacts. *Icarus* 63: 45–68

Drake, M. J. 1983. Geochemical constraints on the origin of the Moon. *Geochim. Cosmochim. Acta* 47: 1759–67

Durisen, R. H., Scott, E. H. 1984. Implications of recent numerical calculations for the fission theory of the origin of the Moon. *Icarus* 58: 153–58

Ganapathy, R., Anders, E. 1974. Bulk composition of the Moon and Earth estimated from meteorites. *Proc. Lunar Sci. Conf., 5th*, pp. 1181–1206

Gault, D. E., Quaide, W. L., Oberbeck, V. R. 1968. Impact cratering mechanics and structures. In *Shock Metamorphism of Natural Materials*, ed. B. M. French, N. M. Short, pp. 87–99. Baltimore: Mono

Gingold, R. A., Monaghan, J. J. 1979.

Binary fission in damped rotating polytropes. *Mon. Not. R. Astron. Soc.* 188: 39–44

Goldreich, P. 1966. History of the lunar orbit. *Rev. Geophys.* 4: 411–39

Goldreich, P., Ward, W. R. 1973. The formation of planetesimals. *Astrophys. J.* 183: 1051–61

Greenberg, R. 1982. Planetesimals to planets. In *Formation of Planetary Systems*, ed. A. Brahic, pp. 515–69. Toulouse, Fr: Lepadues Ed.

Harris, A. W. 1977. An analytical theory of planetary rotation rates. *Icarus* 31: 168–74

Harris, A. W., Kaula, W. M. 1975. A coaccretional model of satellite formation. *Icarus* 24: 516–24

Harris, A. W., Ward, W. R. 1982. Dynamical constraints on the formation and evolution of planetary bodies. *Ann. Rev. Earth Planet. Sci.* 10: 61–108

Hartmann, W. K., Davis, D. R. 1975. Satellite-sized planetesimals. *Icarus* 24: 504–15

Hartmann, W. K., Phillips, R. J., Taylor, G. J., eds. 1986. *Origin of the Moon*. Houston: Lunar Planet. Sci. Inst. 781 pp.

Hashimoto, A. 1983. Evaporation metamorphism in the early solar nebula—evaporation experiments on the melt FeO-MgO-SiO$_2$-CaO-Al$_2$O$_3$ and chemical fractionations of primitive materials. *Geochem. J.* 17: 111–45

Hayashi, C., Nakazawa, K., Nakagawa, Y. 1985. Formation of the solar system. In *Protostars and Planets II*, ed. D. C. Black, M. S. Matthews, pp. 1100–53. Tucson: Univ. Ariz. Press

Hofmeister, A. M. 1983. Effect of a Hadean terrestrial magma ocean on crust and mantle evolution. *J. Geophys. Res.* 88: 4963–83

Holmes, N. C., Radousky, H. B., Moss, M. J., Nellis, W. J., Henning, S. 1984. Silica at ultrahigh temperature and expanded volume. *Appl. Phys. Lett.* 45: 626–27

Horedt, G. P. 1985. Late stages of planetary accretion. *Icarus* 64: 448–70

JANAF Thermochemical Tables. 1971. *Publ. No. NSRDS-NBS37*. Washington, DC: Natl. Bur. Stand. 1141 pp.

Jeanloz, R., Thompson, A. B. 1983. Phase transitions and mantle discontinuities. *Rev. Geophys. Space Phys.* 21: 51–75

Kaula, W. M. 1979. Thermal evolution of Earth and Moon growing by planetesimal impacts. *J. Geophys. Res.* 84: 999–1008

Kaula, W. M. 1980. The beginning of the Earth's thermal evolution. In *The Continental Crust and its Mineral Deposits. Geol. Assoc. Can. Spec. Pap. No. 20*, pp. 25–34

Kaula, W. M., Beachey, A. E. 1986. Mechanical models of close approaches and collisions of large protoplanets. See Hartmann et al 1986, pp. 567–76

Kieffer, S. W. 1975. Droplet chondrules. *Science* 189: 333–40

Kieffer, S. W. 1977a. Impact conditions required for formation of melt by jetting in silicates. See Roddy et al 1977, pp. 751–69

Kieffer, S. W. 1977b. Sound speed in liquid-gas mixtures: water air and water-steam. *J. Geophys. Res.* 82: 2895–2904

Kieffer, S. W. 1979. Thermodynamics and lattice vibrations of minerals. 1. Mineral heat capacities and their relationships to simple lattice vibrational models. *Rev. Geophys. Space Phys.* 17: 1–19

Kieffer, S. W. 1982. Dynamics and thermodynamics of volcanic eruptions: implications for the plumes on Io. In *Satellites of Jupiter*, ed. D. Morrison, pp. 647–723. Tucson: Univ. Ariz. Press

Kinslow, R., ed. 1970. *High Velocity Impact Phenomena*. New York: Academic. 579 pp.

Kipp, M. E., Melosh, H. J. 1986. Origin of the Moon: a preliminary numerical study of colliding planets. *Lunar Planet. Sci. XVII*, pp. 420–21

Kreutzberger, M. E., Drake, M. J., Jones, J. H. 1986. Origin of the Earth's Moon: constraints from alkali volatile trace elements. *Geochim. Cosmochim. Acta* 50: 91–98

Krieger, F. J. 1967. The thermodynamics of the magnesium silicate/magnesium-silicon-oxygen vapor system. *Memo. RM-5337-PR*, Rand Corp., Santa Monica, Calif.

Lambeck, K. 1980. *The Earth's Variable Rotation*. Cambridge: Cambridge Univ. Press

Lambeck, K. 1986. Banded iron formations. *Nature* 320: 574

Lecar, M., Aarseth, S. J. 1986. A numerical simulation of the formation of the terrestrial planets. *Astrophys. J.* 305: 564–79

Lin, C. C., Shu, F. H. 1966. On the spiral arms of disk galaxies. II. Outline of a theory of density waves. *Proc. Natl. Acad. Sci. USA* 55: 229–34

Lucy, L. B. 1977. A numerical approach to the testing of the fission hypothesis. *Astron. J.* 82: 1013–24

Lynden-Bell, D., Pringle, J. E. 1974. The evolution of viscous disks and the origin of nebula variables. *Mon. Not. R. Astron. Soc.* 168: 603–37

Lyzenga, G. A., Ahrens, T. J. 1980. Shock temperature measurements in Mg$_2$SiO$_4$ and SiO$_2$ at high pressure. *Geophys. Res. Lett.* 7: 141–44

Lyzenga, G. A., Ahrens, T. J., Mitchell, A.

C. 1983. Shock temperatures of SiO_2 and their geophysical implications. *J. Geophys. Res.* 88 : 2431–44

Melosh, H. J., Sonett, C. P. 1986. When worlds collide: jetted vapor plumes and the Moon's origin. See Hartmann et al 1986, pp. 621–42

Mizuno, H., Boss, A. P. 1985. Tidal disruption of dissipative planetesimals. *Icarus* 63 : 109–33

Nakazawa, K., Komuro, T., Hayashi, C. 1983. Origin of the Moon: capture by gas drag of the Earth's primordial atmosphere. *The Moon and the Planets* 28 : 311–27

Newsom, H. E. 1984. The lunar core and the origin of the Moon. *Eos, Trans. Am. Geophys. Union* 65 : 369–70

Ohtani, E. 1985. The primordial terrestrial magma ocean and its implications for stratification of the mantle. *Phys. Earth Planet. Inter.* 38 : 70–80

O'Keefe, J. D., Ahrens, T. J. 1977. Impact-induced energy partitioning, melting and vaporization on terrestrial planets. *Proc. Lunar Sci. Conf., 8th*, pp. 3357–74

O'Keefe, J. D., Ahrens, T. J. 1982. The interaction of the Cretaceous/Tertiary extinction bolide with the atmosphere, ocean and solid Earth. See Silver & Schultz 1982, pp. 103–20

Öpik, E. J. 1972. Comments on lunar origin. *Ir. Astron. J.* 10 : 190–238

Orphal, D. L., Borden, W. F., Larson, S. A., Schultz, P. H. 1980. Impact melt generation and transport. *Proc. Lunar Planet. Sci. Conf., 11th*, pp. 2309–23

Rigden, S. M., Ahrens, T. J. 1981. Impact vaporization and lunar origin. *Lunar Planet. Sci. XII*, pp. 885–87 (Abstr.)

Ringwood, A. E. 1966. Chemical evolution of the terrestrial planets. *Geochim. Cosmochim. Acta* 30 : 41–104

Ringwood, A. E. 1979. *Origin of the Earth and Moon.* New York: Springer-Verlag. 295 pp.

Ringwood, A. E. 1986a. The making of the Moon. *Lunar Planet. Sci. XVII*, pp. 714–15 (Abstr.)

Ringwood, A. E. 1986b. Composition and origin of the Moon. See Hartmann et al, pp. 673–98

Roddy, D. J., Pepin, R. O., Merrill, R. B., eds. 1977. *Impact and Explosion Cratering.* Houston: Lunar Sci. Inst. 1299 pp.

Ruff, O., Schmidt, P. 1921. Die dampfdrucke der oxyde des siliciums, aluminiums, calciums und magnesiums. *Z. Anorg. Allg. Chem.* 117 : 172–90

Ruskol, E. L. 1960. Origin of the Moon. I. *Sov. Astron. AJ* 4 : 657–68

Ruskol, E. L. 1982. Origin of planetary satellites. *Izvestiya Earth Phys.* 18 : 425–33

Safronov, V. S. 1966. Sizes of the largest bodies falling onto the planets during their formation. *Sov. Astron. AJ* 9 : 987–91

Safronov, V. S. 1969. Evolution of the protoplanetary cloud and formation of the Earth and planets. *NASA TT F-677*

Silver, L. T., Schultz, P. H., eds. 1982. *Geological Implications of Impacts of Large Asteroids and Comets on the Earth.* GSA Spec. Pap. No. 190. Boulder, Colo: Geol. Soc. Am.

Smith, B. A., and 39 others. 1986. Voyager II in the Uranian system: imaging science results. *Science* 233 : 43–64

Stanyukovich, K. P. 1950. Elements of the physical theory of meteors and the formation of meteor craters. *Meteoritika* 7 : 39–62

Stevenson, D. J. 1981. Models of the Earth's core. *Science* 214 : 611–19

Stevenson, D. J. 1983a. The nature of the Earth prior to the oldest known rock record (the Hadean Earth). In *Origin and Evolution of the Earth's Earliest Biosphere*, ed. J. W. Schopf, Ch. 2. Princeton, NJ: Princeton Univ. Press

Stevenson, D. J. 1983b. Anomalous bulk viscosity of two-phase fluids and implications for planetary interiors. *J. Geophys. Res.* 88 : 2445–53

Stevenson, D. J. 1984a. Lunar origin from impact on the Earth: is it possible? *Conf. Origin of the Moon Abstr. Vol., LPI Contrib. No. 540*, p. 60

Stevenson, D. J. 1984b. Composition, structure and evolution of Uranian and Neptunian satellites. In *Uranus and Neptune*, NASA Conf. Publ. No. 2330, pp. 405–23

Stevenson, D. J., Lunine, J. I. 1986. Mobilization of cryogenic ice in outer solar system satellites. *Nature* 323 : 46–48

Stevenson, D. J., Harris, A. W., Lunine, J. I. 1986. Origins of satellites. In *Satellites*, ed. J. Burns, pp. 39–88. Tucson: Univ. Ariz. Press

Stewart, G. R., Wetherill, G. W. 1986. New formulas for the evolution of planetesimal velocities. *Lunar Planet. Sci. XVII*, pp. 827–28 (Abstr.)

Taylor, S. R. 1986. The origin of the Moon: geochemical considerations. See Hartmann et al 1986, pp. 125–44

Taylor, S. R., Jakes, P. 1974. The geochemical evolution of the Moon. *Proc. Lunar Sci. Conf., 5th*, pp. 1287–1305

Thompson, A. C., Stevenson, D. J. 1983. Two-phase gravitational instabilities in thin disks with applications to the origin of the Moon. *Lunar Planet. Sci. XIV*, pp. 787–88 (Abstr.)

Thompson, A. C., Stevenson, D. J. 1986. Two-phase gravitational instabilities in thin disks with application to the origin of the Moon. Submitted for publication

Tillotson, J. H. 1962. Metallic equations of state for hypervelocity impact. *Rep. GA-3216*, General Atomic Div., General Dynamics, San Diego, Calif.

Walker, J. C. G., Zahnle, K. J. 1986. Lunar nodal tide and distance to the Moon during the Precambrian. *Nature* 320: 600–2

Walker, J. C. G., Klein, C., Stevenson, D. J., Walter, M. R. 1983. Environmental evolution of the Archean early Proterozoic Earth. In *Origin and Evolution of the Earth's Earliest Biosphere*, ed. J. W. Schopf, pp. 32–40. Princeton, NJ: Princeton Univ. Press

Wänke, H., Dreibus, G. 1986. Geochemical evidence for formation of the moon by impact induced fission of the proto-Earth. See Hartmann et al 1986, pp. 649–72

Ward, W. R., Cameron, A. G. W. 1978. Disk evolution within the Roche limit. *Lunar Planet. Sci. IX*, pp. 1205–7 (Abstr.)

Warren, P. H. 1985. The magma ocean concept and lunar evolution. *Ann. Rev. Earth Planet. Sci.* 13: 201–40

Weidenschilling, S. J., Greenberg, R., Chapman, C. R., Herbert, F., Davis, D. R., et al. 1986. Origin of the Moon from a circumterrestrial disk. See Hartmann et al 1986, pp. 731–62

Wetherill, G. W. 1975. Possible slow accretion of the Moon and its thermal and petrological consequences. In *Origins of Mare Basalts*, pp. 184–88. Houston: Lunar Sci. Inst.

Wetherill, G. W. 1980. Formation of the terrestrial planets. *Ann. Rev. Astron. Astrophys.* 18: 77–113

Wetherill, G. W. 1985. Occurrence of giant impacts during the growth of the terrestrial planets. *Science* 228: 877–79

Wetherill, G. W., Cox, L. P. 1984. The range of validity of the two-body approximation in models of terrestrial planet accumulation. I. Gravitational perturbations. *Icarus* 60: 40–55

Wetherill, G. W., Cox, L. P. 1985. The range of validity of the two-body approximation in models of terrestrial planet accumulation. II. Gravitational cross sections and runaway accretion. *Icarus* 63: 290–303

Wetherill, G. W., Stewart, G. R. 1986. The early stages of planetesimal accumulation. *Lunar Planet. Sci. XVII*, p. 939 (Abstr.)

Wood, J. A. 1986. Moon over Mauna Loa: a review of hypotheses of formation of Earth's Moon. See Hartmann et al 1986, pp. 17–55

Wood, J. A., Mitler, H. E. 1974. Origin of the Moon by a modified capture mechanism, or half a loaf is better than a whole one. *Proc. Lunar Sci. Conf., 5th*, pp. 851–53

EXPERIMENTAL AND THEORETICAL CONSTRAINTS ON HYDROTHERMAL ALTERATION PROCESSES AT MID-OCEAN RIDGES

W. E. Seyfried, Jr.

Department of Geology and Geophysics, University of Minnesota, Minneapolis, Minnesota 55455

INTRODUCTION

The discovery of hydrothermal fluids issuing from vents on actively spreading mid-ocean ridges spectacularly confirmed the long suspected and often inferred existence of submarine geothermal systems (Edmond et al 1979a,b, von Damm et al 1985a,b, Michard et al 1984). The chemistry, isotopic composition, and temperature of hot-spring fluids provide incontrovertible evidence of extensive reaction between seawater-derived hydrothermal fluid and basalt (or compositional analogues) at temperatures as high as 350–400°C and pressures of approximately 400–500 bars, conditions that locate the reaction zone in close proximity to subseafloor magma chambers (Sleep et al 1983, von Damm et al 1985a, Bischoff & Rosenbauer 1984). In comparison with seawater chemistry, ridge-crest hydrothermal fluids are distinctly acidic, depleted in Mg and SO_4, and enriched in $SiO_{2(aq)}$, Ca, H_2S, and a wide range of other species (particularly heavy and base metals). Oxygen and hydrogen isotopic compositions of end-member hydrothermal fluid at 21°N East Pacific Rise (Craig et al 1980), for example, indicate high-temperature alteration of basalt (diabase, gabbro) at fluid/rock mass ratios as low as 0.5 (Bowers & Taylor 1985).

The chemistry of hot-spring fluids alone, however, is insufficient to unambiguously establish the mechanisms of hydrothermal alteration and

H^+ generation, the specific nature of reactants, and the responsiveness of alteration processes to changes in pertinent physical and chemical variables for all or a portion of the submarine geothermal system. Information of this sort can be obtained only by integrating appropriate field observations with results of rock-fluid interaction experiments and temperature-, pressure-, and composition-dependent solution-mineral equilibria. Here we apply these different approaches to obtain a comprehensive model for hydrothermal alteration processes at mid-ocean ridges.

EXPERIMENTAL MODELS OF BASALT-SEAWATER INTERACTION

Basalt-seawater interaction studies have contributed significantly to our understanding of ridge-cast hydrothermal processes (Bischoff & Dickson 1975, Hajash 1975, Mottl & Holland 1978, Mottl et al 1979, Seyfried & Bischoff 1979, 1981, Seyfried & Mottl 1982, Rosenbauer & Bischoff 1983, Seyfried & Janecky 1985). Experiments have now been performed at temperatures from 70 to 500°C, at pressures from 400 to 1000 bars, and at fluid/rock mass ratios from 0.5 to 125. These studies typically entail monitoring the change in fluid chemistry during reaction of NaCl-rich fluids with powdered basalt (basalt glass, crystalline basalt, and diabase), followed by an analysis of alteration products after the experiment. The accuracy of fluid chemical data from the experiments, however, is greatly influenced by the type of experimental equipment utilized. Both "cold seal" and on-line sampling (flexible reaction cell; Seyfried et al 1987) equipment have been employed. These differ in several important ways; in particular, the flexible-cell system permits multiple fluid samples to be taken from an experiment at conditions (T, P), thus avoiding quench effects. Quenching can cause dissolution or precipitation of minerals and have a large effect on solution chemistry, especially pH.

Experiments have documented quite clearly the relative mobility of elements during hydrothermal alteration of basalt. As emphasized by Mottl (1983a), three principal mechanisms are important here: (a) depletion from the rock, (b) solution-mineral equilibria, and (c) kinetic effects. The extent to which each of these affects an element during hydrothermal alteration of basalt depends critically on temperature, pressure, relative abundance and nature of primary and alteration phases, and fluid/rock mass ratio. For example, under an appropriate set of conditions, some elements, typically present in trace to moderate concentrations in primary igneous minerals and noncrystalline mesostasis, are variably incompatible in alteration phases and partition effectively and in some cases entirely into the fluid. The concentration that these elements achieve in the fluid is a function only of their concentration in unaltered primary phases and of the effective fluid/rock mass ratio. B, Li, Rb, and to a lesser degree K behave in this manner (Seyfried et al 1984). Major elements, in contrast, are most often controlled by the solubility of key alteration phases; Mg as smectite or chlorite, Ca as anhydrite or clinozoisite, and Na as albite are examples. Kinetic effects characterized by the metastable formation or sluggish precipitation of a mineral clearly play a role in basalt alteration processes, even at relatively high temperatures. The metastable existence of anhydrite and the recalcitrant precipitation and growth kinetics of quartz under acidic conditions can cause wide departures from behavior predicted by mineral solubility relations.

Thus, experimental studies of basalt-fluid interactions are best suited to establish cause-and-effect relationships, but if adequate geologic constraint is available, experiments can be designed to model alteration processes in natural geothermal systems and reproduce almost exactly the chemistry of hot-spring fluids and by inference the chemistry and (to some degree) the mineralogy of alteration assemblages. Owing to excellent field data, this has been accomplished for portions of the submarine geothermal system at mid-ocean ridges [in particular, at 21°N, East Pacific Rise (von Damm et al 1985a)] and likely will be extended to other areas as well, including marginal basins, such as the Guaymas Basin in the Gulf of California (von Damm et al 1985b).

Direction and Magnitude of Chemical Exchange for Basalt- and Seawater-Dominated Systems at Low to Moderate Temperatures of Alteration (150–350°C)

Seawater alteration of basalt under hydrothermal conditions is characterized by Mg removal from solution and formation of Mg-rich secondary phases. The extent to which this occurs, however, is dependent on the seawater/rock mass ratio (Seyfried & Mottl 1982) and temperature. At 150°C, and probably for even lower temperatures, Mg removal from seawater is complete at seawater/rock mass ratios as high as 50 in relatively short periods of time; temperature affects the rate but not the magnitude of the reaction.

Removal of Mg from seawater occurs in the form of $Mg(OH)_2$ components in secondary silicates leaving H^+ behind in solution, as follows:

$$3Mg^{++} + 4SiO_{2(aq)} + 4H_2O = \overbrace{Mg_3Si_4O_{10}(OH)_2}^{\text{Talc}} + 6H^+, \qquad (1)$$

where talc is used as an analogue for more compositionally complex chlorites and smectites, which are the phases that actually form during the experiments either by direct precipitation or, more typically, by replacement of primary minerals in basalt, such as calcic feldspar, pyroxenes, and olivine. Thus, the extent of overall reaction is governed both by H^+ production via Mg metasomatic processes and by H^+ titration via silicate hydrolysis reactions. As long as the solution has a relatively high concentration of Mg, the rate of H^+ production exceeds the rate of H^+ consumption, and thus an acid pH results (Figure 1). At low to moderate temperatures (approximately 150–350°C) and seawater/rock mass ratios less than 50 ("basalt dominated"), the acid-generating mechanism is short lived, and Mg in solution is effectively and entirely replaced by Ca and, at very low seawater/rock mass ratios, by Na and K owing to dissolution of primary silicate phases (Figure 2).

Under rock-dominated conditions and low to moderate temperatures, heavy and base metals are mobile only during early-stage reaction (maximum rate of H^+ production), while SiO_2 is maintained in solution at concentrations at or near quartz saturation [although this depends greatly on the composition and texture of the fresh basalt (Figure 2)].

Neutralization of solution acidity occurs with the elimination of Mg; thus, heavy and base metals are rendered immobile, while the alkali (Na, K) and alkaline earth (Mg, Ca) elements are controlled to varying degrees by secondary mineralization (Mottl & Holland 1978, Hajash 1975, Seyfried & Bischoff 1981). At temperatures above 150°C, an important sink for Ca is anhydrite precipitation, the formation of which is responsible for the conspicuous decrease in dissolved SO_4 during the experiment shown in

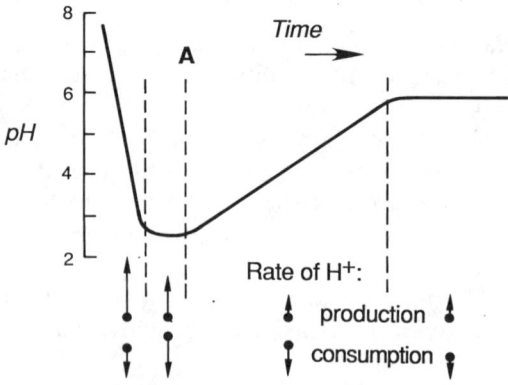

Figure 1 Idealized profile of pH vs time for rapidly heated seawater undergoing reaction with basalt. Length of arrows indicates the rates of H^+ production via Mg removal and H^+ consumption via silicate hydrolysis reactions. The time period to the left of "A" represents seawater-dominated conditions; to the right, rock-dominated conditions. Reprinted with permission of Pergamon Press Ltd. (Seyfried & Mottl 1982).

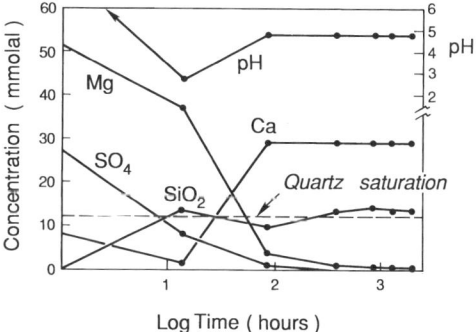

Figure 2 Concentrations of Ca, Mg, SiO$_2$, SO$_4$, and pH (measured at 25°C) in seawater during reaction with diabase at 300°C, 500 bars, and seawater/rock mass ratio of 10. Reprinted with permission of Pergamon Press Ltd. (Seyfried & Bischoff 1981).

Figure 2. In effect, anhydrite formation limits dissolved Ca concentrations in solution and often precludes formation of Ca-bearing silicates, especially at low to moderate temperatures, where the solubility of these phases is relatively high.

At seawater/rock mass ratios greater than 50, Mg removal from solution is less than complete owing to the inadequate abundance of cations in basalt to exchange with Mg; that is, Mg uptake is balanced with respect to electrical charge in solution by leaching of Ca, Na, and K from the rock, since these species are relatively abundant in fresh basalt/diabase and are incompatible with the structure of chlorite or chlorite/smectite, the dominant alteration products of basalt at temperatures from approximately 250–450°C and at high seawater/rock mass ratios. Under these conditions, dissolved Ca is controlled by anhydrite solubility, whereas Na and K concentrations in solution are limited only by their concentrations in the rock. In contrast, Al and Fe are stable in sheet silicate structures and are not extensively replaced by Mg (this is less true of Fe than Al), as evidenced by the relatively high Fe concentrations in solution during reaction of basalt with seawater at fluid/rock ratios greater than 50 (Seyfried & Mottl 1982). Thus, with increasing seawater/rock ratio, the period of H$^+$ production is prolonged and the effectiveness of H$^+$ consumption reduced, culminating at a seawater/rock ratio of approximately 50 in more-or-less continuously acidic conditions (Figure 1) and in a breakdown of all but the most acid-resistant secondary minerals (Seyfried & Mottl 1982).

Solution-Mineral Equilibria at Elevated Temperatures and Pressures

MAGNESIUM METASOMATISM   We have emphasized that the presence of Mg in solution has a dominating effect on pH, even at temperatures as low as

70°C. Perhaps a better way to illustrate this is by means of a reaction involving formation of clinochlore and clinozoisite and anorthite and requisite aqueous species including Mg, namely

$$Mg^{++} + 0.8\overbrace{CaAl_2Si_2O_8}^{\text{Anorthite}} + 0.2SiO_2 + 2.0H_2O$$
$$= 0.2\underbrace{Mg_5Al_2Si_3O_{10}(OH)_8}_{\text{Clinochlore}} + 0.4\underbrace{Ca_2Al_3Si_3O_{12}(OH)}_{\text{Clinozoisite}} + 2H^+, \quad (2)$$

as a function of temperature and pressure (Figure 3A). Ca and Al are

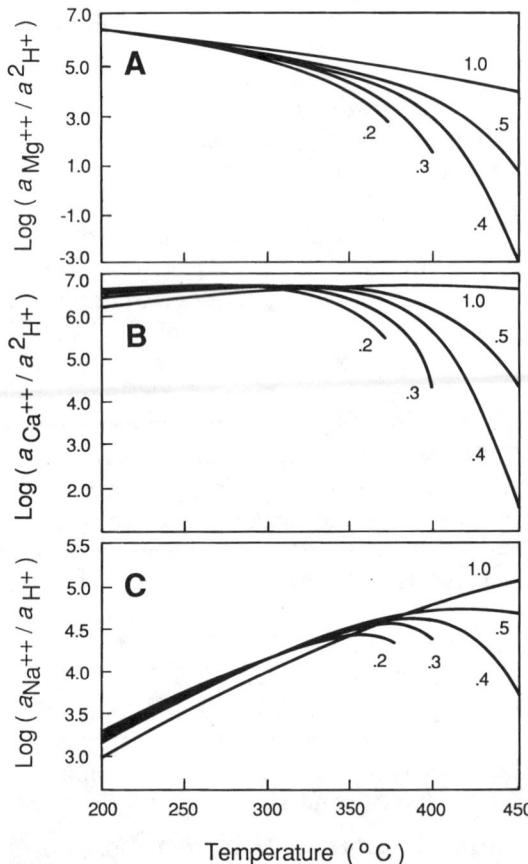

Figure 3 Log activity quotient (log K) for (*A*) Mg, (*B*) Ca, and (*C*) Na fixation reactions (see text) vs temperature at constant pressure (in kilobars). Thermochemical data used for constructions are from Helgeson et al (1978).

conserved in the reaction; thus reaction (2) models more accurately high-temperature than low-temperature alteration processes, since basalt alteration at low temperatures generates Ca-rich fluids because of the relatively high solubility of hydrous Ca-bearing silicates under these conditions.

The curves shown in Figure 3A (also 3B and 3C) were calculated by assuming an activity of unity for minerals and water, and they may differ from the actual positions for naturally occurring and experimentally derived analogues; that is, stoichiometric clinozoisite and mixed-layer chlorite/smectite are common alteration products of high-temperature basalt alteration experiments (Seyfried & Janecky 1985), whereas epidote solid solutions and chlorite (ripidolite) more accurately represent Ca- and Mg-rich secondary phases in nature (Mottl 1983b). The relative positions of the curves, however, which are of primary importance to us here, are most probably not affected significantly by slight to moderate compositional changes, especially for high-temperature, relatively low-pressure conditions where the physical nature (density) of the solvent water dominates mineral solubility relations (Helgeson & Kirkham 1976, Helgeson et al 1978). Thus, the relative stability of the end-member minerals may be used to evaluate the effects of temperature and pressure on cation fixation and, correspondingly, H^+ production.

The continuous decrease in $a_{Mg^{++}}/a_{H^+}^2$ (Figure 3A) indicates a distinct thermodynamic potential for Mg fixation, chloritization of feldspar, and production of acidity with increasing temperature; this is precisely what was observed during the experiments—the higher the temperature, the greater the rate and magnitude of Mg removal from solution (Seyfried & Mottl 1982). Thus, theoretical data explain and confirm the reactivity of Mg during hydrothermal alteration of basalt. *The implication of this is clear; it is highly unlikely that seawater-derived Mg will penetrate completely the sharply increasing grade of hydrothermal metamorphism characterizing the submarine geothermal system at mid-ocean ridges.* The observed mineralogy (chlorite, quartz rich) and chemistry (Mg rich, Ca poor) of metabasalts thought to represent alteration products of basalt within relatively shallow levels of the oceanic crust, which characterize downwelling limbs of subseafloor convection cells (Mottl 1983b), are consistent with this interpretation. Chlorite formation and Mg enrichment in these rocks is due as much to Mg reactivity in response to increasing temperature as it is to seawater chemistry and seawater/rock mass ratio.

The extremely reactive nature of Mg during hydrothermal alteration of basalt requires modification of seawater chemistry to model experimentally deep-seated, high-temperature alteration processes. Thus, recent basalt alteration experiments designed to simulate processes responsible for the chemistry of axial hot-spring fluids have utilized fluids of seawater ionic strength, but ones that were chemically distinct from seawater (Seyfried & Janecky 1985, Rosenbauer & Bischoff 1983)—in particular, free of Mg and SO_4. An important objective of these so-called evolved seawater experiments was to evaluate the effectiveness of H^+ production by Ca and Na metasomatic processes. Ridge-crest hot-spring fluids are invariably acidic, and Ca and Na are the only dissolved species in seawater-derived hydrothermal fluids capable of forming hydrous alteration phases.

CALCIUM AND SODIUM METASOMATISM At temperatures greater than approximately 350°C, the relatively simple mechanism of alteration described above (that is, where Mg exchange is the driving force for basalt alteration) is superseded by more complex alteration processes owing to temperature-dependent changes in the solubility of hydrous secondary silicates. Ca and Na, for example, which are relatively "soluble" at low temperatures, exhibit an increasing tendency with increasing temperature to become "fixed" in solid phases (Mottl & Holland 1978, Seyfried & Janecky 1985). Thus, in a manner similar to that for Mg, removal of these species from solution lowers solution pH, provided the precipitated cation resides in a hydrous alteration phase. Indeed, at sufficiently high temperatures, the fluid need not contain Mg to generate acidity (Seyfried & Janecky 1985), and therefore the relationship between rock- and seawater-dominated systems and solution pH (see Seyfried & Mottl 1982) is valid only for temperatures less than 350°C.

In contrast to the behavior of Mg, the fixation of Ca, illustrated here by clinozoisite-anorthite equilibria (Figure 3B), i.e.

$$\underbrace{3CaAl_2Si_2O_8}_{\text{Anorthite}} + Ca^{++} + 2H_2O = \underbrace{2Ca_2Al_3Si_3O_{12}(OH)}_{\text{Clinozoisite}} + 2H^+, \quad (3)$$

fails to show continuous prograde behavior with increasing temperature. In fact, close evaluation of the change in $a_{Ca^{++}}/a_{H^+}^2$ reveals a slight but nevertheless significant increase with increasing temperature and pressure. The temperature at which this occurs, however, decreases with decreasing pressure. Thus, Ca fixation can result in generation of acidity only if a sufficiently high temperature is attained to cause a decrease in log K of the reaction [reaction (3)]. This is especially significant in light of the chemistry of fluids recharging the submarine geothermal system, which are most likely Ca rich and Mg poor because of relatively low-temperature basalt alteration and seawater modification at shallow levels within the crust. Thus, Ca fixation most probably provides the primary means by which high-temperature hydrothermal fluids become acidic.

Solution chemistry, in particular SiO_2 concentration, can play an important role in modifying mineral solubility relations (Seyfried & Janecky 1985). Acid attack of primary aluminosilicates and of SiO_2-rich interstitial crystalline and noncrystalline phases in basalt often is characterized by nonstoichiometric dissolution and formation of fluids relatively rich in SiO_2 compared with quartz saturation under similar conditions (Hemley et al 1980). This behavior can be attributed to sluggish kinetics of quartz nucleation and growth at low-pH conditions, which results in enhanced stability of SiO_2-rich alteration phases, such as albite. Na replacement of Ca in plagioclase, however, permits Ca to form hydrous alteration phases, such as clinozoisite, and to generate acidity as follows:

$$Na^+ + 2\overbrace{CaAl_2Si_2O_8}^{\text{Anorthite}} + 2SiO_2 + H_2O$$
$$= \overbrace{NaAlSi_3O_8}^{\text{Albite}} + \overbrace{Ca_2Al_3Si_3O_{12}(OH)}^{\text{Clinozoisite}} + H^+. \quad (4)$$

Thus, Na metasomatism contributes directly to acidity production provided sufficient SiO_2 is available to stabilize albite. The more albite that forms, the more anorthite (plagioclase) is recycled into clinozoisite (epidote) and the more H^+ is released to solution. In effect, at constant temperature and pressure, high SiO_2 activities in solution result in low a_{Na^+}/a_{H^+} values, whereas the opposite is true for low SiO_2 activities [reaction (4)].

Because SiO_2 is released to solution more effectively from glassy than from nonglassy basalts, high-temperature hydrothermal alteration of basalts of variable crystallinity is characterized by a very different direction and magnitude of chemical exchange for those elements most diagnostic of alteration processes (in particular, Na and Ca but also Sr, since Sr occupies similar sites as Ca in primary and secondary alteration phases and thus is affected by similar constraints during hydrothermal alteration of basalt). Experimental data show, for example, that hydrothermal alteration of diabase generally results in relatively low-Ca (Ca fixation), high-Sr fluids (Figure 4), with SiO_2 concentrations at or below quartz saturation (Seyfried & Janecky 1985). In contrast, alteration of glassy rocks (crystalline basalt with intercrystalline glass and basalt glass) produces Na for Ca exchange (albitization and/or formation of Na amphibole) and H^+ generation by Na metasomatism and Ca recycling from plagioclase to clinozoisite [reaction (4)]. Ca release to solution often occurs during glass alteration, however, especially at low to moderate temperatures, while Sr

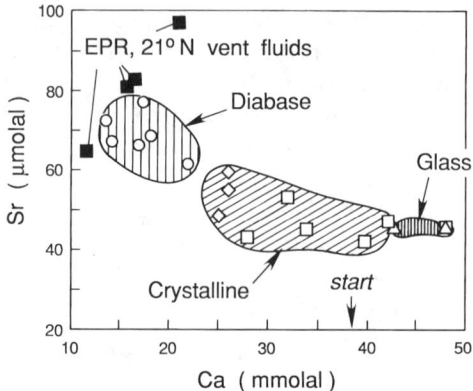

Figure 4 Dissolved Ca (mmolal) and Sr (μmolal) concentrations from basalt alteration experiments at 350–425°C, 400 bars, and fluid/rock mass ratios of 0.5–1.0 (see Seyfried & Janecky 1985). Hydrothermal alteration of diabase (○), crystalline basalt (◇, □) and basalt glass (△) yield different fluid chemistries (characterized here by Ca and Sr) owing to different textural and mineralogic conditions, which affect SiO_2 concentrations in solution and the stability of secondary phases. "Start" refers to the initial Ca and Sr concentrations in the fluid used for the experiment. Ca and Sr data for 21°N, East Pacific Rise hot springs (von Damm et al 1985a) are shown for comparison purposes (see text).

is retained in solid phases, most probably substituting for Mg in hydrous sheet silicate minerals (Figure 4).

Na metasomatism dominates chemical exchange and H^+ production during hydrothermal alteration of relatively SiO_2-rich systems, whereas Ca metasomatism dominates alteration processes for relatively SiO_2-poor systems. Owing to the importance of temperature on the direction of these reactions, however, the need for SiO_2 for Na fixation can be offset significantly by temperature effects. For example, temperature- and pressure-dependent changes in a_{Na^+}/a_{H^+} (log K) for reaction (4) are distinctly different than for the Ca and Mg metasomatic processes discussed previously. In particular, the continuous increase in log K with increasing temperature to approximately the critical point of water (Figure 3C) indicates a distinct tendency for albite to form from solutions ascending from higher to lower temperature environments; that is, Na fixation as illustrated by reaction (4) is characterized by retrograde reactivity (Giggenbach 1984). Moreover, considering this behavior along with the relatively constant temperature-dependent behavior of the equilibrium constant for Ca metasomatism [Figure 3B; reaction (3)], we can conclude that a thermodynamic drive exists to fix Na relative to Ca for cooling geothermal fluids.

End-member hydrothermal fluids at 21°N, East Pacific Rise are characterized by relatively low-Ca, high-Sr concentrations (von Damm et al

1985a); such concentrations suggest that H^+ is produced by Ca fixation [reaction (3)] and that nonglassy basalt (diabase or gabbro) is altered at relatively high temperatures, low pressures, and SiO_2 concentrations in solution at or below quartz saturation (Figure 4). These fluids also reveal slight but significant Na fixation, although the extent of this is complicated somewhat by processes responsible for Cl removal (von Damm et al 1985a, Seyfried et al 1986). If Cl is normalized to seawater, then hot-spring fluids at 21°N, East Pacific Rise show Na gains relative to seawater Na, consistent with the notion of Ca fixation suggested above. However, if this same procedure is applied to 13°N, East Pacific Rise hot-spring fluids, where dissolved Cl is approximately 30% greater than in seawater (Michard et al 1984), then a large Na depletion and Ca enrichment is found, which suggests that Na-fixation and Ca-leaching reactions dominate subseafloor alteration processes at this locality. It is unlikely that this is caused by variations in the composition and/or texture of basalt source rocks or temperature at the primary site of alteration, and most probably it indicates operation of still another process influencing Ca for Na-exchange reactions, this being a concentration condition.

As illustrated by reactions (3) and (4), Ca for Na exchange is controlled by the following equilibria:

$$\underbrace{CaAl_2Si_2O_8}_{\text{Anorthite}} + 2Na^+ + 4SiO_2 = \underbrace{2NaAlSi_3O_8}_{\text{Albite}} + Ca^{++}. \tag{5}$$

Since divalent Ca is more extensively complexed with Cl in aqueous solutions at elevated temperatures and pressures than is monovalent Na (Bowers et al 1987, Berndt et al 1987, and references therein), the effect of increasing Cl concentration in solution is that reaction (5) shifts to the right, fixing Na as albite and increasing the Ca concentration in solution. Thus, the processes that have contributed to the increase in total dissolved Cl concentration in solution at 13°N, East Pacific Rise may have also played a key role in influencing the relative direction and magnitude of Ca and Na exchange. At present, this finding is consistent with theoretical data for a necessarily small and simplified portion of the CaO-Na_2O-SiO_2-Al_2O_3-H_2O system. This system, however, needs to be expanded to include FeO and Fe_2O_3 and activity-concentration relations for plagioclase and epidote solid solutions, so that solubility relations of phases that actually represent alteration products of basalt in the submarine geothermal system can be rigorously evaluated.

Heavy-Metal Mobility

Experiments have shown that Fe and Mn mobility during hydrothermal alteration of basalt is directly related to measured pH (Mottl et al 1979,

Rosenbauer & Bischoff 1983, Seyfried & Janecky 1985), which, as noted, is a sensitive function of temperature, pressure, and solution chemistry. Temperature and pressure, however, are of primary importance here, since for a fluid of relatively constant salinity, these parameters define fluid density and dielectric constant, which dominate mineral solubility relations and aqueous dissociational equilibria. Thus, fluid pH, a function of temperature- and pressure-dependent mineral solubility relations, should correlate with fluid density, chiefly a function of the temperature- and pressure-dependent physical nature of the solvent water. Indeed, Seyfried & Janecky (1985) have recognized just such a relationship and have shown that fluid density and pH correlate linearly ($R^2 = 0.9$), with the variation due entirely to basalt chemistry (primarily the abundance and reactivity of primary SiO_2-bearing minerals and correspondingly dissolved SiO_2 concentration).

The approach taken here has been to emphasize the role of pressure on idealized mineral solubility relations and on activities of aqueous species to account for changes in pH and cation exchange reactions. The role of pressure, however, can also be expressed in terms of dissociational equilibria; that is, decreasing the pressure at elevated temperatures favors formation of aqueous complexes whose dissociation is decreased most strongly by the pressure drop (Hemley et al 1986). Since HCl is affected by this more strongly than the alkali chlorides (Helgeson & Kirkham 1976), for example, a thermodynamic potential exists for cation fixation and transformation of alkali chloride complexes to HCl; that is, the total acidity of the system increases. Of course, HCl dissociates on cooling and contributes H^+ to solution to produce low measured pH values. Thus, mineral hydrolysis reactions and dissociational equilibria act together to increase the total acidity of relatively high-temperature, high-salinity hydrothermal fluids. The relative contribution of each to fluid pH depends critically on temperature, pressure, and rock and fluid chemistry.

The linear relation shown to exist for fluid density and pH can be extended to include Fe and Mn, since results of basalt alteration studies (Mottl et al 1979, Seyfried & Janecky 1985) show that log Fe and log Mn correlate linearly with pH and, correspondingly, fluid density. Data from Seyfried & Janecky (1985), for example, reveal regression coefficients (R^2) of 0.91 and 0.96 for log Fe and log Mn plotted against pH, respectively. The increased abundance of these species in solution with decreasing fluid pH (as measured at 25°C) and decreasing fluid density results from enhanced solubility of Fe and Mn minerals in H^+(HCl)-rich fluids.

Using density data for seawater from Bischoff & Rosenbauer (1985) and Fe, Mn, and pH data from numerous basalt (crystalline basalt and diabase) alteration experiments at temperatures of 350–425°C, pressures of 375–400 bars, and fluid/rock mass ratios of 0.5–1.0, we can establish a general relation for Fe and Mn solubility and solution pH for basalt alteration as a function of temperature and pressure, since temperature and pressure

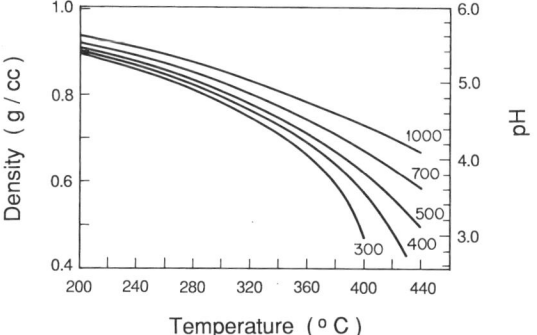

Figure 5 Fluid density and pH (measured at 25°C) vs temperature at constant pressure. The correlation between density, temperature, and pressure (in bars) is for 3.2% NaCl (seawater analogue) and from Bischoff & Rosenbauer (1985). The correlation between pH, temperature, and pressure was determined from basalt alteration experiments utilizing diabase and crystalline basalt at 350–425°C, 375–400 bars, and fluid/rock mass ratios of 0.5–1.0 (Seyfried & Janecky 1985).

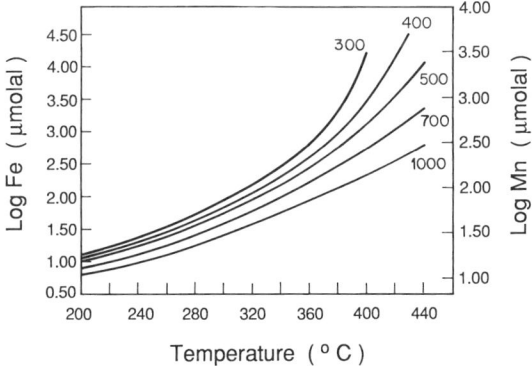

Figure 6 Log Fe and log Mn concentrations (μmolal) in solution from basalt alteration experiments described in Figure 5 as a function of temperature and pressure (in bars) (see text and Seyfried & Janecky 1985). The correlation between dissolved Fe and Mn concentrations and temperature and pressure was indirectly obtained from a correlation between these elements and pH (Mottl et al 1979, Seyfried & Janecky 1985) and from the correlation between density (Bischoff & Rosenbauer 1985) and pH noted in Figure 5 (see text).

define fluid density under the assumption of a constant salinity (Figures 5 and 6). These data clearly illustrate what was inferred earlier from mineral solubility relations or from dissociational equilibria (Hemley et al 1986); that is, at temperatures greater than 350°C, pressure plays an increasingly important role in silicate (basalt) alteration processes by constraining cation (alkali and alkaline chloride) fixation reactions and H^+ (HCl) production, thereby controlling the solubility of key Fe- and Mn-bearing oxide and silicate phases and permitting Fe and Mn concentrations in solution to achieve conspicuously high values in low-density fluids.

The increase in concentration of Fe and Mn and decrease in pH with decreasing density cannot continue indefinitely owing to the limited transport capacity of gaslike fluids. Thus, maximum dissolved metals and minimum pH (maximum HCl concentrations) presumably occur for a compositionally homogeneous fluid just prior to the intersection of the two-phase (low-salinity vapor, high-salinity brine) boundary of seawater. At the two-phase boundary, higher metal concentrations are expected for the brine phase, although appreciable solute exists as well in the so-called vapor phase (Bischoff & Rosenbauer 1984) and metal concentrations may be appreciable.

Ridge-crest hydrothermal fluid at 21°N, East Pacific Rise (von Damm et al 1985a) is characterized by relatively high concentrations of heavy metals (especially Fe and Mn) and low solution pH (Table 1). As in the experiments, Fe and Mn correlate with pH. Furthermore, multiple linear regression analysis of experimental data for Fe, Mn, and pH permits a

Table 1 Measured dissolved concentrations of Fe, Mn, and SiO_2, pH, calculated fluid density, temperature, and pressure for end-member hydrothermal fluids at 21°N, East Pacific Rise

	Measured[a]				Calculated[b]		
Vent[c]	Fe (μmolal)	Mn (μmolal)	SiO_2 (mmolal)	pH	Density (g cm^{-3})	Temperature (°C)	Pressure (bars)
HG	2429	878	15.6	3.3	0.525	395	320
SW	750	699	17.3	3.6	0.579	385	340
OBS	1664	960	17.6	3.4	0.544	400	380
NGS	871	1002	19.5	3.8	0.620	385	400

[a] From van Damm et al (1985a).
[b] Calculated from multiple linear regression analysis of data used for Figures 5 and 6 and quartz solubility data from Fournier (1983) calculated for seawater using effective density of 0.965 V^{-1} (Bischoff & Rosenbauer 1985).
[c] Abbreviations: HG, Hanging Gardens; SW, Southwest; OBS, Ocean Bottom Seismometer; NGS, National Geographic Smoker.

relation to be developed for fluid density in terms of these parameters (Figures 5 and 6, Table 1). Thus, Fe, Mn, and pH data for hot-spring vents at 21°N, East Pacific Rise can be used to estimate fluid density at the site of alteration by direct comparison with experimental results. Fluid densities of 0.525, 0.579, 0.544, and 0.620 are calculated for HG, SW, OBS, and NGS vent fluids, respectively (Table 1).

Fluid density, of course, can be used to estimate temperature or pressure of alteration in the subsurface provided one or the other of these parameters can be reliably constrained. Von Damm et al (1985a), for example, assumed a maximum temperature of approximately 350°C and fluid-quartz equilibrium to establish pressure at the site of upflow for the different hot-spring vents at 21°N. NGS vent fluid, which apparently has cooled conductively from 350 to 273°C gives a subseafloor depth of 3.5 km, while HG, OBS, and SW vent fluids give subseafloor depths of 0.5, 2.0, and 1.5 km, respectively. If we consider the close proximity of vents on the seafloor, especially NGS, OBS, and SW, the relatively large variation in depth of reaction is unsettling. However, by making use of fluid density determined from analysis of Fe, Mn, and pH data (Table 1) and quartz solubility as a function of effective fluid density (i.e. the density of free water), temperature, and pressure (Fournier 1983), we can use the measured SiO_2 concentrations of the different hot-spring fluids to define both temperature and pressure at a site of alteration, precluding assumptions other than quartz saturation. Results of this analysis (Figure 7) show that the reaction zone for all of the 21°N, East Pacific Rise hot-spring vent fluids occurs at subseafloor depths that range from 0.7 to 1.5 km and at

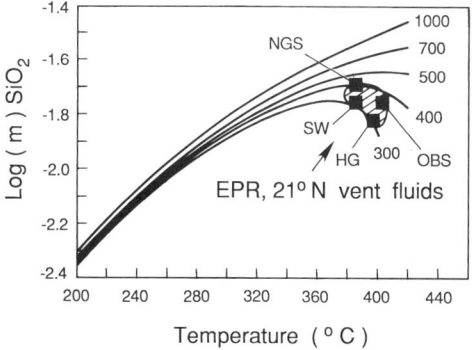

Figure 7 Log SiO_2 concentrations (molal) in solution in equilibrium with quartz as a function of temperature, pressure (in bars), and effective density (Fournier 1983). Dissolved SiO_2 concentrations in 21°N, East Pacific Rise hot-spring vent fluids (von Damm et al 1985a) are superimposed on the diagram to illustrate the conditions of alteration giving rise to these fluids.

temperatures of $395 \pm 10°C$. Thus, if subseafloor temperatures are permitted to exceed 350°C, a rigorous assessment of the SiO_2 solubility data gives rise to relatively similar conditions of alteration for all the hot-spring vents. Again, at temperatures greater than 350°C, relatively slight pressure changes have a profound effect on fluid density and correspondingly dissolved concentrations of heavy metals, SiO_2, and fluid pH. Subseafloor temperatures in the vicinity of 400°C at 21°N, East Pacific Rise, however, require heat loss from ascending fluids in amounts greater than those obtained by simple adiabatic decompression to account for the near 350°C measured temperatures. Either vent temperatures have been underestimated by 30–40°C, or else heat has been lost by conduction or subseafloor mixing processes.

The use of data from bulk-rock alteration experiments to calibrate physicochemical conditions for ridge-crest hot-spring fluids requires that the chemistry and mineralogy of experimentally derived alteration is analogous to that characterizing actual subseafloor hydrothermal systems. Manganese, for example, often shows a high degree of variability in experimental and natural hydrothermal systems owing to its presence in fresh and altered basalt in variable and minor amounts, and thus dissolved Mn is not as reliable an indicator of alteration conditions as is dissolved Fe, SiO_2, and pH. As noted previously, however, it is essential that the solubility of key alteration phases (epidote and plagioclase solid solutions) inferred to characterize subseafloor alteration processes be experimentally determined to assess accurately ridge-crest hydrothermal processes and to constrain the results of bulk-rock alteration experiments, where these phases may or may not have formed in requisite amounts.

Dissolved concentrations of Zn and Cu observed during experiments are not simple functions of pH; instead, they appear to depend on H_2S and Fe concentrations in solution, on temperature, and especially on pressure at high temperatures. Seyfried & Janecky (1985) have suggested that high Fe concentrations act to constrain Cu and Zn by stabilizing Fe-Zn- and Fe-Cu-rich phases such as Fe-rich sphalerite and chalcopyrite, respectively, while sulfide alteration phases are also stabilized by high dissolved H_2S concentrations. These relations are not entirely consistent with chemical data for hot-spring fluids at 21°N, East Pacific Rise (von Damm et al 1985a). Clearly a rigorous assessment of heavy-metal concentrations, especially for Cu and Zn, in ridge-crest hydrothermal fluids in comparison with analogous data from experiments at appropriate temperatures and pressures is needed to establish unambiguously the effect of temperature, pressure, solution chemistry, and basalt/diabase chemistry and texture on heavy-metal mobility during hydrothermal alteration of the oceanic crust at mid-ocean ridges.

CONCLUSIONS

Experimental and theoretical data can be used to constrain hydrothermal alteration processes in the submarine geothermal system at mid-ocean ridges. These data indicate, for example, that seawater chemistry, basalt/diabase chemistry and texture, and temperature and pressure dictate the direction and magnitude of chemical exchange, the chemistry and mineralogy of alteration phases, and the mechanism of H^+ production, which is so critical to the solubility of heavy and base metals.

At low temperatures, seawater Mg dominates alteration processes and provides the driving force for chemical exchange by precipitating as chlorite and smectite and providing H^+ to solution to leach relatively "soluble" and abundant elements from basalt. At temperatures less than 350°C, Mg removal from solution is balanced electrically by addition to solution of Ca, Na, and K from basalt, and a near-neutral pH results. An acid pH is possible only if alteration occurs at a sufficiently high seawater/rock mass ratio to preclude complete Mg removal from solution by cation exchange reactions. At 300°C, a seawater/basalt mass ratio of approximately 50 satisfies this condition.

The continuous prograde nature of Mg-uptake reactions makes it unlikely that seawater-derived Mg can penetrate completely the sharply increasing metamorphic grade at mid-ocean ridges and provide a source of acidity for ridge-crest hydrothermal fluids. However, at temperatures greater than 350°C, Ca metasomatic reactions characterized by Ca fixation and clinozoisite formation can generate acidity at virtually all fluid/rock mass ratios (and especially at relatively low pressures). In contrast, Na-fixation reactions reveal a distinct retrograde tendency and SiO_2 dependence and may contribute to chemical exchange and H^+ production in cooling geothermal systems or systems having relatively high SiO_2 concentrations in solution. Na- and especially Ca-fixation reactions, rather than Mg-fixation reactions, provide the most likely mechanism by which high-temperature ridge-crest hydrothermal fluids become acidic and capable of mobilizing heavy metals.

Experimentally derived Fe, Mn, pH, and density data constrain the temperature and pressure of formation of ridge-crest hydrothermal fluids when it is assumed that these fluids are in equilibrium with quartz. For example, experimental calibrations of Fe, Mn, pH, and SiO_2 data from hot springs at 21°N, East Pacific Rise indicate that end-member hydrothermal fluids from each of the hot-spring vents formed at temperatures of 395 ± 10°C and pressures from 330 to approximately 400 bars. The relatively high temperatures are consistent with theoretical controls on Ca-fixation reactions, especially considering pressure, but require processes other than adiabatic decompression to cool the fluids to the measured temperatures of near 350°C.

In contrast to the behavior of Fe and Mn, dissolved concentrations of Zn and Cu during experiments are not simple functions of pH; they appear to depend on H_2S and Fe concentrations in solution, on temperature, and especially on pressure at high temperatures, perhaps in response to solubility control by Fe-rich sphalerite and chalcopyrite.

ACKNOWLEDGMENTS

I wish to thank Mike Mottl, Dave Janecky, and Pat Shanks, for their thoughtful, helpful, and stimulating discussions, which contributed directly or indirectly to many aspects of this paper. I would also like to acknowledge Michael Berndt, Jeffrey Seewald, and Teresa S. Bowers for their comments and suggestions, which greatly improved the manuscript. I am grateful to my wife Carol for proofing earlier versions of the paper, an effort of considerable magnitude considering the presence of two young children wishing to "help."

Part of this paper was written while on sabbatical at the U.S. Geological Survey, Reston, Va. Funds were provided by NSF grants OCE-8400676 and OCE-8542276.

Literature Cited

Berndt, M. E., Seyfried, W. E. Jr., Beck, J. W. 1987. Hydrothermal alteration processes at mid-ocean ridges: experimental constraints from Ca and Sr exchange reactions. *J. Geophys. Res.* In press

Bischoff, J. L., Dickson, F. W. 1975. Seawater-basalt interaction at 200°C and 500 bars: implications for the origin of seafloor heavy metal deposits and regulation of seawater chemistry. *Earth Planet. Sci. Lett.* 25: 385–97

Bischoff, J. L., Rosenbauer, R. J. 1984. The critical point and two-phase boundary of seawater, 200–500°C. *Earth Planet. Sci. Lett.* 68: 172–80

Bischoff, J. L., Rosenbauer, R. J. 1985. An equation of state for hydrothermal seawater (3.2% NaCl). *Am. J. Sci.* 285: 725–63

Bowers, T. S., Taylor, H. P. 1985. An integrated chemical and stable isotope model of the origin of mid-ocean ridge hot spring systems. *J. Geophys. Res.* 90: 12,583–12,606

Bowers, T. S., Campbell, A. C., Spivack, A. J., Edmond, J. M. 1987. Chemical controls on the composition of vent fluids at 13°–11°N and 21°N, East Pacific Rise. *J. Geophys. Res.* In press

Craig, H., Welhan, J. A., Kim, K., Poreda, R., Lupton, J. E. 1980. Geochemical studies of the 21°N EPR hydrothermal fluids. *Eos, Trans. Am. Geophys. Union* 61: 992 (Abstr.)

Edmond, J. M., Measures, C., McDuff, R. E., Chan, L. H., Collier, R., et al. 1979a. Ridge crest hydrothermal activity and the balances of the major and minor elements in the ocean: the Galapagos data. *Earth Planet. Sci. Lett.* 46: 1–18

Edmond, J. M., Measures, C., Mangum, B., Grant, B., Sclater, F. R., et al. 1979b. On the formation of metal-rich deposits on ridge crests. *Earth Planet. Sci. Lett.* 46: 19–30

Fournier, R. O. 1983. A method of calculating quartz solubilities in aqueous sodium chloride solutions. *Geochim. Cosmochim. Acta* 47: 579–86

Giggenbach, W. F. 1984. Mass transfer in hydrothermal alteration systems. *Geochim. Cosmochim. Acta* 48: 2679–93

Hajash, A. 1975. Hydrothermal processes along mid-ocean ridges: an experimental investigation. *Contrib. Mineral. Petrol.* 53: 205–26

Helgeson, H. C., Kirkham, D. H. 1976. Theoretical prediction of the thermodynamic behavior of aqueous electrolytes at high pressures and temperatures: III. Equation of state for aqueous species at infinite dilution. *Am. J. Sci.* 276: 97–240

Helgeson, H. C., Delany, J. M., Nesbitt, H. W., Bird, D. K. 1978. Summary and critique of thermodynamic properties of rock-forming minerals. *Am. J. Sci.* 278-A: 1–229

Hemley, J. J., Montoya, J. W., Marinenko, J. W., Luce, R. W. 1980. General equilibria in the system Al_2O_3-SIO_2-H_2O and some implications for alteration mineralization processes. *Econ. Geol.* 75: 210–29

Hemley, J. J., Cygan, G. L., D'Angelo, W. M. 1986. Effect of pressure on ore mineral solubilities under hydrothermal conditions. *Geology* 14: 377–79

Michard, G., Albarede, F., Michard, A., Minster, J. F., Charlou, J. L., et al. 1984. Chemistry of solutions from the 13°N East Pacific Rise site. *Earth Planet. Sci. Lett.* 67: 297–308

Mottl, M. J. 1983a. Hydrothermal processes at seafloor spreading centers: application of basalt-seawater experimental results. In *Hydrothermal Processes at Seafloor Spreading Centers*, ed. P. A. Rona, K. Bostrom, L. Laubier, K. L. Smith, pp. 199–224. New York: Plenum. 798 pp.

Mottl, M. J. 1983b. Metabasalts, axial hot springs and the structure of hydrothermal systems at mid-ocean ridges. *Geol. Soc. Am. Bull.* 94: 161–80

Mottl, M. J., Holland, H. D. 1978. Chemical exchange during hydrothermal alteration of basalt by seawater. I. Experimental results from major and minor components of seawater. *Geochim. Cosmochim. Acta* 42: 1103–115

Mottl, M. J., Holland, H. D., Corr, R. F. 1979. Chemical exchange during hydrothermal alteration of basalt by seawater. II. Experimental results for Fe, Mn, and sulfur species. *Geochim. Cosmochim. Acta* 45: 869–84

Rosenbauer, R. J., Bischoff, J. L. 1983. Uptake and transport of heavy metals by heated seawater: a summary of the experimental results. In *Hydrothermal Processes at Seafloor Spreading Centers*, ed. P. A. Rona, K. Bostrom, L. Laubier, K. L. Smith, pp. 177–97. New York: Plenum. 798 pp.

Seyfried, W. E. Jr., Bischoff, J. L. 1979. Low temperature basalt interaction with seawater: an experimental study at 70°C and 150°C. *Geochim. Cosmochim. Acta* 43: 1937–47

Seyfried, W. E. Jr., Bischoff, J. L. 1981. Experimental seawater-basalt interaction at 300°C, 500 bars, chemical exchange, secondary mineral formation and implications for the transport of heavy metals. *Geochim. Cosmochim. Acta* 45: 135–47

Seyfried, W. E. Jr., Janecky, D. R. 1985. Heavy metal and reduced sulfur transport during subcritical and supercritical hydrothermal alteration of basalt: influence of fluid pressure and basalt composition and crystallinity. *Geochim. Cosmochim. Acta* 49: 2545–60

Seyfried, W. E. Jr., Mottl, M. J. 1982. Hydrothermal alteration of basalt by seawater under seawater-dominated conditions. *Geochim. Cosmochim. Acta* 46: 985–1002

Seyfried, W. E. Jr., Janecky, D. R., Mottl, M. J. 1984. Alteration of the oceanic crust: implications for geochemical cycles of lithium and boron. *Geochim. Cosmochim. Acta* 48: 557–69

Seyfried, W. E. Jr., Berndt, M. E., Janecky, D. R. 1986. Chloride depletions and enrichments in seafloor hydrothermal fluids: constraints from experimental basalt alteration studies. *Geochim. Cosmochim. Acta* 50: 469–75

Seyfried, W. E. Jr., Janecky, D. R., Berndt, M. E. 1987. Rocking autoclaves for hydrothermal experiments II: the flexible cell system. In *Experimental Hydrothermal Techniques*. ed. H. Barnes, G. Ulmer. New York: Wiley-Interscience. In press

Sleep, N. H., Morton, J. L., Burns, L. E., Wolery, T. J. 1983. Geophysical constraints on the volume of hydrothermal flow at ridge axes. In *Hydrothermal Processes at Seafloor Spreading Centers*, ed. P. A. Rona, K. Bostrom, L. Laubier, K. L. Smith, pp. 53–70. New York: Plenum. 798 pp.

von Damm, K. L., Edmond, J. M., Grant, B., Measures, C. I., Walden, B., Weiss, R. F. 1985a. Chemistry of submarine hydrothermal solutions at 21°N, East Pacific Rise. *Geochim. Cosmochim. Acta* 49: 2197–2220

von Damm, K. L., Edmond, J. M., Measures, C. I., Grant, B. 1985b. Chemistry of submarine hydrothermal solutions at Guaymas Basin, Gulf of California. *Geochim. Cosmochim. Acta* 49: 2221–39

TECTONICS OF THE SOUTHERN AND CENTRAL APPALACHIAN INTERNIDES

Robert D. Hatcher, Jr.

Department of Geological Sciences, University of Tennessee-Knoxville, Knoxville, Tennessee 37996-1410

INTRODUCTION

The Appalachian orogen was constructed along the ancient Precambrian continental margin of eastern North America by a series of compressional events that began in the Ordovician and episodically spanned much of the Paleozoic era. The processes of accretion of suspect and exotic terranes, together with terrane collision and ultimately completion of the Wilson cycle by continent-continent collision, resulted in construction of the Appalachian orogen (Figure 1). These paleogeographic elements are represented by the various lithotectonic units visible today in the southern and central Appalachians (Figure 2). These subdivisions represent the late Precambrian rifted margin and early Paleozoic platform that were formed directly upon North American Grenvillian basement (Cumberland/Allegheny Plateau, Valley and Ridge, and western Blue Ridge), the offshore North American slope-rise assemblage (Hamburg klippe, southern Appalachian eastern Blue Ridge and Inner Piedmont, and higher thrust sheets of the Maryland–northern Virginia Piedmont), and the exotic Carolina, or Avalon, and Wilmington terranes. The Carolina terrane contains the Alleghanian high-grade metamorphic core. These elements form the basis for this analysis and review of the present state of knowledge of the tectonic history of the southern and central Appalachian internides.

The Appalachian orogen has been the subject of study by geologists at least since the early nineteenth century. Early fundamental concepts, such as the geosynclinal theory of mountain building (Dana 1873, Hall 1883)

Period	Ma.	Orogeny	Tectonic Event
Jurassic	144		Rifting and Opening of Atlantic
Triassic	208		
Permian	245 / 286	ALLEGHANIAN	Collision with Africa
Carb. (Pa 320 / Miss 360)			
Devonian		ACADIAN	Terrane Accretion
Silurian	408 / 438		
Ordovician	505	TACONIC / PENOBSCOT	A-Subduction Obduction Arc Collision/Accretion
Cambrian	570	AVALONIAN	Subduction Volcanic Arc Generation
Precambrian	1000–1200	GRENVILLE	Rifting and Opening of Iapetos

Figure 1 Geologic time scale showing the principal orogenic episodes and tectonic events affecting the rocks of the southern and central Appalachians.

and Willis' (1893) hypotheses of fold mechanisms, were based upon observations of Appalachian rocks. King (1970) described the Appalachians as "the most elegant on earth," but also cautioned that this region is "full of guile, and its geology has aroused controversies as acrimonious as any of those in our science." Controversies have and will continue to rage about the history and origin of structures and other features within the Appalachians. The southern and central Appalachians have had their share of lengthy debates. Controversies have raged for decades over the nature of the Martic line, Cartersville fault, Brevard fault zone, Glenarm Series, Talladega slate belt, Evington Group/James River synclinorium, and Baltimore Gabbro. New controversies, such as those concerning the nature and extent of the Blue Ridge–Piedmont thrust sheet and the locations and characteristics of cryptic sutures, continue to arise as new data become available for interpretation.

SUBDIVISIONS

The southern and central Appalachians may be subdivided in different ways. Subdivisions based upon stratigraphic and lithotectonic units are

TECTONICS OF APPALACHIAN INTERNIDES 339

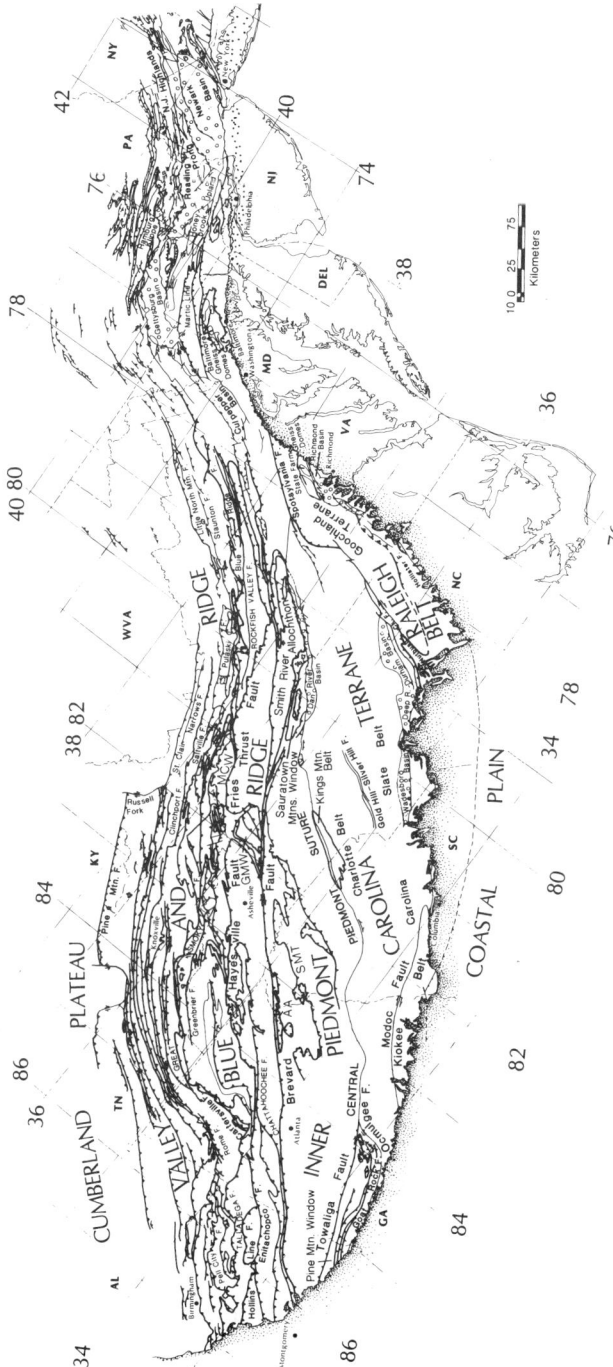

Figure 2 Principal subdivisions and major faults of the southern and central Appalachians. Abbreviations as follows: AA—Alto allochthon; SMT—Six Mile thrust; GMW—Grandfather Mountain window; MCW—Mountain City window. Edge of the Coastal Plain is shown by a stipple pattern. Triassic-Jurassic basins are indicated by open circles.

the most useful and understandable. In addition, there may be other features, such as the prominent gravity gradient that traverses the length of the southern and central Appalachians, that may have a direct bearing upon the tectonic framework of the orogen. The gravity gradient may represent the buried edge of the ancient Precambrian continental margin from which the external and internal basement massifs of the Blue Ridge and Piedmont were plucked and transported northwestward onto the craton aboard thrust sheets.

Basement Massifs

The Proterozoic Grenvillian crystalline rocks form the autochthonous North American reference frame, since they serve as the basement upon which many of the late Precambrian and younger stratigraphic packages that ultimately became involved in the Appalachian orogenies were deposited. Basement also formed during each tectonic cycle within the evolving Appalachian orogen. New continental crust formed in one orogeny may serve as basement for deposition of sediments and volcanic rocks and is then recycled in subsequent orogenies (Osberg 1978, Hatcher 1983, 1986). Consequently, basement produced during the earlier Taconic orogeny and related events may have been reactivated and consolidated with its cover during the Acadian orogeny. Acadian basement is thus generated, and it too may have been reactivated and involved in the late Paleozoic Alleghanian event. The net result is an increase in the total amount of continental crust.

Major Boundaries

Several major boundaries are present within the southern and central Appalachians. These boundaries are marked by large faults and reflect major changes in the character of tectonostratigraphic units. The western edge of the crystalline Appalachians corresponds roughly to the eastern portion of the foreland. There is no single fault or fault zone that can be realistically identified as the boundary between the metamorphic core and the unmetamorphosed foreland, because the major boundary faults with large displacements do not correspond to principal changes in metamorphic grade. Rock units involved in the deformation of the western edge of the metamorphic core and the eastern edge of the foreland behave similarly, with the principal difference being in the nature of the rocks being transported. An abrupt transition exists from the lower Paleozoic sedimentary and metasedimentary rocks involved in the Valley and Ridge thrust sheets to the late Precambrian sedimentary and metasedimentary rocks and some basement in the Blue Ridge. Metamorphic overprint and penetrative strain, characteristic of the more internal parts of mountain chains, extends into the Valley and Ridge foreland in Georgia and Virginia. Moreover, rocks of the frontal Blue Ridge in Tennessee are virtually unmetamorphosed and uncleaved.

HAYESVILLE-FRIES FAULT

A more significant boundary, the Hayesville-Fries fault, occurs in the central Blue Ridge of North Carolina and Georgia (Figures 2, 3a). This boundary separates sequences to the west that were deposited on continental basement and that contain rare mafic/ultramafic rocks and granitoid from sequences to the east that contain abundant mafic and ultramafic rocks. The eastern sequence contains little continental basement and may have been deposited in part on oceanic crust (Hatcher 1978). Most of the Paleozoic granitoids occur in this eastern zone. The Hayesville fault is premetamorphic along much of its extent, but it has been reactivated and overthrust again, particularly toward the north. In the central Appalachians, another boundary may exist southeast of the Fries fault close to where it joins the Blue Ridge thrust in southwestern Virginia.

The Piney Branch mafic-ultramafic complex of northern Virginia (Drake & Morgan 1981) was transported in a higher thrust sheet that may correlate with the Hayesville sheet in the southern Appalachians. The eastern lithotectonic package extends in the southern Appalachians from the eastern Blue Ridge through the Inner Piedmont to the Kings Mountain Belt (Figure 2). The Inner Piedmont contains several syn- to late metamorphic thrust sheets and a metasedimentary-metavolcanic rock assemblage.

BREVARD FAULT ZONE

The Brevard fault zone has long been recognized as a major crustal break (Jonas 1932, King 1955, Reed & Bryant 1964, Bryant & Reed 1970) and has more recently been interpreted as a continental suture (Bird & Dewey 1970, Odom & Fullagar 1973, Rankin 1975). However, the similarity of stratigraphic assemblages on both sides of the Brevard fault zone and the lack of other features (e.g. ultramafic rocks) common to sutures clearly indicate that it cannot be a suture.

The Brevard fault zone ends in North Carolina southwest of the Sauratown Mountains window (Figure 2). The Inner Piedmont likewise ends in this area, except for the erosional remnant of a higher thrust sheet (Smith River allochthon) that passed over the Sauratown Mountains window and presently resides in the Hayesville-Fries thrust sheet of the Blue Ridge in North Carolina and Virginia.

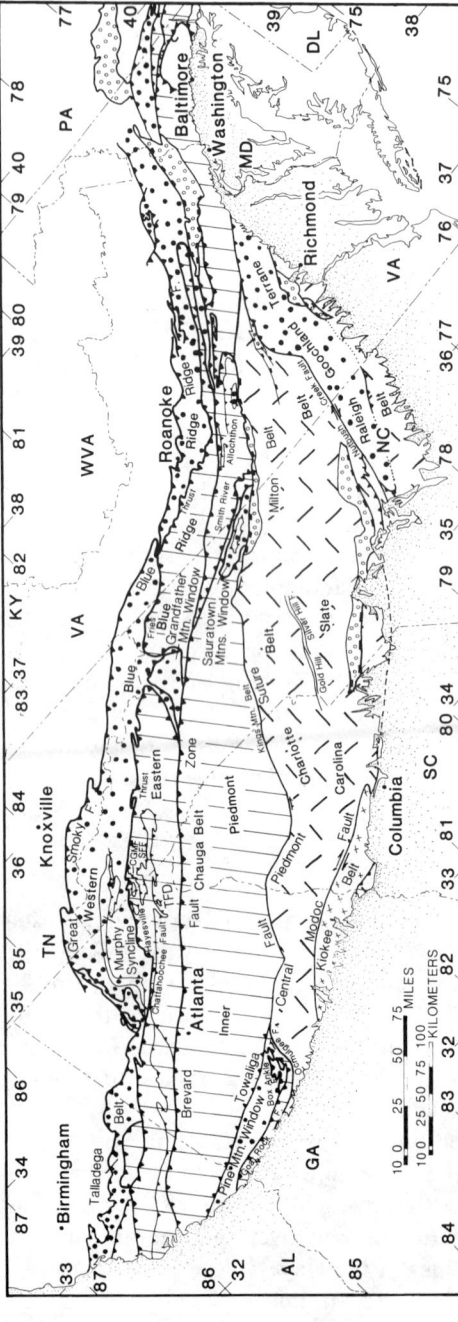

Figure 3a Major structural features, boundaries, and "belts" of the southern and central Appalachian internides. Dot pattern is the allochthonous North American rifted margin containing rocks ranging from middle Proterozoic (Grenville basement) to late Precambrian to Devonian. Most rocks in this zone are late Precambrian to Early Cambrian. The striped pattern represents the Piedmont terrane of Williams & Hatcher (1983), which is composed mostly of late Precambrian(?) to early Paleozoic rocks. The short diagonal line pattern represents the Late Cambrian volcanic and volcaniclastic rocks of the Avalon (Carolina) terrane. The X pattern in the Kiokee belt represents the medium-grade core of the Alleghanian orogen. Rocks in the Kiokee belt are mostly remobilized Avalon terrane rocks. The open-circle pattern represents rocks of Mesozoic basins. The edge of the Coastal Plain is lightly stippled. Abbreviations: SFF, Shope Fork fault; TFD, Tallulah Falls dome; CGMF, Chunky Gal Mountain fault.

Inner Piedmont

There are a number of higher thrust sheets farther south in the Inner Piedmont. One example is the Six Mile thrust sheet in northwestern South Carolina and northeastern Georgia (Griffin 1974). It, together with the Alto allochthon (a large isolated remnant of the Six Mile thrust sheet adjacent to the Brevard fault zone in northeastern Georgia and northwestern South Carolina), transported sillimanite-grade rocks onto rocks of garnet to staurolite grade in the Chauga belt. The northwestern edge of the Alto allochthon is retrograded by deformation within the Brevard fault zone, which puts the time of emplacement of the Alto allochthon at probably more than 360 Myr ago, the time of equilibration of Rb and Sr isotopes in the Brevard fault zone.

Structures in the Inner Piedmont, Chauga belt, and eastern and western Blue Ridge verge toward the northwest. Folds are overturned toward the northwest and are recumbent to reclined in the eastern Blue Ridge, Chauga belt, and Inner Piedmont. There is a zone of near-vertical folds and some east vergence in the central-eastern Blue Ridge of North Carolina and Georgia near the Chattahoochee fault–Shope Fork fault complex in this area (Figure 3a). Toward the southeast, structures steepen in the Inner Piedmont in the vicinity of the Kings Mountain Belt, where they become nearly vertical.

WINDOWS IN THE BLUE RIDGE–PIEDMONT THRUST SHEET

In the western Blue Ridge are several large windows, some of which appear to be classic simple windows such as the Tuckaleechee Cove, Wear Cove, and Cades Cove window complex. These have recently been shown to have developed above a footwall duplex structure (Hatcher et al 1986a). The Hot Springs window is the type eyelid window (Oriel 1950, Boyer & Elliott 1982). The Mountain City and Limestone Cove windows and related structures in the western Blue Ridge are complex duplex structures beneath the Blue Ridge thrust sheet and involve the Cambrian platform sequence (Boyer & Elliott 1982, Diegel & Wojtal 1985). The Grandfather Mountain window is one of the largest windows in the southern Appalachians. It involves basement, late Precambrian metasedimentary rocks, and the Lower Cambrian platform succession (Bryant & Reed 1970). This structure also has been interpreted as a duplex involving the basement-cover succession (Boyer & Elliott 1982). The Shooting Creek and Brasstown Bald windows of the central Blue Ridge in North Carolina and Georgia are smaller structures that involve folding and erosional exhumation of the Hayesville thrust sheet, but they may also be related to structure in the platform rocks beneath. New seismic reflection data (Hatcher et al 1986b) indicate that duplexing in the platform rocks at depth had a profound effect upon late open-fold structures expressed as domes in the Blue Ridge crystalline thrust sheet. This may be a general mechanism for dome formation in all orogenic terranes.

The Sauratown Mountains window exposes basement and cover beneath the same thrust sheets that are present in the eastern Blue Ridge and Inner Piedmont. This window is framed by the Forbush thrust, which appears to be premetamorphic and may be the southeastern equivalent of the Hayesville-Fries thrust. An inner window framed by a later brittle thrust exposes rocks inside that are similar to those above the thrust (Hatcher et al 1986a).

The Pine Mountain window (Schamel & Bauer 1980, Sears & Cook 1984) is a large structure in central Georgia and Alabama exposing basement and a thin cover that may be correlative with the Cambrian platform stratigraphy (Chilhowee Group, Shady Dolomite, Rome Formation). This is a very complex structure; its eastern end is closed by the premetamorphic Box Ankle thrust, and it is flanked on the northwest by the postmetamorphic Towaliga fault, which contains mylonites retrograded from sillimanite-grade to garnet-grade assemblages. The Pine Mountain window is flanked to the southeast by the Goat Rock–Bartletts Ferry fault, which is in part synmetamorphic and in part later. The Goat Rock fault terminates as a statically annealed ductile shear near the east end of the Pine Mountain window (Hatcher et al 1986a).

Central Piedmont Suture

The central Piedmont suture is exposed from near the Georgia-Alabama line to central Virginia. It separates rocks of the Inner Piedmont and eastern Blue Ridge block, which have North American affinities, from an exotic terrane, the volcanic island-arc association of the Carolina (Avalon) terrane. Rocks of the Carolina terrane are markedly different in character and origin from those to the west, and the suture is recognized primarily upon stratigraphic criteria. At its southwestern end, the central Piedmont suture coincides with the Ocmulgee fault, a large fault that may be traced from central Georgia (Hooper & Hatcher 1986), and it becomes coincident with the northeastern extent of the Towaliga fault across central Georgia. It is then traceable across central South Carolina into the Kings Mountain belt along the North Carolina–South Carolina border (Figure 3a). The fault is folded from central South Carolina into the Kings Mountain belt in North Carolina such that units from the Avalon terrane are intermixed with units of the North American affinity Kings Mountain belt sequence

(Hatcher 1983, Horkowitz 1984). Earlier studies suggested the Kings Mountain belt sequence is composed of a volcanic component and a more North American platform–related component (Horton & Butler 1977). The boundary is traceable northward until it passes into the Danville Triassic Basin. It resurfaces once more in northern Virginia (Spotsylvania fault?) and possibly is truncated by faults of the Permian eastern Piedmont fault system.

The Carolina suspect terrane is a late Precambrian to Cambrian volcanic-volcaniclastic sedimentary-rock terrane characterized by the presence of an early Paleozoic European and African fauna, partially the Middle Cambrian *Paradoxides* trilobite. Its characteristics and possible correlatives were recently summarized by Kish & Black (1982) and Secor et al (1983). The Carolina terrane does not resurface again in the central Appalachians once it is truncated and passes beneath the Coastal Plain near Richmond, Virginia. The Chopawamsic terrane in northern Virginia (Pavlides 1981) contains rocks similar to those of the Carolina slate belt (the eastern low-grade part of the Carolina terrane).

In the eastern Piedmont of Virginia, North and South Carolina, and Georgia is a zone where several large faults occur along with a high-grade metamorphic overprint. These faults were originally called the eastern Piedmont fault system by Hatcher et al (1977) and were recognized using a combination of surface geologic and aeromagnetic data. They formed during the late Paleozoic Alleghanian orogeny and flank the antiformal Kiokee and Raleigh belt culminations (Figure 2).

The Raleigh belt in North Carolina is much more complex than the Kiokee belt in South Carolina and Georgia. The Raleigh belt consists of a thrust sheet which carried the Carolina slate belt (Avalonian) rocks over a continental basement-and-platform assemblage (Farrar 1984, 1985). This thrust probably formed during the middle Paleozoic. It was later overprinted and refaulted by Alleghanian metamorphism and faults of the eastern Piedmont fault system. Ophiolitic material rests upon the thrust along the western edge of the Raleigh belt as several dismembered remnants of an ophiolitic melange (Kite & Stoddard 1984). The continental basement-and-cover terrane is part of a suite of possible North American Grenvillian rocks that were metamorphosed in the early Paleozoic, or earlier, and moved to their present positions later. This assemblage was termed the Goochland terrane by Glover et al (1983) and Farrar (1984). It contains granulite facies assemblages that formed during either the Paleozoic or Precambrian.

The Carolina slate belt and Raleigh belt are not exposed at the surface in the central Appalachians. In northern Virginia, rocks of the eastern Blue Ridge rest against the Atlantic Coastal Plain.

Another possibility for an exotic terrane is the Wilmington complex in northern Delaware (Drake et al 1986). It consists of a high-grade assemblage of metasedimentary and metaplutonic rocks of unknown origin that contrast markedly with the rocks immediately west.

Several internal basement massifs occur as the Baltimore gneiss domes and related basement-cored structures (e.g. the Honey Brook Upland) in the central Appalachians (Muller & Chapin 1984, Crawford & Hoersch 1984). These have a cover of late Precambrian and early Paleozoic metasedimentary rocks that is different from that flanking the basement massifs in the Blue Ridge farther west. Hopson (1964) addressed the stratigraphic problems between the Glenarn Series and the basement, but more recently it was recognized that a higher thrust exists within these rocks (Drake et al 1986). The only similar structures to the south are the basement-cored State Farm gneiss domes in the Goochland terrane. The Grenvillian State Farm Gneiss (Glover & Tucker 1979, Mose & Nagel 1982) is a felsic orthogneiss similar to that of the Baltimore gneiss domes.

The Reading Prong is an extensive basement massif complex that occurs in eastern Pennsylvania and northern New Jersey. This complex contains a Paleozoic cover that is again unlike that occurring on the basement rocks farther to the southwest in the southern and central Appalachians as well as that in New England (Drake 1970, 1978). It consists of a thin platform assemblage of carbonate and clastic rocks. This may reflect fundamental differences in the nature of the orogen and in the late Precambrian continental margin from the southern Appalachians to New England.

MAJOR FAULTS

Reference has already been made to many of the large faults in the southern and central Appalachians, and it has been observed that these faults serve as major tectonic boundaries within the crystalline Appalachians (Figures 2, 3a). A definite sequencing of deformation occurred in this part of the Appalachians that proceeded from the most internal parts of the mountain chain in the eastern Blue Ridge and Inner Piedmont outward into the more distal flanks. The Valley and Ridge foreland fold-and-thrust belt and western Blue Ridge were deformed last in the southern and central Appalachians along with the Kiokee-Raleigh belt. This contrasts with the deformation plan in New England, which appears to be west (Taconic dominant) to east (Alleghanian overprinted). In addition, a number of large faults in the internides have experienced a history of multiple reactivation. Probably the best example of this, and the most complex, is the Brevard fault zone, where evidence exists for formation of mylonites in the earliest deformational events affecting the southern Appalachians. The Taconic orogeny and related events produced equilibration of lead isotopes in mylonites about 480 Myr ago (Sinha & Glover 1978), and Rb-Sr isotopes were equilibrated about 360 Myr ago (Odom & Fullagar 1973) during the formation of retrograde mylonites. Recent studies of silica mobility in Brevard fault zone ultramylonite followed by Rb/Sr whole-rock dating have produced an age of 273 Myr for the retrogressive ductile deformation (A. K. Sinha, written communication, 1986). The Brevard fault zone also has a later chaotic brittle zone along its northwestern boundary (Hatcher 1971), termed the Rosman fault by Horton (1982). This structure probably formed during the Alleghanian and is responsible for bringing up pieces of platform carbonate and shale from beneath the thrust sheet. This may indicate that the Blue Ridge–Piedmont thrust sheet was largely emplaced before this last movement occurred on the Brevard fault zone. Segments of other major faults, such as the Hayesville, were not reactivated, but they may have been reactivated along other segments, as was the Fries fault (Hatcher 1978, Hatcher & Odom 1980). Reactivation of faults doubtlessly is related to the degree of folding of fault surfaces during subsequent events as well as to the relationships of the faults to the newly applied stress fields.

Several of the large windows in the southern Appalachians, such as the Grandfather Mountain, Sauratown Mountains, and Pine Mountain windows and the Raleigh belt–Goochland terrane, provide interesting vertical sections through a major part of the crust and expose faults that formed at different times. The Grandfather Mountain window, for example, probably is framed by Alleghanian faults, and these faults juxtapose rocks of higher metamorphic grade outside the window onto rocks of lower metamorphic grade inside (Bryant & Reed 1970). In addition, a higher thrust sheet (Hayesville-Fries), probably inactive at the time of Alleghanian faulting, is present in the area immediately to the southwest and northeast of the Grandfather Mountain window and merges with the Alleghanian faults framing the window along the northeast side (Figure 3a).

The Sauratown Mountains window exposes younger postmetamorphic faults in its internal parts and an older premetamorphic fault (Hayesville equivalent?) forming the frame of the window, which separates rocks of the North American rifted margin and basement from rocks of the Precambrian North American slope, rise, and possibly ocean floor (Hatcher 1980, Hatcher et al 1986a). The Pine Mountain belt is flanked by major faults, which again separate the North American basement/cover assemblage inside from offshore North American rocks of the Piedmont. But here the window is flanked by faults of markedly different ages that exhibit much greater complexity. A pre- or synmetamorphic folded thrust is present at its eastern end; faults that dip more steeply flank the north (Towaliga) and south (Goat Rock–Bartletts Ferry) sides of the window, but these were formed at different times, which indicates that they are not the same fault surface. Immediately east of the Pine Mountain belt window is the Ocmulgee fault (Hooper & Hatcher 1986), which in this area is the central Piedmont suture juxtaposing Carolina terrane rocks against the North American rocks. Studies in the vicinity of the Ocmulgee fault (Hooper & Hatcher 1986) indicate that its motion was right-lateral.

The eastern Piedmont fault system consists of the Modoc and Augusta faults flanking the Kiokee belt, and the Nutbush Creek, Hollister, and Hylas faults flanking the Raleigh belt–Goochland terrane farther north. All are Alleghanian faults that have a right-lateral strike-slip component (Bobyarchick 1981) as well as a thrust component in places. Movement on the Modoc fault in South Carolina is probably best documented (Dennis & Secor 1986, Secor et al 1986), with right-lateral motion involving a maximum incremental displacement of less than 10 km having occurred there. It has also been proposed, based upon a study of shear-sense indicators, that the last movement on the Brevard fault zone was right-lateral (Bobyarchick 1984).

The large thrust sheets of the central Appalachians and the Martic overthrust had a principal component of movement toward the continental craton. The same is true for basement massifs farther north, including the Reading Prong and the New Jersey and Hudson highlands of Pennsylvania, New Jersey, and southeastern New York.

METAMORPHISM

Most of the rocks in the southern and central Appalachian Blue Ridge and Piedmont were metamorphosed to various degrees during several Paleozoic events, with the exception of the frontal thrust blocks in the Blue Ridge in Tennessee and possibly northern Virginia (Figure 3b). Even in these areas, an increase in the illite crystallinity index and anchizone metamorphism is present. Metamorphism affected the rocks of the crystalline southern and central Appalachians at times ranging from 510 Myr ago to about 460 Myr ago at different places during the Taconic or possibly the Penobscot event (Drake et al 1986). The Taconic event is thought to be strongly diachronous based upon the first appearance of Ordovician clastic wedges in the foreland, since these occur much earlier in the southern and Maritime Appalachians (Llanvirnian) than in the central Appalachians (Caradocian) (Shanmugam & Lash 1982). At about 360 Myr ago, the Late Devonian Acadian event affected the rocks in the southern and central Appalachians, although it is not clearly separable from the earlier

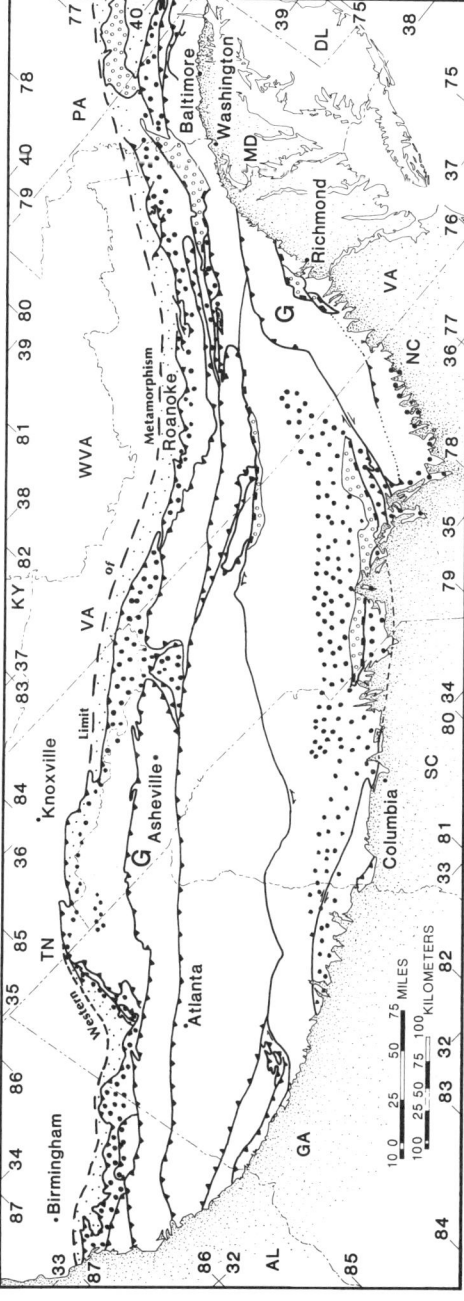

Figure 3b Metamorphic zones of southern and central Appalachians. Large-dot pattern represents greenschist facies metamorphism. Small-dot pattern is anchizone metamorphism along the western edge of the internides. Unpatterned areas in the central and eastern internides are zones of amphibolite facies metamorphism, with the G indicating local zones of Paleozoic granulite facies metamorphism. Other patterns defined in Figure 2. Modified from Glover et al (1983) and other published sources.

event(s). Thermal processes related to the Alleghanian event at about 315–267 Myr ago (Secor et al 1986) affected the rocks principally in the eastern Piedmont. Some evidence of an Alleghanian low-grade metamorphic overprint exists in other parts of the southern and central Appalachians.

The highest grades of Paleozoic metamorphism were reached in the south-central Blue Ridge of North Carolina and nearby Georgia (Hatcher & Butler 1979, Absher & McSween 1985), in the Goochland terrane (Glover et al 1983, Farrar 1984, 1985), and in the Inner Piedmont, where the metamorphic grade reached the sillimanite zone. Uncertainty exists about the timing of metamorphism in the Goochland terrane, since arguments may be made favoring both Grenvillian and Paleozoic ages for the high-grade metamorphism there.

Several zones of lower metamorphic grade also exist in the southern and central Appalachian internides. Low-grade greenschist facies metamorphism occurs in the Carolina slate belt (eastern Carolina terrane) extending from Georgia into Virginia. In the Kings Mountain belt the metamorphic grade is lower to middle amphibolite facies, contrasting with sillimanite grade in the Inner Piedmont and Charlotte belt on either side. Garnet- to staurolite-grade metamorphism affected the rocks of the Chauga belt, whereas in the adjacent Blue Ridge and Piedmont the metamorphic grade is in the kyanite- to sillimanite-grade zones. Metamorphism in the Blue Ridge decreases systematically westward to lower greenschist facies and the anchizone, and it was concluded by Carpenter (1970) to constitute a Barrovian series. The Murphy syncline (Figure 3a) involves some greenschist facies assemblages as a result of postmetamorphic folding of isograd surfaces.

Abundant evidence exists for polyphase metamorphism of the internal parts of the Appalachian metamorphic core (Figure 3c). Overgrown garnets and other minerals that postdate deformed fabrics containing prograde deformed minerals indicate that metamorphism of the internal parts of the southern and central Appalachians was not produced by a single event. Probably the best textural evidence supporting this conclusion is from the Brevard fault zone, where Roper & Dunn (1973) suggested there are at least two prograde events affecting these rocks. Dabbagh (1975) argued similarly for part of the eastern Blue Ridge. The resolution of metamorphic events by geochronological techniques is still not very clear, but there is strong evidence that the western parts of the orogen were metamorphosed primarily during the Taconic or Penobscottian event (Butler 1972, Dallmeyer 1975), and that the more internal parts were overprinted by high-grade Acadian metamorphism (Dallmeyer & Hatcher 1985, Dallmeyer et al 1986). Arguments can also be made that all the metamorphism in the southern and central Appalachians is Taconic

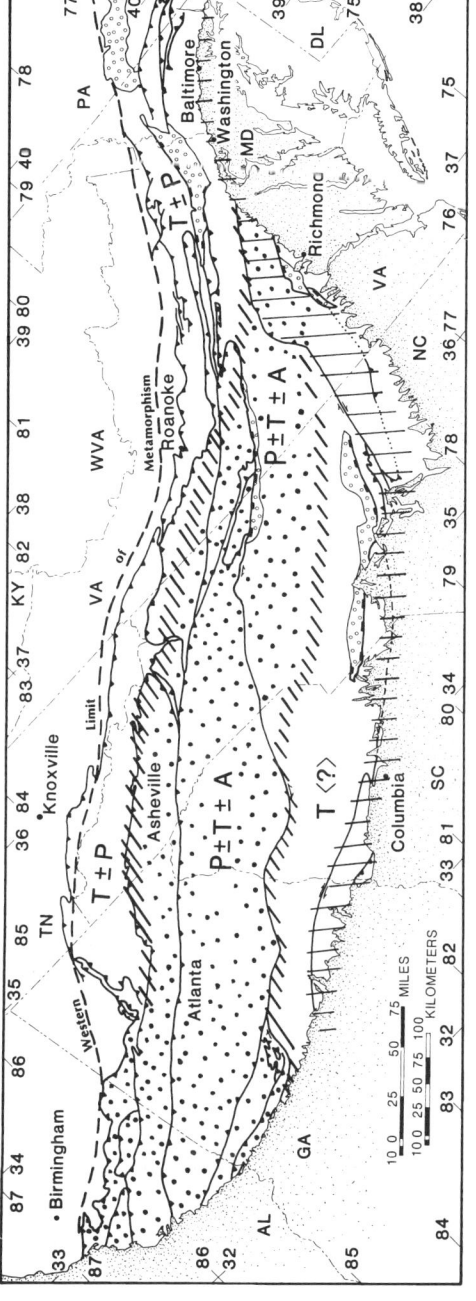

Figure 3c Timing of metamorphism in the southern and central Appalachians. Abbreviations as follows: T—Tacoric; P—Penobscottian; A—Acadian. Diagonal stripe pattern along the southeastern edge of the exposed orogen is the known distribution of the Alleghanian overprint. Other patterns are as defined in Figure 2. Uncertainties in timing indicated by ± and (?). The zone of T(?) represents an event in the Avalonian rocks that occurred about at the time of Taconic metamorphism; however, if the docking of the Avalon terrane did not occur until later, the metamorphism affecting these rocks at 450–480 Myr could not have been the North American Taconic event.

without any Acadian overprint (Drake et al 1986); however, such a conclusion makes it difficult to explain the overgrown garnets and other mineral assemblages that may be indicative of a second progressive high-grade metamorphic event (Figure 3c). Additional evidence may be drawn from some of the plutons, which appear to have been foliated during a later metamorphic event but not deformed by the earlier foliation contained in the rocks that they cut.

Alleghanian metamorphism reached the kyanite and sillimanite zones of the amphibolite facies. It is restricted to a relatively narrow belt (Kiokee-Raleigh belt) in the eastern Piedmont flank of the southern and central Appalachians (Snoke et al 1980, Farrar 1985).

PLUTONIC ACTIVITY

Plutons covering a wide range of compositions were intruded into the rocks of the southern and central Appalachians from late Precambrian until late Paleozoic times (Figure 3d). A suite of late Precambrian alkalic plutons, the Crossnore plutonic-volcanic sequence (Rankin 1976), was intruded into the North Carolina-Virginia Blue Ridge about 650–710 Myr ago (Odom & Fullagar 1984). This suite of plutons and volcanic rocks has been interpreted by Rankin (1976) as representing continental tholeiites formed during rifting of the North American margin, believed by Odom & Fullagar (1984) to have occurred 690–570 Myr ago. However, there is some question about whether the plutons may actually be related to the volcanic assemblage.

Several 500- to 600-Myr-old plutons that formed during the Cambrian occur in the Piedmont and nearby Blue Ridge. These include the Elkahatchee Quartz monzonite in Alabama (Russell 1978), the Henderson Gneiss in North and South Carolina (Odom & Fullagar 1973), the Leatherwood Granite in Virginia (Conley & Henika 1973), and the Occoquan pluton in northern Virginia (Russell 1978). There is also a suite of plutons, mostly granitic but also including some mafic plutons, that formed in the 400–460 Myr interval (Fullagar 1971). These were intruded mostly into the Inner Piedmont, but a few may occur farther to the east. A suite of earlier granitic plutons associated with the Carolina terrane was intruded during the Early Cambrian and late Precambrian at about 550 to 700 Myr ago. The Flat River complex is an example of a typical Eocambrian subvolcanic plutonic complex (McConnell & Glover 1982). A suite of Devonian plutons that intruded at about 360 to 400 Myr ago occurs in the eastern Blue Ridge (Butler 1972), the western Inner Piedmont (Harper & Fullagar 1981), and the Carolina terrane of the central Pied-

TECTONICS OF APPALACHIAN INTERNIDES 353

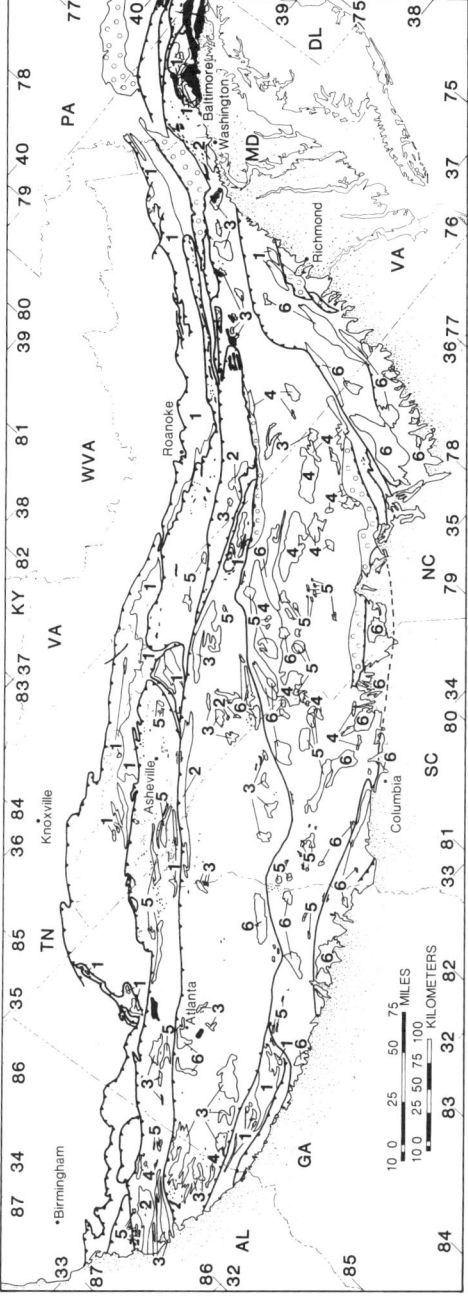

Figure 3d Plutons, Grenville basement rocks, and major boundaries of the southern and central Appalachians. 1—Grenville basement; 2—Cambrian to Early Ordovician (Penobscottian?) plutons; 3—Late Cambrian to Middle Ordovician (Taconic?) plutons; 4—late Precambrian to Cambrian (Avalonian?) plutons; 5—Silurian-Devonian (Acadian?) plutons; 6—Carboniferous-Permian (Alleghanian?) plutons. Ultramafic rocks are indicated in solid black. Other patterns are as defined in Figure 2.

mont (Fullagar 1971). Although these plutons are mostly granitic, mafic plutons are also present.

Ultramafic rocks ranging from dunites to peridotites of different compositions to pyroxenites were emplaced in the eastern Blue Ridge, Inner Piedmont, Charlotte belt, and Raleigh belt during the early Paleozoic (Figure 3d). Their occurrences and distribution have been summarized by Misra & Keller (1978) and Misra & McSween (1984). Most, if not all, of the ultramafic rocks of the southern and central Appalachians are metamorphosed. Arguments may be made for both tectonic and igneous emplacement of these bodies (Hatcher et al 1984), although some ophiolitic bodies probably exist in this part of the orogen, e.g. the Laurel Creek complex (Petty 1982, McSween & Hatcher 1985) and the Piney Branch complex (Drake & Morgan 1981).

The Alleghanian plutons are primarily S-granites (S-type muscovite-bearing granites) and are mostly located in the central to eastern Piedmont, although a few occur in the Inner Piedmont in Georgia and in North Carolina (Sinha & Zietz 1982). The westernmost of these Alleghanian plutons are located immediately southeast of the Brevard fault zone near Atlanta, Georgia, and these have ages ranging from 265–325 Myr (Fullagar 1971, Fullagar & Butler 1979, Ellwood et al 1980).

PLATE TECTONIC AND ACCRETIONARY HISTORY —A COMPLETE WILSON CYCLE

The constructional history of the southern and central Appalachian orogen has been discussed and speculated about by a number of geologists. Brown (1970) was the first to attempt a plate tectonic convergence model for any part of the southern or central Appalachians. His model was based primarily upon the development of the Martinsburg basin as a foredeep adjacent to a westward-dipping subduction zone that formed in response to plate convergence. Bird & Dewey (1970) included the southern and central Appalachians in their plate model, but their model was not clearly directly applicable to this part of the mountain chain. Hatcher (1972) proposed that the southern Appalachian orogen was developed above a westward-dipping subduction zone, and that it involved the collision of Africa as the final event producing the Alleghanian deformation in the late Paleozoic and the closing of the ancient Atlantic Ocean. This model was too simple, since it assumed the entire orogen was constructed above a subduction zone that remained east of the presently exposed Appalachians throughout the Paleozoic.

A subsequent plate model by Odom & Fullagar (1973) suggested an intermediate ocean involving the collision of the Carolina slate belt arc with the North American continent by suturing along the Brevard fault zone in the early Paleozoic (probably during the Acadian). Glover & Sinha (1973) were the first to suggest that a suture exists along the Kings Mountain belt–Charlotte belt boundary (Figure 3a). Rankin (1975) suggested that the collision of Africa with North America involved a transported suture that became the Brevard fault zone, and that suturing probably occurred in the late Paleozoic. Hatcher (1978) proposed that the southern Appalachians and parts of the central Appalachians consisted of an irregular rifted continental margin, a peninsula, and a number of smaller microcontinental blocks. The mountain chain developed by subduction and collision, and, during convergence, these fragmented elements were swept into the North American continent.

Hatcher & Odom (1980) suggested that the sequence of evolution of the North American margin began with a rifting phase that was followed by convergence and obduction of an oceanic eastern Blue Ridge segment into North America during the early Paleozoic, probably during the Taconic orogeny. This was followed by suturing of the Carolina slate belt arc (now the Carolina terrane) to North America during the Acadian orogeny, and later by collision with Africa at the end of the Paleozoic (Alleghanian orogeny). Williams & Hatcher (1982, 1983) suggested that a number of accreted suspect terranes exist in the Appalachians, and that the Avalon (Carolina) terrane (Carolina slate belt–Charlotte belt) is a major exotic terrane having no North American affinities. The Piedmont terrane is the eastern Blue Ridge and Inner Piedmont and equivalents to the north, and it probably had its origins in the late Precambrian continental slope, rise, and ocean floor immediately adjacent to North America.

The Talladega terrane occurs between the Piedmont terrane and the foreland in Alabama and Georgia (Figures 2 and 3a). It is now recognized as containing an equivalent of the late Precambrian to Devonian North American foreland stratigraphy. Correlations have recently been made with the North American platform succession, with the original stratigraphic breaks and unconformities of the platform, such as the Middle Ordovician unconformity at the top of the Knox Group, now recognized in the Talladega sequence (Tull 1978, Tull et al 1985).

The accretionary history of the southern and central Appalachian orogen is still not fully understood. However, using the information from the stratigraphic assemblages, the known timing of the major faults, and the times of the plutonic intrusion and metamorphism, we can perhaps reconstruct the plate tectonic history of the southern and central Appalachians (Figure 4).

The rifted margin of North America formed as the Iapetos and Theic-Rheic Oceans opened in the late Precambrian and thus provided an irregu-

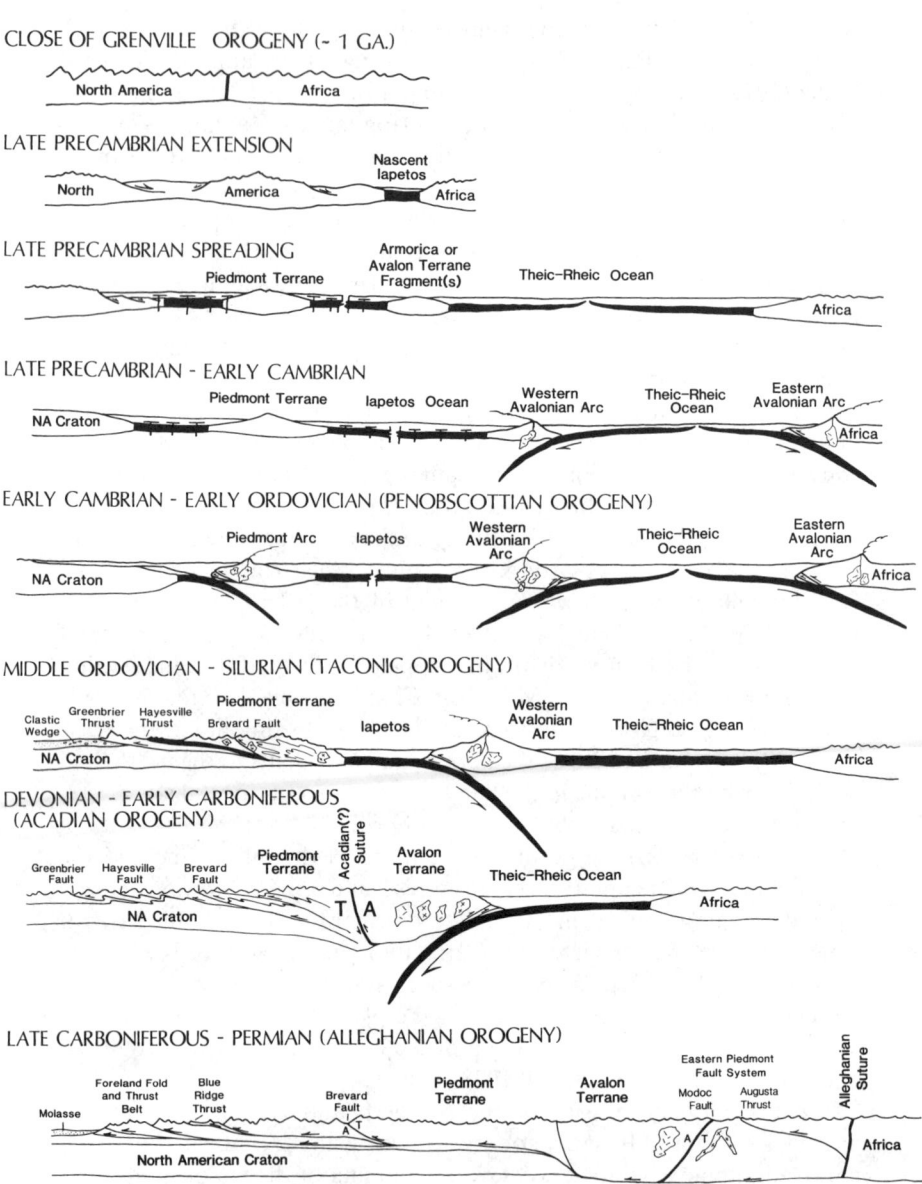

Figure 4 Time sequence diagram showing the development of the Appalachians along a section through the Tennessee-Carolinas-Georgia southern Appalachians (modified from Hatcher & Odom 1980).

lar rift-transform margin in which basins of various depths developed (Thomas 1977, 1983, Rankin 1976, Hatcher & Odom 1980, Williams & Hatcher 1983). Sediments derived from the craton were deposited in block-faulted basins on this rifted margin that influenced the later configurations of thrusts that transported these deposits onto the North American craton. The platform stratigraphic sequence was uninterrupted until Middle Ordovician time. A regional unconformity that developed on top of the Middle Ordovician carbonate bank may be the external vestige of the Taconic/Penobscot orogeny to the east, where thrust sheets loaded the continental margin (Quinlan & Beaumont 1984) and caused subsidence forming a foredeep basin (Sevier basin) about 450 Myr ago in the southern Appalachians. A similar basin formed later, at about 420 Myr ago, in the central Appalachians (Martinsburg basin), which indicates the diachronous nature of the Taconic orogeny. Large faults such as the Hayesville and its equivalents to the north and south formed at this time, and they emplaced the offshore continental slope-rise sediments and some continental and oceanic basement onto the outer portions of the rifted-margin assemblage. The Hamburg klippe and related structures would have also been emplaced at this time (Lash & Drake 1984, Drake et al 1986). This would have been accompanied by emplacement of ultramafic rocks as ophiolitic material, as well as some of the melanges that have been recognized in the southern and central Appalachians in recent years (Drake & Morgan 1981, Drake 1985, Drake et al 1986). Granitic plutons were formed at this time, probably above an eastward-dipping subduction zone. There are relatively few Paleozoic plutons in the western Blue Ridge, which indicates that the subduction zone, represented in remnant form by the Hayesville thrust sheet, probably dipped to the east or southeast.

Later Acadian(?) convergence and suturing of the Avalon (Carolina) terrane to North America produced another suite of granitic and mafic plutons and a large fault that welded the Carolina terrane to the ancient North American margin. This event may have produced a number of plutons above another east-dipping subduction zone and resulted in a metamorphic event that spread across the Inner Piedmont and into the eastern Blue Ridge. This activity may account for the high grades of metamorphism in this area and for the rapid decrease in metamorphic grade immediately east of the central Piedmont suture in the Carolina terrane.

The eastern Piedmont fault system formed later during the final (Alleghanian) collision of North America with Africa; this collision resulted in a detachmant of the ductile-brittle transition zone of the crust, propagating a thrust from the collision zone, which is probably under the Coastal Plain cover (Cook et al 1979, Hatcher & Zietz 1980). The Kiokee-Raleigh belt may be a ramp zone in the thrust (Secor et al 1986). The forces of collision propagated outward into the sedimentary package of the rifted-margin sequence, carrying it onto the continental platform and deforming the remainder of the sediments deposited during the Paleozoic on the western flanks of the Appalachian orogen.

The right-lateral component of strike-slip motion may have resulted from the final stages of the Alleghanian collision, producing an aggregate of a few tens of kilometers of right-lateral strike-slip in most of the southern Appalachians (Secor et al 1986). It may have produced a large amount of strike-slip motion that moved the Goochland terrane (Figure 3a) several hundred kilometers to the south from its initial position in the Washington-Philadelphia area to its present position in North Carolina and Virginia. This is likely if the State Farm gneiss domes are Baltimore-type gneiss domes and if Africa acted as a rigid indenter in this area, producing strike-slip motion like that predicted by Molnar & Tapponnier (1975) for southern Asia.

Beneath the Coastal Plain are several major boundaries that may ultimately prove to be the Africa–North America suture. There is a boundary originally defined by potential field data (Williams & Hatcher 1983, Higgins & Zietz 1983, Horton et al 1984, Nelson et al 1985) that probably is a suture involving the accretion of parts of Africa to southern North America. Another boundary exists under the Coastal Plain and trends northeastward, parallel to the strike of the Appalachians. It also probably represents a suture, because potential field expressions of Appalachian structures appear to be truncated along it. This boundary may define the eastern limit of the Blue Ridge–Piedmont detachment (Hatcher et al 1985). It may also represent a slightly earlier collisional boundary between Africa and North America in the late Paleozoic.

Literature Cited

Absher, R. S., McSween, H. Y. 1985. Granulites at Winding Stair Gap, North Carolina: the thermal axis of Paleozoic metamorphism in the southern Appalachians. *Geol. Soc. Am. Bull.* 96: 647–54

Bird, J. M., Dewey, J. F. 1970. Lithosphere plate–continental margin tectonics and the evolution of the Appalachian orogen. *Geol. Soc. Am. Bull.* 81: 1031–60

Bobyarchick, A. R. 1981. The eastern Piedmont fault system and its relationship to Alleghanian tectonics in the southern Appalachians. *J. Geol.* 89: 335–47

Bobyarchick, A. R. 1984. A late Paleozoic component of strike-slip in the Brevard zone, southern Appalachians. *Geol. Soc. Am. Abstr. With Programs* 16: 126

Boyer, S. A., Elliott, D. 1982. Thrust systems. *Am. Assoc. Pet. Geol. Bull.* 66: 1196–1230

Brown, W. R. 1970. Investigations of the sedimentary record in the Piedmont and Blue Ridge of Virginia. In *Studies in Appalachian Geology: Central and Southern*, ed. G. W. Fisher, F. J. Pettijohn, J. C. Reed Jr., K. N. Weaver, pp. 336–49. New York: Wiley-Interscience

Bryant, B., Reed, J. C. Jr. 1970. Geology of the Grandfather Mountain window and vicinity, North Carolina and Tennessee. *US Geol. Surv. Prof. Pap. 615*. 190 pp.

Butler, J. R. 1972. Age of Paleozoic regional

metamorphism in the Carolinas, Georgia, and Tennessee southern Appalachians. *Am. J. Sci.* 272: 319–33

Carpenter, R. H. 1970. Metamorphic history of the Blue Ridge Province of Tennessee and North Carolina. *Geol. Soc. Am. Bull.* 81: 749–62

Conley, J. F., Henika, W. S. 1973. Geology of the Snow Creek, Martinsville East, Price, and Spray Quadrangles, Virginia. *Va. Div. Min. Resour. Rep. Invest.* 33: 71

Cook, F. A., Albaugh, D. S., Brown, L. D., Kaufman, S., Oliver, J. E., Hatcher, R. D. Jr. 1979. Thin-skinned tectonics in the crystalline southern Appalachians: COCORP seismic reflection profiling of the Blue Ridge and Piedmont. *Geology* 7: 563–67

Crawford, W. A., Hoersch, A. L. 1984. Geology of the Honey Brook Upland, southeastern Pennsylvania. In *The Grenville Event in the Appalachians and Related Topics, Geol. Soc. Am. Spec. Pap. 194*, ed. M. J. Bartholomew, pp. 111–26

Dabbagh, A. E. 1975. *Geology of the Skyland and Dunsmore Mountain quadrangles, western North Carolina.* PhD thesis. Univ. N.C., Chapel Hill. 228 pp.

Dallmeyer, R. D. 1975. $^{40}Ar/^{39}Ar$ ages of biotite and hornblende from a progressively remetamorphosed basement terrane: their bearing on interpretation of release spectra. *Geochim. Cosmochim. Acta* 39: 1655–69

Dallmeyer, R. D., Hatcher, R. D. Jr. 1985. The Alto allochthon: Part 2, Geochronological constraints on tectonothermal evolution. *Geol. Soc. Am. Abstr. With Programs* 17: 86 (Abstr.)

Dallmeyer, R. D., Wright, J. E., Secor, D. T. Jr., Snoke, A. W. 1986. Character of the Alleghany orogeny in the southern Appalachians: Part II. Geochronological constraints on the tectonothermal evolution of the Piedmont in South Carolina. *Geol. Soc. Am. Bull.* 97: 1329–44

Dana, J. D. 1873. On some results of the Earth's contraction from cooling; including a discussion of the origin of mountains, and the nature of the Earth's interior. *Am. J. Sci.* 5: 423–43, 474–75; 6: 6–14, 104–15, 161–72, 304, 381–82. 3rd ser.

Dennis, A. J., Secor, D. T. Jr. 1986. A model for the development of crenulations in shear zones with applications from the southern Appalachian Piedmont. *J. Struct. Geol.* 8: In press

Diegel, F. A., Wojtal, S. F. 1985. Structural transect in SW Virginia and NE Tennessee. In *Field Trips in the Southern Appalachians*, ed. N. B. Woodward, pp. 70–143. Knoxville: Univ. Tenn. Dept. Geol. Sci.

Drake, A. A. Jr. 1970 Structural geology of the Reading Prong. In *Studies of Appalachian Geology: Central and Southern*, ed. G. W. Fisher, F. J. Pettijohn, J. C. Reed Jr., K. N. Weaver, pp. 271–91. New York: Wiley-Interscience

Drake, A. A. Jr. 1978. The Lyon Station–Paulins Kill nappe—the frontal structure of the Musconetlong nappe systems in eastern Pennsylvania and New Jersey. *US Geol. Surv. Prof. Pap. 1023.* 20 pp.

Drake, A. A. Jr. 1985. Tectonic implications of the Indian Run Formation—a newly recognized sedimentary melange in the northern Virginia Piedmont. *US Geol. Surv. Prof. Pap. 1324.* 12 pp.

Drake, A. A. Jr., Morgan, B. A. 1981. The Piney Branch Complex—a metamorphosed fragment of the central Appalachian ophiolite in northern Virginia. *Am. J. Sci.* 281: 484–508

Drake, A. A. Jr., Sinha, A. K., Laird, J., Guy, R. 1986. The Taconic orogen. In *U.S. Appalachians/Ouachitas Orogen, Geol. Soc. Am. Centen. Spec. Vol.*, ed. R. D. Hatcher Jr., G. W. Viele, W. A. Thomas.

Ellwood, B. B., Whitney, J. A., Wenner, D. B., Mose, D., Amerigian, C. 1980. Age, paleomagnetism, and tectonic significance of the Elberton Granite, northeast Georgia Piedmont. *J. Geophys. Res.* 85: 6521–33

Farrar, S. S. 1984. The Goochland granulite terrane: remobilized Grenville basement in the eastern Virginia Piedmont. In *The Grenville Event in the Appalachians and Related Topics. Geol. Soc. Am. Spec. Pap. 194*, ed. M. J. Bartholomew, pp. 215–27

Farrar, S. S. 1985. Tectonic evolution of the easternmost Piedmont, North Carolina. *Geol. Soc. Am. Bull.* 96: 362–80

Fullagar, P. D. 1971. Age and origin of plutonic intrusions in the Piedmont of the southeastern Appalachians. *Geol. Soc. Am. Bull.* 82: 2845–62

Fullagar, P. D., Butler, J. R. 1979. 325 to 265 m.y. old granitic plutons in the Piedmont of the southeastern Appalachians. *Am. J. Sci.* 279: 161–85

Glover, L. III, Sinha, A. K. 1973. The Virgilina deformation, a late Precambrian to Early Cambrian(?) orogenic event in the central Piedmont of Virginia and North Carolina. *Am. J. Sci.* 273-A: 234–51

Glover, L. III, Tucker, R. D. 1979. Virginia Piedmont geology along the James River from Richmond to the Blue Ridge. In *Guides to Field Trips 1–3 for Southeastern Section Meeting*, ed. L. Glover III, J. F. Read, pp. 1–41. Blacksburg: Va. Polytech. Inst. Dep. Geol. Sci.

Glover, L. III, Speer, J. A., Russell, G. S., Farrar, S. S. 1983. Ages of regional metamorphism and ductile deformation in the central and southern Appalachians. *Lithos* 16: 223–45

Griffin, V. S. Jr. 1974. Analysis of the Piedmont in northwestern South Carolina. *Geol. Soc. Am. Bull.* 85: 1123–31

Hall, J. 1883. Contributions to the geological history of the North American continent. *Proc. Am. Assoc. Adv. Sci., 31st, Montreal, 1882*, pp. 31–69

Harper, S. B., Fullagar, P. D. 1981. Rb-Sr ages of granitic gneisses of the Inner Piedmont of northwestern North Carolina and southwestern South Carolina. *Geol. Soc. Am. Bull.* 92: 864–72

Hatcher, R. D. Jr. 1971. Structural petrologic and stratigraphic evidence favoring a thrust solution to the Brevard problem. *Am. J. Sci.* 270: 177–202

Hatcher, R. D. Jr. 1972. Developmental model for the southern Appalachians. *Geol. Soc. Am. Bull.* 83: 2735–60

Hatcher, R. D. Jr. 1978. Tectonics of the western Piedmont and Blue Ridge, southern Appalachians: review and speculation. *Am. J. Sci.* 278: 276–304

Hatcher, R. D. Jr. 1980. Comments on the tectonics of the Piedmont and Blue Ridge near Winston-Salem, North Carolina. In *Geological Investigations of Piedmont and Triassic Rocks, Central North Carolina and Virginia. Rep. CGS 80-B-IV-1–7*, Dep. Energy, Savannah River Lab., Ga.

Hatcher, R. D. Jr. 1983. Basement massifs in the Appalachians: their role in deformation during the Appalachian orogenies. *Geol. J.* 18: 255–65

Hatcher, R. D. Jr. 1986. Basement/cover relationships in the Appalachian-Caledonian-Variscan orogen: mid-Devonian (end of Acadian orogeny) to end of Permian. In *Syntheses of the Caledonide Orogen*, ed. A. L. Harris, D. J. Fettes. London: Geol. Soc. London. In press

Hatcher, R. D. Jr., Butler, J. R. 1979. *Guidebook for Southern Appalachian Field Trip in the Carolinas, Tennessee and Northeastern Georgia. Int. Geol. Correl. Program Project 27, Caledonide Orogen*. 117 pp.

Hatcher, R. D. Jr., Odom, A. L. 1980. Timing of thrusting in the southern Appalachians, USA: model for orogeny? *J. Geol. Soc. London* 137: 321–27

Hatcher, R. D. Jr., Zietz, I. 1980. Tectonic implications of regional aeromagnetic and gravity data from the southern Appalachians. In *The Caledonides in the USA. Va. Polytech. Inst. Geol. Mem. 2*, pp. 235–44

Hatcher, R. D. Jr., Howell, D. E., Talwani, P. 1977. Eastern Piedmont fault system: some speculations on its extent. *Geology* 5: 636–40

Hatcher, R. D. Jr., Hooper, R. J., Petty, S. M., Willis, J. D. 1984. Structure and chemical petrology of three southern Appalachian mafic-ultramafic complexes and their bearing upon the tectonics of emplacement and origin of Appalachian ultramafic bodies. *Am. J. Sci.* 284: 484–506

Hatcher, R. D. Jr., Litehiser, J. J., Zietz, I. 1985. Crustal blocks of the Appalachian region and the eastern limit to the Appalachian detachment. *Eos, Trans. Am. Geophys. Union* 66: 358 (Abstr.)

Hatcher, R. D. Jr., Hooper, R. J., Heyn, T., McConnell, K. I., Costello, J. O. 1986a. Geometric and time relationships of thrusts in the crystalline southern Appalachians. In *Geometry and Mechanisms of Appalachian Thrusting. Geol. Soc. Am. Spec. Pap.*, ed. G. Mitra, S. Wojtal. In press

Hatcher, R. D. Jr., Costain, J. K., Coruh, C., Phinney, R. A., Williams, R. T., Glover, L. 1986b. Tectonic implications of new ADCOH seismic reflection data from the crystalline southern Appalachians. *Proc. Int. Symp. Deep Seism. Reflection Profiling Cont. Lithosphere, 2nd, Cambridge, Engl.* In press

Higgins, M. W., Zietz, I. 1983. Geologic interpretation of geophysical maps of the pre-Cretaceous "basement" beneath the Coastal Plain of the southeastern United States. In *Contributions to the Tectonics and Geophysics of Mountain Chains. Geol. Soc. Am. Mem. 158*, ed. R. D. Hatcher Jr., H. Williams, I. Zietz, pp. 125–30

Hooper, R. J., Hatcher, R. D. Jr. 1986, The Ocmulgee fault: a fundamental tectonic boundary in the Piedmont of central Georgia. *Geol. Soc. Am. Abstr. With Programs* 18: 227

Hopson, C. A. 1964. The crystalline rocks of Howard and Montgomery Counties, Maryland. In *The Geology of Howard and Montgomery Counties, Maryland*, pp. 27–215. Baltimore: Md. Geol. Surv.

Horkowitz, J. P. 1984. Geology of the Philson Crossroads $7\frac{1}{2}$ minute quadrangle, South Carolina—the nature of the boundary separating the Inner Piedmont from the Carolina/Avalon terrane in central-northwestern South Carolina. MS thesis. Univ. S.C., Columbia. 100 pp.

Horton, J. W. Jr. 1982. Geologic map and mineral resources of the Rosman quadrangle, North Carolina. *N.C. Geol. Surv. Map GM185-NE*, scale 1/24,000

Horton, J. W. Jr., Butler, J. R. 1977. Guide to the geology of the Kings Mountain belt in the Kings Mountain area, North Carolina. In *Field Guides for Geological Society of America, SE Section Meet.*, ed. E. R. Burt, pp. 76–143. Winston-Salem: N.C. Geol. Surv.

Horton, J. W. Jr., Zietz, I., Neathery, T. L. 1984. Truncation of the Appalachian Piedmont beneath the Coastal Plain of Alabama: evidence from new magnetic data. *Geology* 12: 51–55

Jonas, A. I. 1932. Structure of the metamorphic belt of the southern Appalachians. *Am. J. Sci.* 24: 228–43. 5th ser.

King, P. B. 1955. A geologic section across the southern Appalachians: an outline of the geology in the segment in Tennessee, North Carolina and South Carolina. In *Guides to Southeastern Geology, Geol. Soc. Am. 68th Ann. Meet. Guideb.*, ed. R. J. Russell, pp. 332–73

King, P. B. 1970. Epilogue. In *Studies of Appalachian Geology: Central and Southern*, ed. G. W. Fisher, F. J. Pettijohn, J. C. Reed Jr., K. N. Weaver, pp. 437–39. New York: Wiley-Interscience

Kish, S. A., Black, W. W. 1982. The Carolina slate belt: origin and evolution of an ancient volcanic arc. In *Tectonic Studies in the Talladega and Carolina Slate Belts, Southern Appalachian Orogen. Geol. Soc. Am. Spec. Pap. 191*, ed. D. N. Bearce, W. W. Black, S. A. Kish, J. F. Tull, pp. 91–97

Kite, L. E., Stoddard, E. F. 1984. The Halifax County complex: oceanic lithosphere in the eastern North Carolina Piedmont. *Geol. Soc. Am. Bull.* 95: 422–32

Lash, G. G., Drake, A. A. Jr. 1984. The Richmond and Greenwich slices of the Hamburg klippe in eastern Pennsylvania—stratigraphy, sedimentology, structure and plate tectonic implications. *US Geol. Surv. Prof. Pap. 1312.* 40 pp.

McConnell, K. I., Glover, L. III. 1982. Age and emplacement of the Flat River complex, an Eocambrian subvolcanic pluton near Durham, North Carolina. In *Tectonic Studies in the Talladega and Carolina Slate Belts, Southern Appalachian Orogen. Geol. Soc. Am. Spec. Pap. 191*, ed. D. N. Bearce, W. W. Black, S. A. Kish, J. F. Tull, pp. 133–44

McSween, H. Y. Jr., Hatcher, R. D. Jr. 1985. Ophiolites(?) of the southern Appalachian Blue Ridge. In *Field Trips in the Southern Appalachians*, ed. N. B. Woodward, pp. 144–70. Knoxville: Univ. Tenn. Dept. Geol. Sci.

Misra, K. C., Keller, F. B. 1978. Ultramafic bodies in the southern Appalachians. *Am. J. Sci.* 278: 389–418

Misra, K. C., McSween, H. Y. Jr. 1984. Mafic rocks of the southern Appalachians: a review. *Am. J. Sci.* 284: 294–318

Molnar, P., Tapponnier, P. 1975. Cenozoic tectonics of Asia: effects of a continental collision. *Science* 189: 419–26

Mose, D. G., Nagel, M. S. 1982. Plutonic events in the Piedmont of Virginia. *Southeast Geol.* 23: 25–39

Muller, P. D., Chapin, D. A. 1984. Tectonic evolution of the Baltimore Gneiss anticlines, Maryland. In *The Grenville Event in the Appalachians and Related Topics. Geol. Soc. Am. Spec. Pap. 194*, ed. M. J. Bartholomew, pp. 127–48

Nelson, K. D., Arnow, J. A., McBride, J. H., Willemin, J. H., Huang, J., et al. 1985. New COCORP profiling in the southeastern United States, Part I: Late Paleozoic suture and Mesozoic rift basin. *Geology* 13: 714–17

Odom, A. L., Fullagar, P. D. 1973. Geochronologic and tectonic relationships between the Inner Piedmont, Brevard zone, and Blue Ridge belts, North Carolina. *Am J. Sci.* 273-A: 133–49

Odom, A. L., Fullagar, P. D. 1984. Rb-Sr whole rock and inherited zircon ages of the plutonic suite of the Crossnore Complex, southern Appalachians and their implications regarding the time of opening of the Iapetos Ocean. In *The Grenville Event in the Appalachians and Related Topics. Geol. Soc. Am. Spec. Pap. 194*, ed. M. J. Bartholomew, pp. 255–63

Oriel, S. S. 1950. Geology and mineral resources of the Hot Springs window, Madison County, North Carolina. *N.C. Geol. Surv. Bull. 60.* 70 pp.

Osberg, P. H. 1978. Synthesis of the geology of the northeastern Appalachians, U.S.A. In *Caledonian-Appalachian Orogens of the North Atlantic Region, Geol. Surv. Can. Pap. 78-13*, ed. E. T. Tozer, P. E. Schenck, pp. 137–47

Pavlides, L. 1981. The central Virginia volcanic-plutonic belt: an island arc of Cambrian(?) age. *US Geol. Surv. Prof. Pap. 1231-A.* 34 pp.

Petty, S. M. 1982. *The geology of the Laurel Creek mafic-ultramafic complex in northeast Georgia: intrusive complex or ophiolite?* MS thesis. Fla. State Univ., Tallahassee. 147 pp.

Quinlan, G. M., Beaumont, C. 1984. Appalachian thrusting, lithospheric flexure and Paleozoic stratigraphy of the eastern interior of North America. *Can. J. Earth Sci.* 21: 973–96

Rankin, D. W. 1975. The continental margin of eastern North America in the southern Appalachians: the opening and closing of the Proto-Atlantic Ocean. *Am. J. Sci.* 275-A: 298–336

Rankin D. W. 1976. Appalachian salients and recesses: Late Precambrian continental breakup and the opening of the Iapetos Ocean. *J. Geophys. Res.* 81: 5605–19

Reed, J. C. Jr., Bryant, B. 1964. Evidence for strike-slip faulting along the Brevard zone in North Carolina. *Geol. Soc. Am. Bull.* 75: 1177–95

Roper, P. J., Dunn, D. E. 1973. Superposed deformation and polymetamorphism, Brevard zone, South Carolina. *Geol. Soc. Am. Bull.* 84: 3373–86

Russell, G. S. 1978. *U-Pb, Rb-Sr and K-Ar isotopic studies bearing on the tectonic development of the southernmost Appalachian orogen, Alabama.* PhD thesis. Fla. State Univ., Tallahassee. 157 pp.

Schamel, S., Bauer, D. T. 1980. Remobilized Grenville basement in the Pine Mountain window. In *The Caledonides in the USA*, ed. D. R. Wones, pp. 313–16. Blacksburg: Va. Polytech. Inst. Dept. Geol. Sci.

Sears, J. W., Cook, R. B. Jr. 1984. An overview of the Grenville basement complex of the Pine Mountain window, Alabama and Georgia. In *The Grenville Event in the Appalachians and Related Topics. Geol. Soc. Am. Spec. Pap. 194*, ed. M. J. Bartholomew, pp. 281–87

Secor, D. T., Samson, S. L., Snoke, A. W., Palmer, A. R. 1983. Confirmation of the Carolina slate belt as an exotic terrane. *Science* 221: 649–51

Secor, D. T. Jr., Snoke, A. W., Bramlett, K. W., Costello, J. O., Kimbrell, O. P. 1986. Character of the Alleghanian orogeny in the southern Appalachians: Part I. Alleghanian deformation in the eastern Piedmont of South Carolina. *Geol. Soc. Am. Bull.* 97: In press

Shanmugam, G., Lash, G. C. 1982. Analogous tectonic evolution of the Ordovician foredeeps, southern and central Appalachians. *Geology* 10: 562–66

Sinha, A. K., Glover, L. III. 1978. U/Pb systematics of zircons during dynamic metamorphism. *Contrib. Mineral. Petrol.* 66: 305–10

Sinha, A. K., Zietz, I. 1982. Geophysical and geochemical evidence for a Hercynian magmatic arc, Maryland to Georgia. *Geology* 10: 593–96

Snoke, A. W., Kish, S. A., Secor, D. T. Jr. 1980. Deformed Hercynian granitic rocks from the Piedmont of South Carolina. *Am. J. Sci.* 280: 1019–34

Thomas, W. A. 1977. Evolution of Appalachian-Ouachita salients and recesses from reentrants and promontories in the continental margin. *Am. J. Sci.* 277: 1233–78

Thomas, W. A. 1983. Continental margins, orogenic belts and intracratonic structures. *Geology* 11: 270–72

Tull, J. F. 1978. Structural development of the Alabama Piedmont northwest of the Brevard zone. *Am. J. Sci.* 278: 442–60

Tull, J. F., Bearce, D. N., Guthrie, G. M., eds. 1985. *Early Evolution of the Appalachian Miogeocline: Upper Precambrian Lower Paleozoic Stratigraphy of the Talladega Slate Belt. Ala. Geol. Soc. Guideb.* 92 pp.

Williams, H., Hatcher, R. D. Jr. 1982. Suspect terranes and accretionary history of the Appalachian orogen. *Geology* 10: 530–36

Williams, H., Hatcher, R. D. Jr. 1983. Appalachian suspect terranes. In *Contributions to the Tectonics and Geophysics of Mountain Chains. Geol. Soc. Am. Mem. 158*, ed. R. D. Hatcher Jr., H. Williams, I. Zietz, pp. 33–53. Boulder, Colo: Geol. Soc. Am.

Willis, B. 1893. The mechanics of Appalachian structure. *US Geol. Surv. 13th Ann. Rep.*, Pt. 2, pp. 211–81

ORGANIC GEOCHEMISTRY OF BIOMARKERS

R. Paul Philp and C. Anthony Lewis

School of Geology and Geophysics, University of Oklahoma, Norman, OK 73019

INTRODUCTION

The classic organic chemical studies by the German chemist Alfred Treibs on the structures of tetrapyrrole-type compounds in crude oils led him to propose a biological origin for crude oils (Treibs 1934, 1936). Treibs based his proposal on structural determinations, made by classical chemical methods, of porphyrins (I) in the oils and their structural relationship with chlorophyll (II) molecules (known to be widely distributed in plants) and compounds with similar structures in animals and bacteria. These observations can be thought of as the beginning of the biomarker concept and precursor/product relationships. Although the idea was initially proposed in the 1930s, it was not widely exploited until major analytical advances occurred in the 1960s and 1970s. In this time period major breakthroughs were made in the coupling of gas chromatographs and mass spectrometer systems and the development of capillary gas chromatography columns.

The ultimate result of the analytical developments, for geochemical purposes, has been the capability to determine trace amounts of components, i.e. biomarkers, in very complex organic extracts derived from geological samples. A biomarker can be best thought of as an organic compound in a geological sample that can be structurally related to its precursor molecule, which occurs as a natural product in a plant, animal, bacteria, spore, fungi, or any other potential source material. (The term biomarker is virtually synonymous with a number of other terms used in the literature, such as chemical fossil and biological marker, but for convenience biomarker is used throughout this article.) The distributions of these compounds permit information on the source, or maturity, of organic matter in a sample to be determined and, in the case of oils, on the extent of biodegradation, the possible migration pathways, and the relative migration distances. It is interesting to note that in the 1970s much of the biomarker work was directed at studies of Recent sediments (Philp et al 1976, Brassell et al 1978). The results and information from those studies have subsequently been invaluable for the application of the biomarker concept to petroleum studies. More recent advances in analytical technology, the continued development of high-performance liquid chromatography systems, and a new generation of mass spectrometers (namely triple-stage quadrupole and hybrid quadrupole/magnetic instruments) are now opening the way for alternative and more sophisticated methods of biomarker determinations.

This article is not an exhaustive review of biomarkers or their applications to various aspects of geochemistry. There have been a number of comprehensive articles of that nature in the relatively recent literature on specific topics such as the geochemistry of steroids (Mackenzie et al 1982), as well as others of a more general nature but still containing detailed reviews of biomarker geochemistry (Mackenzie 1984a, Philp 1985, 1986, Volkman 1986). Instead, we reevaluate early concepts of biomarker geochemistry as a result of new observations and discoveries related to improvements in analytical capabilities. The major areas discussed are (*a*) depositional environments and source-related parameters, (*b*) maturation parameters, (*c*) migration, (*d*) biodegradation, and (*e*) basin modeling.

CORRELATION OF BIOMARKERS WITH DEPOSITIONAL ENVIRONMENTS AND SOURCE MATERIALS

Depositional Environments

Depositional environments play an important role in determining the amount and rate at which organic matter will accumulate in a basin

(Demaison & Moore 1980). The ability to recognize changes in depositional environments through different stratigraphic horizons plays a major role in the evaluation of a basin for petroleum exploration purposes. If a specific set of biomarker parameters can be assigned to a clearly defined depositional environment, it will permit recognition of this type of environment in previously unexplored regions. In many cases the presence of a particular biomarker can be attributed to a specific organism that will grow only in the conditions peculiar to a particular depositional environment, e.g. hypersaline, reducing, freshwater. Alternatively, an unusual biomarker fingerprint, such as the even/odd predominance of n-alkanes, may be associated with a particular environment.

One of the most widely reported classes of biomarkers in the geochemical literature are the n-alkanes. The distributions of n-alkanes are readily determined by gas chromatography alone and were determined for many years prior to the advent of gas chromatograph–mass spectrometer coupling.

The use of n-alkane distributions as source and maturity indicators has been covered extensively in previous articles (Brassell et al 1978, Mackenzie 1984a, Philp 1985), but there are two significant points that need to be mentioned in this article. The first is the increasing number of oil samples that have been reported with an even/odd predominance of n-alkane distributions. Tissot et al (1977) noted in their analysis of 1300 samples of suspected source rocks and crude oils that samples associated with carbonate environments generally showed an even/odd predominance of n-alkanes that can be related to the highly reducing nature of these environments (e.g. Palacas et al 1984). In a highly reducing carbonate environment, even-numbered fatty acid precursors are reduced to produce an even-numbered alkane rather than being decarboxylated and losing a carbon atom. Nishimura & Baker (1986) proposed that in certain cases the n-alkane even/odd predominance may result directly from an input of marine biota. The distributions of the n-fatty acids and alcohols in the same extracts were quite different from those observed for the n-alkanes, eliminating them as possible sources. Based on previous hydrocarbon data for bacteria (Davis 1968) and the fact that Simoneit et al (1980) found an even predominance of n-alkanes in a Chilean paraffin dirt, Nishimura & Baker (1986) speculated that the more likely source for the marked even/odd predominance of n-alkanes (C_{16}–C_{24}) in their samples was from bacteria living under certain marine environmental conditions rather than diatoms.

An even/odd predominance of n-alkanes in the C_{12}–C_{20} range has also been observed recently in a series of surface sediments in the Arabian Gulf (Grimalt et al 1985). The n-alkanes were not derived from pollutant sources, and they are of interest because this appears to be one of the first reports of such a distribution in an apparently well-oxygenated system. The lack of a reducing environment, previously used to explain such distributions, has prompted Grimalt et al (1985) to emphasize an autochthonous origin, possibly bacterial, for these alkanes. The similarity between the observations of Grimalt et al (1985) and Nishimura & Baker (1986) deserves further investigation; such studies may provide additional information on biomarkers in marine organisms, especially bacteria. The recent observations of Grimalt et al (1985), Nishimura & Baker (1986), and the earlier extensive study of alkanes in oils and source rocks by Tissot et al (1974) permit speculation that the even/odd predominance of n-alkanes may thus result from marine organisms as well as specific conditions in the depositional environment.

One of the earliest attempts at recognizing depositional environments from biomarker parameters was based on the relative distribution of the isoprenoids pristane (III) and phytane (IV). In an oxidizing environment, cleavage of the phytol side chain of chlorophyll (II) would be followed by decarboxylation to produce pristane. In a reducing environment, side-chain cleavage of chlorophyll would be followed by reduction and would ultimately produce phytane. The ratio of pristane/phytane has been used extensively for many years, along with other parameters, to provide an indication on the oxidizing or reducing nature of a depositional environment. Didyk et al (1978) showed that the reaction sequences originally proposed for the conversion processes are far more complicated than was originally thought. In view of all the intermediate components involved in the production of pristane and phytane, care needs to be exercised when using pristane/phytane ratios for this purpose. The use of this ratio has been potentially complicated in recent years by the discovery of an alternative source for pristane, namely α-tocopherol (Goossens et al 1984). It has been proposed that the reaction of isoprenoids with H_2S will form thioisoprenoids (Brassell et al 1986), which may in turn remove some of the pristane or phytane normally used in calculating this ratio. Furthermore, additional inputs of certain isoprenoids may result from degradation of the C_{40} isoprenoid ethers present in archaebacteria, which have been extensively investigated in the past few years (Chappe et al 1982, Albaiges et al 1985).

Specific acyclic isoprenoids associated with a particular environment have been reported by Waples et al (1974), who noted that Tertiary sediments deposited in a lagoonal, saline environment possessed relatively high concentrations of the regular C_{25} isoprenoid (V). More recently, ten Haven et al (1986a) observed similarly enhanced concentrations of this compound in sediment samples collected from hypersaline environments. It has been suggested that this C_{25} isoprenoid is associated with a particular organism, as yet unknown, that grows preferentially in saline lagoonal-type environments.

Botryococcane (VI) is a branched hydrocarbon derived from the unsaturated hydrocarbon, botryococcene, which can be directly associated with an organism that will only grow in a specific type of environment. The fresh or brackish water alga, *Botryococcus braunii*, was observed to contain concentrations of botryococcane as high as 70–90% in the senescent phase. The unique occurrence of this compound in *Botryococcus braunii* has been used by Moldowan & Seifert (1980) as evidence that certain oil deposits in Sumatra, Indonesia, were generated principally from prehistoric source material in a fresh or brackish lagoonal-type environment. *B. braunii* exists as two physiologically distinct clonal races, and as such it can contribute both unsaturated hydrocarbons, which are potential precursors of botryococcanes, and long chain *n*-alkanes to freshwater (lacustrine) sediments. A study of biomarkers present in coastal bitumens from the western Otway Basin in Australia revealed the presence of significant concentrations of botryococcane, for the first time, in an Australian crude oil (McKirdy et al 1986). Deposition of *Botryococcus* blooms under anoxic or micro-oxic conditions in deep lakes was proposed to account for the waxy character of the three main bitumen types found in the Otway Basin and their botryococcane content. McKirdy et al (1986) suggested that following the breakup of Australia and Antarctica, such sediments could have accumulated in large meromictic layers in the rift valleys associated with this process. In addition to botryococcane, the coastal bitumens contained relatively high concentrations of 4-methylsteranes previously proposed to be derived from 4-methylsteroids occurring in freshwater dinoflagellates (Robinson et al 1984).

A combination of biomarker and isotopic signatures led Summons & Powell (1986) to suggest the presence of microbial communities containing *Chlorobiaceae*, a green sulfur bacteria, in ancient restricted seas. The biomarker fingerprint consisted of a series of 1-alkyl-2,3,6-trimethyl benzenes (VII) thought to be derived from the aromatic carotenoids of the bacteria. It was proposed that source rocks for oils containing these compounds were deposited under metahaline to hypersaline sulfate- and sulfide-rich, strongly reducing water columns. Such environments are known to be favored by anoxygenic photosynthetic sulfur bacteria. The fact that these alkyl benzenes had an unusually heavy $\delta^{13}C$ value provided additional evidence for their origin from *Chlorobiaceae* via a reverse or reductive Krebs cycle rather than the Benson-Calvin pathway.

A number of compounds recently detected in a detailed geochemical study of a Messianian evaporitic basin in Italy, including the C_{25} isoprenoid hydrocarbon (V), have been proposed as indicators for a hypersaline depositional environment (ten Haven et al 1986a). These compounds include short side-chain steranes and $5\alpha(H)$, $14\beta(H)$, $17\beta(H)$ pregnanes and homopregnanes, and they are thought to be related to certain organisms occurring exclusively in hypersaline environments. A series of C_{31}–C_{35} 22S and 22R hop-17(21)-enes (VIII) were also observed to occur in parallel with the corresponding saturated hopane (IX) series, with both series maximizing at C_{35}. Ten Haven et al (1986b) suggested that the hopene series could possibly be another indicator for a hypersaline environment, and they also noted that it has been observed in a number of other saline environments. Gammacerane (X) has been proposed by a number of workers as an additional indicator of hypersalinity, and it was also observed in the samples studied by ten Haven et al (1986a). The origin of gammacerane is still unclear, but it appears to be indicative of specific depositional environments. Gammacerane was reported by Henderson et al (1969) to occur in extracts of the Green River Shale, and it was proposed that it was derived from tetrahymanol, but this still has not been substantiated. Tetrahymanol, or gammacerane-3β-ol, occurs in ferns and a nonmarine protozoan, *Tetrahymena pyriformis* (Hills et al 1966, Whitehead 1974), and therefore it was originally suggested that gammacerane was a potential nonmarine marker. However, a recent paper by Moldowan et al (1985) suggested a lack of correlation between gammacerane content and depositional environment. High gammacerane indices (gammacerane/C_{30} hopane) may signal hypersaline episodes of source-rock deposition occurring in alkaline lakes as well as in lagoonal carbonate evaporite environments. Fan et al (1984) have shown that a number of oils derived from sediments deposited in a saline environment have inherently higher gammacerane content than those formed from fresh to brackish-type environments. Fan et al (1984) concurred with the earlier proposal by Hills et al (1966) that certain protozoans favoring saline-type environments may be the organisms that contribute significantly to the higher yields of gammacerane in the saline facies oils (Caspi et al 1968). Xianzhang et al (1985), in a study of several terrestrial basins in China, noted that in the Raoyang depression of the North China region, the herbaceous-algal facies source rocks at the center of the depression contained relatively high concentrations of gammacerane. The gammacerane was effectively absent at the edges of the depression, where the predominant source material was from higher plants. In our laboratory we have recently analyzed twenty oils from a variety of depositional environments within China and have shown that samples thought to be produced from freshwater lake environments had concentrations of gammacerane below the limits of detection, whereas those from saline lake depositional environments had gammacerane indices of approximately 1.2. As mentioned above, in their study of hypersaline lake environments, ten Haven et al (1986a) noted the presence of relatively high gammacerane concentrations, as did Rullkötter et al (1984a) in samples from the Upper Cretaceous bituminous chalks from the Ghareb formation (Israel). Ten Haven et al (1986a) pointed out that a careful examination of the literature revealed that the samples examined by Rullkötter et al (1984a) were also from a hypersaline environment.

In addition to specific biomarkers associated with organisms growing in a specific depositional environment, a number of characteristic features can be associated with carbonate source rocks or the oils derived from them. Palacas et al (1984) analyzed a number of oils from the Sunniland Formation of the South Florida Basin that are thought to be derived from the lower Sunniland Limestone, and they assigned a number of features to these oils as characteristic of being derived from carbonate rocks. The oils had a high $Ph/n\ C_{18}$ ratio and a low Pr/Ph ratio indicative of source material being deposited under extremely reducing conditions. The tricyclic terpanes revealed a characteristically high concentration of the C_{23} tricyclic terpane (XI) in all of the oils as well as tricyclic terpanes over the C_{19}–C_{28} range and possibly extending to higher carbon numbers. A number of the oils were characterized by relatively high concentrations of the C_{34} and C_{35} extended hopanes and in many cases by a C_{29}/C_{30} hopane ratio in the 0.8–1.0 range. Ratios such as these are in all probability due to the effects of the strong reducing conditions on the formation mechanisms of the hopanes from their oxygenated precursors. Diasteranes (XII) were observed in these carbonate source oils, although it had been previously proposed that in the absence of clay minerals (i.e. carbonate environments) diasteranes will not be formed (Seiskind et al 1979). Palacas et al (1984) suggested that some, if not all, of the diasteranes present may be the result of microbial activity in a highly reducing hypersaline environment. Similar distributions observed by Zumberge (1984) in the analysis of source rocks from the La Luna Formation, Colombia, were characterized by the abundance of the C_{23} tricyclic terpane, high C_{29}/C_{30} hopane ratios, no diasteranes, and high concentrations of C_{27} steranes along with C_{29} steranes (XIII, $R = C_2H_5$). Some of these peculiarities can be distinguished in oils derived from pre-Ordovician carbonate source rocks (McKirdy et al 1983). In particular, McKirdy et al (1983) observed that the C_{35} extended hopanes were present in higher concentrations than the C_{34} component, and that diasteranes were generally absent or subordinate to nonrearranged steranes. One major difference, however, was the predominance of the C_{28} and C_{29} steranes and the virtual absence of C_{27} steranes in many of these oils.

In a further extension of biomarker fingerprints that can be associated with carbonate environments, Hussler et al (1984) observed that such samples were characterized by high abundances of benzohopanes (XIV; C_{32}–C_{35}) and ring-D aromatized 8,14-secohopanoids (XV; C_{27}–C_{35}). Benzohopanes are probably transformation products of hopanoid precursors that are widespread constituents of the lipidic membranes of prokaryotes. Benzohopanes were present in shale samples, and they were also present in carbonate samples, although in much lower concentrations (Hussler et al 1984). The ring-D aromatic 8,14-secohopanoids do, however, appear to be much more abundant in carbonate-derived crude oils, with the C_{29} and C_{30} compounds being the predominant members of the series. Although the benzohopanes have been proposed to occur at early stages of diagenesis, the 8,14-secohopanoids are presumably formed at a later stage of maturation, possibly as a result of thermocatalytic cleavage of the fragile 8,(14) bond in hopanoid precursors.

Considerable discussion in the recent geochemical literature has been directed at the depositional history of Ordovician age oils in the Michigan Basin. A particularly vexing question is whether or not the oils are actually of Ordovician age or simply trapped in Ordovician age reservoirs. An interesting development in this story is contained in a paper by Foster et al (1986) concerned with the Ordovician Goldwyer Formation, Canning Basin, Australia. Examination of the kerogen from this formation shows that the dominant organism is similar to contemporary *Gloeocapsamorpha prisca*. It was concluded, therefore, that the organism responsible for the kerogen was probably a planktonic cyanobacterium, which bloomed in enormous numbers in a normal marine environment. Ordovician sediments containing either *G. prisca* or oils derived from that source are known to occur in the Baltic Basin (Estonian kukersite), USSR; the Illinois, Michigan, and Williston basins, USA; the Canning and Amadeus basins, Australia; and the Hudson Bay Basin, Canada (Reed et al 1986). Oils and source-rock extracts from these basins have all been shown to be characterized by chromatograms with a pronounced odd/even distribution of alkanes in the C_{17}–C_{19} range and relatively low concentrations of alkanes above C_{20}. The most interesting suggestion proposed by Foster et al (1986) is based on the Ordovician paleogeographic reconstruction of Smith et al (1981). When the Ordovician basins of high source potential are plotted on this map, they occur within a belt approximately 5° north and south of the Early to Mid-Ordovician equator. Deposition would occur in relatively shallow, warm epeiric seas, where nutrients in runoff from surrounding land areas would promote algal growth. Associated bacterial activity could lead to the anoxic conditions necessary for preservation of organic matter. Age differences occurring between the localities, based on conodont studies, may indicate continental drift through the equatorial belt or migration of climatic belts through time.

The results reviewed above have been concerned with hydrocarbon biomarkers, but sulfur-containing biomarkers have also assumed a special significance in the last two or three years for a number of reasons. One is their role as possible indicators of specific depositional environments, and another is the proposal that organic matter in sediments can act as a sink for sedimentary sulfur. In 1974 Ho et al (1974) undertook a detailed study of sulfur compounds in oils. It was observed at that time, and more recently by Radke et al (1982), that a variety of changes occur to the distribution of benzothiophenes (XVI) and methyldibenzothiophenes (XVII) as a result of thermal maturation. An extension of this early work was recently published by Hughes (1984), who compared the distribution of thiophenic organosulfur compounds in oils derived from carbonate sources with those from siliciclastic sources. Carbonate-derived oils in particular contained an abundance of benzothiophenes, an approximately equal distribution among the alkylated dibenzothiophenes, and a distinctive distribution of methyldibenzothiophene isomers. Oils derived from siliciclastic sources not only contained a far lower concentration of the sulfur-containing species but in addition had a significantly different pattern of methyl-dibenzothiophene isomers. Although these distributions are based primarily on source effects, increasing maturity will tend to convert the methyldibenzothiophene isomers to the most thermodynamically favorable distribution. Changes such as these make the distinction between carbonate and siliciclastic oils less well defined, but fortunately such changes do not occur until after the oil generation window has been passed.

Brassell et al (1986) recently suggested a mechanism that may at least in part account for the presence, or origin, of some of the sulfur compounds in geological samples. Identification of isoprenoid thiophenes in Recent sediments has led these authors to propose that organic matter may act as a sink for sedimentary sulfur, which can be incorporated in either an intra- or intermolecular fashion. The latter mode would give rise to sulfur-containing polymeric material within kerogens that could survive into the asphaltene and crude oil stages. Support for the incorporation of sulfur as a cross-linking agent in kerogens comes from some recent pyrolysis gas chromatography work with kerogens using a flame photometric detector to detect sulfur-containing components. Figure 1 shows the sulfur compounds produced by pyrolysis of Kimmeridge and Monterey Shale kerogen as examples. The suggestion of Brassell et al (1986) followed from their identification of isoprenoid thiophenes in a number of relatively recent sediments from the Deep-Sea Drilling Program (DSDP) program. One of the major compounds identified was 3-methyl-2-(3,7,11-trimethyldodecyl)-thiophene (XVIII) and it has been suggested that it may be formed by sulfur incorporation into chlorophyll-derived phytol, archaebacterial phytenes or their diagenetic products. However, a biosynthetic origin for this type of compound cannot be excluded at this time, especially in view of the biogenic origin for dimethylsulfide and more volatile thiophenes.

An analogous pathway to the one described above has been proposed for incorporation of bacterial sulfur into bacteriohopane tetrol (XIX) to form a thienylhopane (XX). This tetrol is a widespread membrane constituent of procaryotes and a surrogate of sterols in bacterial membranes (Valisolalao et al 1984). The bacteriohopane tetrol has been documented as a precursor of the fossil hopanoids (Ourisson et al 1979). Valisolalao et al (1984) proposed that cyclization in the presence of H_2S and possibly a mineral matrix could have led to the formation of the thienylhopane in the early stages of sedimentation.

A recent example of a sulfur compound being used as a depositional environmental indicator is provided by ten Haven et al (1986a) from their studies of hypersaline environments and in particular a marl sample from a Messinian evaporitic basin in Italy. Among the sulfur-containing compounds, the dominant component in the aromatic fraction was an organic sulfur compound, namely 2,3-dimethyl-5-(2,6,10-trimethylundecyl)-thio-

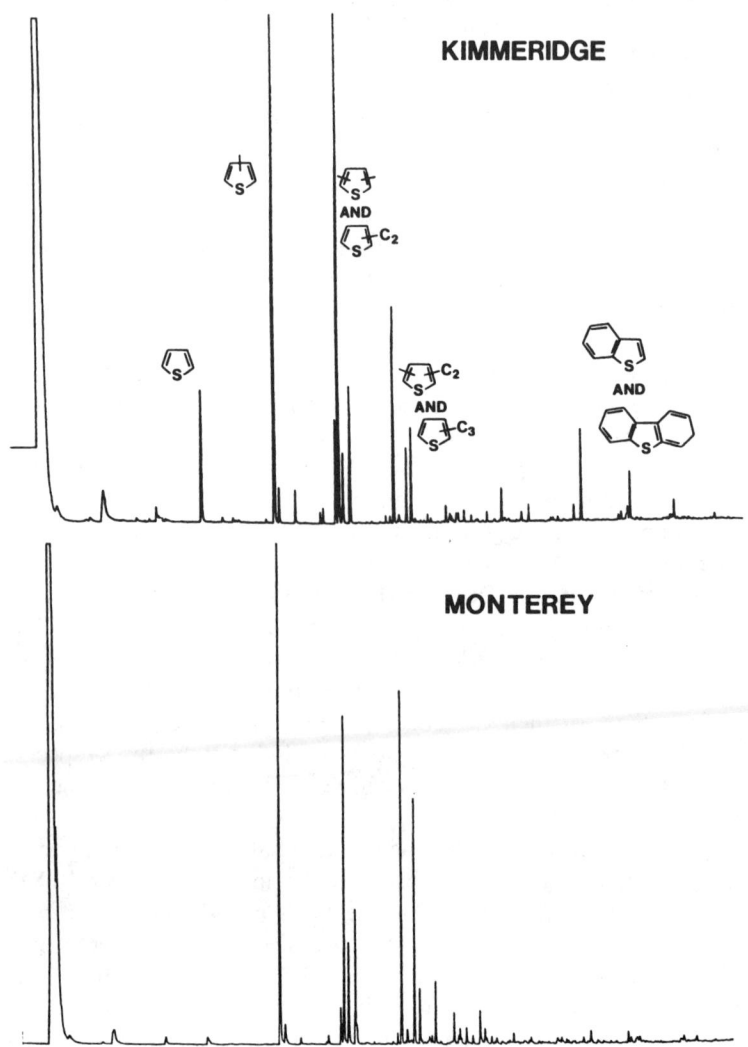

Figure 1 Characterization of various kerogens by pyrolysis-gas chromatography using a flame photometric detector demonstrated the production of various sulfur compounds. The distributions of these compounds vary between different samples, indicating their possible use as depositional environment indicators (A. Bakel, T. Eglinton & R. P. Philp, unpublished results).

phene (XXI). It was speculated that some of the minor uncommon isoprenoid thiophenes resulted from incorporation of sulfur into 2,6,10-trimethyl-7-(3-methylbutyl)-dodecane (XXII) (Sinninghe Damsté et al 1986). Sinninghe Damsté et al (1986) recently identified a series of *n*-alkyl and isoprenoid thiophenes and thiolanes from the marl layer described above. It was proposed that these types of compounds are indicative of hypersaline depositional environments where the sulfur has been incorporated into specific (archae)bacterial and/or algal functionalized components during early diagenesis.

Thus, there are several clearly documented examples where it appears organic matter has been acting as a sink for the sulfur in sediments (Table 1), and the studies by Payzant et al (1983, not described herein) on terpenoid sulfides and sulfoxides in the Athabasca oil sands probably represent another example. As a better understanding of the origin and significance of hydrocarbon biomarkers is obtained, it is clear that our horizons will be broadened and the search for biomarkers containing heteroatoms will be intensified. Results already obtained in the past one or two years show that this will be a most rewarding area. Not only will it lead to the identification of new components, but it will also provide an insight into the origin of sulfur and its incorporation into geological organic matter.

Source Parameters

A number of source parameters based on biomarker distributions have been developed, and one of the most widely used of these parameters

Table 1 Summary of sulfur-containing compounds found in samples of geochemical interest

Compound/compound class	Structure	Reported occurrence	Possible origin	Reference
Thiophenes				
Benzothiophenes		Aromatic fractions of crude oils	Diagenesis/maturation products	Ho et al., 1974
Dibenzothiophenes				
Thioenothiophenes				
Cyclic sulfoxides	a) b)		Oxidation of corresponding sulfides	
Cyclic sulfides	a) b)	Athabasca tar sands	Diagenesis products (?)	Payzant et al., 1983
2,3-Dimethyl-5-(2,6,10-trimethylundecyl)-thiophene		Various hypersaline environments	Cyclization and incorporation of S into 2,6,10-trimethyl-7-(3-methylbutyl)dodecane	ten Haven et al., 1986
30-(2-Methylene thienyl)-17β(H),21β(H)-hopane		Recent sediments/DSDP sediments	Cyclization and incorporation of S into bacteriohopane tetrol	Valisolalao et al., 1984
3-Me-2-(3,7,11-TriMedodecyl)-thiophene				
3-(4,8,12-TriMetridecyl)-thiophene		Recent sediments/DSDP sediments	S incorporation into phytol	Brassell et al., 1986

is based on sterane distributions. The original proposal by Huang & Meinschein (1979) was based on sterol distributions but subsequently extrapolated to steranes; they suggested that C_{29} steranes (XIII, $R = C_2H_5$) were generally associated with a terrestrial input, and that C_{27} steranes (XIII, $R = H$) were associated with a marine input of organic matter to source rocks and their associated oils. However, in the past two or three years significant concentrations of C_{29} steranes have been observed in oil and source rocks thought to be of predominantly marine origin (Palacas et al 1984, Walters & Cassa 1985).

Similarly, an interesting paper by Grantham (1986) has provided evidence that the C_{29} steranes of crude oils may not always be derived from terrestrial source materials. An examination of two types of Oman crude oil, with geological evidence to suggest their generation from Precambrian source rocks (i.e. of an age before land plants evolved), showed a predominance of the C_{29} steranes in the main group of oils, with a very low concentration of the C_{27} and C_{28} homologs. The virtual absence of any rearranged steranes in these oils was reminiscent of the carbonate-source oils described above. Furthermore, the C_{29}/C_{30} hopane ratio of ~ 1, the predominance of the C_{35} extended hopanes over the C_{34} component, and the relatively high tricyclic concentrations are all similar to the observations described above for carbonate-source oils. Since this unusually high abundance of the C_{29} steranes cannot in this case be ascribed to an origin from higher plant sterols, it has been proposed that an alternative source may be primitive algae. Indeed, Patterson (1977) found the C_{29} component to dominate the sterol distributions in brown algae. Walters & Cassa (1985) have noted high concentrations of C_{29} steranes in oils from the Gulf of Mexico that are again believed to be derived from nonterrestrial source rocks, and Volkman et al (1981) have indicated that the dominance of the C_{29} sterol in Recent sediments does not necessarily correspond to a terrestrial contribution to the sediment.

In a further development of the relationship between sterane distributions in geological samples and source materials, Moldowan et al (1985) described the presence of C_{30} steranes as an indicator of oils derived from a marine depositional environment. This was based on the observation that C_{30} steranes were only present in oils known to be derived from marine source rocks. Although the precursor of the C_{30} steranes remains unclear, recent work by Djerassi (1981) has shown that C_{30} sterols are present in marine organisms. Moldowan et al (1985) also proposed that the distribution of monoaromatic steroids is a far more useful parameter for distinguishing marine- vs nonmarine-source oils than the sterane distribution. The distributions of monoaromatic steroids when plotted in a C_{27}–C_{28}–C_{29} ternary diagram provided greater resolution than the corresponding steranes. In particular, the C_{28} monoaromatic steroid content appeared to be greater in marine-source oils. The differences in these distributions can be explained on the basis of the different origins for the steranes and monoaromatic steroids.

MATURATION PARAMETERS

The development of maturity parameters based on the relative proportions of different biomarker stereoisomers has been well documented in several previous publications, and thus we do not discuss them again here in any detail (Mackenzie et al 1982a, Mackenzie 1984a, Philp 1985). In this section, we briefly review some new maturity parameters that have been introduced recently and, perhaps more importantly, discuss anomalies that seem to be appearing currently in some of the more established parameters.

Conversion of a functionalized precursor molecule to its hydrocarbon biomarker generally leads to the molecule having the kinetically preferred configuration, which in many cases is not necessarily the thermodynamically preferred form. Hence, for a given pair of stereoisomers the usual trend is a relative increase in concentration of the more thermally stable isomer with increasing maturity until an equilibrium ratio is obtained. One ratio that has been used extensively in geochemistry is the conversion of the 20R epimer of $5\alpha(H),14\alpha(H),17\alpha(H)$-ethylcholestane (XIII, $R = C_2H_5$) into its corresponding 20S epimer until an equilibrium mixture is obtained with a value of $20S/20R \simeq 1.1$. The majority of values previously reported for this ratio in crude oils have been at or near the equilibrium value. However, anomalous values for the ratio of 0.2–0.3 have been reported in condensates, mainly derived from relatively mature sediments. A recent paper by Strachan et al (1986) suggests that sediments containing coals as the dominant source of hydrocarbons can produce oils with these anomalous sterane epimer ratios if the coals experienced a high heating rate as a result of high geothermal gradients and rapid burial rates. The activation energy necessary for the epimerization reaction is thought to be matrix dependent and is far lower for coals than for shales. Thus, for a sample pair in which the epimerization has progressed further in the shale than the coal, it can be proposed that the heating rate has been high. Conversely, a situation where the epimerization reaction is more advanced in the coal suggests slower heating. A combination of a coal matrix and a high heating rate may therefore result in conditions favorable for petroleum generation but restrict the epimerization reaction to the extent that the petroleum generated has anomalously low sterane epimer ratios (0.2–0.3). An alternative explanation offered by Grantham (1986) is that time is an important factor in isomerization reactions. In young Tertiary basins that have been buried rapidly and have reached a high temperature rapidly, Grantham (1986) proposed that insufficient time has passed for the isomerization to reach equilibrium. A similar observation has been made for the moretane/hopane conversion, which suggests that time constraints also operate on this reaction. In view of the abundance of coal-type material in many of the basins where the oils used by Grantham (1986) occur, it would appear prudent to suggest that the presence of a coal matrix and the associated lower activation energy for the sterane isomerization reaction may also be important factors, as suggested by Strachan et al (1986).

Apparently anomalous values for hopane and sterane maturity indicators have been reported in samples from hypersaline environments (ten Haven et al 1986b). In particular, the premature formation of the $5\alpha(H),14\beta(H),17\beta(H)$ (20R and 20S) steranes and the complete isomerization of the $17\alpha(H),21\beta(H)$-hopanes at the C-22 position were observed. Ten Haven et al (1986b) suggested a possible mechanism for the formation of $14\beta(H),17\beta(H)$ steranes from Δ^7 sterenes, which are known to occur in hypersaline environments along with spirosterenes (Brassell et al 1984). The Δ^7 sterenes can be isomerized to Δ^7, $\Delta^{8(14)}$, and Δ^{14} sterenes, and hydrogenation of these intermediates will produce the most stable configuration, i.e. $14\beta(H),17\beta(H)$ and not $14\alpha(H),17\alpha(H)$ (van Graas et al 1982). These observations are another example of a situation where care must be exercised in the application of steranes and triterpanes as maturity indices. Specific steroidal and hopanoidal precursors obviously can influence strongly the occurrence of stereoisomer distributions at an early stage of diagenesis.

Tetracyclic diterpenoids of the kaurene-phyllocladene type are particularly abundant in leaf resins of conifers belonging to the *Podocarpaceae*, *Araucariaceae*, and *Cupreseaceae*. In recent years a number of reports have appeared concerning their occurrence in oils, source rocks, and brown coals from the Gippsland Basin area in Australia (Philp et al 1981, 1983, Noble et al 1985), where they have been used successfully for the purposes of classifying oils into families and making oil/source-rock correlations. Noble et al (1985) recently studied the effect of thermal maturation upon the relative proportions of epimeric diterpanes by comparing the distribution of compounds in sediments and coals of differing maturity. The two structures studied most extensively in this work were phyllocladane (XXIII) and kaurane (XXIV), both of which exist as the $16\alpha(H)$ and $16\beta(H)$ stereoisomers. It was found in samples of increasing maturity from this region that the proportion of the $16\alpha(H)$ isomers relative to the $16\beta(H)$ isomers decreased. Thermodynamic equilibrium for the interconversion reactions is attained when the $16\alpha(H)/16\beta(H)$ ratio reaches a value of 0.3 for phyllocladanes and <0.1 for kauranes. Equilibrium for the interconversion reactions is reached prior to the onset of oil generation, and thus these parameters may be of use in the lower part of the maturity scale, where the the vitrinite reflectance values are unreliable. At higher levels of maturity, use is made of changes in the dimethylnaphthalene distribution (Alexander et al 1983), and potentially the benzothiophenes may also be of use in this region (Hughes 1984).

Tricyclic terpanes (e.g. XI) were first reported to be present in the Green River Shale by Anders & Robinson (1971), but it was not until more recently that synthetic structural proof for their occurrence was put forward by Aquino Neto et al (1982) and Heissler et al (1984). Since the early studies were generally concerned with tricyclic terpanes in crude oils and not shales, only one major stereochemical series was observed, namely 13β(H),14α(H). However a recent paper by Aquino Neto et al (1986) studied the tricyclic terpanes in a series of immature shales and found that they occurred as a mixture of stereoisomers, with the 13β(H),14α(H) stereoisomers predominating and the 13α(H),14α(H) stereoisomers present in lower concentration. As the maturity level increased into the early part of the oil generation window, the minor series of stereoisomers decreased in concentration to virtually zero, leaving only the 13β(H),14α(H) stereoisomers. Hence this observation introduces another class of biomarkers that may have some potential use as a maturity indicator in the lower part of the maturity scale.

MIGRATION

Migration of petroleum through geological sequences is a poorly understood process, but one that is receiving currently a great deal of attention. A more thorough understanding of migration mechanisms, both primary and secondary, would be extremely useful from an exploration point of view. It would permit predictions to be made of changes to crude oil composition resulting from both processes. Furthermore, studies of compositional changes occurring during primary migration would assist in the assessment of the source-rock expulsion efficiency factor (Leythaeuser et al 1984). Changes in composition may also permit some information to be obtained of relative, and possibly absolute, migration distances. A recent paper by Carlson & Chamberlain (1986) included a short review of previous literature concerned with the compositional changes occurring in certain classes of compounds during migration, and hence these are not discussed here. Changes in composition resulting from migration are further complicated by the observation that compositional changes similar to those resulting from migration may be caused by alternative processes. Such processes include biodegradation, water washing, diffusion, deasphalting, and variations in source material. Major obstacles to the investigation of migration in natural settings have been the lack of adequate and suitable samples and the lack of control over subsurface geologic conditions for interpretation of crude oil composition when determining migrational effects.

Recent years have seen a marked increase in the study of migration using geochemical techniques. Silverman (1965) published a paper on isotopic fractionation in Venezuelan oils, but few additional reports appeared until Seifert & Moldowan (1978, 1981) examined the fractionation of different biomarker classes by migration. Leythaeuser et al

(1983), using shale/sandstone sequences, demonstrated that for C_{15+} alkanes the lower molecular weight components undergo migration more readily than higher molecular weight components and *n*-alkanes migrate faster than branched alkanes. It has been suggested by Leythaeuser et al (1983) that these effects may be produced as a result of interaction with clays (e.g. montmorillonite and illite) and may also be accompanied by apparent isotope fractionation effects.

In an attempt to evaluate the role of clay minerals on geochromatography, Carlson & Chamberlain (1986) calculated the adsorption free energy differences between various chemical compounds and an active solid surface. It was proposed that if differences in certain sterane isomer ratios are a result of migration and not maturation, then these differences will be reflected by differences in the adsorption free energies for the mineral/sterane pairs. In a laboratory study a number of steroid biomarkers were investigated using a montmorillonite clay of increasing water content. The adsorption free energies were found to vary in a systematic way based on carbon number and stereochemistry. In this study the adsorption free energies were used to calculate 253 possible steroid pair concentration ratio variations, and it was concluded that isomers with very similar values would show no migrational enrichments or depletions. One of the values of such an approach is that it enables one to select pairs of isomers that may show large maturational effects but small migrational effects. In this way it is possible to differentiate between these two processes. The published data related only to the laboratory study, and (as stated in the paper) a field comparison of the trends predicted by variations in the adsorption free energy values should provide insight into the validity of the migrational geochromatographic concept.

Using an alternative approach, Bonilla & Engel (1986) studied changes in the chemical composition of crude oils resulting from their simulated migration through clean quartz sand and, in a separate experiment, through a shale sample. Results from the laboratory experiments studying migration through the sand showed that during the course of migration, there was a redistribution of the C_{15+} aliphatic hydrocarbon fraction in favor of the lower molecular weight components with increasing migration distance. In addition, it was found that with increasing migration distance, $\delta^{13}C$ values for the aromatic hydrocarbon fractions remained relatively unchanged, but that the aliphatic and NSO fractions became slightly depleted in ^{13}C with increased migration distance. It was proposed that the almost negligible variation in $\delta^{13}C$ values for the aromatic fractions enhances their value as a parameter for oil/oil and oil/source-rock correlations. The laboratory observations of Bonilla & Engel (1986) on the redistribution of the aliphatic hydrocarbons support numerous observations made during oil migration in natural settings (Vandenbroucke 1972, Mackenzie et al 1983, Leythaeuser et al 1983).

As stated above, migration is still an area that requires a great deal of additional research; however, the advances made in the last few years are extremely significant, and it is clear that many of the effects of migration observed in the field may be reproduced in the laboratory. Estimation of the adsorption free energy has provided a basis for estimating some of the differences in relative migration distances observed for the steranes. The topic is now open to improve and refine the laboratory studies so that the effects of temperature, pressure, mineral matrix, and fluid composition can be predicted and used in data interpretation.

BIODEGRADATION

The effects of biodegradation on crude oil compositions have been studied extensively in the past two decades. Early work by Winters & Williams (1969) demonstrated the preferential removal of n-alkanes during crude oil biodegradation, and this observation was supported by the work of Milner et al (1977). These early studies, utilizing gas chromatography alone, permitted the relative removal rates of normal alkanes and isoprenoids to be documented. However, with the advent of gas chromatography–mass spectrometry, information on the effects of biodegradation on biomarker distributions in crude oils has been published. Alexander et al (1983) presented a table that summarized the relative removal rates of certain component classes based on information available at that time. Connan (1984) published a similar table, and Table 2 herein is a combined version of the data from Alexander et al (1983) and Connan (1984) that includes additional information published since the earlier tables were prepared.

Table 2 attempts to indicate changes that occur with increasing biodegradation, but no attempt has been made to assign the relative extent of biodegradation, since some of the observations reported herein may not be applicable to all situations. For example, in certain situations it has been observed that steranes are degraded prior to terpanes, and in other cases the reverse is true. Likewise, demethylation of the terpanes is not a universal occurrence. A number of other interesting problems still remain to be solved concerning biodegradation. For instance, the fact that 25-norhopanes (XXV) are associated with some biodegraded oils but not others has led to the proposal that these compounds may actually be source indicators or may be formed during biodegradation in Recent sediments. Alternatively, the presence of 25-norhopanes in oils with a predominance of n-alkanes has also been used to support the idea of

Table 2 A summary of the reported effects of biodegradation[a]

Progressive effects of biodegradation	Reference
1. Typical paraffinic oil; abundant n-alkanes	Tissot et al 1977
2. Light end n-alkanes removed	Alexander et al 1983
3. Iso- and anteisoalkanes removed	Alexander et al 1983
4. Alteration and removal of pentacyclic hopane carboxylic acids	Behar & Albrecht 1984
5. $>90\%$ n-alkanes removed	Alexander et al 1983
6. Alkylcyclohexanes, alkylcyclopentanes, alkylbenzenes removed; isoprenoids, naphthalene concentration reduced	Volkman et al 1984
7. Isoprenoids and alkylnaphthalenes removed; selective removal of C_2 naphthalenes	Volkman et al 1984
8. C_{14}–C_{16} bicyclic alkanes removed	Alexander et al 1983
9. $>50\%$ 20R-5α(H),14α(H),17α(H) steranes removed	Alexander et al 1983
10. Relative removal rates for 5α(H),14α(H),17α(H) steranes; $C_{27} > C_{28} > C_{29}$	Seifert et al 1984
11. Preferential removal of C_{30}–C_{35} hopanes and the 22R configuration	Seifert et al 1984 Goodwin et al 1983
12. Preferential removal of C_{27} diasterane	Alexander et al 1983
13. C_{27}–C_{29} hopanes removed with or without demethylation	Seifert et al 1984
14. C_{21}–C_{22} steranes removed	Connan 1984
15. Tricyclic terpanes removed with or without demethylation	Howell et al 1984
16. Alkylated benzenes removed faster than alkylated naphthalenes, which in turn are removed faster than alkylated phenanthrenes; benzo- and dibenzothiophenes also removed at this stage	Williams et al 1986
17. Preferential loss of low molecular weight triaromatic steroids	Wardroper et al 1984
18. Preferential degradation of mono- and triaromatic steroids with 20R configuration	Wardroper et al 1984
19. Low molecular monoaromatic steroids more resistant to biodegradation than high molecular weight monoaromatic steroids	Wardroper et al 1984

[a] This table is a composite of observed effects reported in the recent literature. It is important to emphasize that each case of biodegradation is unique to each sample location and is not necessarily comparable with the effects of biodegradation in another basin. Although an attempt has been made here to list the effects in a relative order based on information in the literature, it should be reiterated that variations in this relative ordering will undoubtedly be observed in some situations.

mixing degraded and nondegraded oils. This is obviously an area that requires further investigation. Despite these observations, it is extremely important to understand the effects of biodegradation, since many of the biomarker maturity and migration parameters also use isomers known to be affected by biodegradation. Extensive biodegradation of crude oils can, under appropriate conditions, reveal the presence of additional resistant biomarkers in a sample. In a heavily biodegraded oil from Greece, Seifert et al (1984) identified the presence of 30-nor-29-methyl-17α(H)-hopane (XXVI) and other members of this family in a homologous series. The presence of such compounds, if widespread in other oils, could be potentially useful for the correlation of severely biodegraded samples.

An additional example of fluctuations in the relative rates of biodegradation is provided by Williams et al (1986), who recently studied some biodegraded Eocene south Texas crude oils. It was observed that regular steranes were not degraded, but that diasteranes were reduced in concentration. Similarly there was a slight decrease in C_{29} and C_{30} hopanes, but no demethylated species were found to occur. The bicyclic sesqui- and diterpanes were degraded irregularly and to a lesser extent than the aromatic hydrocarbons, although it was noted that the degree of biodegradation of the various components varied from one location to another. Hence, this is an example of where the relative rates of biodegradation cited in Table 2 are not followed, and it emphasizes the need for caution in using such an approach to predict relative rates of biodegradation.

BASIN MODELING

The application of biomarkers to problem solving in organic geochemistry has been discussed above. We turn to their use in modeling the thermal history of sedimentary basins. The opposite approach, that of using burial history to predict maturity or hydrocarbon generation, has been the subject of a considerable amount of work (e.g. Karweil 1956, Tissot 1969, Lopatin 1971, Waples 1980, 1984, and references therein); however, these advances are not discussed. This discussion is limited to two isomerization reactions and one aromatization reaction. These are (*a*) the isomerization of C-20

(XXVI)

of 5α-24-ethylcholestane (Figure 2a), (b) the isomerization of C-22 of 17α,21β-bishomohopane (Figure 2b), and (c) aromatization of two C_{29} monoaromatic steroid hydrocarbons (5α and 5β) to a C_{28} triaromatic steroid hydrocarbon (Figure 2c). Considerable interest has been shown in these reactions, not only because they cover a wide range of maturity (e.g. Mackenzie et al 1980, 1981a) but also because their relative rates of change appear to be dependent on the thermal history of the sediment in which they occur (e.g. Mackenzie et al 1982b, Mackenzie & McKenzie 1983, McKenzie et al 1983).

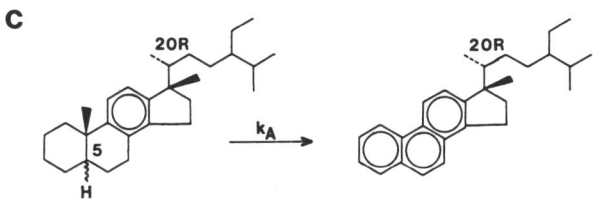

Figure 2 (a) Isomerization reaction of (20R)- to (20S)-5α-24-ethylcholestane. (b) Isomerization reaction of (22R)- to (22S)-17α,21β-bishomohopane. (c) Aromatization of two C_{29} monoaromatic steroid hydrocarbons (5α and 5β) to a C_{28} triaromatic steroid hydrocarbon.

Mackenzie et al (1982b) realized that the relative extent to which each reaction had occurred might be different in different basins, and it was proposed this was due to the basins experiencing different heating rates. Thus, young basins, having a high heating rate (e.g. the Pannonian Basin, Hungary), showed enhanced monoaromatic steroid hydrocarbon aromatization over C-20 sterane isomerization, while older basins, having a low heating rate (e.g. the North Sea), showed the opposite trend (Figure 3).

The reason for these differences can be explained by recourse to the principles behind chemical kinetics. An important factor in determining the rate of any reaction is its rate constant (k). This rate constant has a temperature dependence which is given by the Arrhenius equation (Arrhenius 1889):

$$k = A\ e^{-E/RT},$$

where

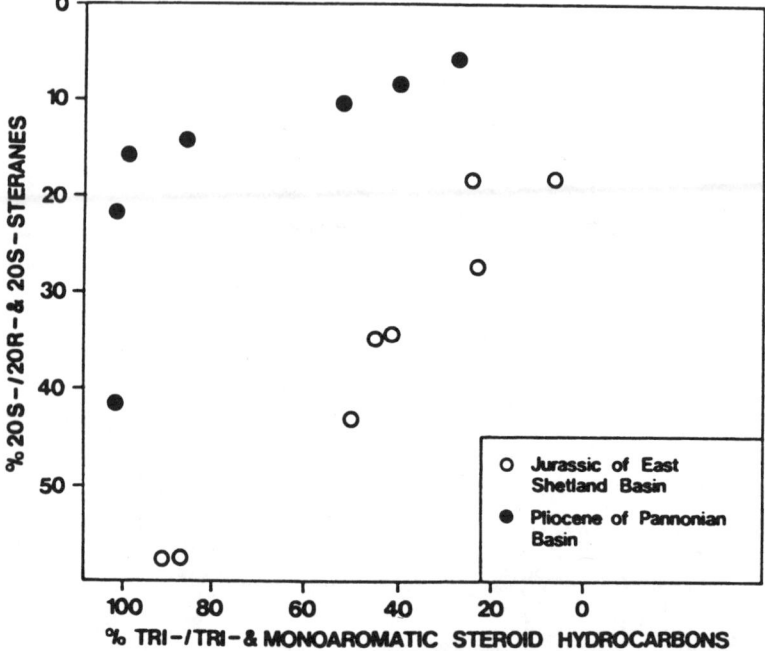

Figure 3 Comparison of the variation of the extent of aromatization of ring-C monoaromatic steroid hydrocarbons relative to the extent of isomerization at C-20 of 5α-24-ethylcholestane between Jurassic shales of the East Shetland Basin, North Sea, and the Pliocene shales of the Pannonian Basin, southeast Hungary. (From Mackenzie 1984b.)

k = rate constant (s^{-1}),

A = preexponential constant (frequency factor; s^{-1}),

E = activation energy (J mol^{-1}),

R = gas constant (8.3143 J K^{-1} mol^{-1}),

T = temperature (K).

A plot of ln k against $1/T$ usually yields a straight line with the intercept and slope giving ln A and $-E/R$, respectively. Thus the temperature dependence of a rate constant is given by the activation energy (E).

To explain the larger rate of steroid aromatization than sterane isomerization in basins of high heat flow, Mackenzie et al (1982b) proposed that the temperature dependence of the aromatization reaction was greater than that for the isomerization reaction. These reactions correspond to lines A and I_s, respectively, shown diagrammatically in Figure 4. It can be seen that at temperatures greater than T_c (i.e. smaller values of $1/T$) the line marked A has larger values of ln k than the line marked I_s. This condition corresponds to that occurring in young basins that are experiencing high heating rates. At temperatures below T_c (i.e. larger values of

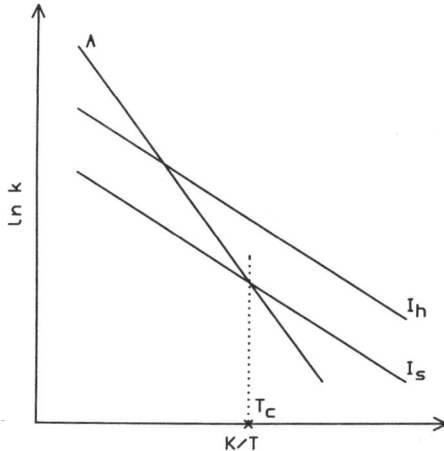

Figure 4 Diagrammatic Arrhenius plot for reactions having different activation energies and preexponential constants. The reactions are represented by lines A, I_s, and I_h, and the point T_c represents the temperature at which the lines for the first two reactions cross. At this temperature the reactions have the same value of rate constant k. Line A has the steepest slope and therefore the reaction it represents has the largest activation energy, while lines I_s and I_h are parallel and therefore represent reactions having the same activation energy but different preexponential constants.

$1/T$), the opposite is true, and this corresponds to old basins experiencing low heating rates. Only at T_c do both reactions have the same value of ln k.

Examination of samples from the Pannonian Basin (Mackenzie & McKenzie 1983) indicated, in addition to the above relationship between steroid aromatization and isomerization, that (a) the temperature dependence of the monoaromatic steroid hydrocarbon aromatization is greater than that for the C-22 bishomohopane isomerization, and (b) that the temperature dependences of the C-20 sterane and the C-22 bishomohopane isomerization reactions are equal but the reactions have different preexponential constants, with that for the latter reaction being larger. A diagrammatic representation of the Arrhenius plot for the C-22 bishomohopane isomerization is shown in Figure 4, where the line is labelled I_h. It is apparent that the lines for the two isomerization reactions are parallel, with the bishomohopane isomerization faster than the C-20 sterane isomerization.

The conclusion to be drawn from (a) above is that in basins with low heating rates, the bishomohopane isomerization is faster than the monoaromatic steroid hydrocarbon aromatization; however, in basins of high heat flow the two reactions have approximately equal rates, because the crossover point occurs at a temperature in the middle of the range typically encountered in this type of basin (Mackenzie & McKenzie 1983).

McKenzie (1978) discussed a model for the development and evolution of sedimentary basins, based on the initial rapid stretching (by a factor β) of a large area of continental lithosphere. In a later paper, McKenzie (1981) used the model to calculate Lopatin's (1971) time-temperature index (TTI). However, since this index has to be calibrated against empirical maturation indices, it was suggested that a better approach would be to test the model directly against maturity indices based on individual organic molecules.

Mackenzie & McKenzie (1983) studied the extent of the reactions (shown in Figure 2) in sediments from the Pannonian Basin (Hungary) and the North Sea (Figure 3). These sediments are thought to have had a relatively simple thermal history and are believed to have experienced little or no uplift (Mackenzie 1984b, and references therein). Thus, Mackenzie & McKenzie (1983) were able to calculate values for the activation energy (E) and preexponential constant (A) that could adequately describe the extent of each reaction in both sedimentary sequences (Table 3).

The method used to calculate the values in Table 3 is described in detail by Mackenzie & McKenzie (1983), and therefore it will not be reiterated here. However, the major kinetic assumptions made should be mentioned. These authors assumed that (a) the substrate [e.g. (20R)-5α-24-ethyl-

Table 3 Estimates of Arrhenius constants for the steroid aromatization and sterane and hopane isomerization reactions

Reaction	A (s^{-1})	E (kJ mol^{-1})
Steroid aromatization	1.8×10^{14}	200
Sterane isomerization at C-20	0.006	91
Hopane isomerization at C-22	0.016	91

cholestane] reacts to form the product [e.g. (20S)-5α-24-ethylcholestane] via a first-order reaction, and (b) the total amount of substrate and product remains constant (i.e. there are no other reactions depleting or adding to either species). The latter assumption appears to be most open to question, since there is some evidence from laboratory thermal studies (Mackenzie et al 1981b, Abbott et al 1984, 1985a,b) and from measurement of the absolute concentrations of biomarkers (Rullkötter et al 1984b) that the absolute amount of these compounds decreases with increasing maturity as a result of thermal breakdown. Mackenzie & McKenzie (1983), however, discussed this point in detail and came to the conclusion that since the model was successful, the assumption was valid.

The Arrhenius parameters thus obtained (Table 3) from basins having relatively simple thermal and subsidence histories can be used to study the timing and extent of uplift in more complex basins. The major assumption when making these investigations is that the same mechanism for the aromatization and isomerization reactions operates in the "simple" as well as the complex basins; however, this assumption is also implicit when calculating the Arrhenius parameters from the "simple" basins. Mackenzie & McKenzie (1983) used these Arrhenius parameters to calculate the timing and the amount of uplift experienced by the Toarcian, Paris Basin shales (France) and the Pliensbachian, Lower Saxony Basin shales (West Germany). For the Paris Basin suite, they stated that it was difficult to calculate reliable estimates of the timing of uplift, although they suggested that it had occurred within 50 m.y. of the stretching event. The amount of uplift could be more reliably estimated and was found to vary from ~0.0 km in the center to ~2.3 km on the eastern margin of the basin. The Lower Saxony Basin, on the other hand, was easier to study, and again the uplift was suggested to have occurred within 50 m.y. of the stretching event. The amount of uplift was found to be in reasonable agreement with that obtained from geological arguments.

Beaumont et al (1985) investigated the Cretaceous Alberta Basin, which was simulated by using a lithostatic flexure model. They were able to show, using the estimates of the Arrhenius parameters previously derived (Table

3; Mackenzie & McKenzie 1983), that the lithostatic flexure model could describe the evolution of the basin. Furthermore, they demonstrated that the paleogeothermal gradient increased with increasing distance from the edge of thrusting, a situation that exists at the present time. This is believed to be due to water being heated at depth, in the faulted foothill area, and then migrating updip, therefore enhancing the thermal gradient in areas to the east.

Mackenzie et al (1985) studied the temperature and burial histories of sediments deposited on the continental margin off Nova Scotia. They obtained good agreement between the predicted and observed present-day bottom-hole temperatures and hence obtained good agreement between the predicted and observed extents of the aromatization and isomerization reactions (Figure 2). These authors were then able to predict subsidence and temperature histories and reaction extents for regions where no samples were examined. They tentatively predicted that oil generation has been occurring in the Verrill Canyon formation (Middle Jurassic to Lower Cretaceous) since the Barremian. Additionally, the top of the oil generation zone presently lies at a depth of 3.8–4.0 km beneath the outer shelf and uppermost slope regions.

Recently, a slightly different approach for calculation of the Arrhenius parameters has been proposed (Strachan et al 1986), one that employs equations used to study nonisothermal kinetics. This approach is, strictly speaking, more correct than that used by Mackenzie & McKenzie (1983); however, the equations are inevitably more complex. It was found (Strachan et al 1986) that there was good agreement between the Arrhenius parameters for the Pannonian Basin, calculated using nonisothermal equations, with those reported by Mackenzie & McKenzie (1983). However, Strachan et al (1986) proposed that different Arrhenius parameters were applicable to shale and coal matrices. This contradicts the conclusions drawn by Mackenzie & McKenzie (1983) from samples from the Mahakam Delta, Indonesia (see also Hoffmann et al 1984). At the present time this problem remains unsolved.

SUMMARY

In the preceding pages we have reviewed some of the areas of biomarker geochemistry that have undergone a number of important reevaluations in the past two or three years. We have not attempted to exhaustively review the field of biomarkers, since this has been done in a number of other recent publications. It was felt that it would be more informative to illustrate how biomarker geochemistry is a rapidly evolving area, one that is changing with advances in analytical technology and with examination of a wider range of samples. It is clear that certain areas, such as sulfur-containing biomarkers, will continue to undergo major developments in the next few years. Other areas, such as maturation and basin-modeling studies, will continue to be refined. There are also areas of geochemistry that could not be covered here that are concerned with the insoluble portion of organic material, such as asphaltenes and kerogens. Developments in these areas parallel to some extent those described above. A combination of results from soluble biomarkers and the insoluble portions of geological samples will greatly enhance the value and utility of biomarker geochemistry.

Literature Cited

Abbott, G. D., Lewis, C. A., Maxwell, J. R. 1984. Laboratory simulation studies of steroid aromatisation and alkane isomerisation. In *Advances in Organic Geochemistry 1983*, ed. P. A. Schenck, J. W. de Leeuw, G. W. M. Lijmbach, pp. 31–38. Oxford: Pergamon

Abbott, G. D., Lewis, C. A., Maxwell, J. R. 1985a. The kinetics of specific organic reactions in the zone of categenesis. *Philos. Trans. R. Soc. London Ser. A* 315: 107–22

Abbott, G. D., Lewis, C. A., Maxwell, J. R. 1985b. Laboratory models for aromatization and isomerization of hydrocarbons in sedimentary basins. *Nature* 318: 651–53

Albaiges, J., Borbon, J., Walker, W. 1985. Petroleum isoprenoid hydrocarbons derived from categenetic degradation of archaebacterial lipids. *Org. Geochem.* 8: 293–97

Alexander, R., Kagi, R. I., Woodhouse, G. W., Volkman, J. K. 1983. The geochemistry of some biodegraded Australian oils. *Aust. Pet. Explor. Assoc. J.* 23: 53–63

Anders, D. E., Robinson, W. E. 1971. Cycloalkane constituents of the bitumens from Green River Shale. *Geochim. Cosmochim. Acta* 35: 661–78

Aquino Neto, F. R., Restle, A., Connan, J., Albrecht, P., Ourisson, G. 1982. Novel tricyclic terpanes (C_{19}, C_{20}) in sediments and petroleums. *Tetrahedron Lett.* 23: 2027–30

Aquino Neto, F. R., Cardoso, J. N., Rodrigues, R., Trindade, L. A. F. 1986. Evolution of tricyclic alkanes in the Espirito Santo Basin, Brazil. *Geochim. Cosmochim. Acta* 50(9): 2069–72

Arrhenius, S. 1889. Über die Reaktionsgeschwindigkeit bei der Inversion von Rohrzucker durch Säuren. *Z. Phys. Chem. (Leipzig)* 4: 226–48

Beaumont, C., Bontilier, R., Mackenzie, A. S., Rullkötter, J. 1985. Isomerization and aromatization of hydrocarbons and the paleothermometry and burial history of Alberta foreland basin. *Am. Assoc. Pet. Geol. Bull.* 69: 546–66

Behar, F. H., Albrecht, P. 1984. Correlations between carboxylic acids and hydrocarbons in several crude oils. Alteration by biodegradation. In *Advances in Organic Geochemistry 1983*, ed. P. A. Schenck, J. W. de Leeuw, G. W. M. Lijmbach, pp. 597–604. Oxford: Pergamon

Bonilla, J. V., Engel, M. H. 1986. Chemical and isotopic redistribution of hydrocarbons during migration: laboratory simulation experiments. In *Advances in Organic Geochemistry 1985*, ed. D. Leythaeuser, J. Rullkötter. In press

Brassell, S. C., Eglinton, G., Maxwell, J. R., Philp, R. P. 1978. Natural background of alkanes in the aquatic environment. In *Aquatic Pollutants*, ed. O. Hutzinger, I. H. van Lelyveld, B. C. J. Zoeteman, pp. 69–86. Oxford: Pergamon

Brassell, S. C., McEvoy, J., Hoffmann, C. F., Lamb, N. A., Peakman, T. M., Maxwell, J. R. 1984. Isomerization, rearrangement and aromatization of steroids in distinguishing early stages of diagenesis. In *Advances in Organic Geochemistry 1983*, ed. P. A. Schenck, J. W. de Leeuw, G. W. M. Lijmbach, pp. 11–23. Oxford: Pergamon

Brassell, S. C., Lewis, C. A., de Leeuw, J. W., de Lange, F., Sinninghe Damsté, J. S. 1986. Isoprenoid thiophenes: novel products of sediment diagenesis? *Nature* 320: 160–62

Carlson, R. M. K., Chamberlain, D. E. 1986. Steroid biomarker-clay mineral adsorption free energies: implications to petroleum migration indices. In *Advances in Organic Geochemistry 1985*, ed. D. Leythaeuser, J. Rullkötter. In press

Caspi, E., Zander, Z. M., Greig, J. B., Mallory, F., Conner, R., Landrey, J. 1968. Evidence for a nonoxidative cyclization of squalene in the biosynthesis of tetrahymanol. *J. Am. Chem. Soc.* 90: 3564–68

Chappe, B., Albrecht, P., Michaelis, W. 1982. Polar lipids of archaebacteria in sediments and petroleums. *Science* 217: 65–66

Connan, J. 1984. Biodegradation of crude oils in reservoirs. In *Advances in Petroleum Geochemistry*, ed. J. Brooks, D. H. Welte, 1: 299–336. London: Academic

Davis, J. B. 1968. Paraffinic hydrocarbons in the sulfate-reducing bacterium *desulfovibrio desulfuricans*. *Chem Geol.* 3: 155–60

Demaison, G. J., Moore, G. T. 1980. Anoxic environments and oil source bed genesis. *Org. Geochem.* 2: 9–31

Didyk, B. M., Simoneit, B. R. T., Brassell, S. C., Eglinton, G. 1978. Organic geochemical indicators of paleoenvironmental conditions of sedimentation. *Nature* 272: 216–22

Djerassi, C. 1981. Recent studies in the marine sterol field. *Pure Appl. Chem.* 53: 873–90

Fan, P., King, J. D., Claypool, G. E. 1984. The characteristics of the biomarker compounds of Chinese crude oils. *Proc. Am. Chem. Soc. Natl. Meet. Hawaii, 1984.* In press

Foster, C. B., O'Brien, G. W., Watson, S. T. 1986. Hydrocarbon source potential of the Goldwyer Formation, Barbwire Terrace, Canning Basin, Western Australia. *Aust. Pet. Explor. Assoc. J.* 26: 142–55

Goodwin, N. S., Park, P. J. D., Rawlinson, A. P. 1983. Crude oil biodegradation under simulated and natural conditions. In *Advances in Organic Geochemistry 1981*, ed. M. Bjorøy et al, pp. 650–58. London: Wiley

Goossens, H., de Leeuw, J. W., Schenck, P. A., Brassell, S. C. 1984. Tocopherols as likely precursors of pristane in ancient sediments and crude oils. *Nature* 312: 440–42

Grantham, P. J. 1986. The occurrence of unusual C_{27} and C_{29} sterane predominances in two types of Oman crude oil. *Org. Geochem.* 9(1): 1–10

Grimalt, J., Albaiges, J., Al-Saad, H. T., Douabul, A. A. Z. 1985. *n*-Alkane distributions in surface sediments from the Arabian Gulf. *Naturwissenschaften* 72: 35–37

Heissler, D., Ocampo, R., Albrecht, P., Riehl, J. J., Ourisson, G. 1984. Identification of long-chain tricyclic terpene hydrocarbons (C_{21}–C_{30}) in geological samples. *J. Chem. Soc. Chem. Commun.* 1984: 496–98

Henderson, W., Wollrab, V., Eglinton, G. 1969. Identification of steranes and triterpanes from a geological source by capillary gas liquid chromatography and mass spectrometry. In *Advances in Organic Geochemistry 1968*, ed. P. A. Schenck, I. Havenaar, pp. 181–207. Oxford: Pergamon

Hills, I. R., Whitehead, E. V., Anders, D. E., Cummins, J. J., Robinson, W. E. 1966. An optically active triterpane, gammacerane in Green River, Colorado, oil shale bitumen. *Chem. Commun.* 1966: 752–54

Ho, T. Y., Rogers, M. A., Drushel, H. V., Koons, C. B. 1974. Evolution of sulfur compounds in crude oils. *Am. Assoc. Pet. Geol. Bull.* 50(11): 2338–48

Hoffmann, C. F., Mackenzie, A. S., Lewis, C. A., Maxwell, J. R., Oudin, J. L., et al. 1984. A biological marker study of coals, shales and oils from the Mahakam Delta, Kalimantan, Indonesia. *Chem. Geol.* 42: 1–23

Howell, V. J., Connan, J., Aldridge, A. K. 1984. Tentative identification of demethylated tricyclic terpanes in nonbiodegraded and slightly biodegraded crude oils from the Los Llanos Basin, Colombia. In *Advances in Organic Geochemistry 1983*, ed. P. A. Schenck, J. W. de Leeuw, G. W. M. Lijmbach, pp. 83–92. Oxford: Pergamon

Huang, W.-Y., Meinschein, W. G. 1979. Sterols as ecological indicators. *Geochim. Cosmochim. Acta* 43: 739–45

Hughes, W. B. 1984. Use of thiophenic organosulfur compounds in characterizing crude oils derived from carbonate versus siliciclastic sources. In *Petroleum Geochemistry and Source Rock Potential of Carbonate Rocks. AAPG Stud. Geol. No. 18*, ed. J. G. Palacas, pp. 181–96

Hussler, G., Connan, J., Albrecht, P. 1984. Novel families of tetra- and hexacyclic aromatic hopanoids predominant in carbonate rocks and crude oils. *Org. Geochem.* 6: 39–49

Karweil, J. 1956. Die Metamorphose der Kohle von Standpunkt der physikalischen Chemie. *Z. Dtsch. Geol. Ges.* 107: 132–39

Leythaeuser, D., Bjorøy, M., Mackenzie, A. S., Schaefer, R. G., Altebäumer, F. J. 1983. Recognition of migration and its effect within two coreholes in shale/sandstone sequences from Svalbard, Norway. In *Advances in Organic Geochemistry 1981*, ed. M. Bjorøy et al, pp. 136–46. London: Wiley

Leythaeuser, D., Mackenzie, A. S., Schaefer, R. G., Bjorøy, M. 1984. A novel approach for recognition and quantification of hydrocarbon migration effects in shale sandstone sequences. *Am. Assoc. Pet. Geol. Bull.* 68: 196–219

Lopatin, N. V. 1971. Temperature and geologic time as factors in coalification. *Izv. Akad. Nauk SSSR Ser. Geol.* 3: 95–106 (In Russian)

Mackenzie, A. S. 1984a. Applications of biological markers in petroleum geochemistry. In *Advances in Petroleum Geochemistry*, ed. J. Brooks, D. H. Welte, 1: 115–214. London: Academic

Mackenzie, A. S. 1984b. Organic reactions as indicators of the burial and temperature histories of sedimentary sequences. *Clay Miner.* 19: 271–86

Mackenzie, A. S., McKenzie, D. 1983. Isomerization and aromatization of hydrocarbons in sedimentary basins formed by extension. *Geol. Mag.* 120: 417–70

Mackenzie, A. S., Patience, R. L., Maxwell, J. R., Vandenbroucke, M., Durand, B. 1980. Molecular parameters of maturation in the Toarcian Shales, Paris Basin, France—I. Changes in the configurations of acylic isoprenoid alkanes, steranes and triterpanes. *Geochim. Cosmochim. Acta* 44: 1709–21

Mackenzie, A. S., Hoffmann, C. F., Maxwell, J. R. 1981a. Molecular parameters of maturation in the Toarcian Shales, Paris Basin, France—III. Changes in aromatic steroid hydrocarbons. *Geochim. Cosmochim. Acta* 45: 1345–55

Mackenzie, A. S., Lewis, C. A., Maxwell, J. R. 1981b. Molecular parameters of maturation in the Toarcian Shales, Paris Basin, France—IV. Laboratory thermal alteration studies. *Geochim. Cosmochim. Acta* 45: 2369–76

Mackenzie, A. S., Brassell, S. C., Eglinton, G., Maxwell, J. R. 1982a. Chemical fossils: the geological fate of steroids. *Science* 217: 491–504

Mackenzie, A. S., Lamb, N. A., Maxwell, J. R. 1982b. Steroid hydrocarbons and the thermal history of sediments. *Nature* 295: 223–26

Mackenzie, A. S., Leythaeuser, D., Schaefer, R. G., Bjorøy, M. 1983. Expulsion of petroleum hydrocarbons from source rocks. *Nature* 301: 506–9

Mackenzie, A. S., Beaumont, C., Boutilier, R., Rullkötter, J. 1985. The aromatization and isomerization of hydrocarbons and the thermal and subsidence history of the Nova Scotia margin. *Philos. Trans. R. Soc. London Ser. A* 315: 203–32

McKenzie, D. 1978. Some remarks on the development of sedimentary basins. *Earth Planet. Sci. Lett.* 40: 25–32

McKenzie, D. 1981. The variation of temperature with time and hydrocarbon maturation in sedimentary basins formed by extension. *Earth Planet. Sci. Lett.* 55: 87–98

McKenzie, D., Mackenzie, A. S., Maxwell, J. R., Sajgó, Cs. 1983. Isomerization and aromatization of hydrocarbons in stretched sedimentary basins. *Nature* 301: 504–6

McKirdy, D. M., Aldridge, A. K., Ypma, P. J. M. 1983. A geochemical comparison of some crude oils from pre-Ordovician carbonate rocks. In *Advances in Organic Geochemistry 1981*, ed. M. Bjorøy et al, pp. 99–107. London: Wiley

McKirdy, D. M., Cox, R. E., Volkman, J. K., Howell, V. J. 1986. Botryococcane in a new class of Australian non-marine crude oils. *Nature* 320: 57–59

Milner, C. W. D., Rogers, M. A., Evans, C. R. 1977. Petroleum transformation in reservoirs. *J. Geochem. Explor.* 7: 101–53

Moldowan, J. M., Seifert, W. K. 1980. First discovery of botryococcane in petroleum. *Chem. Commun.* 1980: 912–14

Moldowan, J. M., Seifert, W. K., Gallegos, E. J. 1985. Relationship between petroleum composition and depositional environment of petroleum source rocks. *Am. Assoc. Petrol. Geol. Bull.* 69(8): 1255–68

Nishimura, M., Baker, E. W. 1986. Possible origin of n-alkanes with a remarkable even-to-odd predominance in recent marine sediments. *Geochim. Cosmochim. Acta* 50: 299–305

Noble, R. A., Alexander, R., Kagi, R. I., Know, J. 1985. Tetracyclic diterpenoid hydrocarbons in some Australian coals, sediments, and crude oils. *Geochim. Cosmochim. Acta* 49: 2141–47

Ourisson, G., Albrecht, P. A., Rhomer, M. 1979. The hopanoids: palaeochemistry and biochemistry of a group of natural products. *Pure Appl. Chem.* 51: 709–29

Palacas, J. G., Anders, D. E., King, J. D. 1984. South Florida Basin—a prime example of carbonate source rocks of petroleum. In *Petroleum Geochemistry and Source Rock Potential of Carbonate Rocks*. *AAPG Stud. Geol. No. 18*, ed. J. G. Palacas, pp. 71–96

Patterson, G. W. 1977. The distribution of sterols in algae. *Lipids* 6: 120–27

Payzant, J. D., Montgomery, D. S., Strausz, O. P. 1983. Novel terpenoid sulfoxides and sulfides in petroleum. *Tetrahedron Lett.* 24(7): 651–54

Philp, R. P. 1985. Biological markers in fossil fuel production. *Mass Spectrom. Rev.* 4(1): 1–54

Philp, R. P. 1986. Geochemistry in the search for oil. *Chem. Eng. News* 64(6): 28–43

Philp, R. P., Maxwell, J. R., Eglinton, G. 1976. Environmental organic geochemistry. *Sci. Prog.* 63(252): 521–47

Philp, R. P., Gilbert, T. D., Friedrich, J. 1981. Bicyclic sesquiterpenoids and diterpenoids in Australian crude oils. *Geochim. Cosmochim. Acta* 45: 1173–80

Philp, R. P., Simoneit, B. R. T., Gilbert, T. D. 1983. Diterpenoids in crude oils and coals of South Eastern Australia. In *Advances in Organic Geochemistry 1981*, ed. M. Bjorøy et al, pp. 698–704. London: Wiley

Radke, M., Welte, D. H., Willsch, H. 1982. Geochemical study on a well in the Western Canada Basin: relation of the aromatic distribution pattern to maturity of organic matter. *Geochim. Cosmochim. Acta* 46(1): 1–10

Reed, J. D., Illich, H. A., Horsfield, B. 1986. Biochemical evolutionary significance of Ordovician oils and their sources. In *Advances in Organic Geochemistry 1985*, ed. D. Leythaeuser, J. Rullkötter. Oxford: Pergamon. In press

Robinson, N., Eglinton, G., Brassell, S. C., Cranwell, P. A. 1984. Dinoflagellate origin for sedimentary 4α-methyl steroids and 5α(H)-stanols. *Nature* 308: 439–41

Rullkötter, J., Aizenshtat, Z., Spiro, B. 1984a. Biological markers in bitumens and pyrolysates of Upper Cretaceous bituminous chalks from the Ghareb Formation, Israel. *Geochim. Cosmochim. Acta* 48: 151–57

Rullkötter, J., Mackenzie, A. S., Welte, D. H., Leythaeuser, D., Radke, M. 1984b. Quantitative gas chromatography–mass spectrometry analysis of geological samples. In *Advances in Organic Geochemistry 1983*, ed. P. A. Schenck, J. W. de Leeuw, G. W. M. Lijmbach, pp. 817–27. Oxford: Pergamon

Seifert, W. K., Moldowan, J. M. 1978. Applications of steranes, terpanes and monoaromatics to the maturation, migration and source of crude oils. *Geochim. Cosmochim. Acta* 42: 77–95

Seifert, W. K., Moldowan, J. M. 1981. Paleoreconstruction by biological markers. *Geochim. Cosmochim. Acta* 45: 783–94

Seifert, W. K., Moldowan, J. M., Demaison, G. J. 1984. Source correlation of biodegraded oils. In *Advances in Organic Geochemistry 1983*, ed. P. A. Schenck, J. W. de Leeuw, G. W. M. Lijmbach, pp. 633–43. Oxford: Pergamon

Seiskind, O., Joly, G., Albrecht, P. 1979. Simulation of the geochemical transformation of sterols: superacid effect of clay minerals. *Geochim. Cosmochim. Acta* 43: 1675–79

Silverman, S. R. 1965. Migration and segregation of oil and gas. In *Fluids in Subsurface Environments. Am. Assoc. Pet. Geol. Mem.*, ed. A. Young, J. E. Galley, 4: 53–65. Tulsa, Okla: AAPG

Sinninghe Damsté, J. S., ten Haven, H. L., de Leeuw, J. W., Schenck, P. A. 1986. Organic geochemical studies of a Messinian evaporitic basin, northern Apennines (Italy). II. Isoprenoids and n-alkyl thiophenes and thiolanes. In *Advances in Organic Geochemistry 1985*, ed. D. Leythaeuser, J. Rullkötter. Oxford: Pergamon. In press

Simoneit, B. R., Crisp, P. T., Rohrback, B. G., Didyk, B. M. 1980. Chilean paraffin dirt—II. Natural gas seepage at an active site and its geochemical consequences. In *Advances in Organic Geochemistry 1979*, ed. A. G. Douglas, J. R. Maxwell, pp. 171–76. Oxford: Pergamon

Smith, A. G., Hurley, A. M., Briden, J. C. 1981. *Phanerozoic Paleocontinental World Maps*. Cambridge: Cambridge Univ. Press

Strachan, M. G., Alexander, R., Noble, R. A., Kagi, R. I. 1986. Constraints on the use of ethylcholestane epimer ratios as maturity indicators for petroleum. Submitted for publication

Summons, R. E., Powell, T. G. 1986. *Chlorobiaceae* in Palaeozoic seas revealed by biological markers, isotopes and geology. *Nature* 319: 763–65

ten Haven, H. L., de Leeuw, J. W., Sinninghe Damsté, J. S., Schenck, P. A., Palmer, S. E., Zumberge, J. E. 1986a. Application of biological markers in the recognition of palaeo hypersaline environments. In *Lacustrine Petroleum Source Rocks*, ed. K. Kelto, A. Fleet, M. Talbot. London: Blackwell. In press

ten Haven, H. L., de Leeuw, J. W., Peakman, T. M., Maxwell, J. R. 1986b. Anomalies in steroid and hopanoid maturity indices. *Geochim. Cosmochim. Acta* 50(5): 853–55

Tissot, B. 1969. Premières données sur les mécanismes et la cinétique de la formation du pétrole dans les sédiments. Simulation du schéma réactionnel sur ordinateur. *Rev. Inst. Fr. Pet.* 24: 470–501

Tissot, B., Pelet, R., Roucaché, J., Combaz, A. 1977. Utilisation des alcanes comme fossiles géochimiques indicateurs des environments géologiques. In *Advances in Organic Geochemistry 1975*, ed. R. Campos, J. Goñi, pp. 117–56. Madrid: ENADIMSA

Treibs, A. 1934. The occurrence of chlorophyll derivatives in an oil shale of the upper Triassic. *Annalen* 517: 103–14

Treibs, A. 1936. Chlorophyll and hemin

derivatives in organic materials. *Angew. Chem.* 49: 682–86

Valisolalao, J., Perakis, N., Chappe, B., Albrecht, P. 1984. A novel sulfur containing C_{35} hopanoid in sediments. *Tetrahedron Lett.* 25(1): 1183–86

Vandenbroucke, M. 1972. Étude de la migration primaire: variation de la composition des extracts de roche à un passage rock/réservoir. In *Advances in Organic Geochemistry 1971*, ed. H. R. von Gaertner, H. Wehner, pp. 547–65. Oxford: Pergamon

van Graas, G., Baas, J. M. A., van de Graaf, B., de Leeuw, J. W. 1982. Theoretical organic geochemistry. I. The thermodynamic stability of several cholestane isomers calculated by molecular mechanics. *Geochim. Cosmochim. Acta* 46: 2399–2402

Volkman, J. K. 1986. Organic geochemical studies of Ace Lake: a saline, meromictic lake in the Vestfold Hills, Antarctica. In *Advances in Organic Geochemistry 1985*, ed. D. Leythaeuser, J. Rullkötter. Oxford: Pergamon. In press

Volkman, J. K., Gillan, F. T., Johns, R. B., Eglinton, G. 1981. Microbial lipids of an intertidal sediment. 2. Sources of neutral lipids in a temperate intertidal sediment. *Geochim. Cosmochim. Acta* 45: 1817–28

Volkman, J. R., Alexander, R., Kagi, R. I., Rowland, S. J., Sheppard, P. N. 1984. Biodegradation of aromatic hydrocarbons in crude oils from the Barrow Subbasin of Western Australia. In *Advances in Organic Geochemistry 1983*, ed. P. A. Schenck, J. W. de Leeuw, G. W. M. Lijmbach, pp. 619–32. Oxford: Pergamon

Walters, C. C., Cassa, M. R. 1985. Regional organic geochemistry of offshore Louisiana. *Gulf Coast Assoc. Geol. Soc. Trans.* 35: 277–86

Waples, D. W. 1980. Time and temperature in petroleum formation: application of Lopatin's method to petroleum exploration. *Am. Assoc. Pet. Geol. Bull.* 64: 916–26

Waples, D. W. 1984. Thermal models for oil generation. In *Advances in Petroleum Geochemistry*, ed. J. Brooks, D. H. Welte, 1: 7–67. London: Academic

Waples, D. W., Haug, P., Welte, D. H. 1974. Occurrence of a regular C_{25} isoprenoid hydrocarbon in Tertiary sediments representing a lagoonal, saline environment. *Geochim. Cosmochim. Acta* 38: 381–87

Wardroper, A. M. K., Hoffmann, C. F., Maxwell, J. R., Barwise, A. J. G., Goodwin, N. S., Park, P. J. D. 1984. Crude oil biodegradation under simulated and natural conditions—II. Aromatic steroid hydrocarbons. In *Advances in Organic Geochemistry 1983*, ed. P. A. Schenck, J. W. de Leeuw, G. W. M. Lijmbach, pp. 605–18. Oxford: Pergamon

Whitehead, E. V. 1974. The structure of petroleum pentacyclanes. In *Advances in Organic Geochemistry 1973*, ed. B. Tissot, F. Bienner, pp. 225–43. Paris: Ed. Tech.

Williams, J. A., Bjorøy, M., Dolcater, D. L., Winters, J. C. 1986. Biodegradation in south Texas Eocene oils—effects on aromatics and biomarker series. In *Advances in Organic Geochemistry 1985*, ed. D. Leythaeuser, J. Rullkötter. Oxford: Pergamon. In press

Winters, J. C., Williams, J. A. 1969. Microbiological alteration of crude oil in the reservoir. *Symp. Pet. Transform Geol. Environ., Am. Chem. Soc. Div. Pet. Chem., New York, Preprints* 14(4): E-22–E-31

Xianzhang, Z., Shuzhen, L., Shunping, M. 1985. Biomarkers as source input indicators in source rocks of several terrestrial basins of China. *Proc. Am. Chem. Soc. Natl. Meet., Miami, 1985.* In press

Zumberge, J. E. 1984. Source rocks of the La Luna Formation (Upper Cretaceous) in the Middle Magdalena Valley, Colombia. In *Petroleum Geochemistry and Source Rock Potential of Carbonate Rocks. AAPG Stud. Geol. No. 18*, ed. J. G. Palacas, pp. 127–34

THERMOMETRY AND BAROMETRY OF IGNEOUS AND METAMORPHIC ROCKS

Steven R. Bohlen and Donald H. Lindsley

Department of Earth and Space Sciences, State University of New York, Stony Brook, New York 11794

INTRODUCTION

Geothermometry and geobarometry refer to the science of inferring the temperatures and pressures at which a rock crystallized from the chemical compositions and physical state (or both) of one or more of the minerals within it. This paper discusses the thermometry and barometry of igneous and metamorphic rocks. Several kinds of reactions can be calibrated for these purposes:

1. *Univariant reactions* (in which all phases have essentially fixed compositions); perhaps the best known is the graphite-diamond transition (a phase change).
1a. *Displaced reactions or equilibria* in which one or more phases have variable compositions, and the variance is therefore greater than one. Such displaced reactions are potentially very useful, for once the effects of the compositional variations are accounted for, they can provide information over a range of pressures and temperatures.
2. *Exchange thermometers*, in which the distribution of two elements between two phases is temperature dependent.
3. *Solvus equilibria*, in which the solubility of a component in a phase is temperature dependent (or less commonly pressure dependent).
4. *Saturation surfaces*, which mark the conditions—including temperature and pressure—under which a melt becomes saturated with a particular phase.

Because most of these reactions are sensitive to both pressure and temperature, we use here the term thermobarometer for the general case.

The application of these methods to thermobarometry involves a number of assumptions, some of which are almost mutually incompatible. (*a*) The basic reaction or equilibrium must be well calibrated, either through experimental or thermochemical data (or both). (*b*) The effects of other components can be accounted for through suitable solution models. (*c*) The minerals in the rock were in equilibrium. (*d*) The equilibrium assemblage has remained unchanged (rare indeed except for some quickly chilled rocks) or can be reconstructed from textural and other information remaining in the rock. Thus we must hope that experimentalists can calibrate in weeks, months, or years reactions or equilibria that nature has millions of years to reset as conditions slowly change from those at which a rock formed to those at the Earth's surface. Each of the types of equilibria listed above has its special problems—and special opportunities—in this regard. In the best of circumstances, different equilibria "lock in" at different conditions, so that one can infer the pressures and temperatures at various stages in the history of a rock—its so-called pressure-temperature-time path.

There are also two fundamentally different approches to calibrating thermobarometers, stemming from the fact that nearly all rocks contain considerably more elements than partake in end-member equilibria: Should one calibrate the simple equilibrium very well and then try to account for the effects of the impurities, or is it better to attack the entire "dirty" system at once? The latter approach might seem inherently superior, until one realizes that it is far more difficult to demonstrate equilibrium in the complex system, and that it is also difficult to isolate the effects of each impurity for purposes of extrapolating to different compositions. At present, both approaches are useful; we are prejudiced toward the former. We should also point out the complementarity of well-designed phase-equilibrium experiments and calorimetric measurements: Rarely is a successful thermobarometer based solely on thermochemical or on experimental data.

In the fall of 1975, an international symposium on geothermometry and geobarometry was held at The Pennsylvania State University. Papers from that conference were published in a special issue of *The American Mineralogist* (Vol. 61, pp. 549–816, 1976). In this brief review we cannot begin to cover all the material that has become available since that time, but we hope to give the flavor of the progress since then. Relatively recent reviews of thermobarometry employing multivariant equilibria in crustal rocks have been published by Essene (1982), Bohlen et al (1983a), and Newton (1983). In this paper, we emphasize the importance of new thermobarometers calibrated since these reviews and of new data on existing thermobarometers. The interested reader is referred to the papers noted above for more complete discussions of thermobarometers not covered in detail here.

UNIVARIANT THERMOBAROMETERS

An often-used type of thermobarometer is one that employs a single univariant curve that is either insensitive to solid solutions or is defined by minerals that occur naturally as end members. Such thermobarometers have the advantage of their simplicity of application: The mere presence of an appropriate phase or evidence from morphological or textural information that an appropriate phase once existed is sufficient to limit pressure (P) or temperature (T). The primary disadvantage is that only limiting P-T information can be inferred directly. Rarely can absolute pressures and temperatures be determined.

Aluminosilicates

There is little doubt that the relative stabilities of andalusite, sillimanite, and kyanite have been used to limit the pressure and temperature conditions in crustal rocks as much or more than any other single thermobarometer. This is, of course, because of their common occurrence in peraluminous rocks worldwide.

So much could be written about the history and development of the aluminosilicates as useful thermobarometers that an entire chapter of this volume would be necessary to cover all significant aspects. However, only a few are mentioned here. The significance of these minerals for elucidating pressure and temperature information was noted in the 1930s by Harker (1932), but useful interpretations were hampered by the emphasis on the role of rock strength in generating "stress" and "antistress" minerals, of which kyanite is an example of the former, and andalusite the latter. Experiments in the 1960s, many of which were unreversed and carried out under nonhydrostatic conditions, suggested that kyanite was not stable much below 10 kbar. This caused many (e.g. Pitcher & Flinn 1965) to consider extreme tectonic scenarios in order to generate 10 kbar or more in rather common, kyanite-bearing crustal rocks. Reliance on unreversed experimental data and failure to evaluate experimental results against field evidence have led to many dead ends in petrology. The problems with the interpretation of the aluminosilicate triple point are, perhaps, the best examples of this. Indeed, based almost solely on field data, and, no doubt, a modicum of geologic intuition, Schuiling (1957) placed the Al_2SiO_5 triple point at 3.5 kbar and 540°C. These values are remarkable considering that they are within 0.5 kbar and 50°C of the currently accepted values.

Based only on reversed experiments in a hydrostatic pressure apparatus, the Al_2SiO_5 triple point is inferred to occur at

6.5 kbar	and	595°C	(Althaus 1967),
5.5 kbar	and	620°C	(Richardson et al 1968),
3.8 kbar	and	500°C	(Holdaway 1971).

When the experimental data are considered in concert with thermochemical data (Anderson et al 1977), it is evident that the triple point of Holdaway best fits the existing data, although some uncertainty (± 50°C and ± 0.5 kbar) remains. The evaluation of Al_2SiO_5 occurrences and independent thermobarometry by field petrologists have also resulted in general agreement on the applicability of the Holdaway triple point.

Quantitative estimates of pressure and temperature are usually deduced from two Al_2SiO_5 minerals coexisting with other minerals that constrain temperature [usually minerals that define a dehydration equilibrium (e.g. Carmichael 1978)]. While this often yields useful results, petrologists are reminded of the potential difficulties that arise from the metastable persistence of one Al_2SiO_5 mineral into the stability field of another (a particular problem for andalusite-sillimanite) and the effects of variable fluid composition on the location of dehydration equilibria. In high-grade metamorphic rocks, the kyanite-sillimanite equilibrium is often employed to limit pressures. Based on the experiments of Richardson et al (1968), the overall uncertainty in the P-T location of this equilibrium is ± 0.5 kbar.

Polymorphs of SiO_2

Because of the common occurrences of SiO_2 polymorphs in rocks, there has long been hope of calibrating extremely useful and widely applicable thermobarometers based on these minerals. This hope has been fostered by the work of Tuttle (1949), who showed that the temperature of the high-low transition in quartz varied inversely with the temperature of formation, and by the work of Dennen et al (1970), who provisionally calibrated the Al content of quartz (in equilibrium with feldspar or some other aluminous phase) as a thermometer. Unfortunately, both practical and theoretical problems hamper the application of these potential thermometers, and workers have been unable to extract useful temperature information from them. Other problems dim the hope of a widely applicable SiO_2 thermobarometer. For example, in volcanic rocks and in experiments, tridymite and cristobalite appear to crystallize well out of their respective fields of stability far more often than not. As a result, few occurrences of these minerals have any relevance to temperature of formation. Still, occurrences that can be documented as having grown within the appropriate stability field can yield useful limiting temperature information, although the effect of pressure on the stabilities of tridymite and cristobalite is poorly constrained (see Tuttle & Bowen 1958). Similarly, the observation of the morphological characteristics of high quartz can also yield useful limiting temperature information (see Cohen & Klement 1967).

The recent discoveries by Smyth (1977), Chopin (1984), and Smith (1984) of coesite in subducted supracrustal rocks and eclogites have generated great interest in the quartz-coesite transition. This equilibrium has been well located as a function of pressure and temperature, as can be seen in Figure 1. In Figure 1, we have noted all available experimental data obtained in the piston-cylinder apparatus for the quartz-coesite transition ($T < 1100°C$). We have drawn the univariant boundary to be consistent with the precise reversals of Bohlen & Boettcher (1982) and to pass through as many of the other brackets as possible. The reversal brackets of Akella (1979) and Mirwald & Massonne (1980) must be reduced in pressure by 0.4–0.8 kbar as a result of recent data on the melting of alkali halides (Bohlen 1984), because the melting curves of various alkali halides were used in those studies for pressure calibration. With this in mind, the quartz-coesite boundary shown in Figure 1 violates only one half-bracket, that of Kitahara & Kennedy (1964) at 1100°C. Such consistency is remarkable considering the various furnace assemblies, experimental techniques, and "pressure corrections" employed.

Despite its rare occurrences, coesite from supracrustal rocks is noteworthy. Based on Figure 1, such rocks must have been buried to depths in excess of 75 km. Perhaps more important, however, is that the coesite survived the return trip to the surface, given the extremely rapid conversion of one polymorph to the other that allows experimental reversals to be obtained at temperatures as low as 300°C! The occurrences of coesite in supracrustal rocks have important consequences for tectonic models of subduction zones, as has been discussed only in part by Gillet et al (1984) and Smith (1984).

MULTIVARIANT THERMOBAROMETERS

Pressure and temperature information is most often inferred from univariant equilibria in which one or more of the phases has appreciable solid solution. These "sliding" or "displaced" equilibria are multivariant in P-T-X space. As a result, the entire thermobarometric assemblage can coexist over a wide range of P and T, which means that quantitative pressure

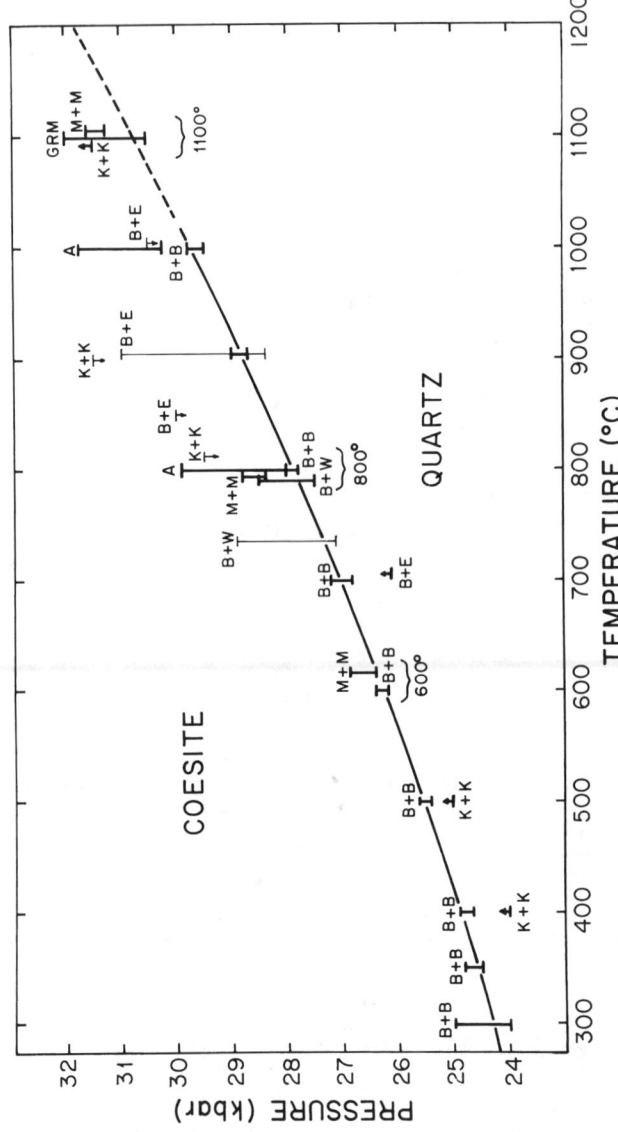

Figure 1 The quartz-coesite equilibrium. Heavy symbols indicate that quartz reacted to coesite and/or coesite reacted to quartz. Light symbols indicate that something other than quartz and/or coesite were used as starting materials. Experimental data indicated as follows: A, Akella (1979); B+B, Bohlen & Boettcher (1982); B+E, Boyd & England (1960); B+W, Boettcher & Wyllie (1968); GRM, Green et al (1966); K+K, Kitahara & Kennedy (1964); M+M, Mirwald & Massonne (1980).

and/or temperature information can be inferred if (*a*) the appropriate equilibrium has been well calibrated, (*b*) the chemical compositions of the phases are known, and (*c*) solution models are available for those minerals with appreciable solid solution. Poorly constrained solution models for the minerals most commonly employed in thermobarometry—plagioclase, pyroxenes, spinels, and, most importantly, garnet—remain as an obstacle to accurate thermobarometry. Recent advancements have improved the situation considerably, but a large proportion of the overall uncertainty in P/T estimates must still be attributed to inaccuracies in the modeling of solution properties.

Thermobarometry in the Upper Mantle

The pressures and temperatures of formation of rocks from the upper mantle can be inferred from equilibria in the model systems MAS (MgO-Al_2O_3-SiO_2) and CMAS (CaO-MAS). Numerous workers have contributed to the experimental delineation of the relative stability fields of plagioclase, spinel, and garnet lherzolite as well as the Al content of ortho- and clinopyroxene coexisting with olivine and spinel or garnet. These include Kushiro & Yoder (1966), MacGregor (1974), Akella (1976), Herzberg (1976, 1978), Herzberg & Chapman (1976), Presnall (1976), Fujii (1977), Danckwerth & Newton (1978), O'Hara & Howells (1978), Haselton (1979), Jenkins & Newton (1979), Perkins & Newton (1980), Benna et al (1981), Dixon (1981), Gasparik (1984), and Gasparik & Newton (1984).

Figure 2 shows the relative stabilities of forsterite-anorthite-pyroxene(s),

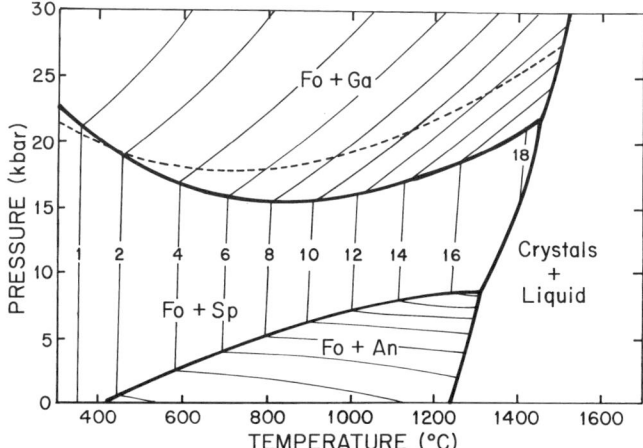

Figure 2 Phase relationships in the CMAS (solid curves) and MAS (dashed curve) systems, modified from Gasparik (1984). Heavy lines indicate univariant reactions; light lines (numerated) indicate the Al content of orthopyroxene.

forsterite-spinel-pyroxene(s), and forsterite-garnet-pyroxene(s) in the model MAS and CMAS systems taken from the recent work of Gasparik (1984). Also shown are the calculated isopleths for the Al content (X_{MgTs}) of orthopyroxene. There are several salient features about these phase relations. As can be seen, the Al content of orthopyroxene is a barometer at low pressures, a thermometer at intermediate pressures, and a thermobarometer (knowledge of either P or T required for determination of the other) at high pressures. The phase boundaries of the stability fields of anorthite and spinel as well as spinel and garnet are marked by strong curvature as a result of changes in pyroxene composition and degree of order in the spinel as a function of temperature and pressure. The curvature of the phase boundaries was first emphasized by Obata (1976), whose early calculations in the CMAS system invalidated previous straight-line extrapolations to lower pressure and temperature. Also evident is the appreciable effect calcium has on the pressure at which garnet becomes stable. It is, perhaps, most important to emphasize that Figure 2 is based on experiments conducted at temperatures of 900°C and above. The boundary separating the anorthite and spinel stability fields is constrained only by half-reversals on the high-pressure side of the reaction, and there are no data that constrain the Al content of orthopyroxene in the anorthite stability field. These data are lacking in part as a result of sluggish reaction kinetics. Nevertheless, the unwary user of these data is cautioned that the

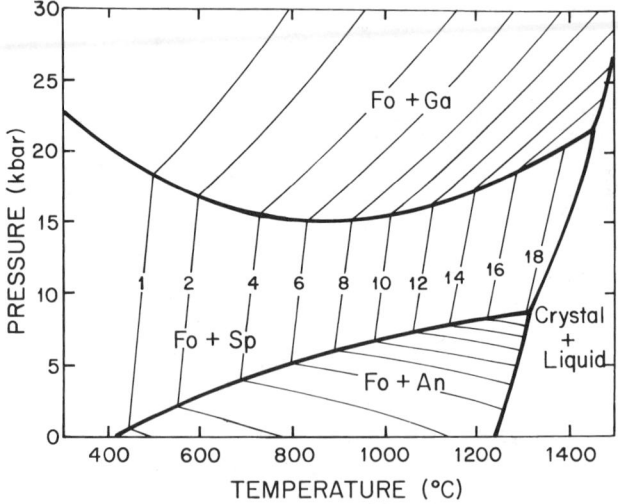

Figure 3 Phase relationships in the CMAS system, modified from Gasparik (1984). Heavy lines indicate univariant reactions; light lines (numerated) indicate the Al content of clinopyroxene.

phase equilibria are unconstrained by experiment at pressures below 8 kbar and temperatures below 900°C. The reaction boundaries and isopleths are dependent on solution models for pyroxene and are likely to be subject to revision as more data are collected.

Features similar to those in Figure 2 can be seen in Figures 3 and 4 (also from Gasparik 1984), which show the Al content (X_{CaTs}) in clinopyroxene for the same assemblages as in Figure 2 and the Al content (X_{MgTs}) of orthopyroxene for plagioclase-saturated assemblages, respectively. As was the case in Figure 2, these diagrams are constrained by data obtained only at relatively high P and T.

Pressure and temperature determinations in mantle rocks require adjustment of the equilibria in Figures 2-4 to account for the effects of small, but important, amounts of Fe and Cr in the natural phases. Even though this can be easily accomplished with existing, but imperfect, solution models, an important remaining task is to experimentally calibrate the effects of Fe and Cr. These effects are of no small consequence, since the primary data used to elucidate the temperature profile of the ancient mantle are obtained by employing the MAS and CMAS systems extrapolated to natural compositions.

The calcium content of olivine coexisting with ortho- and clinopyroxene has gained attention as a potential barometer. The barometer was calibrated initially by Finnerty & Boyd (1978) and has been more precisely

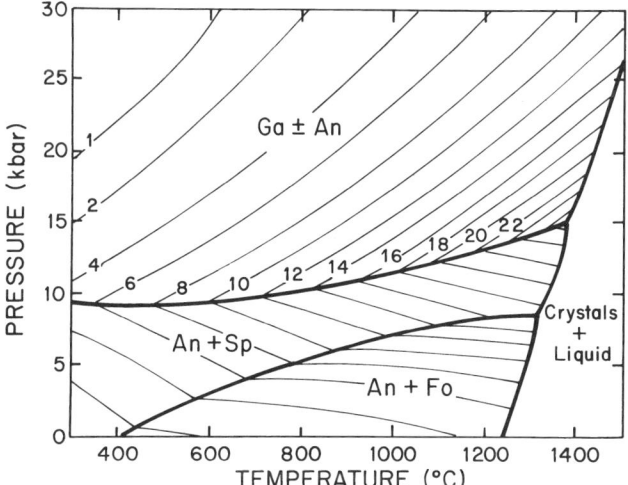

Figure 4 Phase relationships in the CMAS system, modified from Gasparik (1984). Heavy lines indicate univariant reactions; light lines (numerated) indicate the Al content of orthopyroxene in plagioclase-dominated assemblages.

determined by Adams & Bishop (1986). This barometer has considerable potential if extremely accurate analyses of relatively low Ca contents in olivine (0.1–0.2 wt%) can be obtained, and if one assumes that the Ca content of the olivine has not reset.

Thermobarometry in the Crust

A thermobarometer frequently employed in peraluminous crustal rocks is one based on the equilibrium

$$3 \text{ anorthite} = 1 \text{ grossular} + 2 \text{ Al}_2\text{SiO}_5 + 1 \text{ quartz}, \tag{1}$$

wherein coexisting plagioclase-garnet-Al_2SiO_5-quartz defines pressure at a fixed temperature (Ghent 1976, Newton & Haselton 1981). There are numerous problems in the application of this barometer, but it is still widely applied, in part because the barometric assemblage occurs in every pelite. One of the most significant problems is that the end-member equilibrium is known only to within ± 2 kbar, as can be seen in Figure 5. Experiments on this equilibrium are difficult because reaction rates are low and use of H_2O as a flux generates zoisite in the experimental charges. The most tightly constrained reversals of this equilibrium are by Gasparik (1985), but Gasparik's reversals imply a dP/dT slope that is much smaller than that implied by the other experiments. This may be the result of the dissolution of Pb (from the PbO flux) into the anorthite of Gasparik's experiments. The uncertainty in the location of the equilibrium is compounded because the univariant curve must be extrapolated to much lower temperatures and pressures in order to be applied to crustal rocks. In addition, the garnet in the natural assemblages typically contains vanishingly small amounts of Ca along with small, but important, amounts of ferric iron. As a result, it is difficult to determine the mole fraction of grossular component in the garnet. Even if the grossular content can be unambiguously ascertained, the activity of grossular at such extreme dilution is unknown.

Given these problems, use of equilibrium (1) as a barometer is not recommended, despite its popularity. Pressures inferred from it are accompanied by uncertainties of *at least* ± 2 kbar, although most workers apply it as if it were far more accurate.

Recent experimental data on a wide range of equilibria in both peraluminous and metaluminous bulk compositions have provided the basis for numerous thermobarometers. Thermobarometers based on experimental data are in Figure 6, and those that can be calculated with reasonable accuracy are in Figure 7. Both the experimentally based and the calculated equilibria have advantages over equilibrium (1) for constraining P/T information. The equilibria are calibrated within the pressure-tem-

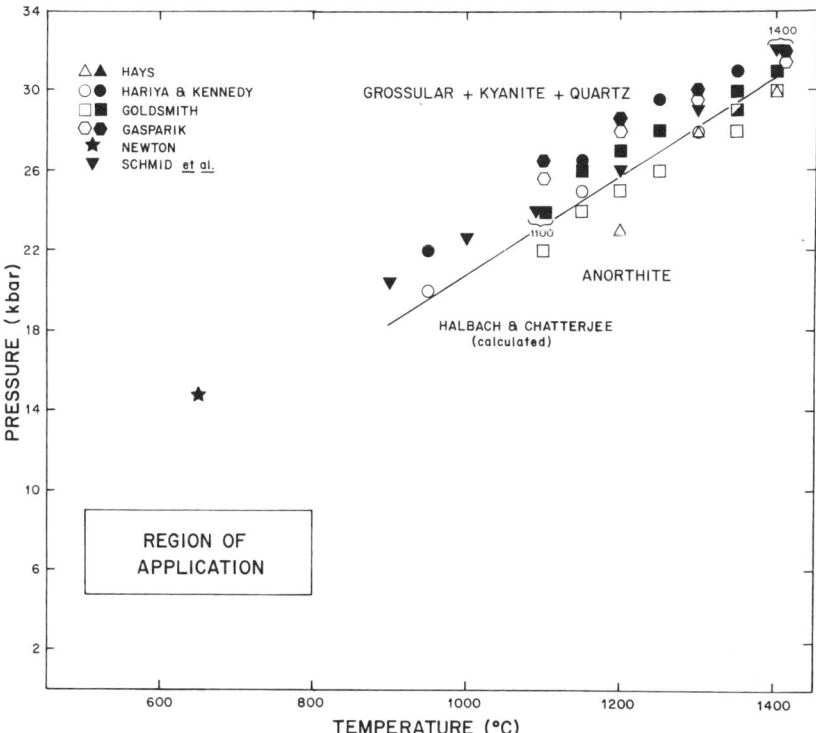

Figure 5 Experimental determinations of the reaction of anorthite to grossular-kyanite-quartz. Note the long extrapolation necessary to apply this equilibrium in rocks. Experimental data by Hays (1967), Hariya & Kennedy (1968), Goldsmith (1980), Gasparik (1985), Newton (1966), and Schmid et al (1978). Calculated boundary by Halbach & Chatterjee (1984).

perature range appropriate for the rocks in which they apply, and therefore long extrapolations are unnecessary. In addition, analyses of minerals for their *major* components are required for application.

Most of the equilibria in Figures 6 and 7 have dP/dT slopes less than 20 bar $°C^{-1}$, making them most suitable for barometry. Assemblages of garnet-rutile-ilmenite-quartz and either Al_2SiO_5 or plagioclase are quite common in peraluminous and metaluminous rocks, respectively. These minerals are the reactants and products of useful barometric assemblages related by

$$3 \text{ ilmenite} + Al_2SiO_5 + 2 \text{ quartz} = 1 \text{ almandine} + 3 \text{ rutile}, \quad (2)$$

$$2 \text{ ilmenite} + 1 \text{ anorthite} + 1 \text{ quartz} = 1 \text{ garnet}(Gr_1Alm_2) + 2 \text{ rutile}. \quad (3)$$

Equilibrium (2) has been tightly reversed by Bohlen et al (1983b), who

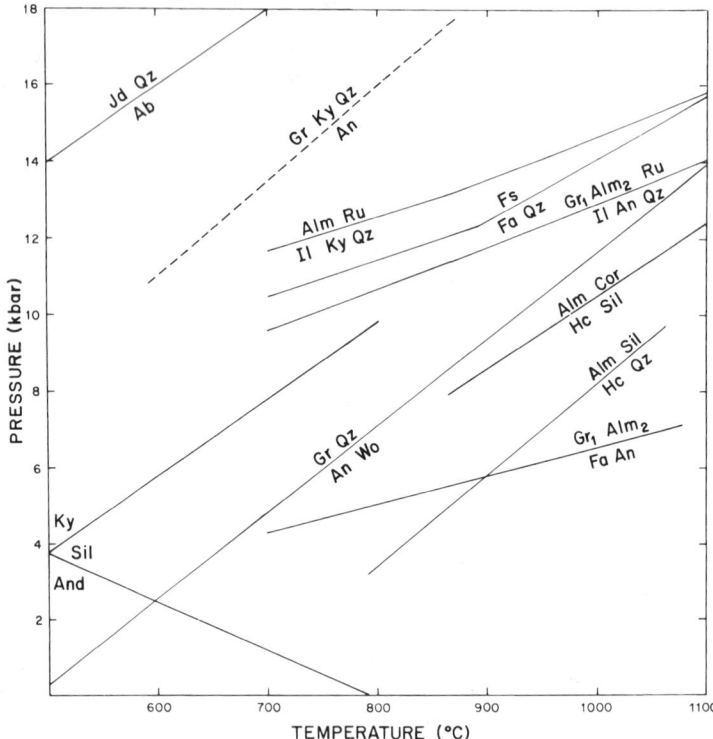

Figure 6 Thermobarometers determined by experiment. Stability of jadeite+quartz is from Holland (1980), and the stability of ferrosilite is from Bohlen et al (1980a,b) and Bohlen & Boettcher (1981). The other equilibria are discussed in the text. The dashed boundary has been extrapolated from data at higher pressure and temperature (see Figure 5).

used the end-member equilibrium as a basis for a P-T-Log K grid from which pressures can be determined easily. A similar P-T-ln K grid has been calculated for equilibrium (3), which has been tightly reversed by Bohlen & Liotta (1986). The attractive aspects of both barometers are that they are quite insensitive to temperature and that only one or two (and at most three) minerals in the requisite assemblage have any appreciable solid solution in natural assemblages. Application of these barometers in numerous terrains throughout the world yields geologically reasonable pressures that agree well with other barometers in terrains where comparisons can be made.

Pressure/temperature information can be deduced in high-grade metamorphic rocks from hercynite equilibria

$$3 \text{ hercynite} + 5 \text{ quartz} = 1 \text{ almandine} + 2 \text{ sillimanite}, \tag{4}$$

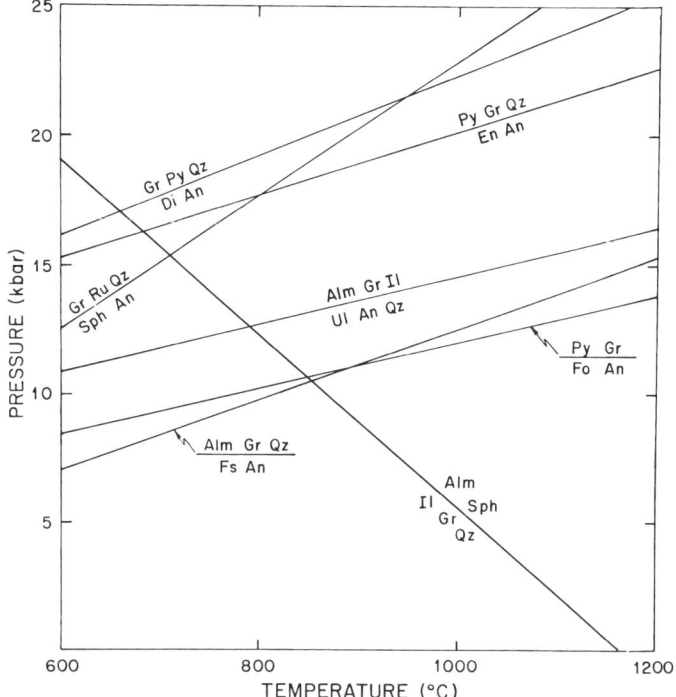

Figure 7 Thermobarometers determined by calculation. See text for discussion and references.

3 hercynite + 3 sillimanite = 1 almandine + 5 corundum, (5)

recently calibrated by Bohlen et al (1986) and S. R. Bohlen (unpublished data), respectively. Reversals of equilibrium (5) locate it between 10.0–10.5 and 12.0–12.5 kbar at 1000 and 1100°C, respectively. Both equilibria are useful thermobarometers in over 20 granulite terrains, although in the case of equilibrium (4) one must be sure that quartz and hercynite-rich spinel are in contact as a necessary requirement for textural equilibration of the assemblage. Application of these thermobarometers can be hampered by complex spinel chemistry for which no reliable solution models are available. Fortunately, most often the spinels are nearly binary in character (hercynite-spinel, hercynite-gahnite), and the activities of hercynite in such spinels can be deduced from the models of O'Neil & Navrotsky (1984).

Grossular-quartz assemblages have been used to limit metamorphic temperatures in numerous terrains, and use of the equilibrium

1 grossular + 1 quartz = 1 anorthite + 2 wollastonite (6)

as a thermometer has been emphasized. However, as can be seen in Figure 6, this equilibrium, experimentally calibrated by Hays (1967), Boettcher (1970), Huckenholz et al (1975), and Windom & Boettcher (1976), has a significant pressure dependence and is as useful a barometer as it is a thermometer. The recent data of Huckenholz et al (1981) calibrating the effect of andradite component in garnet on the equilibrium extend the usefulness of this equilibrium.

Pressures have been inferred in numerous garnet-granulite terrains from five equilibria:

1 fayalite + 1 anorthite = 1 garnet (Gr_1Alm_2), (7)

1 forsterite + 1 anorthite = 1 garnet (Py_2Gr_1), (8)

1 ferrosilite + 1 anorthite = 1 garnet (Alm_2Gr_1) + 1 quartz, (9)

1 enstatite + 1 anorthite = 1 garnet (Py_2Gr_1) + 1 quartz, (10)

1 diopside + 1 anorthite = 1 garnet (Gr_2Py_1) + 1 quartz. (11)

Equilibrium (7) has been calibrated experimentally by Bohlen et al (1983c), while the others have been calculated from experimental and thermochemical data by Johnson & Essene (1982; 8), Bohlen et al (1983c; 9), Newton & Perkins (1982; 10 and 11), and Perkins & Chipera (1985; 10 and 11). These garnet barometers have been widely applied and much discussed (see Bohlen et al 1983a, Newton 1983, Perkins & Chipera 1985). We only comment here that in general these barometers give remarkably consistent and geologically reasonable pressures. These barometers indicate that most granulite terrains record pressures in the range 6–9 kbar with a distinct clustering at 7.5 ± 1 kbar, which suggests similar tectonic processes for the generation of granulites. Problems with accuracy arise when equilibria calibrated for Mg-rich compositions are applied to Fe-rich rocks (and vice versa) as a result, most likely, of inaccuracies in solution models.

Other equilibria useful for barometry and thermobarometry are shown in Figure 7. These have been calculated from a combination of experimental and thermochemical data by Essene & Bohlen (1985). The sphene-bearing equilibria should be useful in elucidating P/T conditions in garnet amphibolites and granulites. Because appropriate phase chemistry data are lacking in the literature, these equilibria have not been "field tested."

EXCHANGE THERMOMETERS

Exchange equilibria involve the distribution of two elements (usually of similar charge and ionic radius) between nonequivalent sites in one mineral (intracrystalline exchange) or between two phases (intercrystalline exchange). The criterion for exchange equilibrium is that the differences in chemical potentials for each component be the same in both phases. The most general form for exchange equilibrium for elements A and B between phases (or sites) 1 and 2 is

$$A + B = B + A$$
(in phase 1) (in phase 2) (in phase 1) (in phase 2)

The equilibrium constant is

$$K = \frac{a_{A(1)} a_{B(2)}}{a^{B(1)} a^{A(2)}} = \frac{X^{A(1)} X^{B(2)}}{X_{B(1)} X_{A(2)}} \cdot \frac{\gamma^{A(1)} \gamma^{B(2)}}{\gamma_{B(1)} \gamma_{A(2)}} = K^D K_\gamma$$

and

$$RT \ln K = \Delta H^{\text{ex}} - T \Delta S^{\text{ex}} + (P-1) \Delta V^{\text{ex}}.$$

If the solutions are ideal ($K_\gamma = 1$), K is replaced by K_D (the ratio of mole fractions), leading to a relatively simple expression for temperature. (Here X is mole fraction, a is activity, T is in kelvins, and "ex" refers to exchange.) It is an ironic measure of our progress in the past dozen years or so that very few exchange thermometers continue to be expressed so simply. Many now incorporate integrated heat-capacity terms for the enthalpy and entropy, and many have expressions for activity coefficients derived from thermodynamic solution models. Because ΔV is usually very small, exchange equilibria are almost exclusively thermometers. The larger the enthalpy of the exchange equilibirum, the greater its temperature dependence and thus its potential for thermometry. Intercrystalline exchange thermometers generally work best if there is little or no solid solution *between* the phases.

The appeal of exchange thermometers is immense: Since most of the phases involved have extensive solid solution, they can apply to rocks with a wide variety of bulk compositions. But they also exact a price: Resetting (retrograde exchange below peak metamorphic or magmatic temperature) occurs very easily in many cases and rarely leaves any textural evidence. Comparison with other thermometers or petrologic intuition is necessary in interpreting the temperatures obtained. Intracrystalline equilibria are especially prone to resetting, for the diffusion paths are but a fraction of a unit cell. As a result, they generally yield a blocking temperature for diffusion rather than a peak temperature. An exception to this rule is the distribution of Al and Si in feldspars, where "high" structural states may be preserved in many igneous rocks and where quantitative data are beginning to become available (Carpenter & Ferry 1984, Carpenter & McConnell 1984, Goldsmith & Jenkins 1985).

Exchange equilibria are not always independent of displaced equilibria. For example, much of the petrologic profession cut its thermodynamic eyeteeth on the formulation of a thermometer based on the exchange of Mg and Fe between olivine and orthopyroxene. We learned about the effects of site multiplicity and whether the $M1$ and $M2$ sites of each of these phases can be considered equivalent. Many papers later, the calibrating experiments were done (Medaris 1968), showing ΔH_{ex} to be small and the system to be of little use as a thermometer. But the addition of quartz to the Fe-rich parts of the system yields a very useful *barometer* based on displacement of the reaction fayalite+quartz = ferrosilite (Bohlen et al 1980a,b). Knowledge of the exchange equilibrium is useful in applying this barometer.

By far the most commonly used elements for exchange thermometry are Fe and Mg, partly because of their abundance in many rocks, but partly also because they substitute so readily for each other in a wide variety of minerals. The exchange of Fe and Mg between biotite and garnet is potentially useful in many metamorphic rocks. The most careful experimental calibration is that of Ferry & Spear (1978), but the effects of other components are so profound that this formulation should be applied only to cases where the phases are nearly pure. The most successful formulations at present appear to be those of Perchuk & Lavrent'eva (1983) and Indares & Martignole (1985). Exchange between garnet and orthopyroxene is useful in granulites and has been calibrated by Harley (1984) and by Sen & Bhattacharya (1984).

Fe-Mg exchange between silicates and oxides is attractive as a thermometer because of the typically high exchange enthalpy. Pyroxene and ilmenite have been calibrated by Bishop (1980), and olivine-ilmenite by Andersen & Lindsley (1981). Recent calibrations of the olivine-spinel exchange include those of Engi (1983) and of Jamieson & Roeder (1984).

One of the more widely used exchange thermometers is based on the coupled exchange of $Fe^{2+} + Ti^{4+}$ for 2 Fe^{3+} between titaniferous magnetite and ilmenite. The exchange is accompanied by an oxidation reaction $4\ Fe_3O_4 + O_2 = 6Fe_2O_3$ that can simultaneously give the oxygen fugacity of the rock. One of the chief benefits of this thermometer is the tendency of the Ti-magnetite to undergo "oxyexsolution" upon cooling: The Fe_2TiO_4 component oxidizes to yield ilmenite lamellae or granules in a Ti-poor magnetite host. The resulting intergrowth must be reintegrated to give the original composition—a nuisance—but the process of oxyexsolution often permits the spinel phase to maintain internal equilibrium upon cooling, so that in many cases there is little or no exchange between it and the *primary* ilmenite. Buddington & Lindsley (1964) gave a graphical form of this thermometer and oxybarometer; since then, there have been a number of thermodynamic treatments. The best of the published versions is that of Andersen & Lindsley (1985); Andersen is now preparing a newer version that utilizes revised data for the oxygen buffers (John Haas, personal communication, 1986).

The assemblage Ti-magnetite+ilmenite+fayalitic olivine+quartz is especially useful because it contains redundant information on oxygen fugacity and therefore can be used as a check on the assumption that the spinel and primary ilmenite have not reset upon cooling (Lindsley et al 1983, Frost et al 1987). It has been generally assumed that oxides in volcanic rocks—especially ash-flow tuffs—have not been reset, and most workers analyze only homogeneous spinels from them. The four-phase equilibrium commonly shows that assumption to be erroneous, and we suggest that oxyexsolved magnetites from those rocks may well give higher temperatures.

Most exchange thermometers involve cations. The oxygen-isotope thermometer is an important exception that uses the temperature-dependent distribution of ^{16}O and ^{18}O between phases. Isotopic exchange has virtually no volume change, so this thermometer is effectively independent of pressure. In general, the more dissimilar the phases, the greater is the exchange enthalpy, so pairs like quartz-magnetite (found in a wide variety of rocks) are most useful (see Friedman & O'Neil 1977).

SOLVUS THERMOBAROMETRY

In this paper we use the term *solvus* in its strict sense to mean the limit of solubility of a component in a given phase. In this usage, the composition of orthopyroxene that coexists with quartz and fayalitic olivine defines a (pressure- and temperature-dependent) solvus. A *miscibility gap* (commonly but unrigorously called a solvus) here is considered to be two solvi along a mutual composition line. If the phases have different structures and thus must be described by two different equations of state (for example, calcite-dolomite or diopside-enstatite), the solvi are nonconvergent. If both phases can be described by the same equation, the solvi converge at a critical, or consolute, point. This usage underscores the important fact that each of the phases across a miscibility gap contains independent thermometric information, because the criterion for *phase* equilibrium (chemical potentials of both components equal in both phases) is more rigorous than that for *exchange* equilibrium (differences in chemical potentials equal). Miscibility gaps are generally strongly temperature dependent and thus serve mainly as thermometers, but if the solid solution has strong excess (nonideal) volume, there will be a pressure dependence as well. Convergent miscibility gaps by definition are nonideal, and the temperature dependence resides in the configurational-entropy and activity-coefficient terms. The solutions in most nonconvergent gaps are also nonideal, but here there may be important contributions from the end-member enthalpy differences as well.

The simplest miscibility gaps are those that are essentially binary such as calcite-dolomite (Goldsmith & Newton 1969, Anovitz & Essene 1982), forsterite-monticellite (Adams & Bishop 1985), and the alkali feldspars (Haselton et al 1983). More complex miscibility gaps include those in the quadrilateral pyroxenes and olivines and in the ternary feldspars. These systems combine elements of exchange equilibrium along with the miscibility gaps. Davidson & Mukhopadhyay (1984) presented a solution model for the miscibility gap in olivines.

Pyroxene thermometry was reviewed by Lindsley (1983); we concentrate here on advances since that time. Important new experimental data on the critical bounding join diopside-enstatite have been obtained by Brey & Huth (1984), Nickel & Brey (1984), and Nickel et al (1985) at high pressures, and by Carlson (1986a,b) at 1 atm. Nickels & Brey presented a solution model for ortho- and clinopyroxenes on this join; their model is satisfactory at high pressures, but as they point out, it is unsatisfactory for much of the data at lower pressures. A solution model that describes all the data is urgently needed.

Lindsley's (1983) two-pyroxene thermometer is graphical. Since then, Davidson has fit the experimental data with quadrilateral solution models, and one can now calculate temperatures analytically (Davidson 1985, Davidson & Lindsley 1985). Both the graphical and analytical thermometers are based on relations among pure Ca-Mg-Fe pyroxenes, and their application to rocks has required projection of natural compositions to the quadrilateral. The projection of Andersen (Lindsley & Andersen 1983) is adequate for relatively low (<10%) amounts of nonquadrilateral components, but it has been shown to yield a serious overcorrection for the effects of Al_2O_3 (Kay & Kay 1983, Andersen & Lindsley 1984). A new projection scheme is being developed by Richard Holmes (University of Michigan) that promises to be more useful for pyroxenes with high alumina contents. Nickel et al (1985) have a projection scheme that works well with their model.

The compositions of coexisting plagioclase and alkali feldspar contain important thermometric information. Most formulations of the two-feldspar thermometer treat it as two separate binaries and obtain temperatures from the partitioning of albite component between them (Stormer 1975, Powell & Powell 1977, Haselton et al 1983). This approach may be valid at very low temperatures, but at higher temperatures the effects of ternary solution become increasingly important (Brown & Parsons 1985). Ghiorso (1983) presented a ternary solution model that accounts for the effects of K in plagioclase and Ca in alkali feldspar, but his model cannot be correct inasmuch as it predicts much more limited ternary solution than is observed in some rocks and in experiments. A revised ternary model (Fuhrman & Lindsley 1986) overcomes this difficulty. A great advantage of the ternary approach is that it allows three temperatures (one for each component) to be calculated rather than one, providing a sensitive test of equilibrium. A serious theoretical drawback to both ternary models is that they treat all feldspars—both triclinic and monoclinic—as a single solution. This assumption cannot be correct, but the available experimental data base is not adequate for fitting the (minimum of) two solutions that would be required. Additional experimental data on the compositions of coexisting feldspars are urgently needed.

A novel barometer has been proposed by Stormer & Whitney (1985), who pointed out that many ash-flow tuffs contain two oxides as well as two feldspars. Since the feldspar thermometer has a rather large pressure correction and the magnetite-ilmenite one does not, one solves for a pressure at which both thermometers give the same temperature.

SATURATION SURFACES

A method of thermometry that is just achieving recognition is the concept of the saturation surface. A saturation surface—the conditions under which a liquid becomes saturated with a given phase—is generally a complex function of temperature, pressure, and bulk composition. Nevertheless, for some rocks, saturation surfaces have been determined for apatite (Watson 1979a), zircon (Watson 1979b, Watson & Harrison 1983), and plagioclase and olivine (Langmuir & Hanson 1981, Glazner 1984). In some cases, these surfaces serve as useful thermometers; in others, their main function is to test whether a given magma would have been saturated with the phase at an independently estimated temperature. This is useful in testing hypotheses involving fractional crystallization of that phase.

ACKNOWLEDGMENTS

Preparation of this work has been suported by National Science Foundation grants EAR 84-16250 to SRB and EAR 84-16254 to DHL.

Literature Cited

Adams, G. E., Bishop, F. C. 1985. An experimental investigation of thermodynamic mixing properties and unit-cell parameters of forsterite-monticellite solid solutions. *Am. Mineral.* 70: 714–22

Adams, G. E., Bishop, F. C. 1986. The olivine-clinopyroxene geobarometer: experimental results in the CaO-FeO-MgO-SiO$_2$ system. *Contrib. Mineral. Petrol.* In press

Akella, J. 1976. Garnet pyroxene equilibria in the system CaSiO$_3$-MgSiO$_3$-Al$_2$O$_3$ and in a natural mineral mixture. *Am. Mineral.* 61: 589–98

Akella, J. 1979. Quartz coesite transition and the comparative friction measurements in piston-cylinder apparatus using talc-alsimag-glass (TAG) and NaCl high-pressure cells. *Neues Jahrb. Mineral. Monatsh.* 5: 217–24

Althaus, E. 1967. The triple point of andalusite-sillimanite-kyanite, an experimental and petrologic study. *Contrib. Mineral. Petrol.* 16: 29–44

Andersen, D. J., Lindsley, D. H. 1981. The correct Margules formulation for an asymmetric ternary solution: revision of the olivine-ilmenite thermometer, with applications. *Geochim. Cosmochim. Acta* 45: 847–53

Andersen, D. J., Lindsley, D. H. 1984. Application of a two-pyroxene thermometer: Correlation of apparent temperature with Al$_2$O$_3$ in augite. *Lunar Planet. Sci. XV, Part 1*, p. 7. Houston: Lunar Planet. Inst. (Abstr.)

Andersen, D. J., Lindsley, D. H. 1985. New (and final!) models for the Ti-magnetite/ilmenite geothermometer and oxygen barometer. *Eos, Trans. Am. Geophys. Union* 66: 416 (Abstr.)

Anderson, P. A. M., Newton, R. C., Kleppa, O. J. 1977. The enthalpy change of the andalusite-sillimanite reaction and the Al$_2$SiO$_5$ phase diagram. *Am. J. Sci.* 277: 585–93

Anovitz, L. M., Essene, E. J. 1982. Phase relations in the system CaCO$_3$-MgCO$_3$-FeCO$_3$. *Eos, Trans. Am. Geophys. Union* 63: 464 (Abstr.)

Benna, P., Bruno, E., Facchinelli, A. 1981. X-ray determination and equilibrium composition of clinopyroxenes in the system CaO-MgO-Al$_2$O$_3$-SiO$_2$. *Contrib. Mineral. Petrol.* 78: 272–78

Bishop, F. C. 1980. The distribution of Fe^{2+} and Mg between coexisting ilmenite and pyroxene with applications to geothermometry. *Am. J. Sci.* 280: 46–77

Boettcher, A. L. 1970. The system CaO-Al$_2$O$_3$-SiO$_2$-H$_2$O at high pressures and temperatures. *J. Petrol.* 11: 337–79

Boettcher, A. L., Wyllie, P. J. 1968. Phase relationships in the system NaAlSiO$_4$-SiO$_2$-H$_2$O to 35 kilobars pressure. *Am. J. Sci.* 267: 875–909

Bohlen, S. R. 1984. Equilibria for precise pressure calibration and a frictionless furnace assembly for the piston-cylinder apparatus. *Neues Jahrb. Mineral. Monatsh.* 1984: 404-12

Bohlen, S. R., Boettcher, A. L. 1981. Experimental investigations and geological applications of orthopyroxene geobarometry. *Am. Mineral.* 66: 951–64

Bohlen, S. R., Boettcher, A. L. 1982. The quartz coesite transformation: a precise determination and the effects of other components. *J. Geophys. Res.* 87: 7073–78

Bohlen, S. R., Liotta, J. J. 1986. A barometer for garnet amphibolites and garnet granulites. *J. Petrol.* 27: In press

Bohlen, S. R., Essene, E. J., Boettcher, A. L. 1980a. Reinvestigation and application of olivine-quartz-orthopyroxene barometry. *Earth Planet. Sci. Lett.* 47: 1–10

Bohlen, S. R., Boettcher, A. L., Dollase, W. A., Essene, E. J. 1980b. The effect of manganese on olivine-quartz-orthopyroxene stability. *Earth Planet. Sci. Lett.* 47: 11–20

Bohlen, S. R., Wall, V. J., Boettcher, A. L. 1983a. Geobarometry in granulites. In *Kinetics and Equilibrium in Mineral Reactions. Adv. Phys. Geochem.*, ed. S. K. Saxena, 3: 141–71. New York: Springer-Verlag

Bohlen, S. R., Wall, V. J., Boettcher, A. L. 1983b. Experimental investigations and geological applications of equilibria in the system FeO-TiO$_2$-Al$_2$O$_3$-SiO$_2$-H$_2$O. *Am. Mineral.* 68: 1049–58

Bohlen, S. R., Wall, V. J., Boettcher, A. L. 1983c. Experimental investigation and application of garnet granulite equilibria. *Contrib. Mineral. Petrol.* 83: 52–61

Bohlen, S. R., Dollase, W. A., Wall, V. J. 1986. Calibration and applications of spinel equilibria in the system FeO-Al$_2$O$_3$-SiO$_2$. *J. Petrol.* 27: In Press

Boyd, F. R., England, J. L. 1960. The quartz-coesite transition. *J. Geophys. Res.* 65: 749–56

Brey, G., Huth, J. 1984. The diopside-enstatite solvus to 60 kbar. *Proc. Int. Kimberlite Conf., 3rd*, ed. J. Kornprobst, 2: 257–64. New York: Elsevier

Brown, W. L., Parsons, I. 1985. Calorimetric and phase-diagram approaches to two-feldspar geothermometry: a critique. *Am. Mineral.* 70: 356–61

Buddington, A. F., Lindsley, D. H. 1964. Iron-titanium oxide minerals and synthetic equivalents. *J. Petrol.* 5: 310–57

Carlson, W. D. 1986a. Reversed pyroxene equilibria in CaO-MgO-SiO$_2$ from 925° to 1175°C at one atmosphere pressure. *Contrib. Mineral. Petrol.* 92: 218–24

Carlson, W. D. 1986b. Pigeonite phase equilibria at atmospheric pressure in CaO-MgO-SiO$_2$. *Eos, Trans. Am. Geophys. Union* 67: 415 (Abstr.)

Carmichael, D. M. 1978. Metamorphic bathozones and bathograds: a measure of post-metamorphic uplift and erosion on a regional scale. *Am. J. Sci.* 278: 769–97

Carpenter, M. A., Ferry, J. M. 1984. Constraints on the thermodynamic mixing properties of plagioclase feldspars. *Contrib. Mineral. Petrol.* 87: 138–48

Carpenter, M. A., McConnell, J. D. C. 1984. Experimental delineation of the CI-II transformation in intermediate plagioclase feldspars. *Am. Mineral.* 69: 112–21

Chopin, C. 1984. Coesite and pure pyrope in high-grade blueschists of the western Alps: a first record and some consequences. *Contrib. Mineral. Petrol.* 86: 107–18

Cohen, L. H., Klement, W. Jr. 1967. High-low quartz inversion: determination to 35 kilobars. *J. Geophys. Res.* 72: 4245–51

Danckworth, P. A., Newton, R. C. 1978. Experimental determination of the spinel peridotite to garnet peridotite reaction in the system MgO-Al$_2$O$_3$-SiO$_2$ in the range 900°–1,100°C and Al$_2$O$_3$ isopleths of enstatite in the spinel field. *Contrib. Mineral. Petrol.* 66: 189–201

Davidson, P. M. 1985. Thermodynamic analysis of quadrilateral pyroxenes. Part 1. Derivation of the ternary nonconvergent site-disorder model. *Contrib. Mineral. Petrol.* 91: 383–89

Davidson, P. M., Lindsley, D. H. 1985. Thermodynamic analysis of quadrilateral pyroxenes. Part II. Model calibration from experiments and applications to geothermometry. *Contrib. Mineral. Petrol.* 91: 390–404

Davidson, P. M., Mukhopadhyay, D. K. 1984. Ca-Fe-Mg olivines: phase relations and a solution model. *Contrib. Mineral. Petrol.* 86: 256–63

Dennen, W. H., Blackburn, W. H., Quesada, A. 1970. Aluminum in quartz as a geothermometer. *Contrib. Mineral. Petrol.* 27: 332–42

Dixon, J. R. 1981. *A spinel lherzolite barometer*. PhD thesis. Univ. Tex., Dallas

Engi, M. 1983. Equilibria involving Al-Cr spinel: I. Mg-Fe exchange with olivine. Experiments, thermodynamic analysis, and consequences for geothermometry. *Am. J. Sci.* 283–A: 29–71

Essene, E. J. 1982. Geologic thermometry and barometry. *Rev. Mineral.* 10: 153–206

Essene, E. J., Bohlen, S. R. 1985. New garnet barometers in the system CaO-FeO-Al$_2$O$_3$-SiO$_2$-TiO$_2$ (CFAST). *Eos, Trans. Am. Geophys. Union* 66: 387 (Abstr.)

Ferry, J. M., Spear, F. S. 1978. Experimental calibration of the partitioning of Fe and Mg between garnet and biotite. *Contrib. Mineral. Petrol.* 66: 113–17

Finnerty, A. A., Boyd, F. R. 1978. Pressure-dependent solubility of calcium in forsterite coexisting with diopside and enstatite. *Carnegie Inst. Washington Yearb.* 77: 713–17

Friedman, I., O'Neil, J. R. 1977. Compilation of stable isotopic fractionation factors of geochemical interest. *US Geol. Surv. Prof. Pap. No. 440-KK*

Frost, B. R., Lindsley, D. H., Andersen, D. J. 1987. Fe-Ti oxide-silicate equilibria: assemblages with fayalitic olivine. In preparation

Fuhrman, M. L., Lindsley, D. H. 1986. An improved, though still imperfect, solution model for ternary feldspars. *Abstr. IMA Meet., Stanford, Calif., 1986*

Fujii, T. 1977. Pyroxene equilibria in spinel lherzolite. *Carnegie Inst. Washington Yearb.* 76: 569–72

Gasparik, T. 1984. Two-pyroxene thermobarometry with new experimental data in the system CaO-MgO-Al$_2$O$_3$-SiO$_2$. *Contrib. Mineral. Petrol.* 87: 87–97

Gasparik, T. 1985. Experimental study of subsolidus phase relations and mixing properties of pyroxene and plagioclase in the system Na$_2$O-CaO-Al$_2$O$_3$-SiO$_2$. *Contrib. Mineral. Petrol.* 89: 346–57

Gasparik, T., Newton, R. C. 1984. The reversed alumina contents of orthopyroxene in equilibrium with spinel and forsterite in the system MgO-Al$_2$O$_3$-SiO$_2$. *Contrib. Mineral. Petrol.* 85: 186–96

Ghent, E. A. 1976. Plagioclase-garnet-Al$_2$SiO$_5$-quartz: a potential geothermometer-geobarometer. *Am. Mineral.* 61: 710–14

Ghiorso, M. S. 1983. Activity/composition relations in the ternary feldspars. *Contrib. Mineral. Petrol.* 87: 282–96

Gillet, Ph., Ingrin, J., Chopin, C. 1984. Coesite in subducted continental crust: *P-T* history deduced from an elastic model. *Earth Planet. Sci. Lett.* 70: 426–36

Glazner, A. F. 1984. Activities of olivine and plagioclase components in silicate melts and their application to geothermometry. *Contrib. Mineral. Petrol.* 88: 260–68

Goldsmith, J. R. 1980. The melting and breakdown reactions of anorthite at high pressures and temperatures. *Am. Mineral.* 65: 272–84

Goldsmith, J. R., Jenkins, D. M. 1985. The high-low albite relations revealed by reversal of degree of order at high pressures. *Am. Mineral.* 70: 911–23

Goldsmith, J. R., Newton, R. C. 1969. *P-T-X* relations in the system $CaCO_3$-$MgCO_3$ at high temperatures and pressures. *Am. J. Sci.* 267-A: 160–90

Green, T. H., Ringwood, A. E., Major, A. 1966. Friction effects and pressure calibration in a piston-cylinder apparatus at high-pressure and temperature. *J. Geophys. Res.* 71: 3589–94

Halbach, H., Chatterjee, N. D. 1984. An internally consistent set of thermodynamic data for twenty-one CaO-Al_2O_3-Si_2-H_2O phases by linear parametric programming. *Contrib. Mineral. Petrol.* 88: 14–23

Hariya, Y., Kennedy, G. C. 1968. Equilibrium study of anorthite under high pressure and high temperature. *Am. J. Sci.* 266: 193–202

Harker, A. 1932. *Metamorphism.* London: Chapman & Hall. 362 pp.

Harley, S. L. 1984. An experimental study of the partitioning of Fe and Mg between garnet and orthopyroxene. *Contrib. Mineral. Petrol.* 86: 359–73

Haselton, H. T. 1979. *Calorimetry of synthetic pyrope-grossular garnets and calculated stability relations.* PhD thesis. Univ. Chicago, Ill.

Haselton, H. T. Jr., Hovis, G. L., Hemingway, B. S., Robie, R. A. 1983. Calorimetric investigation of the excess entropy of mixing in analbite-sanidine solid solutions: lack of evidence for Na,K short-range order and implications for two-feldspar thermometry. *Am. Mineral.* 68: 398–413

Hays, J. F. 1967. Lime-alumina-silica. *Carnegie Inst. Washington Yearb.* 65: 234–39

Herzberg, C. T. 1976. The plagioclase-lherzolite to spinel-lherzolite facies boundary; its bearing on corona structure formation and tectonic history in the Norwegian Caledonides. In *Progress in Experimental Petrology,* ed. G. M. Biggar, D-3: 233–35. London: Natural Environment Res. Counc. Publ.

Herzberg, C. T. 1978. Pyroxene geothermometry and geobarometry: experimental and thermodynamic evaluation of some subsolidus phase relations involving pyroxenes in the system CaO-MgO-Al_2O_3-SiO_2. *Geochim. Cosmochim. Acta* 42: 945–57

Herzberg, C. T., Chapman, N. A. 1976. Clinopyroxene geothermometry of spinel-lherzolites. *Am. Mineral.* 61: 626–37

Holdaway, M. J. 1971. Stability of andalusite and the aluminum silicate phase diagram. *Am. J. Sci.* 271: 97–131

Holland, T. J. B. 1980. The reaction albite = jadeite+quartz determined experimentally in the range 600–1200°C. *Am. Mineral.* 65: 129–34

Huckenholz, H. G., Hölzl, E., Lindhuber, W. 1975. Grossularite, its solidus and liquidus relations in the CaO-Al_2O_3-SiO_2-H_2O system up to 10 kbar. *Neues Jahrb. Mineral. Abh.* 124: 1–46

Huckenholz, H. G., Lindhuber, W., Fehr, K. T. 1981. Stability relationships of grossular + quartz + wollastonite + anorthite. I. The effect of andradite and albite. *Neues Jahrb. Mineral. Abh.* 142: 223–48

Indares, A., Martignole, J. 1985. Biotite-garnet geothermometry in the granulite facies: the influence of Ti and Al in biotite. *Am. Mineral.* 70: 272–78

Jamieson, H. E., Roeder, P. L. 1984. The distribution of Mg and Fe^{2+} between olivine and spinel at 1300°C. *Am. Mineral.* 69: 283–91

Jenkins, D. M., Newton, R. C. 1979. Experimental determination of the spinel peridotite to garnet peridotite inversion at 900°C and 1000°C in the system CaO-MgO-Al_2O_3-SiO_2, and at 900°C with natural garnet and olivine. *Contrib. Mineral. Petrol.* 68: 407–19

Johnson, C. A., Essene, E. J. 1982. The formation of garnet in olivine-bearing metagabbros from the Adirondacks. *Contrib. Mineral. Petrol.* 81: 240–51

Kay, S. M., Kay, R. W. 1983. Thermal history of the deep crust inferred from granulite xenoliths, Queensland, Australia. *Am. J. Sci.* 283-A: 486–513

Kitahara, S., Kennedy, G. C. 1964. The quartz-coesite transition. *J. Geophys. Res.* 69: 5395–5400

Kushiro, I., Yoder, H. S. Jr. 1966. Anorthite forsterite and anorthite-enstatite reactions and their bearing on the basalt-eclogite transformation. *J. Petrol.* 7: 337–62

Langmuir, C. H., Hanson, G. H. 1981. Calculating mineral-melt equilibria with stoichiometry, mass balance, and single-component distribution coefficients. In *Thermodynamics of Minerals and Melts. Adv. Phys. Geochem.,* ed. R. C. Newton, A. Navrotsky, B. J. Wood, 1: 247–71. New York: Springer-Verlag

Lindsley, D. H. 1983. Pyroxene thermometry. *Am. Mineral.* 68: 477–93

Lindsley, D. H., Andersen, D. J. 1983. A two-pyroxene thermometer. *Proc. Lunar Planet. Conf., 13th, Part 2. J. Geophys. Res. Suppl.* 88: A887–A906

Lindsley, D. H., Podpora, C., Frost, B. R. 1983. Experimental calibration of the equilibrium: $Fe_2SiO_4 + 2FeTiO_3 = 2Fe_2TiO_4$

+SiO_2. *Geol. Soc. Am. Abstr. With Programs* 15: 628

MacGregor, I. D. 1974. The system MgO-Al_2O_3-SiO_2: Solubility of Al_2O_3 in enstatite for spinel and garnet peridotite compositions. *Am. Mineral.* 59: 110–19

Medaris, L. G. 1968. Partitioning of Fe^{++} and Mg^{++} between coexisting synthetic olivines and orthopyroxenes. *Am. J. Sci.* 267: 945–68

Mirwald, P. W., Massonne, H.-J. 1980. The low-high quartz and quartz-coesite transition to 40 kbar between 600°C and 1600°C and some reconnaissance data on the effect of $NaAlO_2$ component on the low-quartz-coesite transition. *J. Geophys. Res.* 85: 6983–90

Newton, R. C. 1966. Some calc-silicate equilibrium relations. *Am. J. Sci.* 264: 204–22

Newton, R. C. 1983. Geobarometry of high-grade metamorphic rocks. *Am. J. Sci.* 238-A: 1–28

Newton, R. C., Haselton, H. T. 1981. Thermodynamics of the garnet-plagioclase-Al_2SiO_5-quartz geobarometer. In *Thermodynamics of Minerals and Melts. Adv. Phys. Geochem.*, ed. R. C. Newton, A. Navrotsky, B. J. Wood, 1: 129–45. New York: Springer-Verlag

Newton, R. C., Perkins, D. III. 1982. Thermodynamic calibration of geobarometers for charnockites and basic granulites based on the assemblages garnet-plagioclase-orthopyroxene (clinopyroxene)-quartz with applications to high grade metamorphism. *Am. Mineral.* 67: 203–22

Nickel, K. G., Brey, G. 1984. Subsolidus orthopyroxene-clinopyroxene systematics in the system CaO-MgO-SiO_2 to 60 kb: a re-evaluation of the regular solution model. *Contrib. Mineral. Petrol.* 87: 35–42

Nickel, K. G., Brey, G. P., Kogarko, L. 1985. Orthopyroxene-clinopyroxene equilibria in the system CaO-MgO-Al_2O_3-SiO_2 (CMAS): new experimental results and implications for two-pyroxene thermometry. *Contrib. Mineral. Petrol.* 91: 44–53

Obata, M. 1976. The solubility of Al_2O_3 in orthopyroxenes in spinel and plagioclase peridotites and spinel pyroxenite. *Am. Mineral.* 61: 804–16

O'Hara, M. J., Howells, S. 1978. The enstatite-pyrope geobarometer. In *Progress in Experimental Petrology*, ed. W. S. MacKenzie, D-4: 175–79. London: Natural Environment Res. Counc. Publ.

O'Neil, H. S. C., Navrotsky, A. 1984. Cation distributions and thermodynamic properties of binary spinel solid solutions. *Am. Mineral.* 69: 733–53

Perchuk, L. L., Lavrent'eva, I. V. 1983. Experimental investigation of exchange equilibria in the system cordierite-garnet-biotite. In *Kinetics and Equilibrium in Mineral Reactions*, ed. S. K. Saxena, 3: 199–239. New York: Springer-Verlag

Perkins, D., Chipera, S. J. 1985. Garnet-orthopyroxene-plagioclase-quartz barometry: refinement and application to the English River subprovince and the Minnesota River Valley. *Contrib. Mineral. Petrol.* 89: 69–80

Perkins, D., Newton, R. C. 1980. The compositions of coexisting pyroxenes and garnet in the system CaO-MgO-Al_2O_3-SiO_2 at 900°–1,100°C and high pressures. *Contrib. Mineral. Petrol.* 75: 291–300

Pitcher, W. S., Flinn, G. W., eds. 1965. *Controls of Metamorphism. Geol. J. Spec. Iss. No. 1.* Edinburgh: Oliver & Boyd. 341 pp.

Powell, R., Powell, M. 1977. Plagioclase-alkali feldspar geothermometry revisited. *Mineral. Mag.* 41: 253–56

Presnall, D. C. 1976. Alumina content of enstatite as a geobarometer for plagioclase and spinel lherzolites. *Am. Mineral.* 61: 582–88

Richardson, S. W., Bell, P. M., Gilbert, M. C. 1968. Kyanite-sillimanite equilibrium between 700 and 1500°C. *Am. J. Sci.* 266: 513–41

Schmid, R., Cressey, G., Wood, B. J. 1978. Experimental determination of univariant equilibria using divariant solid-solution assemblages. *Am. Mineral.* 63: 511–15

Schuiling, R. D. 1957. A geoexperimental phase diagram of Al_2SiO_5 (sillimanite, kyanite, and andalusite). *K. Ned. Akad. Wet.* B60: 220–26

Sen, S. K., Bhattacharya, A. 1984. An orthopyroxene-garnet thermometer and its application to the Madras charnockites. *Contrib. Mineral. Petrol.* 88: 64–71

Smith, D. C. 1984. Coesite in clinopyroxene in the Caledonides and its implications for geodynamics. *Nature* 310: 641–44

Smyth, J. R. 1977. Quartz pseudomorphs after coesite. *Am. Mineral.* 62: 828–30

Stormer, J. C. Jr. 1975. A practical two-feldspar geothermometer. *Am. Mineral.* 60: 667–74

Stormer, J. C., Whitney, J. A. 1985. Two-feldspar and iron-titanium oxide equilibria in silicic magmas and the depth of origin of large volume ashflow tuffs. *Am. Mineral.* 70: 52–64

Tuttle, O. F. 1949. The variable inversion temperature of quartz as a possible geologic thermometer. *Am. Mineral.* 34: 723–30

Tuttle, O. F., Bowen, N. L. 1958. Origin of granite in light of experimental studies in the system $NaAlSi_3O_8$-$KAlSi_3O_8$-SiO_2-H_2O. *Geol. Soc. Am. Mem.* 74: 1–153

Watson, E. B. 1979a. Apatite saturation in

basic to intermediate magmas. *Geophys. Res. Lett.* 6: 937–40

Watson, E. B. 1979b. Zircon saturation in felsic liquids: experimental data and applications to trace element geochemistry. *Contrib. Mineral. Petrol.* 70: 407–19

Watson, E. B., Harrison, T. M. 1983. Zircon saturation revisited: temperature and composition effects in a variety of crustal magma types. *Earth Planet. Sci. Lett.* 64: 295–304

Windom, K. E., Boettcher, A. L. 1976. The effect of reduced activity of anorthite on the reaction grossular+quartz = anorthite+wollastonite: a model for plagioclase in the Earth's lower crust and upper mantle. *Am. Mineral.* 61: 889–96

MODELS OF LITHOSPHERIC THINNING

Horst J. Neugebauer

Institut für Geophysik, Technische Universität Clausthal, D-3392 Clausthal-Zellerfeld, Federal Republic of Germany

INTRODUCTION

Lithospheric thinning denotes a geodynamical process that is associated with extensional tectonic phenomena. In plate tectonics the lithosphere is a layer of strength compared with the underlying asthenosphere. Its thickness is commonly defined by the depth to a particular isothermal or by a fraction of the melting temperature of rocks. A typical thickness is of the order 100 km and includes the Earth's crust and parts of the upper mantle. This definition is equivalent to that used in seismology, but it does not necessarily correspond to the elastic thickness derived from flexural deformations in response to loading.

According to its thermal regime, state of stress, and lithology, we expect the lithosphere to contain layers of predominantly brittle and ductile deformation. The oceanic and continental lithospheres have distinctive origins, structures, and compositions. While the age of the oceanic lithosphere is only of order 10^8 Myr, the continental lithosphere contains the oldest rocks known on Earth. The latter has therefore the most comprehensive record of past geodynamical processes.

Lithospheric thinning is expressed by a variety of tectonic and epeirogenic manifestations, such as domal uplifts and hotspot swells (Crough 1983), continental rifts (Morgan & Baker 1983), platform basins (Sleep et al 1980), and plateau uplifts. On the other hand, our understanding of the thinning mechanisms is based on physical concepts in association with observations. Reheating of the lithosphere, buoyantly driven diapiric penetration, or stretching under isostatic conditions are the most frequently discussed models. They implicitly define the nature of the source of thinning.

OBSERVATIONAL CONSTRAINTS

EPEIROGENY   Lithospheric thinning is a time-dependent process. One of the most obvious surface features that has been associated with lithospheric thinning is epeirogenic movement. Domal uplifts at midplate swells and at continental rifts have typical heights of 10^3 m, while depression and subsidence of the lithosphere as recorded by platform basin sediments reach typical values of 4×10^3 m. Subsidence rates deduced from sedimentary records reveal distinct periods of rapid (50–100 m Myr^{-1}) and slow (5–10 m Myr^{-1}) subsidence, and even periods of basin inversion and erosion (Ziegler 1982). On the other hand, a comparison between domal uplift observed at Cenozoic continental rifts with recent crustal movement of the order of 1 mm $\text{yr}^{-1} \triangleq 1000$ m Myr^{-1} suggests the existence of expressed phases of activity for the uplift. One can thus conclude that epeirogenic movements associated with thinning occur episodically over geologic periods of time.

Zones of either uplift or subsidence, such as domes and interior basins, have typical diameters of 500 km and nearly always broad, circular-to-oval geometries. Examples such as Hoggar, Tibesti, and Datur in central North Africa are characteristic of domal uplift areas associated with Cenozoic volcanism (Browne & Fairhead 1983). In addition to basement uplift and volcanism, formation of graben structures is typical for continental rifts such as the Kenya, Baikal, and Rhine rifts. Grabens provide evidence for crustal extension. They usually cross or follow the crest of an uplift area, and their width is of the order of 30–60 km. Examples of interior basins having typical sizes and shapes with well-identified basement graben tectonics are the Paris Basin and the Michigan Basin.

There is good evidence for the existence of domes, rifted domes, and platform basins that are characterized by a minimum unit size and a circular-to-oval standard shape. It is also well known from detailed investigations in the North Sea Basin that there exists a genetic sequence of basin formation based on early domal uplift with volcanism, intermediate extensional graben formation, and mature basin subsidence (Ziegler 1982). Volcanism expands geographically with progression of this sequential development.

Although we are able to identify minimum units of epeirogenic movements, volcanism, and extensional tectonics, it is well known that continental rifts and basins form elongated belts. These belt structures form rather arbitrary patterns such as the U-shaped, 2000-km-long Midcontinent Rift System in North America (Keller et al 1983, Van Schmus & Hinze 1985), the bent Mesozoic North Sea Rift with the Cenozoic Rhine Rift System, or the Cenozoic Afro-Arabian Rift System. Such lineaments of extensional tectonics are considered by many to represent the typical structure that develops prior to continental fragmentation.

However, a closer view of the sedimentology, geophysics, and petrography of these beltlike zones reveals the individual character of small rift or basin segments. In every case, the tectonic history of these belts is the history of all contributing individual substructures. For some well-known areas the typical features listed above for a minimum unit in size and individual development can be revealed from observations. For example, the Upper Rhine Graben is composed of two segments, a southern segment (shown in Figure 1) and a northern segment with

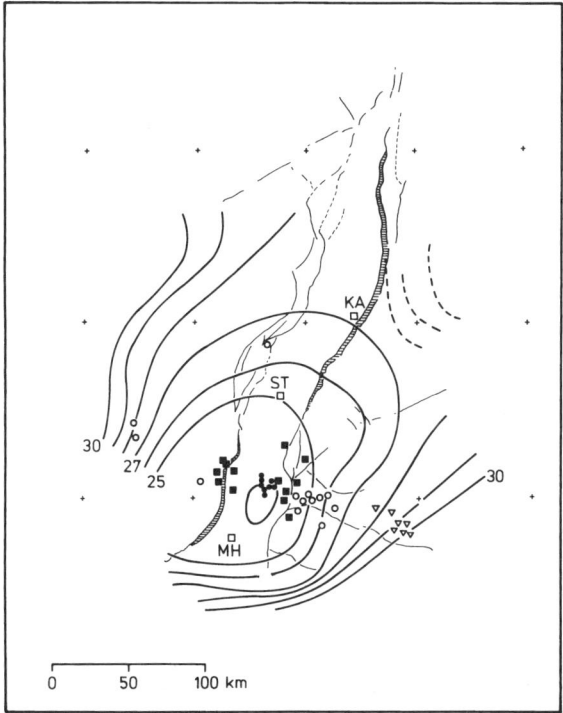

Figure 1 Contour map of the top of the transitional lower crust in kilometers below the surface at the Upper Rhine Rift zone. Graben structure is shown by means of the boundary faults. Stippled area represents amount of horizontal extension in order to reconstruct observed graben wedge subsidence. Dashed lines indicate uncertain data. Volcanic foci of pre-and synrift tectonic age in central position (●, ■) and on the flanks (○, ∇) are all associated with the transitional lower crust. Other notation: ●, Kaiserstuhl volcanism of young age and shallow origin; ∇, Hegau volcanism. Abbreviations as follows: ST, Strasbourg; KA, Karlsruhe; MH, Mulhouse.

characteristic activity in the Eocene and Miocene, respectively. A similar "migration" from north to southeast occurred in the North Sea during the Mesozoic. This suggests that the surface manifestation of lithospheric thinning locally forms a genetic sequence, but that laterally this development happens at different times.

IGNEOUS ACTIVITY   The earliest evidence for lithospheric thinning in terms of the transport of matter and heat over a lithospheric depth range comes from igneous activity. Volcanic activity has been identified at zones of domal uplift and may even have preceded epeirogenic movements. Early phases are observed at a central position; later on, volcanic foci migrate outward over the flanks of rifted domes, while reactivation of the early central sites is frequently observed, especially when large volumes of magma are erupted. Volcanism at mature basin structures is observed less frequently, possibly because of the intrusive deposition of lavas within sediment layers, a process that limits observations.

Volcanism is associated with sites of lithospheric thinning. Although magma ascent through the brittle crust is controlled by stress release along faults, the volcanic eruptions are strongly related to the area of domal uplift or the lower crust structural anomaly (Figure 1).

Barometric depth determinations of the origin of magma segregation suggest that the magma segregates primarily at the lower lithosphere boundary, over a rather long period of time (~ 100 Myr). The eruptions for the southern Upper Rhine Graben are compiled in Figure 2. Generally, the rate of magma ascent must substantially exceed rates of uplift in order to avoid cooling and recrystallization. A brittle mechanical barrier and the occurrence of bimodal volcanism provide arguments that magma is captured in a lower crustal-to-subcrustal position.

GEOPHYSICS   Even more striking evidence on the possible lithological nature of the depth dependence at thinning sites comes from modern seismic sounding surveys. Refraction and reflection studies reveal transition zones with extreme gradients of body-wave velocities and the existence of numerous seismic reflectors in the lower crust underlying domal and rift structures. A lithologic layering has been deduced (Hale & Thompson 1982, Deichmann & Ansorge 1983) based on the best fit to observed seismic sound velocities by rock densities, lithologies, and thermal constraints. O'Reilly & Griffin (1985) suggest a gradational mixture of ultramafic rocks increasing with depth. These arguments are consistent with the view that the lower crust at lithospheric thinning sites is not simply replaced by an upwelling upper mantle, but rather by a layered zone of mafic, ultramafic, and lower crustal rocks. Metamorphic reactions might modify seismic reflection patterns, especially those for old structures.

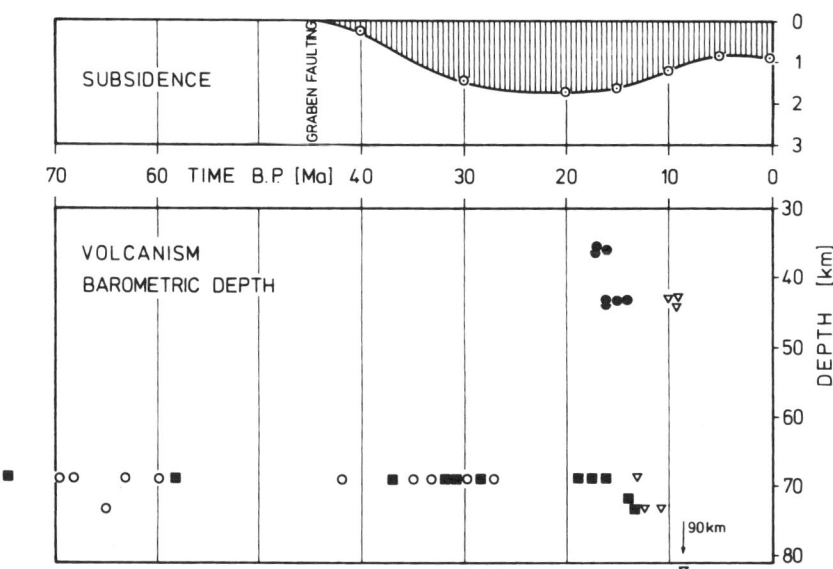

Figure 2 Mean barometric depth of origin of volcanic rocks and the time of eruption for the southern Rhine Rift zone (Neugebauer & Walter 1983). Events older than 70 Myr are not included. Kaiserstuhl volcanism between 10 and 20 Myr corresponds with the inversion of the rift basin. Faulting and basin subsidence with time are shown on top. Geographical positions and definitions of volcanic foci indicated in Figure 1.

The example shown in Figure 1 of a transitional lower crust associated with a rift zone might be understood as a consequence of igneous intrusions from the base of the lithosphere over a period of about 100 Myr (Figure 2), while the tectonic rift event appears to be a mechanical response to the induced stress regime during the middle of this time period. A large volume of magma erupted between 10 and 20 Myr ago over the crest of the dome of transitional lower crust. This volcanism at the Kaiserstuhl was of shallow origin and was accompanied by a period of intense basement uplift that even led to rift-basin inversion, as indicated in Figure 2.

The total heat budget associated with lithospheric-thinning stages is not well constrained by observations. As a result of igneous activity we can assume that magma of a temperature $\sim 900°C$ is penetrating into the crust. Although the time span of activity is long, the efficiency is limited by the volume. This volume will be constrained by the volume of the transitional lower crust as an upper limit. It is thus observed that Cenozoic continental rifts are generally associated with an average increased heat flow of 100 mW m^{-2} (Morgan 1983). On the other hand, measurements from zones marginal to rift basins range about the worldwide continental average of 60 mW m^{-2} (Morgan 1984). This indicates that there is no broad zone of extreme heat flow associated with Cenozoic rifts. The heat flow at mature platform basins approaches the continental average level. Upper limits of paleothermal conditions at basins are detectable from limits on the thermal history of source beds. Generally, the maturation of hydrocarbons restricts the chance for extreme heating events in the past.

Constructive plate boundaries, on the other hand, are assumed to be sites of convective uprise of mantle material, and the associated formation and history of oceanic lithosphere is described as a continuous cooling process (Sclater et al 1980). For comparison the oceanic heat-flow mean is 75 mW m^{-2}, and the anomalies at (for example) the Pacific Rise approach 200 mW m^{-2}. This implies that ocean ridges of 3000-m elevation correspond to an excess heat flow of 125 mW m^{-2}, while rift domes with a mean height of 1000 m exhibit heat-flow anomalies of 40 mW m^{-2}. This factor of three difference provides another constraint on the possible source strength of lithospheric thinning, provided that the nature of mid-oceanic ridges and rift domes is comparable.

Various stages of lithospheric thinning as well as different types of tectonic activity provide variations in the gravity signature. Interpretation of Bouguer gravity anomalies are therefore related to information on typical structural changes. In principle, for the subcrustal part of the lithosphere a density inversion is assumed because of either partial melting, thermal expansion, or the upward movement of lighter rock components, while for the lower crust a density increase is claimed based on the assumption of an uplifted upper mantle or a lithological transition zone due to mixed laminar structures.

Domal basement uplifts with volcanism have typical negative regional Bouguer anomalies. They are explained by a root in the lower lithosphere having a width comparable to that of the dome. The lower crust might have mass excess, but this will be compensated by the deep, dominating mass deficiency. The influence of increasing density contrast within the lower crust is related to a short-wavelength axial gravity high. The more mature the thinned structure becomes, the more the lower crustal mass excess gains influence, which might lead toward a regional positive anomaly. On the other hand, the predominance of mass excess might be supported by a diminishing root within the lower lithosphere. Mature platform basins exhibit only minor anomalies, which indicates an isostatic balance between sediments and a lower crustal mass excess.

It should be pointed out that the tendency toward a reduced regional negative Bouguer anomaly corresponds to an increase in the area of volcanic activity toward the shoulders of rift domes. One might speculate whether the increase of mass excess within the lower crust has a lateral component.

Figure 3 is a comprehensive representation of the three major phases of lithospheric thinning that can be derived from observations. A genetic sequence is suggested in the following order: dome, rift, and basin.

Physical Constraints

The lithosphere is defined according to the specific mode of deformation or physical state that is being considered. Processes like bending, cooling, or extension define the lithosphere by flexural parameters, the apparent thermal state, or the mechanical strength, respectively. Although these characteristic parameters are related to an observed response of the lithosphere, they represent different definitions of the lithosphere.

A common way to define the thickness of the lithosphere is to refer to a thermal interface given by a particular isotherm or by a fraction of the melting temperature. Geotherms can be calculated from specific models of heat production, thermal conductivity, and heat sources, with the determining variable being the surface heat flow. The intersection of these

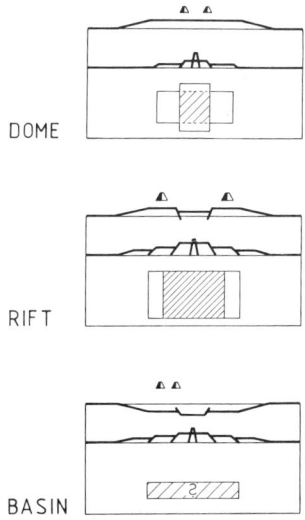

Figure 3 Characteristic stages and structures of lithospheric thinning, as derived from observations. There is good evidence to assume a genetic sequence of dome, rift, and basin. Sources for epeirogenic movements are represented by low-density roots in the lithospheric mantle (stippled areas with variations). Disturbed lower crust is interpreted as a transition zone of higher density with respect to the lower crust. Volcanism spreads from a central position toward a domal position. Volcanic activity in basins is sporadic.

geotherms with a given mantle solidus or even a constant temperature (for example, 1000°C) then defines the lower boundary of the lithosphere (Chapman & Pollack 1977). For a given range of regional heat flow on continents between 40 and 90 mW m^{-2}, the corresponding thicknesses of the lithosphere vary between 200 and 50 km, respectively. Mantle solidus curves for wet conditions are intersected by the above geotherms at a shallower depth and thus define generally thinner lithospheres. The evidence for zones of seismic low velocity and high absorption beneath the oceanic lithosphere and the tectonically active regions on continents have led to the above interpretation of the lithosphere-asthenosphere boundary as the location where melting begins.

The definition of the lithosphere as a layer of strength relative to the underlying asthenosphere depends on typical time spans of loading. The lithosphere behaves elastically on very short time scales, which is evident in its seismic behavior. This is even the case on short time scales or wavelengths of lithospheric loading (Walcott 1973). For longer durations of loading the mechanical strength decreases, which suggests that the lithosphere exhibits viscous behavior also.

Theoretical and experimental rock mechanics (Stocker & Ashby 1973) show that the rate of viscous flow depends on the temperature and pressure conditions, the constitutional and material properties, and the deformation mechanism. Flow laws indicate nonlinear dependence of the strain rate on the thermal and stress state as well. Viscous flow in the lithosphere significantly deforms at a rate larger than 10^{-17} s^{-1} for deviatoric stresses lower than 200 MPa and temperatures higher than 0.5 times the melting temperature of rocks.

Viscous flow, however, is dominated by brittle deformation controlled by normal stresses and fault friction in the uppermost portion of the lithosphere (Brace & Kohlstedt 1980). The highest strength of the lithosphere is thus reached at the brittle-ductile transition, which is determined by parameter sets for both brittle shear and viscous flow. The lithosphere-asthenosphere boundary is not a rheological interface viewed in terms of long-term ductile deformation.

Lithospheric thinning is obviously associated with substantial redistribution of mass in the lower crust and the subcrustal lithosphere, as suggested by gravity anomalies. Is there thus any evidence to associate the above mass heterogeneities with a physical transport mechanism? Support for this view comes from the possible genetic sequence of thinning, the mechanical behavior of the lithosphere, and the existence of igneous activity. In addition, theoretical considerations and experimental investigations of magma segregation and densities provide a suitable physical basis (Stolper et al 1981).

Several authors have suggested that mafic-to-ultramafic magmas are more compressible than common mantle rocks and minerals (Stolper et al 1981, Rigden et al 1984). Thus, at pressures lower than 7 GPa (~200 km) the densities of mafic and ultramafic magmas become drastically lower than the standard values given by the Preliminary Reference Earth Model (PREM), as demonstrated in Figure 4 (Dziewonski & Anderson 1981). Therefore, during melting in the lithospheric mantle, magma escape from the rock matrix becomes more probable with decreasing depth. On the other hand, cooling and/or recrystallization due to a decreasing rate of ascent or in response to a high-strength mechanical barrier will provide a density increase at the depth level of consolidation. Magma upwelling will even become suppressed by the diminishing density contrast in comparison with crustal rocks.

QUANTITATIVE MODELS

Stretching Models

The tectonic mode of normal faulting, the upward bending of structural interfaces, and the corresponding increase of heat flow have all been explained as manifestations of lithospheric extension. If extension is interpreted as stretching in the homogeneous plastic sense, one would expect the lithosphere to become thinner. Stretching models assume that the extending lithosphere is in isostatic balance and hence that mass must

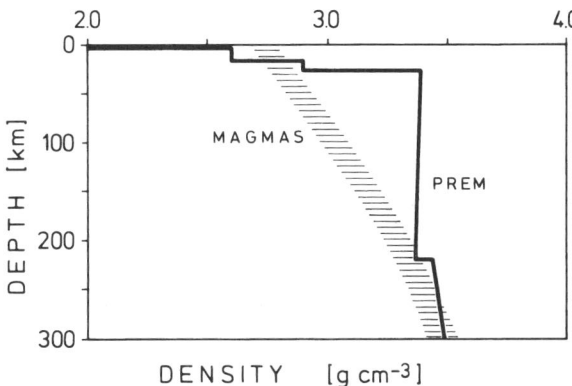

Figure 4 Densities of mafic and ultramafic magmas [after Stolper et al (1981) and Rigden et al (1984)] and distribution of mean upper-mantle densities given by the Preliminary Reference Earth Model (PREM) (after Dziewonski & Anderson 1981).

be moving upward in the asthenosphere beyond the initial depth of the lithosphere-asthenosphere boundary. Thus heat is transferred from the underlying asthenosphere by stretching in the solid state. Model sources are assumed to be either active upwelling of the asthenosphere or horizontal drag within the lithosphere while the asthenosphere behaves passively. The principal concepts are demonstrated in Figure 5a; the model is one dimensionally depth dependent.

Lachenbruch & Sass (1978) presented a kinematic model concept of a steadily extending lithosphere. Their lithospheric layer is defined by a temperature condition at its base. It is assumed to be mechanically and thermally in a steady state. The model thickness of the lithosphere is maintained uniform because the rate of accretion of either crystalline material or, alternatively, solidifying liquid basalt at its base will compensate for the modeled thinning. The rates of extension and of accretion or underplating are assumed to remain relatively uniform. Buoyant plumes, horizontal drag in the lithosphere, and penetrative convection of the asthenosphere are all discussed as possible sources providing appropriate lithosphere-asthenosphere boundary conditions for the models.

The thermal steady-state representation of lithospheric stretching allows one to predict surface heat flow in relation to lithospheric thickness, extension rate, and basal heat flux. The model concept has been used to express data from heat-flow regimes of the Basin and Range province of western North America in terms of extension and thinning by stretching.

A more simplified stretching model of lithospheric thinning has been

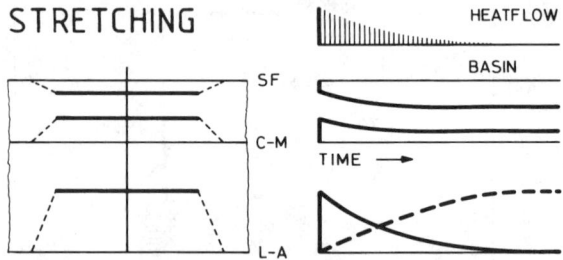

Figure 5a Principles of lithospheric thinning by instantaneous stretching. Structural changes are at the left; variations with time are at the right. The model is only depth dependent. Abbreviations as follows: SF, surface; C–M, crust-mantle interface; L–A, lithosphere-asthenosphere boundary. The model is one dimensional, and it leads to subsidence of the surface only. The dashed line accounts for the assumed accretion or solidification of material with time.

suggested by McKenzie (1978). In this model it is assumed that a thermally equilibrated lithospheric layer is suddenly extended by a factor of β. The stretched layer is assumed to be conserved under isostatic compensation throughout. The results are a subsidence of the surface layer and a passive upwelling of hot asthenosphere because the temperature of the material remains unchanged during stretching. Subsequent cooling of the hot asthenosphere causes subsidence by contraction as the temperature disturbance decays (Figure 5a).

This structurally isostatic concept provides initial subsidence and heat flow for a prescribed extension coefficient. Predictions of heat flow by cooling and subsidence due to contraction are possible as functions of time. Calculations of subsidence and heat flow with time demonstrate that they strongly vary for an extension coefficient $\beta \leq 4$. For larger values of β (up to 100) the predicted surface heat flow and subsidence are insensitive to further variations of β.

Jarvis & McKenzie (1980) introduced finite extension rates and concluded that the previous assumption of instantaneous stretching gives reasonable results for periods of stretching less than 20 Myr. On the other hand, it has been found from applications of McKenzie's model that for several regions the thermal subsidence is much greater than that predicted from stretching models. In order to increase the modeled heat input during extension, Sclater et al (1980) and Royden & Keen (1980) proposed that stretching should thin the subcrustal lithosphere more than the crustal layer. It seems that stretching models are very adaptable to particular observations. However, in terms of a general mechanism for lithospheric thinning, they fail because of the narrow restrictions of the concept when compared with typical observations. The strict equivalence between vertical thinning and horizontal stretching guarantees that no sign of lithospheric thinning exists prior to the onset of stretching, no matter whether one assumes instantaneous or time-dependent stretching. Thus, domal uplift associated with volcanism and even rifted domes are explicitly excluded by such a concept. This is a contradiction that even applies to the North Sea Basin, where thinning has been described by stretching; broad domal uplift, graben formation, and vigorous volcanism are well documented for the initial phase before basin subsidence (Ziegler 1982).

Large nonuniform stretching over a finite distance, made necessary by thermal constraints on such a model, will require the regional exchange of matter, which suggests convective flow processes.

The exclusion by stretching models of any preextension phases of thinning necessitates extreme nonuniform stretching or, equivalently, high-strength heat sources. Therefore these models suggest that lithospheric thinning is a rapid process.

Reheating Models

Reheating hypotheses treat the anomalous heat flow at the surface associated with lithospheric thinning as a temperature or heat-flow increase at the base of the lithosphere. It has been suggested that the source is either within or below the lithosphere and might correspond to a mantle plume or hotspot (Figure 5b). These models describe transient or steady-state solutions of thermal conduction into a lithospheric plate, moving over a stationary source, beneath or within the model layer.

Birch (1975) modeled the surface manifestations of a hotspot by a formulation of conductive heating of a lithosphere moving relative to a point source. The steady-state solution is attained under boundary conditions in which it is assumed that the source heats up the underlying mantle as well. The results indicate that the shape and size of the induced anomaly in terms of heat flow and internal temperature are critically controlled by the assumed velocity. This interpretation tends to overestimate the heat source, which is required to be about 10^{-4} of the total heat flow of the Earth. Gass et al (1978) discussed this approach on the basis of variable lithospheric thickness, while Sandwell (1982) expanded the model representation to three dimensions and a distributed stationary source. Nakiboglu & Lambeck (1985) added transient effects into the concept of reheating.

A variation of the reheating mechanism has been considered by Crough & Thompson (1976). They modeled thinning by monitoring the upward movement of a specific isotherm that corresponds to a physical lower boundary of the lithosphere indicating melting. Spohn & Schubert (1982), on the other hand, adopted an explicit energy conservation criterion by considering partial melting at the moving lithosphere-asthenosphere boundary in response to advective heat sources.

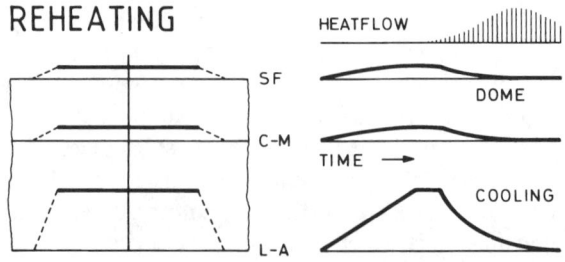

Figure 5b Principles of lithospheric thinning by reheating of the lithosphere due to conduction or source injection. Scheme analogous to that of Figure 5a. Heating-cooling sequence causes preferentially episodic doming because of the absence of mass transport.

The conductive reheating models of lithospheric thinning predict surface uplift due to thermal expansion and thinning of the layer base by thermal erosion that occurs immediately with the increase of heat flux at depth. Modeled surface heat flow, however, is delayed by some tens of millions of years for reasonable model parameters, and models that include interaction between the source and a moving lithosphere may have the rate of uplift and thinning limited by the velocity at which the interaction can occur.

Clearly, thermal conduction of rocks is slow, and thus this process represents the slowest possible way to transfer heat into the lithosphere. In order to be consistent with a lithosphere thinned by up to about 80%, a substantial increase of heat flux by a factor of 8–10 is necessary. The extreme source strength required implies extremely rapid rates of thinning, by definition of the concept. In other words, reheating allows large amounts of thinning only for extremely high source strengths. This is frequently the reason why predicted heat flow is overestimated. The presence of a molten asthenosphere in an elevated upstream position would cause severe gravitational instabilities by density inversion that would rise faster than any thermal perturbation (Neugebauer et al 1983).

Source Injection Models

Conductive reheating of the lithosphere is the slowest possible way to transfer heat and thus leads to overestimation of the assumed sources. However, a more highly efficient reheating mechanism would allow the source strength to be reduced to more reasonable values. Lachenbruch & Sass (1978) suggested such a model on the basis of homogeneous stretching of a lithospheric layer, where the extension is accommodated by intermittent intrusion of dikes that are homogeneously distributed throughout the layer. Although vertical convective transport is achieved with the source model, no solid-state convective return flow is induced because the space that allows asthenosphere uprise is created by lateral stretching. One would expect this approach to be most efficient for the representation of high heat-flow anomalies as a function of extension. The model, however, is critically dependent upon the poorly constrained injection rate. If we assume random magma entry into the lithosphere, then thinning can be calculated by means of a simple energy conservation argument (Withjack 1979). The model produces surface uplift and rapid thinning comparable to observations of the Hawaiian arch, and a fractured lithosphere is not required.

Birch (1979) considered a model in which the lithospheric mantle is detached from the base of the crust. He assumed an initially elongated conduit connecting the underlying asthenosphere with the base of the continental crust. He further assumed that the lithospheric mantle layer is denser than the asthenosphere. Since the conduit pressure is greater than the lithostatic pressure along a weakened crust-mantle boundary, the model describes crustal material being driven horizontally away from the conduit. The loss of the cold mantle boundary would cause increased heat flow and uplift.

The generally assumed density inversion for the asthenosphere, however, is a critical assumption in the modern view of magma compressibility, and thus the driving mechanism might be questioned. The above concepts provide higher strength than nonconductive reheating sources, but all the additional specifications of the source are arbitrary and not part of an expanded self-consistent theory or based on observations. Emerman & Turcotte (1983) solved stagnation flow models for the impingement of a hot cylindrical mantle plume fluid against a cold lithospheric half-space. Their results raise doubts about the ability of mantle plumes to thin the lithosphere without the aid of a preweakened lithosphere.

Both conductive reheating models and modifications using presumed penetrative source characteristics convincingly demonstrate the need for incorporation of dynamic control of thinning processes into the models.

Diapiric Penetration Models

The previously discussed models considered stretching and reheating responsible for lithospheric thinning. Observations suggest that lithospheric thinning is a dynamic-thermal process that involves the transport of both matter and heat. Models incorporating both are rare because of the complexity. There are only a few quantitative treatments of these problems that require the use of numerical techniques.

The motion of a diapir is driven by buoyancy forces due to the difference in densities between the diapir body and the adjacent medium (Figure 5c).

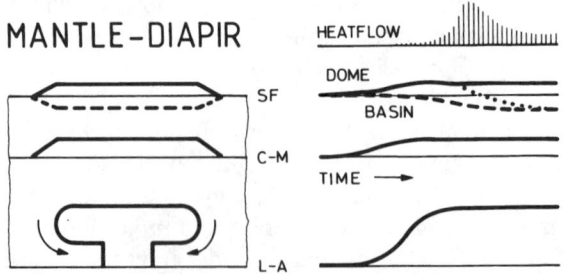

Figure 5c Principles of lithospheric thinning by the uprise of a mantle diapir. Scheme analogous to that of Figure 5a. Models are two dimensional. Dome and/or basin structures might be induced as a consequence of upwelling mantle material at C–M and L–A and of the mechanical properties of the crust (Neugebauer 1983).

Woidt & Neugebauer (1980) have shown by means of a two-dimensional numerical model of large deformations of rocks that the developing interface instability of a two-layer model with inverted density stratification is controlled by structural and mechanical parameters. Thereafter the shapes and sizes of rising diapir structures are controlled by the viscosity ratio between the diapir and the medium and by the ratio of layer thicknesses between overburden and source, respectively. The uprise velocity of a diapir is a function of the density contrast and the absolute viscosity of the host rock layer that is penetrated by the diapir. The growth of a diapir begins with a sluggish initial phase, followed by rapid exponential growth. It ends with stagnation and consolidation due to the fixed boundary at the top.

The dynamic diapir concept has been applied to the geotectonic problems of rifting and plateau uplift by Neugebauer (1983) and Neugebauer et al (1983), respectively. One major conclusion of these investigations was that diapiric uprise rates between 5–10 km Myr^{-1} might be attained by inverted density contrasts between 0.0001 and 0.1 g cm^{-3} with corresponding mantle-lithosphere viscosities between 10^{18} and 10^{22} Pa s. This implies that the lithosphere-asthenosphere system is very sensitive to mechanical and gravitational conditions besides the additional influence of the temperature. Furthermore, it has been shown that the dynamic crustal response to diapir invasion into the lower lithosphere is either synchronous updoming for high-strength crust or stretching in the sense of necking for low-strength crust.

Bridwell & Potzick (1981) considered the coupled thermomechanical continuum representation. However, they were only able to model a short period of diapiric movement starting from a given structure because of the numerical limitations to small deformations.

Diapir models for lithospheric thinning easily overestimate surface uplift and surface heat flow because of an overly efficient transport of matter and heat whenever it is assumed that a single diapir is responsible for the lithospheric thinning. The assumption of a single diapiric event might also be in conflict with the long-term tectonic development at thinning sites and with the observed periods of volcanic activity.

An alternative, quantitative approach to diapiric penetration with melting was performed by Emerman & Turcotte (1984). They assumed that a body of a given shape might melt its way through a solid medium. An unbounded conversion of potential energy into thermal energy is found to be coupled onto a critical size of a sphere. Although Emerman & Turcotte adopted a complex physical approach, the assumption of a given shape for a diapir eliminates the control of size and shape of density instabilities by structural and mechanical parameters of the model. The idea that lithospheric thinning is accomplished by strong secondary convection, which is enhanced by the presence of a weakened horizontal layer, has been considered by Yuen & Fleitout (1985). This mechanism is suggested in response to an initiating diapiric uprise of material.

Igneous Intrusion Model

The most instructive information on lithospheric thinning is provided by igneous volcanic and plutonic activity. Such activity indicates transport of matter and heat through the lithosphere over the total period of lithospheric thinning, a period that is obviously much longer than the tectonic surface manifestations because of the existence of a domal basement uplift associated with volcanism. The rate of magma ascent is high enough to prevent cooling and recrystallization before consolidation. The bimodal character and depth-temperature ranges of magma generation indicate two typical source regions that interact (Wyllie 1981). One is the lithospheric mantle, and the other is the lower crust; the deeper source clearly affects the shallower one. While the lower crustal transition is interpreted by a mixed mafic and ultramafic lithology, the lithospheric mantle source region is related to an increase in temperature. In addition to this vertical scale, there also exist characteristic patterns in the horizontal plane at different depth levels. The foci of volcanism coincide with the predominantly circular areas of domal uplift and the congruent lower crustal lithologic transition zone (Figure 1).

Early volcanism is related to the central position of the region of epeirogenic movement and congruently to that of the transitional crust-mantle boundary. From here, volcanic activity might spread outward despite the control of fault tectonics (Eaton 1982, Neugebauer & Walter 1983). Volcanic activity revives again at a later period at the previous central position. These clusters of igneous extrusive and intrusive activity are of limited diameter (~ 500 km). Phases of epeirogenic uplift are probably associated with periods of igneous activity.

Quantitative models of the ascent and emplacement of igneous intrusions under upper-mantle conditions have been proposed by Reuther et al (1985) and Neugebauer & Reuther (1987). The calculations are based on a representation of the heat-transport equation comprising convection, conduction, shear heating, and adiabatic temperature changes. The thermal equation is coupled with the Navier-Stokes equation. Variable viscosity is assumed to be temperature dependent. In Figure 6, a model box covering a depth range of 160 km and a width of 640 km is shown. The temperature varies according to geotherm representations between 0°C and 1200°C from top to bottom, respectively. Viscosity varies across the vertical axis of the model according to the temperature. The range of viscosity variation has been taken to be three to seven orders of magnitude. In order to approximate a finite volume of source material that has accumulated at depth with inverted density, a closed compositional phase boundary has been introduced. The "melt volume" can be assumed to be different in size and in depth of origin for each calculation. The model represented in Figure 6 shows the sequential uprise of an initially elliptical source region from an original depth of 140 km driven by a constant density contrast of 0.1 g cm^{-3}. The temperature field during uprise and emplacement of the intrusion is contoured by isotherms with 200°C intervals.

The intrusion of the body into the lithosphere reaches its highest rate of ascent within the lithospheric mantle. For instance, a minimum viscosity of 10^{18} Pa s in the lower section allows a maximum rate of the order of 100 km Myr^{-1}. This rate also represents the limiting minimum rate for efficient convective transport of heat exceeding the influence of conductive cooling. The viscosity increase, which is three orders of magnitude for the model shown in Figure 6, that occurs with shallower depth and is a function of the temperature causes a rapid decrease of the rate of ascent; this in turn leads to an emplacement of the intrusion at a lower crustal position between 20–30 km depth. These models demonstrate that the lithosphere is heated by the diapiric intrusions passing through it. When the intruding body approaches shallower depths and therefore a higher viscosity contrast, its diameter is no longer optimum with respect to achieving the maximum possible rate of uprise. Thus, the adjustment to the optimum rate of uprise leads to a splitting of the intruding volume into smaller volumes that become flatter and thinner. Accretion due to intrusion is therefore made possible by the thermal potential that is carried upward, and this accretion is controlled by the gradational change of the mechanical properties of lithospheric rocks. Besides the described processes of the emplacement of intrusions at shallow depths, the lower lithosphere additionally critically controls the size of the source volume of ascending material. Only small assumed source volumes are able to pass the lower lithosphere as a single intrusion.

Beyond a critical size of the source region, the diapiric structures undergo splitting into optimum diameter substructures at the assumed source depth. This phenomenon can be observed from the numerical models in addition to the later emplacement at shallower depth. The process is dependent on the aspect ratio between the dimensions of the source region and of the overlying lithospheric layer for given viscosities and densities. Intrusive transport of magmatic material is therefore limited to small volume sizes at the source region depth by the mechanical and structural conditions of the lithosphere. The emplacement of these intrusions leads to further subdivision into tiny lenses, as demonstrated in Figure 6. The latter process results from the increasing viscosity contrast between the intrusion and the surrounding rock during uprise at shallow depth.

The intrusive uprise of buoyantly driven melt volumes through the lithosphere causes pipe structures in terms of excess heat, which are shown in Figure 6 by the uplifted isotherms. This "convectively" heated part of

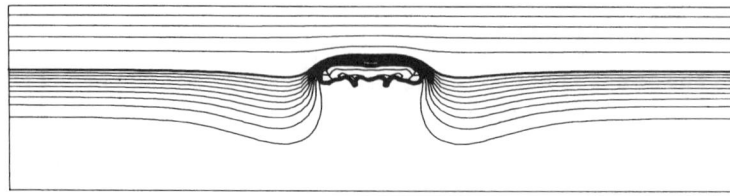

Model 5332 / Timestep 90

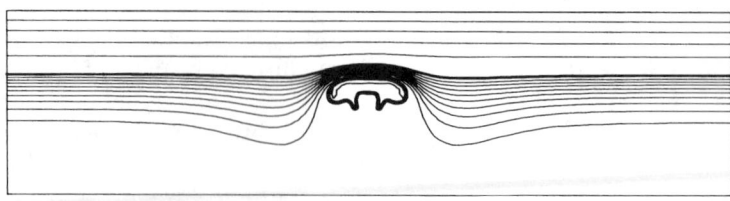

Model 5332 / Timestep 75

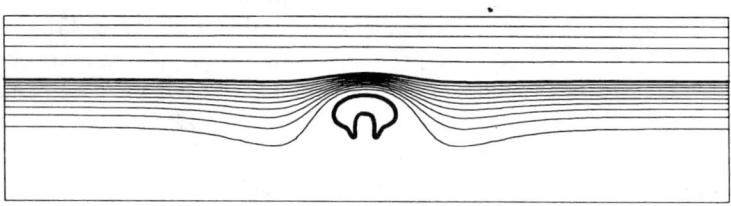

Model 5332 / Timestep 60

Figure 6 (*above* and *facing*) Model calculations on the uprise of a finite volume of "magma" with inverted density under upper-mantle conditions. Intrusion starts at 140 km depth as a horizontally oriented elliptic body. Isotherms are contoured in 200°C intervals between top (0°C) and bottom (1200°C), with the lower temperature interval between 1000 and 1200°C resolved by 20°C steps. For further explanation of the model, see text.

the lithosphere causes thermal expansion and mechanical weakening of the lithosphere before conductive cooling. Conductive heat loss with time to the adjacent parts of the structure will widen the diameter of the pipe and thus its implied effects over the total depth range of the intrusive transport. The induced surface heat flow by emplaced intrusions is of moderate to low amounts and is delayed because of the conductive predominance of transport in the remaining crustal top layer.

Future calculations will not only consider the focusing effect of pre heated lithosphere for subsequent events, but also the influence of the frequency of subsequent intrusive events on the lateral extension of the

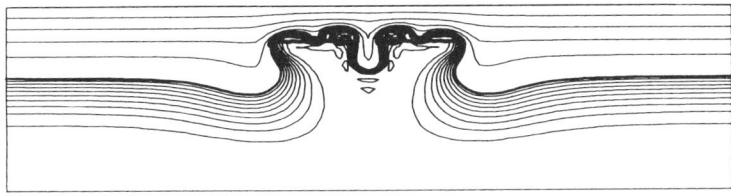

Model 5332 / Timestep 180

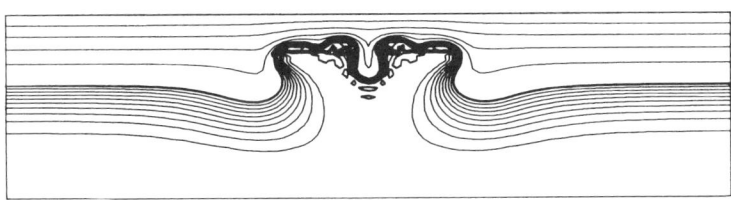

Model 5332 / Timestep 165

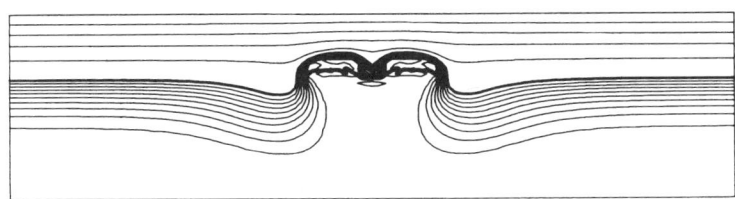

Model 5332 / Timestep 105

induced thermal pipe. One might argue that multiple intrusions will cause the accumulation of small volumes of melt provided by a finite source region during uprise within a wide thermal pipe in the lithosphere. The emplacement and consolidation of this material in a lower or subcrustal position will build up a transitional lithologic layer having a diameter that corresponds to the thermal anomaly throughout the underlying lower lithosphere.

A group of igneous intrusions passing through the lithospheric mantle over a geological period of time might be considered to be a self-consistent system of lithospheric thinning (Figure 7). Beginning with a central, initial igneous event, any productive source will thin the lithosphere appropriately by both conductive heating and the intrusive exchange of material. Thinning is thus immediately apparent over the total subcrustal thickness of the lithosphere according to the initial event. However, the diameter of the thinned part of the lithosphere should be very small at this early stage of thinning. As thinning and extension increase, the problem becomes horizontally directed (Figure 7). The intrusive penetration of matter and heat allows thinning to the degree and the rate that the limitations of the magma source allow. For a limited source productivity, heat transport will dominate lithospheric thinning and determine the tectonic consequences in terms of the dome-to-basin genetic sequence. For ascending convection from the mantle, material transport will dominate, and the large amount of lithosphere replaced by mantle material will be associated with the fragmentation of the lithosphere and the subsequent formation of oceanic structures.

Based on the observations and the small-intrusion model, lithospheric

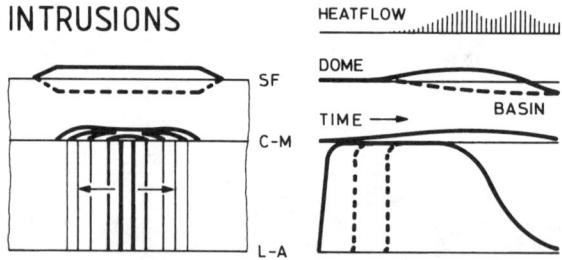

Figure 7 Principles of lithospheric thinning by igneous intrusions. Scheme analogous to that of Figure 5a. The initial state is a central thermal pipe over the subcrustal lithosphere that is achieved by the intrusive transport of matter from the L–A position to the C–M position. Later intrusive events travel the same distance and widen the region of thinning into the horizontal direction. As the intrusions consolidate in the lower crust, episodic events lead to an increase in heat flow and regional epeirogenic movements.

thinning is a rather slow process. In particular, the initial phase associated with domal uplift and occasional rifting can cover well-documented time spans between 50 to 100 Myr without substantial extension. Regional phases of episodic uplift or subsidence are due to repeated reheating by later intrusions. The oval-to-circular shape of the thinned lithosphcre in the initial phase is most likely controlled by the energetically preferred and dynamically induced thermal pipe and its growth with time. The congruence of areas of epeirogeny, heat anomalies, and exchange of matter indicated by different gravitational signatures is implied in the igneous intrusion model. Finally, the limitation of extension of the lithospheric mantle and crustal layer to a confined unit area of standard size agrees with the observations at individual domal and basin structures. The clear identification of these units along rift and basin belts reveals how each unit contributes to a regional tectonic system of lithospheric thinning.

CONCLUSIONS

Magmatic processes provide ample evidence for the joint transport of matter and heat through the lithosphere. Igneous activity is frequently observed long before, as well as in association with, significant tectonic manifestations of extension at the brittle upper crust. Therefore, igneous processes are most likely the signatures of lithospheric thinning, which appears to be a rather slow process with short-term episodes of increased activity. Based on these and additional theoretical arguments, it can be concluded that lithospheric thinning is associated with dome, rift, and basin structures that represent stages of the thinning process and thus form a genetic sequence.

Quantitative models are adopted in order to describe lithospheric thinning by means of its controlling parameters. This necessity for simplification has led to a tendency to represent lithospheric thinning by a single principal parameter, such as extension, temperature, or buoyancy forces. In addition, models have been restricted to selected stages or phenomena of thinning. However, both the models and the observations demonstrate convincingly that lithospheric thinning is controlled by several interdependent parameters. Thus, adequate representation of this complexity allows both a more self-consistent approximation and a correlation of observations with model predictions.

ACKNOWLEDGMENTS

The results presented here also reflect the efforts of many of my students, whose work is gratefully acknowledged. My views on the dynamic-thermal processes involved in lithospheric processes have been deeply influenced by valuable discussions with Drs. D. S. Chapman, A. H. Lachenbruch, G. Schubert, and D. L. Turcotte. This work was partially supported by the Deutsche Forschungsgemeinschaft.

Literature Cited

Birch, F. S. 1975. Conductive heat flow anomalies over a hot spot in a moving medium. *J. Geophys. Res.* 80: 4825–36

Bird, P. 1979. Continental delamination and the Colorado Plateau. *J. Geophys. Res.* 84: 7561–71

Brace, W. F., Kohlstedt, D. L. 1980. Limits on lithospheric stress imposed by laboratory experiments. *J. Geophys. Res.* 85: 6248–52

Bridwell, R. J., Potzick, C. 1981. Thermal regimes, mantle diapirs and crustal stresses of continental rifts. *Tectonophysics* 73: 15–32

Browne, S. E., Fairhead, J. D. 1983. Gravity study of the Central African Rift system: a model of continental disruption. The Ngaoundere and Abu Gabra Rifts. *Tectonophysics* 94: 187–203

Chapman, D. S., Pollack, H. N. 1977. Regional geotherms and lithospheric thickness. *Geology* 5: 265–68

Crough, S. T. 1983. Hotspot swells. *Ann. Rev. Earth Planet. Sci.* 11: 165–93

Crough, S. T., Thompson, G. A. 1976. Numerical and approximate solutions for lithospheric thickening and thinning. *Earth Planet. Sci. Lett.* 31: 397–402

Deichmann, N., Ansorge, J. 1983. Evidence for lamination in the lower continental crust beneath the Black Forest (southwestern Germany). *J. Geophys.* 52: 109–18

Dziewonski, A. M., Anderson, D. L. 1981. Preliminary reference Earth model. *Phys. Earth Planet. Inter.* 25: 297–356

Eaton, G. P. 1982. The Basin and Range Province: origin and tectonic significance. *Ann. Rev. Earth Planet. Sci.* 10: 409–40

Emerman, S. H., Turcotte, D. L. 1983. Stagnation flow with a temperature-dependent viscosity. *J. Fluid Mech.* 127: 507–17

Emerman, S. H., Turcotte, D. L. 1984. Diapiric penetration with melting. *Phys. Earth Planet. Inter.* 36: 276–84

Gass, I. G., Chapman, D. S., Pollack, H. N., Thorpe, R. S. 1978. Geological and geophysical parameters of mid-plate volcanism. *Philos. Trans. R. Soc. London Ser. A* 288: 581–96

Hale, L. D., Thompson, G. A. 1982. The seismic reflection character of the continental Mohorovičić discontinuity. *J. Geophys. Res.* 87: 4625–35

Jarvis, G. T., McKenzie, D. P. 1980. Sedimentary basin formation with finite extension rates. *Earth Planet. Sci. Lett.* 48: 42–52

Keller, G. R., Lidiak, E. G., Hinze, W. J., Braile, L. W. 1983. The role of rifting in the tectonic development of the midcontinent, USA. *Tectonophysics* 94: 391–412

Lachenbruch, A. H., Sass, J. H. 1978. Models of an extending lithosphere and heat flow in the Basin and Range province. In *Cenozoic Tectonics and Regional Geophysics of the Western Cordillera, Geol. Soc. Am. Mem. 152*, ed. R. B. Smith, G. P. Eaton, pp. 209–50

McKenzie, D. P. 1978. Some remarks on the development of sedimentary basins. *Earth Planet. Sci. Lett.* 40: 25–32

Morgan, P. 1983. Constraints on rift thermal processes from heat flow and uplift. *Tectonophysics* 94: 277–98

Morgan, P. 1984. The thermal structure and thermal evolution of the continental lithosphere. *Phys. Chem. Earth* 15: 107–93

Morgan, P., Baker, B. H., eds. 1983. *Processes of Continental Rifting, Dev. Geotectonics*, Vol. 19. Amsterdam: Elsevier. 680 pp.

Nakiboglu, S. M., Lambeck, K. 1985. Thermal response of a moving lithosphere over a mantle heat source. *J. Geophys. Res.* 90: 2985–94

Neugebauer, H. J. 1983. Mechanical aspects of continental rifting. *Tectonophysics* 94: 91–108

Neugebauer, H. J., Reuther, C. 1987. Intrusion of igneous rocks—physical aspects. *Geol. Rundsch.* 76: In press

Neugebauer, H. J., Walter, R. 1983. Volcanism, lithospheric thinning and tectonic implications. *Terra Cognita* 3(2–3): 117

Neugebauer, H. J., Woidt, W.-D., Wallner, H. 1983. Uplift, volcanism and tectonics: evidence for mantle diapirs at the Rhenish Massif. In *Plateau Uplift*, ed. K. Fuchs et al, pp. 381–403. Berlin: Springer-Verlag. 411 pp.

O'Reilly, S. Y., Griffin, W. L. 1985. A xenolith-derived geotherm for Southeastern Australia and its geophysical implications. *Tectonophysics* 111: 41–63

Reuther, C., Neugebauer, H. J., Christensen, U. 1985. Diapiric intrusions and the transport of matter and heat. *IASPEI Assem., 23rd, Tokyo*

Rigden, S. M., Ahrens, T. J., Stolper, E. M. 1984. Densities of liquid silicates at high pressure. *Science* 266: 1071–74

Royden, L., Keen, C. E. 1980. Rifting process and thermal evolution of the continental margin of eastern Canada determined from subsidence curves. *Earth Planet. Sci. Lett.* 51: 343–61

Sandwell, D. T. 1982. Thermal isostasy: response of a moving lithosphere to a distributed heat source. *J. Geophys. Res.* 87: 1001–14

Sclater, J. G., Jaupart, C., Galson, D. 1980. The heat flow through oceanic and continental crust and the heat loss of the Earth. *Rev. Geophys. Space Phys.* 18: 269–311

Sleep, N. H., Nunn, J. A., Chou, L. 1980. Platform basins. *Ann. Rev. Earth Planet. Sci.* 8: 17–34

Spohn, T., Schubert, G. 1982. Convective thinning of the lithosphere: a mechanism for the initiation of continental rifting. *J. Geophys. Res.* 87: 4669–81

Stocker, R. L., Ashby, M. F. 1973. On the rheology of the upper mantle. *Rev. Geophys.* 11: 391–497

Stolper, E., Walker, D., Hager, B. H., Hays, J. F. 1981. Melt segregation from partially molten source regions: the importance of melt density and source region size. *J. Geophys. Res.* 86: 6261–71

Van Schmus, W. R., Hinze, W. J. 1985. The midcontinent rift system. *Ann. Rev. Earth Planet. Sci.* 13: 345–83

Walcott, R. I. 1973. Structure of the Earth from glacio-isostatic rebound. *Ann. Rev. Earth Planet. Sci.* 1: 15–37

Withjack, M. 1979. A convective heat transfer model for lithospheric thinning and crustal uplift. *J. Geophys. Res.* 84: 3008–22

Woidt, W.-D., Neugebauer, H. J. 1980. Finite element models of density instabilities by means of bicubic spline interpolation. *Phys. Earth Planet. Int.* 21: 176–80

Wyllie, P. J. 1981. Plate tectonics and magma genesis. *Geol. Rundsch.* 70: 128–53

Yuen, D. A., Fleitout, L. 1985. Thinning of the lithosphere by small-scale convective destabilization. *Nature* 313: 125–28

Ziegler, P. A. 1982. *Geological Atlas of Western and Central Europe*. Amsterdam: Elsevier. 130 pp.

ARCHITECTURE OF CONTINENTAL RIFTS WITH SPECIAL REFERENCE TO EAST AFRICA

B. R. Rosendahl

Project PROBE, Department of Geology, Duke University, Durham, North Carolina 27706

INTRODUCTION

The Earth's continental skin has been cracked many times in many places during the last several billion years. The resulting rifts number in the hundreds (Burke 1978, Ramberg & Neumann 1978a, Easton 1983, Milanovsky 1983, Ramberg & Morgan 1984), and they occur on virtually every continent. Although the actual land area occupied by rift structures is relatively small (less than 2% of the surface area of continents), at least 3,000,000 words on 10,000 pages in 700 papers have been written on the subject. If colligated, such a body of literature would be comparable to a set of Britainnica encyclopedia. There are many reasons for this attention, but current interests revolve around two main points: (*a*) Continental rifts are a precursor of ocean basins, and (*b*) much of the petroleum yet to be located is either directly or indirectly related to rift structures.

The vastness of the rift literature combined with manuscript limitations make it inappropriate to undertake a full review of rifts here. Even a restriction to African rifting could result in 50 printed pages of citations. Clearly, another approach is needed. The approach taken here is to focus on the morphology and structure of East African rift basins from the perspective of a rifting theme that is being developed by the author and his associates. We then explore some of the more salient implications of the theme, testing them against prevailing doctrines, sentiments, or opinions. The end product is a tightly focused review of African rift morphology and structure packaged within a single architectural theme. It might be noted that this approach automatically eliminates from discussion those topics that do not bear directly on the structural expressions of rifting (e.g. certain aspects of petrogenesis).

NOMENCLATURE FOR RIFTS

The nomenclature of the East African Rift (and of continental rifts in general) can be confusing and misleading. For example, most references to the East African Rift strictly apply only to the Kenya Rift, not the entire rift proper. The use of the word "rift" can refer either to something as large as the entire East African System or to features as small as the Lake Albert trough. The term "rift-valley" can describe something either larger than Lake Malawi or smaller than a football field. Before progressing further, it is useful to establish a more natural, "generic" nomenclature than those available to date. A multitiered scaling terminology is adopted here, building around a theme of blocks, within units, within zones, within branches, within systems. Throughout this discussion, comparisons within or between rift systems are made only at comparable generic scales.

System and Branch Scales

The largest rift scale to be dealt with here is the "system." Most workers agree that at least several rift systems have been operative in Africa since the Permian, but it is not always clear which rift elements belong to which system. The problem is illustrated in Figure 1, which shows that the Tertiary Rift System in East Africa apparently overprints a number of older rift structures. These older structures are mostly erosional remnants of Permo-Triassic to Jurassic Karroo troughs. The incompleteness of exposures, combined with poor age control and lack of subsurface information, makes it difficult to define the system(s). For this reason, the system scale is applied herein only to Cenozoic rifting in Africa.

The branch scale is the first-level subdivision of a rift system. It has long been recognized that the Tertiary System of East Africa is divided into Eastern and Western branches (Gregory 1896). Most workers have attributed the bifurcation to a deflection of rifting stresses around the Tanganyika Shield (e.g. McConnell 1967). It is significant that the division into branches coincides with marked differences in certain rifting traits. Most notable is the extreme difference in the quantity of volcanic effluents; the Eastern Branch is one of the "wettest" continental rift branches in the world, whereas the Western Branch is among the "driest" (Mohr 1982). In this regard, the contrast between these branches is as great as that between any two branches from any two rift systems. It is likely that the

ARCHITECTURE OF RIFTS 447

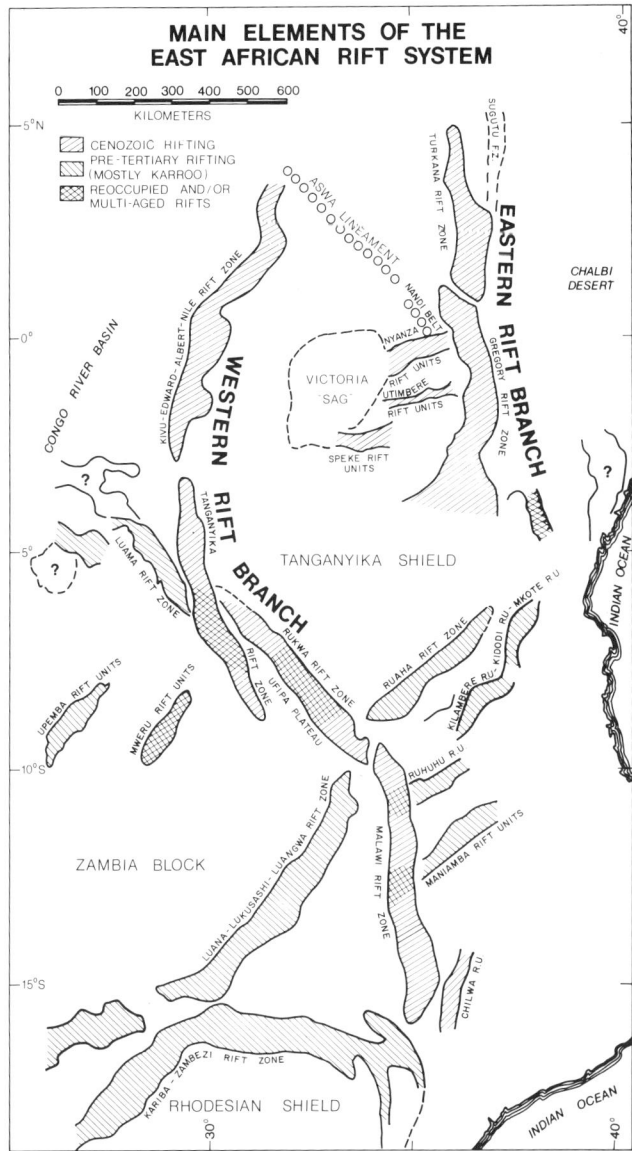

Figure 1 Map portfolio. Figure 1a (*above*) shows the main rift elements, especially zones. Note that the Malawi Zone is associated with the Western Branch, which overprints pre-Tertiary zones. Zonal terminations usually coincide with major kinks, offsets, or bends. Figure 1b shows sedimentary and volcanic cover. Figure 1c presents an interpretation of the inferred rift units throughout East Africa (see text for explanation of terminology). Figure 1d shows a possible trailing plate margin configuration for East Africa.

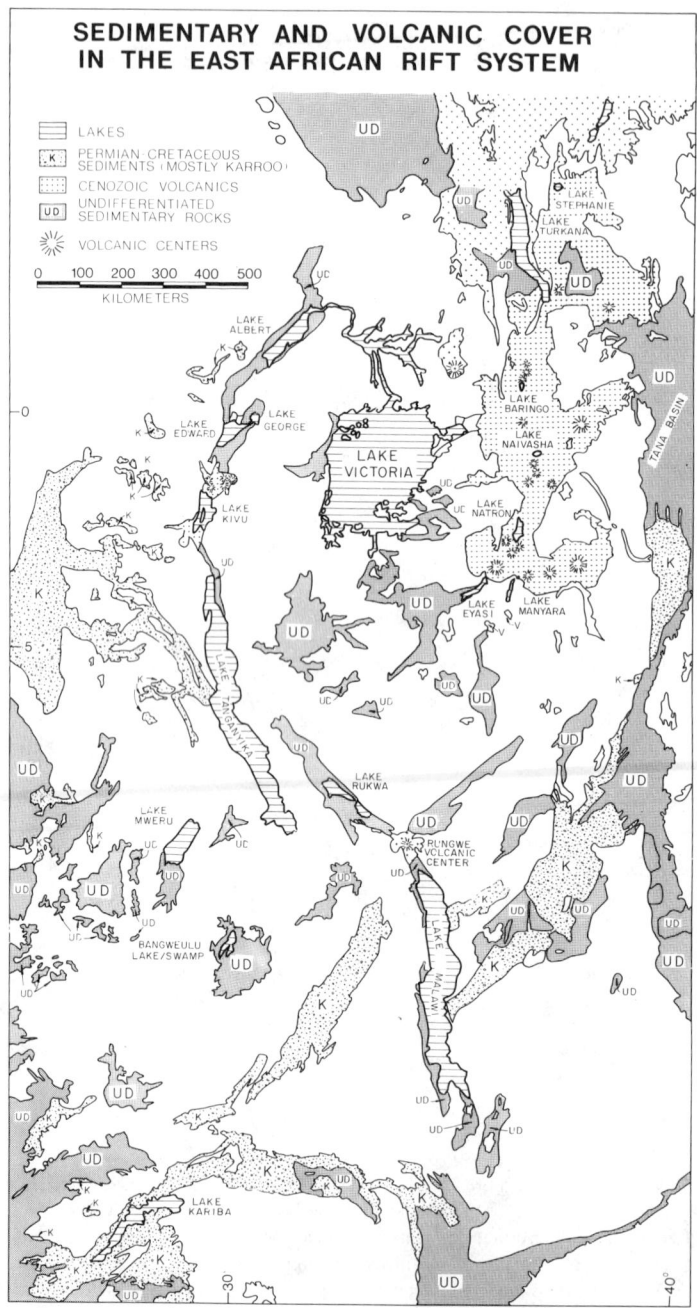

Figure 1b

ARCHITECTURE OF RIFTS 449

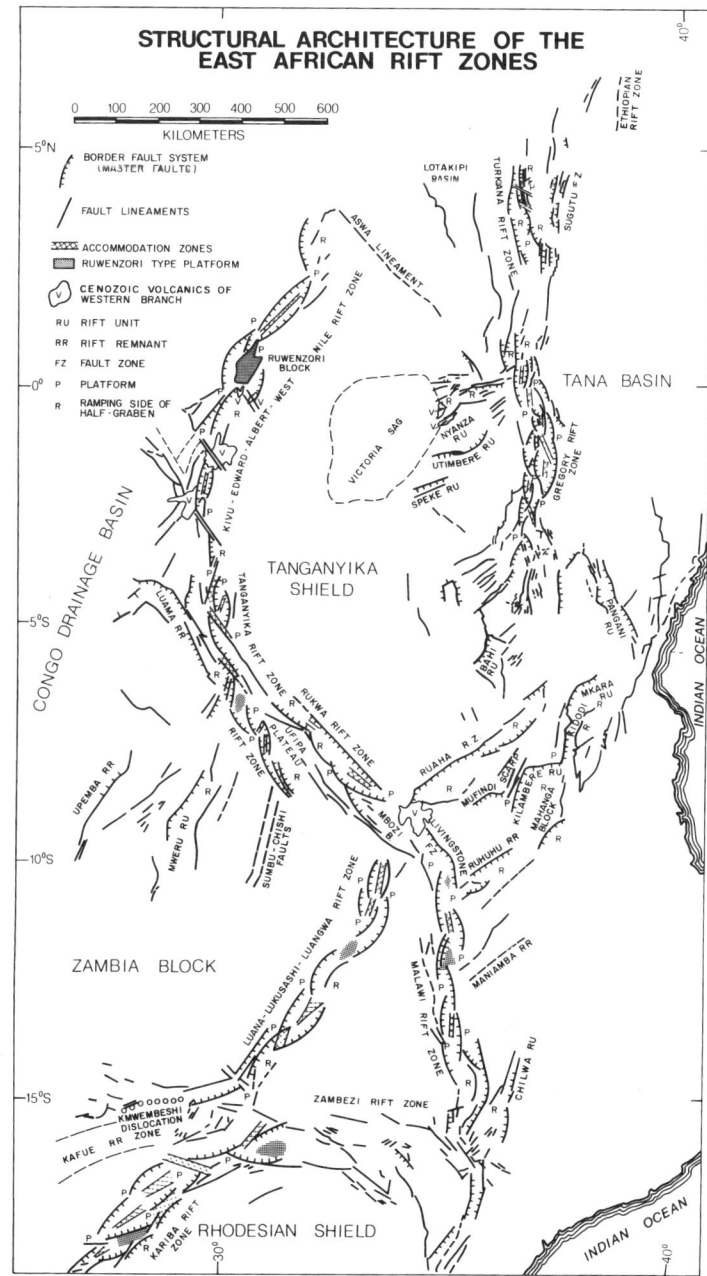

Figure 1c

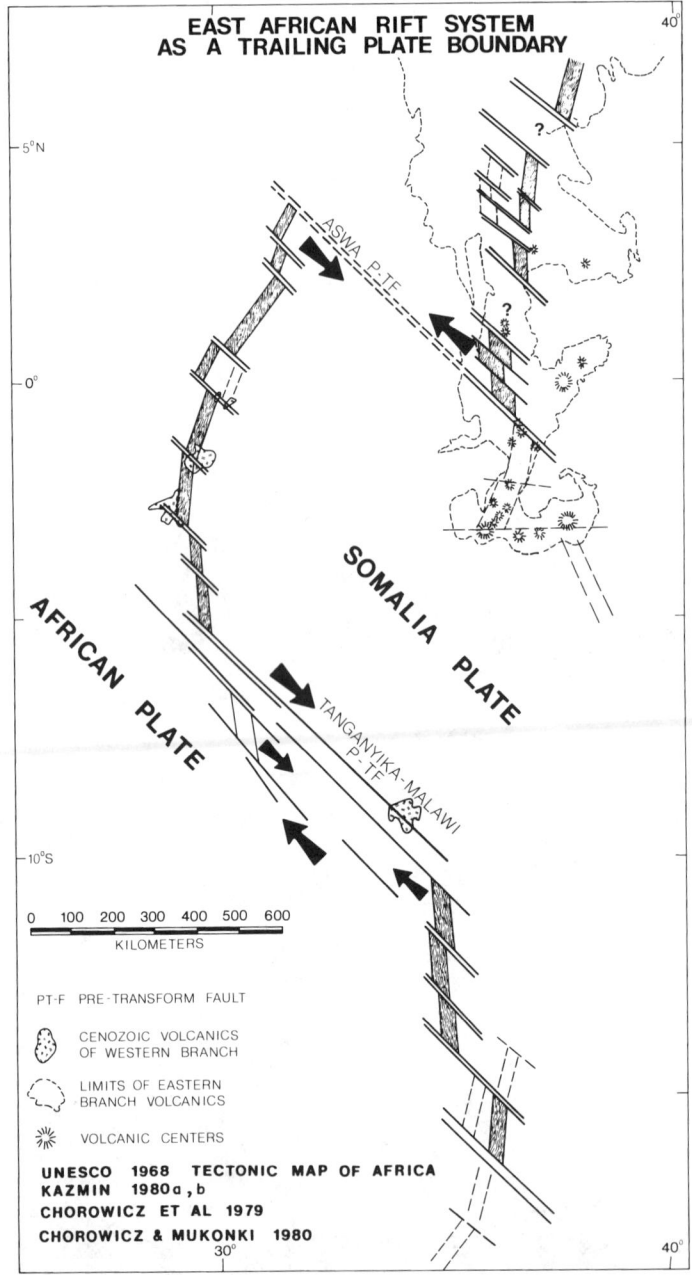

Figure 1d

difference in abundance is even greater if rift-related plutonic rocks also are considered (e.g. see Ramberg & Morgan 1984).

The Malawi Rift Zone (Figure 1) forms the "trunk" of the East African rift tree and might be expected to display affinities with both branches. This does not seem to be true; thus far, our studies of the Malawi Rift Zone show that it is very similar to the other Western Branch rift zones, especially Lake Tanganyika. Hence, the Malawi Zone is included here as part of the Western Branch.

Several of the pre-Cenozoic rift remnants shown in Figure 1 could be grouped into branches, but it is difficult to prove strict genetic, or even temporal, links. Possibly the Luano, Lukusashi, Luangwa, and Ruaha rift zones (Dixey 1937a, Drysdall & Kitching 1963, de Swardt et al 1965, Drysdall & Weller 1966, Vail 1967) belong to a rift branch that originates from a bifurcation of the Kariba Rift Zone (Molyneux 1909, Guernsey 1951, Gair 1959, Tavener-Smith 1960, Drysdall & Weller 1966, Chapman & Pollack 1977, Fairhead & Henderson 1977). The other limb would be the Central Zambezi Branch (Dixey 1944a, de Swardt 1965, Vail 1967). This situation might be analogous to the bifurcation of the Tertiary African Rift System around the Tanganyika Shield.

Rift-Zone Scale

Rift branches are subdivided into rift zones in ways that are morphologically obvious but mechanically poorly understood. Usually the zonal boundaries coincide with offsets, kinks or major changes in the trends of adjacent rift zones, all of which may be associated with centers of volcanism. Such changes are obvious in Figure 1a but less so in Figure 1c. The fact that rift zones worldwide generally have lengths of 500–700 km argues that the division into zones has some genetic validity, but it appears the relationship is neither as simple nor as direct as some believe. We return to this issue in a later section. The Cenozoic rift zones of East Africa are named below, along with listings of key citations. Locations are provided in Figure 1.

MALAWI RIFT ZONE   The Malawi Rift Zone includes all of Lake Malawi proper and the Shire Valley at the south end of the lake. The zone terminates at the north end in the Rungwe volcanic province (Andrew & Bailey 1910, Thiele & Wilson 1915, Dixey 1929, 1937b,c,d, 1944a, 1945a,b, 1956, 1959, Harkin 1960, Garson 1965, Lister 1967, Bloomfield 1966, 1968, Vail 1967, von Herzen & Vacquier 1967, Girdler & Sowerbutts 1970, Carter & Bennett 1973, Andrew 1974, Yairi 1977, Muller & Forstner 1973, Crossley & Crow 1980, Kaufulu et al 1980, Crossley 1982, Rosendahl & Livingstone 1983, Ebinger et al 1984, Ebinger & Rosendahl 1986).

RUKWA RIFT ZONE The Rukwa Rift Zone extends from the Rungwe volcanic province in the south, northwestward through the Rukwa trough and into the Karema gap to the east of Lake Tanganyika (Harvey & Teale 1933, Poussin 1935, Stockley 1938, Holmes 1944, 1965, McConnell 1950, Quennell 1951, 1960b, Spurr 1954, Spence 1954a,b, Harpum 1955a,b, Harkin 1960). The northern end of the Rukwa Zone is rather ill defined, and the connection to the Tanganyika or Luama rift zones is unclear.

TANGANYIKA RIFT ZONE The Tanganyika Rift Zone consists of Lake Tanganyika plus the plain of the Ruzizi at the north end of the lake (Cornet 1905, Holmes 1916, Krenkel 1925, Teale 1932, Veitch 1935, Poussin 1935, Willis 1936, Dixey 1945a,b, 1956, 1959, Capart 1949, McConnell 1950, 1972, Wayland 1952, James 1956, Cooke 1957, Sanders 1965, Yairi & Mizutani 1969, Degens et al 1971, Bath 1975, Chorowicz & Mukonki 1980, LeFournier 1980, Patterson 1982, Rosendahl & Livingstone 1983, Lorber 1984, Burgess 1985, LeFournier et al 1985, Rosendahl et al 1986, Sander 1986). Because our structural understanding of the Tanganyika Rift Zone exceeds that of any other African rift zone, many of the examples in this paper are drawn from this zone. In many ways we consider it to be the archetypical continental rift.

ALBERT RIFT ZONE The Albert Rift Zone includes the rifted troughs from Lake Kivu in the south through the West Nile Basin in the north, including Lakes Edward, George, and Albert (Holmes 1916, 1942, Wayland 1921a,b, 1931, Combe 1930, Holmes & Harwood 1932, 1937, Boustakoff 1933, Willis 1936, Dixey 1944a, 1945a,b, 1956, 1959, Davies & Bisset 1947, Lepersonne 1949, 1956, 1970, Davies 1951, McConnell 1951, 1959, Cahen 1953, 1954, Hopwood & Lepersonne 1953, Meyer 1954, Pallister 1954, 1955a,b, Ruhe 1954, Heinzelin 1955, Harris et al 1956, Pallister & Hepworth 1956, Peeter 1959, Sahama & Meyer 1958, Barnes & Hepworth 1961, von Knorring & du Bois 1961, Hepworth 1962, Sahama 1962, Furon 1963, Bishop 1965, 1969, Gautier 1965, 1967, Bishop & Trendall 1967, de Swardt & Trendall 1969, Macdonald 1968, 1969, King 1970, Degens et al 1973, Wong & von Herzen 1974, Stoffers & Hecky 1978, Chorowicz et al 1979, Chorowicz & Mukonki 1980).

GREGORY RIFT ZONE The Gregory (Kenya) Rift Zone extends from approximately Lakes Manyara and Eyasi in the south to Lake Baringo in the north (Gregory 1896, 1920, 1921, Obst 1913, Shackleton 1945, 1951, 1955, 1978, Dixey 1945a, Baker 1958, 1963a,b, 1965, 1970, 1971, McCall 1964, 1968a,b, McConnell 1967, 1972, 1974a,b, 1979, Tobin et al 1969, Girdler & Sowerbutts 1970, King 1970, Baker & Wohlenberg 1971, Baker et al 1971, 1972, 1978, Logatchev et al 1972, 1983, Mohr 1974, 1976, Fairhead & Walker 1974, Fairhead 1976, Logatchev 1978, Long & Backhouse 1976, Maguire & Long 1976, Mohr & Wood 1976, Mboya 1983). It is convenient to include the Nyanza (Kavirondo), Utimbere, and Speke rift units with the Gregory Zone, but the temporal and mechanical relationships of these units to the zone are not well defined. In comparison to the ends of most of the Western Branch rift zones, the terminations of the Gregory Zone are relatively broad and diffuse. The southern end splays into a region called the Tanzania Divergence, and the northern end partially disappears beneath volcanic cover. This makes the distinction between this zone and the Turkana Zone (see below) somewhat arbitrary.

TURKANA RIFT ZONE The Turkana Rift Zone extends from the Lake Baringo area in the south to the Omo River valley at the north end of Lake Turkana (von Hohnel 1894, Murray-Hughes 1933, Fuchs 1934, 1935, 1939, Dixey 1944a, Shackleton 1945, 1978, Pulfrey 1960, Dodson 1963, McCall 1964, Bishop & Trendall 1967, Howell 1968, Butzer & Thurber 1969, Walsh & Dodson 1969, Butzer 1970, Rhemtulla 1970, Baker et al 1971, 1972, Vondra et al 1971, Bowen & Vondra 1973, Mohr 1974, Behrensmeyer 1976, Mohr & Wood 1976, Truckle 1976, Vondra & Bowen 1976, 1978, Cerling & Powers 1977, Savage & Williamson 1978, Yuretich 1979, Bellieni 1981, Hopson 1982, Dunkelman 1986, Johnson et al 1986, Williamson 1986, Williamson et al 1986).

Like the ends of the Kenya Rift Zone, the southern end of the Turkana Zone is poorly constrained by any obvious surface manifestation. Volcanic cover obscures any clear line of demarcation between the northern end of the Kenya Zone and the southern end of the Turkana Zone.

The same sort of zonal distinctions can be applied to pre-Cenozoic rifting in East Africa (Figure 1). Perhaps the best example is the Luangwa Valley, which for most intents and purposes can be treated as a Karroo-aged Lake Malawi minus the water. The Kariba and Zambezi regions also are grouped into zones, although the latter case is not as well defined as the other two.

Rift-Unit Scale

The use of the term "fundamental unit" is reserved for discrete structural basins with typical lengths of 80–160 km and length-to-width ratios of about 2–4 (Reynolds 1984). Fundamental units are the true building blocks of rifts and should not be confused with the other scales described herein. The distinction between the zone and unit scale is made clear by comparing Figures 1*a* and 1*c*. The distinction between units and smaller-scale structure is shown in Figure 2. The various half-graben identified in Figure 2a are the rift units in the northern half of Lake Tanganyika. Other examples of fundamental units are Lakes Albert (Lake Mobutu Sese Seko) and Edward (Lake Idi Amin). Altogether, there are perhaps 70 of these fundamental units in the Tertiary System of East Africa (Figure 1c). A considerable portion of this article is devoted to the subject of fundamental units, particularly the ways in which they link together and the resultant structural expressions. Suffice it to say that typically these units become structurally interconnected in groups of 8 or more units. Such a genetic grouping of fundamental units creates a rift zone.

Rift units are also easily identified in the older rifting events of East Africa (Figure 1c). For example, the Ruhuhu and Maniamba Karroo troughs cutting across the Malawi Rift Zone (Bornhardt 1900, Andrew & Bailey 1910, Stockley 1932, Spence 1954a, Harkin 1955, McKinlay 1965, Vail 1967, Crossley & Crow 1980) are probably remnants of what were originally deeply subsided and well-developed fundamental units. Lake Mweru (de Swardt 1965, Drysdall & Weller 1966) appears to be a remnant of a single, more or less isolated fundamental unit. The Luangwa Valley is a series of well-preserved, Mesozoic-aged fundamental rift units that together formed a rift zone that bears a remarkable resemblance to the Malawi and Tanganyika rift zones (first noted by Dixey 1937a). The same can be said about the Lake Kariba area, which contains a series of fundamental units that link together to form a rift zone along the Zambia-Zimbabwe border. The Cretaceous graben of southern Sudan (J. Lambiase, personal communication) also are fundamental units linked together into a rift zone, although limitations in both rift exposures and subsurface geophysical coverage make the delineation of pre-Tertiary rift zones in Sudan uncertain at the present time. In Figure 1c I have identified about 55 pre-Tertiary rift units, but there are certainly many more that have gone undetected.

Rift Block Scale

The smallest structural unit of concern here is the "block." Sometimes they are expressed as ideal horst or graben, but more often they are tilted or rotated and they often change form or expression along their lengths. Johnson (1930) referred to these features as "tilt blocks, rift blocks, and step blocks"; Cotton (1950) used the term "fault-angle depression." The rotated and tilted blocks observed in the Bay of Biscay (de Charpal et al 1978, Montadert et al 1979) are synonymous with the blocks described herein. Angelier & Colletta (1983) call these structural units "first-order blocks" to differentiate them from still smaller-scale features.

The dimensions of blocks seem to vary from one rift unit to another,

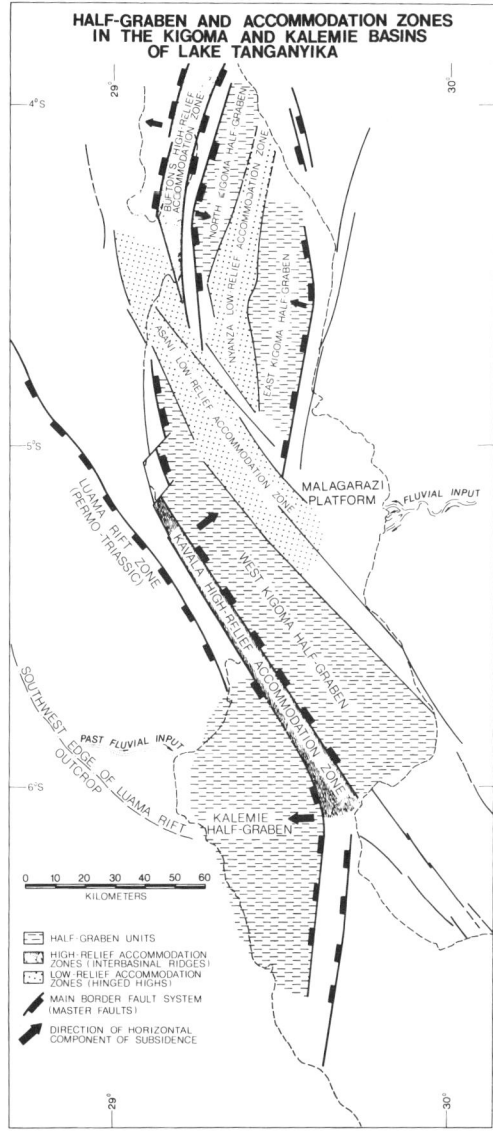

Figure 2 Structure of the northern half of Lake Tanganyika. Figure 2a (*above*) shows the basic architecture of half-graben and accommodation zones, Figure 2b (*over*) the infrastructure. Actual lengths of individual faults in Figure 2b are probably less than shown because of inability to discern small-scale transverse features, but widths and spacings are accurate. Note that low-relief accommodation zones are enclosed by high-relief zones.

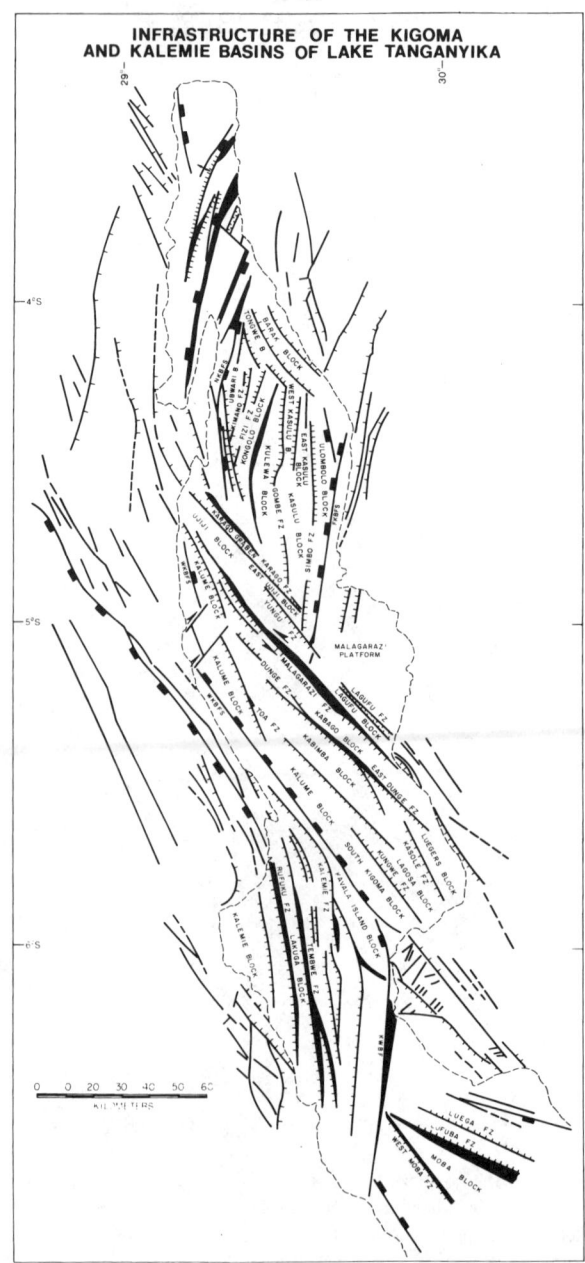

Figure 2b

but widths of 10 km are typical. It is not uncommon for blocks to display apparent length-to-width ratios of up to 10, but these ratios rarely exceed 4 where data density is adequate to delineate transverse block terminations (cf. Aydin & Nur 1982). The problem in deciphering length-width relations of blocks is illustrated by the beautiful block termination shown in a photograph in McGill & Stromquist (1979). Here the Devil's Lane block is terminated by an offset that would be almost impossible to delineate from any subsurface geophysics such as seismics. The significance of blocks lies in the recognition that they represent the infrastructure of rift units, not the fundamental rifting unit of Rosendahl & Livingstone (1983). Good examples of this scale of fracture in African rifts can be found in Ebinger et al (1984) and Figure 2b of this review.

ON THE ARCHITECTURE OF AFRICAN RIFTING

The Historical Perspective

The distinctive morphology of the East African Rift System is, of course, the very thing that attracted researchers to it in the first place. Suess (1891), de Martonne (1897), Gregory (1896, 1921), and de Lapparent (1898) were perhaps the first to recognize the basic structure of the African Rift—bounding normal faults, down-dropped wedge-blocks, broad arching, and horizontal extension. The term rift valley was used by Gregory to denote a long strip of Earth bounded by normal faults. In fairness to his predecessors, both this term and the roots of some of Gregory's ideas are owed to earlier efforts, mainly in the Rhinegraben by workers such as de Beaumont (1827, 1830, 1844, 1847), Suess (1891), de Lapparent (1887, 1898), and Neumayer (1887). H. Cloos (1939) is usually credited with establishing the mechanical validity of the tensile origin of rifts, but again some of the incentive and groundwork for Cloos' clay modeling can be traced back to the ideas espoused by de Beaumont (1827), Suess (1880, 1891), Gregory (1896), de Martonne (1897), de Lapparent (1898), Uhlig (1907, 1912), Obst (1913), Verweke (1913), Krenkel (1922), and Willis (1928, 1936). The term taphrogenesis was used by Krenkel (1922) in ascribing a tensile origin to the East African Rift. Nowadays, the term is essentially synonymous with rifting.

Almost every major geological precept or problem related to the East African Rift can be traced back to these early rift pioneers. These include the present-day controversy between "active" and "passive" rifting mechanisms [e.g. compare de Beaumont (1827) with Suess (1891, 1909)], the issue of fault behaviors at depth, and even the idea that the East African Rift is the outstanding example of a tear between two separating continents (e.g. Krenkel 1922, Willis 1936). Wegener (1912) described the East African

Rift System as the initial stage of continental breakup. Those who doubt how much we owe our early predecessors would benefit from a reading of the syntheses of Suess (1891, 1909), Gregory (1896, 1920), and Krenkel (1922, 1925).

Holmes (1944, 1965) synthesized much of the earlier work and formulated not only the basic picture of the East African Rift that is used today, but also many of the research problems that subsequent workers have pursued. No listing of African rift pioneers would be complete without mention of Dixey, McConnell, and Baker. Much of what is known about the geomorphology of the Western Branch, especially south of Lake Kivu, is owed to Dixey (1944a,b, 1945a,b, 1956, 1959). McConnell has published a long succession of papers dealing mostly with pre-existing structures and their influences on subsequent rifting (e.g. McConnell 1967, 1972, 1974a,b, 1978). Baker, of course, has played the modern keynote role in unraveling the chronology of the Eastern Branch, particularly the Gregory Rift Zone (e.g. Baker 1970, 1971, Baker & Wohlenberg 1971, Baker et al 1971, 1972, 1978).

Data recently acquired by the author and his associates provide an opportunity to update and to some extent revise the morphologic picture of rifting that has emerged from the above studies. We begin with the cross-sectional form of African rift units.

Cross-Sectional Morphology

The traditional morphologic picture of the African rift is one of approximate bilateral symmetry and strong two-dimensionality. This picture is not so much incorrect as it is misleading. Our studies in East Africa (Rosendahl et al 1982a,b, Rosendahl & Livingstone 1983, Patterson 1982, Ebinger et al 1984, Lorber 1984, Reynolds 1984, Reynolds & Rosendahl 1984, Sander et al 1986) demonstrate that asymmetry is the rule, not the exception; the archetypical cross-sectional form of African rift zones is a scalene to obtuse triangle. The steep side represents the major bounding fault system, and the ramping side is characterized by monoclines, steps, or flexures (Rosendahl et al 1986). Bally (1982) has made a similar deduction for the rift zones that he has examined. Hereafter, we use the terms "half-graben" and "full graben" to distinguish asymmetric cross-sectional forms from approximately symmetric cases.

It might be surprising to some workers that half-graben geometries predominate in African rift zones, but the existence of such forms is not an entirely new discovery. Careful readings of Gregory (1921), Krenkel (1925), Dixey (1929, 1956), Baker (1970), Baker et al (1972), Logatchev (1978), Crossley & Crow (1980), and Mohr (1982), among others, reveal that the existence of asymmetric geometries in East African rift zones has been known for many years. Half-graben forms in other rift environments have been described or implied by Florensov (1969), Mueller (1970), Illies (1971), de Charpal et al (1978), Kumarapeli (1978), Harding & Lowell (1979), Bally (1980, 1982, 1983), Bally & Snelson (1980), Weiblen & Morey (1980), Chenet & Montadert (1981), Effimoff & Pinezich (1981), Chenet et al (1982), Nunn (1982), Gibbs (1984), Lillie (1984), and Skilbeck & Lennox (1984). As noted by Bally (1982), the key to recognizing the preponderance of half-graben morphologies in rift zones lies with proper seismic reflection coverage. Hence, it should come as no surprise that some petroleum companies made the above discovery long ago.

Curiously, most experimental modeling of rift structures, whether done with clay (H. Cloos 1928, 1931, 1939, Wunderlich 1957, Oertel 1965, E. Cloos 1955, 1961, 1968, Elmohandes 1981), centrifuges (Ramberg 1963, 1967, 1971, 1978, Mulugeta 1985), wax (Oldenburg & Brune 1972, McGill & Stromquist 1979), cement (Freund & Merzer 1976), glass bottles (Bahat 1979, Bahat & Rabinovitch 1980), plaster (Sales 1976), or more theoretical means (Bott 1976, 1981, 1982a,b, Bott & Kusznir 1979, Withjack 1979, Artyushkov 1981, Neugebauer & Temme 1981, Bott & Mithen 1983, Keen 1985, Rowley & Sahagian 1986, among many others) usually emphasize full graben forms, or at least deformational symmetry. Sometimes this reflects a bias with regard to which "runs" an author chooses to emphasize. Other times it is an outgrowth of boundary constraints, materials, or other experimental parameters, assumptions, or limitations. In cases where the experimental full graben are both "real" and dominant, they usually occur as an end product, or at least late in the experimental rifting cycle. When available, the initial or early fracture expressions are often asymmetric. This is especially apparent in the studies of Stewart (1971), Ramberg (1971), Sales (1976, and shown in Stewart 1978), McGill & Stromquist (1979), and Mulugeta (1985).

Because half-graben seem to be the "type" morphology of African rift zones, whereas youthful ocean basins display grossly symmetric forms, we can conclude that the evolution from continental rift to oceanic spreading center must involve a progression toward symmetric cross-sectional morphologies. This conclusion is supported by our observation (see below) that full graben are a result of the linking together of half-graben (i.e. half-graben must exist first for linking to occur). A way around this conclusion is to postulate that the East African rift zones are not a precursory stage of successful spreading (cf. Fairhead & Browne 1981), but this is difficult to accept. Our group has now examined data from many rift zones that achieved partial or complete success, and in every case where subsurface data are adequate to resolve original geometry, half-graben forms predominate. This supports a similar contention by Bally (1982). We can take this to mean that regardless of their ultimate fate (successful spreading or eventual fossilization), there is nothing morphologically peculiar about the East African rift zones. It also means that half-graben morphologies must evolve into full-graben forms if successful spreading is to be achieved. Before pursuing the details of this evolution, we need a firmer understanding of the dimensional parameters of rifting.

Dimensionality of Rifting: Branch and Zone Scales

A few workers have given special attention to the issue of the third dimension of rifting (e.g. Illies 1972, Bally 1982, Gibbs 1984), but the matter usually has received only passing reference in the African rift literature. One reason is that active normal faults scar the landscape to a much greater extent than transverse fault systems, which are dominated by strike-slip motions. An exception may occur where volcanic features are associated with inferred cross-cutting structures (e.g. Fairhead 1980). A more fundamental reason is that relatively little is known about the length dimensions of rifting outside of a few well-surveyed areas, such as the North Sea and Rhinegraben. (The subsurface reflection data from these areas are largely proprietary.)

Virtually all geophysical and most dynamic models assume that rifts are strongly two dimensional on the larger scales, especially for the system, branch, and zone designations used here (e.g. Girdler et al 1969). Nonetheless, even on these larger scales the length dimension is not limitless. Consider that the distinction made here between Eastern and Western rift branches or between different rift zones implies the existence of terminations along strike, be they loss or change of surface expression, crosscutting structures, or offsets. Perhaps the most impressive terminations in the Africa System are those at the northern end of the Western Branch, in the vicinity of the West Nile Basin (Pallister 1971), and at the southern end of the Eastern Branch, somewhere in north-central Tanzania (James 1956, Dundas 1965, Pallister 1965). The Aswa mylonite belt [also termed the Aswa shear belt, Aswa Lineament, Aswa Dislocation Zone, and Assousa Lineament (Hepworth & MacDonald 1966, Hepworth 1966, Almond 1969, Macdonald 1968, de Swardt & Trendall 1969, King 1970, McConnell 1972, Mohr 1974a)] can be connected with the Nandi fault (Shackleton 1951) and the Mozambique front (Sanders 1965) to form a quasi-continuous connection between the two branches (Figure 1). The studies by Wohlenberg & Bhatt (1972) are taken by McConnell (1972) as additional evidence for an interconnection.

Some authors have suggested the existence of a transform fault offset along the Aswa trend (Chorowicz & Mukonki 1980, Kazmin 1980a,b). The use of transform faults to explain meanders and offsets in continental rifts is an established procedure (e.g. King & Zietz 1971, Chase & Gilmer 1973). However, in the case of the Aswa connection it raises a host of complications. Some of these problems include (a) the age of the Aswa Lineament (Precambian); (b) the great length of the proposed offset (500–600 km); (c) the paucity of seismicity aligned along the proposed offset (e.g. Shudofsky 1985); (d) the scarcity of surface expression, especially recent dislocations, along the proposed offset; (e) the continuation of Eastern Branch rift morphology south of the proposed transform-rift intersection; and (f) the apparent continuation of transform-aligned structures past the rift-branch intersections. For example, the work of Vail (1972), McConnell (1972), and Browne & Fairhead (1983) points to a probable continuation of the Aswa Lineament northwestward, possibly along the Abu Gabra Rift to Darfur, Sudan. McConnell (1972) continues this lineament southeastward past the intersection with the Eastern Branch to the Pangani Rift (termed the Pangani-Aswa Lineament or Pangani–Nandi–Aswa–Jebel Marra line by McConnell) and suggests a further continuation to the Lindi fault zone of Kent et al (1971). We might take note that the volcanic centers of Kilimanjaro and Elgon lie along the lineament as defined by McConnell (1972).

Evidently, the Aswa structures predate the East African Rift System, extend beyond its intersections with the Western and Eastern branches, and cross at least the Eastern Branch without creating any noticeable lateral displacement. It is possible that a transform fault has rejuvenated the central portion of this trend, but if so it is remarkably asymptomatic. A more plausible proposal is that the offset region between the two branches is a transform waiting to be born, not one that has actually happened yet to any significant degree. The term "pretransform" might be more appropriate for such cases.

Transform faults also have been used to connect rift zones. For example, both Chorowicz & Mukonki (1980) and Kazmin (1980a) have proposed a major transform fault zone connecting the north end of Lake Malawi and the middle section of Lake Tanganyika. In effect, these authors are using the Rukwa Rift Zone and the southern half of the Tanganyika Rift Zone to connect the adjacent rift zones. Under this tectonic scheme, the creation of both the Rukwa trough and the southern half of Lake Tanganyika presumably would be due to pull-apart mechanisms. Chorowicz & Mukonki (1980) also terminate the Malawi Rift Zone at its south end with another large transform fault (termed the Zambezi Lineament by these authors). In an earlier paper, Chorowicz et al (1979) connected Lakes Tanganyika and Kivu with what is implied to be an imperfect, 40-km-long transform fault. Recently, Williamson (1986) proposed a Mesozoic transform offset in the southern part of Lake Turkana. These hypothesized "zone-offsetting" transforms might be analogous to those proposed for the Mid-Continent Rift System (or Keweenawan System) in the southeastern corner of Nebraska (e.g. Van Schmus & Hinze 1985).

The continental transform perspective of three-dimensionality treats rifts as the geometric equivalent of the ridge-transform-ridge geometries of oceanic spreading systems. The natural outcome of such a philosophy is a breakup model such as that shown in Figure 1d. We might consider this to be the possible boundary arrangement in East Africa if rifting continues. Although Figure 1d uses the rift-unit concept discussed below, this sort of model still oversimplifies and understates the true architecture of African rifts. This brings us to the subject of the geometry of half-graben units.

Geometry of Simple Fundamental Units

As stated above, rift zones are composed of fundamental units. The intrinsic geometry of an isolated unit is a half-graben whose major bounding fault is arcuate in plan view (Rosendahl et al 1986). The ideal half-graben is depicted in Figure 3. Anisotropy and heterogeneity of prerift structural

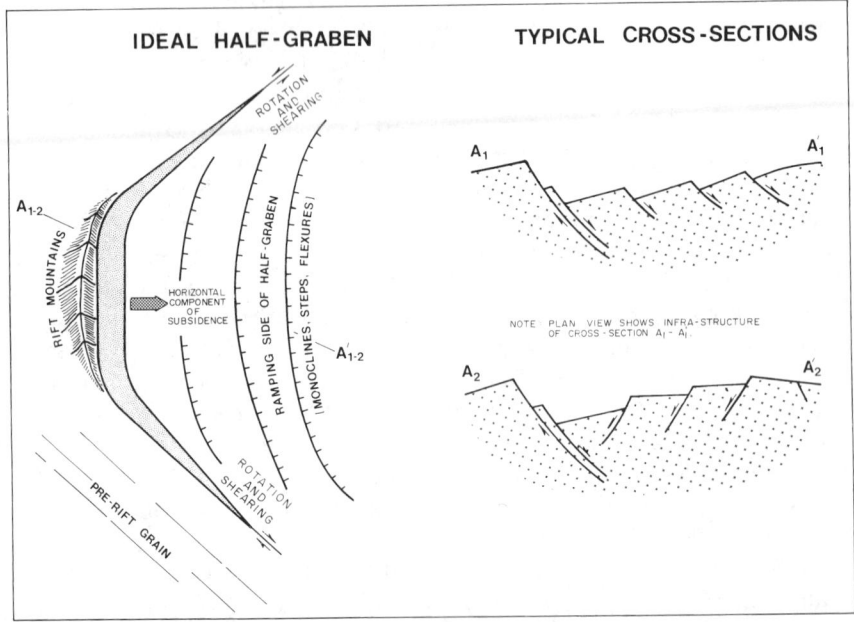

Figure 3 Plan view and hypothetical cross sections of an ideal half-graben. Note that the geometry of subsidence produces oblique-slip faulting along ends of units. Isolated half-graben tend to display synthetically faulted infrastructure (i.e. top cross section).

fabrics make it unlikely that the ideal case will occur in nature. Also, the linking together of units to form a rift zone creates structural modifications to the original geometry, as discussed below.

The closest approximations to the ideal case seem to occur where linking interferences are minimal. This happens at the ends of rift zones, where the terminating fundamental unit is linked only on one side, and where half-graben units alternate polarities along the strike of the rift zone without overlapping. Four seismic sections across relatively simple half-graben units in the Lake Tanganyika and Malawi Rift Zones are shown in Figure 4. Although the basic expressions are the same in these examples, the fracture spacing of the infrastructure is markedly different, ranging from 2 to 15 km. This is about the range we observe throughout the African rift zones, and it is close to the 5–15 km width range reported by Angelier & Colletta (1983) for other regions. De Charpal et al (1978) quote a range of 10–30 km for block spacing in the Bay of Biscay, but examination of their data and other proprietary seismics indicates that this range may be too high by as much as a factor of two. It should be noted that very high-resolution seismics (e.g. echo-sounder data) tend to show smaller fracture spacings, but the dimensions of the major blocks remain the same.

Inspection of the profiles shown in Figure 4 also reveals that the variation in spacing of internal faults on any given profile (i.e. the width of blocks) is considerably less than the variation range between different rift units. For example, the ranges on lines 204, 216, and 222 from Lake Tanganyika are 2–4, 4–8, and 1–6 km, respectively. The block width ranges on lines 814, 834–934, and 817 from Lake Malawi are 7–12, 9–15, and 8–12 km, respectively. Many simple half-graben seem to show a tendency toward decreasing internal fault spacings moving toward the ramping sides of the half-graben. In some cases, the decrease in spacings follows exponential decay relations.

It can be reasoned that relatively simple half-graben also should occur during the initial development of a rift zone, provided that the fundamental units of a rift zone are not strictly contemporaneous. This means that the occurrence of isolated half-graben units should be an early stage of evolution of rift zones.

The seismic profiles in Figure 4 all show infrastructures that are mainly synthetically faulted with respect to the master border fault system. The picture is one of tilted blocks rotated in the same direction along faults that parallel the border fault systems, rather than the alternating horst and graben morphology often associated with continental rifts. The picture is very reminiscent of the tilted "pack-of-cards" geometries of Le Pichon & Sibuet (1981), Wernicke (1981), Bally (1982), Wernicke & Burchfiel

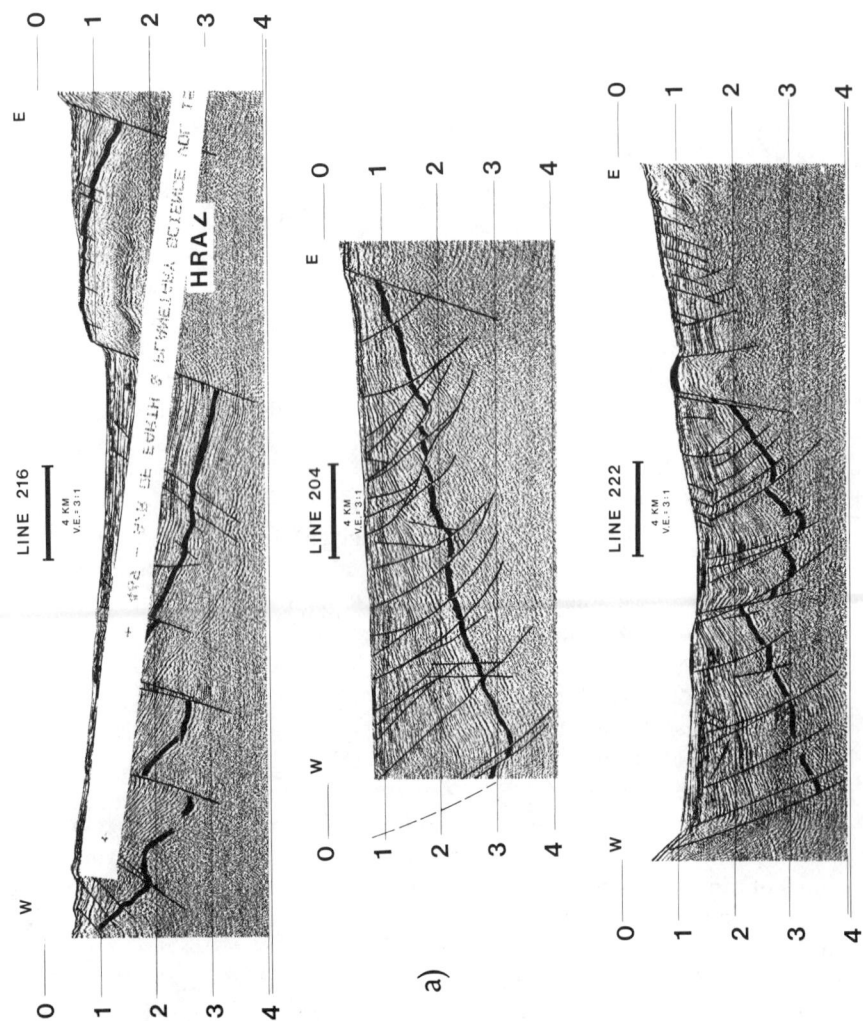

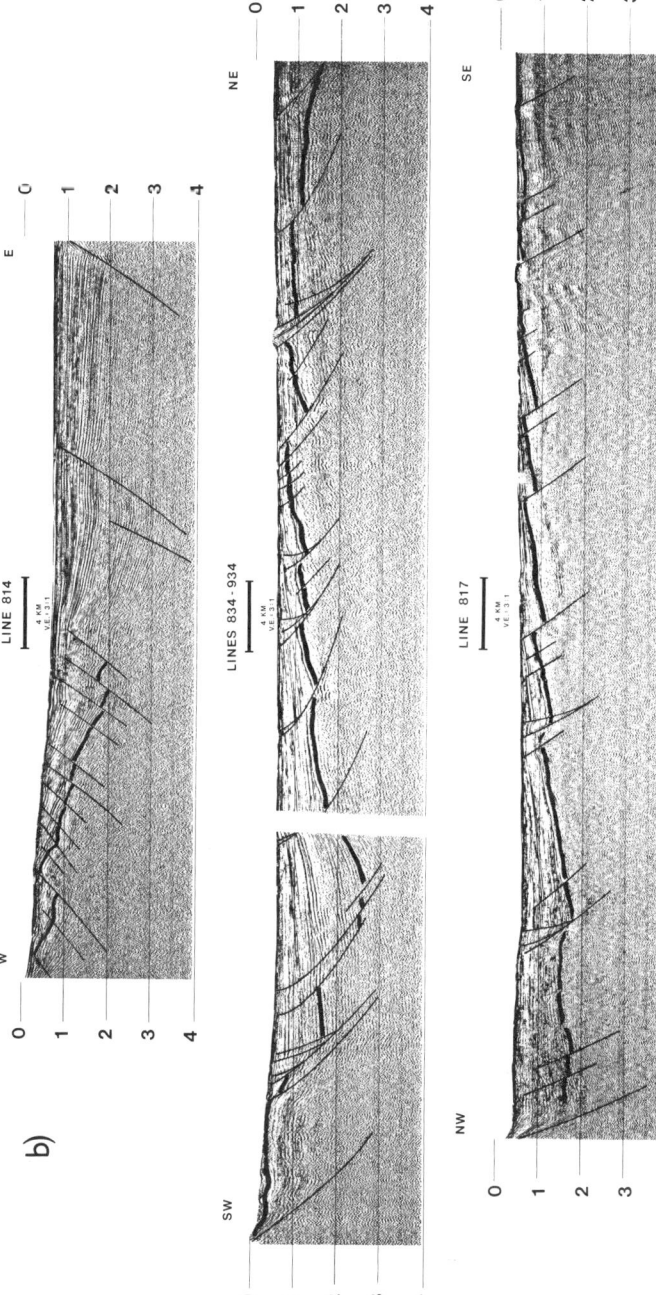

Figure 4 Interpreted multichannel seismic profiles across relatively simple half-graben. Lines 216, 204, and 222 are from the Kalemie, Nyanza-Lac, and Mpulungu half-graben of Lake Tanganyika, respectively. Note the dominance of synthetic infrastructure in these half-graben compared with arrangements where half-graben overlap and face one another (e.g. Figure 7a). Lines 814, 834-934, and 817 are from the Karonga, Nkotakota and Nkhata provinces of Lake Malawi, respectively. All profiles represent 24-fold sections. Tanganyika lines are migrated; Malawi lines are not. See Rosendahl et al (1986) for Tanganyika line locations, and Ebinger et al (1984) for Malawi province locations. Heavy solid line represents acoustic basement, which usually can be taken as the base of the synrift.

(1982), and Angelier & Colletta (1983). It is also similar to the Bay of Biscay morphology described by de Charpal et al (1978). Our studies to date indicate that this geometry is typical of relatively simple, isolated half-graben. It is not necessarily the case, however, for half-graben that link together in ways that create structural interferences (see later discussion).

The sinuosity of the Western Branch of the East African Rift System has been recognized by virtually every worker since Gregory's time, and it is now universally attributed to a deflection of rifting around the Tanganyika Shield. A similar proposition has been made by Van Schmus & Hinze (1985) to explain a lateral offset in the Mid-Continent Rift System that Chase & Gilmer (1973) attributed to a transform fault. In contrast to this large-scale sinuosity, the arcuate forms of half-graben units in plan view are a relatively new development. Agreement on the causes of the curvature is divided between three schools of thought within our research group. One school argues that the true geometry is an orthorhombic fracture pattern, such as that produced in clay models by Oertel (1965) and convincingly explained by Reches (1978, 1983). The asymmetric half-graben form is attributed to unequal development of the four sets of faults that make up the orthorhombic geometry, and the arcuate character of the border fault systems is a consequence of trackline aliasing (Figure 5). The rhomb-shaped basins shown by Thompson & Burke (1974) in the Basin and Range Province could be cited as another example of the same process. The second school also uses trackline aliasing but begins with composite pull-apart fracture geometries such as those proposed by Aydin & Nur (1982). The fact that the infrastructure of rift units tends to be strongly linear, including the individual faults that comprise border fault systems, lends credence to both of these schools of thought. The third school argues that the arcuate forms are real, albeit perhaps not as simple or stylized as indicated in Figure 3. The support for this case comes from both the observational and fault mechanical fronts. Normal faults that curve in map view have been observed in virtually all extensional environments, including the North Sea (e.g. Ziegler 1978), the Gulf of Suez (Robson 1971), the Ethiopian Rift (Figure 1c; Juch 1980), the Gregory Rift (Baker et al 1972), the Rhinegraben (Hirlemann 1974), and the Basin and Range (Wright & Troxel 1966). Such patterns also have been reproduced in laboratory model experiments (e.g. Figure 2 of Bahat 1979, Figure 7 of McGill & Stromquist 1979). Satellite imagery suggests that the prevalence of curvilinear fault patterns has been grossly understated. Another line of evidence comes from a consideration of what happens to bounding rift faults with depth. If these faults flatten with depth, as suggested by many workers (e.g. Proffett 1977, Bally et al 1981, Harding 1984), then they should curve in map view as well. These three schools of

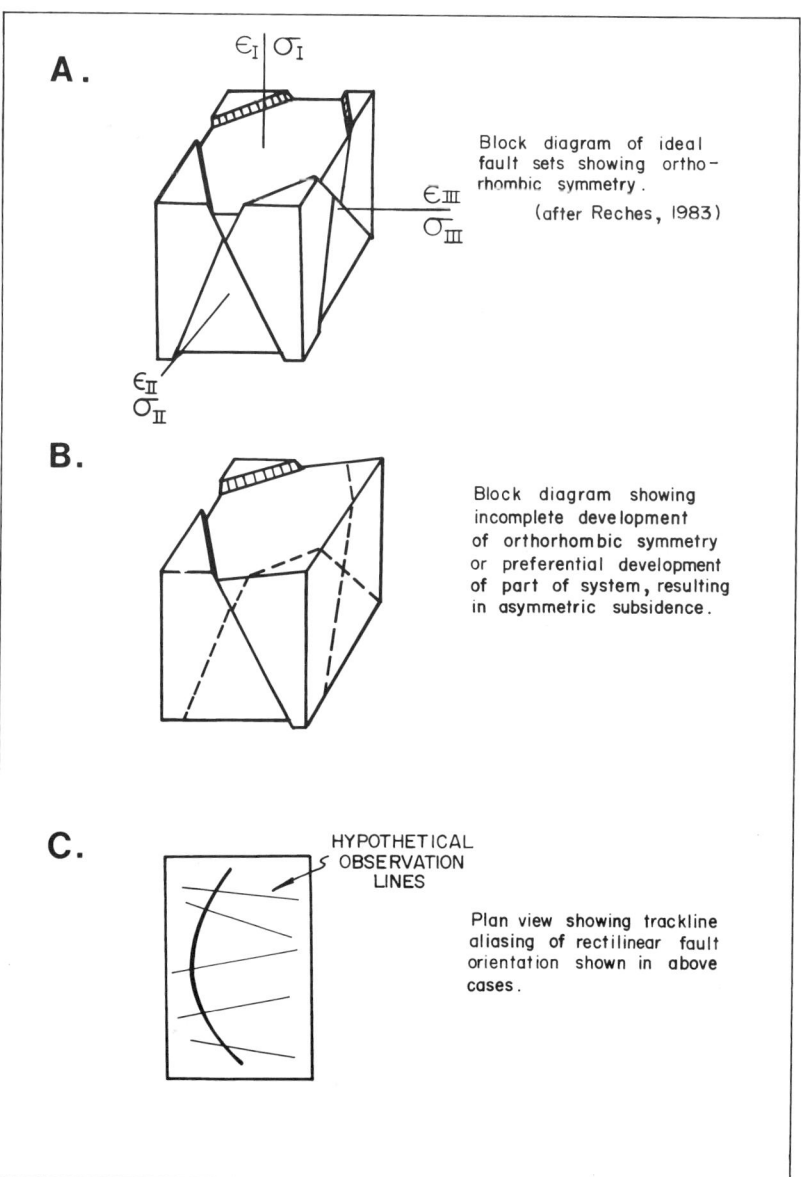

Figure 5 Creation of apparently curvilinear border faults by trackline aliasing of original orthorhombic fracture pattern.

thought are not mutually exclusive, and it is probable that all are operative to various extents in East Africa.

An important consequence of the geometry shown in Figure 3 is that strain expressions within half-graben units are spatially variable. In the ideal case, subsidence is at a maximum adjacent to the geometric center of the border fault, where the displacement is most nearly dip-slip. Toward the arcuate ends of half-graben, the total throw on the border faults decreases as the relative proportion of strike-slip motion increases. Hence, the ends of half-graben ought to be regions of rotation and oblique shearing more than regions of subsidence and pure extension. It is likely that the warping and tilting noted by Thompson & Burke (1974) near the ends of elongate basins in the Basin and Range Province are due to the same process. Much of the strike-slip faulting described in the North Sea rift zones (e.g. Ziegler 1982) also could be interpreted in the same vein. In real half-graben, the depocenters usually are not centered geometrically because half-graben are usually canted toward one end or the other. In other words, not only are half-graben asymmetric in cross section, but they also often plunge along strike as well (Burgess et al 1986). Both anisotropy of prerift structures and the orientation of the applied stress with respect to these structures may be causes of strike asymmetry. We suspect a more general cause is the interaction, or interference, of adjacent half-graben.

Geometries of Linked Half-Graben Units

The key to understanding African rift morphology and structure, including three-dimensionality on the scale of rift zones, pertains to the ways in which half-graben link together to make rift zones. Our group has recognized many different linking arrangements, but all these can be viewed as variations of the half-graben theme of rifting (cf. Rosendahl et al 1986). For the sake of discussion, it is convenient to group these variations into the families and cases shown in Figures 6a–e.

One linking family is created when two half-graben oppose each other, either in overlapping or nonoverlapping modes. A sensible progression of geometric possibilities is depicted in Figure 6 (cases A–F) along with idealized cross-sectional variations for selected cases.

Cases A–C in Figure 6a show half-graben units that overlap and face each other. If the subsiding half-graben units are contemporaneous, then a rather severe space problem seems to be created in the overlapping area. Given that subsidence involves a component of horizontal motion directed toward the ramping sides of half-graben, the facing half-graben must compete for the same piece of terrain. This can result in an antiformal

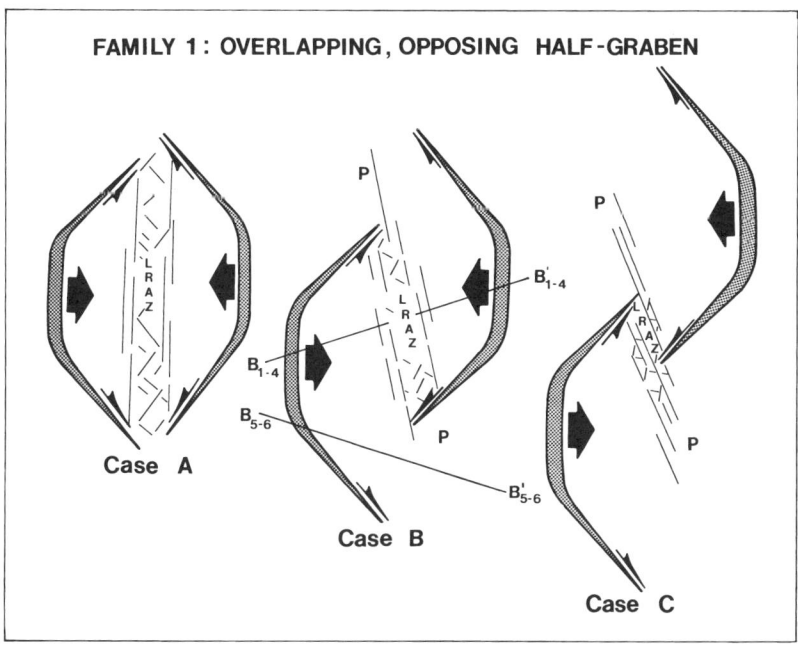

Figure 6 Examples of linked half-graben grouped into families. Opposing geometries create either low-relief or high-relief accommodation zones (LRAZ and HRAZ), depending upon the extent of overlap [compare 6a (*above*) and 6b (*over*)]. Strike-slip accommodation zones (SSAC) may be the intermediate case. Figure 6c shows a family of similar polarity half-graben. Hypothetical cross sections of cases B and E are shown in Figures 6d and 6e.

welt away from which the two half-graben subside in opposite directions (sections B_1 through B_4, Figure 6d). We have termed these welts "hinged highs," or "low-relief accommodation zones." Experimentally produced examples of hinged highs can be seen in the top one third of Figure 108 in Ramberg (1967).

The difference between sections B_1, B_2, and B_3 of Figure 6d relates to the cross-sectional orientations of the internal faults relative to the border fault systems. Based upon the characteristics described above for simple half-graben, one would suspect that the double synthetic arrangement shown in section B_1 should predominate. In actuality, all arrangements have been observed (e.g. Rosendahl et al 1986, Burgess et al 1986, Sander 1986). A seismic example of each is given in Figure 7. We suspect that the development of antithetic infrastructure is an offshoot of interference geometries such as those developed in family 1. Otherwise, this fracture

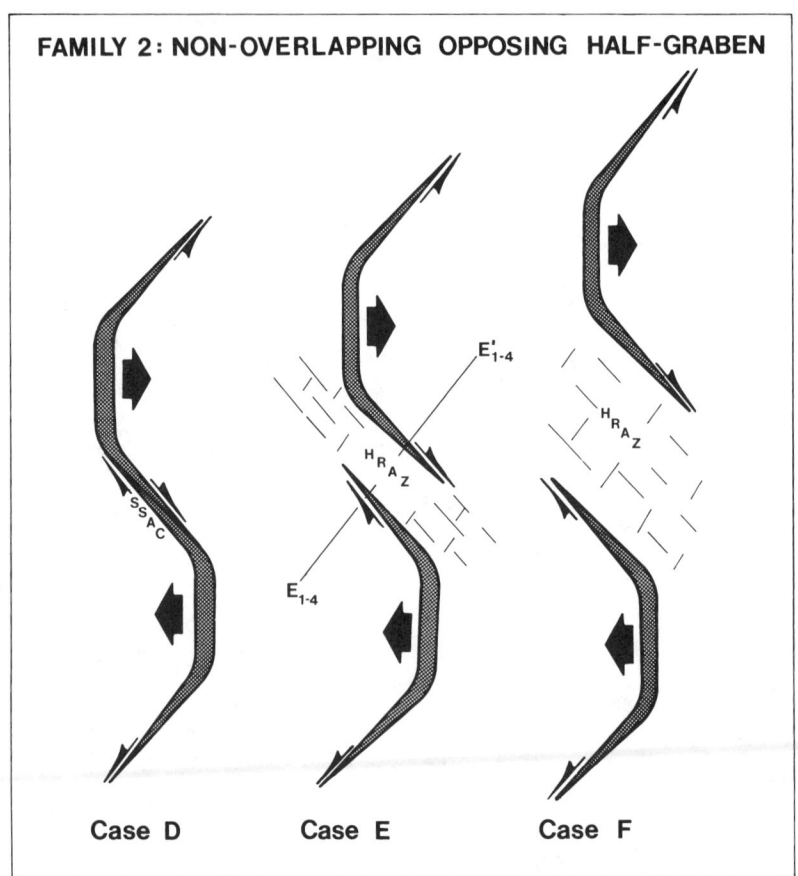

Figure 6b

arrangement should be more prevalent in simple half-graben than it appears to be.

All three of these sections (B_1 through B_3) show what appear to be full graben, but it should be emphasized that this form is a hybrid of facing half-graben geometries, not a primary archetypical morphology. We have yet to find full graben in African rift units that cannot be explained in this fashion, nor any that do not display some type of antiformal welt. We do not deny that "true" full graben can exist, such as those shown in the clay models of H. Cloos (1939) or in profile 4 of Milanovsky (1972), but we suspect they are rare in East African rift zones. The same may be true for other continental rifts, and we believe that most full graben are actually a

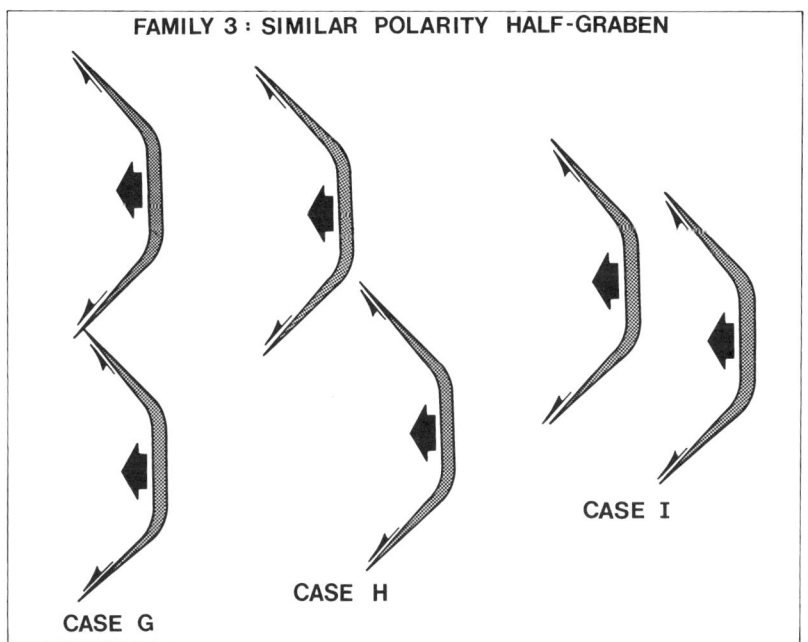

Figure 6c

result of the facing geometries described above. I draw the reader's attention to Figure 6a in Eaton (1982), which looks suspiciously like section B_2, and to Figure 4 of Ghignone & Andrade (1969), which may be an example of section B_1 of Figure 6d. Figure 4 in Wu (1986), which shows the Albuquerque Basin of the Rio Grande Rift, also could be interpreted as facing half-graben separated by a low-relief accommodation zone.

The seismic profiles shown in Figure 7a indicate that subsidence of hinged highs is retarded relative to the depocenters of the facing half-graben. Nonetheless, they do eventually subside. Discussion of the evolution of these highs is taken up in a later section of this paper.

There is no compelling reason why the subsidence of the two facing or adjacent half-graben should be contemporaneous or equal. This suggests another group of variations, exemplified by section B_4. Figure 7a (line 206) shows a seismic profile from Lake Tanganyika that displays a B_4 geometry. Note that subsidence along the "new" border fault system on the east is creating an antiformal welt that is still strongly asymmetric. One outgrowth of this arrangement is that internal faults associated with creation of the original border fault system on the west are being "pulled over" by the

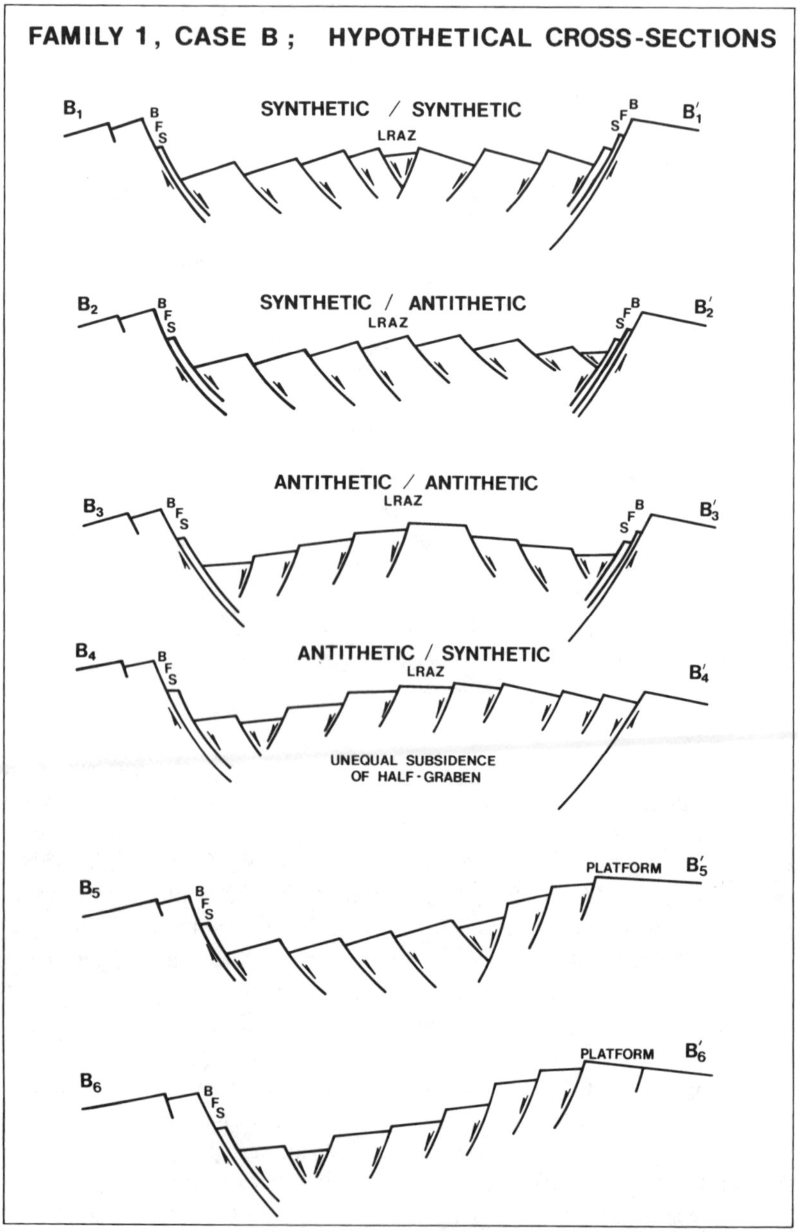
FAMILY 1, CASE B ; HYPOTHETICAL CROSS-SECTIONS
B1
B1'
SYNTHETIC / SYNTHETIC
LRAZ
BFS
B2
B2'
SYNTHETIC / ANTITHETIC
LRAZ
B3
B3'
ANTITHETIC / ANTITHETIC
LRAZ
B4
B4'
ANTITHETIC / SYNTHETIC
LRAZ
UNEQUAL SUBSIDENCE
OF HALF-GRABEN
B5
B5'
PLATFORM
B6
B6'
PLATFORM

Figure 6d

ARCHITECTURE OF RIFTS 473

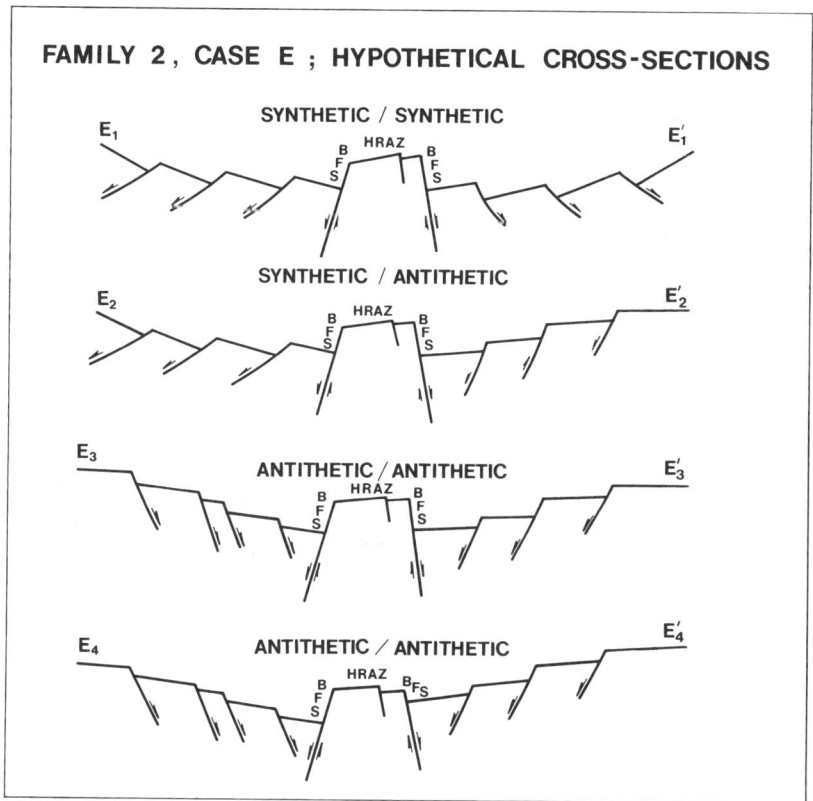

Figure 6e

new subsidence on the east. This is similar to a situation described by Gibbs (1983), and it is an alternative explanation to that of Proffett (1977) for how normal faults can change into reverse faults. Apparently, significant subsidence cannot occur along both border fault systems at the same time. At present, the activity is centered along the new border fault system.

Do morphologies such as those shown in sections B_1 through B_3 evolve from equal, simultaneous subsidence along the facing border fault systems, or rather do they evolve from subsidence that "teeter-totters" between the border faults? This question is tantamount to asking if section B_4 is a precursor of geometries such as those shown in sections B_1 through B_3. Our seismic data from the African rift lakes seem to favor the subsidence

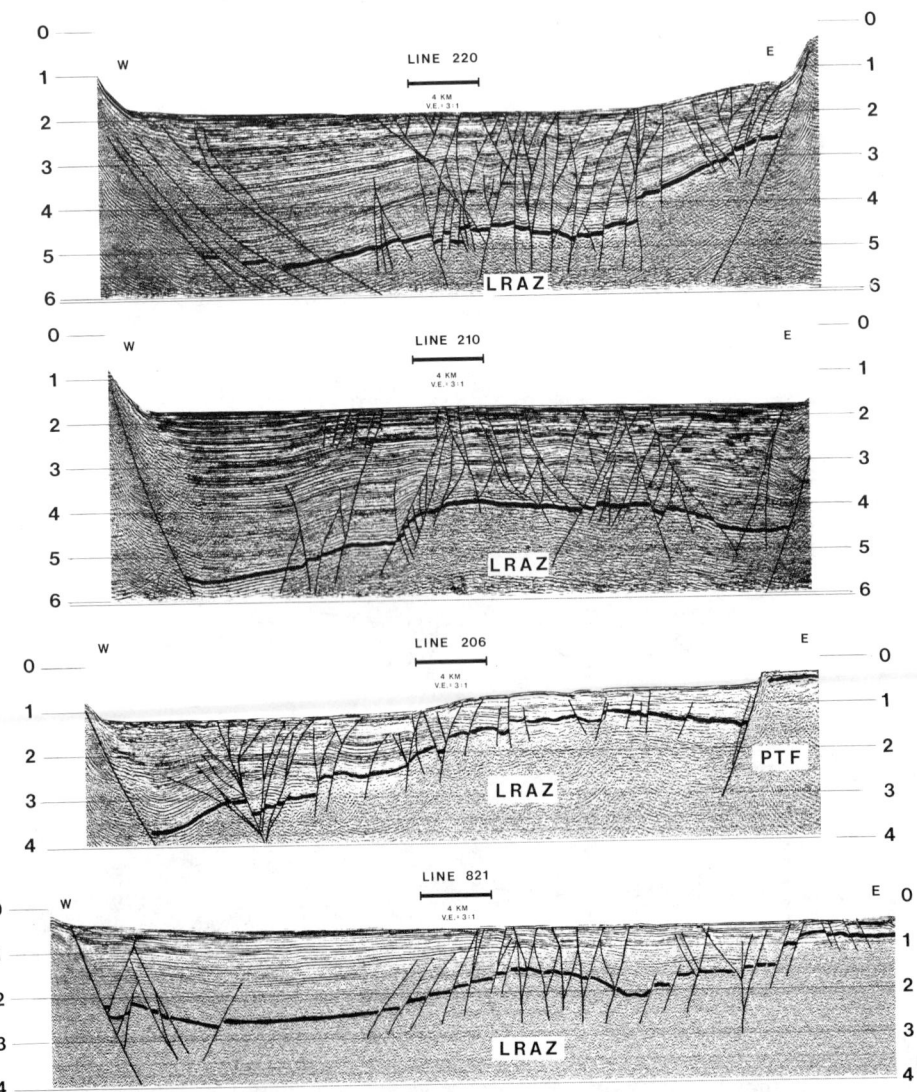

Figure 7a

ARCHITECTURE OF RIFTS 475

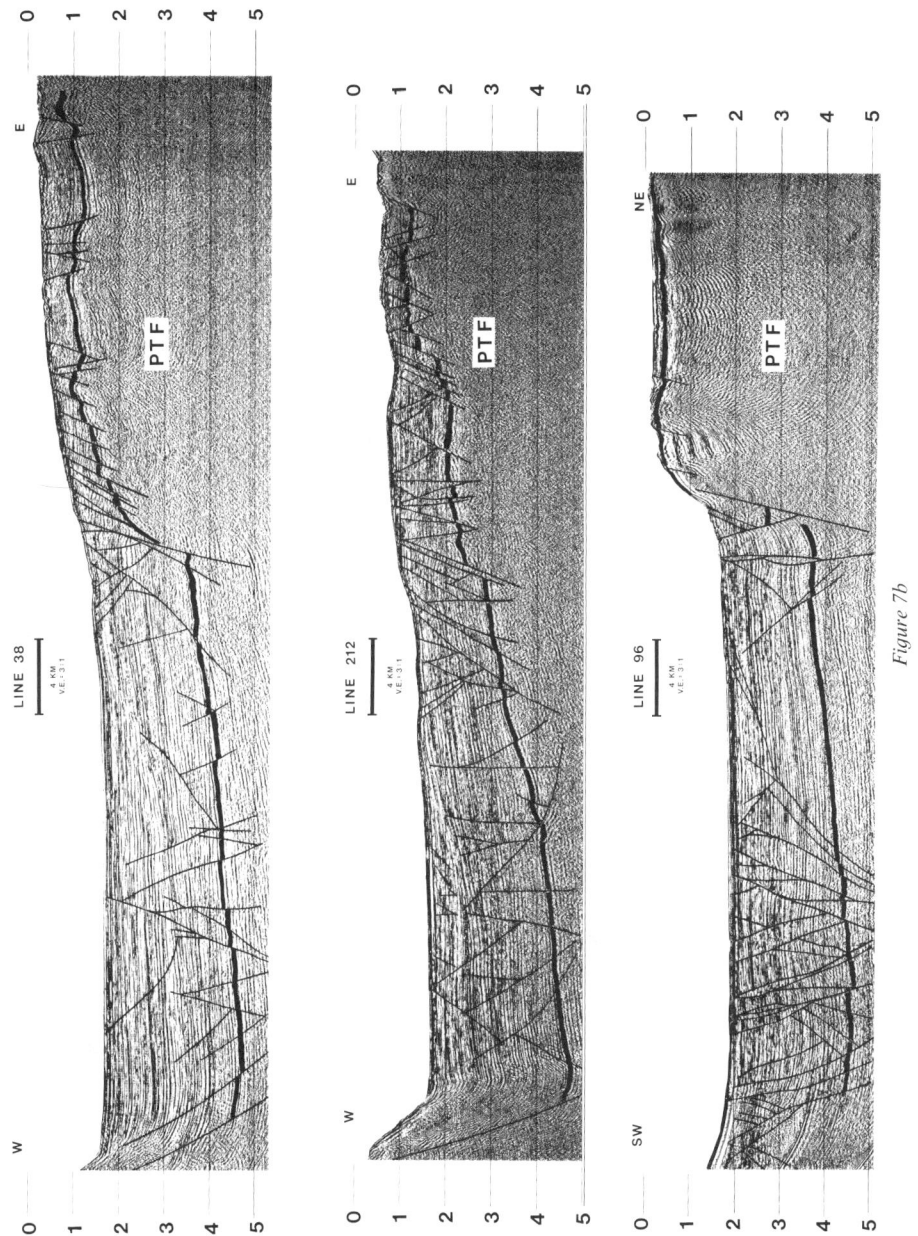

Figure 7b

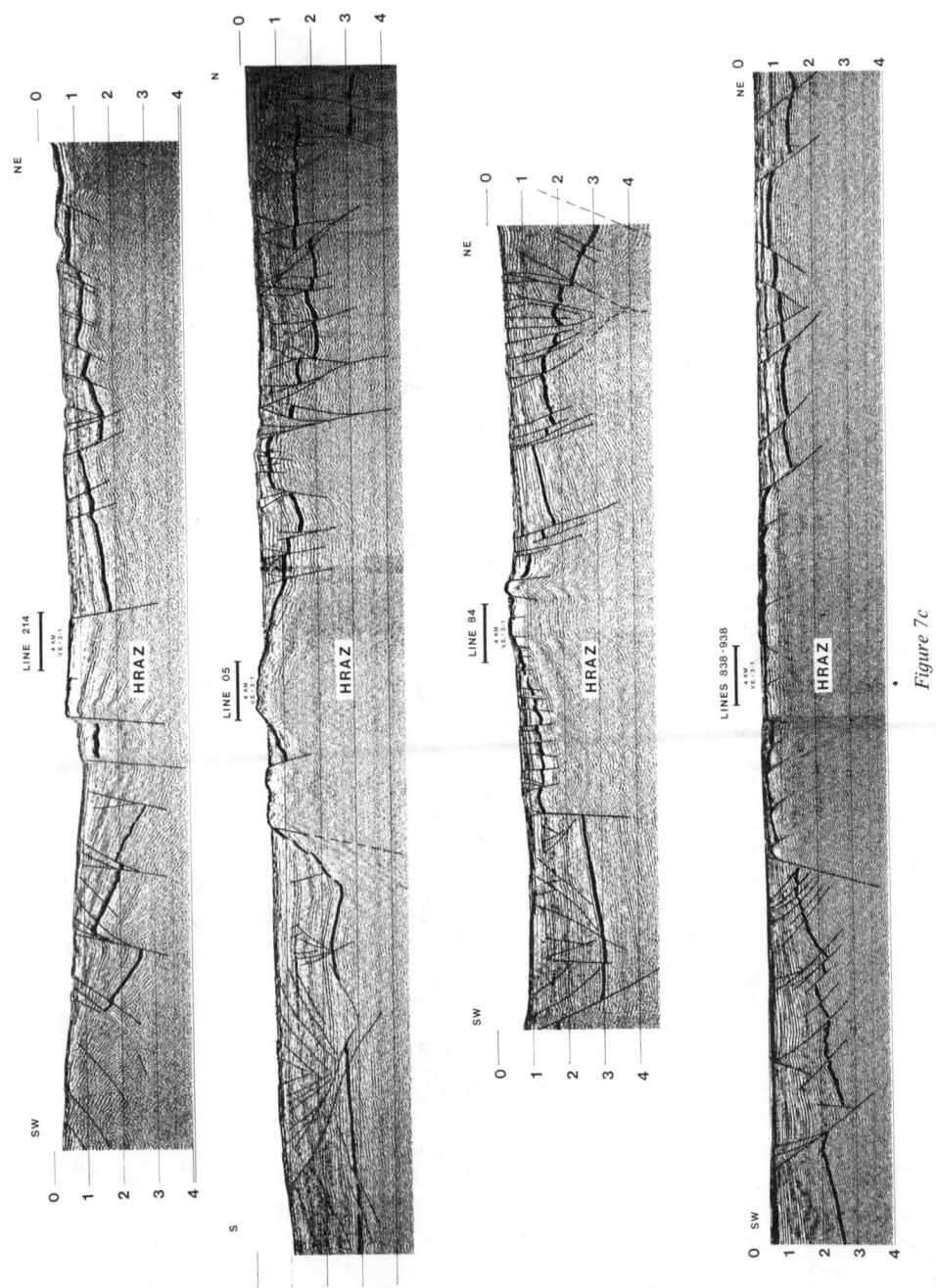

Figure 7c

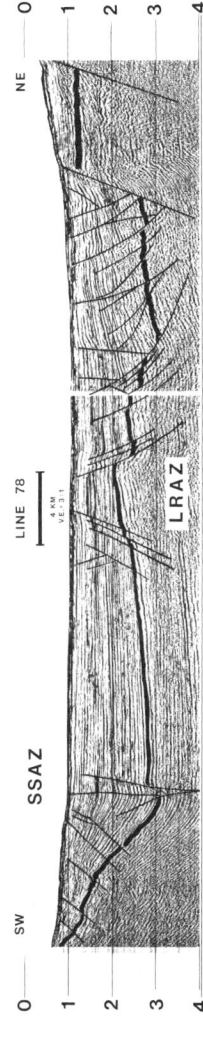

Figure 7 Interpreted multichannel seismic profiles showing actual examples of case B, E, and D geometries. Figure 7*a* shows several types of low-relief accommodation zones (LRAZ), or hinged highs, Figure 7*b* shows examples of platforms (PTF), Figure 7*c* shows examples of high-relief accommodation zones (HRAZ), and Figure 7*d* (*above*) shows a possible example of a strike-slip accommodation zone (SSAZ). Compare to expressions of simple, isolated half-graben (e.g. Figure 4). Lines 220, 210, 206, 38, 212, 96, 214, 05, 84 and 78 are from Lake Tanganyika. Lines 821 and 838–938 are from Lake Malawi. Correlations of seismic lines to cross-sectional geometries in Figures 6*d* and 6*e* are as follow: line 220 to case B_1; 210 and 821 to B_3; 206 to B_4; 38, 212, and 96 to B_5 and B_6; 214 and 05 to E_1; 84 and 838–938 to E_2. Line 78 correlates to case D of Figure 6*b*. All profiles represent 24-fold sections. All Tanganyika lines except 05 are migrated; Malawi lines are not. Note that profile 05 is mainly dip-line, whereas others cross accommodation zones at angles in excess of 30°. See Rosendahl et al (1986) for Tanganyika line locations.

teeter-totter (e.g. Burgess et al 1986, Sander 1986), suggesting that rifting energy there is insufficient for simultaneous development of two facing, overlapping half-graben.

Sections B_5 and B_6 show two hypothetical variations in the morphology across the nonoverlapping regions of facing half-graben. Several actual seismic examples are provided in Figure 7b. Generally, these nonoverlapping areas display a half-graben form bounded by a platform on the ramping side. The platform-basin transitions can range from relatively gradual to abrupt (compare lines 212 and 38, Figure 7b), depending upon the proximity to a border fault "tail." The difference between this geometry and those of simple half-graben (e.g. Figures 3 and 4) is mainly a matter of degree. Also, platforms tend to be associated with larger areas of subdued rift shoulder topography.

Clearly, the amount of overlap between the facing half-graben in cases A–C of Figure 6a has a strong influence on such things as the areal size of platforms, the length and trend of hinged highs, and the ways in which the horizontal stress components of subsidence are relieved along the hinged highs. For example, negligible platform development, longitudinal hinged highs, and pronounced axial compression are more likely for case A than C. The latter geometry is more prone to show large platforms and oblique-trending hinged highs characterized by shearing. Another variable is the relative positioning of the two facing half-graben. Where oblique arrangements occur, the geometries tend to become extremely complicated, especially with regard to the expressions of hinged highs.

Cases E and F in Figure 6b are examples of geometries in which adjacent half-graben display opposite polarities but do not overlap to any significant extent. The "backbone" between the two half-graben is a relatively unsubsided region termed an "interbasinal ridge" by Rosendahl et al (1986) and a "high-relief accommodation zone" by Reynolds (1984) and Burgess et al (1986). We like to view these zones as remnants of prerift rock that have been excluded from significant synrift subsidence by the particular arrangement of the bounding fault systems.

Some of the possible morphologic variations for case E are shown in sections E_1 through E_4 (Figure 6e). The top three sections show various combinations of synthetic and antithetic infrastructures; the bottom section shows a double antithetic arrangement where subsidence of the two half-graben is unequal. Seismic examples of sections E_1 and E_2 are provided in Figure 7c. This type of linking geometry is easily recognized in other continental rifts [e.g. compare section E_1 to Figure 4 in Ghignone & Andrade (1969); section E_2 to Figure 2, profile 5, in Milanovsky (1972) and to line B74–21M in Skilbeck & Lennox (1984); section E_3 to Figure

12 in Okaya & Thompson (1985)]. Geometries such as those shown in cases D–F in Figure 6b also could explain the switching asymmetry described by Mueller (1970) for the Rhinegraben.

The differences between cases E and F of Family 2 pertain to the relative widths, areal extents, and perhaps the elevations of the high-relief accommodation zones between adjacent half-graben. Hypothetically, broad zones such as case F should be subjected to the least subsidence, but the possibility of developing pull-apart terrains atop these zones may obscure this relationship. We return to this issue in the discussions concerning preexisting fabrics and mechanisms of rifting.

Case D represents the transition between overlapping and non-overlapping geometries. The "backbone" between the adjacent half-graben still should be treated as an accommodation zone (because it continues to accommodate the motions generated by opposing subsidence), but the terms low- and high-relief accommodation zones no longer apply, *sensu stricto*. Figure 7d shows a seismic profile that may display this type of linking arrangement. The expression of the accommodation zone is a vertical boundary across which the regional dip changes markedly. Such boundaries may display "flower structure" (Harding 1985), indicative of strike-slip faulting. For this reason, we term these zones "strike-slip accommodation zones."

Cases G through I show linking geometries in which half-graben units face in the same directions. Unless the half-graben in case G are very different in age, it would be difficult to distinguish this geometry from a single, relatively isolated half-graben, especially with synthetically faulted infrastructure. Case I demonstrates another way that structural platforms can be created. A profile taken across either half-graben to the rhomboidal-shaped platform between the units would look much like the sections shown in Figure 7b.

Hypothetically, the linking couplets shown in Figure 6 can be combined with each other in virtually any conceivable mode. In reality, certain linking relationships tend to be repeated during the creation of African rift zones. Burgess et al (1986) have noted that geometries that generate low-relief accommodation zones generally are terminated by high-relief geometries. In other words, case A through C arrangements (Figure 6a) are enclosed along strike by case E or F geometries. The best example of this is the northern half of the Lake Tanganyika Rift Zone (Figure 2). This figure also points out the direct relationship between kinks or doglegs in rifts and accommodation zones. Thus far, our studies have shown that every major kink in an African rift lake correlates to a high-relief or strike-slip accommodation zone. However, not all high-relief or strike-slip accommodation zones are expressed as kinks.

APPLICATIONS AND IMPLICATIONS OF ARCHITECTURAL THEME

Relationships to Prerift Fabrics

There has been much written about the roles of preexisting structure on the development of rift patterns (e.g. Dixey 1956, Holmes 1965, McConnell 1972, 1974a, 1978, King 1970, 1978a,b, Sykes 1978, Mohr 1982, and many other authors). Most of this literature can be divided into two schools of thought. Both contend that rifts follow earlier trends, but they disagree on the genesis of the relationship. The "opportunistic" school argues that African rifts follow antecedent grains because it is mechanically convenient to do so (e.g. Dixey 1956, King 1970). This is the classic "zone-of-weakness" doctrine that has so dominated tectonic thought since Suess's time. The specification of "mechanical convenience" is not universally agreed upon, but the work of Handin (1969) suggests an angular relation of about 25° between the direction of extensional stress and the direction of antecedent lines of weakness. Angular differences greater than this presumably force extensional fractures to adopt paths that ultimately cross existing fabrics. The other school argues for genetic causality, and McConnell (1972, 1974a,b, 1978) is the most outspoken proponent of this position.

It is worth examining the role of antecedent structures in the context of the scales and architectural patterns described here. The first point we should make is that one's perspective on the quality of the coincidence between antecedent structure and rifting is very much scale dependent. On the scale of rift branches, there can be little disagreement with Sykes (1978), who contends that rifting follows the more recent tectonic pulses. Certainly the deflection of the Western Branch around the Tanganyika Shield illustrates the strong effect at this scale of rifting.

A strong correlation to preexisting fabric also occurs on the scale of rift blocks, a point raised by Mohr (1982) and verified by our seismic studies (e.g. Patterson 1982, Lorber 1984). In brief, the orientation of the infrastructure follows the local foliation of the country rock, except where foliations intersect at high angles.

The correlation of preexisting fabric to rift zones and rift units is much more complicated and, as might be deduced from Figure 8, depends largely upon which elements of rift units are considered. The orientations of high-relief accommodation zones almost always follow prerift structural trends very closely, whereas border fault systems usually do not. Low-relief accommodation zones (i.e. hinged highs) may or may not follow antecedent foliation. Figure 8 from Lake Tanganyika portrays these patterns

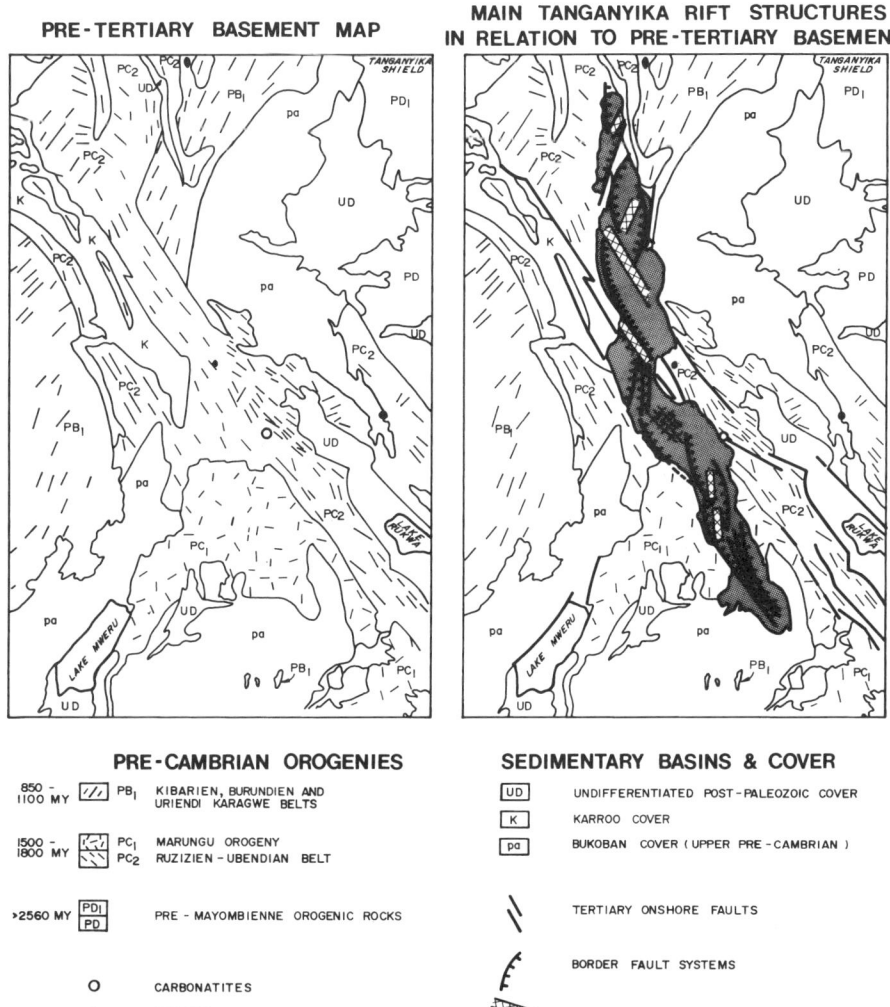

Figure 8 Relationship of Tanganyika rift structures to preexisting basement type and trends. Note the strong correlation between orientation of high-relief accommodation zones and the grain of the basement rocks. Also note that border fault systems, especially in the northern half of the zone, tend to cut across basement fabric. The overall zonal trend seems to follow the Ruzizien-Ubendian and Kibarien-Burundien belts.

quite clearly. Too few strike-slip accommodation zones are known to draw any generalities on their relation to previous structure.

The above discoveries suggest that the features that terminate rift units along strike are also those that follow preexisting grain and display the most shearing. This is equivalent to saying that the termination of rift units occurs by strike-slip faulting along prerift zones of weakness. Border fault systems "take-off" from high-relief accommodation zones and eventually cut across preexisting fabric. The data shown in Figure 8 yield the distinct impression that border fault systems are nature's way of connecting high-relief accommodation zones, not the other way around. Certainly, it is difficult to draw any other conclusion with regard to the Kavala Island and Burton's interbasinal ridges, which are mechanically coupled by the border fault systems along the western edges of the Kigoma and Nyanza-Lac half-graben (Figure 2).

Are transverse fault systems created in nature to transfer motions between offset normal faults, as implied by Bally (1982) and Gibbs (1984), or do border fault systems develop between, and in response to, discrete zones of shear, as suggested herein? The question goes beyond semantics because it strikes at the very heart of the mechanisms of rifting... if we are correct in our analysis, then the world's classic, archetypical rift should be treated initially as the product of pull-apart tectonics.

Because low-relief accommodation zones originate from geometries such as those shown in Figure 6a, it is not surprising that their relationships to preexisting fabric are complicated. On the one hand, they are part of the infrastructure of fundamental units and might be expected to behave like blocks. On the other hand, their overall orientations are strictly constrained by the asymmetric subsidence of the facing half-graben. Likely, the individual fracture patterns of hinged highs will follow older grains, but the overall trends usually will not.

Volcanism

Volcanism is the true "wild card" of rifting, especially in terms of relative abundances, compositions, and timing with respect to structural expressions. Consider that the Eastern and Western branches of the East African System are certainly genetically related, yet one branch is almost devoid of rift-related volcanics, whereas the other is virtually buried in them (Figure 1b). Also, it is now reasonably well established that volcanism can occur at any stage of rift evolution (Mohr 1982). Nothing in the architectural theme presented here explains the differences between the Eastern and Western branches, but the model may shed some light on the issue of where volcanism may first occur in relatively "dry" rift zones.

Within the Western Branch, rift-related volcanics are known to occur in the Rungwe belt between the Malawi and Rukwa rift zones, at the south end of the Albert Zone in the Kivu region, and between the Kivu and Edward rift units (Figures 1b and 1c). All of these known occurrences can be placed on either high-relief accommodation zones between half-graben units or on the pre–transform faults between rift zones. We also are reasonably certain that volcanic cones have been built on two high-relief accommodation zones in Lake Tanganyika. Hence, all known or inferred volcanics along the Western Branch seem to be associated with the boundaries between half-graben, not along the main portions of border faults or within units. This is in keeping with the structural setting of these zones, which suggests that high-relief accommodation zones should offer easier magmatic pathways to the surface than border fault systems, particularly if the latter sole out at relatively shallow depths. Reynolds (1984) believes that this association applies to other continental rift zones, and he cites the Cat's Hill volcanic center in the Rio Grande Rift as an example.

The regular spacing of volcanic centers along the Kenya (Gregory) and Ethiopian (Figure 1c) rift zones (average about 43 km according to Mohr & Wood 1976) and Turkana Rift Zone (about 50 km according to Dunkelman 1986) could be interpreted in the same vein as above. The occurrence of transverse volcanic trends in the southern Kenya Zone (Figure 1d) also could be associated with accommodation-zone volcanism (Figure 1c). However, the volcanic "wild card" reappears with regard to the only definitive Eastern Branch test of this correlation. Our recent seismic surveys of the Turkana Rift Zone suggest that North, Central, and South islands, along with an unnamed volcanic plug between Central and South islands, occur in the middle of half-graben units (Dunkelman 1986). This is exactly the opposite of what seems to occur along the Western Branch rift zones, and it suggests that egress of magmas in "wet" rift zones must be governed by different mechanical rules. It is interesting to note that volcanism in the Oslo graben (e.g. Ramberg & Larsen 1978) occurs in association with both a high-relief accommodation zone (Reynolds 1984) and in the middle region of the southern half-graben. B. T. Larsen (personal communication) believes that the latter occurrences predate the former.

Active or Passive Rifting?

There are nearly as many proposed mechanisms of rifting as there are rift researchers. Most recent reviewers of this subject (e.g. Fischer 1975, Ramberg & Neumann 1978b, Baker & Morgan 1981, Bott 1981, 1982a,b, Illies 1981, Mohr 1982, Morgan & Baker 1983, Turcotte & Emerman 1983, Ramberg & Morgan 1984) have divided propositions into what are now known as the "active" and "passive" modes of rifting (Şengör & Burke

1978). According to the proponents of active rifting, rifts are a tensile response to doming, arching, and/or uplift on a regional scale (e.g. Gregory 1896, 1921, H. Cloos 1939, Dainelli 1943, Holmes 1944, 1965, Quennell 1960a, Gass 1970, 1973a,b, Le Bas 1971, Burke & Whiteman 1973, Lowell et al 1975, Burke 1976a,b, 1977, Bailey 1978, Ramberg & Spjeldnaes 1978).

Variations and elaborations on the active theme concern such issues as the role and timing of volcanism, the causes of doming (e.g. compression of the African plate versus plumes and hotspots), the scales of doming, episodicity, and the interrelations of aulacogens, failed arms, and triple junctions. One interesting African variation is that raised by Le Bas (1971) and Gass (1975), who use different domes to make different parts of the East African Rift System. Mohr (1982) argues that the number of these variations are so great that they render the active mechanism useless as a general theorem of rifting.

The passive mechanism argues that subsidence, not uplift, is the first expression of rifting, and that any doming that occurs is a consequence of later thermal events (e.g. Baker et al 1972, Hutchinson & Engels 1972, Blundell 1978, de Charpal et al 1978, King 1978a,b, Illies 1978, Ramberg & Larsen 1978, Chapin 1979, Faller & Soper 1979). Most passive rifters agree that subsidence is an expression of stretching and thinning of the lithosphere induced mainly by horizontal plate separations. Variations in the passive mode pertain to the physics of lithospheric attenuation and the roles and timing of thermal events, including volcanism.

Some workers take a more noncommittal stance by discussing subsidence and uplift in a quasi-contemporaneous sense (e.g. Logatchev 1978), or else they place them after initial episodes of volcanism (e.g. Williams 1978). Although many workers often favor one or the other modes, they seem willing to accept the premise that different modes are possible for different rifts (e.g. Baker & Morgan 1981). However, the argument goes beyond geography because some authors disagree on individual rifted systems [e.g. Hutchinson & Engels (1972) vs Lowell et al (1975) on the Red Sea; Baker et al (1972) vs Gass (1970) on the Gregory Rift].

Certainly the architectural patterns described here are symptoms of the causative agents of African rifting, but are the symptoms unique enough to diagnose the type of malady? Does the architectural theme constrain the above models? The problem we face is that the symptoms mainly relate to brittle fracture, whereas the mechanisms also involve the ductile portion of the plates. As de Charpal et al (1978) point out, behavior in the brittle crust should not be used to measure necking in the underlying ductile zone. Thinning in the latter can both precede and exceed that in the former. In view of these difficulties, we limit our discussion to a few key points.

The first concerns the issue of subsidence, shearing, and high-relief accommodation zones. Relations to preexisting fabric seem to argue that subsidence is a by-product of shearing along high-relief accommodation zones, not the other way around. If so, then perhaps we ought to concentrate our search for rift mechanisms on situations or conditions that give rise to relatively large-scale intraplate shear stresses. This would argue for a variation of the passive rifting theorem for the same reasons that the Dead Sea rift is classified as passive.

The second point to consider is that nothing in the synrift histories of the African rift zones we have studied requires regional uplift or doming concurrent with extensional deformation. Some of our seismic profiles show unconformities and even some evidence of subaerial erosion, but these events can be explained adequately by facing half-graben geometries with the subsidence teeter-totter, or by climatic changes. It is not so easy to discount the possibility of early or prerupture uplift. Unlike the Bay of Biscay, the rotated fault blocks in the African rift lakes do not display enough internal stratigraphy to prove the existence of prerift sags. Indeed, the lack of such internal stratigraphies could be used to support the active-rifting theorem. Acoustic basement in Lakes Tanganyika, Turkana, and Malawi has the same basic seismic character and is probably the acoustic signature of deeply eroded, prerift crystalline rocks mantled with soils, perisols, laterites, etc (Burgess et al 1986). If the hiatuses across these boundaries prove to be large, as we suspect they will, then the possibility of regional, prerift uplift cannot be discounted.

The third point concerns the issue of volcanism. It has been suggested or implied by some workers (e.g. Williams 1978) that volcanism is the first symptom of rifting, at least in some rift zones. Unfortunately, this is almost impossible to prove. On the other side of the coin, it is certain that subsidence has preceded significant volcanism in the Malawi, Tanganyika, and Albert rift zones. If volcanism and doming are contemporaneous, this relation might argue that subsidence precedes uplift.

The uncertainty regarding the chronology, sequencing, and timing of rifting events in East Africa is deeper than might be ascertained from the active-passive controversy alone. The situation also is further clouded by a certain degree of circularity. For example, those who are especially impressed with regional erosional surfaces (e.g. Dixey 1956) tend to be proponents of active rifting and date the onset of rifting from one or another of these surfaces. In this fashion Dixey placed the initiation of the Malawi Zone in the Miocene. Crossley & Crow (1980), on the other hand, date the onset of Malawi fault activity at the Pliocene. Where volcanics are associated with rifting, such as in the Kenya Rift Zone, it is tempting to correlate the onset of rifting with the first appearance of some particular volcanic event (e.g. Williams 1978). It might be said that the rifting sequence one derives for East Africa is strongly influenced by the criteria one uses to define the onset of rifting.

In Figure 9 I have attempted to construct an evolutionary scheme of rifting that is compatible with our seismic results from African rift zones and with the major points raised in this paper. Some of the more attractive features of previously published models also have been retained. The work of Lowell & Genik (1972) needs specific mention because the model they derived for the southern Red Sea has strongly influenced the conceptualization of the last two stages of Figure 9. It might be noted that Lowell & Genik's study contains many of the elements that now, 15 years later, are the highlights of the more popular extensional models.

Comments on the Development of Passive Margins

When researchers attempt to fit conjugate margins back together, the backtracking usually begins with magnetic anomalies and ends with matching some bathymetric contour (e.g. Talwani & Eldholm 1977). The final fits are never truly clean, and they often leave overlaps, gaps, or misfits that resemble continental rift zones. Another complication of passive-margin rift sequences relates to dip; some appear to dip toward the adjacent continents, whereas others dip away. In some places this change in dip direction occurs over relatively short distances along the same margin. Several authors have addressed these specific problems [initially Lowell & Genik (1972) and more recently Bally (1982)], but satisfactory explanations have proven elusive. We think the architectural theme described here provides some elegant, yet simple answers, and we use the Tanganyika Rift Zone as a case in point.

The first point is that there is little yet to fit back together with regard to the Tanganyika Rift Zone, or to any of the other rift zones south of Ethiopia. The maximum extension we compute across any Tanganyika fundamental unit is 8%, and values of 5–7% are more typical. Although stretching and consequent separation in the ductile portion of the lithosphere could be much greater (cf. Wright & Troxel 1966), it clearly has not reached a stage of significant lithospheric separation. This means that if ocean basins such as the Red Sea and the Norwegian-Greenland Sea began as Tanganyika-like rift zones, then closing the basins by removing the oceanic crust should result in overlaps and misfits that resemble the

Figure 9 Possible stages in the evolution of continental rifts. Stages correspond to cross-sectional snapshots of a hypothetical rift basin. Volcanism is treated as a "wild card" in the rifting process because it can occur at any stage. Note that abandonment also can happen at any stage.

POSSIBLE STAGES IN THE EVOLUTION OF RIFTS

STAGE 1: STRETCH TROUGH (ST)
INITIAL STRETCHING AND THINNING OF CL, CREATING RELATIVELY BROAD SURFACE DEPRESSION. UPPERMOST CRUST MAY SHATTER, BUT FAULT THROWS GENERALLY BELOW RESOLUTION LIMITS OF SEISMICS. TYPICAL AMOUNT OF SUBSIDENCE IS 1.0 TO 1.5 KM. SEDIMENT INFILL DOMINATED BY FLUVIAL OR EOLIAN SANDS SHOWING MUCH REWORKING. SYSTEM WILL APPEAR TRANSGRESSIVE WITH VARIABLE AND COMPLICATED INPUT PATTERN. COMPARE TO STAGE 1 OF KAZMIN (1980 a).

STAGE 2: JUVENILE HALF-GRABEN (JHG)
DEVELOPMENT OF ASYMMETRIC GRABEN COMPOSED OF FAULT BLOCKS THAT TILT MAINLY IN THE SAME DIRECTION. OBFS DIVIDES REGION OF SHOULDER UPLIFT FROM REGION OF CONTINUED SUBSIDENCE, BUT ACTUAL SUBSIDENCE MORE OR LESS EQUALLY DISTRIBUTED AMONG FAULTS THAT DEFINE TILTED BLOCKS. TYPICAL NUMBER OF ACTIVE BLOCKS RANGE FROM 3 TO 6, WITH TYPICAL WIDTH OF 10 KM. REGION OF ACTIVE SUBSIDENCE GENERALLY NARROWER THAN WIDTH OF ST. STRATIGRAPHY CONTROLLED BY WEDGE-BLOCK STYLE OF DEPOSITION OF FLUVIAL CLASTICS, WHICH CAN ENTER FROM EITHER SIDE OF HALF-GRABEN. SEDIMENT LITHOLOGY DEPENDENT ON POSITION WITHIN WEDGE BLOCK AND DISTANCE FROM SOURCE. ANGULAR UNCONFORMITY OFTEN EXISTS BETWEEN STAGE 1 AND STAGE 2, SUGGESTING THAT ONSET OF BLOCK ROTATION IS RELATIVELY ABRUPT. ONSET OF STAGE 2 PROBABLY RELATES TO CREATION OF DETACHMENT ZONE THROUGH CL.

STAGE 3: MATURE HALF-GRABEN (MHG)
DEVELOPMENT OF STRONGLY ASYMMETRIC GRABEN WITH MOST OF SUBSIDENCE TAKEN UP ALONG OBFS. CONTINUED UPLIFT AND ROTATION OF SHOULDER USUALLY CREATES BACK-TILTED RIFT MOUNTAINS, RESULTING IN CESSATION OF SIGNIFICANT SEDIMENT INPUT FROM STEEP SIDE OF HALF-GRABEN. MOTIONS ALONG INTERNAL FAULTS SLOW OR CEASE, BUT OVERALL RATE OF SUBSIDENCE MAY ACCELERATE. SUBSIDENCE GENERALLY OUTPACES SEDIMENTATION, EXCEPT WHERE AXIAL DELTAS OCCUR WITH SWITCH FROM DISTRIBUTED TO LOCALIZED MASTER FAULT SYSTEMS. WEDGE-BLOCK STYLE OF DEPOSITION GIVES WAY TO CONSTRUCTION OF BROAD FANS ORIGINATING FROM RAMPING SIDE OF HALF-GRABEN.
COMPARE TO WERNICKE (1981).

STAGE 4: FACING HALF-GRABEN
ACTIVATION OF FBFS CONCURRENT WITH RETARDATION OR CESSATION OF SUBSIDENCE ALONG OBFS. GEOMETRY EVENTUALLY PRODUCES APPARENT FULL-GRABEN WITH AXIAL HIGH, OR "HINGE ZONE". CENTRAL REGION MAY BECOME A FREE BLOCK. BACK-TILTED RIFT MOUNTAINS CAN DEVELOP ON BOTH SIDES OF BASIN, WHICH RESULTS IN CUTTING OFF FLUVIAL SEDIMENT SUPPLY AND CREATION OF DEEP, STARVED LAKES. ONSET OF ACTIVITY OF FBFS USUALLY MARKED BY ANGULAR UNCONFORMITY AND CREATION OF A NEW GENERATION OF FAULTS. SEDIMENTATION DOMINATED BY BIOGENIC RAIN AND LOCAL BORDER FAULT FANGLOMERATES.
COMPARE TO STAGE III OF LOWELL AND GENIK (1972).

STAGE 5: SUBSIDENCE TEETER-TOTTER
ALTERNATION OF BOUNDARY FAULT ACTIVITY BETWEEN FBFS AND OBFS. POSSIBLY SEVERAL TIMES. ALTERNATIONS APPEAR TO BE RELATIVELY SUDDEN, RESULTING IN CREATION OF ANGULAR UNCONFORMITIES. ABANDONED SITE OF AFG MAY BE UPLIFTED AND ERODED, SOMETIMES EXTENSIVELY. OVERALL PATTERN IS A TEETER-TOTTER, ALBEIT WITH A COMPLICATED AND SPATIALLY CHANGING HINGE LINE. MOST OF FB TO RISING ASTH. IF ASTH RISES ALONG MAJOR DETACHMENTS, PLUME MAY SPLIT INTO EDDIES. DEPOSITIONAL PATTERNS DURING THIS STAGE MAINLY CONTINUATION OF PREVIOUS STAGE.

STAGE 6: JUVENILE OCEANIC RIFT BASIN (JORB)
DEVELOPMENT OF ONE OR MORE JORB. DRAWING SHOWS SITUATION OF SIMULTANEOUSLY ACTIVE JORB SEPARATED BY FB. CENTRAL BLOCK CAN BECOME COMPLETELY ISOLATED IF BOUNDED BY TRANSFORMS IN OTHER PLANE. IFB PROBABLY UPLIFTED RELATIVE TO LEVEL IN STAGE 4, RESULTING IN REDEFORMATION OF BLOCK AND EROSION AND REWORKING OF SEDIMENT PILE ON TOP OF FB.
COMPARE TO STAGE IV OF LOWELL AND GENIK (1972).

STAGE 7: SUCCESSFUL SPREADING SYSTEM (SSS)
CONTINUED OPENING OF ONE JORB AND EVENTUAL ABANDONMENT OF OTHER(S). SITUATION DEPICTED HERE CAN OCCUR AT ANY TIME FROM STAGE 4 ONWARD. SITUATION DEPICTED HERE SHOWS SMALL ARB, BUT FEATURES CAN BE LARGE, THICKLY SEDIMENTED BASINS UNDERLAIN BY BROAD ZONE OF AL. MECHANICALLY, FB BECOMES ATTACHED TO ONE OF PLATES BUT GEOPHYSICALLY AND PETROLOGICALLY IT WILL APPEAR AS A CONTINENTAL "MARSHMALLOW" RIDDLED WITH RIFT-TYPE STRUCTURES. IF BOTH JORB REMAIN ACTIVE FOR LENGTHY PERIOD, IFB MAY BECOME A MICROCONTINENT.

AFG = APPARENT FULL-GRABEN
AL = ACCRETED LITHOSPHERE
ARB = ABANDONED RIFT BASIN
ASTH = ASTHENOSPHERE
BB = BACK BASIN
CL = CONTINENTAL LITHOSPHERE
CM = CONTINENTAL "MARSHMALLOW"
FB = FREE BLOCK
FBFS = FACING BORDER FAULT SYSTEM
IBFS = INCIPIENT BORDER FAULT SYSTEM
IFB = ISOLATED FREE BLOCK
JHG = JUVENILE HALF-GRABEN
JORB = JUVENILE OCEANIC RIFT BASIN
MHG = MATURE HALF-GRABEN
OBFS = ORIGINAL BORDER FAULT SYSTEM
PTF = PLATFORM
RL = REFERENCE LEVEL
RS = RAMPING SIDE OF HALF-GRABEN
SSS = SUCCESSFUL SPREADING SYSTEM
ST = STRETCH TROUGH

Tanganyika Rift Zone. We might say that continental rift zones are the "leftovers" of backtracking plates to their times of opening (but not to their times of rifting).

If the Tanganyika Rift Zone were to evolve into a successful oceanic rift, it could not possibly do so in any symmetric conjugate sense. Indeed, it is impossible to produce biaxial symmetry by splitting the map of Tanganyika (Figure 8) along strike. Bally (1982) flirted with the same conclusion, but the "fitting-together" problem is even more complicated than he imagined because the original fracture patterns are usually not orthorhombic. Let us continue by imagining the hypothetical conjugate margin that would result from the spreading of Figure 8. For the same reason that the fracture margins of high-relief accommodation zones are possible loci of early volcanism, they are also the preferred track of an incipient plate boundary. In a sense, the plate boundary partially exists in the vicinity of the high-relief accommodation zones. The next step is connecting these incipient boundaries together. The connections almost certainly must occur within the half-graben units, either along low-relief accommodation zones or border fault systems. Where both elements exist (e.g. cases A–C, Figure 6a), mechanical decoupling is probably easier along low-relief accommodation zones. Where the connection must occur through a simple half-graben, the border fault system is the convenient choice. Even so, there are no compelling reasons why the interconnections need to follow the surficial trace of the border fault systems, especially if the border fault systems flatten out with depth and intersect the ductile zone within the upper 15–20 km of crust. Figure 10 shows two possible geometries after successful splitting. Simple cross-sectional examples at various positions along the margins of Figure 10b are provided in Figure 10c. These hypothetical cross sections are comparable to those postulated for various passive margins, and they also provide a ready explanation for the variability observed along most passive margins.

Which elements of the fundamental unit become spreading centers and which transform faults? Our intuition would argue that high-relief accommodation zones should be the prototransforms, and a rather compelling case could be derived for situations where the orientation of the stress field remains relatively constant from the onset of rifting through successful splitting. If the stress field changes orientation significantly at the time of successful splitting and our suppositions regarding volcanism are correct, then it is conceivable that border fault systems occasionally could evolve into transforms rather than spreading centers. The breakup scheme in Figure 10 assumes the former case, but it is obvious that either case would result in very asymmetric conjugate margins.

A related issue concerns the proposition that oceanic ridge-ridge transforms can be extended across passive margins and into continental lineaments. This idea began with Wilson's (1965) suggestion that mid-ocean ridge offsets should mimic changes in continental margin orientations. The idea gained popularity from efforts such as Hayes & Ewing (1970) and Arens et al (1971). Extensions into continental zones of weakness have been proposed by Snelgrove (1967), Girdler (1968), Garson & Mitchell (1970), Fail et al (1970), Fuller (1971), Garson & Krs (1976), Fletcher et al (1978), and Sykes (1978), to name just a few. The above discussion, along with Figure 10, suggests that such extensions are possible only where the following circumstances prevail: (*a*) High-relief accommodation zones follow preexisting structural grains of regional extent; and (*b*) high-relief accommodation zones evolve into transforms, which maintain the same orientation. This requires that the stress field not undergo any significant deviation between the continental rifting phase and the spreading phase.

Given these circumstances, especially the unlikelihood of the last condition, it is difficult to accept that a majority of ridge-ridge transforms maintain pole-of-rotation trajectories across hundreds of kilometers of continent. The usual case is probably closer to what Reynolds (1984) has described and what I have shown here in Figure 10. This implies that transform continuations such as those shown in Figures 5 and 7 of Garson & Krs (1976) may be incorrect.

If ridge-ridge transform faults ultimately owe their existences to high-relief accommodation zones (interbasinal ridges), which is probably the general case, then we could conclude that the geometry of trailing plate edges is largely implanted at the onset of the continental rifting phase. Presumably, this applies to such parameters as transform-fault spacing and perhaps even spreading-center placements. It seems likely that the regularities in spacings described by Schouten & Klitgord (1982a,b) are inherited from the regularities in spacings of high-relief accommodation zones. Another consequence of the breakup theme described herein is the prediction that seismic traverses parallel to passive margins ought to show basin-swell-basin morphology. The first-order wavelength should be the average distance between high-relief accommodation zones—about 120 km for the East Africa System. The strike profiles from offshore Brazil (e.g. Asmus & Guazelli 1981) seem to support such a prediction, as do many proprietary lines from petroleum companies.

Concluding Statement

I have molded this review around the idea that African rift morphology has a measure of architectural stability, repeatability, and even predictability with regard to certain parameters. The key to understanding the architectural patterns lay in recognizing the various ways in which rift

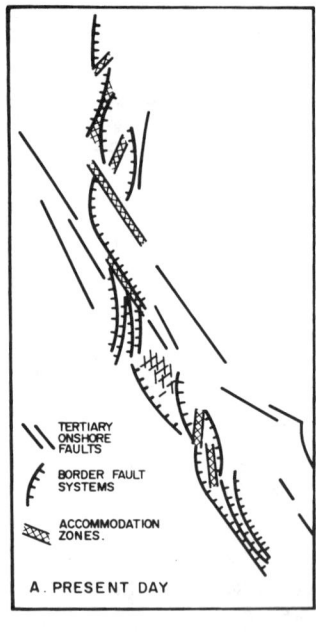
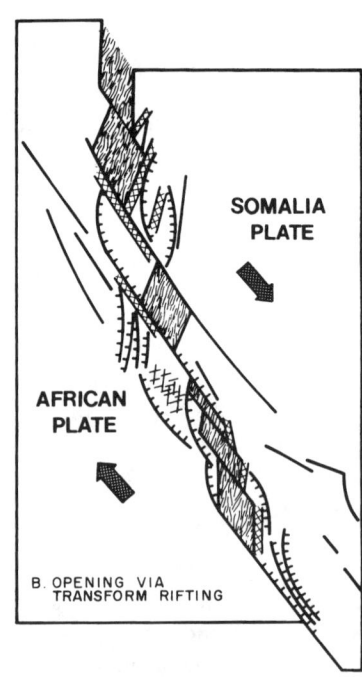

a)

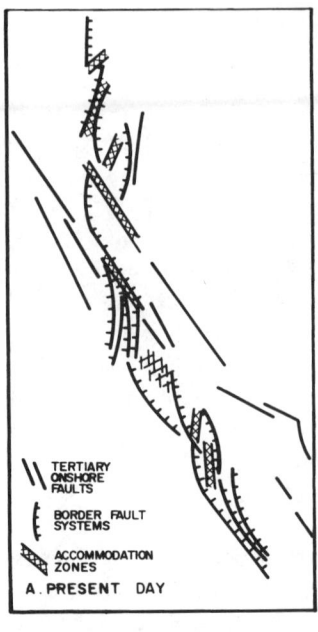
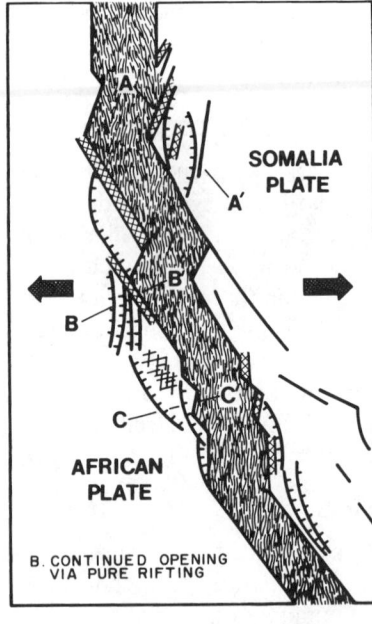

b)

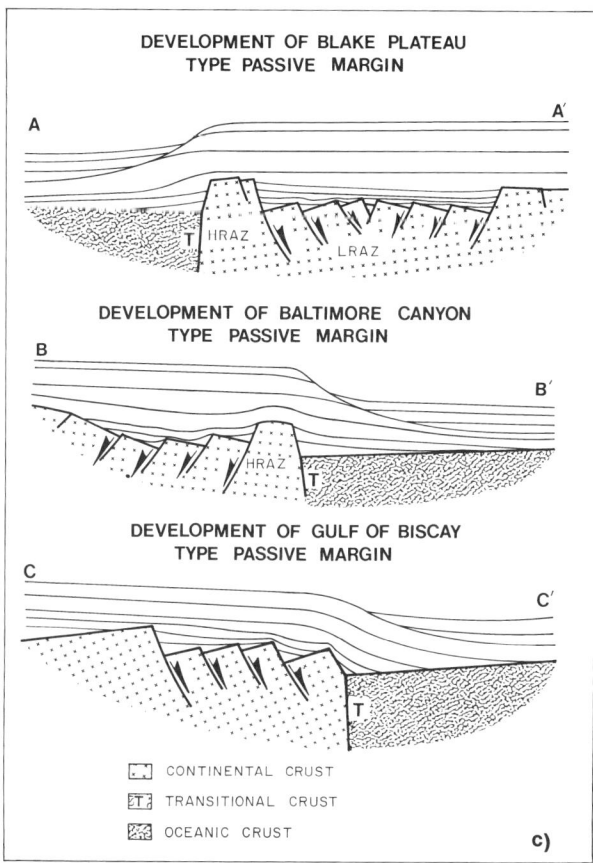

Figure 10 Hypothetical splitting of the Tanganyika Rift by transform (Figure 10*a*) and pure orthogonal (Figure 10*b*) rifting. Figure 10*c* shows passive-margin settings along the traverses shown in Figure 10*b*. Note that the profiles in Figure 10*c* also apply to Figure 10*a*.

units tend to link together. The impact of this geometric theme on our understanding of continental rifting, in Africa and elsewhere, could be very significant. Given what is at stake, perhaps it is fitting to close this paper on a cautionary note. I use the same admonition that Suess used to close his famous treatise on African rifts (Suess 1891):

> In all representations of this kind we have, however, to guard against the assumption of a general arrangement of any kind: indeed, considering the almost incomprehensible variety of the occurrences, a systematic search for regularity is not without danger, because the inquiring mind is so easily led astray from the path of sound synthesis.

Acknowledgments

The original data used in this manuscript were acquired with the financial support of the following petroleum companies: Agip, Amoco, Arco, Conoco, Eason, Esso, Marathon, Mobil, Pecten, Pennzoil, Petrofina, Shell Internationale, and Texaco. The World Bank also provided financial assistance. Many people helped acquire and process the seismic data. A special debt of gratitude is owed to J. McGill and D. Scott, whose tireless efforts made data acquisition possible. I am also indebted to J. Buck, T. Davis, G. Howell, K. Kaczmarick, R. Roessler, W. Schwede, and T. Specht for data processing. Some of the new ideas and concepts described herein are an outgrowth of discussions with my graduate students, especially C. Burgess, T. Dunkelman, P. Lorber, D. Reynolds, S. Sander, and D. Scott. Considerable background research for this manuscript was conducted by C. Blankenship, and assistance in manuscript preparation was provided by M. Bradley, C. Eubank, T. Marchant, and D. Tyler. Figures were drafted by L. Wood. Critical reviews were kindly provided by M. Bishop, T. Dunkelman, J. Flannery, T. Johnson, D. Scott, J. Versfelt, and W. Wheeler.

Special thanks go to the governments of Burundi, Kenya, Malawi, Mozambique, Tanzania, Zaire, and Zambia for issuing research permits and licenses and for providing technical and logistical help.

Literature Cited

Almond, D. C. 1969. Structure and metamorphism of the Basement Complex in Northeast Uganda. *Overseas Geol. Miner. Resour.* 10: 146–63

Andrew, A. R., Bailey, T. E. 1910. The geology of Nyasaland. *Q. J. Geol. Soc. London* 66: 198–252

Andrew, E. M. 1974. Gravity survey of Malawi: field work and processing. *Inst. Geol. Sci. London Rep.* 74: 1–36

Angelier, J., Colletta, B. 1983. Tension fractures and extensional tectonics. *Nature* 301: 49–51

Arens, G., Detteil, J. R., Valery, B., Damotte, B., Montadert, L., Patriat, P. 1971. The continental margin of the Ivory Coast and Ghana. *Inst. Geol. Sci. London Rep.* 70: 61–78

Artyushkov, E. V. 1981. Mechanisms of continental riftogenesis. *Tectonophysics* 73: 9–14

Asmus, H. E., Guazelli, E. 1981. Descricão sumária das estruturas da margem continental Brasileira e das áreas oceanicas e continentais, adjacentes. *CENPES/PETROBRAS, Serie Projecto REMAC No. 9*, PetroBras, Rio de Janeiro

Atti Conv. Lincei. 1980. *Geodynamic Evolution of the Afro-Arabian Rift System No. 47. Int. Meet., Rome, 1979.* Rome: Accad. Naz. Lincei

Aydin, A., Nur, A. 1982. Evolution of pull-apart basins and their scale independence. *Tectonics* 1: 91–105

Bahat, D. 1979. Interpretation on the basis of Hertzian theory of a spiral carbonatite structure at Homa Mountain, Kenya. *Tectonophysics* 60: 235–46

Bahat, D., Rabinovitch, A. 1980. On the initiation of the East African Rift System: a quantitative analysis. See Atti Conv. Lincei 1980, pp. 25–33

Bailey, D. K. 1978. Continental rifting and mantle degassing. See Neumann & Ramberg 1978, pp. 1–13

Baker, B. H. 1958. Geology of the Magadi area. *Geol. Surv. Kenya Rep.* 42: 1–81

Baker, B. H. 1963a. Geology of the area south of Magadi. *Geol. Surv. Kenya Rep.* 61: 1–27

Baker, B. H. 1963b. Geology of the Baragoi area. *Geol. Surv. Kenya Rep.* 53: 1–74

Baker, B. H. 1965. The Rift System in Kenya. See Drake & Loupekine 1965, pp. 82–84

Baker, B. H. 1970. The structural pattern of the Afro-Arabian rift system in relation to plate tectonics. *Philos. Trans. R. Soc. London* 267: 383–91

Baker, B. H. 1971. Explanatory note on the structure of the southern part of the African rift system. See UNESCO 1971, pp. 543–48

Baker, B. H., Morgan, P. 1981. Continental rifting: progress and outlook. *Eos, Trans. Am. Geophys. Union* 62: 585–86

Baker, B. H., Wohlenberg, J. 1971. Structure and evolution of the Kenya rift valley. *Nature* 229: 538–42

Baker, B. H., Williams, L., Miller, J., Fitch, F. 1971. Sequence and geochronology of the Kenya rift volcanics. *Tectonophysics* 11: 191–215

Baker, B. H., Mohr, P. A., Williams, L. A. J. 1972. The geology of the Eastern Rift System of Africa. *Geol. Soc. Am. Spec. Pap. No. 136.* 66 pp.

Baker, B. H., Crossley, R., Goles, G. G. 1978. Tectonic and magmatic evolution of the southern part of the Kenya Rift Valley. See Neumann & Ramberg 1978, pp. 29–50

Bally, A. W. 1980. Basins and subsidence—a summary. In *Dynamics of Plate Interiors. Geodyn. Ser.*, ed. A. W. Bally, P. Bender, T. McGetchin, R. Walcott, 1: 5–20. Washington, DC: Am. Geophys. Union/Geol. Soc. Am.

Bally, A. W. 1982. Musings over sedimentary basin evolution. *Philos. Trans. R. Soc. London* 305: 325–38

Bally, A. W., ed. 1983. *Seismic Expression of Structural Styles—A Picture and Work Atlas*, Vol. 2. Tulsa: Am. Assoc. Pet. Geol. 290 pp.

Bally, A. W., Snelson, S. 1980. Realms of subsidence. *Can. Soc. Pet. Geol. Mem.* 6: 9–94

Bally, A. W., Bernoulli, D., Davis, G. A., Montadert, L. 1981. Listric normal faults. *Oceanol. Acta* 4: 87–101

Barnes, J. W., Hepworth, J. V. 1961. The mineral resources of Uganda. *Bull. Uganda Geol. Surv.* 4: 1–89

Bath, M. 1975. Seismicity of the Tanzania region. *Tectonophysics* 27: 353–79

Behrensmeyer, A. K. 1976. Lothagam Hill, Kanapoi, and Ekora: a general summary of stratigraphy and faunas. See Coppens et al 1976, pp. 163–70

Bellieni, G. E. 1981. Oligocene transitional tholeiitic magmatism in North Turkana (Kenya): comparison with coeval Ethiopian volcanism. *Bull. Volcanol.* 44: 411–27

Bishop, W. W. 1965. Quaternary geology and geomorphology in the Albertine rift valley, Uganda. In *International Studies on the Quaternary*, ed. H. E. Wright, E. G. Frey, pp. 293–321. Boulder, Colo: Geol. Soc. Am.

Bishop, W. W. 1969. Pleistocene stratigraphy in Uganda. *Geol. Surv. Uganda Mem.* 10: 1–128

Bishop, W. W., ed. 1978. *Geological Background to Fossil Man: Recent Research in the Gregory Rift Valley, East Africa.* Buffalo: Toronto Univ. Press

Bishop, W. W., Trendall, A. F. 1967. Erosion surfaces, tectonics, and volcanic activity in Uganda. *Q. J. Geol. Soc. London* 122: 389–420

Bloomfield, K. 1966. A major east-northeast dislocation zone in central Malawi. *Nature* 211: 612–14

Bloomfield, K. 1968. The pre-Karroo geology of Malawi. *Geol. Surv. Malawi Mem.* 5: 1–166

Blundell, D. J. 1978. A gravity survey across the Gardar igneous province, southwest Greenland. *J. Geol. Soc. London* 135: 545–54

Bornhardt, W. 1900. Zur Oberflachergestaltung und Geologie Deutsch-ost-Afrikas. In *Deutsch-ost-Afrika*. Berlin: Reiner. 595 pp.

Bott, M. H. P. 1976. Formation of sedimentary basins of graben type by extension of the continental crust. *Tectonophysics* 36: 77–86

Bott, M. H. P. 1981. Crustal doming and the mechanism of continental rifting. *Tectonophysics* 73: 1–8

Bott, M. H. P. 1982a. The mechanism of continental splitting. *Tectonophysics* 81: 301–9

Bott, M. H. P. 1982b. Origin of the lithospheric tension causing basin formation. *Philos. Trans. R. Soc. London* 305: 319–24

Bott, M. H. P., Kusznir, N. J. 1979. Stress distribution associated with compensated plateau uplift structures with application to the continental splitting mechanism. *Geophys. J. R. Astron. Soc.* 56: 451–59

Bott, M. H. P., Mithen, D. P. 1983. Models of rifting: mechanism of graben formation—the wedge subsidence hypothesis. *Tectonophysics* 94: 11–22

Boustakoff, N. 1933. Le coude du systeme de fractures du graben central Africain au lac Kivu et sa ramification dans la cuvette Congolaise. *Bull. Soc. Belge Geol. Paleontol. Hydrol.* 43: 80–85

Bowen, B. E., Vondra, C. F. 1973. Stratigraphical relationships of the Plio-Pleistocene deposits, East Rudolf, Kenya. *Nature* 242: 391–93

Browne, S. E., Fairhead, J. D. 1983. Gravity study of the central African rift system: a model of continental disruption. See Morgan 1983, pp. 187–204

Burgess, C. F. 1985. *The structural and strati-*

graphic evolution of Lake Tanganyika: a case study of continental rifting. MS thesis. Duke Univ., Durham, N.C. 46 pp.

Burgess, C. F., Rosendahl, B. R., Sander, S., Burgess, C. A., Lambiase, J., et al. 1986. The structural and stratigraphic evolution of Lake Tanganyika: a case study of continental rifting. In *The Triassic System of the Eastern United States. Am. Assoc. Pet. Geol. Spec. Publ.* In press

Burke, K. 1976a. Development of graben associated with the initial ruptures of the Atlantic Ocean. *Tectonophysics* 36: 93–111

Burke, K. 1976b. The Chad Basin: an active intra-continental basin. *Tectonophysics* 36: 197–206

Burke, K. 1977. Aulacogens and continental breakup. *Ann. Rev. Earth Planet. Sci.* 5: 371–96

Burke, K. 1978. Evolution of continental rift systems in the light of plate tectonics. See Ramberg & Neumann 1978a, pp. 1–9

Burke, K., Whiteman, A. J. 1973. Uplift, rifting, and the break-up of Africa. See Tarling & Runcorn 1973, pp. 735–55

Butzer, K. W. 1970. Contemporary depositional environments of the Omo Delta. *Nature* 226: 425–30

Butzer, K. W., Thurber, D. L. 1969. Some Late Cenozoic formations of the lower Omo Basin. *Nature* 222: 1135–38

Cahen, L. 1953. *Esqisse Tectonique du Congo Belge a l'echelle der 1/3,000,000.* Brussels: Comm. Geol. Min. Colon.

Cahen, L. 1954. *Geologie du Congo Belge.* Liege: Vaillant-Carmanne

Capart, A. 1949. Exploration hydrobiologique de Lac Tanganyika resultats scientifiques. *Bull. Inst. R. Sci. Nat. Belg.* 2: 1–16

Carter, G., Bennett, J. 1973. The geology and mineral resources of Malawi. *Geol. Soc. Am. Bull.* 84: 3–14

Cerling, T. E., Powers, D. W. 1977. Paleorifting between the Gregory and Ethiopian Rifts. *Geology* 5: 441–44

Chapin, C. E. 1979. Evolution of the Rio Grande Rift—a summary. In *Rio Grande Rift: Tectonics and Magmatism*, ed. R. E. Reicher, pp. 1–5. Washington, DC: Am. Geophys. Union

Chapman, D. S., Pollack, H. N. 1977. On the regional variation of heat flow, geotherms, and lithospheric thickness. *Tectonophysics* 38: 279–96

Chase, C. G., Gilmer, T. H. 1973. Precambrian plate tectonics: the midcontinent gravity high. *Earth Planet. Sci. Lett.* 21: 70–78

Chenet, P. Y., Montadert, L. 1981. Rifting tectonics on Galicia, Portugal and N. Biscay margin. See *Papers Presented to the Conference on the Processes of Planetary Rifting* 1981, pp. 47–50

Chenet, P. Y., Montadert, L., Gairaud, H., Roberts, D. 1982. Extension ratio measurements on the Galicia, Portugal, and Northern Biscay continental margins: implications for evolutionary models of passive continental margins. *Am. Assoc. Pet. Geol. Bull.* 34: 703–15

Chorowicz, J., Mukonki, M. N. B. 1980. Lineaments anciens, zones transformantes recentes et geotectonique des fosses de l'est African, d'apres la teledetection et la microtectonique. *Rep.*, Mus. R. Afr. Cent., Tervuren Belg., Dep. Geol. Mineral.

Chorowicz, J., Mukonki, M. N. B., Pottier, Y. 1979. Evidence of horizontal compression linked with the opening of the East African rifts (Western branch), in the ridge between the Kivu and the Tanganyika lakes. *C. R. Somm. Seances Soc. Geol. Fr.* 5–6: 231–34

Cloos, E. 1955. Experimental analysis of fracture patterns. *Geol. Soc. Am. Bull.* 66: 241–56

Cloos, E. 1961. Bedding slips, wedges, and faulting in layered sequences. *C. R. Soc. Geol. Finlande* 33: 105–22

Cloos, E. 1968. Experimental analysis of Gulf Coast fracture patterns. *Am. Assoc. Pet. Geol. Bull.* 52: 420–44

Cloos, H. 1928. Über antithetische Bewegungen. *Geol. Rundsch.* 19: 246–51

Cloos, H. 1931. Fliessen und Brechen in der Erdkurste und im geologischen Experiment, plastiche Massen. *Wiss. Tech.* 1: 1–6

Cloos, H. 1939. Hebung, Splatung, Vulkanismus. *Geol. Rundsch.* 30: 403

Combe, A. D. 1930. Volcanic areas of Bunyaruguru and Fort Portal. *Ann. Rep. Geol. Surv. Uganda*, Vol. 16

Cooke, A. 1957. Observations relating to Quartenary environments in east and southern Africa. *Geol. Soc. S. Afr. Bull.* 60: 1–73

Coppens, Y., Howell, F. C., Isaacs, G. L., Leakey, R. E. F., eds. 1976. *Earliest Man and Environments in the Lake Rudolph Basin.* Chicago: Univ. Chicago Press

Cornet, A. 1905. Les dislocations du bassin du Congo. 1. Le Graber de l'Upemba. *Bull. Geol. Soc. Belge* 32: 338–39

Cotton, C. A. 1950. Tectonic scarps and fault valleys. *Geol. Soc. Am. Bull.* 61: 717–57

Crossley, R. 1982. Sedimentation in an active mid-plate continental trough: the Malawi Rift, Africa. *Int. Congr. Sedimentol., 11th,* 11: 1–35

Crossley, R., Crow, M. J. 1980. The Malawi Rift. See Atti Conv. Lincei 1980, pp. 77–87

Dainelli, G., ed. 1943. *Geologia dell'Africa Orientale*, Vol. 3. Reale Accad., Italy

Davies, K. A. 1951. The Uganda section of the Western Rift. *Geol. Mag.* 88: 377–85

Davies, K. A., Bisset, C. 1947. The geology and mineral deposits of Uganda. In *Mineral Resources*, 45: 161–80. London: Imperial Inst.

de Beaumont, E. 1827. Observations geologiques. *Ann. Mines* 2: 5–82

de Beaumont, E. 1830. Memoire: annales des sciences naturelles. *Ann. Sci. Nat.*, Vol. 18–19

de Beaumont, E. 1844. Note sur le rapport qui existe le refroidissement progressif de la masse du globe terrestre et celui de la surface. *C. R. Acad. Sci.* 19: 1327

de Beaumont, E. 1847. Note sur les systemes de montagnes les plus anciens de l'Europe. *Bull Soc. Geol. Fr. 2nd. Ser.* 4: 864–991

de Charpal, O., Guennoc, P., Montadert, L., Roberts, D. G. 1978. Rifting, crustal attenuation and subsidence in the Bay of Biscay. *Nature* 275: 706–11

de Lapparent, A. 1887. Conference sur le sens des mouvements de l'ecoree terrestre. *Bull. Soc. Geol. Fr. 3rd Ser.* 15: 220

de Lapparent, A. 1898. Soulivements et Affaissements. *Rev. Questions Sci.* 14: 5–33

de Martonne, E. 1897. Die hydrographie des oberen Nilbeckers. *Z. Ges. Erdkd.* 32: 315

de Swardt, A. M. J. 1965. Rift faulting in Zambia. See Drake & Loupekine 1965, pp. 105–14

de Swardt, A. M. J., Trendall, A. F. 1969. The physiographic development of Uganda. *Overseas Geol. Miner. Resour.* 10: 207–88

de Swardt, A. M. J., Garrard, P., Simpson, J. G. 1965. Major zones of transcurrent dislocation and superposition of orogenic belts in parts of central Africa. *Geol. Soc. Am. Bull.* 76: 86–102

Degens, E. T., von Herzen, R. P., Wong, H. K. 1971. Lake Tanganyika: water chemistry, sediments, and geological structure. *Naturwissenschaften* 58: 229–40

Degens, E. T., von Herzen, R. P., Wong, H. K., Deuser, W. G., Jannasch, H. W. 1973. Lake Kivu: structure, chemistry and biology of an East African rift lake. *Geol. Rundsch.* 62: 245–77

Dixey, F. 1929. *Annual Report of the Geological Survey Department of Nyasaland Protectorate for the Year 1928*. Zomba: Geol. Surv. Malawi

Dixey, F. 1937a. The geology of part of the Luangwa Valley, northeastern Rhodesia. *Q. J. Geol. Soc. London* 93: 52–74

Dixey, F. 1937b. The pre-Karroo landscape of the Lake Nyasa region, and a comparison of the Karroo structural directions with those of the rift valley. *Q. J. Geol. Soc. London* 93: 77–93

Dixey, F. 1937c. The early Cretaceous and Miocene peneplains of Nyasaland, and their bearing on the age and development of the rift valley. *Geol. Mag.* 74: 49

Dixey, F. 1937d. *Nyasaland Protectorate: Annual Report of the Geological Survey Department for the Year 1936*. Zomba: Geol. Surv. Malawi. 21 pp.

Dixey, F. 1944a. African landscape. *Geogr. Rev.* 4: 457

Dixey, F. 1944b. The geomorphology of northern Rhodesia. *Trans. Geol. Soc. S. Afr.* 7: 9

Dixey, F. 1945a. On the faulting of rift valley structures. *Geol. Mag.* 82(3): 377–85

Dixey, F. 1945b. Erosion and tectonics in the east African rift system. *Q. J. Geol. Soc. London* 102: 339–88

Dixey, F. 1956. The East African Rift System. *Supp. Bull., Overseas Geol. Miner. Resour., No. 1*. 77 pp. London: H. M. Stationery Off.

Dixey, F. 1959. Vertical tectonics in the East African rift system: 20th Int. Geol. Congress. *Assoc. Afr. Geol. Surv.*, pp. 359–75

Dodson, M. H. 1963. Geology of the south Horr area. *Geol. Surv. Kenya Rep.* 60: 52–60

Drake, C. L., Loupekine, I. S., eds. 1965. *Report of the UMC/UNESCO Seminar on the Great African Rift System*. Nairobi: Nairobi Univ. Coll. 145 pp.

Drysdall, A. R., Kitching, J. 1963. A reexamination of the Karroo succession and fossil localities of part of the upper Luangwa valley. *Mem. Geol. Surv. N. Rhod.*, Vol. 1. Salisbury, Rhod.

Drysdall, A. R., Weller, R. K. 1966. Karroo sedimentation in northern Rhodesia. *Trans. Geol. Soc. S. Afr.* 69: 39–69

Dundas, D. L. 1965. Review of rift faulting in Tanzania. See Drake & Loupekine 1965, pp. 95–102

Dunkelman, T. J. 1986. *The structural and stratigraphic evolution of Lake Turkana, Kenya, as deduced from a multichannel seismic survey*. MS thesis. Duke Univ., Durham, N.C. 64 pp.

Easton, R. M. 1983. Crustal structure of rifted continental margins: geological constraints from the Proterozoic rock of the Canadian Shield. *Tectonophysics* 94: 371–90

Eaton, G. P. 1982. The Basin and Range Province: origin and tectonic significance. *Ann. Rev. Earth Planet. Sci.* 10: 409–40

Ebinger, C. J., Rosendahl, B. R. 1986. Tectonic model of the Malawi Rift, Africa. *Tectonophysics*. In press

Ebinger, C. J., Crow, M. J., Rosendahl, B. R., Livingstone, D. A., LeFournier, J. 1984. Structural evolution of Lake Malawi, Africa. *Nature* 308: 627–29

Effimoff, I., Pinezich, A. R. 1981. Tertiary structural development of selected valleys based on seismic data: Basin and Range Province, northeastern Nevada. *Philos. Trans. R. Soc. London* 300: 435–42

Elmohandes, S. 1981. The central European graben system: rifting imitated by clay modeling. *Tectonophysics* 73: 69–78

Fail, J. P., Montadert, L., Delteil, J. R., Valery, P., Patriat, P. h., Schlich, R. 1970. Prolongation des zones de fractures de l'Ocean Atlantique dans le Golfe de Guinee. *Earth Planet. Sci. Lett.* 7: 413–19

Fairhead, J. D. 1976. The structure of the lithosphere beneath the Eastern rift, East Africa, deduced from gravity studies. *Tectonophysics* 30: 269–98

Fairhead, J. D. 1980. The intra-plate volcanic centers of North Africa and their possible relation to the East African rift system. See Atti Conv. Lincei 1980, pp. 45–48

Fairhead, J. D., Browne, S. E. 1981. Rifting within the African continent: a model for continental disruption? See *Papers Presented to the Conference on the Processes of Planetary Rifting* 1981, pp. 67–71

Fairhead, J. D., Henderson, N. B. 1977. The seismicity of southern Africa and incipient rifting. *Tectonophysics* 41: 19–26

Fairhead, J. D., Walker, P. 1974. A detailed gravity study of the crustal structure associated with the Kavirondo Rift Valley, East Africa. See Atti Conv. Lincei 1980, pp. 99–109

Faller, A. M., Soper, N. J. 1979. Paleomagnetic evidence for the origin of the coastal flexure and dyke swarm in central East Greenland. *J. Geol. Soc. London* 136: 737–44

Fischer, A. G. 1975. Origin and growth of basins. See Fischer & Judson 1975, pp. 47–79

Fischer, A. G., Judson, S., eds. 1975. *Petroleum and Global Tectonics*. Princeton, NJ: Princeton Univ. Press. 212 pp.

Fletcher, J. B., Sbar, M., Sykes, L. 1978. Seismic trends and travel-time residuals in eastern North America and their tectonic implications. *Geol. Soc. Am. Bull.* 89: 1656–76

Florensov, N. A. 1969. Rifts of the Baikal mountain region. *Tectonophysics* 8: 443–56

Freund, R., Merzer, A. M. 1976. The formation of rift valleys and their zigzag fault patterns. *Geol. Mag.* 113: 561–68

Fuchs, V. E. 1934. The geological work of the Cambridge expedition to the East African lakes, 1930–31. *Geol. Mag.* 71: 97–112

Fuchs, V. E. 1935. The Lake Rudolf rift valley expedition, 1934. *Geol. J.* 86: 114–42

Fuchs, V. E. 1939. The geological history of the Lake Rudolf Basin, Kenya Colony. *Philos. Trans. R. Soc. London* 229: 219

Fuller, A. O. 1971. South Atlantic fracture zones and lines of old weakness in South Africa. *Nature* 231: 84–85

Furon, R. 1963. *Geology of Africa*. London: Oliver & Boyd. 377 pp.

Gair, H. S. 1959. The Karroo system and coal resources of the Gwembe District, north-east section. *Bull. Geol. Surv. N. Rhod. No. 1*. 86 pp.

Garson, M. S. 1965. Summary of present knowledge of the rift system in Malawi. See Drake & Loupekine 1965, pp. 92–94

Garson, M. S., Krs, M. 1976. Geophysical and geological evidence of the relationship of Red Sea transverse tectonics to ancient fractures. *Geol. Soc. Am. Bull.* 87: 169–81

Garson, M. S., Mitchell, A. H. G. 1970. Transform faulting in the Thai Peninsula. *Nature* 228: 45–47

Gass, I. G. 1970. The evolution of volcanism in the junction area of the Red Sea, Gulf of Aden and Ethiopian rifts. *Philos. Trans. R. Soc. London* 267: 369–81

Gass, I. G. 1973a. The Red Sea Depression: causes and consequences. See Tarling & Runcorn 1973, pp. 779–86

Gass, I. G. 1973b. East African Rifts. *Earth Sci. Rev.* 9: 281–82

Gass, I. G. 1975. Magmatic and tectonic processes in the development of the Afro-Arabian Dome. In *Afar Depression of Ethiopia*, ed. A. Pilger, pp. 10–18. Stuttgart: E. Schweiz Verlagsbuchhandlung

Gautier, A. 1965. Relative dating of peneplains and sediments in the Lake Albert Rift area. *Am. J. Sci.* 263: 537–47

Gautier, A. 1967. New observations on the later Tertiary and early Quaternary in the Western Rift: stratigraphic and paleontological evidence. In *Background to Evolution in Africa*, ed. W. W. Bishop, J. D. Clark, pp. 73–88. Chicago: Univ. Chicago Press

Ghignone, J. I. l., Andrade, G. d. 1969. General geology and major oil fields of Reconcavo Basin, Brazil. In *Geology of Giant Petroleum Fields. Am. Assoc. Pet. Geol. Mem.*, ed. M. T. Halbouty, 14: 337–58

Gibbs, A. D. 1983. Balanced cross-section construction from seismic sections in areas of extensional tectonics. *J. Struct. Geol.* 5: 153–60

Gibbs, A. D. 1984. Structural evolution of extensional basin margins. *J. Geol.* 141: 609–20

Girdler, R. W. 1968. Drifting and rifting of Africa. *Nature* 217: 1102–6

Girdler, R. W., Sowerbutts, W. T. C. 1970.

Some recent geophysical studies of the rift system in East Africa. *J. Geomagn. Geoelectr.* 22: 153–63

Girdler, R. W., Fairhead, R. W., Searle, R. C., Sowerbutts, W. T. C. 1969. Evolution of rifting in Africa. *Nature* 224: 1178–82

Gregory, J. W. 1896. *The Great Rift Valley.* London: John Murray. 424 pp.

Gregory, J. W. 1920. The African rift valleys. *Geogr. J.* 66: 31

Gregory, I. W. 1921. *The Rift Valleys and Geology of East Africa.* London: Seeley, Service. 479 pp.

Guernsey, T. D. 1951. A summary of the provisional geological features of Northern Rhodesia. *Colon. Geol. Miner. Resour.* 1: 121–51

Handin, J. 1969. On the Coulomb-Mohr failure criterion. *J. Geophys. Res.* 74: 5343–48

Harding, T. P. 1984. Graben hydrocarbon occurrences and structural style. *Am. Assoc. Pet. Geol. Bull.* 68: 333–62

Harding, T. P. 1985. Seismic characteristics and identification of negative flower structures, positive flower structures, and positive structural inversion. *Am. Assoc. Pet. Geol. Bull.* 69: 582–600

Harding, T. P., Lowell, J. D. 1979. Structural styles, their plate-tectonic habitats, and hydrocarbon traps in petroleum provinces. *Am. Assoc. Pet. Geol. Bull.* 63: 1016–58

Harkin, D. A. 1955. The geology of the Songwe-Kiwira coalfield, Rungwe District. *Geol. Surv. Tanganyika Bull.* 27: 12–13

Harkin, D. A. 1960. The Rungwe volcanics at the northern end of Lake Nyasa. *Geol. Surv. Tanganyika Mem.* 2: 169–72

Harpum, J. R. 1955a. Recent investigations in pre-Karroo geology in Tanganyika. *Assoc. Serv. Geol. Afr. Rep.*, pp. 165–215

Harpum, J. R. 1955b. Further comments on the age relationships of the Konse, Buanji, and Ukinga series of Tanganyika. *Assoc. Serv. Geol. Afr. Rep.*, pp. 217–18

Harris, N., Pallister, J., Brown, J. 1956. Oil in Uganda. *Geol. Surv. Uganda Mem.* 9: 1–33

Harvey, E., Teale, E. 1933. A physiographical map of Tanganyika Territory. *Geogr. Rev.* 23: 401–13

Hayes, D., Ewing, M. 1970. Northern Brazilian Ridge and adjacent continental margin. *Am. Assoc. Pet. Geol. Bull.* 54: 2120–51

Heinzelin, J. 1955. Le fosse tectonique sous le parallele d'Ishargo. In *Mission J. de Heinzelin de Braucort*, pp. 1–150. Brussels: Congo Belge Inst. Parcs Nat.

Hepworth, J. V. 1962. The relative ages of plateau and plain in West Nile district as indicated by Quaternary erosion surfaces. *Rech. Geol. Surv. Uganda* 1957–8: 37–45

Hepworth, J. V. 1966. A photogeological recognition of ancient orogenic belts in Africa. *Q. J. Geol. Soc. London* 491: 253–92

Hepworth, J. V., MacDonald, R. 1966. Orogenic belts of the northern Uganda basement. *Nature* 210: 726–27

Hirlemann, G. 1974. The structural setup of the Ribeauville fracture field, Haut-Rhin, France: an example of tectonic fragmentation (block-faulting) on the western border of the Rhine Graben. See Illies & Fuchs 1974, pp. 172–76

Holmes, A. 1916. Notes on the structure of the Tanganyika–Nile Rift Valley. *Geogr. J.* 48: 149–59

Holmes, A. 1942. A suite of volcanics from southwest Uganda containing kalsilite. *Mineral. Mag.* 26: 197

Holmes, A. 1944. *Principles of Physical Geology.* London: Nelson. 532 pp. 1st ed.

Holmes, A. 1965. *Principles of Physical Geology.* London: Nelson. 1288 pp. 2nd ed.

Holmes, A., Harwood, H. F. 1932. Petrology of the volcanic fields east and southeast of Ruwenzori, Uganda. *Q. J. Geol. Soc. London* 88: 370–442

Holmes, A., Harwood, H. F. 1937. The volcanic area of Bufumbira, part II. The petrology of the volcanic field of Bufumbira, southwest Uganda, and of other parts of the Birunga Field. *Geol. Surv. Uganda Mem.* 3: 1–300

Hopson, A. J. 1982. In *Lake Turkana: A Report on the Findings of the Lake Turkana Project, 1972–1975*, pp. 1–6. London: Overseas Dev. Off. 1616 pp.

Hopwood, A. T., Lepersonne, J. 1953. Preseice de formations d'age Miocene inferieur dans le fosse tectonique du lac Albert et de la Basse Semliki (Congo Belge). *Bull. Geol. Soc. Belge* 77: 83–113

Howell, F. C. 1968. Omo research expedition. *Nature* 219: 567–72

Hutchinson, R. W., Engels, G. G. 1972. Tectonic evolution in the southern Red Sea and its possible significance to older rifted continental margins. *Geol. Soc. Am. Bull.* 83: 2989–3002

Illies, J. H. 1971. Der Oberrheingraben. *Fridericiana* (*Univ. Karlsruhe*) 9: 17–32

Illies, J. H. 1972. The Rhine Graben System—plate tectonics and transform faulting. In *Geophysical Surveys*, 1: 27–60. Dordrecht: D. Reidel

Illies, J. H. 1978. Two stages in Rhinegraben rifting. See Ramberg & Neumann 1978a, pp. 172–76

Illies, J. H. 1981. Mechanism of graben formation. *Tectonophysics* 73: 249–66

Illies, J. H., Fuchs, K., eds. 1974. *Approaches to Taphrogenesis.* Stuttgart: E. Schweiz Verlag

James, T. C. 1956. The nature of rift faulting in Tanganyika. In *East-Central Reg. Comm. Geol. Meet., Comm. Tech., Afr. South of Sahara,* 1: 81–84

Johnson, D. 1930. Geomorphic aspects of rift valleys. *Int. Geol. Congr. S. Afr., 15th,* 2: 354–73

Johnson, T. C., Halfman, J. D., Rosendahl, B. R., Lister, G. S. 1986. Climatic and tectonic effects on sedimentation in a rift-valley lake: evidence from high-resolution seismic profiles, Lake Turkana, Kenya. *Geol. Soc. Am. Bull.* In press

Juch, D. 1980. Tectonics of the southeastern escarpment of Ethiopia. See Atti Conv. Lincei 1980, pp. 407–18

Kaufulu, Z., Urba, E., White, T. 1980. Age of the Chiwondo Beds, Northern Malawi. *Ann. Transvaal Mus.* 33: 1–8

Kazmin, V. 1980a. Transform faults in the East African Rift System. See Atti Conv. Lincei 1980, pp. 65–73

Kazmin, V. 1980b. Geodynamic control of rift volcanism. *Geol. Rundsch.* 69: 757–69

Keen, C. E. 1985. The dynamics of rifting: deformation of the lithosphere by active and passive driving forces. *Geophysics* 80: 95–120

Kent, P., Hunt, J., Johnstone, D. 1971. The geology and geophysics of coastal Tanzania. *Inst. Geol. Sci. Geophys. Rep.* 6: 101

King, B. C. 1970. Vulcanicity and rift tectonics in East Africa. In *African Magmatism and Tectonics,* ed. T. N. Clifford, I. G. Gass, pp. 263–83. Edinburgh: Oliver & Boyd

King, B. C. 1978a. Structural and volcanic evolution of the Gregory Rift Valley. See Bishop 1978, pp. 29–54

King, B. C. 1978b. A comparison between the older (Karroo) rifts and the younger (Cenozoic) rifts of eastern Africa. See Ramberg & Neumann 1978a, pp. 347–50

King, E., Zietz, I. 1971. Aeromagnetic study of the midcontinent gravity high of central United States. *Geol. Soc. Am. Bull.* 82: 2187–2207

Krenkel, E. 1922. *Die Bruchzonen Ostafrikas.* Berlin: Gebr. Borntraeger. 184 pp.

Krenkel, E. 1925. *Geologie Afrikas,* Vol. 1. Berlin: Gebr. Borntraeger. 210 pp.

Kumarapeli, P. 1978. The St. Lawrence paleo-rift system: a comparative study. See Ramberg & Neumann 1978a, pp. 367–84

Le Bas, M. 1971. Pre-alkaline volcanism, crustal swelling, rifting. *Nature Phys. Sci.* 229: 3027–31

LeFournier, J. 1980. Depots de preouverture de l'Atlantique Sud. Comparaison avec la sedimentation actuelle dans la branche occidentale de rifts Est-Africans. *Rech. Geol. Afr.* 5: 127–30

LeFournier, J., Chorowicz, J., Thouin, C., Balzer, F., Chenet, P. Y., et al. 1985. The Lake Tanganyika Basin: tectonic and sedimentary evolution. *Ac. C. R. Acad. Sci. Paris* 2: 1053–57

Lepersonne, J. 1949. Les grands traits de la geology du Kasai occidental et l'origine secondaire du diamant. *Bull. Geol. Soc. Belge* 58: 284–98

Lepersonne, J. 1956. Les Aplanessements d'erosion du nord-est du Congo Belge et des regiones voisines. *Mem. 8, Acad. R. Belg. Cl. Sci. Nat. Med.* 4: 108

Lepersonne, J. 1970. Revision of the fauna and the stratigraphy of the fossiliferous localities of the Lake Albert–Lake Edward rift (Congo). *Mus. Midden-Afr. Geol. Wet. Ann. Reeks* 67: 171–207

Le Pichon, X., Sibuet, J. C. 1981. Passive margins: a model of formation. *J. Geophys. Res.* 86: 3708–20

Lillie, R. J. 1984. Tectonic implications of subthrust structures revealed by seismic profiling of Appalachian-Ouachita orogenic belt. *Tectonics* 3: 619–46

Lister, L. 1967. Erosion surfaces in Malawi. *Rech. Geol. Surv. Malawi* 7: 15–28

Logatchev, N. A. 1978. Main features of evolution and magmatism of continental rift zones in the Cenozoic. See Ramberg & Neumann 1978a, pp. 351–66

Logatchev, N. A., Beloussov, V. V., Milanovsky, E. E. 1972. East African rift development. *Tectonophysics* 15: 71–81

Logatchev, N. A., Zorin, Y. A., Rogozhina, V. A. 1983. Cenozoic continental rifting and geologic formations (as illustrated by the Kenya and Baikal rift zones). *Geotectonics* 17: 83–92

Long, R., Backhouse, R. 1976. The structure of the western flank of the Gregory Rift, part 2: the mantle. *Geophys. J. R. Astron. Soc.* 44: 677–88

Lorber, P. M. 1984. *The Kigoma Basin of Lake Tanganyika: acoustic stratigraphy and structure of an active continental rift.* MS thesis. Duke Univ., Durham, N.C. 76 pp.

Lowell, J., Genik, G. 1972. Sea floor spreading and structural evolution of the southern Red Sea. *Am. Assoc. Pet. Geol. Bull.* 56: 247–59

Lowell, J., Genik, G., Nelson, T., Tucker, P. 1975. Petroleum and plate tectonics of the southern Red Sea. See Fischer & Judson 1975, pp. 129–53

Macdonald, R. 1968. "Charnockites" in the West Nile District of Uganda: a systematic study in the Groves' type area. *Int. Geol. Congr., 22nd,* 13: 227–49

MacDonald, R. 1969. *The Atlas of Uganda.* Entebbe: Dep. Land Surv. 2nd ed.

Maguire, P. K. H., Long, R. E. 1976. The structure on the western flank of the Gregory Rift (Kenya), Part 1: the crust. *Geophys. J. R. Astron. Soc.* 44: 661–75

Mboya, B. 1983. The genesis and tectonics of the N.E. Nyanza rift valley, Kenya. *J. Afr. Earth Sci.* 1: 315–20

McCall, G. J. H. 1964. Geology of the Sekern area. *Geol. Surv. Kenya Rep.* 65: 1–84

McCall, G. J. H. 1968a. Silali, another major caldera volcano in the rift valley of Kenya. *Proc. Geol. Soc. London* 1644: 267–68

McCall, G. J. H. 1968b. The five caldera volcanoes of the central rift valley, Kenya. *Proc. Geol. Soc. London* 1647: 54–59

McConnell, R. B. 1950. Outline of the geology of Ufipa and Ubende. *Geol. Surv. Tanganyika Bull.* 19: 1–62

McConnell, R. B. 1951. Rift and shield structure in East Africa. *Int. Geol. Congr., 18th,* 14: 199–207

McConnell, R. B. 1959. Outline of the geology of the Ruwenzori Mountains: a preliminary account of the results of the British Ruwenzori expedition 1951–1952. *Overseas Geol. Miner. Resour.* 7: 245–68

McConnell, R. B. 1967. The East African Rift System. *Nature* 215: 578–81

McConnell, R. B. 1972. Geological development of the rift system of eastern Africa. *Geol. Soc. Am. Bull.* 83: 2549–72

McConnell, R. B. 1974a. Evolution of taphrogenic lineaments in continental platforms. *Geol. Rundsch.* 63: 389–430

McConnell, R. B. 1974b. Taphrogenic lineaments and plate tectonics. See Illies & Fuchs 1974, pp. 43–52

McConnell, R. B. 1978. Further data on the alignment of basic igneous intrusive complexes in southern and eastern Africa. *Trans. Geol. Soc. S. Afr.* 81: 225–26

McConnell, R. B. 1979. A Precambrian origin for the protorift dislocation belt of eastern Africa? See Atti Conv. Lincei 1980, pp. 35–43

McGill, G. E., Stromquist, A. W. 1979. The Grabens of Canyonlands National Park, Utah: geometry, mechanics, and kinematics. *J. Geophys. Res.* 84: 4547–63

McKinlay, A. C. M. 1965. The coalfields and the coal resources of Tanzania. *Geol. Surv. Tanzania Bull.* 3: 1–82

Meyer, A. 1954. Une formation a Collenia dans la region de Niangara. *Bull. Geol. Soc. Belge* 62: 213–16

Milanovsky, E. E. 1972. Continental rift zones: their arrangement and development. *Tectonophysics* 15: 65–70

Milanovsky, E. E. 1983. Major stages of rifting evolution in the Earth's history. *Tectonophysics* 94: 599–607

Mohr, P. A. 1974. Mapping of the major structures of the African Rift System. *Smithsonian Astrophys. Obs. Spec. Rep.* 361: 1–85

Mohr, P. A. 1976. ENE-trending lineaments of the African rift system. In *Proc. Int. Congr. New Basement Tectonics, 1st,* 5: 327–36. Salt Lake City: Geol. Assoc. Utah

Mohr, P. A. 1982. Musing on continental rifts. In *Continental and Oceanic Rifts. Geodyn. Ser.,* ed. G. Pálmason, 8: 293–309. Washington, DC: Am. Geophys. Union

Mohr, P. A., Wood, C. A. 1976. Volcano spacings and lithospheric attenuation in the Eastern Rift of Africa. *Earth Planet. Sci. Lett.* 33: 126–44

Molyneux, A. J. C. 1909. The Karroo system in Northern Rhodesia. *Q. J. Geol. Soc. London* 65: 408

Montadert, L., Roberts, D. G., De Charpal, O., Guennoc, P. 1979. Rifting and subsidence of the northern continental margin of the Bay of Biscay. In *Initial Reports of the Deep Sea Drilling Project,* 48: 1025–60. Washington, DC: US Govt. Print. Off.

Morgan, P., ed. 1983. *Processes of Continental Rifting. Tectonophysics,* Vol. 94. 210 pp.

Morgan, P., Baker, B. H. 1983. Introduction: processes of continental rifting. See Morgan 1983, pp. 1–10

Mueller, St. 1970. Geophysical aspects of graben formation in continental rift systems. In *Graben Problems,* ed. J. H. Illies, St. Mueller, pp. 27–37. Stuttgart: E. Schweiz Verlagsbuchandlung

Muller, G., Forstner, U. 1973. Recent iron ore formation in Lake Malawi, Africa. *Mineral. Deposita* 8: 278–90

Mulugeta, G. 1985. Dynamic models of continental rift valley systems. *Tectonophysics* 113: 49–73

Murray-Hughes, R. 1933. Notes on the geological succession, tectonics, and economic geology of the western half of Kenya Colony. *Geol. Surv. Kenya Rep.* 3: 1–8

Neugebauer, H. J., Temme, P. 1981. Crustal uplift and the propagation of failure zones. *Tectonophysics* 73: 33–51

Neumann, E. R., Ramberg, I. B., eds. 1978. *Petrology and Geochemistry of Continental Rifts: Volume One of the Proceedings of the NATO Advanced Study Institute on Paleorift Systems. NATO Adv. Study Inst.,* Vol. 36. Dordrecht: Reidel

Neumayer, M. 1887. *Erdgeschichte,* pp. 326–31. Leipzig

Nunn, J. A. 1982. Subsidence and temperature histories for Jurassic sediments in the North Gulf Coast: a thermal-mechanical model. In *The Jurassic of the Gulf Rim. Ann. Res. Congr. Soc. Abstr.*, ed. D. G. Bebout, et al, 3: 82

Obst, E. 1913. Der ostliche Abschnitt der grossen ostafrikanischen Storungszone. *Mitt. Geogr. Ges. Hamburg* 28: 187

Oertal, G. 1965. The mechanism of faulting in clay experiments. *Tectonophysics* 2: 343–93

Okaya, D. A., Thompson, G. A. 1985. Geometry of Cenozoic extensional faulting: Dixie Valley, Nevada. *Tectonics* 4: 107–25

Oldenburg, D. W., Brune, J. N. 1972. Ridge transform fault spreading pattern in freezing wax. *Science* 178: 301–4

Pallister, J. W. 1954. Erosion levels and laterite in Buganda Province, Uganda. *Int. Geol. Congr., 19th*, 21: 193–99

Pallister, J. W. 1955a. Notes on the northern termination of the western rift. *Rech. 1953, Geol. Surv. Uganda*, pp. 49–52

Pallister, J. W. 1955b. A revised lexicon of Uganda stratigraphy. *Geol. Surv. Uganda*, pp. 1–15

Pallister, J. W. 1965. The rift system in Tanzania. See Drake & Loupekine 1965, pp. 86–91

Pallister, J. W. 1971. The tectonics of East Africa. See UNESCO 1971, pp. 511–42

Pallister, J. W., Hepworth, J. V. 1956. Notes on mylonite and rift faulting in Uganda. *CCTA East-Central Reg. Comm. Geol. Conf., Dar-es-Salam*, pp. 95–97

Papers Presented to the Conference on the Processes of Planetary Rifting. 1981. *Meeting, December 3–5, 1981, Napa Valley, Calif.*

Patterson, M. B. 1982. Structure and acoustic stratigraphy of the Lake Tanganyika Rift Valley: a single-channel seismic survey of the lake north of Kalemie, Zaire. MS thesis. Duke Univ., Durham, N.C. 89 pp.

Peeter, L. 1959. Traits generaux de la geomorphologie et de la genese du bassin du lac Kivu. *Bull. Geol. Soc. Belge* 83: 66–75

Poussin, J. V. 1935. Un graben transversal Tanganyika Rukwa au Tanganyika Territory. *Bull. Geol. Soc. Belge* 59: 330–34

Proffett, J. M. Jr. 1977. Cenozoic geology of the Yerington district, Nevada, and implications for nature and origin of Basin and Range faulting. *Geol. Soc. Am. Bull.* 88: 247–66

Pulfrey, W. P. 1960. Shape of the sub-Miocene erosion level in Kenya. *Geol. Surv. Kenya Bull.* 3: 1–18

Quennell, A. M. 1951. The Luapa goldfield, Tanganyika Territory. *Mineral. Mag.* 85: 341–47

Quennell, A. M. 1960a. The rift system and the East African swell. *Proc. Geol. Soc. London* 1581: 78–86 (Abstr., with discussion)

Quennell, A. M. 1960b. Summary of the geology of Tanganyika, part 2: geological map. *Geol. Surv. Tanganyika Mem.* 1: 1–6

Ramberg, H. 1963. Experimental study of gravity tectonics by means of centrifugal models. *Bull. Geol. Inst. Univ. Uppsala* 42: 1–97

Ramberg, H. 1967. *Gravity, Deformation, and the Earth's Crust.* New York: Academic. 214 pp.

Ramberg, H. 1971. Dynamic models simulating rift valleys and continental drift. *Lithos* 4: 259–76

Ramberg, H. 1978, Experimental and theoretical studies related to thrust nappes. *Geol. Soc. Am. Abstr. With Programs* 10: 475

Ramberg, I. B., Larsen, B. T. 1978. Tectomagnetic evolution. In *The Oslo Paleorift. Norges Geol. Unders. Bull.*, ed. J. A. Dons, B. T. Larsen, 335: 55–73

Ramberg, I. B., Morgan, P. 1984. Physical characteristics and evolutionary trends of continental rifts. *Tectonics* 7: 165–216

Ramberg, I. B., Neumann, E. R., eds. 1978a. *Tectonics and Geophysics of Continental Rifts.* Dordrecht: Reidel

Ramberg, I. B., Neumann, E. R. 1978b. Paleorift systems—introduction. See Neumann & Ramberg 1978, pp. xix–xxvii

Ramberg, I. B., Spjeldnaes, N. 1978. The tectonic history of the Oslo Region. See Neumann & Ramberg 1978, pp. 188–93

Reches, Z. 1978. Analysis of faulting in three-dimensional strain field. *Tectonophysics* 47: 109–29

Reches, Z. 1983. Faulting of rocks in three-dimensional strain field. *Tectonophysics* 95: 133–56

Reynolds, D. J. 1984. Structural and dimensional repetition in continental rifting. MS thesis. Duke Univ., Durham, N.C. 175 pp.

Reynolds, D. J., Rosendahl, B. R. 1984. Tectonic expressions of continental rifting. *Eos, Trans. Am. Geophys. Union* 65: 1116 (Abstr.)

Rhemtulla, S. 1970. A geological reconnaissance of south Turkana. *Geol. J.* 136: 61–73

Robson, D. A. 1971. The structure of the Gulf of Suez (Clysmic) Rift, with special reference to the eastern side. *Q. J. Geol. Soc. London* 127: 247–76

Rosendahl, B. R., Livingstone, D. A. 1983. Rift lakes of East Africa—new seismic data and implications for future research. *Episodes: Q. J. Int. Union Geol. Sci.* 1983: 14–19

Rosendahl, B. R., Ebinger, C., Patterson, M. P., Livingstone, D. A. 1982a. Structural styles during the early stages of continental fragmentation. *Eos, Trans. Am. Geophys. Union* 63: 1117 (Abstr.)

Rosendahl, B. R., Ebinger, C., Patterson, M. P., Livingstone, D. A. 1982b. Stratigraphic styles during the early stages of continental fragmentation. *Eos, Trans. Am. Geophys. Union* 63: 1117 (Abstr.)

Rosendahl, B. R., Reynolds, D. J., Lorber, P. M., Burgess, C. F., McGill, J., et al. 1986. Structural expressions of rifting: lessons from Lake Tanganyika, Africa. In *Sedimentation in the African Rifts. Q. J. Geol. Soc. London Spec. Publ. GSSP No. 23.* In press

Rowley, D. B., Sahagian, D. 1986. Depth-dependent stretching: a different approach. *Geology* 14: 32–35

Ruhe, R. V. 1954. Erosion surfaces of Central African interior high plateaus. *Congo Belge Inst. Nat. Étud. Agron. Publ. Sci.* 59: 1–40

Sahama, T. H. G. 1962. Petrology of Mt. Nyiragongo. *Trans. Edinburgh Geol. Soc.* 19: 1–28

Sahama, T. H. G., Meyer, A. 1958. The volcano Nyiragongo—a progress report. 2. *Congo Belge Inst. Parcs Nat.* 85 pp.

Sales, J. K. 1976. Model studies of continental rifting. *Abstr. Programs, Am. Geophys. Union* 8: 1083 (Abstr.)

Sander, S. 1986. *The geometry of rifting in Lake Tanganyika, East Africa.* MS thesis. Duke Univ., Durham, N.C. 46 pp.

Sander, S., Burgess, C. F., Rosendahl, B. R., Ebinger, C. J. 1986. Structure of Lakes Tanganyika and Malawi. *Eos, Trans. Am. Geophys. Union* 66: 364 (Abstr.)

Sanders, L. D. 1965. Geology of the contact between the Nyanza shield and the Mozambique belt in western Kenya. *Geol. Surv. Kenya Bull.* 7: 1–45

Savage, R. J. G., Williamson, P. G. 1978. The early history of the Turkana Depression. See Bishop 1978, pp. 375–94

Schouten, H., Klitgord, K. D. 1982a. The memory of the accreting plate boundary and the continuity of fracture zones. *Earth Planet. Sci. Lett.* 59: 225–66

Schouten, H., Klitgord, K. D. 1982b. Early Mesozoic Atlantic reconstructions from sea-floor-spreading data. *Eos, Trans. Am. Geophys. Union* 63: 307 (Abstr.)

Şengör, A. M. C., Burke, K. 1978. Relative timing of rifting and volcanism on Earth and its tectonic implications. *Geophys. Res. Lett.* 5: 419–21

Shackleton, R. M. 1945. Geology of the Nyeri Area. *Min. and Geol. Dept., Geol. Surv. Kenya* 12: 1–26

Shackleton, R. M. 1951. A contribution to the geology of the Kavirondo Rift Valley. *Q. J. Geol. Soc. London* 106: 345–92

Shackleton, R. M. 1955. Pleistocene movements in the Gregory' Rift Valley. *Geol. Rundsch.* 43: 257–63

Shackleton, R. M. 1978. Geological map of the Ologesailie Area, Kenya. See Bishop 1978, p. 175

Shudofsky, G. N. 1985. Source mechanisms and focal depths of East African earthquakes using Rayleigh-wave inversion and body-wave modelling. *Geophys. J. R. Astron. Soc* 83: 563–614

Skilbeck, M. J., Lennox, C. G., eds. 1984, *The Seismic Atlas of Australian and New Zealand Sedimentary Basins.* Sydney: Earth Resour. Found. 301 pp.

Snelgrove, A. K. 1967. *Geohydrology of the Indus River, West Pakistan.* Hyderabad: Sind. Univ. Press. 200 pp.

Spence, J. 1954a. The chalcedonic sandstones exposed in the Makutupora-Manyoni section of the Central Railway Line, Tanganyika. *Geol. Surv. Tanganyika Bull.* 1: 34–39

Spence, J. 1954b. The geology of the Galula coalfield, Mbeya district. *Geol. Surv. Tanganyika Bull.* 1: 1–24

Spurr, A. M. M. 1954. Notes on the formation of laterite in the Mbozi area, Mbeya district. *Geol. Surv. Tanganyika Bull.* 1: 30–34

Stewart, J. H. 1971. Basin and Range structure: a system of horsts and grabens produced by deep-seated extension. *Geol. Soc. Am. Bull.* 82: 1019–44

Stewart, J. H. 1978. Basin-Range structure in western North America: a review. In *Tectonics and Regional Geophysics of the Western Cordillera. Geol. Soc. Am. Mem. No. 152,* ed. R. B. Smith, pp. 1–31

Stockley, G. M. 1932. The geology of the Ruhuhu coalfields, Tanganyika Territory. *Q. J. Geol. Soc. London* 88: 610–22

Stockley, G. M. 1938. The geology of parts of the Tabora, Kigoma, and Ufipa districts, northwest Lake Rukwa. *Geol. Surv. Tanganyika Bull.* 20: 1–32

Stoffers, P., Hecky, R. E. 1978. Late Pleistocene–Holocene evolution in the Kivu-Tanganyika basin. *Int. Assoc. Sedimentol. Spec. Publ. No. 2,* pp. 43–55

Suess, E. 1880. Über die vermeintlichen Schwankungen einzelner Theile der Erdoberflaeche. *Verh. K. Geol. Reichsanst. Wien.* 180 pp.

Suess, E. 1891. Die Brüeche des ostlichen Afrikas. *Denkschr. Akad. Wiss. Wien* 553: 1–580

Suess, E. 1909. *The Face of the Earth,* Vol. 3. Wien: F. Tempsky. 316 pp.

Sykes, L. R. 1978. Intraplate seismicity, reactivation of pre-existing zones of weakness, alkaline magmatism, and other tectonism postdating continental fragment-

ation. *Rev. Geophys. Space Phys.* 16: 621–88
Talwani, M., Eldholm, O. 1977. Evolution of the Norwegian-Greenland Sea. *Geol. Soc. Am. Bull.* 88: 969–99
Tarling, D. H., Runcorn, S. K., eds. 1973. *Implications of Continental Drift to the Earth Sciences.* London: Academic
Tavener-Smith, R. 1960. The Karroo system and coal resources of the Gwembe district, south-west section. *Bull. Geol. Surv. N. Rhodesia No. 4.* 84 pp.
Teale, R. 1932. The Eastern shore of Lake Tanganyika and its vicinity. *Ann. Rep. Geol. Surv. Tanganyika,* pp. 22–24
Thiele, E. O., Wilson, R. C. 1915. Portuguese East Africa between the Zambesi River and the Sabi River. *Geogr. J.* 45: 16–39
Thompson, G. A., Burke, D. B. 1974. Regional geophysics of the Basin and Range Province. *Ann. Rev. Earth Planet. Sci.* 2: 213–38
Tobin, D. G., Ward, P. L., Drake, C. L. 1969. Microearthquakes in the rift valley of Kenya. *Bull. Seismol. Soc. Am.* 80: 2043–46
Truckle, P. H. 1976. Geology and Late Cenozoic lake sediments of the Suguta Trough, Kenya. *Nature* 263: 380–83
Turcotte, D. L., Emerman, S. H. 1983. Mechanisms of active and passive rifting. See Morgan 1983, pp. 39–50
Uhlig, C. 1907. Der sogenannte groesse ostafrikanische Graben zwischen Magad (Natron-See) und Lawa ja Mwerie (Manjara-See). *Geogr. Z.* 13: 478–505
Uhlig, C. 1912. Beitrage zur Kenntniss der Geologie und Petrographie Ost-Afrikas. *Centralbl. Mineral.* 1912: 565
UNESCO. 1971. *Tectonics of Africa.* Paris: UNESCO
UNESCO. 1978. *International Tectonic Map of Africa.* Paris: UNESCO
Vail, J. R. 1967. The southern extension of the East African Rift System and related igneous activity. *Geol. Rundsch.* 57: 601–14
Vail, J. R. 1972. Geological reconnaissance in the Zalingei and Jebel Marra areas of western Darfur Province, Sudan. *Bull. Sudan Geol. Surv. Dep.* 19: 1–50
Van Schmus, W. R., Hinze, W. J. 1985. The midcontinent rift system. *Ann. Rev. Earth Planet. Sci.* 13: 345–83
Veitch, A. 1935. Evolution of the Congo Basin. *Geol. Soc. Am. Mem.* 3: 91–98
Verweke, L. 1913. Die Entstehung des Mittelrheintales und der mittelrheinischen Gebirge. *Mitt. Ges. Erdkd. Kolon. Strassburg,* Vol. 4
Vondra, C. F., Bowen, B. E. 1976. Plio-Pleistocene deposits and environments: East Rudolf, Kenya. See Coppens et al 1976, pp. 73–93
Vondra, C. F., Bowen, B. E. 1978. Stratigraphy, sedimentary facies and paleoenvironments, East Lake Turkana, Kenya. See Bishop 1978, pp. 395–414
Vondra, C. F., Johnson, G., Bowen, B. E. 1971. Preliminary stratigraphical studies of the East Rudolf Basin, Kenya. *Nature* 231: 245–48
von Herzen, R. P., Vacquier, V. 1967. Terrestrial heat flow in Lake Malawi, Africa. *J. Geophys. Res.* 72: 4221–26
von Hohnel, R. 1894. *The Discovery by Count Teleki of Lakes Rudolf and Stefanie.*
von Knorring, O., du Bois, C. G. B. 1961. Carbonatitic lava from Fort Portal area in western Uganda. *Nature* 192: 1064–65
Walsh, J., Dodson, R. G. 1969. Geology of northern Turkana: degree sheets 1, 2, 9, and 10. *Geol. Surv. Kenya Rep.* 82: 1–42
Wayland, E. J. 1921a. Some account of the geology of the Lake Albert Rift Valley. *Geogr. J.* 50: 344–59
Wayland, E. J. 1921b. A general account of the geology of Uganda by the geologist. *Rep. Geol. Dep. Uganda,* pp. 8–20
Wayland, E. J. 1931. Summary of the progress for the years 1919–1929. *Geol. Surv. Uganda*
Wayland, E. J. 1952. The study of past climates in tropical Africa. *Proc. Pan-Afr. Congr. Prehist., 1st,* pp. 56–66
Wegener, A. 1912. Die Entstehung der Kontinente. *Geol. Rundsch.* 3: 276–92
Weiblen, P. W., Morey, G. B. 1980. A summary of the stratigraphy, petrology, and structure of the Duluth Complex. *Am. J. Sci.* 208: 88–133
Wernicke, B. 1981. Low-angle normal faults in the Basin and Range Province: nappe tectonics in an extending orogen. *Nature* 291: 645–48
Wernicke, B., Burchfiel, B. C. 1982. Modes of extensional tectonics. *J. Struct. Geol.* 4: 105–15
Williams, L. A. J. 1978. The volcanological development of the Kenya Rift. See Neumann & Ramberg 1978, pp. 101–21
Williamson, P. G. 1986. A Late Mesozoic transform fault system in the southeast Turkana Depression, North Kenya: implications for the early development of the African Rift. *Nature.* In press
Williamson, P. G., Raynolds, G. G., Savage, R. J. G., O'Connell, R. J. 1986. Early rift sedimentation in the Turkana Basin, North Kenya. *J. Geol. Soc. London.* In press
Willis, B. 1928. The Dead Sea problem: rift valley or ramp valley. *Geol. Soc. Am. Bull.* 39: 452–90

Willis, B. 1936. *East African Plateaus and Rift Valleys*. Washington, DC: Carnegie Inst. 358 pp.

Wilson, J. T. 1965. A new class of faults and their bearing on continental drift. *Nature* 207: 343–47

Withjack, M. 1979. A convective heat transfer model for lithospheric thinning and crustal uplift. *J. Geophys. Res.* 84: 3008–22

Wohlenberg, J., Bhatt, N. V. 1972. Report on airmagnetic surveys of two areas in the Kenya Rift Valley. *Tectonophysics* 15: 143–49

Wong, H. K., von Herzen, R. P. 1974. A geophysical study of Lake Kivu, East Africa. *Geophys. J. R. Astron. Soc.* 37: 371–89

Wright, L. A., Troxel, B. W. 1966. Shallow-fault interpretation of Basin and Range structure, southwestern Great Basin. In *Gravity and Tectonics*, ed. K. A. DeJong, R. Scholter, pp. 397–407. New York: Wiley

Wu, Z. 1986. Shallow structure of the southern Albuquerque Basin (Rio Grande Rift), New Mexico, from COCORP seismic reflection data. In *Reflection Seismology—the Continental Crust. Geodyn. Ser.*, ed. M. Barazangi, L. Brown, 14: 293–304. Washington, DC: Am. Geophys. Union

Wunderlich, H. G. 1957. Bruche und Graben im tektonischen experiment. *Geol. Paleontol. Union* 11: 477–98

Yairi, K. 1977. Preliminary account of the lake-floor topography of Lake Malawi in relation to the formation of the Malawi Rift Valley. In *Prelim. Rep Afr. Stud. Nagoya Univ., 2nd*, ed. K. Suiva, pp. 51–69

Yairi, K., Mizutani, S. 1969. Fault system of the Lake Tanganyika rift at the Kigoma area, western Tanzania. *J. Earth Sci. Nagoya Univ.* 17: 71–96

Yuretich, R. F. 1979. Modern sediments and sedimentary processes in Lake Rudolf (Lake Turkana) eastern rift valley, Kenya. *Sedimentology* 26: 313–31

Ziegler, P. A. 1978. North Sea rift and basin development. See Ramberg & Neumann 1978a, pp. 249–77

Ziegler, P. A. 1982. Geological atlas of western and central Europe. *Shell Int. Pet. Maatsch.* 5: 1–130

MARINE MAGNETIC ANOMALIES—THE ORIGIN OF THE STRIPES

C. G. A. Harrison

Division of Marine Geology and Geophysics, Rosenstiel School of Marine and Atmospheric Science, University of Miami, 4600 Rickenbacker Causeway, Miami, Florida 33149

Introduction

The discovery that the ocean floor could be dated by the presence of lineated magnetic anomalies has surely been one of the most surprising, and productive, discoveries in the Earth sciences, even in the whole of science. When Vine & Matthews published their paper (1963), the availability of data to support their contention that the ocean floor recorded the reversal history of the Earth's magnetic field was poor. Other people had suggested the same thing before (Girdler & Peter 1960, Schmalz 1961, Morley 1981). The concept of seafloor spreading had also been suggested many times in the past (e.g. Holmes 1931, Hess 1962, Dietz 1961). But within a few years, evidence had been collected from almost all of the major ocean basins of the Earth to support the idea that reversals of the Earth's field were recorded by oceanic crustal rocks as they upwelled at the mid-oceanic ridges and moved away at speeds from 1 to 10 cm yr^{-1} (Heirtzler et al 1968). Thus if the reversal pattern could be dated, it would be possible to arrive at an estimate of the age of much of the ocean floor.

At about the same time that the marine magnetic anomalies were being studied, the reversal time scale for the Earth's magnetic field over the past few million years was being established with an increasing precision, largely as a result of work done at the United States Geological Survey in Menlo Park and at the Australian National University (Cox et al 1963, McDougall & Tarling 1963). This work followed on earlier, less precise dating of reversals in Europe and Iceland (Rutten & Wensink 1959, Roche 1953).

In this review I examine all the relevant information concerning the magnetization of the oceanic crust in order to answer the question implied in the title of this chapter—Where are the magnetized rocks that cause the Vine-Matthews lineations? In order to do this, it is necessary to look at a variety of data. Some data come from the rocks sampled either by dredging or by drilling. These samples are limited in vertical extent. For instance, the deepest drilled hole has only penetrated just over 1 km into the oceanic crust, and in the worst possible place for studying unoriented cores (on the equator). Although a case can be made for dredging having recovered rocks from deeper in the crust, we are never quite sure that their magnetic properties have not been altered in some way during the processes that brought them to the surface. The same thing can be said about ophiolites, which however have been important in modeling the tectonic nature of oceanic crustal emplacement. Of equal importance is the information that we can obtain from the study of the magnetic anomalies themselves, which give an idea of the directions and intensities of magnetization. Recently, observations of the field at satellite altitude have added a further dimension to the study of magnetization intensity; these observations may prove useful in our understanding of the origin of the stripes, although the stripes themselves cannot be seen at the elevations of the satellites used so far.

Early Models of Lineations

Vine & Matthews (1963) used a model of the magnetized portion of the ocean basins that was about 15 km thick. This was the depth at which they estimated the rocks would be above the Curie temperature of magnetite and so incapable of carrying any remanent magnetization. This model neglected much evidence available at that time that the oceanic crust was considerably thinner (e.g. Hill 1957). The implication was that some of the magnetic signal must come from the mantle. This topic is discussed in greater detail later on. Vine & Wilson (1965) modeled the anomalies over the Juan de Fuca Ridge using a thickness of 8 km, essentially all of the oceanic crust, and also used a model in which the magnetization was confined to layer 2 of the oceanic crust. Vine (1966) used a layer only 1.5 km thick, representing layer 2, on the assumption that the rocks of layer 3 are either gabbro or serpentinite. Gabbro, because of its slow cooling and hence large crystal size, was assumed to have a low magnetization intensity, whereas serpentinite, if magnetized, would have acquired its magnetization at some considerable time subsequent to the surface layers of the oceanic crust, and so would be unable to contribute to the strength of the magnetic anomalies.

Further analysis of magnetic anomalies observed over the Reykjanes

Ridge crest, southwest of Iceland, by Talwani et al (1971) resulted in a model consisting of a layer only 0.5 km thick. In the absence of topographic expression, it is usually possible to change the amplitude of magnetic anomalies either by changing the magnetization or by changing the thickness of the magnetized layer. For a lineated structure, the following equation governs the amplitude of the magnetic field (F) as a function of the amplitude of magnetization (M) of a certain wave number k having an upper surface at a depth d and a thickness of t (Schouten & McCamy 1972):

$$F = 2\pi CM\{\exp(-kd) - \exp[-k(t+d)]\}. \tag{1}$$

The quantity C takes into account the geometry of the situation (Earth's field direction, direction of magnetization, etc, which are assumed to be constant over the area of an individual survey), while expression $2\pi\{\exp(-kd) - \exp[-k(t+d)]\}$ is known as the Earth filter. If kt is small, this can be approximated to

$$F = 2\pi CM \exp(-kd) \cdot kt. \tag{2}$$

This equation indicates that doubling the thickness is equivalent to halving the magnetization intensity. Where the equation breaks down, at high values of kt, other effects (such as the finite width of the zone between normally and reversely magnetized zones) will also be important. Equation (2) demonstrates that it is impossible to determine both the intensity of magnetization and the thickness of the magnetized layer from profiles running perpendicular to the lineated magnetic anomalies. Talwani et al (1971) solved the problem by running profiles parallel to the lineations and over the center of the magnetized blocks. They then assumed that the magnetization along the profile was constant, and that the variations in the field, which correlated well with the small-scale topographic variations, were in fact caused by the topographic variations themselves. Since they were in effect comparing the magnetization of the rock with that of seawater (zero), they were able to determine the magnetization of the upper portion of the magnetized layer. This then enabled them to determine the thickness of the magnetized layer by using Equation (2). Their result was that for crust that was 7.5 m.y. old, a magnetization of 12 A m^{-1} had to be used, and that for young crust within the central anomaly, the magnetization had to be 30 A m^{-1}. The thickness of the crust necessary to give the correct amplitudes of the lineated magnetic anomalies using these magnetization values was about 0.5 km; This result implied that the source was entirely within the extrusive layer of the oceanic crust, consisting of pillow lavas and massive lava flows. No contribution was necessary from the sheeted dike complex or from the intrusive layer that forms the traditional layer 3 of the oceanic crust.

In view of work to be discussed below, where it is shown that the magnetized layer is considerably thicker than that in the model suggested by Talwani et al (1971), it might be useful to consider whether their modeling technique biased their results to give too high a value of the magnetization. One assumption that Talwani et al made was that the topography was lineated perpendicular to the direction of their profiles, or in this case, lineated in the direction of spreading. It is more likely that if topographic lineations exist, they will be parallel to the ridge crest, or perpendicular to the direction used in the model. If we assume that the topography is not lineated, it is possible to estimate what sort of errors might arise from the method of modeling used by Talwani et al (1971). Suppose that a conical shape is responsible for a magnetic anomaly. If the ship track does not go directly over the top of the cone, the apparent height, width, and average depth will all be different from their actual values, whereas the anomaly amplitude might easily be almost as large as if the track went over the top. Suppose that the cone has a height of 0.5 km and a minimum depth of 0.5 km. (These figures are appropriate for the Reykjanes Ridge.) Suppose, further, that the ship's track shows a minimum depth of 0.75 km. The width of the feature will then be $\sqrt{3}/2$ that of the original cone, and so the estimated volume, if it is assumed that the ship went directly over a cone of the appropriate height, would only be 3/8 that of the original cone. In addition, since the average depth of the fictitious cone is somewhat greater, the magnetization necessary to explain the observed anomaly will also be greater. Calculations show that for the situation described above, the magnetization of the fictitious cone will be about four times larger than that of the real cone in order to give the same maximum anomaly over the center of the fictitious cone.

Inversions of Magnetic Anomalies to Obtain Source Functions

It is possible to take a magnetic anomaly record and perform an inversion to determine the magnetization variation necessary to give the observed anomaly. The first method was that suggested by Bott (1967), who used a block model of the crust and obtained the magnetization within each block by inversion of a matrix. The direction of magnetization within each block was assumed, as were the depths of the upper and lower surfaces of each block. A much faster method is to use Fourier analysis. Each Fourier component of the field can be modeled by a Fourier component of magnetization by multiplication with the Earth filter (Schouten & McCamy

1972), which is shown in Equation (1). Suggestions that Fourier analysis of magnetic anomalies was a useful technique were made by Howe (1940), Vestine & Davids (1945), Heirtzler & Le Pichon (1965), Solov'yev (1962), and Gudmundsson (1966, 1967). Details of the method are found in Schouten (1971) and Schouten & McCamy (1972). Resolution is limited at the low and high wave-number ends of the spectrum because the inverse Earth filter becomes infinite at zero and at infinite wave number. Hence, it is usual to cosine taper the magnetic field spectrum at each end before performing the inversion.

When inversions are done in this manner, it is frequently found that it is possible to obtain a model of magnetization that looks like the box model to be expected from the reversal pattern of the Earth's magnetic field.

When the boundaries between oppositely polarized zones are close together, a situation brought about by slow spreading and rapid reversals of the field, the fact that the emplacement process is not infinitely narrow will cause a reduction in the intensity of the magnetic field. It is possible to calculate the value of this effect using the following system. It is thought that to a first approximation the reversal process is a Poisson process, in which the probability of a reversal occurring in a short time interval is approximately constant. In fact it does vary, as was shown by McFadden & Merrill (1984), but the rate of variation is fairly slow, such that if we are studying crust over an age range of 10 m.y. or less, the reversal rate looks constant. The power spectrum of a magnetization that goes from A to $-A$ randomly according to a Poisson distribution is given by

$$P(k) = \frac{2A^2 r}{\pi(4r^2 + k^2)}, \tag{3}$$

where r is the average number of reversals per kilometer of crust (equal to r'/s, where r' is the average number of reversals per million years and s is the spreading rate in kilometers per million years), and k is the wave number in radians per kilometer (Blakely 1979). In order to model the emplacement process, we suppose that the reversal record of magnetization in the ocean floor is convolved with a Gaussian function,

$$\frac{1}{\sqrt{2\pi} t} \exp(-x^2/2t^2),$$

where t is the standard deviation of the Gaussian function in kilometers. In order to determine the power spectrum of the resulting magnetization, we multiply the power spectrum of the reversal record by the power spectrum (the square of the Fourier transform) of the Gaussian function, which is given by (Bath 1974)

$$\exp(-k^2 t^2) = g(k). \quad (4)$$

Finally, in order to determine the field spectrum, it is necessary to multiply by the square of the Earth filter, which we shall call $e(k)$. The final field power spectrum is then

$$F(k) = P(k)g(k)e(k). \quad (5)$$

We compare this with the field spectrum with no Gaussian emplacement process:

$$F'(k) = P(k)e(k). \quad (6)$$

The comparison is done by determining the relative amplitude ratios of the two power spectra over the wave numbers of interest:

$$R = \left[\int_{k_1}^{k_2} F(k) \, dk \bigg/ \int_{k_1}^{k_2} F'(k) \, dk \right]^{1/2}. \quad (7)$$

Using a depth of 5 km, a thickness of 6 km, and with $r = 5$ m.y.$^{-1}$, $s = 10$ km m.y.$^{-1}$, $k_1 = 0.05$ ($\lambda_1 = 125$ km), $k_2 = 1.0$ ($\lambda_2 = 6.3$ km), we obtain a value of $R = 0.823$. Figure 1 shows the various spectra. This reduction in amplitude of only 18% is not very significant. In order to have a larger effect, the spreading rate would have to be smaller, the reversal rate greater, or the wave-number range over which measurements take place shifted to larger wave numbers. We conclude that the emplacement mechanism does not seriously degrade the strength of the signal. However, a second effect, that of finite extrusion events, can produce noise (Blakely 1979, Matthews & Bath 1967, Harrison 1968, Schouten & Denham 1979).

Inversion Results

Harrison (1981, Table 3) summarized data concerning the magnetization of oceanic rocks that had been gathered from inversion studies. Some of these studies were done on very young crust. If crust younger than 4 m.y. is omitted from the calculation on the basis that its magnetization has not decayed sufficiently to be representative of the normal ocean basins, the average magnetization from eight different inversions is 8.14 A m^{-1} when normalized to a 0.5-km-thick layer and corrected to a value appropriate for an equatorial magnetizing field. If this is adjusted to an average latitude of 30° (by multiplying by 1.32) and to a thickness of normal oceanic crust (6 km) by dividing by 10 (see Table 1 and discussion in Harrison 1981), the average magnetization becomes 1.1 A m^{-1}. This value does not take

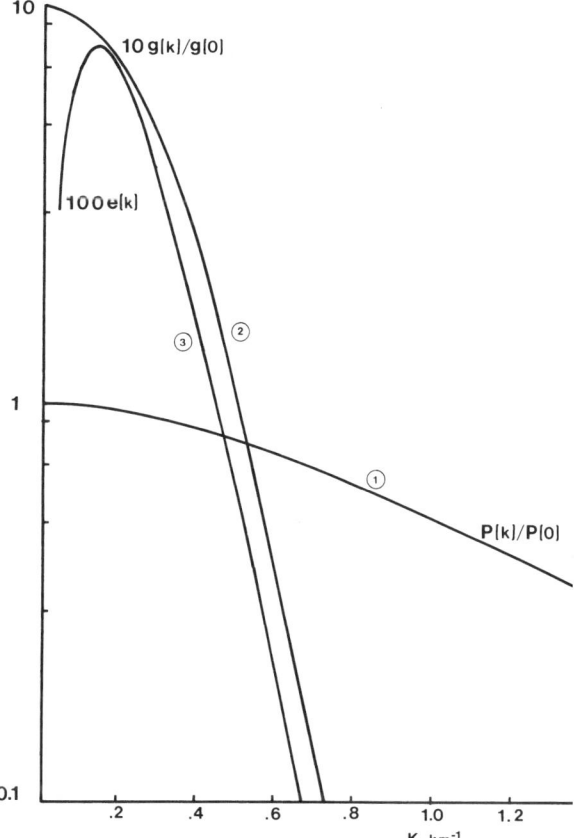

Figure 1 Spectra and filters. Curve 1 is the normalized spectrum for a Poisson square wave produced by a reversal rate of 5 m.y.$^{-1}$ and a spreading rate of 10 km m.y.$^{-1}$. Curve 2 is 10 times the normalized power spectrum of the Gaussian filter, with a standard deviation of 3 km. Curve 3 is 100 times the Earth filter for a body whose depth is 5 km and thickness is 6 km.

into account any magnetization that is parallel to the direction of the lineation, and so it is a minimum value if it is used to compare with results obtained by direct measurements on oceanic rocks.

Skewness Calculations

Skewness calculations have been done in order to study the direction of magnetization in the lineated blocks, with an eye toward making some paleomagnetic determinations. These calculations depend on knowing the reversal history of the field during the time under consideration, and also on having a sufficiently rapid spreading rate such that the reversal boundaries are not too close together (Harrison 1987). The method depends on the fact that as the direction of magnetization in the vertical plane perpendicular to the lineations is rotated, the phase angle of each spectral component of the field is shifted by an equal amount. This results in the shape of the magnetic anomalies varying with the phase angle. For zero phase angle, the anomalies are positive above normally magnetized crust, whereas for a phase angle of 180° (equivalent to an east-west lineated spreading center at the equator), negative anomalies occur above normally magnetized crust. An example of the change in the appearance of magnetic anomalies as the phase angle is varied is shown in Figure 2. Larson & Chase (1972) used the method of phase shifting to determine the direction of magnetization in three sets of lineations in the western Pacific and were able to calculate a paleomagnetic pole appropriate for the age of the anomalies (M1 to M10), whose age is now accepted to be 120 to 132 m.y. (Harland et al 1982).

However, the corollary of these calculations on phase shifting is that if

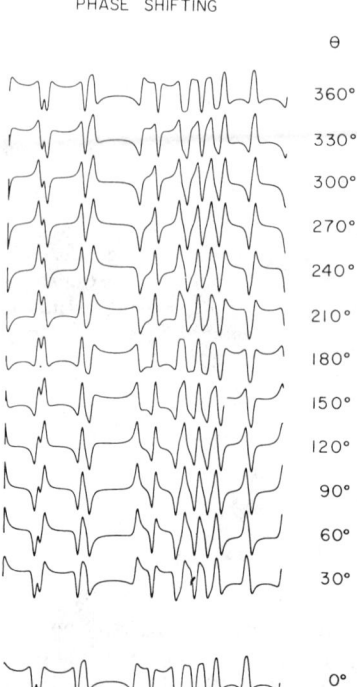

Figure 2 Phase shifting of magnetic anomalies. A set of magnetic anomalies is phase shifted through 360° in 30° increments. After a phase shift of 180°, the anomalies are inverted from the 0° phase-shifted anomalies. From Schouten & Cande (1976).

the direction of magnetization is known, then it is possible to determine something about emplacement history from studying the observed phase shift of the anomalies. In addition, if the same anomalies from each side of a single ridge crest are both available, then a comparison may be made between the phase of each set of anomalies, and information about emplacement mechanisms may be obtained.

Cande & Kristoffersen (1977) analyzed pairs of anomalies of the same age on either side of the Mid-Atlantic Ridge, one pair north of the Azores-Gibraltar Ridge, and one pair south. The anomalies were numbers 31–34, with an age between 61 and 83 m.y. They found that there was a large skewness discrepancy between the pairs of anomalies. Part of the discrepancy is due to the fact that some skewness is caused by the direction of the present-day core field at the location of the anomalies, which is different for each set of anomalies. But even after this is corrected for, there was an anomalous skewness of about $80 \pm 20°$ for each pair. This skewness discrepancy can be explained by an outward rotation of $40°$ (away from the ridge crest) for each member of a pair. An alternative explanation is that the boundary between normally and reversely magnetized crust is highly nonvertical and runs through a thick section of magnetized material.

Weissel & Hayes (1972) found a skewness discrepancy in the southeast Indian Ocean. Anomalies between 5 and 20 on either side of the ridge crest exhibited a skewness discrepancy of $20°$ after having been corrected for the relative movement of the two sets of anomalies since formation. This is equivalent to an outward rotation of $10°$ for each set of anomalies. Cande (1976) also studied the anomalous skewness of magnetic anomalies in the Pacific. He found an anomalous skewness for magnetic anomalies 27 to 32 of about $10–15°$. The evidence came from two sources. The first was that the skewness of anomalies in the southwest Pacific and the North Pacific did not agree with the relative positions of these portions of the same plate. In other words, the locations of possible paleomagnetic pole positions calculated using information from the two sets of anomalies did not agree. Although there is some possibility of relative movement between the two sets of anomalies since they were formed, for instance along the Louisville Ridge or along the Eltanin fracture zone, the amounts of motion along these features either are not in the right direction to explain the anomalous skewness or would have had to have been much too large. The second piece of evidence was from a comparison of the southwest Pacific anomalies with anomalies from the southeast Pacific, which were formed at the same ridge crest. After allowance for the spreading since the sets of anomalies were formed, anomalous skewness of the same amount was found.

The anomalous skewness, just like that in the Atlantic and in the southeast Indian Ocean, was equivalent to what would be caused by outward rotation of magnetized blocks. However, Cande (1976) points out that the evidence from fast-spreading ridge crests is that outward rotation has been observed, but only by a few degrees and not the 12–14° required by the anomalous skewness. An alternative is that the boundary between uniformly magnetized blocks is not perpendicular. This view has not been a very popular one, for the following reasons. The emplacement mechanism for oceanic crust seems to imply that the zone of injection of oceanic crust around the spreading center is fairly narrow. This is borne out by deep submersible studies of the mid-oceanic ridges (Ballard & van Andel 1977) and by analysis of ophiolite complexes, where the relative number of cooling surfaces of dikes within the sheeted dike complex seems to indicate a very narrow zone of injection (Kidd 1977). Another prediction of the Kidd model is that the transition from pillow lavas or ponded lavas to sheeted dikes should be very rapid, an observation borne out by drilling at Deep Sea Drilling Project Site 504B south of the Costa Rica Rift, where the boundary between complete lava sequences to 100% dikes occurred over a vertical depth of only 209 m (Anderson et al 1982). This model of very narrow intrusion, when coupled with the supposition that the magnetized layer was entirely within seismic layer 2 (i.e. the lava and sheeted dike portions) of the oceanic crust and possibly only in the lava portion (the top 0.5 km), suggested that nonvertical boundaries are unlikely to be an important modifying feature of the magnetic signature.

Cande (1976) suggested a third way of producing anomalous skewness, which was that the behavior of the Earth's field itself was responsible. For instance, if there were numbers of undetected reversal events within the reversal time scale, but concentrated toward the ends of periods of constant polarity, then the desired direction of anomalous skewness would be produced. Another mechanism was suggested by Cande (1976)—a pattern of magnetic field behavior produced from a coupled disk dynamo—in which oscillations in the intensity of the dynamo build up through a period of constant polarity until a reversal is achieved (Robbins 1977). This model will give anomalous skewness in the right direction and by the right amount. However, there seems to be little paleomagnetic evidence for such behavior.

Another suggestion was made by Cande & Kent (1976), who proposed a two-layer model for the magnetic source layer. The upper layer, consisting of the surface lava flows of layer 2A and possibly the sheeted dike complex, has sharp, approximately vertical boundaries. The lower layer, consisting of the intrusive part of the crust (equivalent to seismic layer 3), has reversal boundaries that slope downward away from the spreading center. This is the shape of the boundary that might be expected if the intrusive zone is cooled by conductive heat flow from its surface. The lower layers pass through the blocking temperature (a temperature below the Curie temperature of the magnetic minerals at which the magnetization is frozen in) at a later time than the upper layers, and so they become magnetized permanently when they are further away from the ridge crest. Simple cooling calculations by Cande & Kent (1976) show that the distance from the ridge crest where the base of the crust passes through the 500°C isotherm is 45 km for a spreading rate of 5 cm yr^{-1}. A recalculation using the same parameters suggests that the correct distance is closer to 76 km. For the 600° isotherm, the correct distance is 50 km rather than their value of 30 km. These revised figures are closer to those given by the more complex model of Sleep (1975), which was used by Kidd (1977) to model the boundary between opposite polarities in the gabbro layer. Sleep's model (1975) allows for a number of effects that make the calculation of position of isotherms more complicated. It also calls for an even smaller slope of isotherms with distance from the ridge crest, at least for the 5 cm yr^{-1} spreading model. In this model, the 600° isotherm is 2 km from the surface of the crust (or essentially 1 km deep into layer 3) at a distance of 15 km from the ridge crest. However, he did not allow for hydrothermal removal of heat from the surface rocks. If it is assumed that this 2-km depth refers to depth within layer 3, then this 15-km distance is to be compared with the 10-km distance calculated using the simple model of Cande & Kent (1976). Much of this difference may be due to the fact that Sleep allowed for the release of the latent heat of freezing, which causes a slower cooling rate. In any case, it appears likely that a process causing a sloping magnetization acquisition boundary in the lower crust occurs, unless hydrothermal circulation succeeds in cooling off the whole of the oceanic crust very quickly. Opinions differ as to whether this is possible. What remains to be seen is whether such a process of sloping boundaries in layer 3 can cause the large anomalous skewness seen in the Atlantic Ocean anomalies.

Another problem with the rotational explanation for anomalous skewness is that actual rotations of material below the surface may be in the opposite direction to that necessary to explain the anomalous skewness. In the model of formation of the lava and dike layer of the crust proposed by Cann (1974), the deeper lava flows are rotated such that they are dipping toward the zone of intrusion. This results from assuming that the vertical cross-sectional shape of the lava flows is triangular. Average rotations toward the zone of intrusion vary with parameters such as the zone of intrusion and the actual cross-sectional shape of the flows. It is possible, however, that little rotation is produced if the lavas are equally thick throughout their width (Harrison 1974). Rotations achieved in the Cann model can be as great as 40°.

Additional Model Calculations

Further arguments for a magnetic model calling for layer 3 magnetization have been presented by Blakely (1976) and Blakely & Lynn (1977). Blakely noticed that the transition width between crust of opposite polarity, as measured by the distance between minimum and maximum values of the anomaly on either side of the transition, showed an increasing value with the age of the anomaly. After having corrected for depth differences and for details of the reversal time scale at each specific anomaly, he arrived at a transition width that increased from about 4 km for very young anomalies to about 20 km for anomalies aged about 80 m.y. His explanation was that the anomalies were caused by a two-layer crust. The upper layer had a strong magnetization, which became largely demagnetized over a time scale of tens of millions of years, but which had a very small transition width. The lower layer was less strongly, but more stably, magnetized and had a large transition width caused by the slope of the blocking temperature isotherm. Blakely assumed that the upper layer was the layer of lava flows, whereas the lower layer was that of the sheeted dikes. It seems more reasonable, in light of the arguments put forward by Cann (1974) and Kidd (1977), that the upper layer might be a combination of lava flows and dikes, whereas the lower layer might be the intrusive layer equivalent to seismic layer 3.

Another set of modeling experiments was done that confirmed the idea of magnetized oceanic crust being quite thick. Blakely & Lynn (1977) analyzed the youngest set of anomalies over four different ridge crests that were spreading at rates that varied by a factor of 10. The procedure was as follows. For each area, the observed profiles are deskewed (or reduced to the pole). A set of theoretical profiles is then calculated for each area under the assumption of zero skewness. These profiles are calculated using the known time scale and the known spreading rate for each area. The reversal boundaries are modeled by using a step function (Heaviside function) convolved with a Gaussian function, and for each member of the set the standard deviation of the Gaussian function is varied from zero to some appropriate finite value. The field generated by such a magnetization distribution is then differentiated with respect to distance in order to pick out the maximum and minimum of the anomaly associated with the Brunhes-Matuyama polarity boundary. The distance between these extrema is called ΔX. This can then be correlated with the standard deviation of the Gaussian function σ. The transition width is defined as 4σ, which is the region within which 95.4% of the change from normal to reversed magnetization takes place. The same transition is studied for a series of observed profiles from the different regions. Each profile is differentiated in order to determine ΔX, which is then used to calculate 4σ from the relationship established for the models. The values of 4σ are then averaged for all the profiles in each area. The results are shown in Figures 3 and 4, where it can be seen that the transition width is linearly related to the spreading rate. The results for the slowest spreading area, the Reykjanes Ridge, could not be treated in the same manner, since the transition between the Bruhnes and Matuyama intervals is disturbed by the very low spreading rate. The estimates for this ridge were obtained by comparison of the observed profiles with a number of computer-generated profiles using different values of σ.

The preferred explanation of Blakely & Lynn (1977) for this phenomenon is to call upon sloping reversal boundaries in the same way as was done by Kidd (1977) and Cande & Kent (1976). The faster the spreading, the smaller will be the slope of any isotherm within layer 3, and so the greater the transition width.

Further statistical arguments have been presented by Blakely (1979) to argue for a substantial contribution to the magnetic anomalies from layers deeper than layer 2A (the extrusive lava layer). The normal noise associated

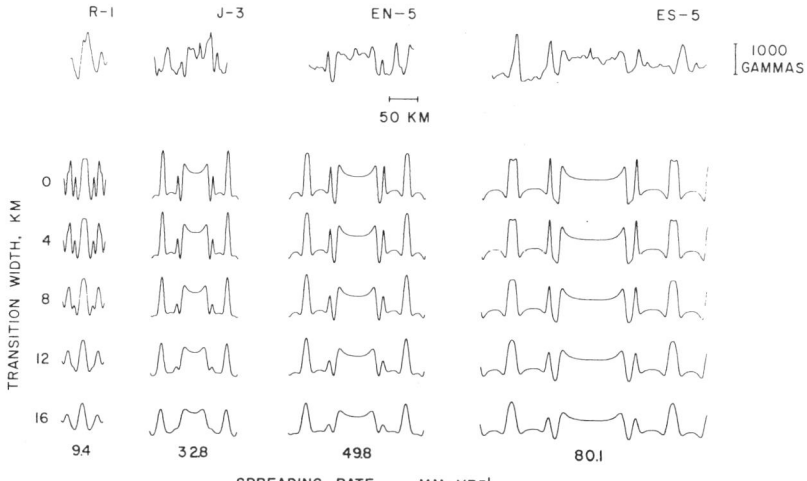

Figure 3 Comparison of data (at the top) and computer-generated magnetic anomalies with different transition widths. The spreading rate is shown along the bottom. The four areas are the Reykjanes Ridge (R-1), the Juan de Fuca Ridge (J-3), the East Pacific Rise at 14°N (EN-5), and the East Pacific Rise at 20°S (ES-5). From Blakely & Lynn (1977).

with the emplacement of lava flows was shown to disrupt the anomalies to a greater extent than commonly seen, if all of the signal is presumed to come from the top 0.5 km of the oceanic crust.

Additional evidence for decay of the magnetization of the upper layers of the oceanic crust is offered by Blakely (1983). He performed spectral analyses of a series of three groups of magnetic anomalies from the Pacific Ocean of about equal spreading rate but of different age, running from the axial zone out to an age of 65–70 m.y. He showed that the ability of the recording process to record short-wavelength (or short time scale) features of the input (in this case, the reversal pattern of the magnetic field) became less the older the set of anomalies.

The results of Blakely & Lynn (1977) may be used to estimate some details about the relative strength of the magnetization in the upper and lower layers of the oceanic crust. If a simple half-plate cooling law is used to determine the width over which the reversal takes place, it is possible to generate the series of curves shown in Figure 5. The value of f shows the fraction of the vertically integrated magnetization intensity that is held in the upper layer, for which it is assumed that there is no smearing out. The horizontal axis in this diagram shows the distance it is necessary to go to capture a certain fraction of the transition zone. (This fraction is shown on the ordinate scale.) Curve 1 shows the fraction predicted for the Gaussian model of Blakely & Lynn (1977), which for a spreading rate of 80 km m.y.$^{-1}$ had a standard deviation of 2.5 km. It can be seen that although the detailed agreement between the Gaussian curve and the others is poor, it seems difficult to explain the observations by a model in which none of the magnetization is in the upper layer, or by a model in which more than 60% of the magnetization is in the upper layer. If, however, the rate of cooling is significantly increased (by, for instance, hydrothermal circulation), then it is possible to match the Gaussian curve

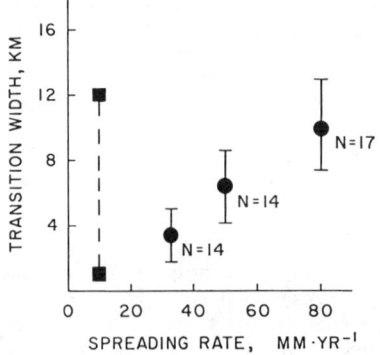

Figure 4 Transition width as a function of spreading rate. Dots are the mean values of the transition widths, and the error bars are equal to four standard errors of the mean. The numbers beside each point show the number of determinations of the transition width made at each location. From Blakely & Lynn (1977).

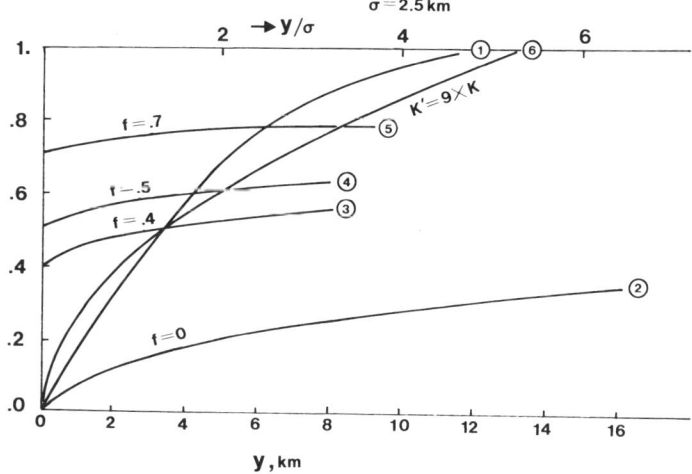

Figure 5 Distance necessary to capture a certain fraction of the transition. The bottom horizontal axis is distance. The top horizontal axis is distance divided by standard deviation, which in the case for old crust is 2.5 km (Blakely & Lynn 1977). Curve 1 is that calculated for the model of Blakely & Lynn. Curves 2 to 5 are calculated assuming that a portion *f* of the magnetization is carried in the upper layer (layer 2) for which there is no smoothing out, and that the rest is carried in layer 3, which cools according to a normal conductive cooling calculation. Curve 6 is for a crust that carries no magnetization in layer 2 and in which the cooling of layer 3 is enhanced by a thermal diffusivity that is nine times greater than that used to calculate curves 2 to 5. This results in a cooling that is three times faster. Such enhanced cooling might be produced by hydrothermal circulation, which eventually penetrates deeply into layer 3.

more exactly with a simple cooling model in which little of the magnetization comes from the upper layer. Such a curve is shown in Figure 5 (curve 6), where the thermal diffusivity has been increased by a factor of nine.

Deep-Tow Studies of Magnetic Anomalies

Observations of the magnetic field close to the ocean bottom have been made for at least two decades. The closeness to the source allows these observations to record in much greater detail some of the processes that occur in the formation of lineated magnetic anomalies. It is possible to invert deeply towed field observations to obtain source functions while allowing for the variable vertical displacement of the sensor (Parker & Klitgord 1972) and for the variable depth of the ocean floor (Parker 1973).

Klitgord et al (1975) describe the results from a series of deep-towed surveys made over a number of ridge crests. One major conclusion that they reached was that at the observation elevation above the oceanic crust that was used (normally only a few hundred meters or less), much of the field signal at wavelengths less than 3 km was caused by topographic effects. It is possible to calculate the relative strength of the signal coming from the upper crust (layer 2) compared with that coming from the lower crust (layer 3) for a lineated system of magnetic anomalies generated by a reversal process that is part of a Poisson distribution (Harrison 1976).

The total power can be written down as

$$\int_{k=0}^{\infty} P(k)\,dk, \tag{8}$$

where k is the wave number and P is the power. This may be integrated into expressions containing the sine and cosine integrals (Gradshteyn & Ryzhik 1980) or auxiliary functions of these integrals for which approximate solutions are available (Abramowitz & Segun 1965). However, since not all wave numbers are available in the observed signal as a result of finite length at the small wave-number end and of noise at the high wave-number end, it is preferable to determine the power within certain wave-number intervals, which has been done numerically. Results are shown in Table 1, where calculations have also been done for observations made at the ocean surface. These results demonstrate that for deep-tow studies, layer 3 contributes about the same power as layer 2, for equal magnetization, despite the fact that it is three times as thick. For surface-tow studies, however, the contribution from each layer is about in proportion to the thickness of the layer. These calculations have been done to point out that although there is an advantage in measuring the field close to the bottom in terms of increased resolution, this goes along with a disadvantage—the uppermost layer is likely to be overemphasized in the resulting models. This is in fact what has happened in the interpretation

Table 1 Power of signal from lineated magnetic structures[a]

Calculation number	Depth to top (km)	Depth to bottom (km)	Thickness (km)	Power (nT^2)
1	0.5	2.0	1.5	31,049
2	2.0	6.5	4.5	32,235
3	3.5	5.0	1.5	3167
4	5.0	9.5	4.5	8808

[a] Calculations 1 and 2 are for deep tow. Calculations 3 and 4 are for surface tow. Calculations 1 and 3 are for layer 2. Calculations 2 and 4 are for layer 3. Magnetization is 1 A m^{-1} in each case. Reversal rate is 5 m.y.$^{-1}$. Spreading rate is 40 km m.y.$^{-1}$. Minimum and maximum wave numbers in summation are 0.05 and 1.9 km^{-1} for deep tow and 0.03 and 1.0 for surface tow.

of deep-tow anomalies, which in most cases have been interpreted using a 0.5-km-thick layer, following Talwani et al (1971).

Klitgord et al (1975) noticed a remarkable decrease in magnetization amplitude on going from zero-aged crust out to crust of a few million years in age. Decreases by factors of 2 or 3 were noted for all six sections of the mid-oceanic ridge in their study. Between 4 and 5 m.y. a number of magnetization intensity measurements were made, resulting in the information shown in Table 2. The quantity in the last column of Table 2 has been corrected for the following effects. Since the intensity of the Earth's field varies with latitude and since thermoremanent magnetization is directly proportional to magnetizing field intensity, the results have been corrected to equatorial field values. Because of the lineated nature of the structures responsible for the magnetic anomalies, magnetizations running horizontally parallel to the lineation have no effect on the magnetic anomalies. Addition of the appropriate component has been made. The mean value of magnetization is 7.90 A m^{-1} for a 0.5-km-thick layer. We desire to estimate the magnetization that would be required in a thicker layer, namely one that is as thick as the oceanic crust (about 6 km). The power spectrum for a Poisson reversal process in crust of depth 0.5 km and thickness 0.5 km gives a total power between wave numbers of 0.05 and 1.9 km^{-1} of 6209 nT^2 for a magnetization of 1 A m^{-1}. If the thickness is increased to 6 km, the power increases to 115,900 nT^2. The ratio of the amplitudes is 4.32, indicating that the average magnetization may be reduced by this ratio on going from a 0.5-km-thick crust to 6-km-thick crust. This results in a magnetization of 1.83 A m^{-1}. This value appears to be a more reasonable magnetization, if it is compared with magnetizations of samples recovered by the Deep Sea Drilling Project (DSDP). A further discussion may be found in Harrison (1976, 1987).

Table 2 Magnetization calculated from deep-tow field values

Location	Magnetization (A m^{-1})	Corrected magnetization (A m^{-1})
Juan de Fuca	17.26	11.49
Pacific Antarctic	11.89	7.58
Gorda	9.45	7.11
East Pacific	6.82	9.28
Costa Rica	4.04	4.05
Mean		7.90
Standard Error		1.23

Macdonald (1977) analyzed deep-towed profiles of the Mid-Atlantic Ridge in the region of the FAMOUS expedition using an inversion involving a 0.5-km-thick magnetized layer. By studying topographic uplifts, he was able to determine intensities of magnetization and found very high values (average 20 A m^{-1}) in the inner floor of the rift valley, decaying very rapidly with age to values of about 5 A m^{-1} for crust aged a few million years. Because the magnetization of the topographic feature is compared with that of water, little ambiguity results in determining the strength of magnetization, and in effect the results of Macdonald (1977) are similar to those of Talwani et al (1971) in this respect. Atwater & Mudie (1973) also used a similar method with deep-tow data. Using the values of magnetization so obtained, Macdonald calculated that in order to explain the surface magnetic anomalies, a thickness of about 0.7 km would have to be used. This thickness was appropriate for crust of age less than 5 m.y. But the common assumption with all of these methods is that magnetizations measured on surface features are continued downward into the crust. If the surface features have stronger magnetizations than the material below, this method becomes invalid. [See also the comments on Talwani et al (1971) above.]

Studies of the transition width between different polarity material were also done by Macdonald (1977). He arrived at widths that varied between 1.0 and 8.0 km (average 2.8 km). Because of the bias toward the upper layers of oceanic crust produced by using deep-towed data, it is unlikely that these figures represent the real transition widths, which would probably be somewhat larger if the whole oceanic crust were to be given a uniform weighting in the determination.

Another very peculiar result obtained by Macdonald (1977) was the presence, in some inversions, of reversely magnetized material right in the center of the rift valley. Although Macdonald claimed that this was a necessity of the inversion, it is likely that it can be removed by a number of factors. For instance, suppose that the magnetized layer were considerably thicker than the 0.5 km used in the inversion, except at the rift-valley floor, where the high temperatures beneath the crust would cause rocks to be above their Curie temperature or where the process of crustal formation has not yet produced crust of normal thickness. This would result in the central area looking less magnetic than the surrounding areas when interpreted using crust of uniform thickness. In addition, the speed at which the deep tow is moved through the water is slow (1.5 km hr^{-1}). This means that a signal of magnetic field caused by the diurnal variation would look like one that has a spatial wavelength of 36 km. The inverse Earth filter used to change fields into magnetizations is very large for deep-towed observations at this wavelength and would give a magnetization signal of a sufficient magnitude to distort the true crustal magnetization. For instance, a 50-nT signal with this wavelength, if interpreted with a slab of depth 0.75 km and a thickness of 0.5 km, would give a magnetization with amplitude 1.1 A m^{-1}.

Another important observation—one having to do with the emplacement width of the magnetic source layer—was made by the deep tow. Macdonald et al (1983) performed a two-dimensional measurement of the magnetic field close to the bottom across a polarity boundary on the East Pacific Rise. They also measured the polarity of small features on the ocean floor, such as pillow lavas, using a magnetic gradiometer mounted on the research submersible ALVIN. Finally, they were able to measure intensities of magnetization by modeling calculations, and these intensities agreed with direct measurements made on pillow lavas sampled by ALVIN. The intensities of magnetization averaged out to be 7.5 ± 2.5 A m^{-1}, neglecting some features of very low magnetization. These low-magnetization features were found between zones of clearly positive and negative magnetization and were presumed to have been magnetized during the process of the reversal of the Earth's magnetic field, when it is usually of low intensity. This magnetization intensity, observed for crust about 0.7 m.y. old, is more like the result for zero-age crust when allowance is made for the latitudinal effect on the dipole field strength. The boundary between the surface polarity change was displaced from that found by inversion of the deep-towed magnetic anomalies. Along the traverse where the surface polarity change was most precisely defined, the separation between the two boundaries was 0.5 km. Macdonald et al (1983) assumed that this separation was due to the fact that lava flows travel a finite distance from their point of origin, spilling over onto older lava flows. This results in a sloping reversal boundary (Harrison & Stieltjes 1977).

Another two-dimensional modeling procedure was done by Miller & Hey (1986), who were studying propagating rifts. They claim that the correct thickness of the magnetized layer is about 0.9 km, based on a correlation between magnetization values calculated for a 1-km-thick layer and actual observations made on rocks recovered from dredge hauls in the area. The assumption is made that the magnetization of the layer does not vary vertically until the bottom of the layer is reached. For their 1-km-thick layer, Miller & Hey found that the total variation in magnetization was over 24 A m^{-1} (from positively magnetized regions to negatively magnetized regions).

The Long-Wavelength Component

Information concerning the magnetization of the oceanic crust may also be sought in the long-wavelength component of the magnetic field. An attempt to do this was made by Harrison & Carle (1981), who studied three very long total-field profiles observed at the ocean surface. They found that an inversion of the magnetic field after having removed the field believed to come from the core produced a rms magnetization between 0.81 and 1.78 A m^{-1} for a 6-km-thick layer. These magnetizations are minimum values, since they were calculated assuming lineated structures, with no magnetization assigned to the direction of lineation. Harrison & Carle also pointed out that the shape of the power spectrum of the field was not what would be expected from lineated structures in the crust coupled with a core field. Shure & Parker (1981) noted that the shape of the power spectrum of the field may have been caused by the incomplete lineation of the crustal structures, but they did not tackle the problem of the high magnetizations directly. The east-west Pacific profile discussed by Harrison & Carle (1981) was 11,000 km long and ran along the 35°S latitude line, approximately. The lineations in this region are predominantly north-south. At 35°S one would expect a horizontal north-south field to be about 71% of the vertical field. After having allowed for this unmeasurable horizontal component, we find that the rms magnetization for the profile is 3.0 A m^{-1}. This is probably an overestimate because of the factor discussed by Shure & Parker (1981), but it is unlikely to be an overestimate by more than 50%, which gives a corrected value of 2 A m^{-1}.

One way out of the problem of studying magnetizations that may not be perfectly lineated, using single profiles, is instead to use large-scale magnetic field observations such as those provided by the magnetic field satellite MAGSAT (Langel et al 1982). Several studies have been done, some in which equivalent source solutions have been made for the whole Earth, and some in which smaller patches of the oceanic crust have been studied.

Arkani-Hamed & Strangway (1985a,b) have come up with an ingenious method for calculating the susceptibility of the crust, assuming that all of the crustal magnetic anomalies are caused by susceptibility variations and that the direction of the magnetizing field is in the centered dipole field direction (established from the first-degree spherical harmonic coefficients). One problem that they did not solve was that of removing negative susceptibility values. The region of the Cretaceous quiet zone in the North Atlantic is obvious on the susceptibility map and is marked by a susceptibility contrast with the surrounding regions of about 0.012 SI units for a crustal thickness of 50 km. Recalculating the magnetization using the strength of the field and a crustal thickness of 6 km produces a value of 4 A m^{-1}. Harrison et al (1986) produced a rough estimate of the minimum rms magnetization for the whole of the oceanic crust (assuming a layer thickness of 6 km) and arrived at a figure of 1.1 A m^{-1}. This result is unreasonably small, since it does not take into account the annihilator and produces both positive and negative magnetizations.

Harrison et al (1986) also studied a smaller area in the northwest Pacific Ocean and came up with an rms magnetization of about 4 A m^{-1} after having applied enough annihilator to remove negative magnetizations. This is necessary because the simple inversion scheme used produces magnetizations that average out to zero. During the inversion, the direction of magnetization for each element of the crust is assumed to be magnetized along the direction of the Earth's field. Induced magnetizations should then all be positive. Remanent magnetizations do not necessarily need to be positive. But considerations of the reversal history of the Earth's magnetic field during the time of formation of oceanic crust in existence today, and of the averaging effect of the high altitude of the satellite, show that the satellite should see only crust of zero or positive magnetization, since there are no long periods of time when the field was reversed. The assumption made in the inversion process was that the crust was uniformly 6 km thick. Portions of the area have elevated plateaus, with presumably much thicker crust, so that the true average magnetization is probably somewhat less than 4 A m^{-1}.

In another study done on the northern and equatorial Atlantic Ocean basin, consisting of a much larger area than that studied in the Pacific, Hayling & Harrison (1986) arrived at an average magnetization of about 4 A m^{-1}, again after having adjusted the magnetization to be everywhere positive or zero by use of an annihilator. There is some evidence that part of the signal is due to the periods of long normal polarity of the Earth's field during the Cretaceous. These were specifically studied by LaBrecque & Raymond (1985), who used a magnetization of 15 A m^{-1} in a 0.5-km-thick layer. Correction to a layer 6 km thick, and for the fact that their modeled anomalies have a smaller amplitude than the observed anomalies, gives a magnetization of 2.5 A m^{-1}. The magnetization model of the northern and equatorial Atlantic produced by Hayling & Harrison (1986) is shown in Figure 6. There is some evidence that this magnetization model is real. Areas with very thick deposits of sediment should have almost zero magnetization because the thick sediment blanket causes the temperature of the underlying crust to be above the Curie temperature. Hayling (1986) has shown that this should be the case for the Laurentian, Amazon, and Congo cones. All of these areas have very small magnetizations in Figure 6. Inspection of Figure 6 shows that the Cretaceous quiet zones are well marked by magnetization stripes of total amplitude of about 6 A m^{-1} on either side of the ridge crest separating North America from Africa. The background magnetization in this region of the North Atlantic is about 3 A

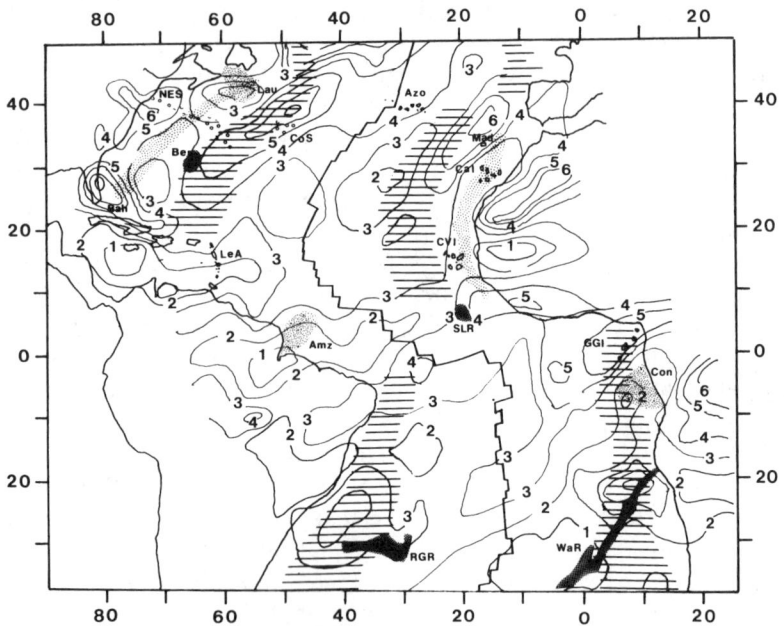

Figure 6 Inversion of MAGSAT magnetic anomaly data to determine a magnetic model for the northern and equatorial Atlantic Ocean basin. The thickness of the magnetic layer was assumed to be 6 km. An annihilator was applied to remove negative magnetizations. The coastlines and the Mid-Atlantic Ridge are marked by heavy lines. The magnetization contours are in A m^{-1}. Very low magnetizations are seen over the Laurentian (Lau), Congo (Con), and Amazon (Ama) cones, where thick sediments have probably caused the crust to lie below the Curie isotherm. The elongated stripes of high magnetization (maxima between 6 and 7 A m^{-1}) on either side of the ridge separating North America from Africa may be caused by the Cretaceous normal polarity interval. The derivation of the magnetization map is described in great detail by Hayling & Harrison (1986), who also identify the meanings of the other symbols.

m^{-1}, which indicates that the magnetization contrast between the normally magnetized normal polarity zones and the surrounding crust is about 3 A m^{-1}, similar to the value found by LaBrecque & Raymond (1985).

It is possible that the MAGSAT data do not accurately record the long-wavelength component of the magnetic field. LaBrecque et al (1985) studied a region in the northwestern Pacific Ocean in which they compared sea-surface data with MAGSAT data. The sea-surface data were bandpassed between wavelengths of 700 and 1900 km and upward continued to an altitude of 400 km, and these were compared with MAGSAT data bandpassed through the same wavelength band. Although the general appearance of the anomalies is roughly similar, the upward-continued sea-surface data show anomalies of considerably larger amplitude. We have done a correlation of the two data sets. The field at rectangular grid points separated by 556 km was read for each field type, with a total of 70 points used in all. The correlation coefficient was only 0.577, and the satellite anomaly values tended to be about half the upward-continued surface values. It is therefore possible that the satellite anomalies are not accurately recovering the true amplitude of the long-wavelength component of the crustal field. If this is the case, then the magnetization values calculated from MAGSAT data may represent minimum values.

Apart from the apparently good correlation between the calculated Cretaceous quiet zone and the observed anomalies (LaBrecque & Raymond 1985, Hayling & Harrison 1986), there is some doubt as to the source of other long-wavelength anomalies observed at satellite altitudes. One clear case of a correlation between a structure and a magnetic anomaly is shown by the Mid-Pacific Mountains (LaBrecque & Cande 1984, Harrison et al 1986). The anomaly appears to be caused by a remanent magnetization acquired when these seamounts were formed at a lower latitude than they occupy today. Two-dimensional models performed by LaBrecque & Cande (1984) show that in order to produce the observed anomaly, it is necessary to have a lineated structure of cross section 200×4 km^2 and magnetized by an amount 6.6 A m^{-1} [twice the value used by LaBrecque & Cande (1984) in order to produce an anomaly of the correct amplitude].

Hayling & Harrison (1986) studied equivalent source models for the long-wavelength crustal field in the northern and equatorial Atlantic (Figure 6). Most of the geologic features in this region show no associated magnetic anomalies apart from the ones already discussed. It can be shown that in some cases, the presence of thick crust beneath elevated platforms should give distinct magnetic anomalies if remanent magnetization of several amperes per meter is assumed for the extra thickness of the crust, compared with the surrounding regions. The absence of such anomalies argues against such high magnetization values and also constrains the susceptibility (and hence induced magnetization) to be fairly small. The absence of strong remanent magnetization may be due in part to the acquisition of magnetization during several polarity epochs.

Studies of satellite-elevation magnetic anomalies have revealed several things. Firstly, it is possible to obtain an estimate of magnetization by studying features like the Cretaceous quiet zone in the Atlantic. Estimates of magnetization vary between 1 and 6 A m^{-1} for the long-wavelength component of magnetization. Secondly, some features show expected magnetic anomalies, such as the Mid-Pacific Mountains, the Cretaceous quiet zone in the North Atlantic, and regions of very low magnetization (such as the three sedimentary cones in the Atlantic), whereas other features display no magnetic anomalies at all, which constrains magnetization values to be less than 1 A m^{-1}, possibly as a result of formation over several polarity reversals.

Direct Measurements of the Magnetic Properties of Oceanic Rocks

Direct measurements of the magnetic properties have been made on many samples of the oceanic basement. The samples fall into three different categories. Each category has its advantages and disadvantages. Dredged rocks were the first type studied. The main disadvantage of these rocks is that they do not sample the oceanic crust uniformly. Because of sediment cover, they are collected preferentially over young oceanic crust. They also tend to be sampled at fault scarps, which offer good opportunities for dredging. This in turn is an advantage because these fault scarps reveal lower layers of the oceanic crust not often available to the driller (Bonatti et al 1971, 1975). Rocks drilled during the Deep Sea Drilling Project (DSDP) and its continuation, the Ocean Drilling Project (ODP), cover a much wider range of ages than do the dredged rocks. In contrast to dredged samples, they probably come from a broader range of tectonic environments. But in general only the upper layers of the oceanic crust have been sampled. Although some gabbros and serpentinites have been sampled by DSDP, by far the majority of rocks studied have been from the upper part of the classical seismic layer 2 (pillow lavas and massive ponded lava flows). Recently, hole 504B penetrated through the pillow lava layer into the sheeted dike complex. Finally, ophiolites have allowed us to sample deeply into the oceanic crust. However, there are some doubts as to the applicability of ophiolites as models of crust produced at midocean ridges. Some people believe that they represent back-arc spreading because of geochemical similarities with back-arc basalts. In addition ophiolites can be magnetically changed during the process of obduction, which thus limits their applicability in determining the magnetic properties of oceanic crust. Levi et al (1978) suggest the use of four criteria that ophiolites must satisfy before they can be considered good magnetic analogues of oceanic crust. Although their argument is somewhat circular, requiring the magnetic properties of the ophiolites to approximate what is known about the oceanic crust from direct measurements, we adopt them in this discussion. The four properties refer only to the upper portions of layer 2, since this was the only oceanic rock analogue reached by drilling available for comparison by Levi et al (1978). The four properties are (a) that the Curie temperature should be between 100 and 450°C, and that for Curie temperatures greater than 200°C the saturation magnetization versus temperature curve should be irreversible; (b) that the natural remanent magnetization (NRM) should have an intensity greater than 0.1 A m^{-1}; (c) that the ratio of remanent to induced magnetization should be greater than unity; and (d) that the alteration of the extrusives should not exceed greenschist-facies metamorphism. These criteria have the effect of limiting the discussion to results from the Oman and Troodos ophiolites (Hayling & Harrison 1986).

Table 3 shows a compilation made from all possible sources of material. For the basalt layer (representing the classical seismic layer 2 of the oceanic crust) there are major differences in the values of the NRM. If most of this layer is considered to be metamorphosed, then the magnetization

Table 3 Magnetization of oceanic rocks (from Hayling & Harrison 1986)

Rock	N	NRM (A m^{-1})	N	Induced magnetization (A m^{-1} in 40,000 nT)	References[a]
Basalt					Ref.
(1) Ophiolites	208	0.342	208	0.3369	1
(2) DSDP	122	2.64	122	0.4702	1, 2
(3) Dredged	309	5.37	309	0.1632	2
(4) Metamorphosed	16	0.0122	16	0.0070	3, 4, 5, 6
Gabbro					
(5) Ophiolites	257	0.478	257	0.1008	7, 8, 9
(6) Unaltered	31	0.621	31	0.4029	1, 3, 6, 10
(7) Metamorphosed/ cataclastic	47	0.894	47	0.5934	6, 11
(8) Serpentinized	6	0.48	6	0.1536	1
Serpentinized peridotites					
(9) Ophiolites	38	6.03	38	0.4852	7, 11
(10) Dredged/ drilled	28	4.15	27	0.8873	1, 3, 5, 6

[a] References
 1. Dunlop & Prevot 1982
 2. Lowrie 1974
 3. Opdyke & Hekinian 1967
 4. Luyendyk & Melson 1967
 5. Irving et al 1970
 6. Fox & Opdyke 1973
 7. Vine & Moores 1972
 8. Luyendyk & Day 1982
 9. Banerjee 1980
 10. Kent et al 1978
 11. Beske-Diehl & Banerjee 1979

becomes very small and in fact insignificant in producing magnetic anomalies. If, on the other hand, the dredged samples are taken as representative throughout this 1.5-km-thick layer, then the average magnetization of the 6-km-thick normal ocean crust, if we neglect any contribution from layer 3, would be 1.3 A m^{-1}. Since dredged samples are heavily biased toward young rocks, which may not have suffered the large decline in magnetization noted by several workers (Irving et al 1970; see also Klitgord et al 1975, Johnson & Atwater 1977, Atwater & Mudie 1973, Macdonald 1977), it would appear unwise to rely heavily on the dredged sample data. The decline in magnetism as a function of age is due to oxidation of titanomagnetite to titanomaghemite (Dunlop & Hale 1977). In order to arrive at the most reasonable figure for layer 2, we have taken 70% of the DSDP average and 10% of the other three averages to arrive at a figure of 2.42 A m^{-1} over this 1.5-km-thick layer (Table 4). If brecciated layers represent a significant thickness of layer 2, then this value should be reduced.

Bleil & Petersen (1983) have documented in some detail the variation in the magnetization of samples recovered by DSDP as a function of their age. They show a fall in magnetization from 5.1 A m^{-1} for samples aged

Table 4 Ocean crust magnetization models

#	Layer (thickness, km)	Remanent magnetization (A m^{-1})	Induced magnetization (A m^{-1} in 40,000 nT)	Fractions (from Table 3)
1	2 (1.5)	2.42	0.380	0.1 from (1)
				0.7 from (2)
				0.1 from (3)
				0.1 from (4)
2	3 (4.5 or 3.78)	0.618	0.313	0.25 from (5)
				0.25 from (6)
				0.25 from (7)
				0.25 from (8)
3	Serpentinite (0.72) in layer 3	5.09	0.686	0.5 from (9)
				0.5 from (10)
4	Serpentinite (1.5) below gabbro	5.09	0.686	0.5 from (9)
				0.5 from (10)
Model A 1+2		1.07	0.330	Short wavelength
Model B 1+2+3		1.60	0.375	
Model C 1+2+3+4		2.88	0.546	Long wavelength

less than 1 m.y. to 1.1 A m^{-1} for rocks aged between 10 and 30 m.y. There were also fluctuations in intensity prior to 30 m.y. ago. The results of their analysis are given in Table 5. These data of Bleil & Petersen (1983) refine the pattern found by Harrison (1976) using a smaller data set. Raymond & LaBrecque (1986) explain these variable intensity phenomena by proposing a model consisting of two components. The first is a thermoremanent magnetization (TRM) that decays exponentially to 16% of its original value with a time constant of 5 m.y. This decay causes the decrease in magnetization shown by Bleil & Petersen and many other authors as one moves away from the ridge crests. The second component is an acquisition of chemical remanent magnetization (CRM), again with a time constant of 5 m.y. The CRM is acquired in the direction of the ambient field and grows to an asymptotic value that is 64% of the original TRM, or four times the TRM after it has decayed to its smallest value. The effect of this CRM is negligible when the reversal rate is rapid, because many different polarity changes are recorded. However, when the reversal rate is very slow, then the CRM acquired can substantially add to or subtract from the decayed TRM, producing a large or small magnetization. This, Raymond & LaBrecque suppose, is the cause of the large magnetization seen during the time of the Cretaceous normal polarity interval and of the generally decreasing magnetization during the Tertiary. The slow fall of magnetization during the Tertiary is caused by the general increase in the rate of reversals during this time interval (McFadden & Merrill 1984). Raymond & LaBrecque (1986) also show that their model can explain the anomalous skewness of magnetic anomalies that has been discussed above. Raymond & LaBrecque's model is described by the following equation:

Table 5 DSDP magnetizations (from Bleil & Petersen 1983)

Age (m.y.)	N	Geometric mean (A m^{-1})	Geometric standard error, range (A m^{-1})	
0–1	5	5.06	4.15	6.17
1–5	14	2.70	2.34	3.12
5–10	7	1.89	1.35	2.65
10–30	14	1.13	0.97	1.31
30–50	15	1.89	1.56	2.29
50–70	12	2.61	2.21	3.08
70–90	8	2.84	2.25	3.58
90–110	8	4.13	3.16	5.40
110–130	6	4.20	3.36	5.25
130–160	4	1.79	1.32	2.42

$$A(T) = M_{trm}R(T)[1+P\exp(-T/\lambda)] + M_{crm}$$
$$\times \frac{1}{\lambda}\int_0^T R(t)\exp[(t-T)/\lambda]\,dt, \tag{9}$$

where M_{trm} is the TRM intensity, M_{crm} is the CRM intensity, $R(T)$ is the geomagnetic polarity time scale expressed as a normalized square wave, P is the ratio of TRM that decays to that which does not and is set at 5, λ is the time constant for TRM decay and CRM acquisition and is set at 5 m.y., and $A(T)$ is the magnetization for crust aged T m.y.

If the Raymond & LaBrecque (1986) model is correct, it should be possible to distinguish between the two remanence carriers in rocks of carefully chosen age using normal rock magnetic techniques unless the magnetic mineralogy and grain size and shape of the magnetic minerals carrying the two remanences are exactly the same.

Layer 3 is represented by the gabbro entries in Table 3. The relatively stable magnetization in oceanic gabbros is not what one would expect from these coarse-grained igneous rocks. However, Davis (1981) has shown that isotropic gabbros from the upper part of layer 3 have thin magnetite rods that have grown within plagioclase crystals. These rods should have the magnetic stability of single-domain grains. Banerjee (1984) has suggested that the lower cumulate gabbro layer does not contribute to the magnetic anomalies because the relatively unmetamorphosed gabbros at this depth have relatively low magnetizations. We note that the arithmetic mean magnetizations of the unmetamorphosed and metamorphosed gabbros in Kent et al (1978) and the magnetizations for these rock types presented in Table 3 are not very different. We therefore prefer to consider the whole of layer 3 as a possible source for marine magnetic anomalies. The timing of the magnetization is an important consideration. It is possible that metamorphism and serpentinization take place sufficiently late that the magnetization acquired during these processes could have no effect on the magnetic field. However, the possibility of rapid cooling in this layer produced by hydrothermal circulation cannot be ruled out. We have therefore used equal weights for all four entries in Table 3 to arrive at an average magnetization for this layer. Notice that since the magnetizations do not vary greatly, most mixtures will give approximately the same average magnetization, which is not true for layer 2. The real question that must be solved is the amount of serpentinized peridotite that is included in layer 3. Although many samples of serpentinite have been recovered by dredging, it is thought that these are due to the uplift of lower crustal or mantle material along the fault scarps that are the usual targets of dredging operations. In order to obtain the fairly large remanent magnetizations for these rocks shown in Table 3, it is necessary for the serpentinization process to be carried out nearly to completion (Saad 1969, Hatherton 1967). This means that P-wave velocities will be very low, in the region of 5 km s^{-1}. Seismic data reanalyzed by modern techniques (Spudich & Orcutt 1980) reveal regions in layer 3 that show a low-velocity region of S-wave velocities. No P-wave low-velocity region was seen, meaning that an increase in the Poisson ratio had occurred. Since serpentinized peridotites have the highest Poisson ratio of all candidate rocks from layer 3 (Christensen 1978), it therefore seems possible that the low S-wave velocities are caused by a mixture of serpentinized peridotite with normal layer 3 constituents such as gabbro or metagabbro. Other rocks such as plagiogranites and trondhjemites, thought to be minor constituents of layer 3, have low Poisson ratios due to their relatively high quartz content. An arithmetic mean velocity of a mixture of 86% gabbro ($V_p = 7.2$ km s^{-1}, $V_s = 3.8$ km s^{-1}, $\sigma = 0.31$) and 16% serpentinite ($V_p = 5.05$ km s^{-1}, $V_s = 2.45$ km s^1, $\sigma = 0.35$) will have a V_p roughly of 6.9 and a V_s of 3.6 km s^{-1}, values that match the low V_s velocity zone of Spudich & Orcutt (1980). [Velocities were taken from Christensen (1978), and σ is the Poisson ratio.] However, it seems unlikely that the percentage of serpentinite within layer 3 could be higher than 16%. There is another possibility, which is that there is a layer of serpentinite between the gabbros of layer 3 and the ultramafic rocks of the mantle (Lewis & Snydsman 1977). A process whereby progressive serpentinization of mantle material takes place with time scales of several tens of millions of years would explain the low-velocity zone seen by Lewis & Snydsman (1977) and also the observation made by Goslin et al (1972) that layer 3 shows a progressive thickening with age, by about 1.5 km. The possibility of a serpentinite layer has recently been confirmed by sophisticated measurements made along a transect in the Atlantic Ocean basin (NAT Study Group 1985). Unless these serpentinization processes occur very rapidly, they will not be capable of contributing to the lineated magnetic anomalies. The crustal-thickening process of Goslin et al (1972) occurs with a time scale of 18 m.y., which is long enough to be unimportant for lineated anomalies but possibly important for long-wavelength anomalies at satellite altitudes. The time scale for observations of the NAT Study Group (1985) appears to be much longer (many tens of millions of years). This long time scale would invalidate this mechanism as a source for even the longest wavelengths of marine magnetic anomalies. We have therefore included several possibilities for models in Table 4, some of which are appropriate for the longer wavelength satellite-elevation anomalies.

In Table 4 we have produced three models. Model A consists just of entries 1 and 2 and is appropriate for short-wavelength seafloor-spreading anomalies. Model B is adjusted to have 16% serpentinite in layer 3, whose total thickness remains 4.5 km. For this model to be important for short wavelength anomalies the serpentinization process must occur within about 10 to 20 km of the axis. Model C is for longer wavelength anomalies, because in addition to serpentinite within layer 3, there is also a lower layer of serpentinite of thickness 1.5 km. One immediate observation that can be made is that induced magnetization is fairly unimportant for generating magnetic anomalies in these types of material. However, it is possible that some of the remanent magnetization may be viscous remanent magnetization, which acts in some ways as a strong induced magnetization, since it is aligned in the direction of the present Earth's field. Although many studies have been done on the acquisition of viscous magnetization and viscous remanent magnetization in oceanic rocks (e.g. Dunlop 1983, Lowrie & Kent 1976, Moskowitz & Banerjee 1981), the full meaning of this research is still unclear, partly because of the very large variation in the ability of oceanic rocks to acquire these types of magnetization. Another complicating factor is that the acquisition of viscous types of magnetization depends heavily on temperature, such that viscous magnetization acquisition rates can be more than doubled by heating to only modest temperatures, such as 50°C (Dunlop & Hale 1977, Smith 1984). It appears clear, however, that the measured magnetizations, if they contain viscous components, may be different than the magnetization available to produce the short-wavelength magnetic anomalies. The difference may be positive or negative, depending on the conditions of storage of the sample between collection and measurement.

Another potentially important factor that has not been considered in arriving at the values shown in Table 4 is the frequent presence of reversals in the pillow lava section of layer 2 (i.e. layer 2A). This has been noted by a number of people and occurs in many of the holes that have penetrated significant thicknesses of this layer. For instance, in holes 332A (penetration 320 m into crust), 332B (penetration 560 m), and 333A (penetration 320 m) there are reversals of inclination down each hole (Hall 1976). Another peculiar feature about the inclinations in these holes is that they do not correspond with the axial dipole field inclination for the location (Harrison & Watkins 1977). In hole 395A (penetration 570 m) there are also reversals, although in this core the inclinations are close to those expected from axial dipole fields (Johnson 1978). In hole 396B (penetration 240 m) Petersen (1979) states that "owing to changes in polarity... the intensity of magnetization integrated over the whole length of the drill hole is much too low to explain the amplitude of the marine magnetic anomaly." Holes 417A and 417D appear to have a more coherent inclination picture, but each of these penetrated less than 200 m of basalt

(Bleil & Smith 1980). In hole 418A there is one reversal in the 550-m section sampled (Levi 1980).

Denham & Schouten (1979) have pointed out that using statistical models of emplacement it is easy to show that in slow-spreading material it is likely that reversals will be seen in vertical section, whereas for fast-spreading ridges this is not necessarily the case. The critical parameters are the standard deviation of emplacement, which depends both on the standard deviation of dike injection and the distance that lavas flow from their point of origin, the rate at which individual emplacement events take place, the rate of spreading, and the rate of reversals of the Earth's field. This suggests that there may be different models appropriate for slow- and fast-spreading sections of the oceanic crust.

The hole that has penetrated farthest into oceanic crust was unfortunately drilled close to the equator, rendering it impossible to determine reversals of polarity from inclination data. Hole 504B penetrated 1076 m into basement, including 300 m of the sheeted dike complex, where drilling terminated on Leg 83 of DSDP (Anderson et al 1982). A review of all magnetic work done on basement samples in this hole has been given by Smith & Banerjee (1986).

Smith & Banerjee (1986) show that the magnetic properties within this 1-km hole can be divided up into three sections, which correspond with different styles of emplacement. The upper 0.6 km consists of extrusive material with properties much like those that other workers have found. Remanent magnetization was on average 6 A m^{-1}, and the magnetic carrier was titanomagnetite (average formula $Fe_{3-x} Ti_x O_4$, with $x = 0.6$), which had been oxidized at low temperature to a single-phase titanomaghemite. This low-temperature oxidation causes the Curie temperature to increase from 160°C to 300–400°C.

The transition unit consists of a region where both lava flows and dikes are mixed and is about 0.2 km thick. The relative thinness of this zone confirms models from ophiolite complexes suggesting that the emplacement scattering width of the dikes above the magma chamber is quite small (Kidd 1977). The remanent magnetization of this zone is a factor of 10 less than that for the lava layer, largely as a result of the alteration that has occurred. This alteration has produced greenschist-facies metamorphism. The magnetic carrier in this case is pure magnetite (Curie temperatures of about 580°C). However, microprobe work indicates that the magnetic minerals have large quantities of ulvospinel, which should reduce the Curie temperature drastically. Exsolution into a magnetite-rich phase and a titanium-rich phase could have occurred, but it is not evident on polished thin sections; thus, any such change must have occurred with a submicroscopic length scale. This suggests that deuteric oxidation, the common way in which exsolution is produced, was not the cause. Rather, Smith & Banerjee (1986) suggest that low-temperature oxidation occurred to titanomagnetite with the normal oceanic value of $x = 0.6$. This was followed by a higher temperature alteration phase, which was responsible for the submicroscopic exsolution and the greenschist-facies metamorphism. During this phase of metamorphism, much magnetite was replaced with silicates, although some secondary magnetite was also produced.

In the bottom section, consisting almost entirely of dikes, it is supposed that the second phase of moderate-temperature alteration took place; this phase would have caused magnetite Curie temperatures, but much less metamorphism in the bulk rock due to the lower permeability of the material. This means that there has been much less dissolution of magnetite, and consequently the average magnetization is about 2 A m^{-1}. If these average magnetizations are used to predict an average magnetization for layer 2, in which the lower dike layer is assumed to continue until a total thickness of 1.5 km has been built up, the figure arrived at is 3.3 A m^{-1}. This is fairly close to the value of the model we have predicted (model A in Table 4). Both of these values have been calculated assuming no reversals in vertical section. The work done on drill holes (in the Atlantic at least) suggests the frequent presence of reversals in vertical section, meaning that the average magnetizations in Table 4 are too high. There may in fact be no net contribution to the magnetic anomalies from layer 2A, especially for slow-spreading material.

Using the information from direct measurements of oceanic crustal rocks, we can put together a magnetic model of the oceanic crust. Such a model is shown in Figure 7. This may be compared with other models of the magnetic nature of the oceanic crust that have been presented previously. Kent et al (1978) had a stronger magnetization for layer 2A and a much weaker magnetization for layer 2B, based on the supposition that layer 2B was metamorphosed. It appears likely that the strongly metamorphosed rocks of layer 2 are confined to the fairly thin transition zone between layer 2A and 2B (Smith & Banerjee 1986). The magnetization for layer 3 suggested by Kent et al (1978) is approximately the same as the value we have suggested (i.e. 1 A m^{-1}). Their model had no deeper serpentinite layer. The average magnetization for their model assuming a total thickness of 6 km is 1.13 A m^{-1} after the pillow basalt layer has weathered to its lower magnetization state. Dunlop & Prevot (1982) have proposed a model in which the layer 2A and 2B magnetizations are somewhat lower and the layer 3B results are considerably lower than the ones we show in Figure 7. They do show the possibility of elevated magnetization in the ultramafic cumulate layer. If this layer is excluded and their results normalized to a 6-km-thick crust, the average magnetization

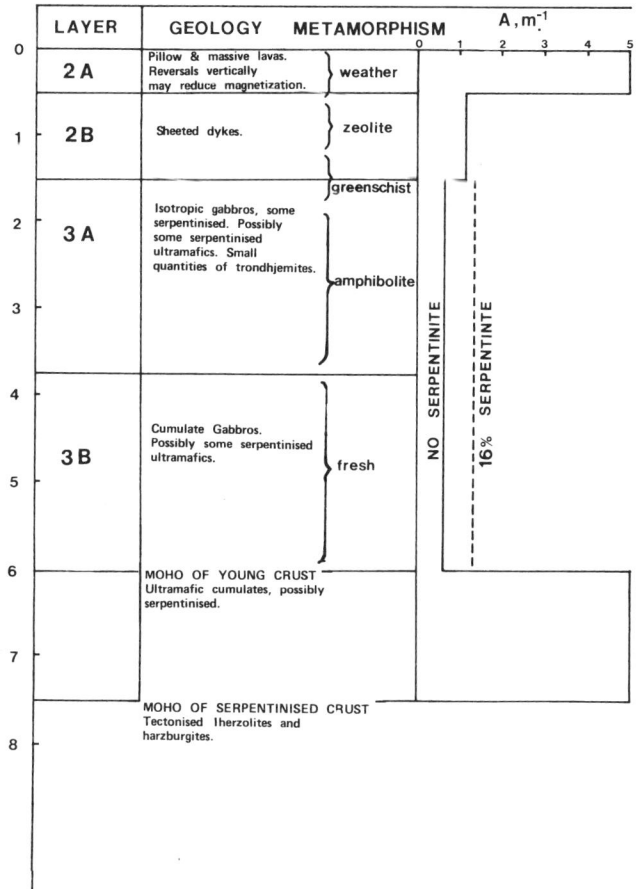

Figure 7 Model for the oceanic crust. The left-hand column shows the location of the classical seismic layers and depths (in kilometers). Modern methods of seismic data analysis suggest that the only sharp discontinuity in seismic velocity is at the Moho. Nevertheless, the old seismic names are important distinguishing marks. The geology is based on dredged, drilled, and ophiolite samples. The values for the magnetization of the various layers are discussed in the text. The dashed line for layer 3 is the magnetization value assuming that 16% of this layer is composed of serpentinized ultramafic rocks. The two Mohos are those when the crust has solidified from the magma chamber, and when it is very young. Subsequent serpentinization of mantle material causes this material to gain magnetization and to lose seismic velocity, resulting in it being transferred from mantle to crust. This is a possible explanation for the thickening of the crust with age seen in the Atlantic (Goslin et al 1972). The lower Moho is placed below this 1.5-km-thick layer, and it will be appropriate for crust that is several tens of millions of years old.

of their model is 0.808 A m^{-1}. Banerjee's (1984) model has a lower magnetization for layers 2B and 3A and zero magnetization for layer 3B for reasons discussed above. His average magnetization, again normalized to a 6-km-thick crust, is 0.708 A m^{-1}.

Discussion and Conclusions

From what has been presented in this paper, it seems obvious that the source of the lineated magnetic anomalies must reside in most of the oceanic crust. The problem of having the causative layer restricted to the pillow lavas of layer 2A can be summarized as follows:

1. The magnetization of the rocks from layer 2A is insufficient to give the required size of the magnetic anomaly. Inversions of the field require average magnetizations over the 6-km-thick oceanic crust of greater than 1 A m^{-1}, whereas direct measurements of layer 2A rocks would give an average magnetization of only 0.42 A m^{-1} when averaged over the 6-km-thick oceanic crust. If reversals in vertical section are as prevalent as they seem to be in drilled sections, then even this value will be too high, requiring more of the source to come from below layer 2A.

2. Many indirect experiments have been done that suggest a contribution from layer 3. One observation showed that a destruction of the magnetization of the upper layers had occurred with time. In another experiment, the nature of the sloping boundary between zones of opposite polarity for crust created at different spreading rates was measured. The results confirmed that much of the signal has come from the lower layers of the crust. Other arguments were offered by Blakely (1979) for a thick source layer.

3. The presence of anomalous skewness suggests that a sloping boundary between crust of opposite polarity—one of considerable horizontal extent—is required. The alternative explanation of large-scale rotations seems to go against tectonic arguments and direct observations of approximately flat-lying lava flows. One unresolved problem that needs to be tackled is the rate of cooling of the material of the lower crust. Many models have been produced where it is assumed that all cooling takes place by conduction. But it is well known that significant quantities of heat (perhaps a preponderate amount) are removed by circulation of water through young oceanic crust. This will have the effect of making the slope of the blocking temperature isotherm steeper than for the case when purely conductive cooling takes place. Further work on this important topic needs to be carried out.

Calculation of the magnetization of the oceanic crust done from satellite data reveals that larger magnetizations are necessary to explain the signal coming from the Cretaceous normal polarity interval than for the Vine-Matthews lineations seen in ocean surface lineations. One possible cause for this is magnetization of a layer of serpentinized mantle lying at the base of the crust. The timing of the magnetization of this layer is less critical than for magnetizations necessary to explain the short-wavelength lineations because of the large lateral extent of the Cretaceous normal polarity material in the North Atlantic Ocean basin.

The preferred model is shown in Figure 7. There are still many uncertainties, however. For instance, we need to known the statistics of the possibility of going through reversals vertically, especially for fast-spreading sections of the crust. The nature of the lower oceanic crust is still uncertain, especially as to its degree of metamorphism. The amount of serpentinization of normal gabbro needs to be better determined, and an especially important task is to determine the amount of ultramafic material trapped in the oceanic crust and how much of this material is serpentinized. The timing of magnetization in serpentinized and metamorphosed material also needs to be studied. Finally, the role of viscous magnetization effects, especially at high temperatures, needs to be further investigated.

ACKNOWLEDGMENTS

My research has been supported by the National Aeronautics and Space Administration, Geodynamics Program. I thank Richard Blakely for providing an original figure and Carol Raymond for providing a preprint. I have benefited from discussions with Huang Qilin, Nancy Wittpenn, and Kjell Hayling. Contribution from the University of Miami, Rosenstiel School of Marine and Atmospheric Science.

Literature Cited

Abramowitz, M., Segun, I. A. 1965. *Handbook of Mathematical Functions*. New York: Dover. 1046 pp.

Anderson, R. N., Honnorez, J., Becker, K., Adamson, A. C., Alt, J. C., et al. 1982. DSDP hole 504B, the first reference section over 1 km through layer 2 of the oceanic crust. *Nature* 300: 589–94

Arkani-Hamed, J., Strangway, D. W. 1985a. Lateral variations of apparent magnetic susceptibility of lithosphere deduced from MAGSAT data. *J. Geophys. Res.* 90: 2655–64

Arkani-Hamed, J., Strangway, D. W. 1985b. Correction to "Lateral variations of apparent magnetic susceptibility of lithosphere deduced from MAGSAT data." *J. Geophys. Res.* 90: 4655–57

Atwater, T., Mudie, J. D. 1973. Detailed near-bottom geophysical study of the Gorda rise. *J. Geophys. Res.* 78: 8665–86

Ballard, R. D., van Andel, T. H. 1977. Morphology and tectonics of the inner rift valley at lat. 36°50′N on the Mid-Atlantic Ridge. *Geol. Soc. Am. Bull.* 88: 507–30

Banerjee, S. K. 1980. Magnetism of the oceanic crust—evidence from ophiolite complexes. *J. Geophys. Res.* 85: 3557–66

Banerjee, S. K. 1984. The magnetic layer of the ocean crust—how thick is it? *Tectonophysics* 105: 15–27

Bath, M. 1974. *Spectral Analysis in Geophysics*. New York: Elsevier. 563 pp.

Beske-Diehl, S., Banerjee, S. K. 1979. An example of magnetic properties as indicators of alteration in ancient oceanic lithosphere—the Othris ophiolite. *Earth Planet. Sci. Lett.* 44: 451–62

Blakely, R. J. 1976. An age-dependent, two-layer model for marine magnetic anomalies. In *The Geophysics of the Pacific Ocean Basin and its Margins, Geophys. Monogr. Ser.*, 19: 227–34. Washington, DC: Am. Geophys. Union

Blakely, R. J. 1979. Random crustal magnetization and its effect on coherence of short-wavelength marine magnetic anomalies. *Earth Planet. Sci. Lett.* 46: 43–48

Blakely, R. J. 1983. Statistical averaging of marine magnetic anomalies and the aging of the oceanic crust. *J. Geophys. Res.* 88: 2289–96

Blakely, R. J., Lynn, W. S. 1977. Reversal transition widths and fast-spreading centers. *Earth Planet. Sci. Lett.* 33: 321–30

Bleil, U., Petersen, N. 1983. Variations in magnetization intensity and low-temperature titanomagnetite oxidation of ocean floor basalts. *Nature* 301: 384–88

Bleil, U., Smith, B. 1980. Paleomagnetism of basalts, Leg 51, In *Initial Reports of the Deep Sea Drilling Project*, ed. T. Donnelly, J. Francheteau, W. Bryan, P. Robinson, M. Flower, M. Salisbury, 51–53: 1351–61. Washington, DC: US Govt. Print. Off.

Bonatti, E., Honnorez, J., Ferrara, G. 1971. I. Ultramafic rocks. Peridotite-gabbro-basalt complex from the equatorial Mid-Atlantic Ridge. *Philos. Trans. R. Soc. London Ser. A* 268: 385–402

Bonatti, E., Honnorez, J., Kirst, P., Radicati, F. 1975. Metagabbros from the Mid-Atlantic Ridge at 06°N: contact-hydrothermal-dynamic metamorphism beneath the axial valley. *J. Geol.* 83: 61–78

Bott, M. H. P. 1967. Solution of the linear inverse problem in magnetic interpretation with application to oceanic magnetic anomalies. *Geophys. J. R. Astron. Soc.* 13: 313–23

Cande, S. C. 1976. A palaeomagnetic pole from Late Cretaceous marine magnetic anomalies in the Pacific. *Geophys. J. R. Astron. Soc.* 44: 547–66

Cande, S. C., Kent, D. V. 1976. Constraints imposed by the shape of marine magnetic anomalies on the magnetic source. *J. Geophys. Res.* 81: 4157–62

Cande, S. C., Kristoffersen, Y. 1977. Late Cretaceous magnetic anomalies in the North Atlantic. *Earth Planet. Sci. Lett.* 35: 215–24

Cann, J. R. 1974. A model for oceanic crustal structure developed. *Geophys. J. R. Astron. Soc.* 39: 169–87

Christensen, N. I. 1978. Ophiolites, seismic velocities and oceanic crustal structure. *Tectonophysics* 47: 131–57

Cox, A., Doell, R. R., Dalrymple, G. B. 1963. Geomagnetic polarity epochs and Pleistocene geochronology. *Nature* 198: 1049–51

Davis, K. E. 1981. Magnetite rods in plagioclase as the primary carrier of stable NRM in oceanic floor gabbros. *Earth Planet. Sci. Lett.* 55: 190–98

Denham, C. R., Schouten, H. 1979. On the likelihood of mixed polarity in oceanic basement drill cores. In *Deep Drilling Results in the Atlantic Ocean: Ocean Crust, Maurice Ewing Ser.*, ed. M. Talwani, C. G. A. Harrison, D. E. Hayes, 2: 160–65. Washington, DC: Am. Geophys. Union

Dietz, R. S. 1961. Continent and ocean basin evolution by spreading of the sea floor. *Nature* 190: 854–57

Dunlop, D. J. 1983. Viscous magnetization of 0.04–100 μm magnetites. *Geophys. J. R. Astron. Soc.* 74: 667–87

Dunlop, D. J., Hale, C. J. 1977. Simulation of the long-term changes in the magnetic signal of the oceanic crust. *Can. J. Earth Sci.* 14: 716–44

Dunlop, D. J., Prevot, M. 1982. Magnetic properties and opaque mineralogy of drilled submarine intrusive rocks. *Geophys. J. R. Astron. Soc.* 69: 763–802

Fox, P. J., Opdyke, N. D. 1973. Geology of the oceanic crust: magnetic properties of oceanic rocks. *J. Geophys. Res.* 78: 5139–54

Girdler, R. W., Peter, G. 1960. An example of the importance of natural remanent magnetization in the interpretation of magnetic anomalies. *Geophys. Prospect.* 8: 474–83

Goslin, J., Beuzart, P., Francheteau, J., Le Pichon, X. 1972. Thickening of the oceanic layer in the Pacific Ocean. *Mar. Geophys. Res.* 1: 418–27

Gradshteyn, I. S., Ryzhik, I. M. 1980. *Tables of Integrals, Series and Products*. Corrected and enlarged edition prepared by A. Jeffrey. Transl. by Scr. Tech. New York: Academic. 1160 pp.

Gudmundsson, G. 1966. Interpretation of one-dimensional magnetic anomalies by use of the Fourier-transform. *Geophys. J. R. Astron. Soc.* 12: 87–97

Gudmundsson, G. 1967. Spectral analysis of magnetic surveys. *Geophys. J. R. Astron. Soc.* 13: 325–37

Hall, J. M. 1976. Major problems regarding the magnetization of oceanic crustal layer 2. *J. Geophys. Res.* 81: 4223–30

Harland, W. B., Cox, A. V., Llewellyn, P. G., Pickton, C. A. G., Smith, A. G., Walters, R. 1982. *A Geologic Time Scale*. Cambridge: Cambridge Univ. Press. 131 pp.

Harrison, C. G. A. 1968. Formation of magnetic anomaly patterns by dyke injection. *J. Geophys. Res.* 73: 2137–42

Harrison, C. G. A. 1974. Tectonics of mid-ocean ridges. *Tectonophysics* 22: 301–10

Harrison, C. G. A. 1976. Magnetization of the oceanic crust. *Geophys. J. R. Astron. Soc.* 47: 257–84

Harrison, C. G. A. 1981. Magnetism of the oceanic crust. In *The Sea*, ed. C. Emiliani, 7: 219–39. New York: Wiley

Harrison, C. G. A. 1987. The crustal field. In *Geomagnetism*, ed. J. A. Jacobs. London: Academic. In press

Harrison, C. G. A., Carle, H. M. 1981. Intermediate wavelength magnetic anomalies over ocean basins. *J. Geophys. Res.* 86: 11,585–99

Harrison, C. G. A., Stieltjes, L. 1977. Faulting within the median valley. *Tectonophysics* 38: 137–44

Harrison, C. G. A., Watkins, N. D. 1977. Shallow inclinations of remanent magnetism in Deep Sea Drilling Project igneous cores: geomagnetic field behavior or postemplacement effects? *J. Geophys. Res.* 82: 4869–77

Harrison, C. G. A., Carle, H. M., Hayling, K. L. 1986. Interpretation of satellite elevation magnetic anomalies. *J. Geophys. Res.* 91: 3633–50

Hatherton, T. 1967. A geophysical study of Nelson-Cook strait region, New Zealand. *N. Z. J. Geol. Geophys.* 10: 1330–47

Hayling, K. L. 1986. Heat flow and magnetization in the oceanic crust, and the possibility to determine a zero level for magnetic models derived from satellite magnetic anomalies. *Eos, Trans. Am. Geophys. Union* 67: 264 (Abstr.)

Hayling, K. L., Harrison, C. G. A. 1986. Magnetization modelling in the North and equatorial Atlantic Ocean using MAGSAT data. *J. Geophys. Res.* 91: 12,423–43

Heirtzler, J. R., Le Pichon, X. 1965. Crustal structure of the mid-ocean ridges. 3. Magnetic anomalies over the Mid-Atlantic Ridge. *J. Geophys. Res.* 70: 4013–33

Heirtzler, J. R., Dickson, G. O., Herron, E. M., Pitman, W. C. III, Le Pichon, X. 1968. Marine magnetic anomalies, geomagnetic field reversals, and motions of the ocean floor and continents. *J. Geophys. Res.* 73: 2119–36

Hess, H. H. 1962. History of ocean basins. In *Petrologic Studies: A Volume to Honor A. F. Buddington*, ed. A. E. J. Engel, H. L. James, B. F. Leonard, pp. 599–620. Boulder, Colo: Geol. Soc. Am.

Hill, M. N. 1957. Recent geophysical exploration of the ocean floor. *Phys. Chem. Earth* 2: 129–63

Holmes, A. 1931. Radioactivity and Earth movements. *Geol. Soc. Glasgow Trans.* 18: 559–606

Howe, H. H. 1940. Height of magnetic anomalies. *Trans. Am. Geophys. Union* 1940: 309–11

Irving, E., Robertson, W. A., Aumento, F. 1970. The Mid-Atlantic Ridge near 45°N. VI. Remanent intensity, susceptibility, and iron content of dredged samples. *Can. J. Earth Sci.* 7: 226–38

Johnson, H. P. 1978. Paleomagnetism of igneous rock samples—DSDP Leg 45. In *Initial Reports of the Deep Sea Drilling Project*, ed. W. G. Melson, P. D. Rabinowitz, et al, 45: 387–96. Washington, DC: US Govt. Print. Off.

Johnson, H. P., Atwater, T. 1977. Magnetic study of basalts from the Mid-Atlantic Ridge, lat. 37°N. *Geol. Soc. Am. Bull.* 88: 637–47

Kent, D. V., Honnorez, B. M., Opdyke, N. D., Fox, P. J. 1978. Magnetic properties of dredged oceanic gabbros and the source of marine magnetic anomalies. *Geophys. J. R. Astron. Soc.* 55: 513–37

Kidd, R. G. W. 1977. The nature and shape of the sources of marine magnetic anomalies. *Earth Planet. Sci. Lett.* 33: 310–20

Klitgord, K. D., Huestis, S. P., Mudie, J. D., Parker, R. L. 1975. An analysis of near-bottom magnetic anomalies: sea-floor spreading and the magnetized layer. *Geophys. J. R. Astron. Soc.* 43: 387–424

LaBrecque, J. L., Cande, S. C. 1984. Intermediate-wavelength magnetic anomalies over the Central Pacific. *J. Geophys. Res.* 89: 11,124–34

LaBrecque, J. L., Raymond, C. A. 1985. Sea-floor spreading anomalies in the MAGSAT field of the North Atlantic. *J. Geophys. Res.* 90: 2565–75

LaBrecque, J. L., Cande, S. C., Jarrard, R. D. 1985. Intermediate-wavelength magnetic anomaly field of the north Pacific and possible source distributions. *J. Geophys. Res.* 90: 2549–64

Langel, R., Ousley, G., Berbert, J., Murphy, J., Settle, M. 1982. The MAGSAT mission. *Geophys. Res. Lett.* 9: 243–45

Larson, R. L., Chase, C. G. 1972. Late Mesozoic evolution of the western Pacific Ocean. *Geol. Soc. Am. Bull.* 83: 3627–44

Levi, S. 1980. Paleomagnetism and some magnetic properties of basalts from the Bermuda Triangle. In *Initial Reports of the Deep Sea Drilling Project*, ed. T. Donnelly, J. Francheteau, W. Bryan, P. Robinson, M. Flower, M. Salisbury, 51–53: 1363–78. Washington, DC: US Govt. Print. Off.

Levi, S., Banerjee, S. K., Beske-Diehl, S., Moskowitz, B. 1978. Limitations of ophiolite complexes as models for the magnetic layer of the oceanic lithosphere. *Geophys. Res. Lett.* 5: 473–76

Lewis, B. T. R., Snydsman, W. E. 1977. Evidence for a low velocity layer at the base of the oceanic crust. *Nature* 266: 340–44

Lowrie, W. 1974. Oceanic basalt magnetic properties and the Vine and Matthews hypothesis. *J. Geophys.* 40: 513–36

Lowrie, W., Kent, D. V. 1976. Viscous remanent magnetization in basalt samples. In *Initial Reports of the Deep Sea Drilling Project*, ed. R. S. Yeats, S. R. Hart, et al, 34: 479–84. Washington, DC: US Govt. Print. Off.

Luyendyk, B. P., Day, R. 1982. Paleomagnetism of the Samail ophiolite, Oman: 2. The Wadi Kadir gabbro section. *J. Geophys. Res.* 87: 10,903–17

Luyendyk, B. P., Melson, W. G. 1967. Magnetic properties and petrology of rocks near the crest of the Mid-Atlantic Ridge. *Nature* 215: 147–49

Macdonald, K. C. 1977. Near-bottom magnetic anomalies, asymmetric spreading, oblique spreading, and tectonics of the Mid-Atlantic Ridge near lat. 37°N. *Geol. Soc. Am. Bull.* 88: 541–55

Macdonald, K. C., Miller, S. P., Luyendyk, B. P., Atwater, T. M., Shure, L. 1983. Investigation of a Vine-Matthews magnetic lineation from a submersible: the source and character of marine magnetic anomalies. *J. Geophys. Res.* 88: 3403–18

Matthews, D. H., Bath, J. 1967. Formation of magnetic anomaly pattern of mid-Atlantic ridge. *Geophys. J. R. Astron. Soc.* 13: 349–57

McDougall, I., Tarling, D. H. 1963. Dating of polarity zones in the Hawaiian Islands. *Nature* 200: 54–56

McFadden, P. L., Merrill, R. T. 1984. Lower mantle convection and geomagnetism. *J. Geophys. Res.* 89: 3354–62

Miller, S. P., Hey, R. N. 1986. Three dimensional magnetic modelling of a propagating rift, Galapagos 95°30′W. *J. Geophys. Res.* 91: 3395–3406

Morley, L. W. 1981. An explanation of magnetic banding in ocean basins. In *The Sea*, ed. C. Emiliani, 7: 1717–19

Moskowitz, B. M., Banerjee, S. K. 1981. A comparison of the magnetic properties of synthetic titanomaghemites and some oceanic basalts. *J. Geophys. Res.* 86: 11,869–82

NAT Study Group. 1985. North Atlantic transect: a wide-aperture, two-ship multichannel seismic investigation of the oceanic crust. *J. Geophys. Res.* 90: 10,321–41

Opdyke, N. D., Hekinian, R. 1967. Magnetic properties of some igneous rocks from the Mid-Atlantic Ridge. *J. Geophys. Res.* 72: 2257–60

Parker, R. L. 1973. The rapid calculation of potential anomalies. *Geophys. J. R. Astron. Soc.* 31: 447–55

Parker, R. L., Klitgord, K. D. 1972. Magnetic upward continuation from an uneven track. *Geophysics* 37: 662–68

Petersen, N. 1979. Rock- and paleomagnetism of basalt from Site 396B, Leg 46. In *Initial Reports of the Deep Sea Drilling Project*, ed. L. Dmitriev, J. R. Heirtzler, 46: 357–62. Washington, DC: US Govt. Print. Off.

Raymond, C. A., LaBrecque, J. L. 1986. Magnetization of the oceanic crust: TRM or CRM? *J. Geophys. Res.* In press

Robbins, K. G. 1977. A new approach to sub-critical instability and turbulent transitions in a simple dynamo. *Math. Proc. Cambridge Philos. Soc.* 82: 309

Roche, A. 1953. Sur l'origine des inversions de l'aimantation constantées dans les roches d'Auvergne. *C.R. Acad. Sci. Paris.* 236: 107–9

Rutten, M. G., Wensink, H. 1959. Geology of the Hvalfjordur-Skorradalur area (southwestern Iceland). *Geol. Mijnbouw* 21: 172–81

Saad, A. H. 1969. Magnetic properties of ultramafic rocks from Red Mountain, California. *Geophysics* 34: 974–87

Schmalz, R. F. 1961. A case for convection. *Miner. Ind. (University Park, Pa.)* 30(5): 1–8

Schouten, H., Cande, S. C. 1976. Palaeomagnetic poles from marine magnetic anomalies. *Geophys. J. R. Astron. Soc.* 44: 567–75

Schouten, H., Denham, C. R. 1979. Modelling the oceanic magnetic source layer. In *Deep Drilling Results in the Atlantic Ocean: Ocean Crust, Maurice Ewing Ser.*, ed. M. Talwani, C. G. A. Harrison, D. E. Hayes, 2: 151–59. Washington, DC. Am. Geophys. Union

Schouten, H., McCamy, K. 1972. Filtering marine magnetic anomalies. *J. Geophys. Res.* 77: 7089–99

Schouten, J. A. 1971. A fundamental analysis of magnetic anomalies over ocean ridges. *Mar. Geophys. Res.* 1: 111–44

Shure, L., Parker, R. L. 1981. An alternative explanation for intermediate-wavelength magnetic anomalies. *J. Geophys. Res.* 86: 11,600–8

Sleep, N. H. 1975. Formation of oceanic crust: some theoretical constraints. *J. Geophys. Res.* 80: 4037–42

Smith, B. M. 1984. Magnetic viscosity of some doleritic basalts in relation to the interpretation of the oceanic magnetic anomalies. *Geophys. Res. Lett.* 11: 213–16

Smith, G. M., Banerjee, S. K. 1986. The magnetic structure of the upper kilometer of the marine crust. *J. Geophys. Res.* In press

Solov'yev, O. A. 1962. Use of frequency method for the determination of some parameters of magnetic bodies. *Izv. Sib. Otd. Akad. Nauk SSSR Geol. Geofiz.* 2: 122–25

Spudich, P., Orcutt, J. 1980. Petrology and porosity of an oceanic crustal site: results from wave form modeling of seismic refraction data. *J. Geophys. Res.* 85: 1409–33

Talwani, M., Windisch, C. C., Langseth, M. G. Jr. 1971. Reykjanes Ridge crest: a detailed geophysical study. *J. Geophys. Res.* 76: 473–517

Vestine, E. H., Davids, N. 1945. Analysis and interpretation of geomagnetic anomalies. *Terr. Magn. Atmos. Electr.* 50: 1–36

Vine, F. J. 1966. Spreading of the ocean floor: new evidence. *Science* 154: 1405–15

Vine, F. J., Matthews, D. H. 1963. Magnetic anomalies over oceanic ridges. *Nature* 199: 947–49

Vine, F. J., Moores, E. M. 1972. Model for the gross structure, petrology, and magnetic properties of oceanic crust. *Geol. Soc. Am. Mem.* 182: 195–205

Vine, F. J., Wilson, J T. 1965. Magnetic anomalies over a young oceanic ridge off Vancouver Island. *Science* 150: 485–9

Weissel, J. K., Hayes, D. E. 1972. Magnetic anomalies in the Southeast Indian Ocean. In *Antarctic Oceanology II: The Australian-New Zealand Sector, Antarct. Res. Ser.*, 19: 165–96. Washington, DC: Am. Geophys. Union

STRESS NEAR THE SURFACE OF THE EARTH

D. I. Gough and W. I. Gough

Department of Physics, University of Alberta, Edmonton, Alberta T6G 2J1, Canada

INTRODUCTION

Over most of the Earth's surface earthquake hypocenters are concentrated in the depth range 0–15 km—that is, in the upper half of the crust (Chen & Molnar 1983). The exception is the subduction zones, where cold lithosphere is underthrust to greater depths. As a first approximation, the top 10–15 km of the crust can be regarded as a region of nearly elastic response to stress, with brittle fracture under sufficient stress difference. This is the depth range discussed in this review. The lower crust appears to be characterized by some form of ductile response to stress. In large areas of the upper crust, the stress field is related to large-scale tectonic processes and represents the tractions acting on the plate concerned. However, some decoupled crustal blocks are known and are noted herein. Caution is necessary in generalizing the stress in a region to the plate that contains it.

Recent reviews of crustal stress are given by Ranalli & Chandler (1975), Wyss (1977), McGarr & Gay (1978), and Gay (1980). In the upper crust one principal stress is generally within 20° of the vertical except beneath very rough topography, and this "vertical" principal stress σ_z is usually within 10% of the *lithostatic pressure* $\int \rho g \, dz$ due to overlying rock (McGarr & Gay 1978). The horizontal principal stresses generally increase with depth and may be less or greater than the vertical pressure. Within a few hundred meters of the surface, the rock mass is divided into blocks by numerous fractures, including large and small *faults* on which shear displacements can be identified, and *joints* on which no shear movement can be seen. If the ambient stress is comparable to the strengths of these fractures, its component magnitudes and orientation will vary and may not represent the general crustal stress field.

In a classic analysis of geological faulting, Anderson (1951) showed that the type of faulting is simply related to the orientation of the stress tensor. With principal stresses $\sigma_1 > \sigma_2 > \sigma_3$, normal faults arise with σ_1 vertical, strike-slip faults with σ_2 vertical, and thrust faults with σ_3 vertical; the fault plane makes an angle less than 45° with σ_1 and contains σ_2 in each case. These familiar facts appear in several parts of this review and are recapitulated here for convenience.

The stress field in crustal rocks can be investigated by means of direct measurement methods, which include the strain-relief overcoring and hydraulic-fracturing techniques, or through several techniques that describe the orientation of the stress tensor but not its magnitudes. These include the use of earthquake mechanisms, of recent movements on faults, of a class of shear fractures in the walls of boreholes known as breakouts, and of the orientation of contemporary igneous intrusions. These methods and their limitations are outlined next.

STRAIN-RELIEF MEASUREMENTS

Measurements of this kind are usually made in a mine or quarry. A strain sensor is located in a borehole, one tunnel diameter or more from the wall, and it records strains when the rock containing the borehole is isolated by overcoring with a bit of larger diameter. These strains are combined with elastic moduli of the rock to give stress components. A sensor contains one or more rosettes of three strain gauges. The types in common use are discussed by McGarr & Gay (1978). Most types require measurements in two or three nonparallel holes to determine the stress tensor, and where there are small-scale variations of rock properties, several measurements may be required along each borehole. The technique demands a high level of skill and involves assumptions in the corrections applied for the effects of modification of the stress field by the borehole. Unfortunately, other techniques involve equally serious assumptions, and thus strain-relief measurements form an indispensable part of the data set. Stephenson & Murray (1970), Herget (1973), and Gay (1975) describe meticulous measurements with doorstopper-type sensors. In a study of stress in near-surface rocks, de la Cruz & Raleigh (1972) used five methods to determine the stress tensor and found consistent directions but widely varying magnitudes.

Overcoring measurements in mines and tunnels, at depths of 500 meters or more, most closely represent the general stress field in the crust and have special importance. In surface exposures the stress may be strongly affected in magnitudes and orientation by nearby faults and joints (Engelder & Sbar 1977). Strain relaxation in a granite outcrop has been shown to be related to the opening of microfractures (Engelder et al 1977). Tullis (1977) has considered conditions for valid strain-relief studies of residual stress.

STRESS MEASUREMENTS BY HYDRAULIC FRACTURE

The use of hydraulic pressure in boreholes to fracture rock began as a technique in oil production, and it is now a major source of knowledge of the state of stress in the upper crust. The physics of the process has been elucidated in studies by Hubbert & Willis (1957), Scheidegger (1962), Kehle (1964), Fairhurst (1964), and Haimson & Fairhurst (1967), among others. Current techniques and methods of interpretation are collected in a volume of workshop proceedings edited by Zoback & Haimson (1982). The pressure in fluid in a short section of the borehole, between packers, is increased until a fracture is produced. The recorded pressures, together with flow rates into the section, are used to estimate the stress field.

Provided that the least principal stress σ_3 is approximately horizontal and the rock is isotropic and cannot be penetrated by the fluid, Hubbert & Willis (1957) showed that a vertical cylindrical hole modifies the stress field so that the first-cycle breakdown pressure is

$$p_b = \tau_0 + 3S_h - S_H - p, \tag{1}$$

where S_H and S_h are the greater and lesser horizontal "principal" stresses, respectively, and the rock has pore-fluid pressure p and tensile strength τ_0. (Strictly, S_H and S_h are principal stresses only if the third is exactly vertical.) Immediately after breakdown the pressure falls to p_c, where

$$p_c = p_b - \tau_0. \tag{2}$$

The tensile strength τ_0 can be estimated from (2), a method that is sometimes preferred over laboratory measurement (Bredehoeft et al 1976). In subsequent cycles the breakdown pressure, with the existing fracture, is p_c. In any cycle, when pumping ceases, the pressure pauses as it falls at the *instantaneous shut-in pressure* (*ISIP*). Assuming that the fracture formed normal to σ_3 and is just held open by the ISIP, we have

$$S_h = \sigma_3 = \text{ISIP}. \tag{3}$$

The least principal stress is most directly determined by hydraulic fracturing. If the pore pressure p can be measured in a stem test, then S_H can be found from (1). Finally, it is assumed that the vertical "principal" stress is equal to the lithostatic pressure:

$$S_v = \int \rho g \, dz. \tag{4}$$

If the rock is isotropic and σ_3 is vertical, then the fracture will be vertical and its azimuth can be found by means of an inflatable impression sleeve, or by forcing a suspension of a plastic into the fracture and using an acoustic borehole televiewer.

This outline of the theory gives the principles of the method but ignores complications from fluid penetration into the rock or from fracture propagation. Such effects, and the pressure changes observed with various orientations of the stress tensor, are discussed by Hickman & Zoback (1982) and by other authors in Zoback & Haimson (1982).

The formation of a vertical fracture requires σ_3 horizontal, so that the stress orientation can generally be determined by hydraulic fracture in regions of normal and strike-slip faulting but not in a thrust-faulting stress field, where σ_3 is vertical. The expected horizontal fractures are difficult to verify. Bredehoeft et al (1976) reported horizontal hydraulic fractures at depths less than 120 m.

Stress estimates from hydraulic fracture have special value because they reveal the stress field at depths of hundreds to thousands of meters in undisturbed rock far from stress relief by near-surface weathering. Stress magnitudes from this technique can sometimes be combined with stress orientations from breakouts or from earthquake mechanisms.

INVESTIGATION OF STRESS THROUGH EARTHQUAKE MECHANISMS

A lithospheric earthquake is generated in failure, under shear stress, of some area A of an existing or new fracture. Figure 1a shows a part FF' of a fault plane in a stress field with principal stresses σ_1, σ_3 so oriented that slip will occur on FF' if $(\sigma_1 - \sigma_3)$ grows large enough. If sudden slip occurs, compressional waves are radiated from a source that at distances large compared with the source dimensions is a *double couple* corresponding to reduction of shear stress across FF' and also across the auxiliary plane GG' perpendicular to FF'. The intersection of planes FF' and GG' is parallel to the intermediate principal stress σ_2. The near-field and far-field radiations are discussed by Brune (1970). The double couple arises because the compressional radiation pattern has four lobes separated by nodal planes FF' and GG', and it conserves zero angular momentum.

STRESS NEAR THE SURFACE OF THE EARTH 549

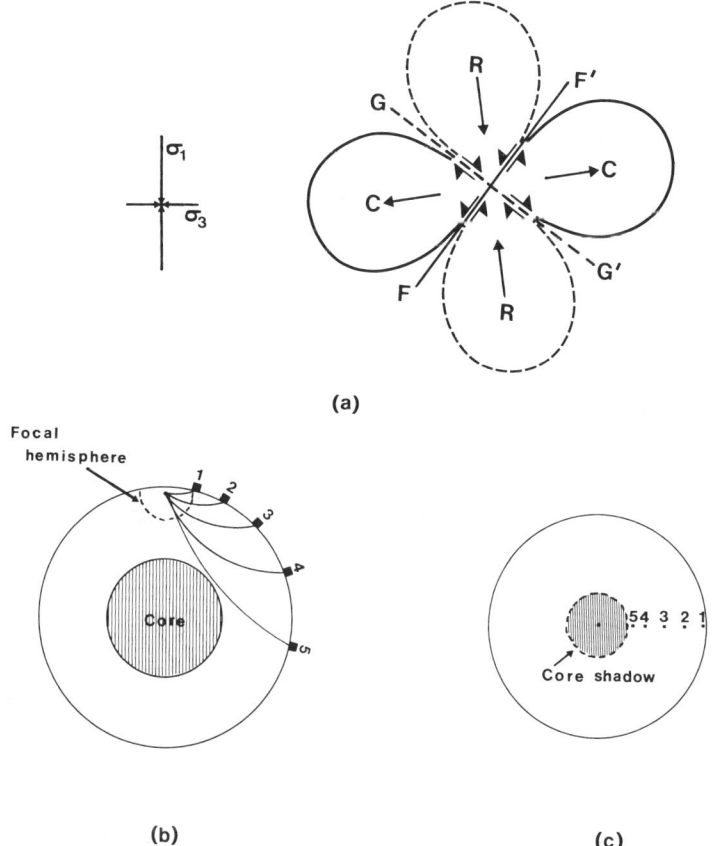

Figure 1 (*a*) The double-couple source mechanism. *CC*, compressive first arrivals; *RR*, rarefactive first arrivals; *FF'*, fault plane; *GG'*, auxiliary plane. (*b*) Rays for *P* waves outside the core. (*c*) Intersections of the rays to the five stations in (*b*) with the lower focal hemisphere.

Initial compressions are radiated into the quarter spaces marked *C* in Figure 1*a*, and initial rarefactions into those marked *R*. The radiation pattern from an earthquake is conveniently represented on the lower half of the *focal sphere* centered at the hypocenter, or on a projection of that hemisphere on its horizontal diametral plane (Figure 1*b*) to give a *mechanism solution*. For detectors very near the epicenter, the upper hemisphere is projected. Distant observatories record as first arrivals compressional or *P* waves, which start downward from the source and return to the surface, because of the general increase of velocity with depth in the Earth (Figure 1*b*). Consequently, rays to nearer observatories pass through the lower hemisphere near its edge, and those to far stations closer to its center. Rays to observatories pass through the focal hemisphere at points determinable from the velocity-depth relation in the Earth and can be used to identify the compressive and rarefactive quarter spaces and nodal planes. In Figure 2 the three classic fault types discussed by Anderson (1951) are shown at the left, focal hemisphere diagrams (with the quarter spaces receiving compressional first arrivals blackened) in the center, and stress orientations at the right. A limitation of "beach-ball" diagrams like those in Figure 2 as indicators of earthquake mechanisms is that the fault plane and auxiliary plane cannot be distinguished unless the first motion of shear waves can be established. For major, shallow earthquakes, which generate surface waves of large amplitudes, long-period S waves can often be so used. Alternatively, aftershock locations may serve to identify the fault plane. Beach-ball diagrams like those in Figure 2 may represent either a single large earthquake recorded by many observatories or a number of

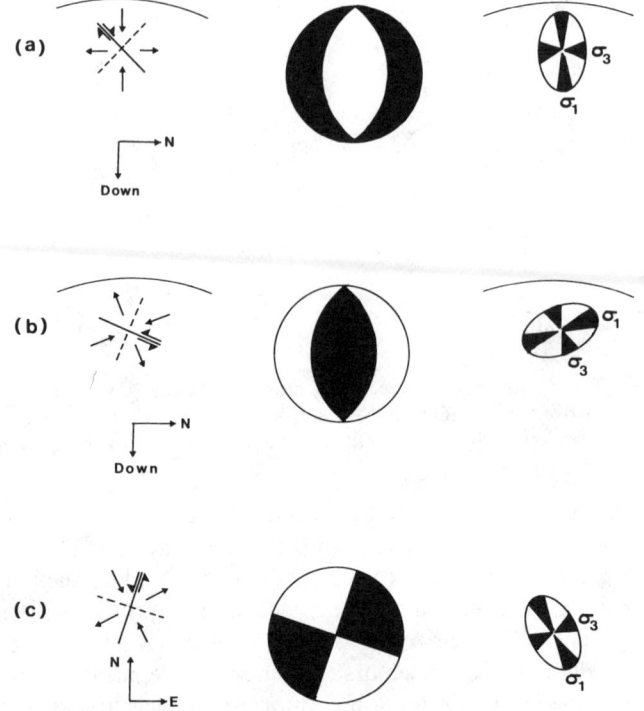

Figure 2 (*Left*) Source configuration, (*center*) source mechanism plots ("beach-balls"), and (*right*) inferred orientations of greatest and least principal stresses for (*a*) a normal-faulting earthquake, (*b*) a thrust-faulting earthquake, and (*c*) a strike-slip-faulting earthquake.

similar small events radiating through the upper focal hemisphere to a local array of detectors to yield a *composite mechanism*.

Earthquake mechanisms thus give an unambiguous indication of the orientation of the stress tensor in the earthquake source region, and they may indicate normal, thrust, or strike-slip fault displacements (Figure 2). The precision of stress orientation is limited, as indicated in Figure 2, especially where the earthquake activated an existing fracture. In general, the greatest principal stress σ_1 lies somewhere in the two quarter spaces receiving rarefactive first motions (McKenzie 1969).

Certain stress magnitudes can be estimated if a source model is assumed. The *seismic moment* M_0 of an earthquake in which an average sudden displacement u occurs on a fault of area A is

$$M_0 = \mu u A, \tag{5}$$

where μ is the rigidity modulus (Aki 1966). The *stress drop* $\Delta\tau$, or the change of shear stress across the fault during the earthquake, is

$$\Delta\tau = \frac{\mu u}{CW} = \frac{M_0}{CAW}, \tag{6}$$

where W is half of the smallest dimension of the fault, and C is a form factor dependent on the fault geometry and direction of slip (Madariaga 1977). For a circular fault of radius r, we have $W = r$ and $C = 16/7\pi$ (Keilis-Borok 1959), so that

$$\Delta\tau = \mu u \frac{7\pi}{16r} = \frac{7M_0}{16r^3}. \tag{7}$$

A circular fault is often assumed for simplicity. For some well-instrumented shallow earthquakes, u can be estimated from a surface rupture and A from aftershock locations, so that both M_0 and $\Delta\tau$ can be estimated from source-region field data. The spectra of seismograms can also be used to estimate M_0 and r for distant earthquakes (Wyss & Hanks 1972). The stress drop $\Delta\tau$ can then be found from (7).

Teleseismic data from earthquakes of magnitude 5 or greater thus yield information primarily about the orientation of the fault and the principal stresses, and to a lesser extent about the area of the fault on which displacement occurred, the magnitude of the displacement, and the stress drop. The total shear stress τ across the fault is not given by seismic data and is always larger than $\Delta\tau$ and often much larger. If the dynamic friction on the fault remains near the static value during sliding, it can be shown that $\Delta\tau \ll \tau$ and radiation efficiency is low. If the dynamic friction drops, either through partial melting or through heating of pore water, the radiation efficiency may rise and $\Delta\tau$ may approach τ (Sibson 1977). Brune (1970) has pointed out that during failure the particle velocity V_{par} at the fault is related to shear-wave velocity V_s by

$$V_{par} = \frac{\tau V_s}{\mu}. \qquad (8)$$

In principle, if V_{par} is recorded by a strong-motion seismometer near the fault plane, a lower limit can be set to τ. In this way Brune found that California earthquakes indicated $\tau \simeq 10$ MPa at the faults concerned. Stress drops $\Delta\tau$ lie in the range 0.1–10 MPa in many lithospheric earthquakes. Near-surface stress information from earthquake mechanisms is considered later in this review.

STRESS ORIENTATIONS FROM OBSERVATION OF ACTIVE FAULTS

The crust of the Earth is everywhere divided by faults, or fractures on which shear displacement has occurred. At a given time most of these faults are inactive, and those that are active may be undergoing displacement in a quite different stress field from that in which they were formed. For example, the Zambezi Rift, in which Lake Kariba lies, is bounded by active normal displacements (σ_1 vertical) of faults formed in Precambrian time in a strike-slip stress field (σ_2 vertical) that then produced intense folding about vertical axes (Drysdall & Weller 1966, Gough & Gough 1970). Another example is the Rocky Mountain Trench–Tintina fault in the Canadian Cordillera, north of 51° N, which was the site of hundreds of kilometers of strike-slip displacement during Mesozoic time, when it formed the boundary between the North American craton and accreting lithospheric blocks, but is now undergoing normal displacement to form the northeast wall of the Trench. Once formed, a fault may be reactivated in different stress fields at different times. Only where very recent structures are displaced can a fault be used to infer the orientation of the present stress field. Young watercourses, man-made structures such as roads, or postglacial varved sediments might serve such a purpose; other examples will occur to the reader.

As an example of stress orientations by this method, structures across the San Andreas fault show recent right-lateral strike-slip displacements indicative of a stress field with σ_2 vertical and σ_1 oriented NNE-SSW, in agreement with mechanisms of earthquakes on and near the fault. Again, many faults in east Texas show recent normal displacements, downthrown to the southeast, which imply a stress field with σ_1 vertical and σ_2 along the strike of the faults (Zoback & Zoback 1980). A concordant stress orientation is inferred from the azimuths of borehole breakouts there, as reported by Brown et al (1980) and interpreted by Gough & Bell (1982).

STRESS ORIENTATIONS FROM IGNEOUS INTRUSIONS AND VOLCANIC VENTS

Where the least principal stress σ_3 is horizontal, rising magma will tend to form and fill fractures to produce dikes at right angles to σ_3. Dikes will therefore strike parallel to the greater horizontal compression S_H, which may be σ_1 (in a strike-slip stress field) or σ_2 (in a normal stress field). The flank vents of some multiple volcanoes are concentrated near a straight line running through the main vent, which indicates the dike feeding them. If the satellite vents are active, their azimuth gives that of S_H.

Nakamura et al (1977) have shown that some 16 systems of multiple vents in the Aleutian Arc are aligned in the northwest-southeast quadrants, parallel to the velocity vector of the Pacific plate as it is thrust under the North American and Asian plates (Figure 3). As the dikes feeding the vents must be generally vertical, σ_3 is horizontal and at right angles to the direction of underthrusting. This implies a strike-slip stress orientation, with σ_2 vertical, rather than a thrust-faulting stress field, which would produce horizontal sheet intrusions without alignment of the vents. Behind

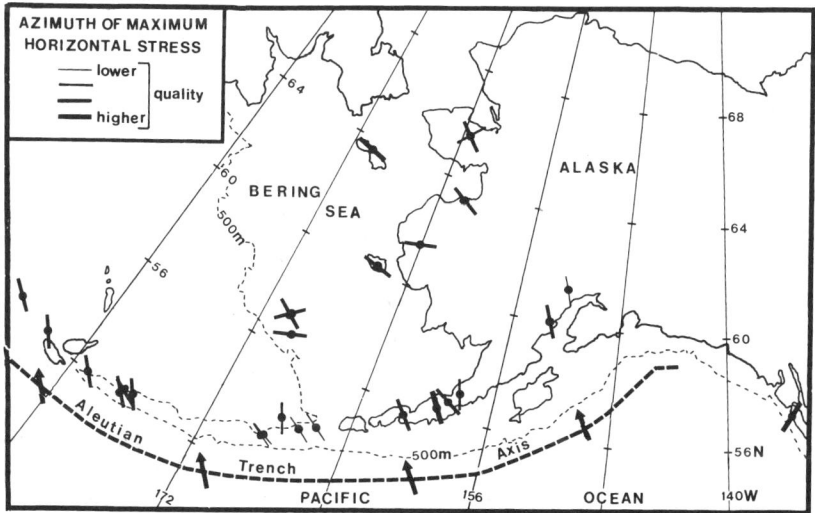

Figure 3 Azimuths of maximum horizontal stress inferred from orientations of satellite volcanic vents in the Aleutian Arc and Alaska. After Nakamura et al (1977).

the arc, in Alaska, vents tend to be aligned east-west, which Nakamura et al interpret as meaning that $\sigma_2 = S_H$ is east-west, σ_3 north-south, and σ_1 vertical (a normal-faulting, extensional stress field).

Dike swarms are common in exposed Precambrian shields and probably lie along S_H and at right angles to σ_3 in the stress field at the time of intrusion. Horizontal sheet intrusions may similarly indicate thrust-faulting paleostress fields, but unfortunately they offer no orientation of the horizontal principal stresses. An advantage of igneous intrusions over ancient faults as paleostress indicators is that the intrusive rock can often be dated.

BOREHOLE BREAKOUTS AS INDICATORS OF STRESS ORIENTATION

Breakouts are fractures that develop spontaneously in the walls of boreholes, such as oil wells, in the Earth's upper crust. They produce an elongation of the cross section of the hole, which is readily detected by the four-arm dipmeter used in the hydrocarbon industry to determine the attitudes of sedimentary beds intersected by a well. In the sedimentary basin of western Canada, it has been noticed that breakouts elongate the cross sections of wells in a northwest-southeast azimuth, not only within a well but also between wells distributed throughout the Alberta basin (Cox 1970). Babcock (1978) showed that the alignment extended over some 10^5 km^2 of the Alberta basin and was independent of the lithology of the rock and of the attitude of the sedimentary beds.

The phenomenon has been explained by Bell & Gough (1979) and independently by Hottman et al (1979) as being caused by concentration, near the walls of the borehole, of a crustal stress field with unequal horizontal principal stresses. The breakouts elongate the hole section by shear fracturing in the direction of the lesser horizontal principal stress S_h. Northwest-southeast elongations would thus indicate a northeast-southwest orientation of the greater horizontal compression S_H. Figure 4 shows breakout azimuths in the western Canadian sedimentary basin.

The development of breakouts as shear fractures, in the amplified stress difference near the borehole wall, has been discussed by Gough & Bell (1982) in terms of brittle fracture theory. Consider first a vertical cylindrical hole of radius a in an extended medium under uniaxial horizontal stress S. At a point a distance r from the axis of the hole on a radius making angle θ with S, the stress components are given by the Kirsch (1898) equations, which are quoted in textbooks such as Timoshenko & Goodier (1951) in the following form:

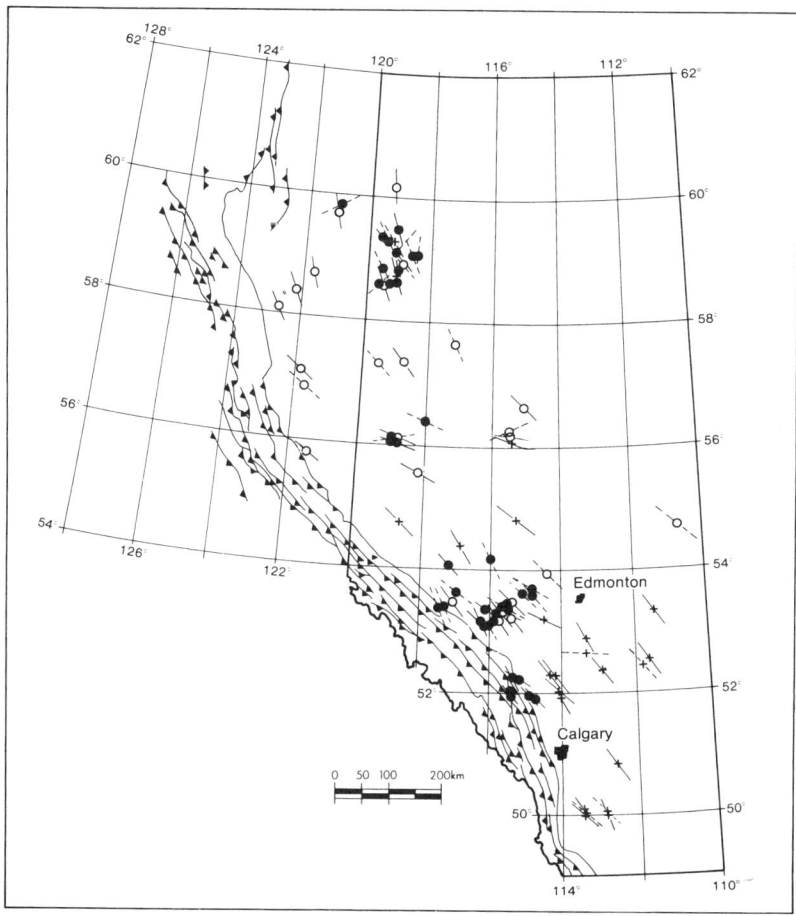

Figure 4 Azimuths of breakouts in 94 oil wells in the western Canadian sedimentary basin. After Fordjor et al (1983). Results are shown from Babcock (1978), crosses; Gough & Bell (1981), circles; and Fordjor et al (1983), dots. Solid lines show mean azimuths from five or more breakouts in the well; broken lines show means from four or less breakouts. Babcock's azimuths follow his use in showing majority and minority means.

$$\sigma_r = \frac{S}{2}\left\{1 - \frac{a^2}{r^2} + \left(1 + \frac{3a^4}{r^4} - \frac{4a^2}{r^2}\right)\cos 2\theta\right\},$$

$$\sigma_\theta = \frac{S}{2}\left\{1 + \frac{a^2}{r^2} - \left(1 + \frac{3a^4}{r^4}\right)\cos 2\theta\right\}, \tag{9}$$

$$\tau_{r\theta} = -\frac{S}{2}\left\{1 - \frac{3a^4}{r^4} + \frac{2a^2}{r^2}\right\}\sin 2\theta.$$

At the hole wall ($r = a$), the radial normal stress σ_r and the tangential shear stress $\tau_{r\theta}$ vanish, as they must at a free surface, and we have

$$\sigma_\theta = S - 2S \cos 2\theta. \tag{10}$$

In a biaxial stress field with horizontal principal stresses S_H and S_h, with $S_H > S_h$, the radial normal and tangential shear stresses still vanish at the wall of the hole, leaving the tangential normal stress

$$\sigma_\theta = S_H + S_h - 2(S_H - S_h) \cos 2\theta. \tag{11}$$

When $\theta = 90°$ or $270°$, σ_θ takes its maximum value of

$$\sigma_{\theta max} = 3S_H - S_h, \tag{12}$$

and as $\sigma_r = 0$ at the wall, Equation (12) gives the stress difference there. Figure 5a represents the stresses that produce a breakout by shear fracture, near P and Q, and Figure 5b illustrates the near-hole amplification of the stress difference for various values of S_H/S_h far from the hole. Large

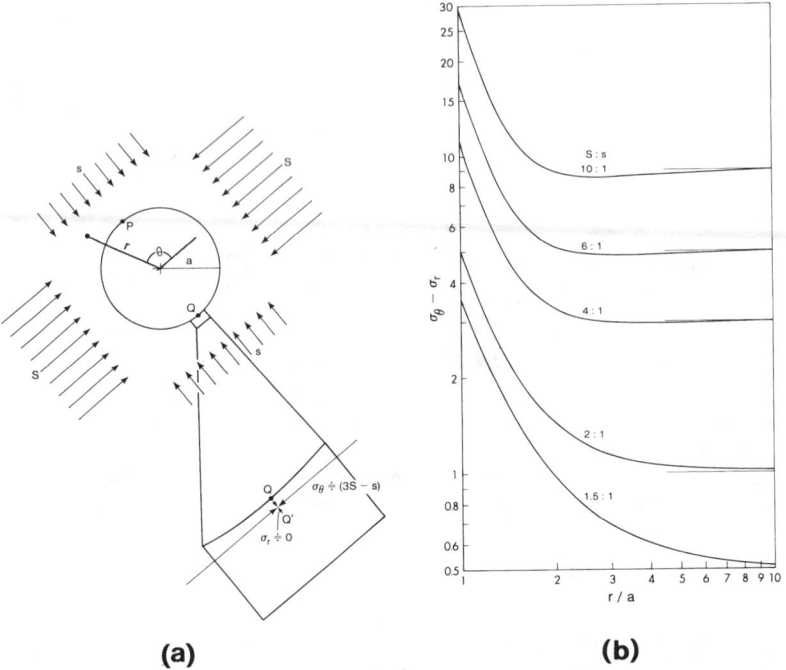

Figure 5 Amplification of stress difference near a cylindrical hole in a medium under biaxial stress with $S > s$. (a) Breakouts form by shear fracture near P and Q. (b) The stress difference $(\sigma_\theta - \sigma_r)$ as a function of r/a along the line PQ in (a). Curves are shown for five values of the stress ratio S/s far from the hole. Both scales are logarithmic. After Gough & Bell (1982).

amplification factors arise because the radial normal stress σ_r vanishes at the wall. Gough & Bell (1982) have applied Mohr-Coulomb failure criteria to the initial fracture.

For observations of crustal stress orientation, the essential point is that breakouts elongate a borehole in the direction of S_h and so at right angles to S_H. This prediction of the hypothesis has been tested in four stress provinces against determinations of stress orientations by means of overcoring at outcrops, hydraulic fracture, recent movements on faults, and earthquake mechanisms, and it has passed all such tests at the time of this writing. A detailed study has been made by Fordjor et al (1983) of the statistics of breakouts in the western Canadian sedimentary basin.

THE NEAR-SURFACE STRESS FIELD

The distribution of stress observations over the Earth's surface is far from ideal. Strain relaxation measurements at depths greater than 500 m are highly concentrated in deep mines subject to rock bursts, notably in Archean shield rocks of South Africa and Canada, with a few deep observations and many near-surface observations in the United States, Europe, and Australia. Hydraulic fracture measurements are concentrated in the United States. Stress information from earthquake mechanisms refers mainly to regions of high seismicity, and so to plate margins. Intraplate earthquakes are of special interest, but few are large enough to allow good mechanism solutions. The stress field in the oceanic crust is known mainly from earthquakes at active plate boundaries. Breakouts have so far yielded orientation data mainly in hydrocarbon-bearing sediments in North America, but they could potentially enlarge the data set for other continents and continental margins.

McGarr & Gay (1978) have collected observations of stress magnitudes at depths greater than 100 m, mainly from Canada and South Africa with some from Australia and the US. For the vertical pressure, their Figure 3 shows that most values lie within 20% of the lithostatic stress or overburden pressure given by Equation (4), but with large deviations. The highest known value is more than double the lithostatic stress and comes from a shear zone in the Canadian shield, at a depth of 1.7 km (Herget et al 1975). McGarr & Gay (1978) examined another common assumption—that one principal stress is vertical—and found that in southern Africa most sites have one principal stress within 30° of the vertical.

In considering the horizontal stresses, a useful standard of comparison is Heim's rule, namely

$$S_{\text{hor}} = S_v = \int \rho g \, dz, \tag{13}$$

from which the term "lithostatic stress" takes its origin. In an elastic half-space, the assumption of zero horizontal strain gives

$$S_{\text{hor}} = S_v \frac{n}{1-n} \quad (14)$$

in rock of Poisson's ratio n, but observed horizontal stresses are in most places larger than (14) indicates, probably because rocks yield inelastically in geological time. Generally, Equation (13) approximates better to measured horizontal stresses.

Where the stress field is known, the horizontal "principal" stresses are commonly unequal and vary greatly between regions (Gay 1980). The large regional variation is well shown in two figures reproduced in Figure 6 from McGarr & Gay (1978), presenting results of strain-relief measurements in South Africa and Canada. In both cratons the horizontal stresses increase with depth, but otherwise they differ radically. In South Africa, horizontal stresses much in excess of the lithostatic stress are found in the depth range 100–500 m, and σ_1 is generally horizontal. At greater depths, σ_1 becomes near-vertical and even S_H is almost always less than $\rho g z$, with no obvious preferred orientation. Most of the deeper results come from mines in the Witwatersrand basin, where faulting is mainly normal, in harmony with a vertical σ_1. The high near-surface horizontal stresses may be an effect of removal of overburden by erosion, which has left residual horizontal compressions associated with higher σ_v values before erosion (Voight 1966). Another possibility is that the stresses result from folding or flattening of strata (Gay 1975).

A very different stress field prevails in the shield of Ontario, Canada. Even S_h is larger than the lithostatic value in all results given by Herget and his associates [see McGarr & Gay (1978) for references]. One extremely high value is associated with anomalously high S_v in the shear zone noted above. Thrust faulting, consistent with S_v as σ_3, is dominant in the shield of Ontario, which forms part of the North American Cratonic stress province shortly to be described.

The crustal stress field is best known in North America, where all of the techniques discussed above have been applied. Sbar & Sykes (1973, 1977) used strain-relief measurements, earthquake mechanisms, and recent geological displacements such as "pop-ups" in quarries to reveal very high S_H values in a generally northeast-southwest direction through much of the northeastern United States. Lo (1978) showed that fractures in tunnels and pop-ups in excavations indicated a similar stress field in Ontario. Zoback & Zoback (1980) compiled and reviewed all available stress data, with the exception of breakouts, for the conterminous United States, which

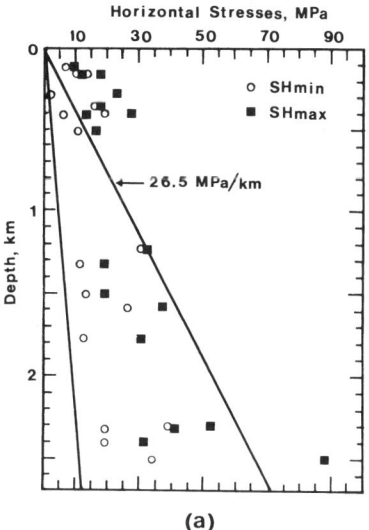

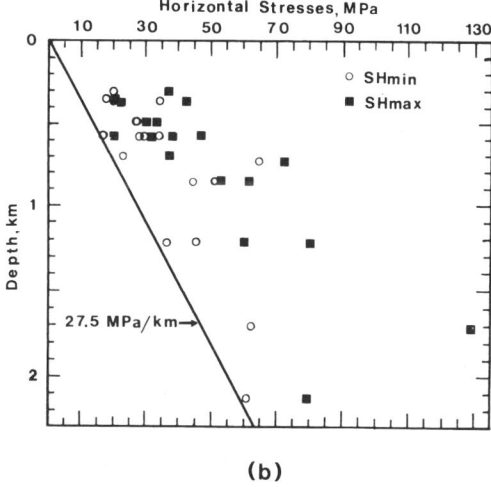

Figure 6 Greatest and least horizontal compressions measured in mines (*a*) in southern Africa and (*b*) in Ontario, Canada. After McGarr & Gay (1978), who give the sources of the data shown.

they divided into 14 stress provinces. In each of these the horizontal "principal" stresses have approximately constant directions and relative magnitudes. The largest such province they named the Mid-Continent stress province, characterized by large S_H in a NE-SW direction. Zoback & Zoback mapped this stress province from the east front of the Rocky Mountains of Colorado and New Mexico to New England and Quebec. Gough & Bell (1981) and Gough et al (1983) have used breakouts in oil wells to show that this stress province extends through the western Canadian sedimentary basin, east of the Rocky Mountains, to the Arctic islands of Canada, and thus it should more properly be named the North American Cratonic stress province. It is shown in Figure 7, with pairs of broad arrows to indicate the direction of S_H from data reported by the authors quoted in this paragraph. Bell et al (1986) have shown that

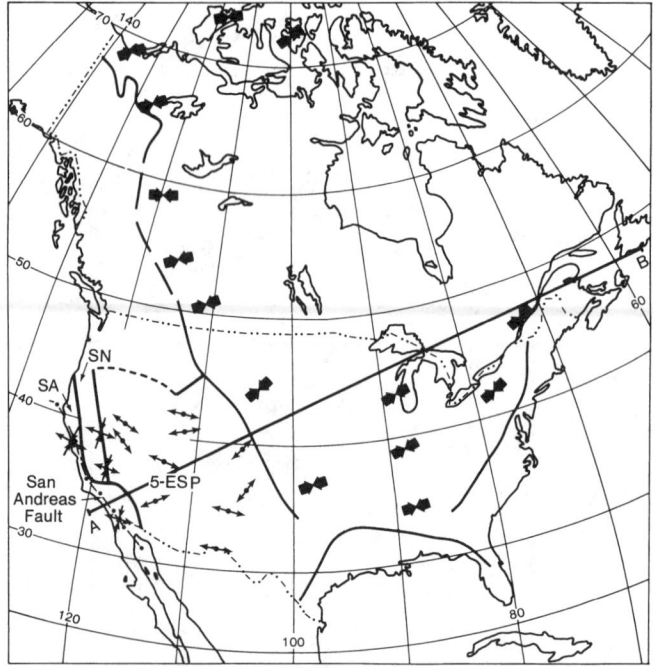

Figure 7 Stress orientations in North America. Pairs of broad arrows show directions of the maximum horizontal compression S_H in the craton. Two-headed arrows indicate directions of the least horizontal compression S_h in five extensional stress provinces (5-ESP). Abbreviations: SN, Sierra Nevada stress province; SA, San Andreas stress province. *AB* marks the great circle of the section in Figure 8. Each arrow symbol represents from 3 to 30 orientation data. After Gough (1984), where sources are listed. Reproduced by permission of Macmillan Journals Ltd.

the cratonic stress field extends to the continental shelf off Atlantic Canada.

The relationship between stress patterns in North America and tractions on the North American plate has been considered by Raleigh (1974), Zoback & Zoback (1980), and Gough (1984), among others. In the San Andreas stress province, near the western plate boundary, the stress orientation is shown mainly by earthquake mechanisms and indicates the right-lateral strike-slip traction on the edge of the plate. The great area of the North American Cratonic stress province indicates traction on the underside of the plate as the probable cause of the crustal stress field. Such a traction would produce the observed stress field if it were directed either to the northeast or to the southwest. Zoback & Zoback (1980) regard it as a result of viscous drag from the underlying asthenosphere as the plate slides southwestward. Gough et al (1983) have noted the alternative possibility that northeastward flow in the mantle transmits shear by viscous drag through the asthenosphere to the base of the lithosphere. The stress field in the craton can be explained by either view. Gough (1984) has argued that the five extensional stress provinces west of the Rocky Mountains (Figure 7), together with other geophysical and petrological evidence, lead to the conclusion that there is a convective upflow in the mantle beneath western North America. Figure 8 is a cartoon illustrating the dynamical situation proposed. The cratonic stress field would then result from northeastward mantle flow under the plate.

In Alaska the stress field reflects the traction of the underthrust Pacific plate on the edge and underside of the North American plate, as shown by Nakamura et al (1977; see Figure 3). Breakouts in northwestern Canada appear to locate the boundary between the Alaskan and North American Cratonic stress provinces (Gough et al 1983).

The study by Zoback & Zoback (1980) makes the important distinction between a plate and stress provinces within a plate. Not all stress provinces necessarily show the stress field of the underlying lithosphere. In North America the Cratonic stress province, the five extensional stress provinces of Zoback & Zoback between the east front of the southern Rockies and the Sierra Nevada, and both the San Andreas and Alaskan stress provinces have stress orientations clearly related to tractions on the plate. The Appalachian and Gulf of Mexico coastal stress fields appear to represent rock masses decoupled from the underlying plate.

Strain-relief measurements (Greiner 1975), earthquake mechanisms (Ahorner 1975), and geological study of the fracture systems (Illies 1975) have contributed to an interesting picture of the stress field in Germany and adjoining countries from the Alps to the coastal plain adjoining the North Sea (Greiner & Illies 1977). Between Miocene time and the present,

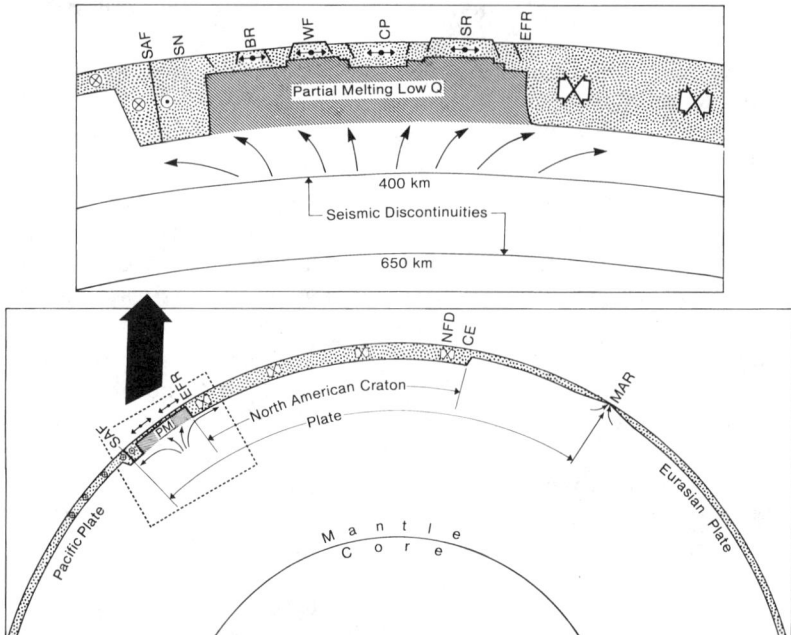

Figure 8 A cartoon section, without radial exaggeration, along the great circle *AB* in Figure 7. The lithosphere is stippled. Stress symbols correspond with those in Figure 7. Abbreviations: SAF, San Andreas fault; EFR, east front of the southern Rocky Mountains; NFD, Newfoundland; CE, continental edge; MAR, Mid-Atlantic Ridge. Inset abbreviations: SN, Sierra Nevada; BR, Basin and Range province; WF, Wasatch Front; CP, Colorado Plateau; SR, southern Rockies. After Gough (1984). Reproduced by permission of Macmillan Journals Ltd.

the direction of S_H has rotated from NNE-SSW to NW-SE. The Upper Rhine Rift opened in the earlier stress field, but earthquakes there now have strike-slip mechanisms indicating maximum compression in the NW-SE quadrants. In the Lower Rhine Rift and coastal plain, the earthquakes show strike-slip-, thrust-, and normal-faulting mechanisms, but the NW-SE direction is dominant for S_H. Strain-relief overcoring measurements show S_H oriented mainly NW-SE in the upper Rheingraben, increasing in magnitude and rotating slightly to NNW-SSE in the Alps. The overall picture is consistent with compression of the crust from the Alpine interplate collision between the Eurasian and African plates. Originally normal faults in the Upper Rhine Valley appear to be moving in strike-slip failure under the present stress field. Ahorner (1975) has compared slip rates from geodetic and seismological data and has shown that 80–90% of the displacement occurs in aseismic creep.

Hast (1973) has reported stress-relief measurements in Scandinavia that show very large horizontal stresses, exceeding twice the lithostatic value, and a linear dependence of mean horizontal stress on depth in the depth range 0–900 m. Other workers in other places discover complexity rather than simplicity. McGarr & Gay (1978, their Figure 8) show maximum shear stress $(\sigma_1 - \sigma_3)/2$ measured in mines in North America and South Africa in the depth range 100 m to 5 km. There is a general increase with depth, but values at a given depth can range through a factor of five. The stronger crystalline rocks support shear stresses in the range 10–30 MPa at depths exceeding 1 km.

In Iceland, stress-relief measurements by Hast (1973) at five stations gave directions of S_H parallel to the rift axis at two stations, perpendicular to the rift at two stations, and in an intermediate direction at one. Hydraulic fracture measurements by Haimson & Voight (1977) in two shallow holes show S_H perpendicular to the rift. Earthquake mechanisms show normal faulting in the rift but strike-slip faulting east and west of it (Klein et al 1977). Clearly the stress pattern near the Mid-Atlantic Rift in Iceland is highly variable between sites less than 100 km apart. This is not surprising, but it carries a warning against generalization from sparse data elsewhere.

Stress drops inferred from earthquakes range from less than 0.1 MPa to over 60 MPa (Hartzell & Brune 1977). Richardson & Solomon (1977) found stress drops between 0.2 and 7 MPa for five intraplate earthquakes of special relevance to intraplate stress. Hanks (1977) reports stress drops determined from 390 earthquakes, mainly shallow events in California but some deeper ones from the Tonga–Kermadec subduction zone. He concluded that the stress drops were nearly independent of source strength and were lognormally distributed, with an average value near 1.0 MPa and two standard deviations at 0.1 and 10 MPa. As remarked above, stress drops $\Delta\tau$ give lower limits to the shear stress τ acting across the fault before slip. Most measured maximum shear stresses exceed 10 MPa at depths of 1 km or more. If shear stresses are larger at midcrustal depths, the present data set is consistent with typical values of 0.01 to 0.1 of the ratio $\Delta\tau/\tau$. Brace (1972) has shown in the laboratory that shear stresses of hundreds of megapascals are required to produce slip on variously prepared surfaces in crustal rocks under confining pressures of 100–500 MPa, corresponding to depths of 3–15 km. In conjunction with the earthquake-inferred stress drops, Brace's results support the view that earthquakes commonly lower the shear stress across the fracture by 1–10% of the value before slip.

Stress measurements are difficult, expensive, and time consuming, but recent developments in the state of knowledge of stress fields in North America and Europe encourage the hope that accumulating stress data will increasingly reveal patterns of tractions acting on the plates. Breakouts should rather quickly extend knowledge of the orientation of the horizontal stresses, at least in sedimentary basins, and can be most useful when combined with hydraulic fracture measurements of the stress magnitudes in the same borehole. First results from breakouts in the oceanic crust are also encouraging (Newmark et al 1984).

ACKNOWLEDGMENTS

We thank many authors cited in this review for supplying us with reprints, preprints, and helpful comments. A. McGarr and N. C. Gay kindly allowed us to reproduce two of the figures from their 1978 review in our Figure 6, and K. Nakamura, K. H. Jacob, and J. M. Davies gave permission for reproduction of Figure 3 from their paper of 1977. We acknowledge permission of Macmillan Journals Ltd. for reproduction of Figures 7 and 8, which originally appeared in *Nature* (Gough 1984). The research of D.I.G. is supported by the Natural Sciences and Engineering Research Council of Canada.

Literature Cited

Ahorner, L. 1975. Present-day stress field and seismotectonic block movements along major fault zones in Central Europe. *Tectonophysics* 29: 233–49

Aki, K. 1966. Generation and propagation of *G* waves from the Niigata earthquake of June 16, 1964. Part 2. Estimation of earthquake moment from the *G* wave spectrum. *Bull. Earthquake Res. Inst. Tokyo Univ.* 44: 73–88

Anderson, E. M. 1951. *The Dynamics of Faulting.* Edinburgh: Oliver & Boyd. 2nd ed.

Babcock, E. A. 1978. Measurement of subsurface fractures from dipmeter logs. *Am. Assoc. Pet. Geol. Bull.* 62: 1111–26

Bell, J. S., Gough, D. I. 1979. Northeast-southwest compressive stress in Alberta: evidence from oil-wells. *Earth Planet. Sci. Lett.* 45: 475–82

Bell, J. S., Todrouzek, A. J., Ervine, W. B. 1986. Offshore in-situ stress regimes in eastern Canada. *RESERVES Conf. Can. Soc. Pet. Geol., 21st, Calgary* (Abstr.)

Brace, W. F. 1972. Laboratory studies of stick-slip and their application to earthquakes. *Tectonophysics* 14: 189–200

Bredehoeft, J. D., Wolff, R. G., Keys, W. S., Shuter, E. 1976. Hydraulic fracturing to determine the regional in situ stress field, Piceance Basin, Colorado. *Geol. Soc. Am. Bull.* 87: 250–58

Brown, R. O., Forgotson, J. M., Forgotson, J. M. Jr. 1980. Presented at Ann. Fall Conf. Soc. Pet. Eng. AIME, 55th, Dallas (Pap. 9269)

Brune, J. N. 1970. Tectonic strain and the spectra of seismic shear waves from earthquakes. *J. Geophys. Res.* 75: 4997–5009

Chen, W.-P., Molnar, P. 1983. Focal depths of intracontinental and intraplate earthquakes and their implications for the thermal and mechanical properties of the lithosphere. *J. Geophys. Res.* 88: 4183–4214

Cox, J. W. 1970. *The high-resolution dipmeter reveals dip-related borehole and formation characteristics.* Presented at Ann. Log. Symp., Soc. Prof. Well Log Anal., Los Angeles

de la Cruz, R. V., Raleigh, C. B. 1972. Absolute stress measurements at the Rangely anticline, northwestern Colorado. *Int. J. Rock Mech. Min. Sci.* 9: 625–34

Drysdall, A. R., Weller, R. K. 1966. Karroo sedimentation in Northern Rhodesia. *Trans. Geol. Soc. S. Afr.* 69: 39–69

Engelder, T., Sbar, M. L. 1977. The relationship between in situ strain relaxation and outcrop fractures in the Potsdam sandstone, Alexandria Bay, New York. See Wyss 1977, pp. 41–55

Engelder, T., Sbar, M. L., Krantz, R. 1977. A mechanism for strain relaxation of Barre granite: opening of microfractures. See Wyss 1977, pp. 27–40

Fairhurst, C. 1964. Measurement of in situ rock stresses with particular reference to hydraulic fracturing. *Rock Mech. Eng. Geol.* 2: 129–47

Fordjor, C. K., Bell, J. S., Gough, D. I. 1983. Breakouts in Alberta and stress in the North American plate. *Can. J. Earth Sci.* 20: 1445–55

Gay, N. C. 1975. In-situ stress measurements in Southern Africa. *Tectonophysics* 29: 447–59

Gay, N. C. 1980. The state of stress in the plates. In *Dynamics of Plate Interiors. Am. Geophys. Union Geodyn. Ser.*, ed. A. W. Bally, P. L. Bender, T. R. McGetchin, R. I. Walcott, 1: 145–54

Gough, D. I. 1984. Mantle upflow under North America and plate dynamics. *Nature* 311: 428–32

Gough, D. I., Bell, J. S. 1981. Stress orientations from oil-well fractures in Alberta and Texas. *Can. J. Earth Sci.* 18: 638–45

Gough, D. I., Bell, J. S. 1982. Stress orientations from borehole wall fractures with examples from Colorado, east Texas and northern Canada. *Can. J. Earth Sci.* 19: 1358–70

Gough, D. I., Gough, W. I. 1970. Load-induced earthquakes at Lake Kariba—II. *Geophys. J. R. Astron. Soc.* 21: 79–101

Gough, D. I., Fordjor, C. K., Bell, J. S. 1983. A stress province boundary and tractions on the North American plate. *Nature* 305: 619–21

Greiner, G. 1975. In-situ stress measurements in southwest Germany. *Tectonophysics* 29: 265–74

Greiner, G., Illies, J. H. 1977. Central Europe: active or residual tectonic stresses. See Wyss 1977, pp. 11–26

Haimson, B. C., Fairhurst, C. 1967. Initiation and extension of hydraulic fractures in rock. *Soc. Pet. Eng. J.* 7: 310–18

Haimson, B. C., Voight, B. 1977. Crustal stress in Iceland. See Wyss 1977, pp. 153–90

Hanks, T. C. 1977. Earthquake stress drops, ambient tectonic stresses and stresses that drive plate motions. See Wyss 1977, pp. 441–58

Hartzell, S. H., Brune, J. N. 1977. Source parameters for the January 1975 Brawley–Imperial Valley earthquake swarm. See Wyss 1977, pp. 333–55

Hast, N. 1973. Global measurements of absolute stress. *Philos. Trans. R. Soc. London Ser. A* 274: 409–19

Herget, G. 1973. First experiences with the C.S.I.R. triaxial strain cell for stress determinations. *Int. J. Rock Mech. Min. Sci.* 10: 509–22

Herget, G., Pahl, A., Oliver, P. 1975. Ground stresses below 3000 feet. *Proc. Can. Rock Mech. Symp., 10th, Kingston,* 1: 281–307

Hickman, S. H., Zoback, M. D. 1982. The interpretation of hydraulic fracturing pressure-time data for in situ stress determination. See Zoback & Haimson 1982, pp. 103–46

Hottman, C. E., Smith, J. H., Purcell, W. R. 1979. Relationship among Earth stresses, pore pressure and drilling problems, offshore Gulf of Alaska. *J. Pet. Technol.* 31: 1477–84

Hubbert, M. K., Willis, D. G. 1957. Mechanics of hydraulic fracturing. *Trans. Am. Inst. Min. Eng.* 210: 153–60

Illies, H. 1975. Recent and paleo-intraplate tectonics in stable Europe and the Rhinegraben rift system. *Tectonophysics* 29: 251–64

Kehle, R. O. 1964. Determination of tectonic stresses through analysis of hydraulic well fracturing. *J. Geophys. Res.* 69: 256–66

Keilis-Borok, V. 1959. On estimation of the displacement in an earthquake source and of source dimensions. *Ann. Geofis. (Rome)* 12: 205–14

Kirsch, G. 1898. Die Theorie der Elastizität und die Bedürfnisse der Festigkeitslehre. *Z. Ver. Deutsch. Ing.* 42: 707

Klein, F. W., Einarsson, P., Wyss, M. 1977. The Reykjanes Peninsula, Iceland earthquake swarm of September 1972 and its tectonic significance. *J. Geophys. Res.* 82: 865–88

Lo, K. Y. 1978. Regional distribution of in situ horizontal stresses in rocks of southern Ontario. *Can. Geotech. J.* 15: 371–81

Madariaga, R. 1977. Implications of stress-drop models of earthquakes for the inversion of stress drop from seismic observations. See Wyss 1977, pp. 301–16

McGarr, A., Gay, N. C. 1978. State of stress in the Earth's crust. *Ann. Rev. Earth Planet. Sci.* 6: 405–36

McKenzie, D. P. 1969. The relation between fault plane solutions for earthquakes and the directions of the principal stresses. *Bull. Seismol. Soc. Am.* 59: 591–601

Nakamura, K., Jacob, K. H., Davies, J. N. 1977. Volcanoes as possible indicators of tectonic stress orientation: Aleutians and Alaska. See Wyss 1977, pp. 87–112

Newmark, R. L., Zoback, M. D., Anderson, R. N. 1984. Orientation of in situ stresses in the oceanic crust. *Nature* 311: 424–28

Raleigh, C. B. 1974. Crustal stress and global tectonics. *Proc. Congr. Int. Soc. Rock Mech., 3rd,* pp. 593–97

Ranalli, G., Chandler, T. E. 1975. The stress field in the upper crust as determined from in situ measurements. *Geol. Rundsch.* 64: 653–73

Richardson, R. M., Solomon, S. C. 1977. Apparent stress and stress drop for intraplate earthquakes and tectonic stress in the plates. See Wyss 1977, pp. 317–31

Sbar, M. L., Sykes, L. R. 1973. Contemporary compressive stress and seismicity in eastern North America: an example of intraplate tectonics. *Geol. Soc. Am. Bull.* 84: 1861–82

Sbar, M. L., Sykes, L. R. 1977. Seismicity and lithospheric stress in New York and adjacent areas. *J. Geophys. Res.* 82: 5771–86

Scheidegger, A. E. 1962. Stresses in the Earth's crust as determined from hydraulic fracturing data. *Geol. Bauwes.* 27: 45–50

Sibson, R. H. 1977. Kinetic shear resistance, fluid pressures and radiation efficiency during seismic faulting. See Wyss 1977, pp. 387–400

Stephenson, B. R., Murray, K. J. 1970. Application of the strain rosette relief method to measure principal stresses throughout a mine. *Int. J. Rock Mech.* *Min. Sci.* 7: 1–22

Timoshenko, S., Goodier, J. N. 1951. *Theory of Elasticity*. New York: McGraw-Hill. 506 pp. 2nd ed.

Tullis, T. E. 1977. Reflections on measurement of residual stress in rock. See Wyss 1977, pp. 57–68

Voight, B. 1966. Beziehung zwischen grossen horizontalen Spannungen in Gebirge und der Tektonik und der Abtragung. *Proc. Congr. Int. Soc. Rock Mech., 1st, Lisbon,* 2: 51–56

Wyss, M., ed. 1977. Special issue, *Stress in the Earth*, of *Pure Appl. Geophys.* 115: 1–458

Wyss, M., Hanks, T. C. 1972. The source parameters of the San Fernando earthquake inferred from teleseismic body waves. *Bull. Seismol. Soc. Am.* 62: 591–602

Zoback, M. D., Haimson, B. C., eds. 1982. *Proc. Workshop XVII on Hydraulic Fracturing Stress Measurements.* U.S. Geol. Surv. *Open-File Rep. 82–1075.* 2 vols.

Zoback, M. L., Zoback, M. D. 1980. State of stress in the conterminous United States. *J. Geophys. Res.* 85: 6113–56

POLAR WANDERING AND PALEOMAGNETISM

Richard G. Gordon

Department of Geological Sciences, Northwestern University, Evanston, Illinois 60201

INTRODUCTION

An important hypothesis of paleomagnetism is that *apparent polar wander*, the motion of the paleomagnetic pole relative to a continent or plate, is caused entirely by plate motion relative to the mesosphere, which is the relatively stronger and slowly deforming mantle beneath the asthenosphere. Recent research has challenged this hypothesis and suggests that polar wandering (i.e. true polar wandering) affects apparent polar wander (APW) paths. The debate about polar wandering has been revived by studies comparing paleomagnetic poles to plate motion relative to hotspots (Morgan 1981). What makes the new work interesting is that the underlying hotspot model makes many predictions that are testable by available and obtainable data. One prediction, that the Hawaiian hotspot has shifted southward over Cenozoic time, has been severely tested and confirmed (Kono et al 1978, Kono 1980, Gordon & Cape 1981, Gordon 1982). Recent tests have mainly, but not entirely, taken the form of a reexamination of existing data in the light of the new model. Future tests can and should be made, however, by obtaining high-quality data from paleomagnetic study of critical time intervals already identified by analysis of existing data.

This paper discusses what polar wandering is and reviews how the available paleomagnetic data and plate reconstructions have been used to investigate past polar wandering. Implications of work to date, possible causes of polar wandering, and promising directions of future research are also discussed.

POLAR WANDERING

Polar motion is the motion of the whole Earth relative to its axis of rotation (hereafter termed "spin axis") (Figure 1). Polar motion consists of two

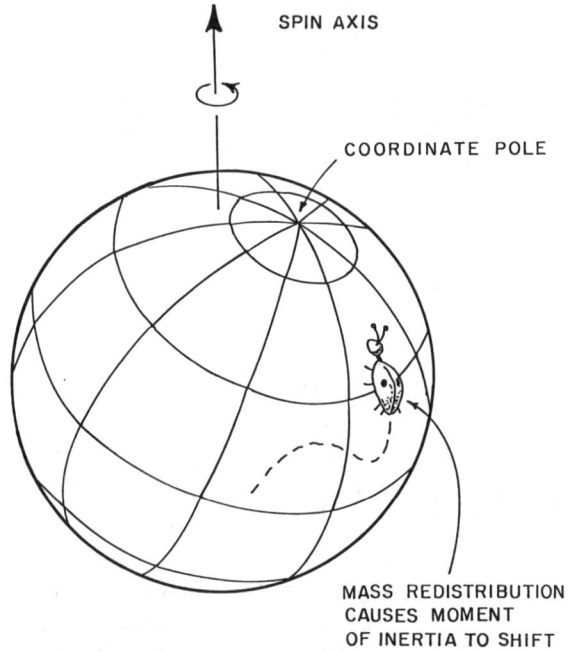

Figure 1 Diagram illustrating polar wandering [from Andrews (1985), after Gold (1955) and Goldreich & Toomre (1969)]. Coordinate system moves with the planet, which moves relative to the spin axis in response to a mass redistribution. Here the change in mass distribution is represented by the motion of a beetle, which causes the greatest axis of nonhydrostatic moment of inertia to shift. Reproduced with permission from the American Geophysical Union.

parts: the *wobble* (annual wobble and Chandler wobble), which is periodic and transient; and *polar wander*, which represents the long-term trend of migration of the Earth (relative to its spin axis). Owing to conservation of angular momentum, the spin axis is fixed and it is the Earth that moves in a reference frame fixed relative to the stars. In fine detail the spin axis departs from the angular momentum vector, but this departure is several hundred times smaller than the polar motion (Gold 1955).

Polar motion is superimposed on the precession and nutation of the Earth's spin axis, which are caused by the gravitational torque exerted on the Earth's equatorial bulge by the Moon and Sun. The dynamics of polar wandering, which is caused by redistribution of mass or angular momentum on or within the Earth (Figure 1) and is unrelated to external torques, are distinctly different from the dynamics of precession and associated nutation. On a perfectly rigid Earth (or an Earth with a very high viscosity throughout its mantle), there would be no polar wandering. Conservation of angular momentum would require the angular momentum vector and the Earth rotation (angular velocity) vector to coincide and remain fixed (aside from the precession and associated nutation due to external torques). However, the Earth deforms and these deformations have two implications for polar wandering. First, it permits polar wandering to occur. Second, the size of the Earth's equatorial bulge is near, but slightly exceeds, what it would be if the Earth were composed of an ideal fluid.

The excess bulge was at one time interpreted as a fossil bulge recording a higher rotation rate from the past (Munk & MacDonald 1960). This long memory of the Earth would imply that the viscosity of the mantle is high enough to prohibit polar wandering. However, an alternative and now widely accepted interpretation of this observation is that the excess (nonhydrostatic) part of the bulge simply reflects the departure of the nonhydrostatic Earth from perfect sphericity (Goldreich & Toomre 1969). That the axis of greatest nonhydrostatic moment of inertia is parallel to the spin axis can be interpreted as showing that the Earth wanders relative to the spin axis so as to align these two axes. In this model, changes in the mass distribution of the Earth might trigger polar wandering.

Polar wandering appears to be occurring now. Astronomical data collected this century by the International Latitude Service (ILS) measure polar motion directly. Superimposed on the wobble, a trend of 70–110 mm yr^{-1} toward eastern Canada ($\sim 75°$W) is observed (Yumi & Wako 1970). Although the tectonic stability of the sites of some of the observatories has been questioned (McKenzie 1972), these results have been widely accepted by the geophysical community (e.g. Dickman 1981, Sabadini & Peltier 1982, Morgan 1981). Moreover, preliminary results from measurements of five years of polar motion using very-long-baseline interferometry support the ILS results (Carter & Robertson 1986). Thus the problem seems not to be whether the pole has wandered, but whether the pole has wandered enough to affect the interpretation of APW paths.

MEASURING POLAR WANDERING WITH PALEOMAGNETIC DATA

Given the observed APW paths of the plates, how can one determine if polar wandering has occurred? Before the acceptance of continental drift, it was recognized that polar wandering should be inferred if APW paths from different continents coincided (Irving 1964). With the acceptance of plate tectonics, however, polar wandering is an unnecessary concept. Polar wandering might nevertheless prove useful, (McKenzie 1972), but there is no consensus on what would demonstrate its usefulness. McKenzie (1972) suggested that it is a useful concept if the motion of the pole relative to any one plate is very much faster than the motion between plates. Many later investigations, however, seem to have taken the less restrictive viewpoint that polar wander has occurred if there has been significant motion of the pole relative to a reference frame external to the plates. This is discussed in more detail below.

Central to the paleomagnetic polar-wandering problem is the development of definitions that permit determination of polar wandering from available and obtainable data. That all paleomagnetic data are derived from plates in relative motion presents an immediate dilemma—how to separate plate motion from polar wandering. Research has focused on three methods: the vector-sum method, the mean-lithosphere method, and the hotspot method.

Vector-Sum Method

To investigate whether polar wandering affects APW paths, McKenzie (1972) suggested the following test. First, APW vectors are converted into equivalent equatorial components of a plate rotation (or angular velocity) vector. Next, each vector is weighted by the area of the plate it represents, and the weighted vectors are summed. A large resultant would show a commonality of motion of the plates relative to the paleomagnetic pole. Though he did not explicitly carry out the test, McKenzie (1972) argued that the general northward motion of the major plates since Cretaceous time suggested that no polar wandering has occurred. McElhinny (1973) applied this test to early Tertiary poles from all the major plates and found that the vector sum was insignificantly different from zero, a result suggesting that little, if any, polar wandering had occurred over Tertiary time. Idnurm (1985) has recently reapplied the vector-sum test to early Tertiary paleomagnetic data and found a small ($4 \pm 2.5°$) polar shift. However, the criterion that the motion of the pole relative to any one plate is very much faster than the motion between the plates (McKenzie 1972) is unsatisfied, which suggests that polar wandering has been unimportant, on average, over Cenozoic time.

Advantages of the vector-sum test are its common sense appeal and its simplicity of application. This simplicity is most apparent when the method is applied to the present plate geometry. Because the number and area of plates change with time, the method is harder to apply to ancient polar wander and plate motions.

Mean-Lithosphere Method

Jurdy & Van der Voo (1974) proposed a polar-wandering test, here termed the mean-lithosphere test, that is similar in intent to the vector-sum test and is based on global relative plate motions. They proposed that global plate motions could be decomposed into random plate motions superimposed on a rigid rotation. They attach a coordinate system (with arbitrary paleolongitude) to the mean paleomagnetic pole from all the plates. Their rigid rotation is found from the mean rotation of the lithosphere in this coordinate system, the "random" motions being the difference between the rotation of an individual plate and the mean rotation. They define true polar wander to be the equatorial component of the mean-lithosphere rotation relative to their paleomagnetic coordinate system.

In principle, results of the mean-lithosphere and vector-sum tests differ. In the mean-lithosphere test, the weight given to an element of plate area is proportional to the sine of the angular distance of the element from the plate's Euler vector relative to the mean lithosphere. In the vector-sum test, equal weight is given to each element of plate area regardless of its distance from the plate's Euler vector. In practice, results of these two tests are similar. Jurdy & Van der Voo's (1974) results agree with McKenzie's (1972) and McElhinny's (1973)—no significant polar wandering has occurred since early Tertiary time, a result that Jurdy & Van der Voo (1975) extended to Early Cretaceous time.

The mean-lithosphere approach can be generalized in a useful way. Implicit in the mean-lithosphere test is the assumption of the existence of a plate-motion reference frame external to any one plate. This reference frame is the unique frame defined by assuming that there has been no net rotation of the lithosphere. Strictly speaking, the test as originally defined did not specify a complete reference frame because Jurdy & Van der Voo (1974) left paleolongitude unspecified. However, rotation of the present-day coordinate system using the rigid rotation they find specifies a complete reference frame to which paleomagnetic poles can be referred, as is done in Figure 2. The paleomagnetic pole never deviates from the North Pole of the mean-lithosphere reference frame by more than a few degrees. Some of the poles differ significantly from the North Pole of the mean-lithosphere reference frame, but the net Cenozoic shift is insignificant, in agreement with the results of Jurdy & Van der Voo (1974).

The mean-lithosphere reference frame can be criticized because it is ad hoc. However, dynamical models of absolute (i.e. relative to the mesosphere) plate motion suggest the mean-lithosphere reference frame may have physical significance because it is kinematically equivalent to Solomon & Sleep's (1974) uniform-drag, no-net-torque model of absolute plate motion (Simpson 1975). In other words, if the coupling between lithosphere and asthenosphere has no lateral variation on the Earth, then the mean-lithosphere reference frame describes the motion of the lithosphere relative to the mesosphere.

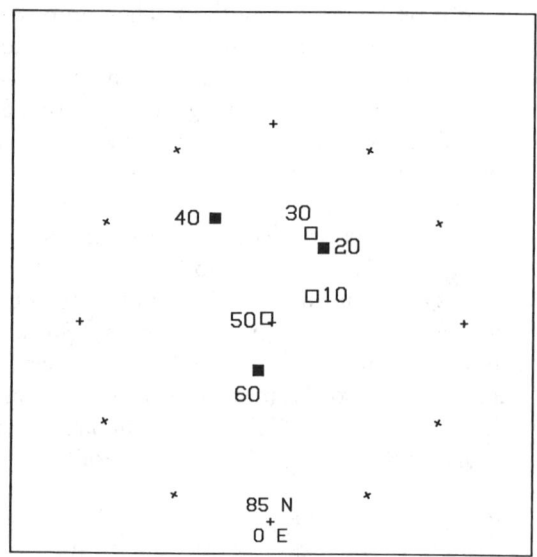

Figure 2 Cenozoic migration of the paleomagnetic pole relative to the mean-lithosphere reference frame, based on the motion of the pole relative to the hotspots determined by Livermore et al (1984) and on the net rotation of the lithosphere relative to the hotspots determined by Gordon & Jurdy (1986). Squares indicate the pole positions; their ages (in millions of years) are given by the adjacent numerals. The 95% confidence circles (not shown) range in radius from 1.0 to 4.9°. The 10, 20, and 40 Ma poles differ significantly (but marginally so) from the North Pole of the mean-lithosphere reference frame, whereas the 30, 50, and 60 Ma poles do not. Equal-area projection.

It seems unlikely, however, that no lateral variations of lithosphere-asthenosphere coupling exist. Continental lithosphere and old oceanic lithosphere are likely to be more strongly coupled to the deeper mantle than young oceanic lithosphere, but it is hard to quantify this variation. Moreover, the torques on downgoing slabs in subduction zones may be balanced in part by drag at the base of plates, which could cause a net rotation of the lithosphere (Solomon & Sleep 1974, Davis & Solomon 1981, 1985, Harper 1986). Thus the uniform-drag, no-net-torque model likely is at best an approximation to the absolute motion of plates.

Moreover, if plate motions are random, as proposed by Jurdy & Van der Voo (1974), the mean-lithosphere reference frame must be an inherently fuzzy concept. That random plate motions relative to the mesosphere should sum exactly to zero would be remarkable. Thus, the mean-lithosphere reference frame has a definite uncertainty associated with it. The Earth now has 7–12 major plates with a root-mean-square velocity of ~ 5 cm yr^{-1}, which suggests that the probable error in the mean velocity of the lithosphere as a whole is ~ 1 cm yr^{-1}, corresponding to an error in angular velocity of $\sim 0.1°$ m.y^{-1}. Accumulated over Cenozoic time, this would imply a probable error in the mean rotation of the lithosphere as a whole of $\sim 6°$. Thus any gradually accumulated shift of the mean-lithosphere reference frame relative to the pole of less than 10–15° (i.e. roughly twice the probable error) should be considered insignificant because this is just the uncertainty in the definition of the mean-lithosphere reference frame itself.

Thus it seems premature to define true polar wander as the shift of the pole relative to the mean-lithosphere reference frame. Rapid motion of the pole relative to the mean-lithosphere frame would be strong evidence that polar wandering has occurred, but slow motion can be explained by other processes. Paleomagnetic practice has shown that it is useful to distinguish between what could be observed (namely APW) and the hypotheses to be tested (in this case polar wandering and net rotation of the lithosphere). At the risk of needless precision, I propose that the motion of the pole relative to the mean-lithosphere reference frame be termed *mean-lithosphere apparent polar wander*, and that the terms "polar wandering" and "true polar wander" are best reserved for the process described by Goldreich & Toomre (1969). This proposal has no effect on Jurdy & Van der Voo's (1974, 1975) conclusion that little polar wandering has occurred, but it does matter when comparing the mean-lithosphere and hotspot reference frames.

Hotspot Method

The current interest in polar wandering stems largely from comparison of paleomagnetic poles to plate-hotspot motions (Morgan 1972, 1981, Hargraves & Duncan 1973). If hotspots have been fixed relative to the paleomagnetic axis, plate-hotspot motions can be used to predict the APW path of a plate. Consider the idealized traces of hotspots on a plate (Figure 3). Furthermore, consider an imaginary hotspot now at the North Pole to which the same hotspot-plate rotation parameters were applied. The "trace" of this imaginary hotspot is precisely the APW path expected if no polar wandering has occurred and if the paleomagnetic pole is aligned with the paleo–spin axis (Figure 3).

This idealization and the observed North America–hotspot motion is used to predict North America's 180–0 Ma APW path in Figure 4. A comparison of high-quality paleomagnetic poles from North America to the predicted North American APW path shows small but significant differences between predicted and observed poles superimposed on a general agreement between the two.

In analogy to the description of the motion of the paleomagnetic pole

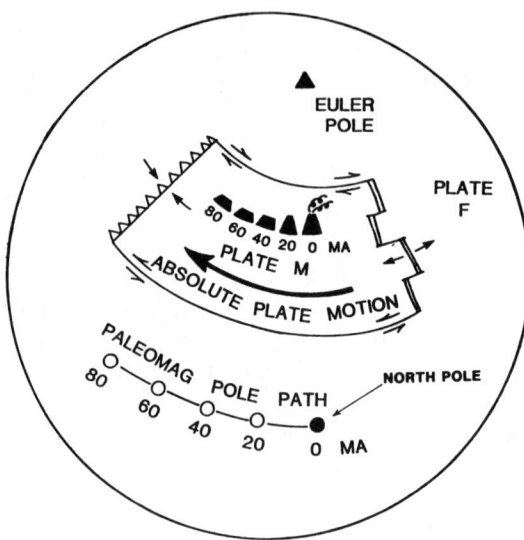

Figure 3 Cartoon illustrating the relationship between hotspot tracks and segments of apparent polar wander paths if the spin axis has been fixed relative to the hotspots. Plate F is fixed relative to the hotspots, while plate M rotates about the Euler pole relative to plate F, the hotspots, and the paleomagnetic axis (assumed to coincide with the spin axis). The hotspot trace on plate M, the transform faults dividing plate M from plate F, and the segment of the paleomagnetic apparent polar wander path all lie on small circles centered on the Euler pole. From Gordon et al (1984). Reproduced with permission of the American Geophysical Union.

relative to the mean-lithosphere reference frame, it seems useful to use the term *hotspot apparent polar wander* to describe the motion of the paleomagnetic pole relative to hotspots. If an individual or group of hotspots moves relative to the paleomagnetic pole, there are several possible explanations, one of which is true polar wander. Here the case is not unlike the classical problem of continental drift versus polar wandering: Is the APW of a hotspot or hotspots due to hotspot drift or polar wander? If these hotspot–paleomagnetic pole shifts reflect polar wandering, then there should be a large common motion of hotspots relative to the pole. On the other hand, motion between hotspots would be reflected in the difference between APW of individual hotspots or widely separated groups of hotspots.

There are at least two important differences between the concepts of mean-lithosphere APW and hotspot APW. First, the mean-lithosphere reference frame is unique. Once the mean-lithosphere APW has been estimated, no further tests are available. On the other hand, it is possible

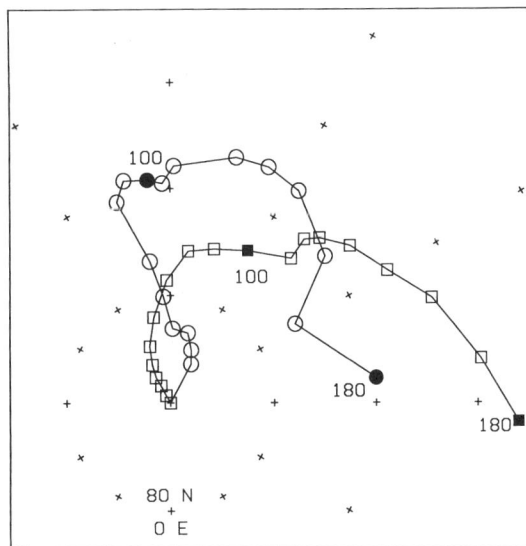

Figure 4 Observed apparent polar wander path of North America (circles) is compared with the apparent polar wander path of North America predicted if the hotspots have been fixed relative to the spin axis (squares). Poles are given every 10 m.y. from 20 to 180 Ma. The 100 and 180 Ma poles are solid; the rest are open. Predicted poles are determined from the North America–hotspot motion model of Morgan (1983). The radii of the 95% confidence circles of the observed poles range from 3 to 10° (Harrison & Lindh 1982a). Differences between observed and predicted poles, when significant, imply motion of the hotspots relative to the paleomagnetic axis. Equal-area projection.

(at least in principle) to observe the APW of individual or groups of hotspots. An individual hotspot cannot have an APW path in the usual sense, but it can have a time history of apparent latitudinal changes. With two or more hotspots, however, it is possible to define a unique APW path. This is advantageous, as hotspot APW paths can be defined independently for widely separated regions. An APW path for one region serves as a testable prediction of the APW path for other regions. A second difference is that the mean-lithosphere frame cannot be defined unless the motions of all the plates are known. This limits its use to a time interval shorter than that to which the hotspot method can be applied.

Paleomagnetic poles are determined from rocks that move not with hotspots but with plates. Therefore the APW path of a group of hotspots, such as those beneath the Pacific plate, must be inferred indirectly. In practice, this means reconstructing sampling sites and their associated paleomagnetic poles or directions to their former positions relative to the hotspots.

Figure 5 (from Morgan 1981) shows a reconstruction relative to hotspots of the continents and of paleomagnetic poles at 60 Ma. The 13 blunt arrows in the figure show paleomagnetic sites for ages ranging from 55 to 65 Ma. The circles show the paleomagnetic poles of these sites rotated to their locations at ~60 Ma. In addition to these poles, a small circle arc is shown to illustrate the locus of poles consistent with the paleo-colatitude of Suiko seamount found from paleomagnetic study of the azimuthally unoriented basalt cores obtained by deep-sea drilling (Kono et al 1978, Kono 1980). This arc is centered on the present location of Hawaii, which is the reconstructed location of Suiko seamount, a 65 Ma seamount in the Hawaiian-Emperor island and seamount chain. The larger circle with the "+" is the average of the 13 poles. Figure 5 includes poles only up to 1973. A similar plot with up-to-date data, and with poles based on limited

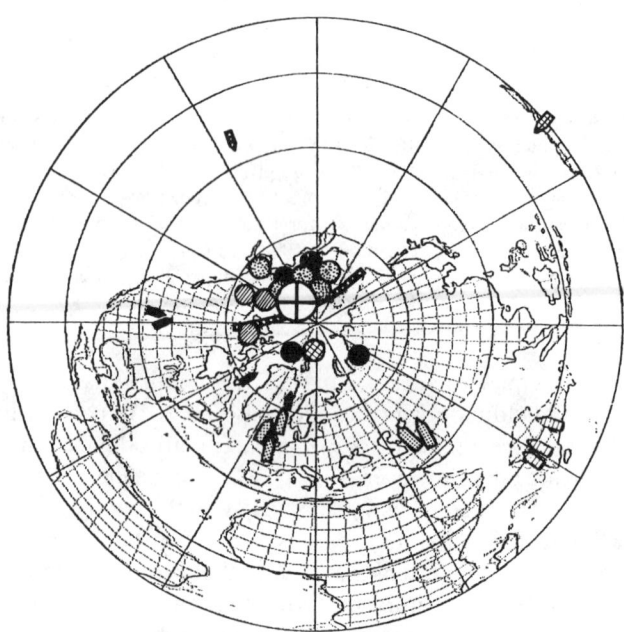

Figure 5 Paleomagnetic poles for 60 Ma as seen in a reference frame fixed relative to the hotspots. The continents, the paleomagnetic sampling sites, and the paleomagnetic poles are rotated to their positions 60 Ma. The sites (arrows) and pole positions (dots) for each continent are pattern coded to illustrate possible systematic effects. The heavy curved line shows the locus of paleomagnetic pole locations consistent with the paleo-colatitude of Suiko seamount inferred from inclination-only paleomagnetic data (Kono 1980). The large circle with the "+" is the average of the site poles. Polar equal-area projection with grid lines every 30°. From Morgan (1981). Reproduced with permission of John Wiley & Sons, Inc.

sampling omitted, would show a similar mean pole position (but closer to the North Pole of the hotspot reference frame) and much less scatter in the individual poles.

Many studies comparing the paleomagnetic and hotspot reference frames have been published in the past few years (Kono et al 1978, Kono 1980, Morgan 1981, Jurdy 1981, Gordon & Cape 1981, Gordon 1982, 1983, Harrison & Lindh 1982b, Sager 1983, Epp et al 1983, Livermore et al 1983, 1984, Andrews 1985, Idnurm 1985). Most of these papers share the conclusion that the hotspots have moved relative to the paleomagnetic pole. There is some disagreement, however, about the size and details of hotspot APW. Instead of extensively discussing the differences between results from different investigators, this review focuses on the results of Livermore et al (1984). Among the various studies, their data appear to be the most up to date, both in their compilation of paleomagnetic poles and in the relative plate motions they adopt. Their study is also the only one that includes data from the major continents and extensive data from the Pacific plate. Moreover, they explicitly modeled the effects of the low-order zonal nondipole paleomagnetic field, which otherwise might be an important alternative to hotspot APW when considering paleomagnetic data from a single plate or from a geographically restricted set of plates. It is also noteworthy that their estimate of hotspot APW is smaller than estimates in other published studies.

Livermore et al (1984) found that the difference between the paleomagnetic axis and the hotspots has typically been $5°$ or less over the past 90 m.y. (Figure 6). From 20 to 80 m.y. ago, the paleomagnetic pole lay (in fixed hotspot coordinates) systematically on the Pacific side of the present spin axis. The confidence limits of all the poles exclude the present spin axis, but a point a few degrees away, for example near $86-87°N$, $160-170°E$, lies within the 95% confidence circles of all the 20–80 Ma poles. Most of these confidence limits for poles for the past 90 m.y. also exclude poles more than $\sim 7-8°$ away from the North Pole of the hotspot reference frame.

Livermore et al (1984) found that the difference was larger (up to $17-19°$) 100–200 Ma. To explain this larger shift they considered three possibilities: (*a*) relatively rapid polar wander occurred before 90 Ma, (*b*) hotspot tracks older than 100 Ma have been misidentified, and (*c*) there are errors in the data. Knowing the extent to which each of these explanations applies is important, as they have important implications for the interpretation of APW paths. Livermore et al (1984) favor explanations (*b*) and (*c*) and suggest that estimates of the size of pre-90 Ma hotspot APW will diminish when improved estimates of the ages and directions of hotspot tracks are available. They point out that the hotspot tracks in the Atlantic

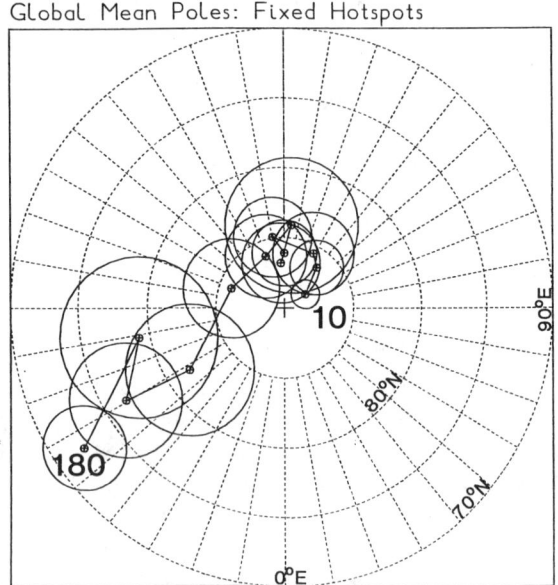

Figure 6 Migration of the paleomagnetic pole (and 95% confidence circles) relative to global hotspots over the past 180 m.y. Average ages of data used to determine a pole are given adjacent to two of the poles. The best-fit pole locations were determined from a dipole-plus-quadrupole field model. From Livermore et al (1984). Reproduced with permission of the Royal Astronomical Society.

are generally clear and well dated for the past 100 m.y., whereas older tracks are identified only on the African continent and are more diffuse, and presumably more subject to error.

However, a case might also be made for polar wandering having caused the apparent polar shift at 90 m.y. The larger errors of the pre–100 Ma reconstructions do not affect the interpretation of a shift that occurred since 100 Ma. Moreover, analysis of data from only the Pacific plate suggests a similar shift between $\sim$ 90 and $\sim$ 80 Ma (Gordon 1983). Analysis of paleomagnetic data from only the North American plate (R. G. Gordon, unpublished analysis) show that the 85–90 Ma pole (in fixed hotspot coordinates) is distinctly different from the 20–80 Ma pole position recognized by Livermore et al (1984) and others. These different interpretations of the significance of Late Cretaceous hotspot APW can and should be tested by determining the net rotation of the lithosphere relative to the hotspots in the Late Cretaceous. If the hotspot–paleomagnetic pole shift is caused by misidentified hotspot tracks, then a large and rapid net rotation of the lithosphere, one paralleling the paleomagnetic pole–hotspot rotation, would be expected from the close agreement of the hotspot and mean-lithosphere reference frames over Cenozoic time (Gordon & Jurdy 1986). If only a small net rotation is measured, it would show that the pole also shifted rapidly relative to the mean-lithosphere reference frame and would support the polar-wandering interpretation.

MOTIONS BETWEEN HOTSPOTS An outstanding problem of analysis of polar wander based on the hotspot method is whether motion between hotspots is slow enough that hotspots provide a useful absolute frame of reference. The key question is how fast hotspots move relative to one another, and this has been a contentious issue. Morgan (1972) argued that the motion between hotspots is ~ 1 cm yr^{-1} or less. Generally this implies that a "mean-hotspot" reference frame can be chosen such that the maximum motion of hotspots relative to the mean-hotspot frame is ~ 5 mm yr^{-1}. Clague & Jarrard (1973) have also found that the hotspots in the Pacific basin move very slowly. Other workers (Molnar & Atwater 1973, Molnar & Francheteau 1975, Pilger 1981) have found, however, that between-hotspot motions can be as large as several centimeters per year. Morgan (1981) and Duncan (1981) have disputed this; their analyses show that the data are consistent with 5 mm yr^{-1} or less motion of Atlantic and Indian Ocean hotspots relative to a mean-hotspot reference frame. Morgan (1983) has furthermore shown that if only the plates bordering the Atlantic (i.e. the African, South American, and Eurasian plates) are considered, only 3 mm yr^{-1} or less motion of any hotspot relative to a mean-hotspot frame is needed. He suggests that errors in the reconstructions of the Indian Ocean are larger than in the Atlantic, and that incorporation of the Indian Ocean data forces faster apparent motion between hotspots than actually occurs.

An alternative interpretation is that motion between nearby hotspots (such as those in the Atlantic) is slow, but that widely separated groups of hotspots (e.g. Atlantic Ocean hotspots versus Indian Ocean hotspots) may move rapidly relative to one another. This must be considered when the problematical motion of the Pacific basin plates relative to other plates is studied. If the widely adopted relative motion circuit (Pacific–Antarctica–Africa) is used to relate the motion of the plates bordering the Atlantic to that of the Pacific basin plates, the motions are consistent with only slow wander between hotspots for the past few tens of millions of years, but they break down for the early Tertiary (Morgan 1981, Duncan 1981, Gordon & Jurdy 1986). The differences are so large that either the hotspots in the Pacific must move as a group at several centimeters per year relative to hotspots in the Atlantic and Indian oceans, or else an additional plate boundary, perhaps between East and West Antarctica or between the

North and South Pacific, is required. Fortunately for the credibility of the hotspot reference frame, paleomagnetic data also require that a now-fossil, early Tertiary plate boundary exist within the present Antarctic plate, the present Pacific plate, or both (Gordon & Cox 1980, Suarez & Molnar 1980).

Motion Between the Mean-Lithosphere and Hotspot Reference Frames

In pioneering studies of polar wandering, large motion between the paleomagnetic axis and the hotspots was found, whereas only insignificant motion between the paleomagnetic axis and the mean-lithosphere (using either the vector-sum method or the mean-lithosphere method) was found (Duncan et al 1972, McElhinny 1973, Hargraves & Duncan 1973, Jurdy & Van der Voo 1974). This observation suggested that while the mean-lithosphere was fixed relative to the spin axis, the mesosphere rotated independently, a process termed *mantle roll* by Hargraves & Duncan (1973). Moreover, in a study of early Tertiary absolute plate motions, Solomon et al (1977) found that the early Tertiary plate motions relative to the mean-lithosphere reference frame were very different from the early Tertiary motions relative to the hotspots.

Later work has shown that the mean-lithosphere and hotspot reference frames are more similar than was previously thought (Jurdy & Gordon 1984, Gordon & Jurdy 1986). Although a recent analysis by Gordon & Jurdy (1986) confirms that the mean-lithosphere frame has rotated over Cenozoic time relative to the hotspot frame, the amount of rotation is only half that found by Hargraves & Duncan (1973). Moreover, the rate of relative rotation of the two reference frames over Cenozoic time is typically $\sim 0.1°$ m.y.$^{-1}$, much less than that found by Solomon et al (1977).

The precise relationship between the paleomagnetic, hotspot, and mean-lithosphere frames is unclear. Within the error limits of the data and analysis, it is just possible that the mean-lithosphere has been locked relative to the paleomagnetic axis over Cenozoic time (Figure 2). However, at the other extreme it is possible that the motion between the hotspot and mean-lithosphere frames is slow compared with the motion of either frame relative to the paleomagnetic axis, and thus to a first approximation the hotspot and mean-lithosphere reference frames move in unison relative to the paleomagnetic axis. This is suggested by the observation that much, perhaps most, of the Cenozoic hotspot APW occurred in the past few million years (Figure 6) at a rate several times faster than the rotation of the mean-lithosphere relative to the hotspots. The truth may lie somewhere between these extremes: The hotspot reference frame, mean-lithosphere reference frame, and paleomagnetic axis may all be in relative motion of comparable magnitude, and the agreement of the mean-lithosphere frame and the paleomagnetic axis 50–60 Ma may be fortuitous. The most certain statement that now can be made is that the paleomagnetic pole–hotspot motion is not less than the motion of the mean-lithosphere reference frame relative to the hotspots.

AXIAL GEOCENTRIC DIPOLE HYPOTHESIS

In most analyses of polar wandering, a working hypothesis is that paleomagnetic poles determined using the dipole formula coincide with the spin axis. Two important ways that the paleomagnetic field may depart from this model are (*a*) if nondipole components of the paleomagnetic field do not average to zero over long intervals of time, and (*b*) if the field is nonaxial. From available data, these hypotheses can be tested in two ways.

First, the relative motion of the plates determined from seafloor magnetic anomalies and traces of transform faults can be combined with globally distributed paleomagnetic data to test whether the time-averaged field is dipolar (Wilson 1970, 1972, Coupland & Van der Voo 1980, Livermore et al 1983, 1984). These tests show a time-averaged nondipole field a few percent as large as the time-averaged dipole field. Thus these effects should not be ignored, especially when paleomagnetic data from a single plate are employed, as has been stressed by Epp et al (1983). With these effects taken into account, the dipole axis is nevertheless found to have moved relative to the hotspots (Livermore et al 1984). Hotspot APW paths show that from ~ 20 to 80 m.y. ago the paleomagnetic pole lay systematically on the Pacific side of the present spin axis (Figure 6). If this were caused by a nonaxial paleomagnetic field, it would be surprising if the recent paleomagnetic field were axial. Yet many analyses have shown that the field averaged over the past 5 m.y. is axial. It thus appears that a nonaxial paleomagnetic field is the less likely explanation for the discrepancies between the paleomagnetic axis and the North Pole of the hotspot reference frame. Nevertheless, from only paleomagnetic data and plate reconstructions, polar wandering cannot be distinguished from a nonaxial paleomagnetic field (Figure 7).

A second type of test arises in those rare cases where nonpaleomagnetic estimates of paleolatitude are available. Unlike the first type of test, the second type of test can distinguish polar wandering from a nonaxial field (Figure 7). Paleolatitudes of sites on the Pacific plate have been determined from two nonpaleomagnetic data types—equatorial sediment facies (Winterer 1973, Clague & Jarrard 1973, van Andel et al 1975, Suarez & Molnar 1980, Gordon & Cape 1981), and the location of the boundary between

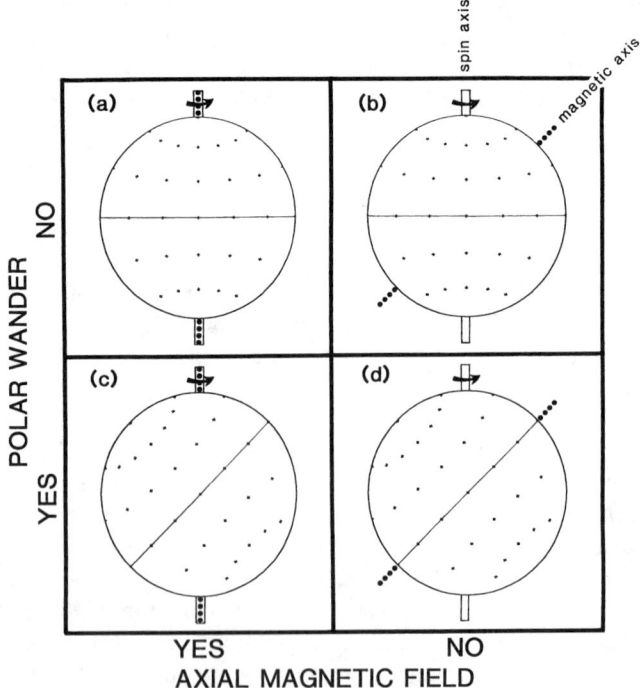

Figure 7 Contrast between polar wandering and a nonaxial paleomagnetic field. In parts (*a*) and (*b*) no polar wandering has occurred, as is shown by the agreement of the spin axis with the north pole of the coordinate system fixed to the Earth. In parts (*c*) and (*d*) polar wandering has occurred, as is shown by the departure of the spin axis from the north pole of the coordinate system fixed to the Earth. In parts (*a*) and (*c*) the paleomagnetic field is axial, as is shown by the agreement of the paleomagnetic axis with the spin axis. In parts (*b*) and (*d*) the paleomagnetic field is nonaxial, as is shown by the departure of the paleomagnetic axis from the spin axis. It is important to note that from paleomagnetic data and plate reconstructions alone, polar wandering [part (*c*)], the main subject of this review, cannot be distinguished from a nonaxial paleomagnetic field [part (*b*)]. Polar wandering and a nonaxial field cannot be distinguished even if the paleomagnetic data come from sites that are well distributed globally. Only with nonpaleomagnetic estimates of paleolatitude can these two cases be distinguished.

the bryozoan-algal and coral-algal carbonate facies (Schlanger & Konishi 1975). Equatorial sediment facies reflect the high biologic productivity of the Pacific Ocean from $\sim 5°S$ to $\sim 5°N$, wherein calcareous deposits very near the equator give way to siliceous deposits to the north and south. Ideally, the age of equatorial transit for a Northern Hemisphere site lying on crust that formed in the Southern Hemisphere can be inferred from a deep-sea drilling core (Winterer 1973, Heezen et al 1973).

The coral-algal/bryozoan-algal carbonate facies boundary is recognized in modern-day to Eocene facies distribution patterns that are apparently controlled by latitude. Within 20° latitude of the equator, shallow-water carbonates are dominated by a coral-algal facies, whereas at latitudes exceeding 30°, coeval carbonate deposits are dominated by a bryozoan-algal facies (Schlanger & Konishi 1975). This observation of the bipolar distribution of the tropical reef facies and the cooler-water bryozoan-rich facies permits a semiquantitative evaluation of paleolatitude, although the estimated paleolatitudes can be biased if there has been a latitudinal shift of surface-water isotherms. However, the biases of each of these paleolatitude indicators presumably are independent. Thus a strong case for a latitudinal shift can be made if these independent measures of paleolatitude agree.

These tests can be applied to the determination of the paleolatitude of the Hawaiian hotspot at the time of the formation of Suiko seamount (65 Ma). Analysis of equatorial sediment facies suggests that Suiko seamount formed 5–10° north of the present latitude of Kilauea (19.6°N) (Gordon & Cape 1981), in good agreement with the paleomagnetically determined paleolatitude of 27°N (Kono 1980). In detail, the difference between northward motion inferred from equatorial sediment facies and from hotspot traces has been very small (1–3°) for the past 40 m.y., but much larger (5–10°) 60 to 70 Ma.

The analysis of the shallow-water carbonate facies distribution further supports the northerly paleolatitude of Suiko and other Emperor seamounts. Samples from four seamounts in the Emperor seamount chain that have been drilled reveal Paleogene shallow-water carbonate sediments of the bryozoan-algal facies recovered from the tops of the edifices (McKenzie et al 1980). In detail there is a gradient from north to south: The northernmost seamount, Suiko, has the greatest proportion of the bryozoan-algal facies, and the southernmost seamount, Koko, has the least, which suggests a warming trend from 65 Ma (the age of Suiko) to 48 Ma (the age of Koko), a pattern echoing that inferred from the equatorial sediment facies. The bryozoan-algal assemblages suggest that the seamounts, especially Suiko, formed in cooler waters than those now occupied by Hawaii and can be explained by a southward shift of the Hawaiian hotspot of 5–10° since the formation of Suiko seamount (McKenzie et al 1980).

One of the puzzling departures of the paleomagnetic field from a dipole is a persistent tendency for poles, when viewed from their sampling sites, to be located to the right-hand side of the dipole (Wilson 1970, 1972). This right-handed tendency, if real, is hard to explain. For example, to explain the observation in terms of the electric currents needed to produce a right-handed magnetic field requires an Earth-air electric current system more than a million times that of currently observed currents. Andrews (1985) has proposed that this right-handed tendency is due to polar wandering combined with a sampling bias weighted heavily toward Europe, where the right-handed tendency is very pronounced. This hypothesis merits further investigation. Possibly some differences between the time-averaged normal and reversed paleomagnetic field can also be explained by polar wandering.

In summary, analyses of globally distributed paleomagnetic data show that the small corrections required to the geocentric dipole formula are too small to alter the conclusion that the hotspots have shifted relative to the paleomagnetic axis, although they show smaller hotspot–paleomagnetic axis motion than do analyses assuming a geocentric dipolar paleomagnetic field. These departures are difficult to explain in terms of a nonaxial field because the field averaged over the past 5 m.y. is axial. Moreover, two different nonpaleomagnetic paleolatitude indicators available from the Pacific plate show a Cenozoic southward shift of the Hawaiian hotspot, which most likely occurred between $\sim$ 40 and 65 Ma. This southward shift is consistent with that inferred from Pacific plate paleomagnetic data (Kono 1980, Gordon & Cape 1981, Gordon 1982, Sager 1983) and predicted by the APW of the hotspots in the Atlantic and Indian oceans. The anomalous tendency for poles, when viewed from their sampling sites, to be located to the right-hand side of the axial geocentric dipole might be explainable in terms of a recent episode of polar wandering.

IMPLICATIONS TO INTERPRETATION OF APW PATHS

Paleo-Reconstructions and Paleo-Velocities

Although the analysis reviewed here suggests that the paleomagnetic pole is an imperfect marker of the lower mantle, the motion between the paleomagnetic pole and the mesosphere is probably small enough that past plate velocities determined from paleomagnetic data should provide good estimates of plate-mesosphere motion. Encouragingly, there is a hint in the data that polar wandering is concentrated in brief intervals (≤ 10 m.y.) that divide long intervals (many tens of millions of years) of negligible polar wandering. There may be a danger in using the angular distance between poles not widely separated in age if an apparently rare episode of polar wandering divides them. Polar wandering might cause APW to be jerky, with episodes of relatively rapid APW marking episodes of polar wandering. Related to this is a possible pitfall whereby poles of close but not identical ages, if separated in time by an episode of polar wandering, record very different positions of the pole relative to the mesosphere. If these poles are from separate continents and are used to guide pre-Jurassic plate reconstructions, they might lead to unreliable relative positions of these continents. It may be possible to determine the boundary between these long quasi-static intervals from data independent from pole positions, as is discussed further in the next section.

Polar Wandering and Polarity Bias

If the reversal time scale is viewed through a sliding window 25 m.y. long, the polarity pattern seen in the window undergoes conspicuous changes in character as the window is moved from the present backward many tens or hundreds of millions of years (Harland et al 1982). The two most recent conspicuous changes occur at 83 and 118 Ma. Since 83 Ma the field has reversed often and symmetrically, spending equal amounts of time in the normal and reversed polarity state. Between 83 and 118 Ma, however, the field remained in the normal state with few, if any, brief scattered intervals of reversed polarity. Looking further back, before 118 Ma and at least as far back as 165 Ma (which marks a decrease in the quality of the record of reversals, being the oldest reversal reliably recorded in seafloor magnetic lineations), the field reversed often and symmetrically.

This pattern, in which the field is characterized by long intervals of *polarity bias*, is thought to typify the reversal time scale throughout geologic time. The length of time characterized by polarity bias (in the cases above, *normal polarity bias* for the interval from 118–83 Ma and *mixed polarity bias* for the other two intervals) is much longer than the time between reversals during the Cenozoic. In the nomenclature of the magnetostratigraphic time scale, these long intervals of polarity bias are now termed *superchrons*, which distinguishes them from *chrons*, which are time intervals of constant polarity between reversals, possibly interrupted by short intervals of opposite polarity (Harland et al 1982).

When the Permo-Carboniferous reversed (PC-R) superchron was first recognized from a detailed magnetostratigraphic study in Australia, Irving (1966) noticed that the base of the superchron recorded a jump in the pole position inferred from the directions of magnetization. This led Irving & Robertson (1969) to postulate that this jump was caused by polar wandering, and that there was a link between reversals (polarity bias) and paleomagnetic directions—namely, that the boundary between polarity bias superchrons coincided with brief intervals of polar wandering.

New data have recently led to a similar hypothesis. Gordon (1983) proposed that the end of the Cretaceous normal (K-N) polarity bias superchron coincided with a 10–15° polar jump inferred from a comparison of Pacific plate paleomagnetic poles and hotspot tracks. Moreover, the

80–40 Ma position of the pole relative to the Pacific hotspots shows only minor wander relative to the abrupt ~ 85 Ma shift. This analysis is consistent with the pole having been quasi-fixed during the Cretaceous-Tertiary-Quaternary mixed (KTQ-M) polarity superchron, after a brief episode of rapid polar wandering near the boundary dividing the K-N superchron from the KTQ-M superchron.

This hypothesis has recently been tested further by an analysis of North American APW and motion relative to the hotspots over the past 200 m.y. (R. G. Gordon, unpublished). A prediction of the polar wander–polarity bias model is that poles from each polarity superchron, when reconstructed to their location in the fixed (Atlantic) hotspot reference frame, should divide by their membership in one of the three polarity superchrons (KTQ-M, K-N, or JK-M) into distinct groups. This prediction is confirmed when high-quality paleomagnetic poles from North America are reconstructed into the hotspot reference frame. Not only is the K-N to KTQ-M polar jump (first observed from independent Pacific plate data) found, but so too is a jump between the JK-M and K-N superchrons. However, the paleomagnetic data are too sparse to be completely convincing. If further work confirms this pattern, a possible implication of the polar wander–polarity bias model is that plate motions determined from paleomagnetic data within a single polarity bias superchron are unbiased, whereas data extending beyond a single polarity superchron may introduce a small bias because of a pole-mesosphere offset of 5–10°.

CAUSES OF POLAR WANDERING

The location of the spin axis is controlled by the location of the axis of greatest nonhydrostatic moment of inertia. Lateral density heterogeneities in the Earth in turn control the location of the inertial axis. The possible role of different sources of lateral heterogeneity has been investigated with two approaches. The first is to model observed lateral heterogeneity and examine whether a particular source of heterogeneity, such as continent-ocean differences, has an appropriate geometric relationship to the spin axis. The second follows from interpretation of the geoid. The geoid connection arises because the same integrals that determine the moment of inertia tensor of the Earth also determine the coefficient of the second-order harmonic of the geopotential. Because anomalous potential is simply related to geoid anomalies, so too must anomalous moment of inertia be simply related to geoid anomalies.

One of the earliest and most widely discussed mechanisms of polar wander is the redistribution of mass caused by continental drift. This idea is now unpopular for two reasons. First, the present arrangement of continents and oceans bears little relationship to the present spin axis. The greatest principal axis of nonhydrostatic moment of inertia calculated from the present arrangement of either continents or oceans does not coincide with the spin axis (Munk & MacDonald 1960). Moreover, the principal axis of earlier distributions of the continents and oceans have positions very similar to those at present (Jurdy 1978). Second, the long-wavelength components of the geoid bear little relationship to the present arrangement of continents and oceans (Chase 1979, Crough & Jurdy 1980). The geoid height differences between continents and oceans are small, perhaps 10 m, and are hard to appraise in the presence of the large anomalies that cut indiscriminately across continent-ocean boundaries (Chase 1985).

Some recent research has focused on glaciation and deglaciation as the mechanism driving polar wander [Dickman 1979, Sabadini & Peltier 1982, Sabadini et al 1982; see a recent paper by Boschi et al (1985) for a more complete set of references]. The focus on deglaciation occurs partly because hotspot APW was negligible 20–80 Ma but rapid over the past 5 to 20 m.y. [Morgan's (1981), Andrews's (1985), and to a lesser extent Harrison & Lindh's (1982b) and Livermore et al's (1984) estimates of hotspot APW suggest that the pole has been wandering rapidly, possibly as fast as $\sim 1.0°$ m.y.$^{-1}$, over the past few million years. Andrews (1985) has found that recent hotspot APW agrees with astronomically observed polar wander, which suggests an average motion of 0.7–1.1° m.y.$^{-1}$ toward $\sim 75°$W (Munk & MacDonald 1960, Yumi & Wako 1970); this result in turn suggests that hotspot APW reflects polar wandering.] It appears that deglaciation can cause polar wandering having the speed and direction observed, with the speed being about right if the Earth's viscosity is assumed to have the widely adopted value of 10^{22} poise (Sabadini & Peltier 1982, Sabadini et al 1982). Deglaciation is an attractive mechanism because it explains the current pattern of polar wander, but rapid hotspot APW is thought by several workers to have occurred during times, such as the Late Cretaceous (Gordon 1983), for which there is no evidence of glaciation. Thus, if the deglaciation mechanism is correct, and if the rapid hotspot APW of earlier times is caused by polar wandering, there presumably is more than one source of density heterogeneity that can drive polar wandering.

An alternative source of heterogeneity is the changing configuration of subduction zones. Among the few present-day tectonic features that have a prominent geoid signal are subduction zones, which have a geoid high generally attributed to the uncompensated excess mass of the cold subducting slab (Chase 1979). Because of the relationship (noted above) between the second-order harmonics of the geopotential and the moment of inertia tensor, changing subduction-zone configurations seem an attractive mechanism for causing polar wandering. Jurdy (1983) has studied whether the change in configuration of subduction zones from the early Tertiary to the present might produce the (hotspot) polar wander observed to have occurred over the same interval. She adopts an empirical scaling relation to calculate the change in the (nonhydrostatic) moment of inertia tensor caused by changing subduction-zone configuration and finds reasonable agreement between observed and predicted polar wandering. However, the scaling between the contributions to the moment of inertia of the subducting slab and other sources are, of necessity, arbitrary, since so little is known about most of the sources. Moreover, although the mechanism matches the overall polar wander since early Tertiary time, analyses of hotspot APW suggest that most of the Cenozoic polar wander has occurred during the past 10 m.y., whereas the subduction geometry modeled by Jurdy (1983) appears to have evolved more or less continuously over $\sim$ 50 m.y.

Using the geoid as a guide, the main controls to the Earth's nonhydrostatic moment of inertia must arise from mantle convection (Chase 1979, 1985). The geoid is dominated by two large highs, one centered in the central Pacific, and a second centered over Africa but running from Greenland across Africa and the Southwest Indian Ridge (Chase 1979, Crough & Jurdy 1980). The Pacific high is the larger of the two highs and apparently controls the present location of the spin axis. If the African high were to become larger than the Pacific high, possibly by changes in the regime of deep mantle convection or by deformation of the core-mantle boundary, rapid polar wander would occur, with Greenland migrating to the equator (Chase 1979). Our ignorance of deep mantle convection processes is probably even greater than our ignorance of polar wander. The smallness of polar wandering over the past 200 m.y. implies that the deep mantle convection pattern responsible for the second-order nonhydrostatic geopotential has been fairly stable (Chase 1985).

Another explanation that has been proposed for the current long-wavelength geoid highs, which in turn are the controls of the present location of the spin axis, is upper mantle overheating. Anderson (1982) has noted that the African high lies over the Mesozoic position of Pangea in Morgan's (1981) plate reconstructions. Anderson (1982) proposed that the insulating effect of thick continental lithosphere of the Pangean assemblage caused heating of the upper mantle, which in turn caused the African geoid high. Anderson (1982) further proposed that a now-dispersed continent existed in the Mesozoic Pacific and provided an insulating effect that caused the Pacific geoid high. Chase (1985) has suggested, alternatively, that stifling of asthenosphere return flow under the Pangean continent by the subduction zones ringing Pangea caused upper-mantle overheating beneath Pangea. Chase (1985) further suggested that the Pacific geoid high is related to the anomalously shallow, volcano-studded Darwin Rise.

SUMMARY

The evidence that there has been enough polar wandering to affect APW paths is suggestive but not entirely convincing. Weighing against the polar-wandering hypothesis is that the net Cenozoic motion of the paleomagnetic pole relative to the mean-lithosphere reference frame is insignificant. Although the coincidence of the paleomagnetic and mean-lithosphere frame 50–60 Ma may be fortuitous, such a chance alignment does not seem likely. Further weighing against the polar-wandering hypothesis is that estimates of the amount of motion of the paleomagnetic pole relative to the hotspots has diminished as higher quality paleomagnetic data and plate reconstructions have become available (Livermore et al 1984). That the size of hotspot APW appears to be about the same as the size of the corrections to the dipole formula necessary to model the time-averaged nondipole field also argues against the usefulness of the concept of polar wandering in the interpretation of APW paths. Lastly, almost all inferences about past polar wandering depend on the hotspot reference frame. Unless the present debate about motion between hotspots is resolved in favor of very slow motions between hotspots, inferences based solely on the hotspot frame are bound to be unconvincing.

On the other hand, the occurrence of polar wandering large enough to affect the interpretation of APW paths is suggested by several lines of evidence: Polar wander measured directly this century occurs at a rate (~ 100 mm yr^{-1}) that, if continued for millions or tens of millions of years, would have a noticeable effect on APW paths. Analyses of paleomagnetic data by several independent groups of investigators find that significant shifts of the paleomagnetic axis relative to the hotspots have occurred in the past 20 m.y. (Morgan 1981, Harrison & Lindh 1982b, Livermore et al 1984, Andrews 1985). These shifts might reasonably be attributed to polar wandering, and if they are as large as found in some of these analyses, they are too large to depend on the choice of a hotspot rather than some alternative reference frame (e.g. the mean-lithosphere reference frame). Polar wandering might be able to explain a long-standing enigma of paleomagnetism, Wilson's (1970, 1972) "right-handed" effect (Andrews 1985). The hotspot-based determination of polar wandering is a testable model. In particular, paleomagnetic data and plate motions of the continents can be used to predict hotspot–paleomagnetic pole motion in the Pacific Ocean basin and vice versa. Independent data from the Pacific plate agree with these predictions (Kono et al 1978, Morgan 1981, Gordon & Cape 1981, Jurdy 1981, Gordon 1982) and support the hypothesis that the hotspots have moved largely in unison relative to the paleomagnetic pole. The limited tests available from nonpaleomagnetic paleolatitude indicators suggest that to a useful approximation the paleomagnetic pole has been tracking the spin axis, and that the apparent polar shifts inferred from paleomagnetic data are not an artifact of nondipole components of the paleomagnetic field or a nonaxial dipole field (McKenzie et al 1980, Gordon & Cape 1981). Moreover, when the nondipole field is explicitly modeled, the inferred shift of the dipole axis relative to the hotspots is barely affected (Livermore et al 1984).

In any event, polar wandering clearly has been small, probably less (perhaps much less) than $\sim 20°$ over the past 200 m.y., and less than $\sim 10°$ over the past 80 m.y. Large polar shifts of 60–90°, as in the model of Goldreich & Toomre (1969), have not occurred over the past 200 m.y. The lack of large-scale polar wandering implies that the Earth's nonhydrostatic moment of inertia has changed little over the past 200 m.y. Viewed over many tens of millions of years, there are good reasons for thinking APW primarily reflects plate motion, with polar wandering being a secondary effect. Separation of polar wandering from plate motion presents a considerable challenge to paleomagnetists because polar wandering has been small enough that it will be difficult to appraise its effect on APW paths.

Directions of Future Research

Two time intervals of critical importance for further paleomagnetic study of polar wandering have been identified from rapid intervals of APW relative to the hotspots. Further study is needed to document APW during these intervals. Does rapid APW occur synchronously on several plates? Do the APW paths of different plates have a large common element over the critical time intervals? Is APW ever fast enough to exclude an explanation in terms of rapid plate motion?

The most recent critical interval is the past 5–20 m.y. Several investigations have concluded that global mean poles less than 10 Ma in average age differ significantly from the present spin axis. Morgan (1981) found that the pole has been moving relative to the hotspots at 100–300 mm yr^{-1} over the past 5 m.y., although the rate found by Livermore et al (1984) is considerably less. Morgan's (1981) rate is fast enough that it would not depend on what reference frame is used to measure polar wandering. Andrews (1985) found that the most recent interval of hotspot APW agrees with the direction and rate of polar wandering inferred from the International Latitude Service data (Yumi & Wako 1970, Dickman 1981).

The second critical interval for study is the time interval ~ 90–80 Ma (across the boundary between the K-N and KTQ-M superchrons, which also divides Santonian from Campanian time). The postulated shift across this boundary (Gordon 1983) is large enough to be measured in a single high-quality paleomagnetic study, and its precise location in a stratigraphic section should be identifiable magnetostratigraphically and biostratigraphically. Thus, the prospects of fixing the timing and the rate of this shift are good. In particular, it should be possible to estimate how closely in time any rapid hotspot APW coincides with the boundary dividing the K-N and KTQ-M superchrons.

Harrison & Lindh (1982b) and Livermore et al (1984) also found very rapid shifts of the paleomagnetic axis relative to the hotspots at ~ 70–100 Ma. Livermore et al (1984) have suggested that the inferred Late Cretaceous motion between the paleomagnetic axis and the hotspots might be due to inaccurate plate-hotspot reconstructions. This should be investigated through continuing programs of dating suspected hotspot traces; Duncan (1984) has already made an important first step toward solving the Late Cretaceous plate-hotspot motion problem. Relative plate motions for Late Cretaceous time are probably known accurately enough to warrant comparisons of the mean-lithosphere and hotspot reference frames. This would serve to test whether the mean-lithosphere more closely coincided with the hotspot or paleomagnetic reference frames during the rapid shift between the paleomagnetic axis and the hotspots. Large discrepancies between the mean-lithosphere and hotspot reference frames might serve to identify errors in reconstructions relative to the hotspots.

ACKNOWLEDGMENTS

This manuscript was largely prepared while I was a visitor at the Department of Earth Sciences, University of Cambridge. I thank the members of the department for their hospitality. I also thank Roy Livermore and Sy Schlanger for helpful comments about the manuscript. This work was supported by NSF grant EAR-8417323 and by an Alfred P. Sloan Foundation Research Fellowship.

Literature Cited

Anderson, D. L. 1982. Hotspots, polar wander, Mesozoic convection, and the geoid. *Nature* 297: 391–93

Andrews, J. 1985. True polar wander: an analysis of Cenozoic and Mezozoic paleomagnetic poles. *J. Geophys. Res.* 90: 7737–50

Boschi, E., Sabadini, R., Yuen, D. A. 1985. Transient polar motions and the nature of the asthenosphere for short time scales. *J. Geophys. Res.* 90: 3559–68

Carter, W. E., Robertson, D. S. 1986. Studying the Earth by very-long-baseline interferometry. *Sci. Am.* 255(5): 46–54

Chase, C. G. 1979. Subduction, the geoid, and lower mantle convection. *Nature* 282: 464–68

Chase, C. G. 1985. The geological significance of the geoid. *Ann. Rev. Earth Planet. Sci.* 13: 97–117

Clague, D. A. Jarrard, R. D. 1973. Tertiary Pacific plate motion deduced from the Hawaiian-Emperor chain. *Geol. Soc. Am. Bull.* 84: 1135–54

Coupland, D. H., Van der Voo, R. 1980. Long-term nondipole components in the geomagnetic field during the last 130 m.y. *J. Geophys. Res.* 85: 3529–48

Crough, S. T., Jurdy, D. M. 1980. Subducted lithosphere, hotspots, and the geoid. *Earth Planet. Sci. Lett.* 48: 15–22

Davis, D. M., Solomon, S. C. 1981. Variations in the velocities of the major plates since the Late Cretaceous. *Tectonophysics* 74: 189–208

Davis, D. M., Solomon, S. C. 1985. True polar wander and plate-driving forces. *J. Geophys. Res.* 90: 1837–41

Dickman, S. R. 1979. Continental drift and true polar wandering. *Geophys. J. R. Astron. Soc.* 57: 41–50

Dickman, S. R. 1981. Investigation of controversial polar motion features using homogeneous International Latitude Service data. *J. Geophys. Res.* 86: 4904–12

Duncan, R. A. 1981. Hotspots in the Southern oceans—an absolute frame of reference for motion of the Gondwana continents. *Tectonophysics* 74: 29–42

Duncan, R. A. 1984. Age progressive volcanism in the New England Seamounts and the opening of the Central Atlantic. *J. Geophys. Res.* 89: 9980–90

Duncan, R. A., Petersen, N., Hargraves, R. B. 1972. Mantle plumes, movement of the European plate, and polar wandering. *Nature* 239: 82–86

Epp, D., Sager, W. W., Theyer, F., Hammond, S. R. 1983. Hotspot–spin axis motion or magnetic far-sided effect? *Nature* 303: 318–20

Gold, T. 1955. Instability of the Earth's axis of rotation. *Nature* 175: 526–29

Goldreich, P., Toomre, A. 1969. Some remarks on polar wandering. *J. Geophys. Res.* 74: 2555–67

Gordon, R. G. 1982. The late Maastrichtian paleomagnetic pole of the Pacific plate. *Geophys. J. R. Astron. Soc.* 70: 129–40

Gordon, R. G. 1983. Late Cretaceous apparent polar wander of the Pacific plate: evidence for a rapid shift of the Pacific hotspots with respect to the spin axis. *Geophys. Res. Lett.* 10: 709–12

Gordon, R. G., Cape, C. D. 1981. Cenozoic latitudinal shift of the Hawaiian hotspot and its implications for true polar wander. *Earth Planet. Sci. Lett.* 55: 37–47

Gordon, R. G., Cox, A. 1980. Paleomagnetic test of the early Tertiary plate circuit between the Pacific Basin plates and the Indian plate. *J. Geophys. Res.* 85: 6534–46

Gordon, R. G., Jurdy, D. M. 1986. Cenozoic global plate motions. *J. Geophys. Res.* 91: 12,389–12,406

Gordon, R. G., Cox, A., O'Hare, S. 1984. Paleomagnetic Euler poles and the apparent polar wander and absolute motion of North America since the Carboniferous. *Tectonics* 3: 499–537

Hargraves, R. B., Duncan, R. A. 1973. Does the mantle roll? *Nature* 245: 361–63

Harland, W. B., Cox, A. V., Llewellyn, P. G., Pickton, C. A. G., Smith, A. G., Walters, R. 1982. *A Geologic Time Scale*. New York: Cambridge Univ. Press. 131 pp.

Harper, J. F. 1986. Mantle flow and plate motions. *Geophys. J. R. Astron. Soc.* 87: 155–71

Harrison, C. G. A., Lindh, T. 1982a. A polar wandering curve for North America during the Mesozoic and Cenozoic. *J. Geophys. Res.* 87: 1903–20

Harrison, C. G. A., Lindh, T. 1982b. Comparison between the hot spot and geomagnetic field reference frames. *Nature* 300: 251–52

Heezen, B. C., MacGregor, I. D., Foreman, H. P., Forristall, G. Z., Hekel, H., et al. 1973. The post-Jurassic sedimentary sequence of the Pacific plate: a kinematic interpretation of diachronous deposits. In *Initial Reports of the Deep Sea Drilling Project*, 20: 725–38. Washington, DC: Govt. Print. Off.

Idnurm, M. 1985. Late Mesozoic and Cenozoic palaeomagnetism of Australia—II. Implications for geomagnetism and true polar wander. *Geophys. J. R. Astron. Soc.* 83: 419–33

Irving, E. 1964. *Paleomagnetism and its Application to Geological and Geophysical Problems*. New York: Wiley. 399 pp.

Irving, E. 1966. Paleomagnetism of some Carboniferous rocks from New South Wales and its relation to geological events. *J. Geophys. Res.* 71: 6025–51

Irving, E., Robertson, W. A. 1969. Test for polar wandering and some possible implications. *J. Geophys. Res.* 74: 1026–36

Jurdy, D. M. 1978. Ridges, trenches, and polar wander excitation. *J. Geophys. Res.* 83: 4989–94

Jurdy, D. M. 1981. True polar wander. *Tectonophysics* 74: 1–16

Jurdy, D. M. 1983. Early Tertiary subduction zones and hot spots. *J. Geophys. Res.* 88: 6395–6402

Jurdy, D. M., Gordon, R. G. 1984. Global plate motions relative to the hot spots 64 to 56 Ma. *J. Geophys. Res.* 89: 9927–36

Jurdy, D. M., Van der Voo, R. 1974. A method for the separation of true polar wander and continental drift, including results for the last 55 m.y. *J. Geophys. Res.* 79: 2945–52

Jurdy, D. M., Van der Voo, R. 1975. True polar wander since the Early Cretaceous. *Science* 187: 1193–96

Kono, M. 1980. Paleomagnetism of DSDP Leg 55 basalts and implications for the tectonics of the Pacific plate. In *Initial Reports of the Deep Sea Drilling Project*, 55: 737–52. Washington, DC: Govt. Print. Off.

Kono, M., Morgan, W. J., scientific staff. 1978. Paleolatitude of Emperor seamounts from DSDP Leg 55. *Eos., Trans. Am. Geophys. Union* 59: 297 (Abstr.)

Livermore, R. A., Vine, F. J., Smith, A. G. 1983. Plate motions and the geomagnetic field—I. Quaternary and late Tertiary. *Geophys. J. R. Astron. Soc.* 73: 153–71

Livermore, R. A., Vine, F. J., Smith, A. G. 1984. Plate motions and the geomagnetic field—II. Jurassic to Tertiary. *Geophys. J. R. Astron. Soc.* 79: 939–61

McElhinny, M. W. 1973. Mantle plumes, palaeomagnetism, and polar wandering. *Nature* 241: 523–24

McKenzie, D. P. 1972. Plate tectonics. In *The Nature of the Solid Earth*, ed. E. C. Robertson, pp. 323–60. New York: McGraw-Hill. 677 pp.

McKenzie, J., Bernoulli, D., Schlanger, S. O. 1980. Shallow-water carbonate sediments from the Emperor Seamounts: their diagenesis and paleogeographic significance. In *Initial Reports of the Deep Sea Drilling Project*, 55: 415–55. Washington, DC: Govt. Print. Off.

Molnar, P., Atwater, T. 1973. Relative motion of hot spots in the mantle. *Nature* 246: 288–91

Molnar, P., Francheteau, J. 1975. The relative motion of "hot spots" in the Atlantic and Indian oceans during the Cenozoic. *Geophys. J. R. Astron. Soc.* 43: 763–74

Morgan, W. J. 1972. Plate motions and deep mantle convection. *Geol. Soc. Am. Mem.* 132: 7–22

Morgan, W. J. 1981. Hotspot tracks and the opening of the Atlantic and Indian Oceans. In *The Sea*, ed. C. Emiliani, 7: 443–87. New York: Wiley

Morgan, W. J. 1983. Hotspot tracks and the early rifting of the Atlantic. *Tectonophysics* 94: 123–39

Munk, W., MacDonald, G. J. F. 1960. *The Rotation of the Earth: A Geophysical Discussion*. London: Cambridge Univ. Press. 323 pp.

Pilger, R. H. Jr. 1981. Plate reconstructions, aseismic ridges, and low-angle subduction beneath the Andes. *Geol. Soc. Am. Bull.* 92: 448–56

Sabadini, R., Peltier, W. R. 1982. Pleistocene deglaciation and the Earth's rotation: implications for mantle viscosity. *Geophys. J. R. Astron. Soc.* 66: 553–78

Sabadini, R., Yuen, D. A., Boschi, E. 1982. Polar wandering and the forced response of a rotating, multilayered, viscoelastic planet. *J. Geophys. Res.* 87: 2885–2903

Sager, W. W. 1983. A Late Eocene paleomagnetic pole for the Pacific plate. *Earth Planet. Sci. Lett.* 63: 408–22

Schlanger, S. O., Konishi, K. 1975. The geographical boundary between the coralalgal and the bryozoan-algal limestone facies: a paleolatitude indicator. *Theme 1, Sedimentologic Indicators, Int. Congr. Sedimentol. 9th, Nice, Fr.*, pp. 187–90

Simpson, R. W. 1975. Relations between a criterion for polar wander and some conditions for absolute plate motion. *J. Geophys. Res.* 80: 4823–24

Solomon, S. C., Sleep, N. H. 1974. Some simple physical models for absolute plate motions. *J. Geophys. Res.* 79: 2557–67

Solomon, S. C., Sleep, N. H., Jurdy, D. M. 1977. Mechanical models for absolute plate motions in the early Tertiary. *J. Geophys. Res.* 82: 203–12

Suarez, G., Molnar, P. 1980. Paleomagnetic data and pelagic sediment facies and the motion of the Pacific plate relative to the spin axis since the Late Cretaceous. *J. Geophys. Res.* 85: 5257–80

van Andel, T. H., Heath, G. R., Moore, T. C. Jr. 1975. Cenozoic history and paleoceanography of the central equatorial pacific. *Geol. Soc. Am. Mem. No. 143.* 134 pp.

Wilson, R. L. 1970. Permanent aspects of the Earth's nondipole magnetic field over upper Tertiary times. *Geophys. J. R. Astron. Soc.* 19: 417–37

Wilson, R. L. 1972. Palaeomagnetic differences between normal and reversed field sources, and the problem of far-sided and right-handed pole positions. *Geophys. J. R. Astron. Soc.* 28: 295–305

Winterer, E. L. 1973. Sedimentary facies and plate tectonics of the equatorial Pacific. *Am. Assoc. Pet. Geol. Bull.* 57: 265–82

Yumi, S., Wako, Y. 1970. Secular motion of the pole. In *Earthquake Displacement Fields and the Rotation of the Earth*, ed. L. Mansinha et al, pp. 82–87. Dordrecht, Neth: D. Reidel.

SUBJECT INDEX

A

Acid rain, 185–86
Advection
  and diagenesis, 157–59
African continental rifts, 445–91
Alaskan stress province, 561
Albert rift zone, 452
Albitization
  of detrital feldspars in sandstones, 143
Aleutian Arc
  and stress orientations, 553
Alexandria, 104
Algae
  eukaryotic, 54–55
Aliphatic hydrocarbons
  redistribution of, 381–82
Alkyl benzenes
  as biomarkers, 368
Alpide collisions
  and sea level, 237
Alpides, 219, 223
  alpinotype and germanotype, 227
  on the ruins of the Cimmerides, 231–32
Alpine-Himalayan ranges
  and Tethys, 213–15, 238
Aluminosilicate mineral stability
  and diagenesis, 152–53
Aluminosilicates
  as thermobarometers, 399–400
Andalusite
  as a thermobarometer, 399
Angular momentum
  conserving, 301
  of the Earth-Moon system, 275–76, 298, 300
  of material ejected from a large impact, 293–94, 298
  of the Moon, 275–76
  and planetary formation, 284
  and protomoons, 310
Annamia, 222, 226
Appalachian internides
  tectonics of, 337–58
Appalachians
  boundaries within, 340–41
  major faults in, 346–48

southern and central
  metamorphism in, 348–52
  plutons in, 352–54
  subdivisions of, 338–41
Apparent polar wander, 567
  hotspot, 574–75, 577–78, 587–89
Apparent polar wander path(s)
  hotspot, 581
  interpretation of, 584–86, 589
  mean-lithosphere, 573
  and polar wandering, 570
Aromatization
  of C_{29} monoaromatic steroid hydrocarbons, 385–88
Arrhenius equation, 386–87
Arrhenius parameters
  in basins, 389
  calculation of, 390
Arrhenius plot
  for C-22 bishomohopane isomerization, 387–88
Asteroidal impacts, 206–8
Asthenosphere
  and lithosphere, 428, 430–34, 571–72
Atmosphere
  of Venus, Earth, and Mars compared, 171–209
  composition and chemistry, 182–92
  origin and evolution, 192–209
  structure and circulation, 173–81
Atmospheric composition
  and biology
    on Venus, Earth, and Mars, 183–85
  and chemical cycles, 185–91
  global changes in
    on Venus, Earth, and Mars, 191–92
Atmospheric formation
  and volatile degassing, 201–6
Atom bomb test
  at Bikini Atoll, 11–13
Australia
  Otway Basin
    biomarkers in, 367–68
Autotrophic carbon fixation
  and the $\delta^{13}C_{org}$ record, 63–68

isotopic biogeochemistry of, 49–52
Autotrophic organisms
  isotopic composition of, 52–56
Axial geocentric dipole hypothesis
  and polar wandering, 581–84, 589–90

B

Bacteria
  anoxygenic photosynthetic, 55
  chemoautotrophic, 55–56
Bacteriohopane tetrol, 373
Barometry
  of igneous and metamorphic rocks, 397–415
Basalt
  and calcium and sodium metasomatism, 324–27
  hydrothermal alteration of, 318–19
  and heavy metal mobility, 327–32
  and interaction with seawater, 318–32
  and magnesium metasomatism, 321–24
Basaltic fire fountains, 89
Basement massifs
  in the Appalachian orogenies, 340
  in the central Appalachians, 346
Basin modeling
  and biomarkers, 384–90
Basins
  and continental rifts form elongated belts, 422–24
  interior
    typical sizes and shapes, 422
  and lithospheric thinning, 426–27
Benzohopanes, 371
Bikini Atoll
  atom bomb tests at, 11–13
Binary accretion
  and the origin of the Moon, 284–85

595

596 SUBJECT INDEX

Biodegradation
  of biomarkers, 382–84
Biology
  and atmospheric composition
    on Venus, Earth, and
    Mars, 183–85
Biomarkers
  and basin modeling, 384–90
  biodegradation of, 382–84
  and depositional environ-
    ments, 364–75
  maturation parameters, 377–
    80
  migration of, 380–82
  organic geochemistry of, 363–
    91
  source parameters, 375–77
  sulfur-containing, 372–76
Bishomohopane isomerization,
  385–88
Blocks
  Tethyan, 218–24
Blue Ridge-Piedmont thrust
  sheet, 347
  windows in, 343–46
Body waves
  and radial models of D'',
    29–33
  and seismic imaging, 116,
    133
  travel times
    and seismic imaging, 117–
      23
    and radial models of the
      core-mantle boundary,
      26–29
Borehole(s)
  breakouts
    and stress orientation, 554–
      57, 564
  and hydraulic fracturing, 547–
    48
Botryococcane
  in biomarkers, 367
Bouguer gravity anomalies, 426
Branch scales
  in the African rift system,
    446, 451, 460–62
Brasstown Bald window
  in the central Blue Ridge,
    343–44
Brevard fault zone, 341, 346–
  47, 348, 355
  prograde events affecting, 350

C

Calaveras fault, 128
Calcium metasomatism
  and geothermal mid-ocean
    ridges, 324–27

Calvin cycle photosynthesis, 51–
  55
  and ribulose-1,5-bisphosphate
    caroboxylase, 51, 53–56,
    65, 68
Carbon
  carbon isotopes
    and biochemical evolution,
      47–69
    isotopic geochemistry of, 56–
      57
  organic
    isotopic record of, 61–62
  organic and carbonate
    sedimentary reservoirs of,
      57–61, 66
Carbonate mineral stability
  and diagenesis, 152
Carbonate rocks
  oils from, 370–72
Carbonates
  isotopic studies of, 144
  sedimentary, 57–58
  siliceous
    and decarbonation reac-
      tions, 59
Carolina suspect terrane, 345
Chemical remanent magnetization (CRM)
  and marine magnetic anoma-
    lies, 531–32
Chert artifacts
  trace-element sourcing for,
    109
Chlorine cycles
  on Venus and Earth, 188–89
Chlorobiaceae
  in ancient restricted seas, 368
Cimmerian continent, 218–22,
  226, 229–31, 238
Cimmeride collisions
  and sea-level changes, 236–37
Cimmeride sutures, 229
Cimmerides
  alpinotype and germanotype,
    227
  orogenic zone of, 232
  and the Tethyside orogenic
    complex, 218
Clastic diagenesis, 141–67
Clasts
  in melt sheets, 253
Clouds
  on Venus, Earth, and Mars,
    173
Cometary impacts, 206–8
Cometary showers
  periodic, 265–67
Compaction
  and diagenesis, 158–59
Continental drift
  and polar wander, 586–87

Continental rifts
  African, 445–91
  and basins form elongated
    belts, 422–24
  see also Rifts
Convection
  and diagenesis, 158
  in the outer core, 38
Convective stability
  of the tropospheres of Venus,
    Earth, and Mars, 174–
    75
Copper
  provenance studies of North
    American, 109
  in ridge-crest hydrothermal
    fluids, 332, 334
Coral-algal/bryozoan-algal
  carbonate facies, 582–
    83
Core-mantle boundary, 25–42
  anelastic properties near
    and seismic constraints,
      35–36
  compositional models for, 39
  and lateral heterogeneity of
    D'', 33–35
  models of dynamics near, 39–
    41
  and radial models of D'',
    29–33
  and seismological constraints,
    25–36, 41
  and a thermal boundary layer,
    36–41
Coriolis force
  on Earth and Mars, 175, 178
  on Venus, 174
Cosmic rays
  and atmospheric N_2, 184
Crassulacean Acid Metabolism
  (CAM), 54
Cratering mechanics, 254–56
Cratering rates, 257, 266
  estimates, 261–63
Cretaceous normal (K-N) polarity bias superchron, 585–
    86, 591
Cretaceous-Tertiary boundary
  and impact-induced ex-
    tinctions, 264–66
Cretaceous-Tertiary impactor,
  207–8
Cretaceous-Tertiary-Quaternary
  mixed (KTQ-M) polarity
  superchron, 586, 591
Crust
  thermobarometry in, 406–10
Crustal stress field, 545–64
Cyanite
  as a thermobarometer, 399
Cyanobacteria, 55

SUBJECT INDEX 597

D

Deglaciation
 and polar wander, 587
Depositional environment(s)
 and biomarkers, 364–75
 hypersaline
  indicators for, 368–69
Diagenesis
 chemical modeling of, 145
 clastic, 141–67
 clay
  of the Gulf Coast Tertiary section, 143
 and hydrocarbons, 165–67
 and isotopic studies, 144–45
 and kerogen, 147–49, 151, 162–63
 and mass balance, 142–45
 and mass transfer, 156–60
 and organic/inorganic interactions, 145–56
 predicting, 164–65
 reaction studies, 143–44
 and redox reactions, 149–51
 and specific basin models, 160–64
Diagenetic processes
 and isotopic compositions of sedimentary carbon species, 57–58, 61
Diapiric penetration models
 of lithospheric thinning, 434–36
Diapirism, 74–75
Diasteranes
 in carbonate source oils, 370–71
Diffusion
 and diagenesis, 156–57
Dipole hypothesis
 axial geocentric, 581–84, 589–90
Disk evolution, 300–7
Domal uplifts
 and lithospheric thinning, 422, 441
Domes
 and lithospheric thinning, 424, 426

E

Earth
 atmosphere of
  compared with Venus and Mars, 171–209
 atmospheric structure and circulation on, 173–81
 and giant impacts, 311

Earth filter
 and marine magnetic anomalies, 507, 509–10
Earthquake hypocenters, 545
Earthquake mechanisms
 and stress investigation, 548–52, 557
Earthquakes
 stress drops inferred from, 563
East Pacific Rise, 319, 326–27, 330–33
Epeirogeny
 and lithospheric thinning, 422–24
Equatorial bulge
 of the Earth, 568–69
Eurasia
 tectonic subdivisions of, 216–18
Evolution
 biochemical
  and carbon isotopes, 47–69
 terrestrial
  and large-scale impact, 263–67
Extinctions
 mass
  in the geologic record, 264–67
  periodic, 266

F

Fault(s)
 in the Appalachians, 346–48
 and half-graben geometries, 471, 473
 Hayesville-Fries, 341
 and stress orientations, 552–53
 transform
  and the African rift system, 460–61
  and the Mendocino Fracture Zone, 19, 22
Fault scarps
 rocks from the oceanic crust at, 528
Fault system(s)
 border
  and rifting, 488
 Piedmont, 345, 348, 357
Fault zones
 Brevard, 341, 346–48, 355
 and seismic imaging, 127–28
Faulting
 and stress, 546
 strike-slip, 233, 236
Fire fountains
 basaltic, 89
Fischer-Tropsch reactions, 68

Fission
 and the origin of the Moon, 282–83
Fossil $\delta^{13}C_{org}$ record, 65
Fossil
 chemical, 364
Fourier analysis
 of magnetic anomalies, 508–9
Fries fault, 341, 347

G

Gabbro(s), 528
 magnetization of, 506
 oceanic, 532
 serpentinization of, 539
Gammacerane
 as a biomarker, 369
Gas(es)
 and pyroclasts, 83–89
 rare
  on Venus, Earth, and Mars, 194–95
Geoarchaeology, 97–111
 history of, 97–99
Geoid
 two large highs on, 588–89
Geology
 of Tethyside blocks, 224
Geomagnetic constraints
 and the outermost core, 38
Geothermal areas
 and seismic imaging, 128–31
Geotherms
 and lithospheric thinning, 427–28
Germany
 stress field in, 561–62
Geysers–Clear Lake region
 and seismic imaging, 128–29
Gondwana-Land, 215–18, 226–27, 229–30, 238
Goochland terrane, 345–48, 358
 metamorphism in, 350
Graben
 full
  in African rifts, 458–60, 470
 half
  in African rifts, 458–60, 462–68
  geometries of linked, 468–79
 and rifting, 488
Grandfather Mountain window
 in the southern Appalachians, 343, 347
Grätz number
 and lava flows, 82
Gravity gradient
 in the Appalachians, 340

SUBJECT INDEX

Great Lakes
  fluctuations in, 106
Greenhouse effect
  on Earth, 192
  on Earth, Mars, and Venus, 172
  on Venus, 174
Gregory rift zone, 452–53

H

Hawaii
  and seismic imaging, 128–29
Hayesville fault, 341, 347
Heat flow
  through the seafloor, 17–18
Heim's rule, 557–58
Herodotus, 104
Holocene coastal change, 102
Hotspot(s)
  Hawaiian
    shift of, 583–84
  plate
    motions between, 579–80, 589
    and polar wandering, 573–80, 585–86, 589, 591
Huanan block, 221–22, 226
Hydraulic fracturing
  and stress measurements, 547–48, 557
Hydrocarbons
  and diagenetic modeling, 165–67
Hydrothermal alteration
  at mid-ocean ridges, 317–34

I

Iceland
  stress-relief measurements in, 563
Igneous activity
  and lithospheric thinning, 424–25, 441
  see also Volcanism; Volcanology
Igneous intrusion model
  of lithospheric thinning, 436–41
Igneous intrusions
  and stress orientations, 553–54
Igneous rocks
  thermometry and barometry of, 397–415
Impact hypothesis
  of lunar origin, 271–312
Impact structures
  characteristics, 246–54
  currently known, 256–61, 266
  formational processes, 254–56

morphology, 246–50
  submarine, 257
  terrestrial, 245–67
Impacts
  large, 206–8
  and orbital injection, 293–98
  physics of, 285–98
  and terrestrial evolution, 263–67
  and vaporization, 288–93, 310
  thermodynamics of, 286–93
Indus
  Holocene evolution of, 104
Instantaneous shut-in pressure (ISIP)
  and hydraulic fracturing, 547
Io
  explosive eruptions on, 80
  pyroclast dispersion on, 83–85
  volatile component of, 276
Iron
  and hydrothermal alteration processes, 327–34
  in the mantle and core, 39
  on the Moon, 284
  as Moon-forming matter, 307–8
Iron Age soils
  in Denmark, 101–2
Isomerization
  in basins
    and steroid aromatization, 385–89
  of C-20 of 5α-24-ethylcholestane, 384–88
  of C-22 of 17α, 21β-bishomohopane, 385–88
Isoprenoids
  C_{25}, 367
  pristane, 366–67
  phytane, 366–67
Isotope(s)
  carbon
    and biochemical evolution, 47–69
  fractionations, 49–52
  in the geologic past, 56–68
  stable
    and provenance studies, 110
Isotopic studies
  and diagenesis, 144–45

J

Japan
  subduction-zone studies in, 130–31
Jovian satellite system, 275–76

K

Kenya rift zone, 453
Kerogen
  and diagenesis, 147–49, 151, 162–63
  formation of, 58–59
  isotopic record of, 61–62
  in Ordovician Goldwyer Formation, 371
  Precambrian, 65–66
  sulfer as a cross-linking agent in, 373
  three types, 147–48
Kilauea volcano
  and seismic imaging, 128–30

L

Latitudinal motion
  of the Tethysides, 232–33
Laumontite cementation
  and volcanogenic sandstones, 143–44
Laurasia
  and the Cimmerian continent, 230
Lightning
  and atmospheric N_2, 183–85
Lithosphere
  and asthenosphere, 428, 430–34, 571–72
  mean-lithosphere test
    and polar wandering, 570–73, 578–79, 589, 591
  oceanic, 236, 426
  oceanic and continental, 421
  viscous flow in the, 428
Lithospheric thinning, 421–41
  diapiric penetration models, 434–36
  and epeirogeny, 422–24
  geophysics of, 424–27
  and igneous activity, 424–25, 441
  igneous intrusion model, 436–41
  reheating models of, 432–33
  source injection models, 433–34
  stretching models of, 429–31
Lithostatic flexure model
  of basins, 389–90
Longitudinal motion
  of the Tethysides, 233
Luama rift zone, 452
Lunar origin
  collisional hypothesis of, 271–312
  ten propositions assessing, 272–75, 309–10
  see also Moon

SUBJECT INDEX 599

Lunar basins, 264
see also Moon

M

Magma(s)
  availability and eruption onset, 73–76
  chamber evolution, 75
  cooling and heating in conduits, 79
  and dike propagation, 75–76
  mafic-to-ultramafic
    and lithospheric thinning, 429
  motion of conduits, 77
  rheology of, 76–77
  segregation
    and the lower lithosphere, 424
  subsurface motion, 76–79
Magma ocean(s)
  on the Moon, 277, 284
  in the postimpact Earth, 292–93, 310
Magnesium
  and geothermal mid-ocean ridges, 333
  metasomatism
    and geothermal mid-ocean ridges, 321–24
  removal from seawater, 319–21
Magnetic anomalies
  deep-tow studies of, 519–23
  inversions of, 508–11, 538
  marine, 505–39
  phase shifting of, 512–13
Magnetic field
  reversal of the Earth's, 22, 505
Magnetic striations
  on the seafloor, 22
Malawi rift zone, 447, 451
Manganese
  and hydrothermal alteration processes, 327–34
Mantle
  convection, 588
  core-mantle boundary, 25–42
  of the Moon and the Earth, 307
  upper
    overheating, 588–89
    thermobarometry in, 403–6
Mantle flow
  and seismic imaging, 134
Margins
  passive
    and rift sequences, 486
Marine magnetic anomalies, 505–39

  early models of lineations, 506–8
  long-wavelength component, 523–28
  and magnetic measurements of oceanic rocks, 528–38
  model calculations, 516–19
  skewness calculations, 511–16, 538
Mars
  atmosphere of
    compared with Earth and Venus, 171–209
  atmospheric structure and circulation on, 173–81
  explosive eruptions on, 80
  and giant-impact effects, 311
  rille-forming eruptions on, 83
Mass balance
  and diagenesis, 142–45
Mass transfer
  and diagenesis, 156–60
Mean lithosphere
  and plate-hotspot reference frames, 580–81
Mean-lithosphere apparent polar wander
  and hotspot apparent polar wander, 574–75
Mean-lithosphere test
  and polar wandering, 570–73, 578–79, 589, 591
Melt formation, 73–74
Melt rocks
  impact, 252–54
Mesosphere
  plate motion relative to, 567, 572
Metamorphic rocks
  thermometry and barometry of, 397–415
Metamorphism
  and isotopic content of sedimentary species, 68
  and isotopic effects in sedimentary carbons, 58–61, 66
  and marine magnetic anomalies, 532, 539
  shock
    and terrestrial impact studies, 250–54
  in the southern and central Appalachians, 348–52
Metasomatism
  calcium and sodium
    and geothermal mid-ocean ridges, 324–27
  magnesium
    and geothermal mid-ocean ridges, 321–24

Meteoric water
  and diagenesis, 159
Meteorites
  CO_2 and N_2 in, 199–200
  interstellar material in, 196
Microfossils
  cellular, 67–68
Mid-ocean ridges
  hydrothermal alteration at, 317–34
Migration routes
  of Tethyside blocks, 224–26
Miller-Urey-type spark discharge syntheses, 68
Mineral/sterane pairs
  and adsorption free energies, 381
Miscibility gap, 413
Monoaromatic steroid hydrocarbon aromatization, 385–88
Moon
  bulk chemistry, 276
  explosive eruptions on, 80
  mass and angular momentum, 275–76
  orbital evolution, 277–78
  origin of
    and binary accretion, 284–85
    chemistry of the impact hypothesis, 307–9
    and capture, 283–84
    collision hypothesis, 271–312
    and fission, 282–83
    numerical simulations of, 298–300
    and planetesimal impacts, 206–8
  primordial high temperatures on the, 277
  rille-forming eruptions on, 83
  trace elements on the, 276–77
  volatile depletion on the, 276–77
  vulcanian explosions on, 85
Mylonites
  in the southern Appalachians, 346–47

N

n-alkanes
  as biomarkers, 365–67, 371
  and crude oil biodegradation, 382–84
  migration of, 381
Natural remanent magnetization (NRM)
  and marine magnetic anomalies, 529

Nile
  Holocene evolution of the, 104
25-norhopanes
  and biodegraded oils, 382–84
North American Cratonic stress province, 560–61
North American-hotspot motion, 573
North China block, 224–25
North China fold belt, 221
Nuclear magnetic resonance (NMR)
  studies of kerogen, 147–48
Numerical simulations
  of an impact origin of the Moon, 298–300
  of planetary formation, 280–82

O

Obsidian artifacts,
  trace-element sourcing for, 108–9
Oceanic crust
  magnetic model of, 536–39
Oceans
  Tethyan, 215, 230–31
Oil-field waters
  organic species in, 145–47, 149, 162
Oort cloud, 265–66
Ophiolites, 528
  magnetization of, 506
Orbital angular momentum vectors
  of Earth and Mars, 173–74
Orbital injection, 310
  and large impacts, 293–98
Orbital evolution
  of the Earth and Moon, 277–78
Ordovician Goldwyer Formation, 371–72
Orogen
  Appalachian, 337–58
Orogenic collage
  Tethyside, 213–38
Orogenic complex
  Cimmeride, 231
Orogenic zone
  Alpide, 232
  of the Cimmerides, 232
Oscillations
  free
    and radial models of the core-mantle boundary, 26, 28–29
    and seismic imaging, 116, 133–34

Oxidation cycles
  on Venus, Earth, and Mars, 189–91

P

Pacific basin plates
  motion of, 579–80, 589–90
Pacific plate, 130–31
  and stress orientations, 553, 561
Pacific trenches, 20–21
Paleobiogeography
  and Tethyside blocks, 224, 226
Paleoclimatology
  and Tethyside blocks, 224, 226
Paleogeomorphology
  reconstructing, 102–8
Paleomagnetism
  and polar wandering, 567–91
  and Tethyside blocks, 224, 226
Pangea
  and Tethys, 214–15, 228, 238
Peridotite
  serpentinized
    in layer 3, 532–33
Permo-Carboniferous reversed (PC-R) superchron, 585
Phase shifting
  of magnetic anomalies, 512–13
Phosphate
  inorganic
    in archaeological sediments, 101
Phosphorus
  and terrestrial biomass, 67
Photosynthesis
  Calvin cycle, 51–55
  C4, 54, 63, 65
  inhibition of, 207
Phytane
  and isoprenoids pristane, 366–67
Phytoliths
  in archaeological sediments, 102
Piedmont
  fault system, 348, 357
  Inner
    plutons in, 352
    metamorphism in, 350
    suture, 344–46
    thrust sheet
      windows in, 343–46
Pine Mountain window, 344, 347
Planet-satellite systems, 275
Planetary formation, 278–82

Planetesimals
  and planetary formation, 278–80, 284–85
Plate
  Pacific, 130–31
  and stress orientations, 553, 561
Plate hotspots
  and mean-lithosphere reference frames, 580–81
  motions between, 579–80, 589
  and polar wandering, 573–80, 585–86, 589, 591
Plate motion(s)
  lithospheric, 130, 134
  and polar wandering, 570–73, 585–86, 590–91
  relative to the mesosphere, 567, 572
Plate tectonics
  and polar wandering, 569–70
  of the southern and central Appalachian orogen, 354–58
  and the Tethys, 214
Plinian eruptions, 86–87
Plutons
  in the southern and central Appalachians, 352–54
Polar wandering
  and axial geocentric dipole hypothesis, 581–84, 589–90
  causes of, 586–89
  and interpretation of apparent wander paths, 584–86, 589
  and the mean-lithosphere test, 570–73, 578–79, 589, 591
  measuring, 569–81, 589
  and paleomagnetism, 567–91
  and plate-hotspot motions, 573–80, 585–86, 589, 591
  and polarity bias, 585–86
  and the vector-sum test, 570–71
Pristane/phytane ratios, 366–67
Provenance
  of Tethyside blocks, 224–26
Provenance studies, 108–10
Ptolemy, 104
Pyroclast dispersion
  in explosive eruptions, 80, 83–90
Pyroclast formation
  and volatile exsolution, 78–79
Pyroclastic flows, 87–88
Pyroclastic surges, 89

SUBJECT INDEX 601

Q

Quartz-coesite equilibrium, 401–2

R

Raleigh belt
 in North Carolina, 345, 347–48
Reading Prong
 basement massif, 346
Redox reactions
 and diagenesis, 149–51
Rheology
 and collisional lunar origin, 285, 293–95
Ribulose-1,5-bisphosphate carboxylase
 and Calvin cycle photosynthesis, 51, 53–56, 65, 68
Rift branches
 divided into rift zones, 451
Rift zones
 Cenozoic
  of East Africa, 447, 451–53
 and linked half-graben, 468–79
 and rift block scale, 454, 457
 and rift-unit scale, 453–54
Rifting
 active and passive, 483–86
 evolutionary scheme of, 486–87
Rifts
 African
  cross-sectional morphology, 458–60
  continental, 445–91
  and lithospheric thinning, 424–26
  nomenclature for, 446–57
  prerift fabrics, 480–82
  and volcanism, 482–83, 485
Rocks
 igneous and metamorphic
  thermometry and barometry of, 397–415
 melt, 252–54
 oceanic
  magnetic properties of, 528–38
 sedimentary
  carbonate and organic carbon in, 57–61, 66
Rocky Mountain Trench–Tintina fault
 and stress orientations, 552
Rotation rates
 of Venus, Earth, and Mars, 174

Rukwa rift zone, 452

S

San Andreas fault, 127–28
 and stress orientations, 552
San Andreas stress province, 561
Satellites
 studies of magnetic anomalies by, 526–28
Saturation surfaces
 and thermometry, 415
Saturnian satellites, 312
Sauratown Mountains window, 344, 347–48
Scandinavia
 stress-relief measurements in, 563
Sea-level changes, 104, 236–37
 and collisions in the Tethysides, 238
Seafloor
 heat flow through, 17–18
 spreading, 505
Seamount(s), 13, 16–17
 Emperor, 583
 Suiko, 576, 583
Seasons
 on Earth and Mars, 173–74
Seawater-basalt interaction, 318–32
Sediments
 archaeological, 99–102
 in the Pacific, 17, 20–21
Seismic imaging
 and body-wave travel times, 117–23
 and fault zones, 127–28
 and medical tomography, 115
 methodology, 116–27
 regional and global studies, 133–34
 and subduction zones, 130–32
 and surface-wave dispersion, 123–27
 and teleseismic methods, 118–20
 three-dimensional, 115–35
Seismic reflection,
 and African rift zones, 459
Seismic studies
 in the Pacific, 17
Seismic waves
 three classes of, 116
Seismological constraints
 and the core-mantle boundary, 25–36, 41
Serpentinite(s), 528
 magnetization of, 506

Serpentinization
 and marine magnetic anomalies, 532–34, 539
Shock metamorphism
 and terrestrial impact studies, 250–54
Shooting Creek window
 in the central Blue Ridge, 343–44
Siderophile patterns
 in the Earth and Moon, 277
Silica (SiO_2)
 and entropy production in a shock event, 288–91
 as a thermobarometer, 400–1
Sillimanite
 as a thermobarometer, 399
Skewness calculations
 of marine magnetic anomalies, 511–16, 538
Smoothed particle hydrodynamics (SPH)
 and numerical simulations of an impact origin of the Moon, 299
Sodium metasomatism
 and geothermal mid-ocean ridges, 324–27
Solar nebula
 capture of gas from, 194–95
 volatile retention in, 195–201
Sound speed
 and large impacts, 304–6
 in seismic studies in the Pacific, 17
Spin angular momentum vectors
 of Earth and Mars, 173–74
Spin axes
 planetary, 282
Spin axis
 of the Earth, 567–69, 577, 581, 586–88, 590
Sterane(s)
 biodegradation of, 382, 384
 and biomarker studies, 376–77
 maturity indicators, 378–79
 mineral/sterane pairs
  and adsorption-free energies, 381
Stereoisomers
 biomarker, 377–80
Steroid aromatization
 and isomerization in basins, 385–89
Strain-relief measurements
 of the Earth, 546–47, 557
Stratospheric temperature
 and zonal winds on Earth and Mars, 175–78

Stress
  and borehole breakouts, 554–57, 564
  and earthquake mechanisms, 548–52, 557
  and igneous intrusions and volcanic vents, 553–54
  and observation of active faults, 552–53
Stress field
  crustal, 545–64
  near-surface, 557–64
  and strain-relief measurements of the Earth, 546–47, 557
Stress measurements
  and hydraulic fracturing, 547–58, 557
Stress provinces, 560–61
Strike-slip motion
  of the Tethysides, 233, 236
Subducting slabs
  and seismic imaging, 133
Subduction zone(s), 236
  Alpide, 237
  and polar wandering, 587–88
  and seismic imaging, 130–32
Subsidence
  and rifting, 484–85
Sulfur-containing biomarkers, 372–76
Sulfur cycles
  on Earth and Mars, 185–88
Sulfur isotopic studies
  and diagenesis, 145, 151
Superchrons, 585–86
Surface waves
  dispersion
    and seismic imaging, 123–27
  and seismic imaging, 116, 133–34
Suture(s)
  central Piedmont, 344–46
  Cimmeride, 229
  Tethyan, 217–19
  and the Tethysides, 229–31, 233
Suture-parallel strike-slip faulting, 233, 236
Sverdrup, Harald, 7–8

T

Tanganyika rift zone, 452, 486, 488
Tectonic subdivisions
  of Eurasia, 216–18
Tectonics
  of the southern and central Appalachian internides, 337–58

  of the Tethysides, 213–38
Tectosilicates
  shock-metamorphism studies of, 250–52
Teleseismic methods
  and seismic imaging, 118–20
Terpanes
  biodegradation of, 382, 384
Tethyan blocks, 218–24
Tethyan orogens, 226–27
Tethyan paradox, 214–15
Tethyan sutures, 217–19
Tethys
  paleobiogeographic and tectonic concept, 213–14
Tethyside blocks
  provenance and migration routes, 224–26
Tethysides
  tectonics of, 213–38
Tetracyclic diterpenoids
  in oils, 379
Texas
  faults in, 552–53
Thermal boundary layer
  and the core-mantle boundary, 36–41
Thermal wind equation
  on Mars and Earth, 178
Thermobarometers, 398
  multivariant, 401–10
  univariant, 399–401
Thermobarometry
  in the crust, 406–10
  solvus, 413–15
  in the upper mantle, 403–6
Thermodynamics
  and collisional lunar origin, 285
  of impacts, 286–93
Thermometers
  exchange
    in geothermometry, 411–13
Thermometry
  of igneous and metamorphic rocks, 397–415
  and saturation surfaces, 415
Thermopylae
  paleogeography of, 105–7
Thermoremanent magnetization (TRM)
  and marine magnetic anomalies, 531–32
Thienylhopane
  as a biomarker, 373
Thiophenic organosulfur compounds
  in oils, 372–73
Thompson, Sir William, 271
Thrust sheet(s)
  of the Appalachians, 348

Blue Ridge-Piedmont, 347
  windows in, 343–46
Tidal dissipation
  in the Earth and Moon, 278
Tigris-Euphrates
  Holocene evolution of, 104
Tonga Trench, 20
Trace-element fingerprinting, 108–9
Trace elements
  on the Moon, 276–77
Tricyclic terpane(s), 370
  in oils, 379–80
Troposphere
  of Venus, Earth, and Mars, 174–75
Tropospheric temperature
  and zonal winds on Earth and Mars, 175–78
Troy
  archaeological geology of, 104–5
Tunguska impact, 207
Turkana rift zone, 453

U

Ultraviolet bombardment
  and atmospheric N_2, 184
Ur
  Sumerian city, 104
Uranian system
  and giant-impact effects, 311–12

V

Vaporization
  in large impacts, 288–93, 310
Vector-sum test
  and polar wandering, 570–71
Venus
  atmosphere of
    compared with Earth and Mars, 171–209
  atmospheric structure and circulation on, 173–81
  explosive eruptions on, 80
  and giant impacts, 311
  loss of water from, 208–9
Viscosity
  artificial, 299
  of lava, 82
  of magma, 76–77
  and the physics of large impacts, 293–95
Viscous flow
  in the lithosphere, 428
Volatile(s)
  degassing
    and atmospheric formation, 201–6

SUBJECT INDEX   603

depletion of
  on the Moon, 276–77
  on Venus, Earth, and Mars, 194
exsolution
  and pyroclast formation, 78–79
  retention of
    by solid grains in solar nebula, 195–201
Volcanic areas
  and seismic imaging, 128–31
Volcanic eruption
  on Venus, 192
Volcanic vents
  and stress orientations, 553–54
Volcanism
  impact-induced, 264
  and lithospheric thinning, 424
  and rifting, 482–83, 485
  see also Igneous activity
Volcanoes
  on Venus, 187
Volcanology
  atmospheric influences, 85–86
effusive eruptions, 80–83
eruption styles, 79–80
magma availability and eruption onset, 73–76
physical, 73–90
and pyroclast dispersion, 83–90
see also Igneous activity

W

Water
  loss of
    from Venus, 208–9
    of Venus and Earth, 195–97
Waves
  body
    and seismic imaging, 116–23, 133
  Love, 123
  quasi-geostrophic
    on Earth and Mars, 178–79
  Rayleigh, 123
  surface
    and seismic imaging, 116, 123–27, 133–34
Wilson cycle
  of the southern and central Appalachians orogen, 354–58
Winds
  on Earth and Mars, 178–79
  zonal
    on Earth and Mars, 175–78
    on Venus, 179–81

Y

Yangtze block, 221, 224, 226
Yellowstone
  and seismic imaging, 128–29

Z

Zambeze Rift
  and stress orientations, 552
Zinc
  in ridge-crest hydrothermal fluids, 332, 334
Zone scales
  in the African rift system, 460–62

CUMULATIVE INDEXES

CONTRIBUTING AUTHORS VOLUMES 1–15

A

Abelson, P. H., 6:325–51
Aki, K., 15:115–39
Alyea, F. N., 6:43–74
Anderson, D. E., 4:95–121
Anderson, D. L., 5:179–202
Andrews, J. T., 6:205–28
Anhaeusser, C. R., 3:31–53
Apel, J. R., 8:303–42
Arculus, R. J., 13:75–95
Armstrong, R. L., 10:129–54
Arnold, J. R., 5:449–89
Axford, W. I., 2:419–74

B

Bada, J. L., 13:241–68
Bambach, R. K., 7:473–502
Banerjee, S. K., 1:269–96
Banks, P. M., 4:381–440
Barnes, I., 1:157–81
Barry, R. G., 6:205–28
Barth, C. A., 2:333–67
Barton, P. B. Jr., 1:183–211
Bassett, W. A., 7:357–84
Bathurst, R. G. C., 2:257–74
Behrensmeyer, A. K., 10:39–60
Benninger, L. K., 5:227–55
Benson, R. H., 9:59–80
Bergström, S. M., 14:85–112
Bhattacharyya, D. B., 10:441–57
Birch, F., 7:1–9
Bishop, F. C., 9:175–98
Black, R. F., 4:75–94
Blandford, R., 5:111–22
Bodnar, R. J., 8:263–301
Bohlen, S. R., 15:397–420
Bonatti, E., 3:401–31
Bottinga, Y., 5:65–110
Brewer, J. A., 8:205–30
Brown, L., 12:39–59
Browne, P. R. L., 6:229–50
Brownlee, D. E., 13:147–73
Bryan, K., 10:15–38
Buland, R., 9:385–413
Bullard, E., 3:1–30
Burdick, L. J., 7:417–42
Burke, D. B., 2:213–38
Burke, K., 5:371–96
Burnett, D. S., 11:329–58

Burnham, C. W., 1:313–38
Burnham, L., 13:297–314
Burns, J. A., 8:527–58
Burns, R. G., 4:229–63; 9:345–83
Burst, J. F., 4:293–318
Busse, F. H., 11:241–68

C

Cane, M. A., 14:43–70
Carpenter, F. M., 13:297–314
Carter, S. R., 7:11–38
Castleman, A. W. Jr., 9:227–49
Champness, P. E., 5:203–26
Chapman, C. R., 5:515–40
Chase, C. G., 3:271–91; 13:97–117
Chave, K. E., 12:293–305
Chen, C.-T. A., 14:201–35
Chou, L., 8:17–33
Clark, D. R., 5:159–78
Clark, G. R. II, 2:77–99
Claypool, G. E., 11:299–327
Cluff, L. S., 4:123–45
Coroniti, F. V., 1:107–29
Crompton, A. W., 1:131–55
Crossey, L. J., 15:141–70
Crough, S. T., 11:165–93
Crutzen, P. J., 7:443–72
Cruz-Cumplido, M. I., 2:239–56
Cunnold, D. M., 6:43–74

D

Daly, S. F., 9:415–48
Damon, P. E., 6:457–94
Decker, R. W., 14:267–92
Dieterich, J. H., 2:275–301
Domenico, P. A., 5:287–317
Donahue, T. M., 4:265–92
Donaldson, I. G., 10:377–95
Drake, E. T., 14:201–35
Duce, R. A., 4:187–228
Durham, J. W., 6:21–42

E

Eaton, G. P., 10:409–40
Eugster, H. P., 8:35–63

Evans, B. W., 5:397–447
Evensen, N. M., 7:11–38

F

Farlow, N. H., 9:19–58
Fegley, B. Jr., 15:171–212
Filson, J., 3:157–81
Fischer, A. G., 14:351–76
Fripiat, J. J., 2:239–56

G

Garland, G. D., 9:147–74
Gay, N. C., 6:405–36
Gibson, I. L., 9:285–309
Gierlowski, T. C., 13:385–425
Gieskes, J. M., 3:433–53
Gilluly, J., 5:1–12
Gingerich, P. D., 8:407–24
Gordon, R. G., 15:567–93
Gough, D. I., 15:545–66
Gough, W. I., 15:545–66
Graf, D. L., 4:95–121
Grant, T. A., 4:123–45
Grey, A., 9:285–309
Grieve, R. A. F., 15:245–70
Gross, M. G., 6:127–43
Grossman, L., 8:559–608
Grove, T. L., 14:417–54
Gueguen, Y., 8:119–44
Gulkis, S., 7:385–415

H

Haggerty, S. E., 11:133–63
Hales, A. L., 14:1–20
Hallam, A., 12:205–43
Hamilton, P. J., 7:11–38
Hanson, G. N., 8:371–406
Hargraves, R. B., 1:269–96
Harms, J. C., 7:227–48
Harris, A. W., 10:61–108
Harrison, C. G. A., 15:505–43
Hart, S. R., 10:483–526; 14:493–571
Hatcher, R. D. Jr., 15:337–62
Hay, W. W., 6:353–75
Heirtzler, J. R., 7:343–55
Helmberger, D. V., 7:417–42
Hem, J. D., 1:157–81
Herron, E. M., 3:271–91

CONTRIBUTING AUTHORS

Hinze, W. J., 13:345–83
Hoffman, E. J., 4:187–228
Holman, R. A., 14:237–65
Holton, J. R., 8:169–90
Housen, K. R., 10:355–76
Howell, D. G., 12:107–31
Hsü, K. J., 10:109–28
Hubbard, W. B., 1:85–106
Hulver, M. L., 13:385–425
Hunt, G. E., 11:415–59
Hunten, D. M., 4:265–92
Huppert, H. E., 12:11–37

J

James, D. E., 9:311–44
Javoy, M., 5:65–110
Jeanloz, R., 14:377–415
Jeffreys, H., 1:1–13
Jenkins, F. A. Jr., 1:131–55
Johns, W. D., 7:183–98
Johnson, M. E., 7:473–502
Johnson, N. M., 12:445–88
Johnson, T. C., 12:179–204
Johnson, T. V., 6:93–125
Jones, D. L., 12:107–31; 14:455–92
Jones, K. L., 5:515–40

K

Kanamori, H., 1:213–39; 14:293–322
Karig, D. E., 2:51–75
Keesee, R. G., 9:227–49
Kellogg, W. W., 7:63–92
Kinzler, R. J., 14:417–54
Kirkpatrick, R. J., 13:29–47
Kistler, R. W., 2:403–18
Koeberl, C., 14:323–50
Komar, P. D., 14:237–65
Koshlyakov, M. N., 6:495–523
Kröner, A., 13:49–74
Ku, T.-L., 4:347–79
Kvenvolden, K. A., 3:183–212; 11:299–327

L

Langevin, Y., 5:449–89
Lay, T., 15:25–46
Leckie, J. O., 9:449–86
Lerman, A., 6:281–303
Lerman, J. C., 6:457–94
Levine, J. S., 5:357–69
Levy, E. H., 4:159–85
Lewis, B. T. R., 6:377–404
Lewis, C. A., 15:363–95
Lick, W., 4:49–74; 10:327–53
Lilly, D. K., 7:117–61
Lindsay, E. H., 12:445–88
Lindsley, D. H., 15:397–420
Lindzen, R. S., 7:199–225

Lion, L. W., 9:449–86
Lister, C. R. B., 8:95–117
Long, A., 6:457–94
Lottes, A. L., 13:385–425
Lowe, D. R., 8:145–67
Lowenstein, J. M., 14:71–83
Lupton, J. E., 11:371–414

M

Macdonald, K. C., 10:155–90
Macdonald, R., 15:73–95
Malin, M. C., 12:411–43
Margolis, S. V., 4:229–63
Mavko, G. M., 9:81–111
McBirney, A. R., 6:437–56; 12:337–57
McConnell, J. C., 4:319–46
McConnell, J. D. C., 3:129–55
McGarr, A., 6:405–36
McKerrow, W. S., 7:473–502
McNally, K. C., 11:359–69
Melosh, H. J., 8:65–93
Mendis, D. A., 2:419–74
Mercer, J. H., 11:99–132
Millero, F. J., 2:101–50
Miyashiro, A., 3:251–69
Mogi, K., 1:63–94
Mohr, P., 6:145–72
Molnar, P., 12:489–518
Monin, A. S., 6:495–523
Morris, S., 14:377–415
Munk, W. H., 8:1–16
Murase, T., 12:337–57
Murchey, B., 14:455–92
Murthy, V. R., 13:269–96
Mysen, B. O., 11:75–97

N

Nairn, A. E. M., 6:75–91
Navrotsky, A., 7:93–115
Naylor, R. S., 3:387–400
Ness, N. F., 7:249–88
Neugebauer, H. J., 15:421–43
Newburn, R. L. Jr., 10:297–326
Nicolas, A., 8:119–44
Nier, A. O., 9:1–17
Nixon, P. H., 9:285–309
Normark, W. R., 3:271–91
Norton, D. L., 12:155–77
Nozaki, Y., 5:227–55
Nunn, J. A., 8:17–33

O

Okal, E. A., 11:195–214
Oliver, J. E., 8:205–30
O'Nions, R. K., 7:11–38
Opdyke, N. D., 12:445–88
Ostrom, J. H., 3:55–77

P

Pálmason, G., 2:25–50
Palmer, A. R., 5:13–33
Park, C. F. Jr., 6:305–24
Parker, R. L., 5:35–64
Pasteris, J. D., 12:133–53
Peltier, W. R., 9:199–225
Philander, S. G. H., 8:191–204
Phillips, R. J., 12:411–43
Philp, R. P., 15:363–95
Pilbeam, C. C., 3:343–60
Pinkerton, H., 15:73–95
Pittman, E. D., 7:39–62
Poag, C. W., 6:251–80
Pollack, H. N., 10:459–81
Pollack, J. B., 8:425–87
Prinn, R. G., 6:43–74; 15:171–212

R

Raitt, W. J., 4:381–440
Rapp, G. Jr., 15:97–113
Reeburgh, W. S., 11:269–98
Ressetar, R., 6:75–91
Revelle, R., 15:1–23
Richet, P., 5:65–110
Richter, F. M., 6:9–19
Ridley, W. I., 4:15–48
Riedel, W. R., 1:241–68
Roden, M. F., 13:269–96
Rodgers, J., 13:1–4
Roedder, E., 8:263–301
Rogers, N. W., 9:285–309
Rosendahl, B. R., 15:445–503
Ross, R. J. Jr., 12:307–35
Rothrock, D. A., 3:317–42
Rowley, D. B., 13:385–425
Rubey, W. W., 2:1–24
Rudnicki, J. W., 8:489–525
Rumble, D. III, 10:221–33
Russell, D. A., 7:163–82

S

Sachs, H. M., 5:159–78
Saemundsson, K., 2:25–50
Sahagian, D. L., 13:385–425
Saleeby, J. B., 11:45–73
Savage, J. C., 11:11–43
Savin, S. M., 5:319–55
Schermer, E. R., 12:107–31
Schidlowski, M., 15:47–72
Schnitker, D., 8:343–70
Schopf, J. W., 3:213–49
Schopf, T. J. M., 12:245–92
Schramm, S., 13:29–47
Schubert, G., 7:289–342
Schumm, S. A., 13:5–27
Schunk, R. W., 4:381–440
Scotese, C. R., 7:473–502
Sekanina, Z., 9:113–45

CHAPTER TITLES VOLUMES 1–15

PREFATORY CHAPTERS

Developments in Geophysics	H. Jeffreys	1:1–13
Fifty Years of the Earth Sciences—A Renaissance	W. W. Rubey	2:1–24
The Emergence of Plate Tectonics: A Personal View	E. Bullard	3:1–30
The Compleat Palaeontologist?	G. G. Simpson	4:1–13
American Geology Since 1910—A Personal Appraisal	J. Gilluly	5:1–12
The Earth as Part of the Universe	F. L. Whipple	6:1–8
Reminiscences and Digressions	F. Birch	7:1–9
Affairs of the Sea	W. H. Munk	8:1–16
Some Reminiscences of Isotopes, Geochronology, and Mass Spectrometry	A. O. Nier	9:1–17
Early Days in University Geophysics	J. T. Wilson	10:1–14
Personal Notes and Sundry Comments	J. Verhoogen	11:1–9
The Greening of Stratigraphy 1933–1983	L. L. Sloss	12:1–10
Witnessing Revolutions in the Earth Sciences	J. Rodgers	13:1–4
Geophysics on Three Continents	A. L. Hales	14:1–20
How I Became an Oceanographer and Other Sea Stories	R. Revelle	15:1–23

GEOCHEMISTRY, MINERALOGY, AND PETROLOGY

Chemistry of Subsurface Waters	I. Barnes, J. D. Hem	1:157–82
Genesis of Mineral Deposits	B. J. Skinner, P. B. Barton Jr.	1:183–212
Order-Disorder Relationships in Some Rock-Forming Silicate Minerals	C. W. Burnham	1:313–38
Low Grade Regional Metamorphism: Mineral Equilibrium Relations	E. Zen, A. B. Thompson	2:179–212
Clays as Catalysts for Natural Processes	J. J. Fripiat, M. I. Cruz-Cumplido	2:239–56
Phanerozoic Batholiths in Western North America: A Summary of Some Recent Work on Variations in Time, Space, Chemistry, and Isotopic Compositions	R. W. Kistler	2:403–18
Microstructures of Minerals as Petrogenetic Indicators	J. D. C. McConnell	3:129–55
Advances in the Geochemistry of Amino Acids	K. A. Kvenvolden	3:183–212
Volcanic Rock Series and Tectonic Setting	A. Miyashiro	3:251–69
Metallogenesis at Oceanic Spreading Centers	E. Bonatti	3:401–31
Chemistry of Interstitial Waters of Marine Sediments	J. M. Gieskes	3:433–53
Multicomponent Electrolyte Diffusion	D. E. Anderson, D. L. Graf	4:95–121
Argillaceous Sediment Dewatering	J. F. Burst	4:293–318
The Uranium-Series Methods of Age Determination	T.-L. Ku	4:347–79
Interactions of CH and CO in the Earth's Atmosphere	S. C. Wofsy	4:441–69
A Review of Hydrogen, Carbon, Nitrogen, Oxygen, Sulphur, and Chlorine Stable Isotope Fractionation Among Gaseous Molecules	P. Richet, Y. Bottinga, M. Javoy	5:65–110

607

Title	Author(s)	Citation
Transmission Electron Microscopy in Earth Science	P. E. Champness	5:203–26
Geochemistry of Atmospheric Radon and Radon Products	K. K. Turekian, Y. Nozaki, L. K. Benninger	5:227–55
Transport Phenomena in Chemical Rate Processes in Sediments	P. A. Domenico	5:287–317
Metamorphism of Alpine Peridotite and Serpentinite	B. W. Evans	5:397–447
Hydrothermal Alteration in Active Geothermal Fields	P. R. L. Browne	6:229–50
Temporal Fluctuations of Atmospheric ^{14}C: Causal Factors and Implications	P. E. Damon, J. C. Lerman, A. Long	6:457–94
Geochemical and Cosmochemical Applications of Nd Isotope Analysis	R. K. O'Nions, S. R. Carter, N. M. Evensen, P. J. Hamilton	7:11–38
Calorimetry: Its Application to Petrology	A. Navrotsky	7:93–115
Clay Mineral Catalysis and Petroleum Generation	W. D. Johns	7:183–98
Geochemistry of Evaporitic Lacustrine Deposits	H. P. Eugster	8:35–63
Geologic Pressure Determinations from Fluid Inclusion Studies	E. Roedder, R. J. Bodnar	8:263–301
Rare Earth Elements in Petrogenetic Studies of Igneous Systems	G. N. Hanson	8:371–406
Depleted and Fertile Mantle Xenoliths from Southern African Kimberlites	P. H. Nixon, N. W. Rogers, I. L. Gibson, A. Grey	9:285–309
The Combined Use of Isotopes as Indicators of Crustal Contamination	D. E. James	9:311–44
Intervalence Transitions in Mixed-Valence Minerals of Iron and Titanium	R. G. Burns	9:345–83
The Role of Perfectly Mobile Components in Metamorphism	D. Rumble III	10:221–33
Phanerozoic Oolitic Ironstones—Geologic Record and Facies Model	F. B. Van Houten, D. B. Bhattacharyya	10:441–57
Applications of the Ion Microprobe to Geochemistry and Cosmochemistry	N. Shimizu, S. R. Hart	10:483–526
The Structure of Silicate Melts	B. O. Mysen	11:75–97
Radioactive Nuclear Waste Stabilization: Aspects of Solid-State Molecular Engineering and Applied Geochemistry	S. E. Haggerty	11:133–63
In Situ Trace Element Microanalysis	D. S. Burnett, D. S. Woolum	11:329–58
Terrestrial Inert Gases: Isotope Tracer Studies and Clues to Primordial Components in the Mantle	J. E. Lupton	11:371–414
Double-Diffusive Convection Due to Crystallization in Magmas	H. E. Huppert, R. S. J. Sparks	12:11–37
Applications of Accelerator Mass Spectrometry	L. Brown	12:39–59
Kimberlites: Complex Mantle Melts	J. D. Pasteris	12:133–53
Theory of Hydrothermal Systems	D. L. Norton	12:155–77
Rheological Properties of Magmas	A. R. McBirney, T. Murase	12:337–57
Cooling Histories from $^{40}Ar/^{39}Ar$ Age Spectra: Implications for Precambrian Plate Tectonics	D. York	12:383–409
Solid-State Nuclear Magnetic Resonance Spectroscopy of Minerals	R. J. Kirkpatrick, K. A. Smith, S. Schramm, G. Turner, W.-H. Yang	13:29–47
Evolution of the Archean Continental Crust	A. Kröner	13:49–74

Oxidation Status of the Mantle: Past and Present	R. J. Arculus	13:75–95
Direct TEM Imaging of Complex Structures and Defects in Silicates	D. R. Veblen	13:119–46
Mantle Metasomatism	M. F. Roden, V. R. Murthy	13:269–96
Petrogenesis of Andesites	T. R. Grove, R. J. Kinzler	14:417–54
Chemical Geodynamics	A. Zindler, S. Hart	14:493–571
Application of Stable Carbon Isotopes to Early Biochemical Evolution on Earth	M. Schidlowski	15:47–72
Thermometry and Barometry of Igneous and Metamorphic Rocks	S. R. Bohen, D. H. Lindsley	15:397–420

GEOPHYSICS AND PLANETARY SCIENCE

Structure of the Earth from Glacio-Isostatic Rebound	R. I. Walcott	1:15–38
Interior of Jupiter and Saturn	W. B. Hubbard	1:85–106
Magnetospheric Electrons	F. V. Coroniti, R. M. Thorne	1:107–30
Theory and Nature of Magnetism in Rocks	R. B. Hargraves, S. K. Banerjee	1:269–96
Geophysical Data and the Interior of the Moon	M. N. Toksöz	2:151–77
Solar System Sources of Meteorites and Large Meteoroids	G. W. Wetherill	2:303–31
The Atmosphere of Mars	C. A. Barth	2:333–67
Satellites and Magnetospheres of the Outer Planets	W. I. Axford, D. A. Mendis	2:419–74
Interaction of Energetic Nuclear Particles in Space with the Lunar Surface	R. M. Walker	3:99–128
Array Seismology	J. Filson	3:157–81
High Temperature Creep of Rock and Mantle Viscosity	J. Weertman, J. R. Weertman	3:293–315
The Mechanical Behavior of Pack Ice	D. A. Rothrock	3:317–42
Mechanical Properties of Granular Media	J. R. Vaišnys, C. C. Pilbeam	3:343–60
Petrology of Lunar Rocks and Implications to Lunar Evolution	W. I. Ridley	4:15–48
Paleomagnetism of Meteorites	F. D. Stacey	4:147–57
Generation of Planetary Magnetic Fields	E. H. Levy	4:159–85
Hydrogen Loss from the Terrestrial Planets	D. M. Hunten, T. M. Donahue	4:265–92
The Ionospheres of Mars and Venus	J. C. McConnell	4:319–46
The Topside Ionosphere: A Region of Dynamic Transition	P. M. Banks, R. W. Schunk, W. J. Raitt	4:381–440
Understanding Inverse Theory	R. L. Parker	5:35–64
Discrimination Between Earthquakes and Underground Explosions	R. Blandford	5:111–22
Composition of the Mantle and Core	D. L. Anderson	5:179–202
Laser-Distance Measuring Techniques	J. Levine	5:357–69
The Evolution of the Lunar Regolith	Y. Langevin, J. R. Arnold	5:449–89
Theoretical Foundations of Equations of State for the Terrestrial Planets	L. Thomsen	5:491–513
Cratering and Obliteration History of Mars	C. R. Chapman, K. L. Jones	5:515–40
Mantle Convection Models	F. M. Richter	6:9–19
The Galilean Satellites of Jupiter: Four Worlds	T. V. Johnson	6:93–125
The Interstellar Wind and its Influence on the Interplanetary Environment	G. E. Thomas	6:173–204
Evolution of Ocean Crust Seismic Velocities	B. T. R. Lewis	6:377–404
The Magnetic Fields of Mercury, Mars, and Moon	N. F. Ness	7:249–88
Subsolidus Convection in the Mantles of Terrestrial Planets	G. Schubert	7:289–342
The Diamond Cell and the Nature of the Earth's Mantle	W. A. Bassett	7:357–84

Title	Authors	Citation
The Magnetic Field of Jupiter: A Comparison of Radio Astronomy and Spacecraft Observations	E. J. Smith, S. Gulkis	7:385–415
Synthetic Seismograms	D. V. Helmberger, L. J. Burdick	7:417–42
Cratering Mechanics—Observational, Experimental, and Theoretical	H. J. Melosh	8:65–93
Heat Flow and Hydrothermal Circulation	C. R. B. Lister	8:95–117
Seismic Reflection Studies of Deep Crustal Structure	J. A. Brewer, J. E. Oliver	8:205–30
Geomorphological Processes on Terrestrial Planetary Surfaces	R. P. Sharp	8:231–61
Origin and Evolution of Planetary Atmospheres	J. B. Pollack, Y. L. Yung	8:425–87
The Moons of Mars	J. Veverka, J. A. Burns	8:527–58
Refractory Inclusions in the Allende Meteorite	L. Grossman	8:559–608
Rotation and Precession of Cometary Nuclei	Z. Sekanina	9:113–45
The Significance of Terrestrial Electrical Conductivity Variations	G. D. Garland	9:147–74
Ice Age Geodynamics	W. R. Peltier	9:199–225
Free Oscillations of the Earth	R. Buland	9:385–413
Long Wavelength Gravity and Topography Anomalies	A. B. Watts, S. F. Daly	9:415–48
Dynamical Constraints of the Formation and Evolution of Planetary Bodies	A. W. Harris, W. R. Ward	10:61–108
Pre-Mesozoic Paleomagnetism and Plate Tectonics	R. Van der Voo	10:191–220
Interiors of the Giant Planets	D. J. Stevenson	10:257–95
Halley's Comet	R. L. Newburn Jr., D. K. Yeomans	10:297–326
Regoliths on Small Bodies in the Solar System	K. R. Housen, L. L. Wilkening	10:355–76
Heat and Mass Circulation in Geothermal Systems	I. G. Donaldson	10:377–95
Magma Migration	D. L. Turcotte	10:397–408
The Heat Flow from the Continents	H. N. Pollack	10:459–81
Strain Accumulation in Western United States	J. C. Savage	11:11–43
Oceanic Intraplate Seismicity	E. A. Okal	11:195–214
Creep Deformation of Ice	J. Weertman	11:215–40
Recent Developments in the Dynamo Theory of Planetary Magnetism	F. H. Busse	11:241–68
Seismic Gaps in Space and Time	K. C. McNally	11:359–69
The Atmospheres of the Outer Planets	G. E. Hunt	11:415–59
Asteroid and Comet Bombardment of the Earth	E. M. Shoemaker	11:461–94
The History of Water on Mars	S. W. Squyres	12:83–106
Tectonic Processes Along the Front of Modern Convergent Margins—Research of the Past Decade	R. von Huene	12:359–81
Tectonics of Venus	R. J. Phillips, M. C. Malin	12:411–43
The Geological Significance of the Geoid	C. G. Chase	13:97–117
Cosmic Dust: Collection and Research	D. E. Brownlee	13:147–73
The Magma Ocean Concept and Lunar Evolution	P. H. Warren	13:201–40
Downhole Geophysical Logging	A. Timur, M. N. Toksöz	13:315–44
The Midcontinent Rift System	W. R. Van Schmus, W. J. Hinze	13:345–83
Triggered Earthquakes	D. W. Simpson	14:21–42
Forecasting Volcanic Eruptions	R. W. Decker	14:267–92
Rupture Process of Subduction-Zone Earthquakes	H. Kanamori	14:293–322
Geochemistry of Tektites and Impact Glasses	C. Koeberl	14:323–50
Temperature Distribution in the Crust and Mantle	R. Jeanloz, S. Morris	14:377–415
The Core-Mantle Boundary	C. J. Young, T. Lay	15:25–46

Physical Processes in Volcanic Eruptions	L. Wilson, H. Pinkerton, R. Macdonald	15:73–95
Three-Dimensional Seismic Imaging	C. H. Thurber, K. Aki	15:115–39
The Atmospheres of Venus, Earth, and Mars: A Critical Comparison	R. G. Prinn, B. Fegley, Jr.	15:171–212
Terrestrial Impact Structures	R. A. F. Grieve	15:245–70
Origin of the Moon—The Collision Hypothesis	D. J. Stevenson	15:271–315
Models of Lithospheric Thinning	H. J. Neugebauer	15:421–43
Marine Magnetic Anomalies—The Origin of the Stripes	C. G. A. Harrison	15:505–43
Stress Near the Surface of the Earth	D. I. Gough, W. I. Gough	15:545–66
Polar Wandering and Paleomagnetism	R. G. Gordon	15:567–93

OCEANOGRAPHY, METEOROLOGY, AND PALEOCLIMATOLOGY

Electrical Balance in the Lower Atmosphere	B. Vonnegut	1:297–312
The Physical Chemistry of Seawater	F. J. Millero	2:101–50
Numerical Modeling of Lake Currents	W. Lick	4:49–74
Features Indicative of Permafrost	R. F. Black	4:75–94
Chemical Fractionation at the Air/Sea Interface	R. A. Duce, E. J. Hoffman	4:187–228
Pacific Deep-Sea Manganese Nodules: Their Distribution, Composition, and Origin	S. V. Margolis, R. G. Burns	4:229–63
Quaternary Vegetation History—Some Comparisons Between Europe and America	H. E. Wright Jr.	5:123–58
The History of the Earth's Surface Temperature During the Past 100 Million Years	S. M. Savin	5:319–55
Photochemistry and Dynamics of the Ozone Layer	R. G. Prinn, F. N. Alyea, D. M. Cunnold	6:43–74
Effects of Waste Disposal Operations in Estuaries and the Coastal Ocean	M. G. Gross	6:127–43
Glacial Inception and Disintegration During the Last Glaciation	J. T. Andrews, R. G. Barry	6:205–28
Synoptic Eddies in the Ocean	M. N. Koshlyakov, A. S. Monin	6:495–523
Influences of Mankind on Climate	W. W. Kellogg	7:63–92
The Dynamical Structure and Evolution of Thunderstorms and Squall Lines	D. K. Lilly	7:117–61
Atmospheric Tides	R. S. Lindzen	7:199–225
The Role of NO and NO_2 in the Chemistry of the Troposphere and Stratosphere	P. Crutzen	7:443–72
The Dynamics of Sudden Stratospheric Warnings	J. R. Holton	8:169–90
The Equatorial Undercurrent Revisited	S. G. H. Philander	8:191–204
Satellite Sensing of Ocean Surface Dynamics	J. R. Apel	8:303–42
Particles Above the Tropopause	N. H. Farlow, O. B. Toon	9:19–58
Nucleation and Growth of Stratospheric Aerosols	A. W. Castleman Jr., R. G. Keesee	9:227–49
The Biogeochemistry of the Air-Sea Interface	L. W. Lion, J. O. Leckie	9:449–86
Poleward Heat Transport by the Oceans: Observations and Models	K. Bryan	10:15–38
Thirteen Years of Deep-Sea Drilling	K. J. Hsü	10:109–28
The Transport of Contaminants in the Great Lakes	W. Lick	10:327–53
Cenozoic Glaciation in the Southern Hemisphere	J. H. Mercer	11:99–132
Oceanography from Space	R. H. Stewart	12:61–82
Pre-Quaternary Sea-Level Changes	A. Hallam	12:205–43
El Niño	M. A. Cane	14:43–70
Carbon Dioxide Increase in the Atmosphere and Oceans and Possible Effects on Climate	C.-T. A. Chen, E. T. Drake	14:201–35

Experimental and Theoretical Constraints on Hydrothermal Alteration Processes at Mid-Ocean Ridges	W. E. Seyfried Jr.	15:317–35

PALEONTOLOGY, STRATIGRAPHY, AND SEDIMENTOLOGY

Origin of Red Beds: A Review—1961–1972	F. B. Van Houten	1:39–62
Mammals from Reptiles: A Review of Mammalian Origins	A. W. Crompton, F. A. Jenkins Jr.	1:131–56
Cenozoic Planktonic Micropaleontology and Biostratigraphy	W. R. Riedel	1:241–68
Growth Lines in Invertebrate Skeletons	G. R. Clark II	2:77–99
Marine Diagenesis of Shallow Water Calcium Carbonate Sediments	R. G. C. Bathurst	2:257–74
The Origin of Birds	J. H. Ostrom	3:55–77
Early Paleozoic Echinoderms	G. Ubaghs	3:79–98
Precambrian Paleobiology: Problems and Perspectives	J. W. Schopf	3:213–49
Adaptive Themes in the Evolution of the Bivalvia (Mollusca)	S. M. Stanley	3:361–85
Biostratigraphy of the Cambrian System—A Progress Report	A. R. Palmer	5:13–33
Paleoecological Transfer Functions	H. M. Sachs, T. Webb III, D. R. Clark	5:159–78
The Probable Metazoan Biota of the Precambrian as Indicated by the Subsequent Record	J. W. Durham	6:21–42
Stratigraphy of the Atlantic Continental Shelf and Slope of the United States	C. W. Poag	6:251–80
Chemical Exchange Across Sediment-Water Interface	A. Lerman	6:281–303
Organic Matter in the Earths Crust	P. H. Abelson	6:325–51
Quantifying Biostratigraphic Correlation	W. W. Hay, J. R. Southam	6:353–75
Volcanic Evolution of the Cascade Range	A. R. McBirney	6:437–56
Recent Advances in Sandstone Diagenesis	E. D. Pittman	7:39–62
The Enigma of the Extinction of the Dinosaurs	D. A. Russell	7:163–82
Primary Sedimentary Structures	J. C. Harms	7:227–48
Paleozoic Paleogeography	A. M. Zeigler, C. R. Scotese, W. S. McKerrow, M. E. Johnson, R. K. Bambach	7:473–502
Archean Sedimentation	D. R. Lowe	8:145–67
Quaternary Deep-Sea Benthic Foraminifers and Bottom Water Masses	D. Schnitker	8:343–70
Evolutionary Patterns of Early Cenozoic Mammals	P. D. Gingerich	8:407–24
Form, Function, and Architecture of Ostracode Shells	R. H. Benson	9:59–80
Metamorphosed Layered Igneous Complexes in Archean Granulite-Gneiss Belts	B. F. Windley, F. C. Bishop, J. V. Smith	9:175–98
Ancient Marine Phoshorites	R. P. Sheldon	9:251–84
The Geological Context of Human Evolution	A. K. Behrensmeyer	10:39–60
Rates of Biogeochemical Processes in Anoxic Sediments	W. S. Reeburgh	11:269–98
Methane and Other Hydrocarbon Gases in Marine Sediment	G. E. Claypool, K. A. Kvenvolden	11:299–327
Sedimentation in Large Lakes	T. C. Johnson	12:179–204
Rates of Evolution and the Notion of "Living Fossils"	T. J. M. Schopf	12:245–92
Physics and Chemistry of Biomineralization	K. E. Chave	12:293–305
The Ordovician System, Progress and Problems	R. J. Ross Jr.	12:307–35

Blancan-Hemphillian Land Mammal Ages and Late Cenozoic Mammal Dispersal Events	E. H. Lindsay, N. D. Opdyke, N. M. Johnson	12:445–88
Patterns of Alluvial Rivers	S. A. Schumm	13:5–27
Amino Acid Racemization Dating of Fossil Bones	J. L. Bada	13:241–68
The Geological Record of Insects	F. M. Carpenter, L. Burnham	13:297–314
Paleogeographic Interpretation: With an Example From the Mid-Cretaceous	A. M. Ziegler, D. B. Rowley, A. L. Lottes, D. L. Sahagian, M. L. Hulver, T. C. Gierlowski	13:385–425
Molecular Phylogenetics	J. M. Lowenstein	14:71–83
Conodonts and Biostratigraphic Correlation	W. C. Sweet, S. M. Bergström	14:85–112
Occurrence and Formation of Water-Laid Placers	R. Slingerland, N. D. Smith	14:113–47
Genesis of Mississippi Valley–Type Lead-Zinc Deposits	D. A. Sverjensky	14:177–99
Coastal Processes and the Development of Shoreline Erosion	P. D. Komar, R. A. Holman	14:237–65
Climatic Rhythms Recorded in Strata	A. G. Fischer	14:351–76
Geologic Significance of Paleozoic and Mesozoic Radiolarian Chert	D. L. Jones, B. Murchey	14:455–92
Geoarchaeology	G. R. Rapp Jr.	15:97–113
Integrated Diagenetic Modeling: A Process-Oriented Approach for Clastic Systems	R. C. Surdam, L. J. Crossey	15:141–70
Organic Geochemistry of Biomarkers	R. P. Philp, C. A. Lewis	15:363–95

TECTONOPHYSICS AND REGIONAL GEOLOGY

Rock Fracture	K. Mogi	1:63–84
Mode of Strain Release Associated with Major Earthquakes in Japan	H. Kanamori	1:213–40
Iceland in Relation to the Mid-Atlantic Ridge	G. Pálmason, K. Saemundsson	2:25–30
Evolution of Arc Systems in the Western Pacific	D. E. Karig	2:51–75
Regional Geophysics of the Basin and Range Province	G. A. Thompson, D. B. Burke	2:213–38
Earthquake Mechanisms and Modeling	J. H. Dieterich	2:275–301
Current Views of the Development of Slaty Cleavage	D. S. Wood	2:369–401
Precambrian Tectonic Environments	C. R. Anhaeusser	3:31–53
Plate Tectonics: Commotion in the Ocean and Continental Consequences	C. G. Chase, E. M. Herron, W. R. Normark	3:271–91
Age Provinces in the Northern Appalachians	R. S. Naylor	3:387–400
Radar Imagery in Defining Regional Tectonic Structure	T. A. Grant, L. S. Cluff	4:123–45
Geochronology of Some Alkalic Rock Provinces in Eastern and Central United States	R. E. Zartman	5:257–86
Aulacogens and Continental Breakup	K. Burke	5:371–96
Paleomagnetism of the Peri-Atlantic Precambrian	R. Ressetar	6:75–91
Afar	P. Mohr	6:145–72
Critical Mineral Resources	C. F. Park Jr.	6:305–24
State of Stress in the Earth's Crust	A. McGarr, N. C. Gay	6:405–36
The North Atlantic Ridge: Observational Evidence for Its Generation and Aging	J. R. Heirtzler	7:343–55
Platform Basins	N. H. Sleep. J. A. Nunn, L. Chou	8:17–33
Deformation of Mantle Rocks	Y. Gueguen, A. Nicolas	8:119–44

Title	Author(s)	Citation
Seismic Reflection Studies of Deep Crustal Structure	J. A. Brewer, J. E. Oliver	8:205–30
Fracture Mechanics Applied to the Earth's Crust	J. W. Rudnicki	8:489–525
Mechanics of Motion on Major Faults	G. M. Mavko	9:81–111
Cordilleran Metamorphic Core Complexes—From Arizona to Southern Canada	R. L. Armstrong	10:129–54
Mid-Ocean Ridges: Fine Scale Tectonic, Volcanic and Hydrothermal Processes Within the Plate Boundary Zone	K. C. Macdonald	10:155–90
Earthquake Prediction	Z. Suzuki	10:235–56
The Basin and Range Province: Origin and Tectonic Significance	G. P. Eaton	10:409–40
Accretionary Tectonics of the North American Cordillera	J. B. Saleeby	11:45–73
Hotspot Swells	S. T. Crough	11:165–93
The Origin of Allochthonous Terranes: Perspectives on the Growth and Shaping of Continents	E. R. Schermer, D. G. Howell, D. L. Jones	12:107–31
Structure and Tectonics of the Himalaya: Constraints and Implications of Geophysical Data	P. Molnar	12:489–518
The Appalachian-Ouachita Connection: Paleozoic Orogenic Belt at the Southern Margin of North America	W. A. Thomas	13:175–99
Earthquakes and Rock Deformation in Crustal Fault Zones	R. H. Sibson	14:149–75
Tectonics of the Tethysides: Orogenic Collage Development in a Collisional Setting	A. M. C. Şengör	15:213–44
Tectonics of the Southern and Central Appalachian Internides	R. D. Hatcher Jr.	15:337–62
Architecture of Continental Rifts With Special Reference to East Africa	B. R. Rosendahl	15:445–503

Annual Reviews Inc.
A NONPROFIT SCIENTIFIC PUBLISHER

4139 El Camino Way
P.O. Box 10139
Palo Alto, CA 94303-0897 • USA

ORDER FORM

Now you can order
TOLL FREE
1-800-523-8635
(except California)

Annual Reviews Inc. publications may be ordered directly from our office by mail or use our Toll Free Telephone line (for orders paid by credit card or purchase order, and customer service calls only); through booksellers and subscription agents, worldwide; and through participating professional societies. Prices subject to change without notice.   ARI Federal I.D. #94-1156476

- **Individuals:** Prepayment required on new accounts by check or money order (in U.S. dollars, check drawn on U.S. bank) or charge to credit card — American Express, VISA, MasterCard.
- **Institutional buyers:** Please include purchase order number.
- **Students:** $10.00 discount from retail price, per volume. Prepayment required. Proof of student status must be provided (photocopy of student I.D. or signature of department secretary is acceptable). Students must send orders direct to Annual Reviews. Orders received through bookstores and institutions requesting student rates will be returned.
- **Professional Society Members:** Members of professional societies that have a contractual arrangement with Annual Reviews may order books through their society at a reduced rate. Check with your society for information.
- **Toll Free Telephone orders:** Call 1-800-523-8635 (except from California) for orders paid by credit card or purchase order and customer service calls only. California customers and all other business calls use 415-493-4400 (not toll free). Hours: 8:00 AM to 4:00 PM, Monday-Friday, Pacific Time.

Regular orders: Please list the volumes you wish to order by volume number.
Standing orders: New volume in the series will be sent to you automatically each year upon publication. Cancellation may be made at any time. Please indicate volume number to begin standing order.
Prepublication orders: Volumes not yet published will be shipped in month and year indicated.
California orders: Add applicable sales tax.
Postage paid (4th class bookrate/surface mail) **by Annual Reviews Inc.** Airmail postage or UPS, extra.

ANNUAL REVIEWS SERIES		Prices Postpaid per volume USA/elsewhere	Regular Order Please send: Vol. number	Standing Order Begin with: Vol. number
Annual Review of ANTHROPOLOGY				
Vols. 1-14	(1972-1985)	$27.00/$30.00		
Vol. 15	(1986)	$31.00/$34.00		
Vol. 16	(avail. Oct. 1987)	$31.00/$34.00	Vol(s). _____	Vol. _____
Annual Review of ASTRONOMY AND ASTROPHYSICS				
Vols. 1-2, 4-20	(1963-1964; 1966-1982)	$27.00/$30.00		
Vols. 21-24	(1983-1986)	$44.00/$47.00		
Vol. 25	(avail. Sept. 1987)	$44.00/$47.00	Vol(s). _____	Vol. _____
Annual Review of BIOCHEMISTRY				
Vols. 30-34, 36-54	(1961-1965; 1967-1985)	$29.00/$32.00		
Vol. 55	(1986)	$33.00/$36.00		
Vol. 56	(avail. July 1987)	$33.00/$36.00	Vol(s). _____	Vol. _____
Annual Review of BIOPHYSICS AND BIOPHYSICAL CHEMISTRY				
Vols. 1-11	(1972-1982)	$27.00/$30.00		
Vols. 12-15	(1983-1986)	$47.00/$50.00		
Vol. 16	(avail. June 1987)	$47.00/$50.00	Vol(s). _____	Vol. _____
Annual Review of CELL BIOLOGY				
Vol. 1	(1985)	$27.00/$30.00		
Vol. 2	(1986)	$31.00/$34.00		
Vol. 3	(avail. Nov. 1987)	$31.00/$34.00	Vol(s). _____	Vol. _____

ANNUAL REVIEWS SERIES	Prices Postpaid per volume USA/elsewhere	Regular Order Please send: Vol. number	Standing Order Begin with: Vol. number
Annual Review of COMPUTER SCIENCE			
Vol. 1	(1986) .... $39.00/$42.00		
Vol. 2	(avail. Nov. 1987)............. $39.00/$42.00	Vol(s). _____	Vol. _____
Annual Review of EARTH AND PLANETARY SCIENCES			
Vols. 1-10	(1973-1982)................. $27.00/$30.00		
Vols. 11-14	(1983-1986)................. $44.00/$47.00		
Vol. 15	(avail. May 1987)............. $44.00/$47.00	Vol(s). _____	Vol. _____
Annual Review of ECOLOGY AND SYSTEMATICS			
Vols. 1-16	(1970-1985)................. $27.00/$30.00		
Vol. 17	(1986) .... $31.00/$34.00		
Vol. 18	(avail. Nov. 1987)............. $31.00/$34.00	Vol(s). _____	Vol. _____
Annual Review of ENERGY			
Vols. 1-7	(1976-1982)................. $27.00/$30.00		
Vols. 8-11	(1983-1986)................. $56.00/$59.00		
Vol. 12	(avail. Oct. 1987)............. $56.00/$59.00	Vol(s). _____	Vol. _____
Annual Review of ENTOMOLOGY			
Vols. 10-16, 18-30	(1965-1971, 1973-1985)........ $27.00/$30.00		
Vol. 31	(1986) .... $31.00/$34.00		
Vol. 32	(avail. Jan. 1987)............. $31.00/$34.00	Vol(s). _____	Vol. _____
Annual Review of FLUID MECHANICS			
Vols. 1-4, 7-17	(1969-1972, 1975-1985)........ $28.00/$31.00		
Vol. 18	(1986) .... $32.00/$35.00		
Vol. 19	(avail. Jan. 1987)............. $32.00/$35.00	Vol(s). _____	Vol. _____
Annual Review of GENETICS			
Vols. 1-19	(1967-1985)................. $27.00/$30.00		
Vol. 20	(1986) .... $31.00/$34.00		
Vol. 21	(avail. Dec. 1987)............. $31.00/$34.00	Vol(s). _____	Vol. _____
Annual Review of IMMUNOLOGY			
Vols. 1-3	(1983-1985)................. $27.00/$30.00		
Vol. 4	(1986) .... $31.00/$34.00		
Vol. 5	(avail. April 1987)............. $31.00/$34.00	Vol(s). _____	Vol. _____
Annual Review of MATERIALS SCIENCE			
Vols. 1, 3-12	(1971, 1973-1982)............ $27.00/$30.00		
Vols. 13-16	(1983-1986)................. $64.00/$67.00		
Vol. 17	(avail. August 1987)........... $64.00/$67.00	Vol(s). _____	Vol. _____
Annual Review of MEDICINE			
Vols. 1-3, 6, 8-9 11-15, 17-36	(1950-1952, 1955, 1957-1958) (1960-1964, 1966-1985)........ $27.00/$30.00		
Vol. 37	(1986) .... $31.00/$34.00		
Vol. 38	(avail. April 1987)............. $31.00/$34.00	Vol(s). _____	Vol. _____
Annual Review of MICROBIOLOGY			
Vols. 18-39	(1964-1985)................. $27.00/$30.00		
Vol. 40	(1986) .... $31.00/$34.00		
Vol. 41	(avail. Oct. 1987)............. $31.00/$34.00	Vol(s). _____	Vol. _____